D0074006

IMPORTANT:

HERE IS YOUR REGISTRATION CODE TO ACCESS
YOUR PREMIUM McGRAW-HILL ONLINE RESOURCES.

For key premium online resources you need THIS CODE to gain access. Once the code is entered, you will be able to use the Web resources for the length of your course.

If your course is using **WebCT** or **Blackboard**, you'll be able to use this code to access the McGraw-Hill content within your instructor's online course.

Access is provided if you have purchased a new book. If the registration code is missing from this book, the registration screen on our Website, and within your WebCT or Blackboard course, will tell you how to obtain your new code.

Registering for McGraw-Hill Online Resources

TO gain access to your McGraw-Hill web resources simply follow the steps below:

1. USE YOUR WEB BROWSER TO GO TO: **http://www.mhhe.com/spencer**
2. CLICK ON **FIRST TIME USER**.
3. ENTER THE REGISTRATION CODE* PRINTED ON THE TEAR-OFF BOOKMARK ON THE RIGHT.
4. AFTER YOU HAVE ENTERED YOUR REGISTRATION CODE, CLICK **REGISTER**.
5. FOLLOW THE INSTRUCTIONS TO SET-UP YOUR PERSONAL UserID AND PASSWORD.
6. WRITE YOUR UserID AND PASSWORD DOWN FOR FUTURE REFERENCE. KEEP IT IN A SAFE PLACE.

TO GAIN ACCESS to the McGraw-Hill content in your instructor's **WebCT** or **Blackboard** course simply log in to the course with the UserID and Password provided by your instructor. Enter the registration code exactly as it appears in the box to the right when prompted by the system. You will only need to use the code the first time you click on McGraw-Hill content.

Thank you, and welcome to your McGraw-Hill online Resources!

*YOUR REGISTRATION CODE CAN BE USED ONLY ONCE TO ESTABLISH ACCESS. IT IS NOT TRANSFERABLE.

0-07-234146-7 SPENCER, EARTH SCIENCE

MCGRAW-HILL
ONLINE RESOURCES

REGISTRATION CODE

regionally-11381530

How's Your Math?

Do you have the math skills you need to succeed?

Why risk not succeeding because you struggle with your math skills?

Get access to a web-based, personal math tutor:

- Available 24/7, unlimited use
- Driven by artificial intelligence
- Self-paced
- An entire month's subscription **for much less** than the cost of one hour with a human tutor

ALEKS is an inexpensive, private, infinitely patient math tutor that's accessible any time, anywhere you log on.

 ALEKS® Mc Graw Hill

Log On for a **FREE** 48-hour Trial

 www.highedstudent.aleks.com

ALEKS is a registered trademark of ALEKS Corporation.

Earth Science

Understanding Environmental Systems

Edgar W. Spencer

Washington and Lee University

McGraw Hill

Boston   Burr Ridge, IL   Dubuque, IA   Madison, WI   New York   San Francisco   St. Louis
Bangkok   Bogotá   Caracas   Kuala Lumpur   Lisbon   London   Madrid   Mexico City
Milan   Montreal   New Delhi   Santiago   Seoul   Singapore   Sydney   Taipei   Toronto

McGraw-Hill Higher Education ⚛

*A Division of The **McGraw-Hill** Companies*

EARTH SCIENCE: UNDERSTANDING ENVIRONMENTAL SYSTEMS

Published by McGraw-Hill, a business unit of The McGraw-Hill Companies, Inc., 1221 Avenue of the Americas, New York, NY 10020. Copyright © 2003 by The McGraw-Hill Companies, Inc. All rights reserved. No part of this publication may be reproduced or distributed in any form or by any means, or stored in a database or retrieval system, without the prior written consent of The McGraw-Hill Companies, Inc., including, but not limited to, in any network or other electronic storage or transmission, or broadcast for distance learning.

Some ancillaries, including electronic and print components, may not be available to customers outside the United States.

 This book is printed on recycled, acid-free paper containing 10% postconsumer waste.

1 2 3 4 5 6 7 8 9 0 QPD/QPD 0 9 8 7 6 5 4 3 2

ISBN 0–07–234146–7

Publisher: *Margaret J. Kemp*
Sponsoring editor: *Thomas C. Lyon*
Developmental editor: *Lisa Leibold*
Executive marketing manager: *Lisa Gottschalk*
Senior project manager: *Rose Koos*
Production supervisor: *Sherry L. Kane*
Media project manager: *Jodi K. Banowetz*
Lead media technology producer: *Judi David*
Design manager: *Stuart Paterson*
Cover/interior designer: *Ellen Pettengell*
Cover image: *Edgar W. Spencer*
Lead photo research coordinator: *Carrie K. Burger*
Compositor: *Interactive Composition Corporation*
Typeface: *10/12 Times Roman*
Printer: *Quebecor World Dubuque, IA*

Inside front cover image: This image was generated from a digital database of land and seafloor elevations on a 2-minute latitude/longitude grid. Assumed illumination is from the west; shading is computed as a function of the east-west slope of the surface with a nonlinear exaggeration favoring low-relief areas. A Mercator projection was used for the world image, which spans 360° of longitude from 270° West around the world eastward to 120° East; latitude coverage is from 80° North to 80° South. The resolution of the gridded data is at least 5-minutes of latitude and longitude. Data points were resampled from 2-minute gridded ocean depths derived from satellite altimetry of the sea surface between 72° N and S, and from 5-minute U.S. Navy data poleward from 72°. Land data were primarily from 30-second gridded data collected from various sources by the National Imagery and Mapping Agency.

Other computerized digital images and associated databases are available from the National Geophysical Data Center, National Oceanic and Atmospheric Administration, U.S. Department of Commerce, Code E/GC3, Boulder, Colorado 80303. www.ngdc.noaa.gov/mgg/image/images.html

Digital image by Dr. Peter W. Sloss, NOAA/NESDIS/NGDC.

The credits section for this book begins on page 501 and is considered an extension of the copyright page.

Library of Congress Cataloging-in-Publication Data

Spencer, Edgar Winston.
 Earth science : understanding environmental systems / Edgar W. Spencer. — 1st ed.
  p.   cm.
 Includes index.
 ISBN 0–07–234146–7 (acid-free paper)
 1. Earth sciences   I. Title.

QE28 .S7   2003
 550—dc21

                                        2002025073
                                        CIP

www.mhhe.com

In grateful appreciation for their help, encouragement, and support, this book is dedicated with love to Liz, Shawn, Shannon, Rich, Tucker, and Jonah.

Brief Contents

unit four

The Solar System and Its Place in the Universe   435

Contents

unit one   Major Elements of Earth Systems   13

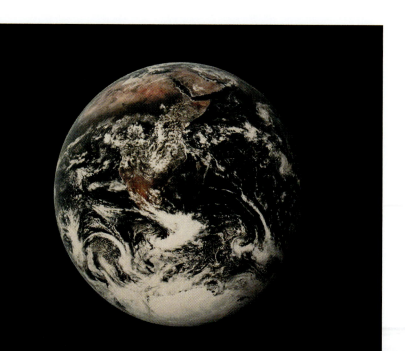

unit three Earth's Physical Climate System   173

part 1   Oceans and Coasts

unit four   The Solar System and Its Place in the Universe   435

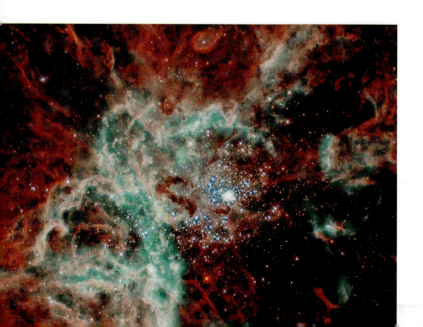

Preface

As we enter the twenty-first century, the need for a better understanding of Earth and the environment in which we live is increasingly apparent. In almost every community, some look to scientists for leadership in explaining the environmental issues society faces on a day-to-day basis. Ultimately, responsibility for making decisions about environmental issues rests with individual citizens and their representatives, few of whom are trained as scientists. Most of the students who enroll in earth science courses do so to satisfy a general education requirement. For many, this will be their only formal college-level science course. This course may be their only opportunity to learn what science is, how scientists go about examining issues, how scientific ideas evolve over time, and why science has been so successful as a way of looking at the world. In addition to providing broad perspectives on science and the environment, earth science courses also establish links between science, economics, politics, and history. For all these reasons, earth science courses are ideal general education courses.

This book is intended for introductory courses that place emphasis on the systems approach to earth science with special attention to the impact these systems have on the environment. It is appropriate for liberal-arts nonmajors with no previous college science or mathematics. By using a systems approach, students gain a better understanding of Earth as a whole, the relationships among processes acting on and inside Earth, and the connections between physical and biological sciences.

OVERVIEW

The primary goals of this book are to provide the background the general student needs to understand the way Earth works, how knowledge of Earth relates to the environmental issues confronting our society, and how scientists go about examining these issues. The unifying focus used to achieve these goals is the study of the processes that govern the evolution of Earth and control the operation of its environmental systems.

It is difficult for students to understand the present condition of Earth unless they are familiar with the processes that cause change. The time frame for these changes may vary from relatively short spans to those that take place slowly over thousands or even millions of years. Thus, the time frame in which natural processes bring about change is an important aspect of Earth systems. Slow processes such as those that cause the gradual shifting of conti-

nents, the rise of mountains, the formation of new sea floor, and the evolution of living organisms have set the stage for the present Earth. Although these barely perceptible processes continue, we are more acutely aware of changes produced over short time intervals. These include the development of atmospheric storms that can have great impact in a matter of days or even hours. Waves and currents can transform the coast by eroding some areas while depositing sediment in others, and earthquakes and volcanic eruptions can cause widespread changes in a day and can transform regions as the effects continue to accumulate over thousands of years.

Throughout human history, people have adjusted to natural conditions. Thousands of years ago, our ancestors moved south as glaciers gradually covered the northern latitudes. Subsequently, as the ice margin receded, some groups followed it north, eventually crossing from Siberia to North America where they began the settlement of the Americas. During most of the million of years that humans have inhabited Earth, their numbers were not great. They lacked the capability of modifying Earth's systems and surface significantly. But human population has grown exponentially in the last hundred years, and since the industrial revolution, humans have made significant changes to the environment. Both the size of the human population and our ability to alter Earth's environment have increased dramatically during the last few decades. We have constructed road systems that are long enough to reach the moon; extracted and used a large percentage of the total amount of oil and gas on Earth within a few decades; dug millions of miles of trenches for pipes; and connected oceans by canal works. We have stripped the soil from vast areas; modified drainage basins in ways that cause floods; and disposed of toxic waste using methods that endanger our drinking water. Many scientists think it is likely that humans are changing the climate of Earth, but at present, we do not know how fast those changes may take place or how extreme they may be. The last of the virgin forests are being cut down rapidly, causing thousands of animal and plant species to approach the edge of extinction. We do not know the extent to which our own survival is tied to that of other species or to the overall health of the ecosystem. These will be among the most challenging issues facing the current generation of students. How well they will be able to deal with these as scientists, business managers, teachers, politicians, and citizens will depend on how well they understand the nature of the problems, the interdependence and operation of natural systems.

Can we manage Earth's systems without damaging our own environment? Can we understand and live in harmony with the processes that govern the environment? Can we learn how the different parts of Earth systems are interrelated well enough to predict the consequences of our own actions? These are questions all informed individuals, especially scientists and engineers, must try to answer. These are the subjects this book explores and clarifies by examining the character of the solid earth, the oceans, the atmosphere, biological processes, and the ways they are interrelated.

This book provides a basis for understanding natural systems and identifies many of the problems we face as we try to avoid the difficulties caused by natural processes such as volcanic eruptions, earthquakes, and landslides, as well as those created by our own actions.

FEATURES

1. **Systems Emphasis.** The systems approach emphasizes the relationships among the great variety of processes that affect Earth and its environment. Although the book provides coverage used in earth science courses, emphasis in the first three units is placed on Earth systems, the connections between geology, oceanography, meteorology, and ecology. The physical systems identified by NASA in a book entitled *Earth System Science—A Closer View* are used as a basis for organization of the systems approach. NASA recognizes three major physical systems, the core-mantle, plate tectonic, and physical climate systems. The close relationships between living organisms and the physical environment are treated through examination of biogeochemical cycles.

2. **Major Ecosystems.** Close connections exist between the study of the physical and biological aspects of Earth's environments. These connections are emphasized in this book. Chapter 14 provides an introduction to the land surface and the relationship between terrestrial geology and ecosystems. Similar emphasis is given to marine environments in chapter 8 and to coastal environments in chapter 10.

3. **The Relationship Between Humans and Their Environment.** Not only do natural processes shape the human environment, human activities have become a major influence on natural systems. This theme runs throughout this book, and special attention is given to the relevance of Earth systems for understanding environmental problems.

4. **Flexible Organization.** The interaction of Earth systems is like a web of interrelationships. Some instructors prefer to begin by looking at Earth as a planet and end by examining details of surface processes. Others prefer to cover topics more familiar to students at the start and end with a large-scale picture of Earth's interior. To accommodate various approaches, this book provides flexible organization. After completion of the Introduction in unit one, the instructor can proceed with other units of the book. However, the author thinks that most students will find later chapters easier to

understand after they are more familiar with the recycling of Earth materials, the importance of time in the operation of Earth processes, and major features of Earth and its interior.

5. **Chapter Guide.** Each chapter begins with a brief introduction that relates the chapter to other parts of the text. The Chapter Guide summarizes the main objectives of the chapter and indicates how the subject matter of the chapter relates to Earth systems.

6. **Key Words.** Key words are identified in bold-faced print.

7. **Case Histories.** Where appropriate, case histories of events or localities are used to illustrate principles and demonstrate the relevance of the subject matter. Most students find the concepts presented with a case-history approach interesting and easy to understand and to remember because they can associate them with real situations.

8. **Data Banks.** Some tables contain files such as soil and climate types that students need to have readily available for reference as they read the text.

9. **Summary Points.** Summary Points provide a quick review of the most important concepts presented in the chapter.

10. **Review Questions.** Review questions are included to help students evaluate their mastery of basic materials presented in each chapter and to prepare for tests.

11. **Thinking Critically.** Thinking Critically questions encourage the student or a class to go beyond the text materials in exploring the relevance of the subject matter to environmental issues.

12. **Units of Measure.** Both English and SI units of measure are included in the text. Precise conversions are used for measurements that are exact, but approximate and rounded-off conversions are used if exact measurements are unknown.

13. **Appendix.** The appendixes provide supplementary information about several important subjects that some instructors may wish to include in their courses.

Appendix A. Units and Conversions. This appendix contains basic information about units of measure and conversions between English and SI units.

Appendix B. Minerals. This appendix explains how to identify minerals by using their physical properties. It contains a chart of physical properties and composition of common rock-forming minerals and common, economically important minerals.

Appendix C. Rock Identification. This section contains additional details about common rocks.

Appendix D. Topographic and Geologic Maps. This appendix contains a brief explanation of how to read topographic and geologic maps.

Appendix E. Star Charts. Students who enjoy identifying stars will find these star charts and information about locating objects in the sky useful.

Appendix F. The Periodic Table of Elements.

14. **Glossary.** The glossary provides brief definitions of the technical terms used throughout the text.

ORGANIZATION

The organization of this book allows instructors to vary the sequence of coverage. It is organized into four units. Unit three, which covers surficial processes, is subdivided into parts that cover the oceans, the atmosphere, and the land surface.

Unit 1 Major Elements of Earth Systems

The introduction defines science and Earth system science, explains how this field came into existence, and provides basic information that students without previous background in earth science will need in order to understand Earth systems. Unit one contains basic information about Earth, the materials that compose it, the time frame in which the processes cause change on Earth, modern ideas about the origin and place of Earth in the solar system, and the model scientists have developed of Earth's interior. This part encourages students to think of Earth as a planet, its place in time and space, and how processes operating deep inside Earth affect the surface and the human environment. Throughout the text, biogeochemical processes and ecology are emphasized in the discussion of Earth environments.

Unit 2 The Plate Tectonic System

This unit explains the operation of the plate tectonic system and the way movements in the outer parts of Earth cause earthquakes and volcanic activity that have such profound impacts on the surface. The changes in the way Earth scientists think about Earth have been revolutionary. By examining the history of the development of plate tectonic theory, this unit gives students a better understanding of how hypotheses evolve to become widely accepted theories.

Unit 3 Earth's Physical Climate System

This unit includes Earth's surface features and the processes that shape the surface.

Part 1 Oceans and Coasts Chapters 8, 9, and 10 examine the oceans, which cover more than 70% of Earth's surface, the character of the ocean basins, the movement of water in the oceans, and the environments in the oceans. These are important components of systems that interact with the atmosphere and land surface. This part also includes the coastal transition zone between the open oceans and the land surface.

Part 2 The Atmosphere Chapters 11, 12, and 13 begin with a chapter that defines the composition and structure of the atmosphere. This introductory chapter includes discussions of atmospheric pollution, greenhouse gases, global warming, and the ozone hole. It also provides modern coverage of El Niño and its effects on weather.

Part 3 The Land Surface Chapters 14 through 20 examine the terrestrial environment. Terrestrial environments are covered in greater detail than transitional or marine environments because the processes that operate on land environments affect humans more directly. The chapter topics—soil, mass movement, streams, ground water, wind, and ice—resemble those found in conventional Earth Science texts. However, the treatment of these topics emphasizes systems and the relationships of these subjects to the environment. Special attention is given to the hydrologic cycle. Water-supply issues are so widespread and of such great importance that a separate chapter is devoted to this topic.

Unit 4 The Solar System and Its Place in the Universe

Unit four examines the place of Earth in the solar system and as part of the universe. These topics, which are first introduced in chapter 4, are developed more fully in unit four for use in courses that have sufficient time for this coverage.

THE SUPPLEMENTS PACKAGE

Online Learning Center

This website, located at www.mhhe.com/spencer, hosts instructor and student tools. The instructor center is password protected and offers topics for classroom discussions, test questions, an image bank, and PowerPoint lecture outlines. Students will find online quizzing, flashcards, additional readings, study tips, and Internet exercises that will enhance the text material and offer a thorough review of the content.

PowerWeb

Included within the Online Learning Center, PowerWeb provides access to a course-specific website, developed with the help of instructors teaching the course, to provide instructors and students with curriculum-based materials, updated weekly assessments, informative and timely world news, refereed web links, and much more. You'll get daily news updates and have access to 5,900 research sources through the Internet's most thorough search tool, Northern Light. This differs from the Online Learning Center in that it extends the learning experience beyond the core textbook content into other subject areas. PowerWeb is designed to supplement the text content by offering more outside readings, research opportunities, and more.

Digital Content Manager

Free to adopters, this CD-ROM contains illustrations, photos, and lecture outlines that can be imported into PowerPoint, as well as other presentation software, to create your own personalized presentation.

Transparencies

One hundred acetate transparencies of key text illustrations are available to qualified adopters.

Classroom Testing Software

Test questions are also available on McGraw-Hill testing software, for use with Macintosh and IBM PC computers.

ACKNOWLEDGMENTS

It is difficult to emphasize how much this book represents the efforts of many people. I am deeply indebted to my students, colleagues, family, friends, reviewers, and the staff at McGraw-Hill who made the publication of this book possible. In addition to individuals, I am most grateful to the administrators of Washington and Lee University who have provided long-standing support and encouragement of the faculty to pursue their professional interests.

My wife has provided help and supported my work on this book for many years. She and our daughter Shawn prepared many of the illustrations. Our daughter Shannon and her husband Rich Wallace offered suggestions about ways to integrate recent work from fields of environmental studies in the text. Kary Smout, director of the Writing-Across-the-Curriculum program at Washington and Lee University, helped me improve the organization and presentation of ideas throughout the book. Cynthia Abelow, Katherine McAlpine, and Jennifer Strawbridge helped me improve the composition of the manuscript. Kate Metznik, Richard Kilby, Greg Bank, Betty Mitchell, Madelyn Miller, and Linda Davis assisted in editing the manuscript and preparing illustrations. I greatly appreciate the help of the people who assisted me in locating figures, as well as those who granted permission for the use of their illustrations. The reviewers provided constructive criticisms and detailed suggestions for content, organization, depth of treatment, presentation of ideas, and new information. The book has been greatly improved through their efforts. Thomas Arny was especially helpful in reviewing the chapters on astronomy. I appreciate the work of the developmental editor, Lisa Leibold, and the production team, including Rose Koos, Carrie Burger, and Stuart Paterson, who transformed the manuscript into a book.

REVIEWERS

Miguel F. Acevedo
University of North Texas

Ray Arvidson
Washington University

DeWayne Backhus
Emporia State University

Ray Beiersdorfer
Youngstown State University

Stephen K. Boss
University of Arkansas

Lawrence W. Braile
Purdue University

Walter J. Burke
Wheelock College

Wayne F. Canis
University of North Alabama

Lindgren L. Chyi
University of Akron

James Collier
Fort Lewis College

William C. Culver
St. Petersburg Junior College

Paul K. Grogger
University of Colorado, Colorado Springs

Jack C. Hall
University of North Carolina at Wilmington

Darrell Kaufman
Northern Arizona University

Michael J. Kirby
Johnson State College

Robert Lawrence
Oregon State University

Keenan Lee
Colorado School of Mines

Michael B. Leite
Chadron State College

Kathleen J. Lemke

Peter F. LeRose
Mount St. Mary College

Judy Ann Lowman
Chaffey College

Constantine Manos
State University of New York at New Paltz

R. A. Mason
Memorial University of Newfoundland

Barry R. Metz
Delaware County Community College

Archie Moore
Southern Louisiana University

Dr. Hallan C. Noltimier
Ohio State University

Donald J. Perkey
University of Alabama in Huntsville

Michael R. Rampino
New York University

Max Reams
Olivet Nazarene University

Godfrey A. Uzochukwu
North Carolina AT&T State University

Anthony J. Vega
Clarion University

Charles Todd Watson
University of the Ozarks

Edgar Spencer
Lexington, Virginia

Meet the Author

Experiences from childhood kindled my interest in nature and eventually led to my career in earth science. I grew up in a small town in southeastern Arkansas and have clear memories of frightening hours watching violent electrical storms during summer months, visiting neighboring towns destroyed by tornadoes, listening to stories about the famous 1927 flood on the Mississippi River, spending many nights camping out as a Boy Scout, fishing on oxbow lakes, and taking float trips down local streams. On a trip west, I was amazed to see the Rocky Mountains rise out of the Great Plains. When I went to college at Washington and Lee University, I knew I wanted to take a course in geology. During that first exposure to the science of the earth, I discovered the scope, application, and fascination of studying earth science. As a senior, I was a field assistant to Dr. Marcellus Stow who was helping prospectors evaluate uranium properties in Montana. We made the Yellowstone Bighorn Research Association camp at Red Lodge our base of operation. There I was introduced to the variety of research interests of geologists from many universities and to the enthusiasm they had for what they were doing. Later I would get to know the YBRA camp well as I studied the structure of the Beartooth Mountains while a graduate student at Columbia University.

During my years at Columbia, I spent a summer on one of the early transatlantic voyages of the R/V Vema. Collecting cores, taking underwater photographs, and making seismic refraction surveys and continuous depth recordings provided a hands-on introduction to oceanography. It was an exciting time at the Lamont Observatory. Evidence that would lead to the development of plate tectonics was being collected and analyzed by an extraordinary group of scientists.

After graduation, I accepted an invitation to join the Washington and Lee faculty where I have had the good fortune of being associated with colleagues who have been enthusiastic teachers, outstanding scientists, and good friends. It has been a special pleasure to work with faculty and students in the Keck Geology Consortium, with members of the Virginia Division of Mineral Resources, and with a number of Washington and Lee alumni college groups. My earlier research continued in Montana, but it gradually shifted to the Central Appalachian Mountains. I've never tired of fieldwork, trying to resolve challenging structural and stratigraphic problems, and making interpretations of complex data. Sabbatical leaves in New Zealand, Australia, Switzerland, Scandinavia, Great Britain, and Greece have provided wonderful opportunities to study geology in many parts of the world. I have especially enjoyed working with students in the field, and in the 1970s developed an introductory course in geology that is centered around work in the field as much as possible. As with many earth science teachers, working closely with undergraduates, following their careers, and sharing their experiences has been richly rewarding.

While a graduate student, I taught at Hunter College and complained to my apartment mate about the content of the textbooks. A few years later, as an editor at T. Y. Crowell, Philip Winsor urged me to write a book. I found that gathering information about a broad range of subjects and looking for new ways to organize and express that information offered a challenging, interesting, and enjoyable way to keep up with the field. Over the years, I have written a number of books, including ones on earth science, structural geology, and, most recently, geologic map interpretation.

Fieldwork in parts of Montana and Virginia that were later designated as wilderness areas is partially responsible for my interest in conservation and the use of geologic information in land use and regional planning. In 1975, I joined a group of individuals who shared an interest in land use planning. We formed a local conservation council that has worked with other organizations and local governments to promote conservation in our community. The insights gained have helped focus my attention on many of the topics discussed in this book.

Professor Spencer was head of the Washington and Lee geology department for over thirty-five years and was named the Ruth Parmly Professor. He is an honorary member of Phi Beta Kappa and the national leadership fraternity Omicron Delta Kappa. He received a National Science Foundation faculty fellowship and research grants from NSF, the American Chemical Society, and the Mellon Foundation. In 1991, he received an Outstanding Faculty Award from the Virginia Council of Higher Education.

The Proven
Earth Science *Learning System*

Systems Emphasis

The systems approach emphasizes the relationships among the great variety of processes that affect Earth and its environment. Emphasis in the first three units is placed on Earth systems: the connections between geology, oceanography, meteorology, and ecology.

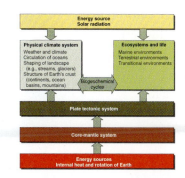

Figure I.5 Earth's Four Major Systems. Incoming solar radiation, gravity, rotation of Earth on its axis, and heat and movements of material deep inside Earth provide the energy to drive Earth's environmental systems.

land surface. The movement of water between the atmosphere, oceans, and land surface, called the hydrologic cycle, is one of the most important concepts of the physical climate system (figure I.7). The **hydrologic cycle** starts with evaporation of water from the surface. The water vapor rises into the atmosphere, condenses, and eventually falls back to Earth as precipitation. Part of this precipitation runs off on the surface as streams or forms glaciers, part infiltrates into the ground, plants use some of this water, and large quantities evaporate and return to the atmosphere. The action of streams, glaciers, and the wind are among the processes in the physical climate system that shape the landscape and modify the climate. This system, along with plate tectonics and biological processes, creates the physical environment at the surface.

Physical processes that operate within Earth systems shape environments at the surface. Study of the relationships between

Figure 14.9 A Taiga Forest. This taiga forest, composed mainly of spruce and fir trees, is located in the mountains north of Yellowstone Park. Similar forests cover vast areas farther north in Canada.

Interaction of ... in Major Earth Systems. ...cesses in each of the ... systems interact, creating ...ts favorable for the exis-...ing organisms. With the ...f energy received from the ...rth is almost a completely ...em. The magnetic field ...rough time. Today, the ...les do not coincide with ...rotation, but these two ...generally been close ...the past, as shown in this

Introduction to Earth System Science **7**

Boreal (High-Latitude) Forests

South of the tundra, winters are also extremely cold, but they are short compared with arctic winters. Summers vary in length. Some are warm; others are cool and wet. When the ice sheets retreated from these areas, they left a low-lying, poorly drained rocky surface with thin soil. Some evergreen trees prosper under these conditions. Fir, spruce, and pine dominate the forests of northern Canada, northern Europe, and parts of Asia, but in northern Siberia, most of the trees are larches. Ecologists refer to these high-latitude forests as **taiga forests** (figure 14.9). Broadleaf deciduous trees, such as oaks and elms, which are so abundant farther south (figure 14.10), are rare in the taiga forests. The wet, mossy, and boggy ground is an ideal habitat for mosquitoes, and the bogs also make an ideal home for moose, mink, and beavers.

Midlatitude Forests

In the midlatitudes, global atmospheric circulation causes the eastern sides of continents to have cold winters and warm summers, and precipitation is evenly distributed throughout the year. These are ideal conditions for a great variety of deciduous trees, such as oaks, hickory, maples, elms, sweet gums, and conifers. Seasonal changes, especially the loss of leaves in the fall and new growth in the spring, are prominent features of the temperate deciduous forests (figure 14.10). Most of these forests lie south of the areas that were glaciated during the last glacial advance that culminated about 18,000 years ago. Undoubtedly, the advancing ice disrupted similar forests that grew farther north during the last interglacial period.

With the changing seasons, numerous migratory birds come and go as they move from summer habitats in Canada to winter feeding grounds in temperate or tropical climates. These birds as well as other animals help disperse seeds of trees and shrubs that

produce their fruit in the autumn and early winter. Large numbers of animal species, including bears, foxes, deer, rabbits, squirrels, skunks, and groundhogs, also populate these forests.

Savannas

The grass-covered plains that define savannas cover large areas in both tropical and temperate regions. Small areas of trees that prosper along rivers and in places where ground water rises dot the

Figure 14.10 A Deciduous Forest. This deciduous forest is in the Blue Ridge Mountains, part of the Central Appalachian Mountains of Virginia and contains a great variety of deciduous trees, including oak, poplar, and maples, as well as some conifers.

Major Ecosystems

Close connections exist between the study of the physical and biological aspects of Earth's environments. These connections are emphasized in chapters 8, 10, and 14.

Chapter Guide

Each chapter begins with a brief introduction that relates the chapter to other parts of the text. The **Chapter Guide** summarizes the main objectives of the chapter and indicates how the subject matter of the chapter relates to Earth systems.

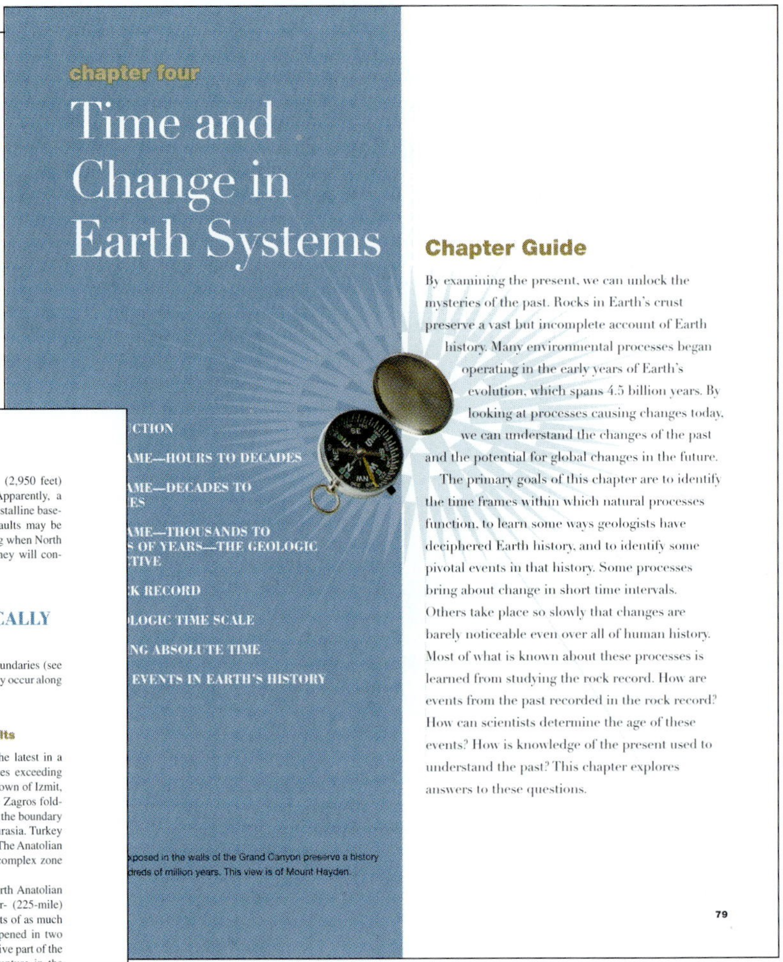

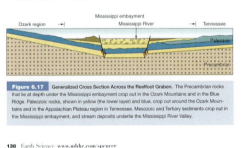

igneous and metamorphic rocks buried 900 meters (2,950 feet) below sediments of the Atlantic Coastal Plain. Apparently, a number of steeply inclined faults have broken the crystalline basement layer into big blocks that are shifting. The faults may be ancient breaks formed as a result of continental rifting when North America and Africa began to separate. How long they will continue to be active remains unknown.

EARTHQUAKES IN SEISMICALLY ACTIVE AREAS

Most seismic activity is concentrated along plate boundaries (see figure I.10). The most dangerous earthquakes generally occur along transform faults and in subduction zones.

Seismic Activity on Great Strike-Slip Faults

Turkey, 1999 At 3:02 A.M. on August 17, 1999, the latest in a series of eleven devastating earthquakes (magnitudes exceeding 6.7) along the North Anatolian fault struck near the town of Izmit, Turkey. This major strike-slip fault extends from the Zagros fold-and-thrust belt in Iran into the Aegean Sea and forms the boundary between the Anatolian Plate (most of Turkey) and Eurasia. Turkey is moving westward relative to Eurasia (figure 6.20). The Anatolian Plate is a relatively small plate located in a highly complex zone between the Africian and Eurasian plates.

The latest sequence of earthquakes along the North Anatolian fault began in 1939. At that time, a 360-kilometer- (225-mile) long segment of the fault ruptured with displacements of as much as 7 meters. The August 17, 1999 earthquake happened in two phases. The first, located at the western end of the active part of the fault that caused a 90-kilometer- (56-mile) long rupture in the ground, lasted 12 seconds and had a magnitude of 7.4. After an 18-second pause, a second major earthquake ruptured a zone 30 kilometers (18 miles) long farther east. This caused an offset of 5 meters (16 feet) across a small country road.

Figure 6.16 Mississippi Valley Earthquakes. The New Madrid earthquake of 1811–1812 (open circles) occurred in a section of the crust, called the Reelfoot rift, that has dropped down as a result of crustal extension. Some geologists think this rift is located over a zone where deep movements are stretching the continental crust of North America. Faults lie buried beneath sediments of the Mississippi embayment of the coastal plain. Epicenters of some recent earthquakes are shown by crosses.

Seismic activity in the region continues today at a level higher than it was prior to 1886. A number of recent earthquakes lie along a line that trends northwest from Charleston, South Carolina, into eastern Tennessee, as shown in figure 6.19. It seems probable that earthquakes originated as a result of movements on faults in

Figure 6.17 Generalized Cross Section Across the Reelfoot Graben. The Precambrian rocks that lie at depth under the Mississippi embayment crop out in the Ozark Mountains and in the Blue Ridge. Paleozoic rocks, shown in yellow (the lower layer) and blue, crop out around the Ozark Mountains and in the Appalachian Plateau region in Tennessee. Mesozoic and Tertiary sediments crop out in the Mississippi embayment, and stream deposits underlie the Mississippi River Valley.

138 Earth Science www.mhhe.com/spencer

---

chapter four

Time and Change in Earth Systems

Chapter Guide

By examining the present, we can unlock the mysteries of the past. Rocks in Earth's crust preserve a vast but incomplete account of Earth history. Many environmental processes began operating in the early years of Earth's evolution, which spans 4.5 billion years. By looking at processes causing changes today, we can understand the changes of the past and the potential for global changes in the future.

The primary goals of this chapter are to identify the time frames within which natural processes function, to learn some ways geologists have deciphered Earth history, and to identify some pivotal events in that history. Some processes bring about change in short time intervals. Others take place so slowly that changes are barely noticeable even over all of human history. Most of what is known about these processes is learned from studying the rock record. How are events from the past recorded in the rock record? How can scientists determine the age of these events? How is knowledge of the present used to understand the past? This chapter explores answers to these questions.

exposed in the walls of the Grand Canyon preserve a history dreds of million years. This view is of Mount Hayden.

79

Case Histories

Case Histories of events or localities are used to illustrate principles and demonstrate the relevance of the subject matter. Students will find the concepts presented with a case-history approach not only interesting but also easy to understand because they can associate them with real situations.

End-of-Chapter Material

Each chapter concludes with **Summary Points,** which provide a quick review of the most important concepts; **Review Questions,** intended to help students evaluate their mastery of basic concepts; **Thinking Critically,** which encourages students to go beyond the text materials in exploring the relevance of the subject matter to environmental issues; and **Key Words** which identify important terms and concepts.

SUMMARY POINTS

1. Weathering refers to the decay and disintegration of rocks where they come in contact with the atmosphere, water, and living organisms. Mechanical weathering processes, such as expansion and contraction caused by solar radiation, crystal growth, unloading, and freezing and thawing, cause rocks to disintegrate. These processes break down the whole rock. Chemical weathering processes, such as oxidation, carbonation, dissolution, and hydrolysis, which act on surfaces, cause rocks to decompose. Because chemical processes act on surfaces, smaller particles are more prone to chemical weathering than larger particles.

2. Rocks and minerals differ greatly in their susceptibility to weathering and erosion.

3. Soil is a product of weathering. Some soils have distinct zonal structure, and others have no zones. In representative zonal soils, the upper A horizon contains organic matter; the B horizon is leached and rich in clay; the C horizon consists of decaying fragments of the bedrock. Among the factors that influence the development of soil are composition of the parent rock, climatic conditions, and slope.

4. Clay plays an especially important role in the value of soil for plant growth. The clay holds and releases nutrients to the roots of plants.

5. Soil degradation is an important environmental problem. Soil erosion, chemical degradation caused by changes in soil chemistry, and physical degradation, such as the compaction of soil by traffic and cattle, may severely damage soils.

6. What are the main causes of soil degradation?
7. How do the soils formed in humid climates differ from those formed in arid climates?
8. What is the role of freezing water in weathering processes?
9. Based on susceptibility to weathering, what minerals would you expect to find in beach sand?

THINKING CRITICALLY

1. How might weathering processes in the continental North America change if Earth's climate experienced a shift toward cooler temperatures? Toward warmer temperatures?
2. Explain why limestone strata commonly form cliffs in arid climates but most cliffs in humid climates are composed of quartz minerals.
3. What steps should individuals take to protect soil on land they own?
4. What steps can a community take to help protect soil?

KEY WORDS

A horizon	hydrolysis
badlands	laterite
bauxite	mechanical weathering
B horizon	O horizon
caliche	oxidation
capillary action	permafrost
carbonation	pH
carbonic acid	regolith
chemical weathering	residual soils
C horizon	sheeting
differential weathering	soil
frost heaving	spheroidal weathering
gossan	subsoil
hard pan	topsoil
hydration	

REVIEW QUESTIONS

1. What common rocks are most susceptible to weathering by dissolution?
2. How does acid rain form?
3. What bedrock conditions help neutralize effects of acid rain in streams?
4. Where is freezing and thawing most effective as a weathering agent?
5. Why does clay play such an important role in soil chemistry?

Chapter Fifteen Weathering and Soil Development **335**

Page Out

Proven. Reliable. Class Tested.

More than 16,000 professors have chosen **PageOut** to create course websites. And for good reason: **PageOut** offers powerful features, yet is incredibly easy to use.

Now you can be the first to use an even better version of **PageOut.** Through class testing and customer feedback, we have made key improvements to the grade book, as well as the quizzing and discussion areas. Best of all, **PageOut** is still free with every McGraw-Hill textbook. And students needn't bother with any special tokens or fees to access your **PageOut** website.

Customize the Site to Coincide with Your Lectures

Complete the **PageOut** templates with your course information and you will have an interactive syllabus online. This feature lets you post content to coincide with your lectures. When students visit your **PageOut** website, your syllabus will direct them to components of McGraw-Hill web content germane to your text or to specific material of your own.

New Features Based on Customer Feedback

- Specific question selection for quizzes
- Ability to copy your course and share it with colleagues or use as a foundation for a new semester
- Enhanced grade book with reporting features
- Ability to use the **PageOut** discussion area or add your own third party discussion tool
- Password-protected courses

Short on Time? Let Us Do the Work

Send your course materials to our McGraw-Hill service team. They will call you for a 30-minute consultation. A team member will then create your **PageOut** website and provide training to get you up and running. Contact your McGraw-Hill representative for details.

PowerWeb–Geology

http://www.dushkin.com/powerweb

Free PowerWeb—Your Professor's Turnkey Solution for Keeping Your Course Current with the Internet!

PowerWeb is a website developed by McGraw-Hill/Dushkin giving instructors and students

- Course-specific materials
- Refereed course-specific web links and articles
- Student study tools—quizzing, review forms, time management tools, web research
- Interactive exercises
- Weekly updates with assessment
- Informative and timely world news
- Access to Northern Light Research Engine (received multiple Editor's Choice awards for superior capabilities from *PC Magazine*)
- Material on how to conduct web research
- Daily news feed of topic specific news

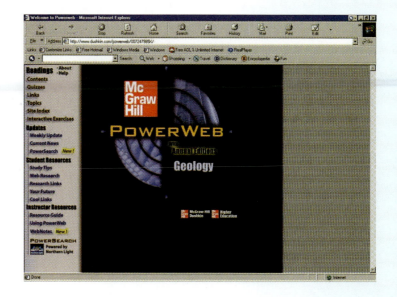

Earth Science

Introduction to Earth System Science

Chapter Guide

In studying the Earth, it is important to be familiar with some of science's characteristics, to use an interdisciplinary approach, and to understand certain Earth systems concepts. This introduction will present these to you.

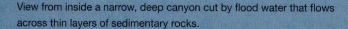

View from inside a narrow, deep canyon cut by flood water that flows across thin layers of sedimentary rocks.

Viewed from the distance of the moon, the astonishing thing about the Earth, catching the breath, is that it is alive. The photographs show the dry, pounded surface of the moon in the foreground, dead as an old bone. Aloft, floating free beneath the moist, gleaming membrane of bright blue sky, is the rising Earth, the only exuberant thing in this part of the cosmos. If you could look long enough, you would see the swirling of the great drifts of white cloud, covering and uncovering the half-hidden masses of land. If you had been looking for a very long, geologic time, you could have seen the continents themselves in motion, drifting apart on their crustal plates, held afloat by the fire beneath. It has the organized, self-contained look of a live creature, full of information, marvelously skilled in handling the sun. (Figure I.1.) (from *The Lives of a Cell* by Lewis Thomas)

Like a spaceship, Earth is essentially self-contained. It has one atmosphere, and its size and natural resources are finite. As the atmosphere, organisms, oceans, and solid Earth interact, changes in one part of the system ultimately affect the entire planet. Early European settlers in North America had a far different view of Earth. To them, it had limitless forests and vast natural resources. When the soil became exhausted in one place, farmers simply moved to another. Streams appeared to have unbounded capacity to remove waste; the oceans were thought to contain an unlimited food supply. Today we recognize that the atmosphere and waters of the ocean belong to no nation; the quality of the environment anywhere depends on the quality of the environment everywhere.

The processes that shape Earth's environment operate as an integrated system (figure I.2). In the past, scientists studied these processes through a number of related, but distinctly separate disciplines, such as geology and biology. Geologists study solid material, Earth's deep interior, and the structure and evolution of

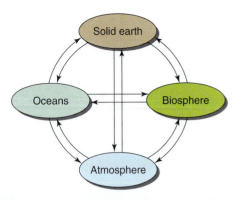

Figure I.2 **The Integrated Earth Systems.** Earth's natural environment is comprised of the solid earth, the oceans, the atmosphere, and the biosphere. These components work synergistically. For example, processes in the solid earth result in emission of volcanic gases that change the composition of the atmosphere. Breakdown of materials in the earth caused by interactions between the atmosphere and rocks provides nutrients that nourish plants and animals of the biosphere. Biological processes involve exchanges between living organisms, the oceans, and the atmosphere. Movements inside the solid earth change the shape of ocean basins. As the position of the ocean basins shift, the climate changes, affecting plants, animals, and oceanic and atmospheric circulation.

Earth's crust. Meteorologists examine the atmosphere; oceanographers study the oceans; and biologists describe and analyze Earth's ecosystems. Gradually, scientists have recognized the importance of transcending these disciplinary boundaries and have begun to integrate knowledge accumulated through these disciplines.

Over the past few decades, three extraordinary developments have convinced the scientific community and the public of the value of interdisciplinary approaches to understanding Earth processes. First, views of the solar system focused attention on Earth as a planet. These views clarified that Earth is unlike any other planet in the solar system and that Earth is the only planet with an environment favorable for the existence of complex life forms, and perhaps for life of any type. John-David Bartoe, a U.S. astronaut, said this on seeing Earth from space:

> As I looked down, I saw a large river meandering slowly along for miles, passing from one country to another without stopping. I also saw huge forests, extending across several boundaries. And I watched the extent of one ocean touch the shores of separate continents. Two words leaped to mind as I looked down on all this, commonality and interdependence. We are one world.

The second development, inspired by Rachel Carson's book *Silent Spring* (1962), occurred when people realized that Earth's environment is fragile and that the web of life on Earth is sensitive to environmental change. Carson pointed out that falcons and pelicans were dying off due to ingestion of DDT, which causes thinning of their eggshells, making them too fragile to survive.

Figure I.1 Photograph of Earth Appearing to Rise over the Surface of the Moon.

Scientific research and environmental activism erupted because of Carson's book.

The third development was a new focus on interdisciplinary approaches to study natural processes. Ecology, geophysics, and geochemistry, which bridge the gap between more conventional fields like geology, physics, biology, and chemistry, have made the most progress in this area.

These three developments have raised the environmental consciousness of people around the world and set the stage for working toward a more integrated approach to studying Earth's environmental systems. To make wise decisions about how to interact with Earth, people must understand the processes that act on Earth and the time frame within which those processes operate.

WHAT IS SCIENCE?

The word **science** comes from the Latin *scientia,* meaning knowledge. It refers to specialized knowledge—the understanding of natural phenomena. It implies a method of study that involves the interaction of observations, experiments, and ideas, called hypotheses, about how things are related or how natural processes operate. This definition emphasizes science as a process of understanding natural phenomena, rather than as fact finding. In his book *The Sciences—Their Origins and Methods* (1967), R. Harre identifies another aspect of science, the knowledge of causes: "Anyone can accumulate facts, scientists accumulate explanations." For example, the average temperature of the atmosphere and the oceans is gradually increasing. Scientists are concerned not only with verifying this observation and what it implies for the future, but also with understanding what is causing this change.

Although the expression *scientific method* is often used, it is difficult to define in any widely acceptable way. The work of most scientists involves solving problems or trying to gain a better understanding of a particular natural process. Although collecting data using established routines is common, many scientists do not give much thought to the approach or method that should guide their steps. Most scientists agree that science involves an interplay of observation, **hypothesis,** and experimentation and that certain qualities characterize good scientific work. These qualities include precise use of words and symbols and care and precision in making observations and conducting experiments. Scientists should examine and interpret the results of their work without bias or undue influence of preconceived notions. Finally, scientists must be willing to face criticism, to consider alternate interpretations, and when necessary, to discard an idea that has been refuted. The history of science contains many examples of prominent scientists who failed to consider alternate interpretations or discard old ideas, but the progress of science depends on these qualities in scientists as a group. The test of good observational and experimental work is that the results can be duplicated by others working independently.

The interaction of observation, experimentation, and hypotheses follows no prescribed pattern. Some of the problems under investigation in fields like astronomy and geology involve such long periods of time and such great distances that experimentation in a laboratory is not possible. Under these conditions, scientists must rely on well-designed observations. In most scientific investigations, it is common for an observation to lead to the formulation of a hypothesis, which, in turn, may suggest an experiment or other observations that may confirm or disprove the hypothesis. Other sequences are not uncommon. The role of the hypothesis is clearly central. A good hypothesis suggests new experiments or observations from which still more facts can be established. Hypotheses may be valuable if they suggest new connections between facts or connections that lead to new experiments. Even if these experiments ultimately show them to be invalid, they serve to move science toward a better understanding of natural phenomena and their causal mechanisms. The test of a good hypothesis is that it explains observations and experiments. The best test of a hypothesis is that it can be used to accurately predict the results of experiments and observations that had not been made at the time the hypothesis was formulated. Many would insist that the best hypothesis is the simplest one that will explain observed phenomena.

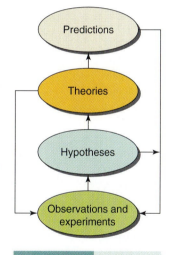

Figure I.3 Interplay of Scientific Study. Scientific studies involve interplay among observations, experiments, hypotheses, and the development of theories. Based on hypotheses and theories, it is possible to make testable predictions. The results of failed tests commonly lead to new hypotheses or suggest the need for more observations and experiments.

Some people use the words *theory* and *hypothesis* as synonyms, but more often the term **theory** represents a more advanced stage of understanding (figure I.3). Theories are conceptual schemes that demonstrate the relationship of several hypotheses to one another or to some common explanation. Thus, the theories of evolution and relativity and the atomic theory of matter are concepts that incorporate, tie together, and provide central explanations for many hypotheses. Successful theories enable scientists to predict the outcome of controlled tests. In geology, the theories that continents move and that continents move as part of large plates are examples of unifying theories. Theories have evolved in each of the special disciplines involved in the study of Earth. It is increasingly clear that ideas coming from the study of fossils, volcanism, the structure of mountains, and other fields are interrelated.

A few natural phenomena occur so invariably and in such a regular manner that scientists refer to them as laws. The law of superposition is an example. This law states that in a sequence of sedimentary rocks that has not been overturned, the layer at the top is younger than layers lower in the sequence.

EARTH SYSTEM SCIENCE

Earth system science is an interdisciplinary approach to understanding the way Earth functions. It is concerned with the present condition of Earth and the evolution of its environments. A **system** is a set or arrangement of things related or connected to one another, forming a single interdependent unit. For example, the Sun, planets, and satellites that revolve around them are called the solar system. In anatomy, a system is a group of organs acting together to perform one of the main bodily functions. The circulatory system and the digestive system are examples. Each of these systems consists of connected organs that interact with one another through natural processes. As used in science, **process** refers to a method of doing something or the steps or operations involved in an action. In Earth system science, a system consists of the materials and the processes involving those materials that play a role in maintaining the present environment or in bringing about changes on Earth.

The National Research Council defines **Earth system science** as "an integrated approach to the study of Earth systems, requiring interdisciplinary investigations of the geology, physics, chemistry, and biology of the whole Earth, because all parts of Earth are interconnected through geological, biological, geophysical, and geochemical processes . . ." (see figure I.2). The objectives of Earth system science are to understand the processes involved in the global Earth system and the ways parts of the Earth system interact; to sustain a sufficient supply of natural resources; to mitigate geological hazards; and to adjust to global and environmental changes. Earth system science is concerned with the evolution of Earth's environment and its present condition.

A system isolated from processes taking place outside the system so that no material enters or leaves is said to be **closed.** Although Earth is largely isolated from other bodies in the solar system, Earth systems are not completely closed (figure I.4). A cold, solid rock located so far out in space that nothing changes on the rock is a closed system. Earth is not such a rock! If matter enters or leaves the system during the operation of the processes that take place within it, the system is **open.** Although Earth is largely isolated from other celestial bodies, Earth has an open system, both in terms of energy and mass transfer. Energy in the form of solar radiation drives many of the physical and biological processes that operate in the atmosphere, the oceans, and near the surface of the land. Particles blown away from the Sun in huge explosions reach Earth's atmosphere where they interact with matter high in the atmosphere and give rise to the aurora. Clearly, matter and energy reach Earth from outside the solar system, and some matter and energy is lost. Gases such as hydrogen and helium are so light they drift off into space, and Earth radiates energy.

In 1997, the Hale-Bopp comet dramatically demonstrated the movement of extraterrestrial matter through the solar system. Impact craters, like Meteor Crater in Arizona, demonstrate that large masses of rock and metal have reached Earth in the past. Much larger craters tens of kilometers in diameter lie buried under sediment beneath Chesapeake Bay and in Yucatán, Mexico. Impacts that formed these craters almost certainly contributed to

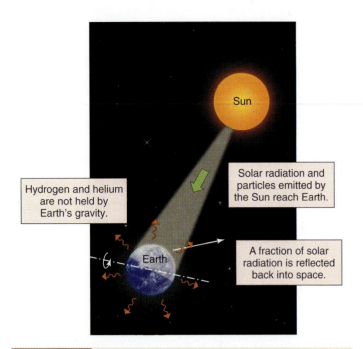

Figure I.4 Open Earth Systems. If Earth were a completely closed system, external sources of energy and matter would not enter the system. Solar radiation is an important source of energy for processes that take place on Earth's surface. Part of this radiation is reflected back into space. Minor amounts of matter reach Earth from the Sun, and hydrogen and helium escape from Earth.

drastic changes in Earth's climate. These climatic changes probably ended the dinosaurs' long reign as the dominant animal group and extinguished or brought many other organisms to the edge of extinction. Such prehistoric events are convincing evidence of the potential for change in Earth systems.

EARTH SYSTEMS

Four major systems operate on and within Earth (figures I.5 and I.6). The **core-mantle system** operates deep inside Earth where a large reservoir of heat drives movements that take place inside and near the surface. Earth's magnetic field originates in the central part of Earth, called the **core,** which is hot liquid metal. Heat from the core softens the overlay mantle, causing the material within the mantle to flow very slowly. The upward movement of mantle material affects the outer layers of Earth, causing them to move. The **plate tectonic system** includes processes involved in the movement of the solid, brittle part of Earth's interior, a layer that extends about 100 kilometers (60 miles) deep. These movements cause new sea floor to form along ridges in the ocean basins, mountain chains to rise, and parts of Earth's surface to become elevated. They are also responsible for most earthquakes and volcanic activity. The **physical climate system** includes all the interactions that occur between the atmosphere, the oceans, and the

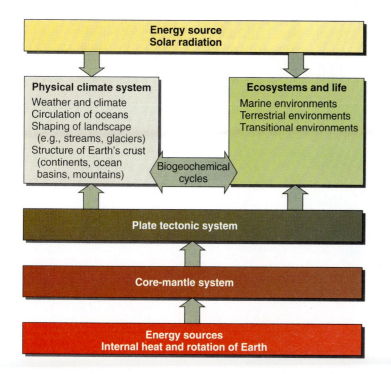

Energy source
Solar radiation

Physical climate system
Weather and climate
Circulation of oceans
Shaping of landscape
   (e.g., streams, glaciers)
Structure of Earth's crust
   (continents, ocean
   basins, mountains)

Ecosystems and life
Marine environments
Terrestrial environments
Transitional environments

Biogeochemical
cycles

Plate tectonic system

Core-mantle system

Energy sources
Internal heat and rotation of Earth

Figure I.5 Earth's Four Major Systems. Incoming solar radiation, gravity, rotation of Earth on its axis, and heat and movements of material deep inside Earth provide the energy to drive Earth's environmental systems.

land surface. The movement of water between the atmosphere, oceans, and land surface, called the hydrologic cycle, is one of the most important concepts of the physical climate system (figure I.7). The **hydrologic cycle** starts with evaporation of water from the surface. The water vapor rises into the atmosphere, condenses, and eventually falls back to Earth as precipitation. Part of this precipitation runs off on the surface as streams or forms glaciers, part infiltrates into the ground, plants use some of this water, and large quantities evaporate and return to the atmosphere. The action of streams, glaciers, and the wind are among the processes in the physical climate system that shape the landscape and modify the climate. This system, along with plate tectonics and biological processes, creates the physical environment at the surface.

Physical processes that operate within Earth systems shape environments at the surface. Study of the relationships between

Figure I.6 Interaction of Processes in Major Earth Systems. Natural processes in each of the major Earth systems interact, creating environments favorable for the existence of living organisms. With the exception of energy received from the Sun, the Earth is almost a completely closed system. The magnetic field changes through time. Today, the magnetic poles do not coincide with the pole of rotation, but these two poles have generally been close together in the past, as shown in this illustration.

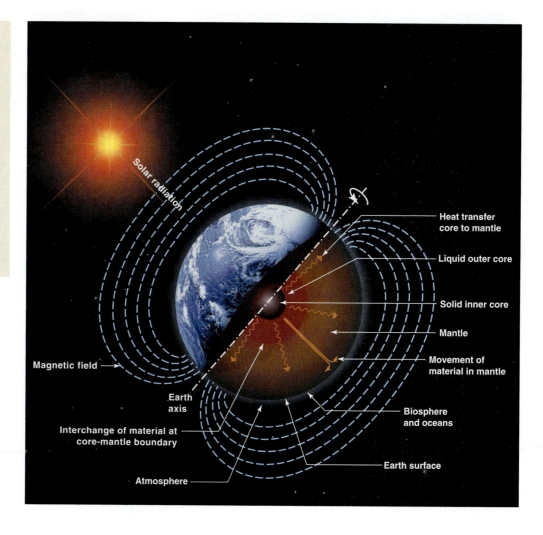

Solar radiation

Heat transfer
core to mantle

Liquid outer core

Solid inner core

Mantle

Movement of
material in mantle

Magnetic field

Earth
axis

Biosphere
and oceans

Interchange of material at
core-mantle boundary

Earth surface

Atmosphere

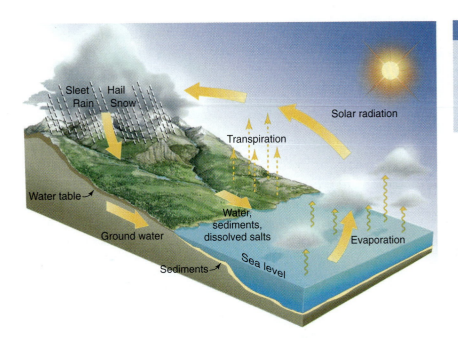

Figure I.7 **The Hydrologic Cycle.** The hydrologic cycle involves the movement of water through the atmosphere, streams, glaciers, and the ground back into the ocean. This cycle is important in the processes that shape Earth and govern the environments of living organisms.

organisms and the environment is the primary interest of ecologists. A community of plants, animals, fungi, and bacteria and their related physical and chemical environment constitute an **ecosystem.** Mountain streams, freshwater lakes, marine swamps, or sand beaches are examples of ecosystems. Earth's surface contains a vast number of ecological niches, each with its own distinctive environment. Many of these terrestrial environments may be conveniently grouped into larger ecological units, called **biomes,** such as tropical rain forests, ice sheets, and hot deserts. Even broader environmental units (figure I.8 and table I.1)—terrestrial, marine, and transitional—form the basis for discussion later in this book.

Materials such as water and carbon dioxide found near Earth's surface move through several Earth systems. As water moves through the hydrologic cycle, it dissolves some materials in the ground and absorbs atmospheric gases. The chemicals in these materials eventually become involved in biological processes. In this way, matter circulates through biological, geological, and atmospheric processes. Carbon dioxide from deep inside Earth enters the atmosphere through volcanic eruptions. Once in the atmosphere, it may be used by plants as they grow. When plants die and begin to decay, carbon dioxide is freed and may become part of the soil. In a sense, carbon dioxide is recycled as it moves through Earth systems. Recycling of matter that involves organisms is a **biogeochemical cycle.**

THE EARTH MODEL

Earth systems operate at different levels on and inside Earth. The classical Earth model devised by geophysicists depicts an Earth composed of a number of concentric shells differing from one another in composition and state (solid, liquid, or gas) (figure I.9). A solid **inner core** composed mainly of an iron alloy lies within a hot, liquid, iron melt that forms the **outer core.** A **mantle** composed mainly of silicate minerals, rich in iron and magnesium, surrounds the outer core. The mantle is solid, but it is so hot that all but the uppermost portion of it flows. One section of the mantle located at a depth of about a hundred kilometers is close to its melting temperature and contains a small amount of molten matter. This partially molten, plastic layer, called the **asthenosphere,** allows the solid brittle layer to move over the underlying mantle. Cooling has produced the solid rock layer known as the **lithosphere.** The lower part of the lithosphere is part of the mantle, but the upper portion, the **crust,** varies from one part of the lithosphere to another. Continental crust has a similar composition to granite; oceanic portions of the crust consist mainly of rocks that have

Table I.1	Earth Environments

Earth has three major groups of ecosystems, each containing numerous smaller systems.

Major Ecosystems	Examples
Marine	Deep sea
	Shallow open ocean
	Continental shelves
Transitional	Marine swamps
	Intertidal zone (beaches, tidal flats, and deltas)
Terrestrial	Hot deserts
	Ice caps and tundra
	Forests
	Freshwater ecosystems (swamps, streams, lakes)

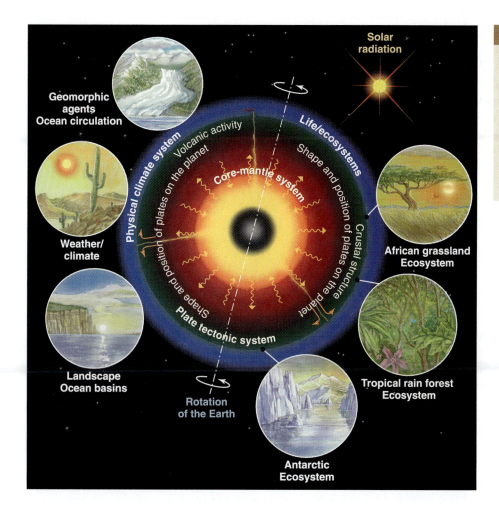

Figure I.8 A Schematic Depicting the Operation of Earth's Systems. The physical climate and ecosystems operate on the surface and are responsible for the great variety of environments. Both the physical climate and ecosystems obtain most of their energy from solar radiation, but Earth's rotation and the plate tectonic system affect them. The plate tectonic system derives energy from Earth's internal heat.

higher iron and magnesium content—a dark rock called basalt—overlain by a thin layer of sediment.

PLATE TECTONICS

The theory of plate tectonics is that several large and relatively thin plate-like units make up the lithosphere and that movements of these plates relative to one another provide a basis for understanding the structure, shape, and history of the lithosphere (figure I.10). Three types of boundaries—divergent, convergent, and fault boundaries—define the plates. Where plates diverge, they spread apart as new oceanic crust forms through volcanic activity. Where plates converge, part of one lithospheric plate may sink beneath another. These zones where one lithospheric plate sinks deep into the mantle are called **subduction zones** (figure I.11). Chains of volcanic islands form over subduction zones in ocean basins, but long mountain systems, such as the Himalayas, form where continents collide along convergent boundaries. Breaks through the lithosphere, known as **transform faults,** form plate boundaries where plates move laterally by one another. The prominent fault zone that runs along the coast of California, the San Andreas fault, is an example of a transform fault.

ENERGY FOR EARTH SYSTEMS

The Sun provides most of the energy required for processes in biogeochemical cycles and in the physical climate system (see figure I.5), which include circulation of the atmosphere and oceans, weather and climate, the movement of water through the environment, and biological processes. Solar energy drives the hydrologic cycle. Heat energy from the core and mantle, combined with the effects of Earth's rotation, provide energy to drive processes in the core-mantle and the plate tectonic systems. Heat from the core, supplemented by the decay of radioactive isotopes in the mantle, keep the mantle so hot that it flows like a hot plastic material. The hot matter rises toward the surface where it breaks and moves the cooler, outer, brittle lithosphere. This drives the plate tectonic system. As the hot, electrically conductive liquid in the core moves, it generates Earth's magnetic field. This field extends far beyond the atmosphere and even affects some of the electromagnetic radiation and particles coming from the Sun. In this way, processes operating deep inside Earth affect the atmosphere and the physical climate system.

The rotation of Earth and the gravitational attraction of material inside and on the surface affect all Earth systems. Everything on and inside Earth rotates as the planet spins on its axis. Because of

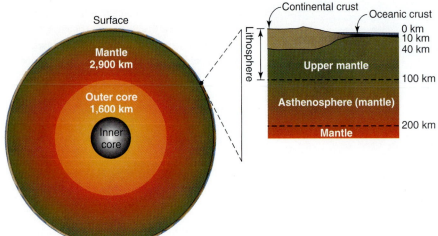

Figure I.9 A Model of Earth's Interior.

this rotation, large masses of air and water flow in curved paths as they move across the surface. Earth's rotation plays an important role in the movement of tides in the oceans and movement in the solid Earth. Rotation is also a factor in the movement of lithospheric plates near the surface and of molten material near the center of Earth.

Gravity causes streams and glaciers to flow downslope, it pulls water into the ground, and it causes water in the oceans to seek a level. Gravity holds the atmosphere to Earth, preventing it from drifting off into space. Gravitational attraction pulled together the matter that formed Earth, and it has affected Earth processes ever since. During the early stages of Earth's evolution, gravity caused molten materials of different density to separate, just as oil floats on top of water. This resulted in the formation of a central iron-rich core of high-density matter surrounded by a less-dense mantle. Even today, when rocks deep in Earth melt, the lower density liquids rise toward the surface, causing volcanic and tectonic activity. If they become more dense than the underlying material, heavy slabs of the outer part of the lithosphere sink back into the mantle.

THE METHODOLOGY OF EARTH SYSTEM SCIENCE

Scientific study of Earth usually begins with observation and description of the character and distribution of the plants and animals that inhabit the surface, the materials that lie on and near the surface, the waters of the ocean, and the atmosphere. Scientists have devoted considerable time to this activity since they first began to investigate natural phenomena. The job of monitoring the environment is more sophisticated today. Investigators now probe the depths of the oceans and the deep interior of Earth. Satellites carry instruments that measure the conditions of the atmosphere and monitor changes that take place on Earth's surface.

Earth system science emphasizes connections between different parts of the planet, changes within systems over time, and the way changes in one part of a system affect other parts (see figures I.2 and I.8). The movement of water on Earth is a good example of such a system. At any given moment, water resides in a number of reservoirs, including the atmosphere, oceans, lakes, streams, ice sheets, underground, and as part of organisms and rocks. Water continually

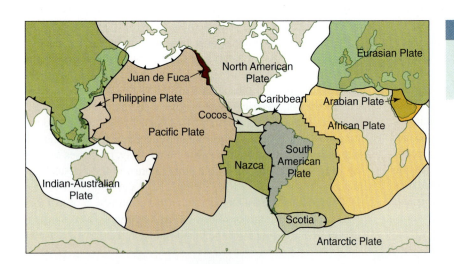

Figure I.10 The Lithospheric Plates. The lithosphere consists of several very large plates which may be subdivided into smaller plates. Many of the smaller plates are not shown.

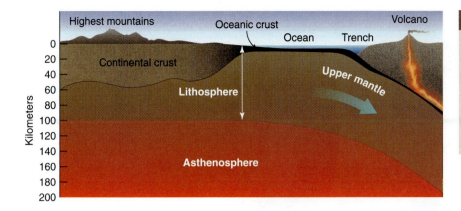

Figure I.11 **Plate Subduction.** This schematic cross section of the outer part of Earth shows a continent and adjacent oceanic crust that are part of the same lithospheric plate. A subduction zone is located where the oceanic crust sinks. Volcanoes form above the subduction zone. The plates move above the asthenosphere, a plastic layer in the mantle that contains a small amount of molten material.

moves among these reservoirs through evaporation, precipitation, stream action, and transpiration.

Analysis of this system may focus on the flow of water from one reservoir to another over time. Analysis of the sequence of reactions in a system is typical of the systems approach. For example, an increase in global temperature causes water in the ocean to expand, resulting in a rise in sea level. Higher temperatures also cause more evaporation from the ocean, increasing cloud cover and precipitation. This leads to more runoff in streams and perhaps even more snowfall. Increased cloud cover reflects incoming solar radiation, which decreases temperature. Thus, a negative feedback in the system may tend to counteract the initial reaction. Examination of the history of similar events in the past is one way to study these interactions. Sediments and rocks that contain a record of past events provide a rich source of much of this information. If no record exists, scientists must rely on their understanding of how and at what rates these chain reactions occur by monitoring the environment. One can observe and measure many of these reactions.

MONITORING THE ENVIRONMENT

Concern that changes in Earth systems might reduce their support of life is a strong motivation for monitoring these systems. Careful monitoring of the environment allows detection of changes and may answer many critically important questions about long-term trends. Is the warming of Earth that started many thousands of years ago accelerating? Are patterns of biomes shifting? Are groups of plants and animals disappearing? Is sea level rising? Are significant changes taking place in air or water quality? Monitoring makes it possible to track changes, avoid or reduce losses resulting from natural catastrophes, and improve understanding of the way Earth systems operate.

Many government agencies are responsible for measuring the quality of air and water, the quantity of water flowing through streams, the health of animal and plant populations, crop production, shifts in the position of the margins of glaciers, and changes in the shoreline. Before the advent of satellites, almost all monitoring of Earth systems took place in restricted locations on the ground or in the oceans.

The most familiar monitoring systems are those that track weather. Weather reports help people plan day-to-day activities and frequently save lives by helping people avoid hurricanes and tornadoes. Extended forecasts help farmers and businesses plan for unexpected conditions that could affect the success of their operations. In making predictions, meteorologists integrate monitoring of ground conditions with data collected at sea, from airplanes, and from satellites.

In recent decades, the technology used to monitor Earth's surface from the air has changed dramatically. Initially, photographers used ordinary cameras to take pictures from airplanes, and these provided the only images available of Earth's surface. Then, cameras designed to take vertical photographs that produced map-like images came into use. Later, color film and film sensitive to infrared light, which detects heat sources, added variety to the types of photographic images available. Recently, major advances have accompanied the development of satellites capable of circling Earth in just hours.

Satellites have dramatically enhanced our ability to monitor the Earth's environment. Many of the most cost-effective monitoring operate from satellites. In a fraction of the time needed to collect data at ground level, satellites can provide information from large areas of Earth's surface, including areas that are remote and inaccessible on the ground. Consequently, satellites provide a practical means of keeping track of many different types of global changes. The movement of icebergs can be tracked, the growth and decay of ice sheets on land can be measured, and seasonal changes in production of plant life in the oceans and on land can be followed (figure I.12). Areas where fires are destroying forests can be identified, and the source of sediments or pollutants entering streams and oceans can be located.

Significant advances in obtaining and interpreting satellite imagery followed the development of digital recording. Today, digital-recording devices provide a wide variety of images of Earth and other planets. The digital-recording devices scan Earth's surface, recording the amount of radiation reaching the scanner in each of a number (commonly seven) of narrow bands in the spectrum, including visible and infrared radiation. Computer processing of these data produces images that show what was recorded from any one or any combination of these bands. The processor assigns a color to each band of radiation, including bands outside

Figure I.12 Monitoring the Environment. Satellites orbiting Earth send back information about the environment at the surface like that illustrated. From left to right, these show sea-surface topography, sea-surface temperature, concentration of chlorophyll and terrestrial vegetation, mean cloud cover, and stratospheric ozone.

the visible part of the spectrum. These images are called *false-color images* because visible colors are used to illustrate invisible radiation, such as infrared or ultraviolet. By selecting the right combination of bands, the processor may produce an image that looks like a color photograph (see figure I.12). Other combinations of bands produce images that emphasize objects on the ground that are radiating certain wavelengths of radiation. For example, because heat sources produce infrared radiation, images emphasizing this part of the spectrum show differences in water temperature. This makes it possible to trace the movement of warm water in the Gulf Stream as it flows across the Atlantic Ocean. Images obtained from bands sensitive to green light show the distribution of plants (see figure I.12). Green colors on the ground appear as shades of red in these false-color images.

Monitoring also provides valuable information about Earth's interior. A network of seismograph stations throughout the world monitors earthquakes. Data collected through this network allow seismologists to identify areas prone to earthquakes and to determine areas affected and possibly in need of aid following major earthquakes.

Most of what scientists know about Earth systems comes from studying the materials that compose the Earth, atmosphere, oceans, and organisms and the processes that involve those materials. In exploring Earth's environmental systems, the first step is examination of the materials involved in these systems. This is followed by study of the major divisions of Earth, changes that have taken place over time, and the processes that govern the interaction of materials with one another and their impacts on people.

KEY WORDS

asthenosphere	crust	lithosphere	process
biogeochemical cycle	Earth system science	mantle	science
biome	ecosystem	open system	subduction zone
closed system	hydrologic cycle	outer core	system
core	hypothesis	physical climate system	theory
core-mantle system	inner core	plate tectonic system	transform faults

Major Elements of Earth Systems

chapter one

The Building Blocks of Earth Materials

The elements that compose the materials on Earth take many forms. These folded rocks contain many of the same elements that occur in plants and animals.

Chapter Guide

What is the nature of matter? This question has intrigued philosophers and scientists for at least 3,000 years. Matter, of which Earth is made, continually changes form and composition. For example, living matter grows and decays, and snow melts. Studying organic and inorganic matter and the processes that cause them to change teaches us about Earth systems. Organic and inorganic matter are extremely varied, and organic matter in particular is highly complex. Earth system science is concerned with tracing the movement of matter and energy through the natural processes that act on Earth. Almost all energy used by Earth processes comes from the Sun or the internal heat and rotational movements of Earth. To understand how matter and energy behave and interact with each other, one must first have a basic understanding of the nature of matter.

INTRODUCTION

Knowledge of materials, how they behave, and how they interact with one another is essential to understanding Earth systems. Plants, animals, minerals, rocks, and water have one thing in common. All consist of just a few elemental particles. These particles are connected in a variety of ways to produce the diverse organic and inorganic matter on Earth. Because the same basic materials exist in organic and inorganic matter (figure 1.1), it is not surprising that natural processes cause materials to change or that these materials move through the environment.

THE NATURE OF MATTER

Scientists define **matter** as anything that occupies space and has mass. Some matter consists of pure substances that have identical composition and exhibit identical properties when they are under the same natural conditions. Although almost all materials, such as rocks, water, and atmospheric gases, are impure, the component parts of those materials contain elements and compounds. **Elements,** the basic unit of most matter on Earth, remain unchanged when they participate in chemical reactions. A single unit of an element is an **atom.**

Although elements do not break down to form other elements in chemical reaction, **radioactive elements** emit particles and break down spontaneously to form other elements. Some radioactive elements break down naturally; others break down only in nuclear accelerators and during atomic explosions. Geochemists use naturally occurring radioactive elements to determine the age of some rocks (see chapter 4).

Modern Models of Atomic Structure

Scientific models are the products of our imagination. They are constructed to help us visualize what we cannot see. The model itself does not exist in nature. The creation of the model is simply an effort to provide a better description of nature. If the model, along with any mathematical formulation, truly reflects reality, it eventually will be validated experimentally. The greatest value of a model, however, lies in its ability to predict unknown behavior. . . . When a model fails to agree with reality, we must modify it or discard it in favor of a better idea. (From F. M. Miller, 1984, p. 163, *Chemistry,* McGraw-Hill)

Atoms are exceedingly small; enlarged 100 million times, an atom would be about the size of a pea, and even at this magnification, the particles that make up atoms would not be visible to the unaided eye. According to modern models of atomic structure, an atom enlarged enough to see its parts would comprise a number of smaller particles characterized by their mass and electrical charge. These basic components of atoms are **protons, electrons,** and **neutrons.** Protons have one positive charge, written $+1$; electrons have one negative charge, written -1; and neutrons have no charge, written 0. Because atoms are so small, physicists use a special unit of measure, the **atomic mass unit,** to describe their mass. One

Figure 1.1 Carbon Occurs in Many Forms. Plant matter and sediments, including the coal shown here, are two of the largest carbon reservoirs on Earth. Carbon also makes up some rare materials such as diamonds and graphite.

atomic mass unit is equal to 1/12 of the mass of the most common type of carbon, carbon-12. Electrons have insignificant mass. Although neutrons behave as though they are combinations of protons and electrons, they are not composed of protons and electrons. A neutron has a mass approximately equal to the combined mass of one electron and one proton and no net electrical charge.

Elements differ from one another in the number of protons in the nucleus, called the *atomic number* of the element (figure 1.2). Chemists have found 90 naturally occurring elements. Only eight main elements (nitrogen, oxygen, iron, magnesium, silicon, aluminum, sodium, and chlorine) make up most matter on Earth, but the distribution of elements among the various major parts of the Earth is dramatically different (figure 1.3). Based on weight, iron is the most abundant element. It is the primary component of Earth's core and is one of the main constituents in Earth's mantle, which also contains magnesium, silicon, and oxygen. The crust consists mainly of oxygen, silicon, aluminum, iron, and calcium.

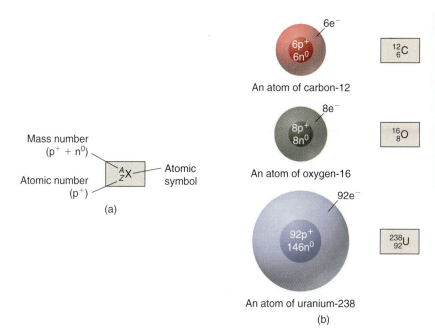

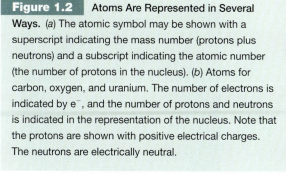

Figure 1.2 **Atoms Are Represented in Several Ways.** (a) The atomic symbol may be shown with a superscript indicating the mass number (protons plus neutrons) and a subscript indicating the atomic number (the number of protons in the nucleus). (b) Atoms for carbon, oxygen, and uranium. The number of electrons is indicated by e^-, and the number of protons and neutrons is indicated in the representation of the nucleus. Note that the protons are shown with positive electrical charges. The neutrons are electrically neutral.

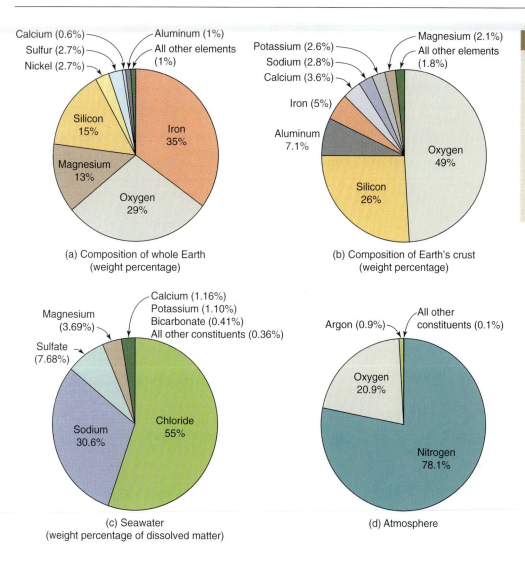

(a) Composition of whole Earth
(weight percentage)

(b) Composition of Earth's crust
(weight percentage)

(c) Seawater
(weight percentage of dissolved matter)

(d) Atmosphere

Figure 1.3 **Composition of Earth Showing the Percentage of Earth's Total Weight Made Up of Each of the Most Abundant Elements.** Compare the composition of (a) Earth as a whole, (b) with that of the crust, (c) seawater, and (d) the atmosphere. All of these are shown as weight percentages. Similar data might be expressed as volume percentages.

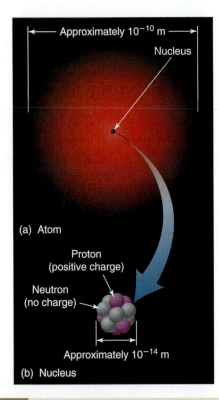

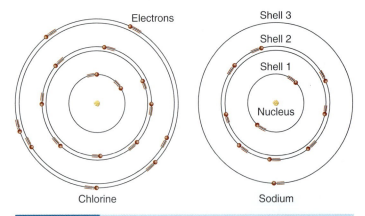

The atmosphere is comprised mainly of nitrogen and oxygen, and the oceans contain salts of sodium, chloride, and a large number of minor elements dissolved in water.

In 1913, the Danish physicist Niels Bohr devised one of the most successful and simple atomic models developed after the discovery of the electron, proton, and neutron. His model depicts the atom as a planetary system in which equal numbers of protons and neutrons lie at the center of the atom and make up its nucleus (figure 1.4 and see figure 1.2). Electrons occupy orbits at specific distances from the nucleus. Some scientists refer to Bohr's model as the *hard-ball model*. In this model, a limited number of electrons occupy each orbit, which are referred to as shells. The first shell can contain no more than two electrons. The second shell has two subshells. The inner subshell may contain two electrons; the outer subshell may contain up to six electrons. The third shell may have three subshells, which may contain two, six, and ten electrons (figure 1.5). The most

Figure 1.4 Model of an Atom. In this representation of an atom, the electrons appear as a fuzzy cloud. Protons and neutrons are present in the nucleus. If the number of electrons is equal to the number of protons, the atom is electrically neutral. Almost all of the atom's mass is concentrated in the nucleus.

Figure 1.5 Schematic Representation of the Bohr Model of Atoms for Chloride and Sodium. The electrons are arranged with two in the first shell, two in the subshell of the second shell, and six in the second subshell of the second shell. Because sodium has one electron in its third shell, it is easy for this electron to be lost, leaving sodium with a net positive charge. Chlorine has seven electrons in its outer shell and is likely to gain an electron, giving it a net negative charge.

Table 1.1	Earth Materials

Subatomic particles combine to form atoms characterized by the number of neutrons and protons in their nucleus. Elements combine in various ways to form compounds. The compounds comprise two major groups. Organic compounds, most of which contain carbon, hydrogen, and oxygen, occur in plants and animals. Inorganic compounds occur as minerals, which are the primary components of the three major rock types—igneous, metamorphic, and sedimentary rocks.

  I. **Subatomic particles**
   (e.g., neutrons, protons [+], and electrons [−] combine in varying numbers to form atoms)

 II. **Atoms and elements**
   (i.e., electrically neutral atoms; ions; radioactive atoms)
   (elements may combine to become ionic, covalent, or metallic compounds)

III. **Molecules of compounds and elements**
   (molecules are the smallest unit of an element or compound; e.g., a molecule of water)
   (may occur in solid, liquid, or gaseous state and in inorganic or organic compounds)

 IV. **Solid inorganic elements and compounds**
   Minerals
   (solid, inorganic elements or compounds; e.g., ice, salt, copper, gold, quartz)
   Unconsolidated solids (sediment) (e.g., soil, mud, silts, sand, gravel)
   Rocks (solid aggregates of one or more minerals)
    Igneous rocks (crystallize from molten matter)
    Volcanic rocks (surface igneous rocks)
    Plutonic rocks (deep igneous rocks)
    Metamorphic rocks (altered rocks)
    Sedimentary rocks (formed by settling)

  V. **Liquids** (magma, lava, pure water, seawater, solutions in the earth)

 VI. **Gases** (air, volcanic gases, magmatic gases)

VII. **Organic compounds (carbohydrates, fats, proteins)**
   (Compounds found in plants and animals. They are composed primarily of carbon, hydrogen, oxygen, and nitrogen. They also contain small amounts of other elements.)

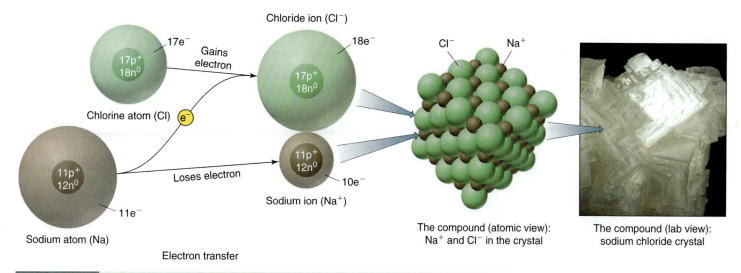

Chloride ion (Cl⁻)

17e⁻   17p⁺ 18n⁰

Gains electron

Chlorine atom (Cl)   e⁻

18e⁻   17p⁺ 18n⁰

Cl⁻   Na⁺

11p⁺ 12n⁰

Loses electron

11p⁺ 12n⁰   10e⁻

Sodium ion (Na⁺)

11e⁻

Sodium atom (Na)

Electron transfer

The compound (atomic view): Na⁺ and Cl⁻ in the crystal

The compound (lab view): sodium chloride crystal

Figure 1.6 **Formation of a Compound Showing Ionic Bonding.** In the first step, an electron from sodium is freed and attaches to chloride. Ions of both atoms are formed. These are attracted to one another and to the compound that is growing. The way the ions fit together creates a cubic crystal lattice. A picture of a crystal of this compound, known as halite or salt, appears at the right.

chemically stable elements are those that have eight electrons in their outer shell. This electron configuration is present in all inert gases (e.g., argon, krypton, and xenon). If the element is electrically neutral, the number of electrons in orbit equals the number of protons in the nucleus. Many elements, other than the inert gases, may gain or lose electrons until they have eight in their outer shell. In the process, these electrically charged atoms, called **ions,** may join with other ions to form compounds (figure 1.6).

Compounds

Two or more elements may combine to form **compounds** (table 1.1). The term **molecule** applies to the smallest unit particle of an element or compound that can have a stable independent existence. For example, a single combination of two atoms of the element hydrogen and one atom of the element oxygen make one molecule of water, H_2O. Most compounds are combinations of metal and nonmetal elements. Some of these compounds result from the attraction between metal atoms such as sodium that have lost one or more electrons, becoming positively charged (Na^+), and nonmetal ions such as chloride. These nonmetal atoms have gained one or more electrons and become negatively charged (Cl^-). Compounds formed in this way are **ionic compounds** (see figure 1.6). Other compounds, **covalent compounds** (figure 1.7), including water and many organic compounds, form through a process in which atoms of metals and nonmetals share electrons in such a way that each atom has eight electrons around its nucleus. In the case of water, two hydrogen atoms share electrons with one oxygen atom, producing H_2O. The third class of compounds, called **native metals,** such as copper (Cu), gold (Au), silver (Ag),

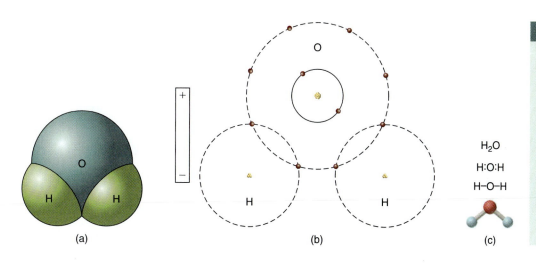

O

+

−

H   H

(a)

O

H   H

(b)

H_2O

H:O:H

H—O—H

(c)

Figure 1.7 Schematic Representations of the Bonding of a Covalent Compound, Water. (a) The two hydrogen atoms are so close to the oxygen atom that they share electrons. (b) Because both hydrogen atoms are located on one side of the molecule, the molecule has more electrons on one side than the other. This gives one side of the molecule a negative charge. The other side is positive. (c) Additional representations of the water molecule.

and semimetals like sulfur (S), consist of atoms of a single element (figure 1.8). Because the atoms are all the same size, they pack together so closely that the electrons are shared among the atoms. With this tight packing, the electrons can move through the metal, making it possible for the metal to carry an electrical current.

STATES OF MATTER ON EARTH

On Earth, matter occurs in three states—solid, liquid, and gas (figure 1.9). Chemists define solids as matter that does not conform to the shape of a container. At the Earth's surface, most solid compounds are brittle. They expand slightly when heated and contract when cooled. Solids also expand or contract in response to changes in the pressure surrounding them, but again these changes are slight. This response to pressure is called **compressibility.** Unlike solids, liquids are not rigid. They flow and assume the shape of the container in which they are placed. Like solids, they do not undergo large changes in volume in response to temperature change, and they are only slightly compressible. Gases are quite different. They lack rigidity, and they expand to fill any container in which they reside. They respond to changes in temperature and pressure and are highly compressible.

Many of the solids that compose the Earth are **crystalline solids.** In such solids, the atoms have an orderly internal arrangement formed during the process of crystallization. This crystalline structure determines the physical properties of the material and the shape a crystal of the solid assumes if it is free to grow (figure 1.10). These properties include density (mass per unit volume), hardness, color, a tendency to break on certain planes, and the

Figure 1.9   Comparison of Some Physical Properties of the Three States of Water. (a) Atoms in ice have a well-defined geometric structure. As ice melts, the crystalline structure breaks down. In the gaseous state, water molecules are independent of one another. Heating ice changes it into liquid form. Additional heat causes the liquid to evaporate and form water vapor. Conversely, cooling water vapor causes liquid to form, and additional cooling produces ice. The amount of heat involved in these transfers is shown. (b) Representation of the molecular structure of ice, water, and steam.

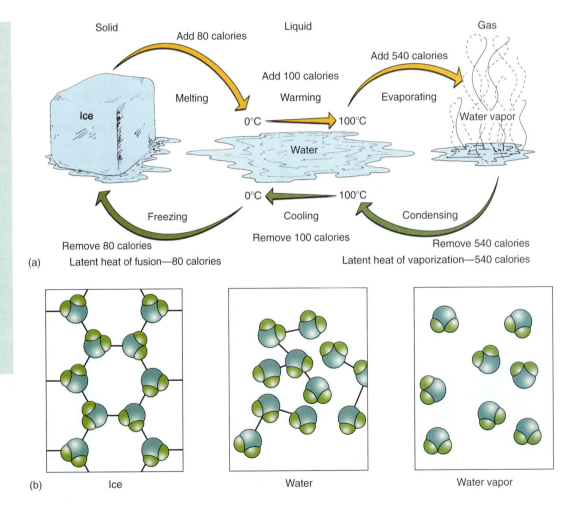

Figure 1.10 Crystalline Solids. Crystals may form where the growth of a crystalline compound is not restricted by the growth of other crystalline solids. The crystals shown here are pyrite (FeS_2).

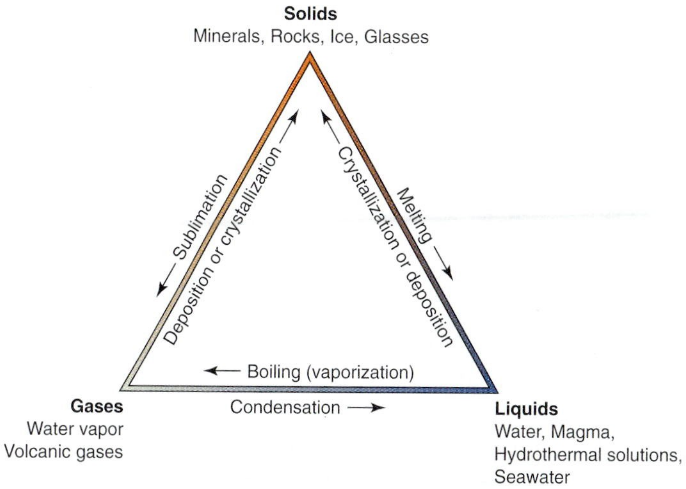

Solids
Minerals, Rocks, Ice, Glasses

Sublimation

Deposition or crystallization

Crystallization

Melting

Crystallization or deposition

← Boiling (vaporization)

Gases
Water vapor
Volcanic gases

Condensation →

Liquids
Water, Magma,
Hydrothermal solutions,
Seawater

Figure 1.11 Diagrammatic Representation of the Changes of State. The end members (gases, liquids, and solids) may undergo changes through the processes indicated on the sides of the triangle.

ability of the solid to conduct electricity and to transmit and bend light rays. Scientists use the physical properties of solid elements and compounds to identify them (see appendix B).

Changes of State

In crystalline solids, the atoms, molecules, and ions are arranged in orderly geometric patterns (see figures 1.6, 1.8, and 1.9). Electrical bonds hold atoms in position within the pattern, referred to as a crystal lattice. As temperature and pressure rise, the structure may yield, and as the temperature reaches the melting point and the solid changes into a liquid, the lattice breaks down and the atoms move freely as a liquid. In noncrystalline solids, such as natural glass, and in liquids and gases, molecules possess no fixed orderly arrangement. The molecules are free to move. The rate of movement depends on temperature and pressure. As temperature increases, molecules move more rapidly and the material flows, but increased pressure constrains the movement. If the temperature rises or the pressure drops, a liquid may change state and become a gas (figure 1.11). In gases, molecules are highly mobile, and they move with little or no restraint. Consequently, most gases consist primarily of empty space. As the temperature of a gas increases, the molecules become more active, and the distance between molecules increases.

Substances change from one state to another primarily as a result of increases or decreases in temperature and pressure. As temperature increases, solids change to liquids through a process called **melting.** Ice melts when the temperature reaches 0°C (32°F) at Earth's surface. Ice is the only solid on Earth with a melting point within the range of temperatures found on Earth's surface. However, as sediment and rock sink into Earth's interior in subduction zones, temperature steadily rises, and at a depth of about 100 kilometers (60 miles), temperature is close to the melting point of many of the common rock-forming minerals in the mantle.

An increase in temperature or a decrease of pressure may cause molecules in liquids to change into gases through a process known as **vaporization.** At Earth's surface where pressure is low, increases in temperature cause liquids to boil. Under certain temperature and pressure conditions, solids may change to gases through the process of **sublimation.** On a cold, but sunny, day in winter, it is possible to see water vapor rise from a snow-covered surface. The process is more obvious when dry ice is taken out of a freezer. Both vaporization and sublimation are reversible. The change of state from gas to liquid is called **condensation.** Most changes from liquid to solid take place as a result of freezing, or **crystallization,** the process in which components of a solid (i.e., atoms, molecules, ions) form a regular, internal, and geometrical pattern (see figures 1.6, 1.8, and 1.9). Gases may also change into solids through crystallization.

EARTH MATERIALS

Although elements and compounds make up all matter, the materials on Earth's surface take a great variety of forms. The atmosphere consists mainly of nitrogen and oxygen with small quantities of many other elements. Water in streams is primarily a compound of hydrogen and oxygen with minor amounts of other elements dissolved from solids on the ground surface. In seawater, sodium and chloride ions are the major constituents dissolved in the water, but small amounts of many other elements are present. Carbon, oxygen, and hydrogen are the most abundant elements in organisms, but trace quantities of many other elements are essential for plants and animals to develop. Organisms extract the elements they require from the water, the atmosphere, and the solids in the ground.

Soil, a mixture of organic matter and the products of decay of the underlying rock, covers most of the land surface (figure 1.12).

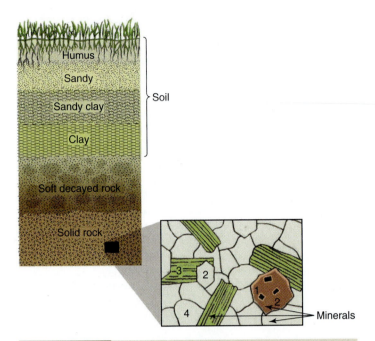

Soil. Rocks or sediment composed of minerals are present everywhere beneath the ground surface. Over long time periods, these minerals usually break down chemically and combine with organic matter to form soil. The character of the soil depends on the composition of the minerals and organisms that live near the surface.

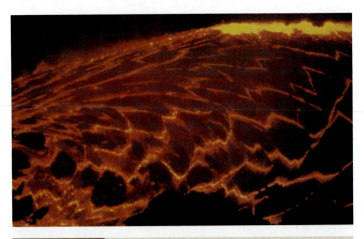

Figure 1.13 A Lava Lake. Molten lava pours out of a lava lake at Kilauea volcano in Hawaii. This lava rose from a magma chamber located deep under the islands, formed a lake, and was draining back into the pipe from which it originally came at the time this photograph was taken.

Although this decay takes place primarily as a result of chemical reactions between the rock and the constituents of the atmosphere, organisms commonly play an important role in these chemical reactions, called **chemical weathering.** The **rock** beneath the soil or exposed where the soil has eroded away consists of an aggregate of naturally occurring inorganic elements or compounds called **minerals** (see figure 1.10). Mineralogists have identified over 2,000 minerals. Each has a definite chemical composition or range of compositions, certain distinct physical properties, and a distinct crystalline structure.

Although all rocks contain minerals, the aggregate of minerals may form in several ways. As temperature of molten matter inside Earth drops, the melt crystallizes and produces **igneous rock.** This crystallization may occur deep inside Earth, or the molten material may rise to the surface (figure 1.13) and form a **volcanic rock.** Other rock types form from fragments of older rocks that are freed by weathering processes at Earth's surface. For example, weathering of rocks that contain quartz produces quartz sand grains that may accumulate on beaches or in sand dunes. Chemical reactions that take place during weathering release ions that may precipitate as solids from solutions or provide the materials organisms use to make shells. Unconsolidated deposits such as sand or shell fragment are referred to as **sediment** (figure 1.14a). If the minerals are compacted, cemented together, or recrystallized, the result is a **sedimentary rock** (figure 1.14b). High temperature and pressure can cause alterations of igneous or sedimentary rock. Rocks formed as a result of these alterations compose the third main class of rocks, called **metamorphic rocks.**

GASES IN EARTH'S ENVIRONMENT

Many of the processes in Earth's environmental systems involve gases that are mobile and move freely through these systems. They are present in the molten matter inside Earth and escape into the rocks around bodies of magma or into the atmosphere through volcanic eruptions. Much of Earth's atmosphere formed in this way. Once in the atmosphere, gases such as oxygen, nitrogen, and carbon dioxide interact with materials and organisms on the surface. These same gases move between the air and upper layers of the oceans where they are involved in marine processes. These processes include respiration by animals, photosynthesis by plants, and decay of organic matter as a result of bacterial action.

Gases in the atmosphere play a variety of roles. Some, such as oxygen, are essential for human existence. Others, such as nitrogen oxides, are harmful to humans, and some, such as ozone and carbon dioxide, have both beneficial and detrimental effects. A layer rich in ozone high in the atmosphere protects organisms by reducing the amount of ultraviolet radiation from the Sun that reaches them. Yet, at ground level, ozone damages living organisms. Similarly, carbon dioxide is essential for plant growth, but high quantities of carbon dioxide in the atmosphere absorb radiation and cause the atmosphere to become warmer—in what is known as the *greenhouse effect* (see chapter 11). Clearly, the processes that govern the production, movement, and reactions of gases are important for understanding Earth's environmental systems and the role people play in changing and protecting them.

LIQUIDS IN EARTH'S ENVIRONMENT

Of the liquids on and inside Earth, **magma** (molten rock) (see figure 1.13) is responsible for the formation of igneous rocks, and it is the source of many of the elements found in the atmosphere, hydrosphere, and biosphere. Magma forms deep inside Earth, where both temperature and pressure are high, and most magma remains at a depth where it slowly cools to form igneous rocks. Hot fluids from magmas invade the surrounding rocks and may even reach the surface. These solutions commonly carry metals and are responsible for the formation of many important mineral deposits. Volcanoes form where magma reaches near the ground surface. Volcanic emissions may consist primarily of gases, but generally some magma flows out on the surface, where it is called **lava** (see chapters 2 and 7). The highly fluid lavas that form the Hawaiian Islands have a temperature in excess of 1,000°C (1,800°F), and may flow great distances across the surface. Although silicon and oxygen are the primary components of magma, these rock melts contain many elements and water.

Water

Water plays a critical role in biological, chemical, and geological processes. The large quantity of water produced by melting rock supports the idea that most of the water on Earth originated through volcanic activity. However, some scientists think that part of the water on Earth may have come from comets that contained large quantities of ice.

Unlike other substances on Earth, water changes state under conditions at the Earth's surface (figure 1.15). Most minerals melt, vaporize, crystallize to form solids, or condense from vapor to form liquids at much higher temperatures and pressures than those normally encountered at the Earth's surface. The behavior of ice under the pressure conditions found near sea level is most familiar. At this pressure, ice begins to melt at 0°C (32°F) to form water, and water vaporizes at 100°C (212°F) to form water vapor. At low pressures, ice passes directly into a vapor state without melting. These changes of state involve consumption or production of energy. Heat energy is needed to cause melting, vaporization, or evaporation. But heat energy is released when water vapor changes into a liquid. These processes are important in the transfer of heat from the tropics to high latitudes (see chapter 12).

Water is both cohesive and adhesive. **Cohesion** is the holding together of like substances, and **adhesion** is the holding of one substance to an unlike one. Water molecules stick together because the bonds between the hydrogen and oxygen hold the water molecules to one another. For this reason, water has surface tension, which means that a water surface behaves as though a thin skin covers it. Surface tension is responsible for the beads of water on any nonabsorbent surface, such as a newly waxed car. Cohesive and adhesive effects cause water to exhibit **capillarity,** the ability of water to move upward against the pull of gravity through small openings. Because of this property, water can move upward

(a)

(b)

Figure 1.14 Sediments and Sedimentary Rocks. (a) Sediments are unconsolidated material generally derived from the breakdown of rocks. The gravel in a stream channel, sand on the beach, sand dunes in deserts, and volcanic ash deposits are examples of sediment. See chapter 2 for additional discussion. (b) Sedimentary rocks generally form in layers called strata.

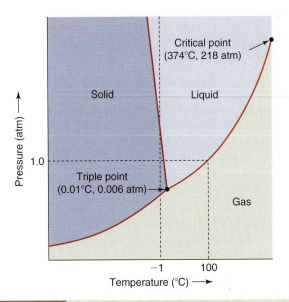

Figure 1.15 Temperature and Pressure Conditions Under Which Water Is Solid, Liquid, or Gas. Note: One atmosphere of pressure is the atmospheric pressure at mean sea level. (1 bar = 100,000 pascals = 14.7 lb/in².) The triple point is the combination of temperature and pressure at which water vapor, water, and ice may coexist.

through a piece of porous paper or creep through the fine pores in plants and in the soil.

At temperatures below 100°C (212°F), water molecules remain in a liquid state. Heat makes the water molecules more active and eventually overcomes the attraction between water molecules. As the bonds between the molecules break, the water changes from a liquid to a gas.

Water is a solvent that is capable of dissolving ionic compounds and polar molecules in which the negative and positive charges are concentrated at ends of the molecular structure. However, it will not dissolve nonpolar substances like those composed of carbon and hydrogen. For this reason, oil, a nonpolar substance, and water will not mix. Oil, which has lower density than water, floats on water. Because of this, the first step in cleaning oil spills is the use of skimming devices designed to remove the layer of oil.

Heat capacity is the quantity of heat, measured in calories, required to increase the temperature of a system by one degree Celsius. Water has high heat capacity relative to most other materials, such as rock or soil at Earth's surface. Between 540 and 600 calories (1,100 and 2,500 joules) are required to evaporate a gram of water. (A **calorie** is the amount of heat needed to raise the temperature of 1 gram [0.002 pounds] of water 1°C. A gram is about the weight of a paper clip.) The high heat capacity of water explains why water is an important energy storage and transfer medium in the atmosphere. Because water has such high heat capacity, the oceans are slower to cool in winter or to warm during summer than adjacent land areas. Therefore, the climate over the ocean is more moderate than that over land.

The high heat capacity of water is also important to organisms because the chemical reactions that take place in organisms produce heat. The organism is protected from the resulting high temperatures by the fact that so much heat is needed to raise the temperature of water. Because so much heat is required to vaporize water, it is a good evaporative coolant and works effectively to reduce high body temperature by evaporating sweat.

Water has a high freezing point, and unlike almost all other substances, it is less dense when it is frozen solid than when it is a liquid. The arrangement of the molecules in ice causes it to be less dense than water. Unlike most substances, the freezing point of water lies within the range of temperatures commonly encountered on Earth. Thus, near Earth's surface, water is present in solid as well as liquid and gaseous states in the atmosphere.

Solutions

Many natural processes involve water. Some of these processes involve the water molecule. Others involve reactions in which hydrogen ions (H^+) and hydroxyl ions (OH^-) form by disassociation of H_2O. Water is a major constituent of organic materials, and it is equally important as the medium for transferring elements from one place to another, both within organisms and inside and on the surface of Earth. For example, minerals in the soil may dissolve in water. The roots of plants pick up the aqueous solution and move it into the plant structure where the plant uses the elements in the construction of new organic molecules (figure 1.16). Water commonly contains and transports dissolved mineral matter formed by the chemical breakdown of minerals. Under favorable conditions, new minerals may form from the solution. Cave formations, petrified wood, and cement in sedimentary rocks originate in this manner (figure 1.17). Aqueous solutions also find their way through streams and ultimately into the ocean where the elements in solution (i.e., ions in water) become part of the salt content of the sea. Water is also the essential energy storage and transport medium of the atmosphere (see chapter 11).

The special nature of the water molecule makes it extremely versatile, allowing it to be the primary component of many solutions and to play a wide variety of roles in Earth processes. Water is an unusual covalent compound in which the distribution of electrons in the molecule is uneven. One end of the molecule acquires a slight positive electric charge and the other, a slight negative charge. Such molecules are called **polar molecules.** Because of the polarity of its molecules, water can dissolve solid ionic compounds by neutralizing the electrical charges that hold atoms together. For example, water molecules surround a small crystal of halite in water. The water molecules become oriented with their positive ends toward negative ions on the outer surface of the salt. Negative ends of water molecules align toward positive ions in the solid (figure 1.18). The attraction between the water molecules and ions in the solid loosens and finally breaks the molecular bonds in the solid, allowing the ions to move into the water. This process, known as **dissolution,** creates what is commonly referred to as a **solution.** The amount of a compound that dissolves in a solution is limited, however, because once water molecules join to an ion,

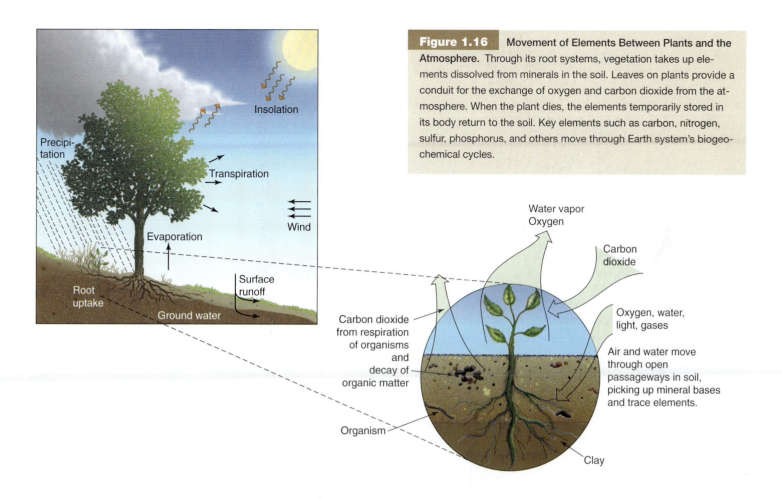

Figure 1.16 **Movement of Elements Between Plants and the Atmosphere.** Through its root systems, vegetation takes up elements dissolved from minerals in the soil. Leaves on plants provide a conduit for the exchange of oxygen and carbon dioxide from the atmosphere. When the plant dies, the elements temporarily stored in its body return to the soil. Key elements such as carbon, nitrogen, sulfur, phosphorus, and others move through Earth system's biogeochemical cycles.

Insolation

Precipitation

Transpiration

Wind

Evaporation

Root uptake

Surface runoff

Ground water

Water vapor Oxygen

Carbon dioxide

Carbon dioxide from respiration of organisms and decay of organic matter

Oxygen, water, light, gases

Air and water move through open passageways in soil, picking up mineral bases and trace elements.

Organism

Clay

(a)

(b)

Figure 1.17 **Rocks Created by the Formation of Solids from Solutions Carrying Elements in Solution.** (a) Petrified wood formed by the deposition of silicon dioxide, quartz, in wood. (b) Deposits composed of calcium carbonate hang down from the roof of a cave.

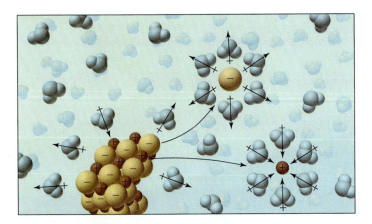

The Dissolution of Sodium Chloride, an Ionic Compound. Water molecules surround the ions in the compound and gradually separate and disperse the ions into the liquid. The negative ends of the water molecules face the positive ions, and the positive ends of the molecules face the negative ions.

they remain attached to that ion in the solution. Thus, the number of potentially active water molecules decreases as more ions from the solid dissolve in the solution.

However, if the temperature of the solution increases, the effectiveness of water in dissolving ionic compounds usually increases because the mobility of the water molecules increases. Consequently, more halite dissolves in hot water than in cold. Eventually, no more of the solid will go into the solution. At this point, the solution is **saturated,** with respect to that particular material. Any additional halite added to the solution does not dissolve. If the temperature decreases, or if some of the water evaporates, ions will regroup once more as the solid, forming what is called a **precipitate.** Many minerals form by this process, known as **precipitation,** which is especially important in the formation of certain minerals in sedimentary rocks. In many arid regions, people obtain salt by evaporating seawater in shallow human-made lakes, called pans. As the water evaporates, the salts eventually precipitate from the solution (figure 1.19).

Scientists commonly represent the combination of ions or the dissolution of compounds by equations such as $[Na^+ + Cl^- \rightleftharpoons NaCl]$. The elements, ions, ionic groups, or compounds shown on one side of the equation combine to form those shown on the other side of the equation. Most of these equations are reversible so the chemical reactions may proceed from left to right, in which case the sodium and chloride ions on the left are combining to form the compound sodium chloride on the right. The arrow pointing from right to left indicates the dissolution of the sodium chloride to form sodium and chloride ions. Which direction the equation proceeds generally depends on temperature, pressure, and on the amounts of the various ions or compounds on each side. In many cases, the reaction proceeds in one direction until an equilibrium condition results. In the preceding equation, an equilibrium condition exists when the combination of sodium and chloride ions

forming salt is equal to the breakdown of salt to form sodium and chloride ions. At that point, the reaction from left to right is balanced by the reaction from right to left. Solids formed in this way may precipitate.

Colloids and Colloidal Suspensions

Although many solutions contain ions, certain mixtures of solids, liquids, or gases, known as colloids, are also common in the environment. A **colloid** is a dispersion of particles of one substance (solid, liquid, or gas) in another substance. Fog, containing water droplets dispersed in air, is a colloid. Colloids may be solids (such as alloys in which one metal is dispersed in another), liquids, or gases. Many colloidal suspensions consist of water-containing particles that are so small they remain suspended. Colloidal suspensions differ from ionic solutions in that the dispersed particles are larger than ions. Consequently, light rays do not generally penetrate through colloids as easily as they do through ionic solutions.

In colloidal suspensions, particles remain in suspension because they have an electrostatic charge on their surface and thus tend to repel one another. If these charges decrease or are neutralized, the colloidal particles come together, often in clumps or aggregates, and sink. Clay deposits in the ocean form in this way. The reaction between suspended clay and salt water may also have adverse environmental effects where seawater mixes with fresh water in rivers near the coast. The deposits of clay may eventually accumulate in the channel, clogging it, and making it necessary to dredge the channel to keep it open.

A Precipitate Example. Salts, primarily sodium chloride, deposited as a result of evaporation in a shallow lake in Death Valley, California.

ORGANIC COMPOUNDS

Every organism lives in an environment of which it is itself a part, along with others of its kind, organisms of other kinds, and many non-living substances such as air, water, and soil. No organism lives alone, and no living thing is sufficient unto itself. . . . Perhaps the most fundamental interdependence of all is illustrated by the fact that all cells require certain sugars that are synthesized . . . only by the green cells of plants. Those green cells themselves are not independent. They require . . . supplies of raw materials from the environment and radiation from the distant sun. (From Simpson et al., 1957, *Life,* Harcourt, Brace)

Sources of the Elements in Organic Compounds

All natural organic compounds—those formed by organisms—contain carbon and form only as products of living organisms (figure 1.20). Despite the small number of elements involved in most organic molecules, a vast number of complex compounds exist in organisms. In addition to carbon, these organic compounds commonly contain water, oxygen, nitrogen, and a number of elements derived from minerals. All of these elements ultimately come from inorganic sources in the environment. Some come from soil; others come from water or the atmosphere. Once incorporated in organic molecules, the molecular structure of the organic molecules may undergo hundreds of modifications as the elements pass from one organism to another in the food chain. When organisms die, the elements that had become parts of organic molecules return to the soil or become sediment that accumulates on the sea floor.

Water Water plays several critical roles in living organisms. Water is the primary source of both hydrogen and oxygen—two elements that, along with carbon, are present in most organic compounds. The water molecules enter into chemical reactions with other substances, producing more complex organic compounds. Water also facilitates other chemical reactions that take place in cells, and it serves as the medium in which nutrients and other substances move through living matter. For example, in trees, water is the main component of

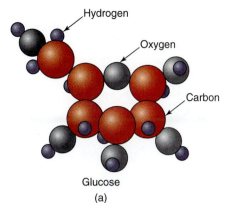

Glucose
(a)

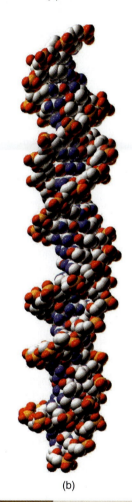

(b)

Figure 1.20 Organic Molecules. Most organic molecules are composed of carbon, hydrogen, and oxygen. These combine in a huge number of variations with minor amounts of other elements to compose organisms. (*a*) Model of the simple sugar glucose and (*b*) the DNA double helix.

sap, and in animals, water is the main component of blood.

Although water is a by-product of chemical reactions that take place inside most animals, especially those adapted for life in deserts, most animals and plants must take in water from their environment. Animals obtain water from food sources and from the sea, streams, or lakes. Plants obtain most of their water from water supplies held in soil.

Carbon Much of the carbon used by organisms comes from the atmosphere, which contains 0.04% carbon dioxide. The atmosphere is the most obvious readily available source of carbon. However, carbon dioxide does not enter most chemical reactions directly. Instead, the carbon in carbon dioxide becomes involved in chemical reactions after first dissolving in water, which it does readily. The reaction between water and carbon dioxide takes place in the atmosphere, producing a weak acid known as carbonic acid.

$$(H_2O + CO_2 \rightarrow H_2CO_3)$$

In the presence of sunlight, green plants take up water from the soil and, through the process of photosynthesis, convert the water and carbon dioxide dissolved in it into simple sugars that become part of the plant matter. For organisms, this process is perhaps the most important of all chemical reactions. Plants store and use the sugar for their energy. Animals feed on the plants, take in the sugars, and use the sugars as raw materials for many complex reactions that lead to the formation of the more elaborate organic molecules.

Oxygen Earth's atmosphere is nearly 21% oxygen, and oxygen is also available in water. Most of the oxygen used by organisms in the manufacture of organic molecules comes from water and water-bearing foods. With few exceptions, plants and animals must have access to oxygen. Only a few bacteria and parasites can live in oxygen-deficient (anaerobic) environments. In green plants, excess oxygen results when carbon dioxide and water combine to form sugar. If oxygen forms faster than the plant can use it, the plant releases oxygen into the surrounding water or atmosphere. This process is responsible for much of the oxygen in Earth's atmosphere. This process is also one source of the oxygen in the upper part of the ocean, where sunlight that penetrates the water makes it possible for algae and cyanobacteria (blue-green algae) to carry on photosynthesis. Without this important source of oxygen, the surface layers of the ocean would be largely devoid of animals.

Mineral Components Organisms also require elements that result from weathering and chemical decomposition of minerals. Most organisms require small amounts of sulfur, potassium, sodium, calcium, magnesium, chlorine, iron, and copper to manufacture more complex organic compounds. In addition, all organisms require some phosphate. Many organisms require even more elements. For example, humans must have small amounts of iodine, manganese, zinc, and fluorine. Trace quantities of still other elements are essential for good health in both plants and animals.

Most of these elements come from the soil or from water in which they are dissolved. As rainwater infiltrates into the soil, it dissolves the elements derived from rock. The water carries the dissolved matter into lakes and finally into the ocean, where the salts slowly accumulate. As a result of these processes, seawater contains most of the elements found on Earth.

Nitrogen Although the atmosphere is nearly 80% nitrogen (N_2), no animals and few plants can use atmospheric nitrogen. Most of the nitrogen organisms use comes from nitrogen-bearing compounds such as ammonia (NH_3) and nitrates (NO_3^-) present in soil. Before most organisms can gain access to nitrogen, some other process must first remove it from the atmosphere. A few types of bacteria and some algae can remove nitrogen from the air, combining it with soil, and lightning bolts fix a minor amount of nitrogen in soil. Some of these bacteria live in the roots of beans and other legumes. Thus, these crops enrich the soil with nitrogen compounds.

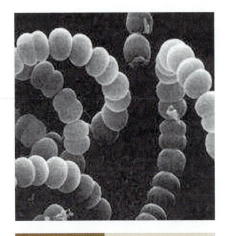

Figure 1.21 A 12,000x Enlargement of a Typical Bacterium, *Listeria monocytogenes.*

Organisms—Their Influence on Other Systems

Not only do living organisms make Earth a unique planet, but plants and animals also influence the composition of seawater, the atmosphere, and the soil. They are also responsible for the formation of some minerals and rocks. Some scientists argue that living organisms regulate the operation of many natural processes, such as the climate, and are responsible for the formation of some minerals and rocks. According to the Gaia hypothesis, first proposed in the 1970s by the British inventor and philosopher James Lovelock, living organisms work together to maintain an environment on Earth at optimum levels. According to this hypothesis, organisms regulate the salinity, temperature, and composition of seawater and the temperature and composition of the atmosphere. In short, living organisms act as one great self-regulating organism.

According to the Gaia hypothesis, which has attracted the attention of many scientists, organisms accomplish this regulatory function in a variety of ways. An obvious influence is the way organisms affect the oxygen and carbon dioxide content of the atmosphere through respiration and photosynthesis. Without life, the atmosphere would contain more carbon dioxide and carbon monoxide and much less nitrogen, oxygen, hydrogen, and methane. Organisms influence their environment in many less obvious ways, and their impact can be significant because organisms exploit optimum environmental conditions by growing more vigorously. If conditions are unsuitable, population size decreases, and under extremely adverse conditions, the organisms die out. Through the interaction of variations in growth rates, population size, and the ability to change the composition of their environment, organisms regulate or at least impact other natural systems, including inorganic materials used in organic compounds.

Major Groups of Organic Compounds

Most organic compounds belong to one of four major groups—carbohydrates; fats; nucleic acids, such as DNA; and proteins (see figure 1.20). Carbohydrates and most fats consist entirely of carbon, hydrogen, and oxygen. Proteins are more complex and contain carbon, hydrogen, oxygen, nitrogen, phosphorus, and small amounts of other elements.

These four major groups of organic compounds combined with water are the basic building blocks of all plants and animals. Plants can manufacture these organic molecules, but animals eat plants to obtain ready-made organic molecules. Once in the animal, the molecules may undergo a great variety of chemical reactions that lead to the production of more complex molecules. Other materials, notably vitamins and hormones (a protein), facilitate these modifications.

The Important Roles of Bacteria

Bacteria are extremely small (10^{-14} liters in volume), single-celled microorganisms, some of which live as individuals, while others live in colonies. With the exception of viruses, bacteria are the smallest of all living things. Bacteria are complex in composition and structure, and they are extremely abundant. Bacteria were among and may have been the first organisms to form on Earth (figure 1.21). Although they are active only in a liquid medium, they occur everywhere and can survive in dry environments for long periods. Some even thrive in oxygen-free environments in which most other life forms die. Unlike algae, bacteria (with the exception of cyanobacteria, the blue-green algae) do not contain chlorophyll and thus do not carry on photosynthesis. Some bacteria are necessary for fermentation; others remove nitrogen from the air; still others cause diseases such as pneumonia and tuberculosis.

Bacteria are involved in many natural processes. One of the most important of these is the breakdown of organic molecules in soft tissues of dead plants and animals. In the absence of bacteria, these tissues break down slowly. As the soft tissues decompose, carbon dioxide is released into the soil, seawater, or the atmosphere. However, some of the carbon remains fixed in animal and plant hard parts, such as shells that contain calcium carbonate ($CaCO_3$). These do not readily decompose. They may accumulate on the sea floor, eventually becoming limestone. In this way, carbon that originates in the atmosphere may eventually accumulate in rock.

TRACING MATTER THROUGH EARTH SYSTEMS— BIOGEOCHEMICAL CYCLES

Although organisms may contain a great variety of elements, a relatively small number are essential, and all of these are involved in movements and transformations as they pass through the environment. These Earth materials are recycled. The path of movement may be as direct as the recycling of water from the ocean to the atmosphere and back, but many cycles are much more complex.

For example, plants take up water and nutrients from the soil and carbon dioxide from the atmosphere. They transform these materials into structural components of plants and return oxygen and water vapor to the atmosphere. Animals eat plants and transform plant matter into animal tissue. At the same time, they exhale carbon dioxide into the atmosphere. Eventually, when plants die, the solid organic compounds of the plant decompose and return nutrients, water, and carbon dioxide to the environment. These processes recycle water, carbon dioxide, and nutrients from the soil. Transfers of matter that involve both biological and geological processes are called **biogeochemical cycles.** These cycles are part of comprehensive cycling that affects the rocks of the crust and mantle and organisms and other materials present on Earth's surface. Water (H_2O), oxygen (O), carbon (C), phosphate (PO_4^{-3}), and nitrogen (N_2) are key materials in Earth systems.

The Oxygen Cycle

Although oxygen is one of the most abundant elements, making up about 45% of the total mass of Earth, most of the oxygen is tied up with silica in minerals in the crust and mantle. The atmosphere is the readily available reservoir of oxygen that supports essential animal life (figure 1.22). Most of this atmospheric oxygen comes from the

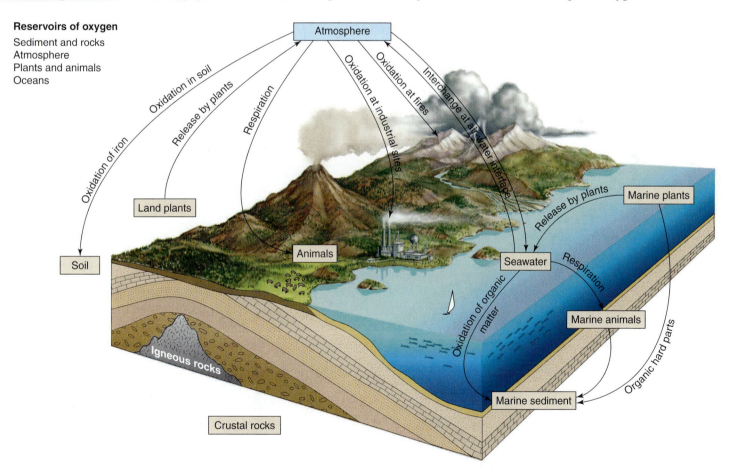

Reservoirs of oxygen
Sediment and rocks
Atmosphere
Plants and animals
Oceans

Figure 1.22 Biogeochemical Cycles for Oxygen. Major reservoirs and processes involved in moving oxygen from one reservoir to another are indicated.

process of **photosynthesis** in which sunlight is taken in by specialized cells, known as *chloroplasts,* in plant leaves. The solar energy is used to drive chemical reactions that can be simplified as follows: carbon dioxide taken from the air, plus water absorbed by plant roots, plus solar energy produces glucose, water, and oxygen.

$$6CO_2 + 12H_2O + \text{solar energy} \rightleftharpoons C_6H_{12}O_6 + 6H_2O + 6O_2$$

Before plants were abundant on Earth, the atmosphere contained very little oxygen, and that was produced when short wavelength (ultraviolet) radiation broke down water vapor.

Oxygen production through photosynthesis, oxygen consumption by animals, and carbon dioxide emission through respiration are perhaps the most familiar of all biogeochemical cycles (figure 1.23 and see figure 1.22). Photosynthesis and respiration are responsible for cycling almost all the oxygen in the atmosphere each year. However, the movement of oxygen through Earth systems is complex. In addition to the reservoirs of oxygen in sediment and rocks and the atmosphere, large quantities are present in plants and animals on land and in the oceans and in the carbon dioxide present in the atmosphere and oceans.

The Carbon Cycle

Carbon is the primary building block of organic matter. It provides a good example of the complex ways in which these key elements move through natural systems. Figure 1.23 depicts the movement of carbon through the earth, atmosphere, oceans, and biosphere. Carbon, in the form of carbon dioxide, enters the atmosphere from volcanoes, from burning hydrocarbons such as oil and gas, from plants, and from respiration of organisms. Carbon dioxide is also released into the environment through decay of dead organic matter and chemical alteration of some rocks. Carbon is removed from the atmosphere by both land plants, such as those in the rain forests, and by the microscopic plants that inhabit the upper layers of the oceans. Carbon dioxide in the atmosphere also combines with water to form a weak acid that falls as precipitation. This rainwater infiltrates the soil where plants pick it up or it runs off eventually into the ocean. Carbon dioxide also gets into the oceans through the exchange of gases between water and air. In the oceans, invertebrate animals such as clams build their shells out of carbon in the form of carbonate. The shells may eventually become part of the sediment on

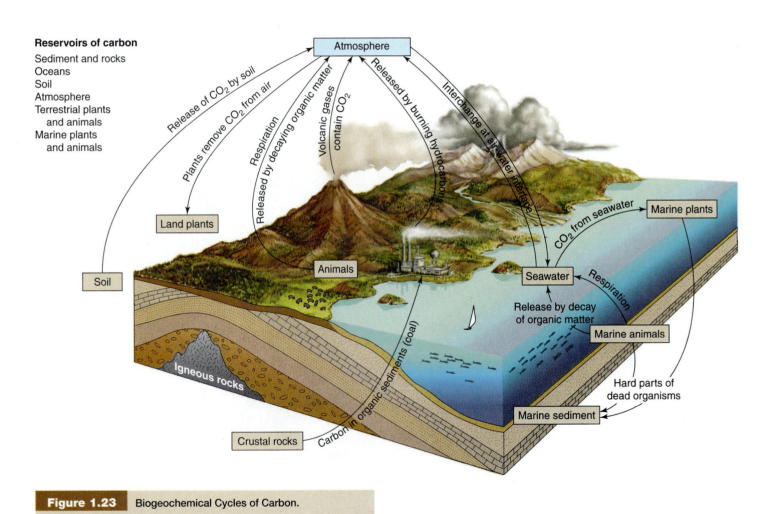

Reservoirs of carbon

Sediment and rocks
Oceans
Soil
Atmosphere
Terrestrial plants
  and animals
Marine plants
  and animals

Figure 1.23 Biogeochemical Cycles of Carbon.

the sea floor where the carbon remains inert as part of the sediment or in rock formed by compaction and solidification of the sediment.

Large quantities of some key elements that move through earth systems are present in natural reservoirs. For carbon, these reservoirs include, in order of total volume, sediments and rocks, the deep oceans, soil, surface layers of the oceans, atmosphere, and biomass (the mass of living organisms) on land and in the oceans. Although the quantity of carbon in sediments and rocks is many millions of times greater than that in all biomasses, the rate at which carbon moves through biological systems is far greater. Scientists have a good understanding of the processes involved in the movement of the key elements from one reservoir to another. Determination of the rate of movement of carbon from one reservoir to another is currently an important topic of research. The impact that increasing carbon dioxide in the atmosphere has on the rate of movement of carbon dioxide into biomass is still under investigation. This knowledge is critical to understanding how much global temperature may rise in the future.

The Phosphorus Cycle

Phosphorus (P), one of the most important elements in biogeochemical processes, originally found its way into Earth systems through the weathering of the mineral apatite, a calcium phosphate, which occurs as a minor constituent of igneous rocks of granitic composition. As granite decomposes, the phosphorus goes into solution and moves through the soil in ground water. Plants take up the phosphorus from soil water, and eventually, animals obtain this critical element as they consume the plant matter (figure 1.24). Phosphorus is essential for plant life, animals, and even microbes. Perhaps most important is phosphorus' role as part of DNA and RNA structures, but it is present in many other organic molecules, and it is an important constituent of bone. Fossil bone beds, formed where vast numbers of animals died in the geologic past, provide one of the most readily available sources of phosphate (PO_4^{-3}).

Replenishing phosphate in the soil is the major use of bone-phosphate and other phosphate deposits. The importance of this use as fertilizer is apparent from studies of crop yields from fertilized versus unfertilized lands. The yield from crops fertilized with phosphate is three to four times greater than it is from

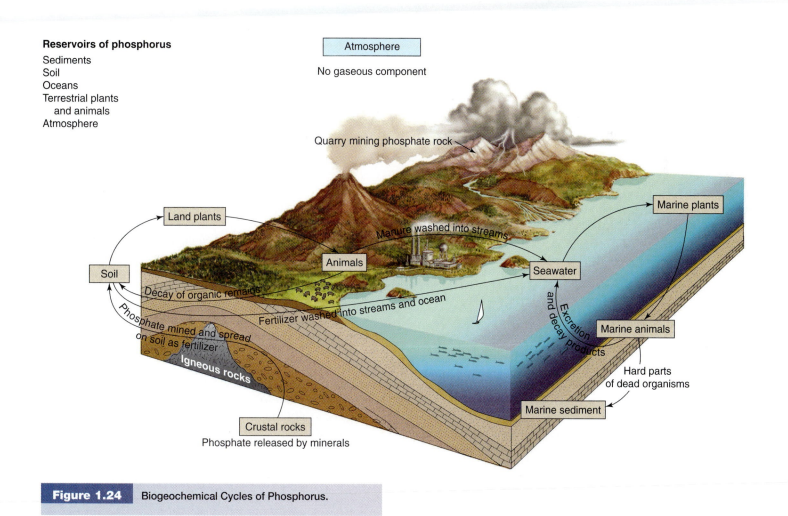

Reservoirs of phosphorus

Sediments
Soil
Oceans
Terrestrial plants
   and animals
Atmosphere

Figure 1.24 Biogeochemical Cycles of Phosphorus.

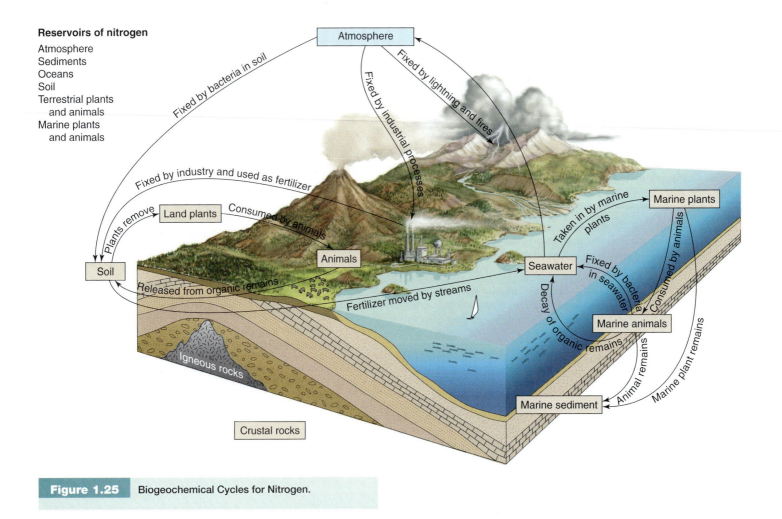

Reservoirs of nitrogen

Atmosphere
Sediments
Oceans
Soil
Terrestrial plants
    and animals
Marine plants
    and animals

Figure 1.25   Biogeochemical Cycles for Nitrogen.

unfertilized crops. Consequently, the demand for phosphate has grown rapidly, raising questions about the long-term supply. A number of agricultural experiment stations are currently seeking ways to conserve the supply of rock phosphate, which is one of a number of important natural resources related to weathering processes.

The Nitrogen Cycle

Although nitrogen (N_2) is the main constituent of the atmosphere, plants cannot obtain the nitrogen they require directly from the atmosphere. Nitrogen must first be altered, a process called *fixation,* to nitrates (NO_3^-) and ammonium (NH_4^+). Atmospheric nitrogen is converted to nitrate by cosmic radiation, fires, and lightning (figure 1.25). These natural processes provide the energy needed to break the strong bonds between nitrogen atoms. Certain bacteria can also break the bonds, particularly bacteria that live in the soil, especially on the roots of peas, alfalfa, beans, and clover and cyanobacteria (blue-green algae) that live in water ecosystems. Despite these natural sources, many plants do not get enough nitrogen. Consequently, industrial processes are used to fix

nitrogen that is used as a fertilizer. this accounts for nearly 30% of the total nitrogen supply available to plants. Animals that eat the plants take in the nitrogen and return it to the environment as urea. The nitrogen is also released when the plants or animals die and decay. Part of the nitrogen in the environment is preserved in minerals that become incorporated in sediments and sedimentary rocks that form a huge reservoir of nitrogen.

Burning hydrocarbons produces nitrogen oxides (NO_x) and sulphur dioxide. These combine with water vapor in the atmosphere and form acids that return to the surface and accumulate in lakes and streams. The increase in acidity reduces crop yields on land and makes lakes acidic. In sunlight, nitrogen oxides and hydrocarbons also react with one another to produce smog. Smog contains some ozone (O_3), which damages plants and can cause human respiratory and nervous problems.

The processes involved in biogeochemical cycles involve the atmosphere, the hydrosphere, and the biosphere. Clearly, humans play an important role in some of these processes that move matter through Earth's environments. These biogeochemical cycles of which humans are involved are part of an even larger-scale cycling seen in the rock cycles discussed in chapter 2.

SUMMARY POINTS

1. The number of protons in the nucleus of atoms defines elements, pure substances that cannot be broken down by chemical means. By sharing electrons or becoming ionized, elements may bond together, forming compounds.
2. Inorganic substances are solid, liquid, or gas, depending on temperature and pressure conditions.
3. Water, the only natural substance found in all three states on Earth, is critically important in the operation of many Earth processes.
4. In addition to its roles in oceans and streams, water is the transfer agent for movement of many elements through the atmosphere, biological processes, and in the crust.
5. Organic compounds are composed primarily of hydrogen, oxygen, and carbon. Nitrogen and phosphorous are minor components. The elements in these compounds originate from inorganic matter.
6. Many elements and compounds of the biosphere (notably silica, phosphate, nitrogen, oxygen, and carbon) move through the lithosphere, and some (notably nitrogen, oxygen, and carbon dioxide) also pass between the atmosphere, biosphere, and oceans, as well as the lithosphere.
7. Matter moves through Earth systems through many processes that may involve changes in state, biological processes, or chemical or biochemical reactions. Melting, crystallization, evaporation, condensation, ingestion, respiration, excretion, sedimentation, and metamorphism are examples of these processes.

REVIEW QUESTIONS

1. By what mechanism does water dissolve minerals?
2. How do compounds with ionic bonding differ from those with covalent or metallic bonding?
3. How does pressure affect the temperature at which most solids melt?
4. In what ways do organic compounds differ from most inorganic compounds?
5. How do colloids differ from ions?
6. What are the main reservoirs of oxygen, nitrogen, phosphates, and carbon in Earth systems?
7. List important properties of water that make it so important in Earth systems.
8. What do organic compounds have in common with one another?

9. What natural conditions cause materials on Earth to change state?

THINKING CRITICALLY

1. Based on what you know about salt, trace the movement of the elements sodium and chloride through Earth systems.
2. What types of compounds dissolve most readily in water? What about this group makes them more likely to dissolve?
3. What are the sources of the elements found in seawater? What natural processes remove elements from seawater?
4. Why is water so important in Earth systems?
5. How would processes now so active on Earth differ if the temperature at the Earth's surface never rose above freezing?
6. What are the characteristics of biogeochemical processes?

KEY WORDS

adhesion	magma
atom	matter
atomic mass unit	melting
biogeochemical cycles	metamorphic rocks
calorie	minerals
capillarity	molecule
chemical weathering	native metals
cohesion	neutrons
colloid	photosynthesis
compounds	polar molecules
compressibility	precipitate
condensation	precipitation
covalent compounds	protons
crystalline solids	radioactive elements
crystallization	rock
dissolution	saturated solution
electrons	sediment
elements	sedimentary rocks
heat capacity	soil
igneous rocks	solution
ionic compounds	sublimation
ions	vaporization
lava	volcanic rock

Minerals and the Rock Cycle

Unusual patterns formed in laminations in a specimen of malachite, a copper carbonate mineral.

Chapter Guide

Rocks appear to be static to the naked eye; however, the process of rock formation is just as dynamic as that of the birth and growth of any living organism. As volcanoes erupt, lava and ash spread across the land surface, old rocks melt beneath the volcano, or new ones form as the melt cools and crystallizes. Country rock around magma undergoes chemical reactions that change its form. At the same time, rivers carry sediment eroded from the surface into the ocean where it sinks to the sea floor and forms new deposits. In other parts of the ocean, the remains of dead organisms settle to the sea floor and form limy oozes. In deserts, salts precipitate in lake beds that dry out as the water evaporates, and sand freed from rocks that contained quartz form dunes that move downwind across the desert floor. The materials created by these daily processes bear the mark of the environment in which they form. The character of the sediment, rocks, and the minerals of which they are composed is the subject of this chapter. Like the elements described in chapter 1, these materials and the minerals that compose the rocks are part of cycles in which they are continually transformed. Geologists refer to these as rock cycles.

INTRODUCTION

Rocks and minerals make up most of the solid materials on Earth. Learning to recognize and identify them is an essential first step toward obtaining an understanding of the physical environment. Discovering how these materials form and how they change as the environment around them varies is equally important. They hold key evidence to many aspects of Earth's history including records of the evolution of life, the formation of mountains, and the movement of continents, as well as past climates and the conditions that caused climatic change. We will examine portions of this history in later chapters. First it is important to examine the character of minerals and the conditions under which various types of rocks form.

MINERALS

Minerals are naturally occurring, solid elements or compounds that have a definite chemical composition, an orderly internal arrangement of atoms, and as a result, have certain definite physical properties (see appendix B). The systematic arrangement of atoms determines the shape that crystals of a particular mineral assume if the mineral is free to grow in an unrestricted space. The crystal shape provides one key to mineral identification (figure 2.1). However, perfectly shaped crystals form only under special circumstances. The perfectly shaped quartz crystals in figure 2.1 formed in open cracks filled with water. Because most minerals grow in spaces between or around other minerals that are growing, their shapes are irregular. The resulting texture commonly resembles an irregular mosaic (figure 2.2 and see figure 1.12).

Chemical composition is a useful way to classify minerals into groups. A few minerals composed of a single element form a class known as *native elements* (figure 2.3); however, most minerals are compounds (table 2.1).

Identifying Minerals

It is possible to identify many of the common minerals by making observations and using simple tests on hand specimens. These include measuring the hardness of the specimen and observing the way it breaks, the color of the specimen and its powder, its luster, and the shape of crystals. A mineral's hardness is measured by scratching it. Hardness is measured on a scale of 1 to 10 and is based on what other minerals will scratch the unknown. A set of minerals is used to make this comparison. Talc, the mineral from which talcum powder is produced, has a hardness of 1, and diamond is assigned a hardness of 10. The entire scale is described in appendix B. To test hardness, it is convenient to use common objects such as a fingernail, which has a hardness of 2.5, a penny (H = 3), a knife blade (H = 5), a piece of glass (H = 6), and a steel file (H = 6.5).

The way a mineral breaks gives information about the arrangement of atoms within the mineral and particularly the strength of bonds between layers of atoms. In a few minerals like quartz (see figure 2.1), the bonds are uniformly strong in all directions. Such minerals commonly break like a piece of glass with a shell-shaped

Figure 2.1   Cluster of Quartz Crystals.

fracture. Other minerals break with smooth surfaces called **cleavages.** The degree of development of cleavages may be perfect or imperfect, and the number and orientation of the cleavages vary among minerals. The single perfect cleavage of mica, the three perfect cleavages at right angles of salt, the mineral halite, and the cleavage of calcite, which produces a rhombohedral-shaped fragment (figure 2.4), are good examples.

Although the color of minerals is one of their more obvious characteristics, color of many is misleading. A small amount of an impurity can cause minerals that are ordinarily colorless to have a wide range of colors. The color of the powder of minerals is much less misleading. Powder, called the *mineral's streak,* is commonly produced by rubbing the mineral across a piece of unglazed porcelain. The way light reflects from a mineral is called its *luster.* Some minerals have a metallic luster; others are nonmetallic. Among minerals with nonmetallic luster, some look like resin, others are silky or look like glass. The shape of crystals is a useful guide to mineral identification, but finding good crystals is rare. Some crystals are cubic; others have the shape of hexagonal prisms or octahedrons. Crystals can be grouped in one of seven forms. Use of these physical properties to identify common minerals is described in appendix B.

Laboratory techniques make it possible to identify less common minerals. These techniques include measurement of density and optical properties. X-ray analysis is used to determine the arrangement of atoms within crystalline structures and to identify the presence of the chemical composition of minerals.

Silicate Minerals

Of the more than 3,000 known minerals, a large proportion are **silicates,** minerals that contain silicon and oxygen. Most of the rocks and minerals near the ground surface contain only a few major elements. Oxygen and silicon combined with aluminum, iron, calcium, sodium, potassium, carbon, sulfur, chlorine, or magnesium

account for about 99% of the weight of materials composing Earth's crust. All silicates contain silicon and oxygen atoms held together by covalent bonds (figure 2.5a). These groups consists of silicon (element #14), which has four electrons in its outer shell, and oxygen (element #8), which has six electrons in its outer shell. Four oxygen atoms cluster around one silicon atom, forming a tetrahedron in which electrons are shared between the oxygen and silicon. The tetrahedron has a net negative charge of minus four $(SiO_4)^{-4}$. Because the silicon-oxygen tetrahedron is not an electrically neutral compound, each tetrahedron must join with other tetrahedra or other ions to become electrically neutral and stable. The combination of positive and negative ions creates this neutral condition in ionic compounds. In covalent compounds, adjacent elements share electrons, so each has the stable configuration (eight electrons in its outer orbit). The negative charges on the silicon-oxygen tetrahedron are neutralized by combinations with other ions or groupings. In most silicates, this neutral condition is achieved in whole or in part by the sharing of one or more oxygen atoms by adjacent tetrahedra.

Mineralogists classify the silicate minerals according to the way the silicon-oxygen tetrahedra are grouped (figure 2.5b). The simplest group includes minerals composed of independent tetrahedral groups not directly linked together, but bound to positive metal ions like magnesium or iron. One such mineral is **olivine**

$(Fe,Mg)_2SiO_4$, one of the most abundant minerals in the mantle. Because their ionic radii are similar, iron and magnesium can substitute for one another in the crystalline structure of olivine. Thus, some olivine is Fe_2SiO_4; some is Mg_2SiO_4, but most olivine contains a mixture of iron and magnesium. Other silicates have double tetrahedra, single or double chains, rings, or sheet structures as shown with common rock-forming mineral groups in figure 2.5b. In these arrangements of the tetrahedra, metal ions commonly fit

Table 2.1	Mineral Groups and Examples of Common Minerals in Each Group
Native elements	Gold, silver, copper, platinum, arsenic, sulfur, diamond, and graphite
Silicates	Quartz, feldspar, kaolinite (a common clay mineral)
Sulfides	Galena, sphalerite, pyrite
Oxides	Hematite, corundum, magnetite
Halides	Halite, fluorite
Carbonates	Calcite, dolomite
Sulfates	Gypsum
Phosphates	Apatite

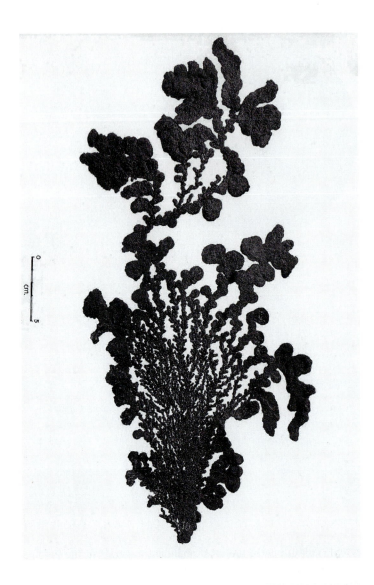

into spaces in the silicate structures. The metals neutralize the net electrical charges on silicon-oxygen tetrahedral groupings and bind the sheets, rings, and chains together. Some groups contain water molecules (H_2O), or hydroxyl groups (OH) with additional oxygen atoms formed by the separation of water into hydroxyl and oxygen atoms. Only one grouping consists of a network structure in which the oxygen atoms of every tetrahedron are shared with adjacent tetrahedra. That is the structure of one of the most common silicate minerals, quartz (SiO_2).

Quartz is especially interesting because it is present in many different rock types and takes many different forms in the environment (figure 2.6 and see figure 2.1). Quartz is a common mineral in many crustal rocks. Where these rocks are exposed at the Earth's surface, they slowly break down as a result of interactions with the atmosphere. Quartz is highly resistant to chemical alteration and it is strong. Thus, quartz is released as decay proceeds, and small pieces of quartz eventually move into streams where they become the sand in sandbars or on beaches. During weathering processes, silica held in other minerals is released into ground-water solutions and eventually into the ocean. While moving through the ground, the silica in the water may replace wood, forming petrified wood, or it may crystallize in cracks or cavities to form quartz crystals like those in figure 2.1. Once the silica reaches the ocean, certain organisms may extract the silica to form their skeletons.

Quartz and olivine are important rock-forming minerals. The feldspar, pyroxene, amphibole, and mica groups are significant for the same reason. **Feldspars** (figure 2.7), the most abundant minerals

(a)

(b)

(c)

Figure 2.4 Minerals with Perfect Cleavage. (a) Mica has one perfect cleavage, (b) halite (salt) has three cleavages at right angles to one another, called cubic cleavage, and (c) calcite has three cleavages that are not at right angles to one another, called rhombohedral cleavage.

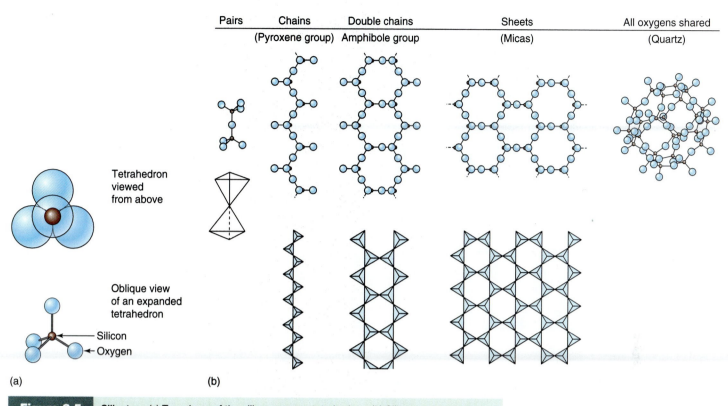

Pairs	Chains	Double chains	Sheets	All oxygens shared
	(Pyroxene group)	Amphibole group	(Micas)	(Quartz)

Tetrahedron
viewed
from above

Oblique view
of an expanded
tetrahedron

— Silicon

← Oxygen

(a)　　　　　(b)

Figure 2.5   Silicates. (a) Two views of the silicon-oxygen tetrahedron. (b) Silicon-oxygen tetrahedra join in a variety of ways. Examples of silicate minerals that have each type of silicon-oxygen linkage are indicated.

in Earth's crust, fall into two subgroups. One of these subgroups contains potassium (K) $KalSi_3O_8$ and is known as potash feldspar. The second subgroup, known as plagioclase feldspars, contains calcium, sodium, or mixtures of these two instead of potassium. Because calcium and sodium have similar ionic radii, they can substitute for one another in the structure. Unlike quartz, feldspar undergoes chemical reactions with constituents of the atmosphere. Clay is an end product of these reactions. For this reason, feldspar is much less common in sedimentary rocks than quartz.

The **pyroxenes** (augite is a member of this group) and **amphiboles** (hornblende is a member of this group) are silicate minerals with silica-oxygen tetrahedra arranged in chains (see figure 2.5). Both of these groups contain magnesium and iron. Some members also contain calcium, sodium, and aluminum.

The sheet arrangement of silica-oxygen tetrahedra characterizes the **mica** group (see figures 2.4 and 2.5b), but different elements bind the sheets together. In the muscovite variety (white mica), potassium and aluminum bind the sheets. In the biotite variety (dark mica), magnesium and iron are also present.

Carbonate Minerals

Like quartz, minerals composed of carbonate ions occur in a great variety of forms. The carbonate ion CO_3^{-2} (figure 2.8), is the essential structural unit of carbonate group minerals. Although it

forms compounds with manganese, iron, barium, strontium, and other positive ions, by far the most abundant carbonates in Earth's crust are those of calcium, $CaCO_3$ (calcite), and a combination of calcium and magnesium, $CaMg(CO_3)_2$ (dolomite).

Calcite   Much of the calcium in the oceans comes from the weathering of calcium-bearing minerals that occur in igneous rocks. Rivers carry the ions produced by weathering into the sea. In locations where evaporation raises the concentration of calcium and carbonate ions to a level at which the water is saturated, these ions combine and form calcium carbonate ($CaCO_3$), the mineral **calcite**. Rocks composed primarily of calcite are called limestone.

Limestones form in both marine and nonmarine waters. Many limestones now exposed on land originally formed in the shallow waters of the continental shelf. Some limestones form by chemical precipitation of calcium carbonate as previously described, but most are composed of the remains of microscopic-sized marine organisms. Many such organisms use calcium carbonate derived from the seawater to construct their shells and other hard parts. Eventually, these hard parts become constituents of sediment. The chalk in the cliffs of Dover, England, consists of billions of minute shells of single-celled plants and animals that lived in a shallow sea over 70 million years ago. As more sediment accumulated on top of the shells, they compacted to form the chalk used in classrooms.

Figure 2.6 **Varieties of Quartz.** Crystals are in the center. Clockwise from upper left corner are petrified wood, calcedony, moss agate, banded jasper, a horn-shaped nodule of chert (flint) covered by chalk, and tiger's eye quartz.

Figure 2.7 Feldspars. (a) Potassium feldspar crystal. (b) Plagioclase, variety labradorite, showing play of colors and striations that are characteristic of plagioclase feldspar.

(a)

(b)

When heated under pressure, limestone becomes a metamorphic rock called *marble*. Because calcium carbonate is soluble in rainwater, rocks containing calcium carbonate may dissolve. The calcium carbonate moves as ions (Ca^{2+} and CO_3^{2-}) in streams or water underground until it reaches the ocean or precipitates as

calcite elsewhere in the crust. Veins of calcite commonly fill cracks in carbonate rocks. Chemical precipitation also results in deposition of calcium carbonate around springs, in cavities, and as coatings on plants bathed in the fine mist of calcium-rich water found near waterfalls. Most cave deposits form from chemical

(a)

(b)

Figure 2.8 Calcium Carbonate. (*a*) Molecular model of a carbonate ion CO_3^{-2}. This ionic grouping is the basic unit in the common rocks limestone and marble, both of which are calcium carbonate ($CaCO_3$). (*b*) The mineral calcite commonly breaks into rhombohedrons, which reflect the internal arrangements of atoms. Calcium carbonate occurs in many ways. Clockwise from upper center are calcite crystals, cross section of a cave formation, a cleavage fragment from a vein of calcite, a cluster of calcite crystals, and crystals that grew on the wall of a cave.

precipitation of calcite from evaporating water. These movements of calcium carbonate from one rock to another are examples of the transformations that are typical of the rock cycle.

THE ROCK CYCLE

Rocks and their component minerals may follow one or more of many paths as they move through the rock cycle (figures 2.9 and 2.10). Some parts of the rock cycle take place inside Earth. Others occur on the surface. Almost all energy for surficial processes comes from the Sun. Solar energy is essential for biogeochemical processes, and it drives the **hydrologic cycle** (see figures I.7, 2.9, and 2.10), which traces the movement of water through Earth systems. The hydrologic cycle involves the evaporation of water from the surface, transfer of water vapor through atmospheric circulation, precipitation of rain and snow, and finally, return of water to the oceans through streams, glaciers, and underground.

Heat from deep in Earth's interior drives processes that operate inside Earth. Movements that are part of the core-mantle system cause plates of lithosphere to move. Along some plate margins, the lithosphere splits and pulls apart, allowing magma to rise, consolidate, and form new sea floor. Sediment accumulates on the

new sea floor as it moves. The plate may or may not include a continent, but in either case, the outer edge of a plate may reach a subduction zone where it sinks. As the plate descends deeper inside Earth, elevated temperature and pressure make the minerals unstable, and they recombine or recrystallize into metamorphic rocks. If subduction proceeds and the temperature continues to rise, the metamorphic rock may melt to form magma. The hot magma expands and rises toward the surface above the subduction zone where it pours out as lava or crystallizes if it remains deep. In some places, continents that are carried along on a subducting plate approach one another. Because all continental crust has lower density than oceanic crust, the continent cannot subside into the mantle. When continents collide, sediments along their margins and the metamorphic and igneous rocks that were in the subduction zone rise and form mountains. In this way, igneous and metamorphic rocks formed thousands of feet inside Earth are subject to erosion.

Once exposed to the atmosphere, the hydrosphere, and the biosphere at the surface, rocks weather—literally, they break down, decay, and decompose. Eventually, plants and organisms grow in the products produced by decomposition and disintegration of rocks. This mixture of organic and inorganic matter is soil. Clay minerals, which form as a result of the chemical decomposition of common rock-forming minerals such as feldspar, are common

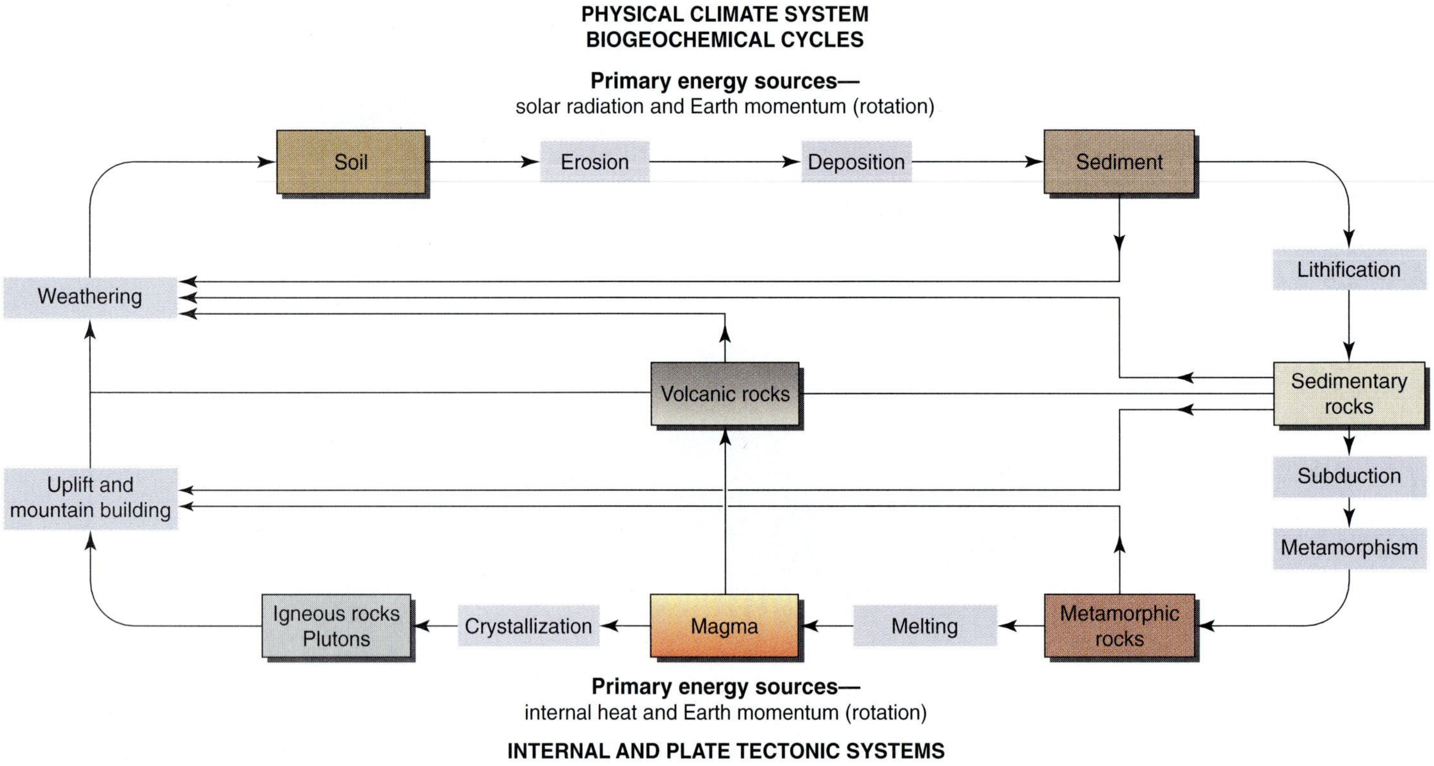

PHYSICAL CLIMATE SYSTEM
BIOGEOCHEMICAL CYCLES

PHYSICAL CLIMATE SYSTEM
BIOGEOCHEMICAL CYCLES

Primary energy sources—
solar radiation and Earth momentum (rotation)

Primary energy sources—
internal heat and Earth momentum (rotation)

INTERNAL AND PLATE TECTONIC SYSTEMS

Figure 2.9 The Rock Cycle. The rock cycle depicts some of the ways rock materials are recycled.

ingredients of soil. Minerals such as quartz, which is highly resistant to chemical decomposition, remain solid after weathering causes less-resistant minerals to decompose. As rock decomposes, the resistant minerals are freed from them. Eventually, rainfall and streams carry the weathering products toward the sea, where they are likely to consolidate and compact as sedimentary rocks along the continents' margins. Progressive changes might stop at this point were it not for the dynamic character of Earth's crust. Plates carrying sediments on the sea floor move into subduction zones, and once again, continents eventually collide, setting the stage for continuation of the cycles.

Although the above processes may proceed at extremely slow rates and some materials may remain unaffected by recycling for hundreds of millions of years, evidence that recycling takes place is all around us. Most of the minerals in modern sediments on beaches originally formed in igneous rocks. All metamorphic rocks represent a second stage in the recycling of matter. These types of cyclic transformations have taken place repeatedly throughout Earth's history. Because of them, few of the rocks originally formed during the earliest part of Earth history remain. However, the recycling is rarely complete, and fragments of the materials formed at various steps in the process remain. They constitute the *rock record*. This record makes it possible to decipher many of the main events of Earth's history, described in chapter 4. For a closer look at the materials involved in the rock cycle, it is appropriate to start with igneous rocks. It is probable that Earth was completely molten early in its history. At that time, all newly formed rocks were of igneous origin.

IGNEOUS ROCKS

Although all **igneous rocks** form from magma, geologists refer to those that solidify at or near Earth's surface as extrusive **volcanic rocks** (figure 2.11), and those that solidify inside Earth as **plutonic rocks,** which are discussed in chapter 7. Volcanic and plutonic rocks differ from one another in texture. Rocks that form by the slow cooling of magma inside the Earth consist of relatively large crystals (figure 2.12). At the surface where cooling takes place quickly, the time needed for minerals to grow into large crystals is not sufficient, and the resulting minerals are small. Individual mineral crystals are commonly so small that one cannot see them without the aid of a microscope, and sometimes the **lava** freezes so fast that crystals cannot grow at all. The atoms do not have sufficient time to develop a crystal form. Under these conditions, a noncrystalline, amorphous substance called **obsidian,** a glass, forms (see figure 2.12*d*).

Most magmas originate in three geological settings: along ridges formed where the sea floor pulls apart; in zones where rocks near the surface sink or are dragged down into the interior; and at isolated places, called **hot spots,** such as Hawaii, where magmas rise from deep in the mantle. Partial melting of the mantle generally produces a magma with a composition of **basalt** (figure 2.13). The

Primary energy sources

Solar radiation   Earth momentum (rotation)

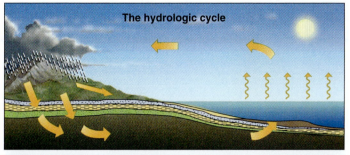

The hydrologic cycle

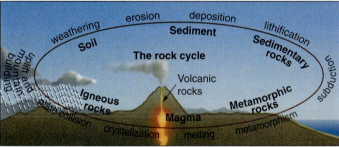

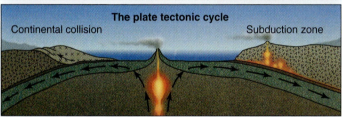

The plate tectonic cycle

Continental collision   Subduction zone

Figure 2.10   Schematic of the Rock Cycle, the Hydrologic Cycle, and Recycling Associated with Plate Tectonics.

sium and *fic* for iron). The composition of rocks in combination with rock texture provides the basis used for rock classification (see figure 2.13) (see appendix C).

Generation of Magma

Both temperature and confining pressure govern the melting of minerals that make up magma. At Earth's surface, where the pressure is one atmosphere (one pascal, or about 14.7 lb/in^2), some minerals melt at temperatures as low as 500° to 600°C (about 900°F); others melt only above 1,500°C (2,700°F). The basaltic lavas that flow out of the volcanoes in Hawaii generally have a temperature in the range of 1,000° to 1,200°C (1,800° to 2,200°F) (see figure 2.11). Deep in the earth, where pressure caused by the weight of the overlying rocks confines the materials, the temperature at which rocks melt is higher than it is at the surface.

In the crust, temperature increases at a rate of about 30°C per kilometer (1.8°F per 100 feet) of depth. (The rate is somewhat higher where cooling igneous rocks or rocks containing radioactive minerals are in the subsurface.) This rate of increase in temperature with depth is the *geothermal gradient.* At this rate, one might expect that temperatures are high enough to melt basalt at a depth of 30 to 40 kilometers (18 to 25 miles), near the base of the continental crust. However, the weight of the overlying rock causes pressure to increase with depth, and increased pressure raises the temperature at which melting occurs. Because temperature and pressure govern the melting point of minerals, magma may be generated by increasing temperature to the melting point or by reducing the pressure on rock that is already close to its melting point.

Most magmas found in the crust belong to one of three main compositional groups (see figure 2.14). Magmas in the oceanic crust have the composition of basalt, and most of the large bodies of igneous rock produced by melting continental crust have granitic composition. Magmas intermediate in composition between granite and basalt produce andesite, the main volcanic rock found over

Figure 2.11   Igneous Rocks Form from Magma. Red hot magma shows through a hole in a lava lake in Hawaii.

mineral olivine, a major constituent of the mantle, commonly crystallizes from these magmas. The active volcanoes located along oceanic ridges also erupt magmas that are products of partial melting of the mantle. The magmas formed in and above subduction zones are distinctly different in composition. These magmas result from partial melting of the sinking lithosphere combined with parts of the upper mantle. Melting of the upper mantle is affected by water derived from the sediment and upper part of the slab. Magmas that originate in this way are much more varied in composition than those that originate deep in the mantle. Because magmas formed in subduction zones originate by the melting of rocks that contain more silica than mantle rock, the magmas may produce a rock called **andesite** (figure 2.14 and see figure 2.13).

All magmas contain some silica; most are basically silicate melts, but they differ in the proportions of silica present. Magmas containing the highest percentage of silica also contain more potassium and aluminum than other rocks. Magmas and rocks that contain lower percentages of silica contain more calcium, magnesium, and iron and are referred to as *mafic* (*ma* for magne-

Chapter Two Minerals and the Rock Cycle **43**

Figure 2.12 Common Igneous Rocks. (a) Coarse-grained granite; (b) rhyolite, showing flow structure; (c) pumice; (d) obsidian; (e) gabbro; (f) basalt; and (g) scoria.

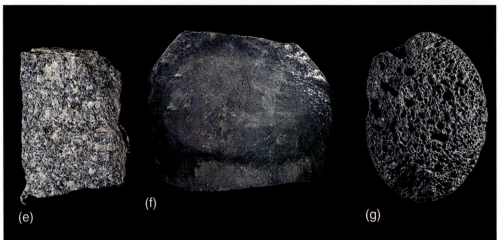

subduction zones. Because these three types are so abundant, the conditions under which they form deserve special consideration. Many of the magmas of basaltic composition, which makes up most of the sea floor, come to the surface from a depth of about 100 to 200 kilometers (60 to 120 miles), where the combination of temperature and pressure is close to that needed to melt the mantle. The magma generated in the mantle results from partial melting of mantle rocks. The magma rises to the surface at hot spots such as Hawaii. Although granite, the main igneous rock found in continental crust, can melt at considerably lower temperatures than basalt and could originate at shallower depths, temperature is not normally high enough for either granitic or basaltic magmas to form in the crust. However, several processes may locally increase temperature or decrease pressure.

High temperature in the outer, solid parts of Earth may increase locally as a result of the decay of radioactive isotopes or by heat rising from greater depths in the mantle. The rise of heat from deep in the mantle and pressure reduction during deformation vary from place to place depending on crustal deformation. Some scientists have suggested that frictional heating may be a factor in subduction zones where the lithosphere sinks beneath another plate. The presence of water in the sinking lithosphere certainly contributes to the melting that occurs there. Local reduction in pressure takes place in the crust as it deforms. Pressure reduction is almost certainly a factor in the generation of magma beneath oceanic ridges where the crust is extended as the sea floor spreads.

Although small quantities of radioactive minerals are present in numerous rocks in Earth's crust, they are far more abundant in granitic rocks located on continents than they are in basalts in oceanic crust. Heat generated during the decay process may build up locally because rock is such a poor conductor. Although concentrations of radioactive minerals may produce sufficient heat to melt

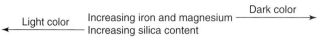

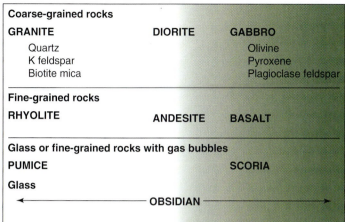

(a)

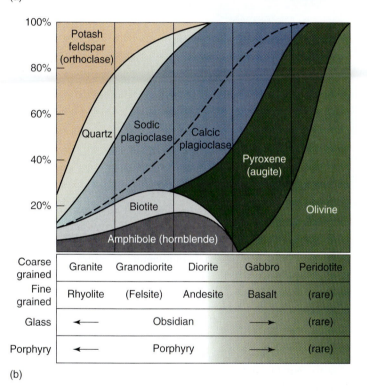

(b)

Figure 2.13 Classification and Mineral Components of Igneous Rocks. (a) Simplified classification of igneous rocks. (b) Mineral components of various common igneous rocks.

rock locally, not enough is known about the amount or the distribution of radioactive minerals in the crust or the mantle to determine how important the process is as a cause of volcanic activity.

Formation of Rocks from Magma

Evidence concerning the change of magma into igneous rock occurs at sites of active volcanism, in deeply eroded mountain ranges

where igneous rocks are exposed at the ground surface, and in the laboratory. In many mountain belts and in the ancient rocks exposed in the interiors of many continents, large bodies of igneous rock formed long ago are present at the surface. Erosion has cut deeply into gigantic **plutons** in the Sierra Nevada Mountains of California. These exposures provide an excellent field laboratory for studying the origin of rocks that form from magmas. In the laboratory, petrologists, who study the origin of rocks, melt samples of rock and make observations as the magma cools and crystallizes under carefully controlled conditions. In Hawaii, geologists study the cooling and crystallization of lava in its natural setting. There, lava may pour into lava lakes, where it cools and solidifies, or it may flow across the surface of the volcanoes directly into the sea.

Crystallization of Lavas in Hawaii

Lava lakes formed in collapsed parts of volcanoes in Hawaii create ideal conditions for studying the cooling and crystallization history of basaltic lava (figure 2.15). Holes are drilled through the solidified crust on lava lakes into the molten magma below. By measuring the temperature in the magma and in the crust as it thickens, volcanologists document the cooling history of the magma. From samples collected in these holes, they see the mineralogy and chemistry of both the magma and the resulting rock as it changes over time.

When lava first entered the lake shown in figure 2.15, it was about 1,200°C (over 2,000°F). The surface cooled rapidly, and as it did, its color changed from yellow or orange to nearly black. A month after the eruption, the rock at the surface was cool enough to pick up, but because rock is such a good insulator, the temperature at a depth of only 3 meters (10 feet) was still about 1,130°C (over 2,000°F). After four years, the crust was still only 15 meters (about 50 feet) thick (figure 2.16).

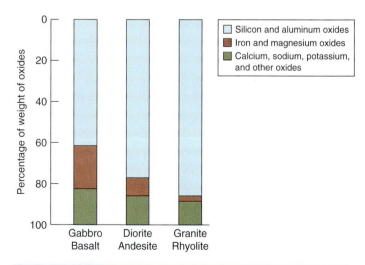

Figure 2.14 Comparison of the Chemical Compositions of Representative Samples of Gabbro and Basalt, Diorite and Andesite, and Granite and Rhyolite.

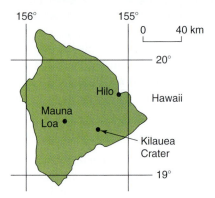

The newly crystallized lava at the surface of a lava lake consists of glassy basalt and **scoria,** a basaltic rock containing many cavities formed by the expansion of gases (see figure 2.12). Because cooling of lava at this air-lava interface is so rapid, crystals do not have time to grow. The crystals in the basalt are too small to be visible throughout the upper 6 meters (20 feet). Grain size gradually increases at greater depths where the slow cooling allows crystals in the magma time to grow. A similar fine-grained border is present around the edges of most igneous rock masses, called **intrusions,** that formed where magma was injected into cool rock. These fine-grained borders form as a result of rapid crystallization of the magma where it comes in contact with cool rock along the edge of the intrusion.

Olivine is the first mineral to crystallize as basaltic lava cools. The first crystals grow surrounded by liquid. As a result, the crystals can grow into perfectly shaped crystals unrestricted by other solids. Slow cooling deep in the interior followed by more rapid cooling at the surface produces a rock containing large crystals in a matrix of finer crystals, called a **porphyry** (figure 2.17). However, because olivine is dense, it usually sinks. A few olivine crystals may freeze into the uppermost part of a flow if it cools rapidly, but by the time most of the melt cools and is viscous enough to hold crystals in place, the olivine has already settled out of the upper part of the lava. The resulting basalt contains little or no olivine. The olivine accumulates in the basalt near the floor of the lava lake, retaining its near-perfect crystal shape as it forms in and sinks through the melt.

The Crystallization of Magma

The process of magma crystallization is not as simple as might be expected from watching water turn to ice. The main difference is that magmas contain an assortment of elements that can combine in different ways to form a variety of minerals. In contrast, water contains only two elements that combine to form only one mineral, ice. In magmas, some minerals begin to crystallize at higher temperatures than others, and some of the first minerals undergo chemical reactions with the remainder of the liquid as the cooling process continues. N. L. Bowen, the first scientist to study these reactions in laboratory experiments, found what was subsequently confirmed in Hawaii, that olivine and a calcium-rich plagioclase feldspar are the first minerals to crystallize in a basaltic magma. As soon as solid crystals of olivine are present in the melt, chemical reactions start to occur between these early-formed crystals and the melt. These reactions lead to the formation of minerals of the pyroxene group. If cooling is slow, all of the olivine may react to form pyroxene, but often some of the original olivine is left with a coating of pyroxene. As the temperature of the melt drops, reactions between the melt and the pyroxene produce new minerals of the amphibole group. These same amphibole minerals may also appear in the melt. At still lower temperatures where biotite mica starts to crystallize, the melt reacts with the amphibole minerals to produce mica. Similar reactions occur between the early-formed feldspar and the melt. The sequence of minerals produced by these reactions are known as Bowen's Reaction Series (figure 2.18). Although basaltic

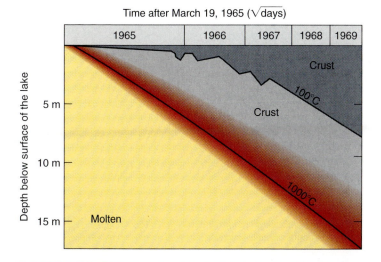

Time after March 19, 1965 (√days)

| 1965 | 1966 | 1967 | 1968 | 1969 |

Depth below surface of the lake

5 m
10 m
15 m

Crust
Crust
100°C
1000°C
Molten

Figure 2.16 Depth to Magma Beneath the Surface of the Lava Lake Formed in Hawaii in 1965.

magma generally lacks enough silica to produce them, quartz and muscovite mica are the last common minerals to crystallize in magmas. A number of other rare and precious minerals, including gold, may also form late in the cooling of magmas of granitic composition. This explains why these minerals commonly occur in vein-like bodies located near the top of large intrusions of granite.

Once igneous rocks are exposed at the ground surface, weathering processes begin to break down the rock and produce the materials that eventually form sediments.

Igneous Intrusions

Although magma chambers filled with magma are inaccessible, solidified plutons that formed thousands of feet below the surface are exposed in young and ancient mountain belts. Erosion has stripped away the rock intruded by the plutons, providing three-dimensional views of these masses. Like roots growing out of the magma, offshoots and feeder systems leading from the plutons toward the volcanoes that once stood above them are plainly visible. The tops of some large plutons in the Sierra Nevada Mountains are still in place, and the deeper interior portions are also exposed, making it possible to study the once-buried magma chambers and their long-since frozen contents. As magma moves closer to the surface, it follows the course of least resistance. This course may be cracks in the rock around the intrusion, called **country rock,** or along contacts between bodies of rock of different types. The country rock may be an older pluton, but often the magma intrudes sedimentary rocks.

SEDIMENT AND SEDIMENTARY ROCKS

The term **sediment** applies to loose, unconsolidated materials like **quartz sand** (figure 2.19), clay, soft muds composed of the shells

of marine organisms, and mixtures of these materials. Most of the inorganic components of sediment originate from weathering processes, but some sediments are composed entirely of organic matter. Freezing and thawing of water in cracks, growth of roots or crystals, and repeated changes in temperature are among the processes that cause rocks to disintegrate mechanically. These mechanical weathering processes often combine with chemical reactions to release minerals from the rocks in which they originate. Once fragments of the rock are released, streams, the wind, or ice may move them away. Chapter 15 provides more detailed information about these weathering processes.

Chemical weathering involves chemical reactions between rocks and the atmosphere or aqueous solutions that infiltrate through the soil and into solid rock. Weathering effects commonly show on the surface of rocks. Because of this, the surface of a rock frequently has a different color from the freshly broken interior. Rocks and minerals differ in their susceptibility to chemical weathering. Some minerals, such as salt, dissolve rapidly in water; others, like quartz, dissolve extremely slowly. As a result of weathering processes, massive rock outcrops disintegrate. Some parts of the rock may dissolve; other parts may undergo chemical reactions or remain as fragments.

Soluble minerals dissolve and become part of the aqueous solutions that move in streams or seep through the ground. Elements carried in these solutions may eventually form compounds that precipitate as vein fillings or deposits on the walls of caves, or they may travel with surface water into the ocean. Marine organisms may pick up elements that reach the sea and use them in the construction of shells and other hard parts. Some minerals undergo chemical reactions that result in the formation of other minerals, such as clay, that are stable at Earth's surface. Other minerals, such as quartz, that do not react chemically with other substances at the surface remain as small grains that get washed into streams and slowly move toward the ocean.

Figure 2.17 Photograph of a Basalt Porphyry. The large crystals of feldspar (about 1 inch across) formed before the much finer-grained dark minerals that compose the basalt.

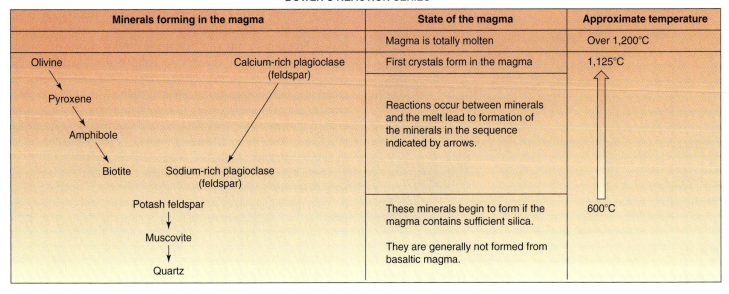

BOWEN'S REACTION SERIES

Minerals forming in the magma		State of the magma	Approximate temperature
		Magma is totally molten	Over 1,200°C
Olivine	Calcium-rich plagioclase (feldspar)	First crystals form in the magma	1,125°C
Pyroxene		Reactions occur between minerals and the melt lead to formation of the minerals in the sequence indicated by arrows.	
Amphibole			
Biotite	Sodium-rich plagioclase (feldspar)		
Potash feldspar		These minerals begin to form if the magma contains sufficient silica.	600°C
Muscovite			
Quartz		They are generally not formed from basaltic magma.	

Figure 2.18 Bowen's Reaction Series. (See text for discussion.)

Types of Sedimentary Materials

Sediments are composed of organic materials, fragmental materials, and precipitates from ionic solutions. Ions are produced by the process of solution, and their chemical reactions release them, leaving them free to move in solutions. Organic matter, including waste products of organisms, solid parts such as shells, and plant remains, all may become part of the sediment. Some sediments, such as coal, are made almost entirely of just such remains of plants and animals. Most fragmental materials are pieces of minerals or rock produced by the disintegration of preexisting rock. Volcanic ash and meteoritic particles fall in this category as well. Unlike most other sediments, they represent new additions of material to the crust. All such fragmental materials are referred to as **clastic.**

The finest clastic materials are called **colloids.** This term refers to a particular size of particles—those in the range of 0.00001 to 0.0000001 centimeters in diameter. Colloidal particles of mud and clay are so small that they cannot be distinguished individually, even with a high-powered optical microscope. If clay is stirred in water until the particles move in suspension through the solution, the heavier and larger aggregates will settle out, but colloidal-sized particles will remain in suspension indefinitely and will move more or less at random through the solution.

Processes of Sedimentation

Many sediments transported in water or air eventually settle. Windblown sand and dust and fragments suspended in streams are examples. Extraterrestrial material, meteoritic dust, falls to Earth through the atmosphere and sinks through ocean water. Clay and other colloids form aggregates in salt water and settle out; shells and other remains of animals or plants living in water settle when the organisms die; and fecal pellets, mainly invertebrate excreta, also settle through water.

Although much of the sediment carried by streams into the sea settles as layers in the shallow water along continental margins, some of this sediment finds its way into the deep sea. In addition, the remains of organisms that thrive in sunlit water near the surface settle to the sea floor. Volcanic ash, windblown dust, and clay colloids may be carried long distances and into deep water by slow but steady oceanic currents. Other sediments are transported into the deep sea as clouds of suspended sediment caused by submarine landslides.

Chemical Precipitation Ocean and lake water usually contain dissolved gases, solids (ions), and liquids. Most natural solutions are not single-component systems; they contain many different ions in solution; so, the chemical reactions involved may be complex. Precipitation occurs when the solution becomes saturated. A solution is said to be saturated with respect to a particular compound when no more of that compound can be dissolved in a given amount of solvent. Because the amount that can be dissolved decreases as the temperature of the solution drops, a solution becomes saturated as it cools. As a result, the ions combine to form solids. Deposits around hot springs form in this way. Evaporation of the solution also causes saturation and brings about precipitation. The deposits of calcium carbonate found along some streams and near waterfalls or cascades precipitate from evaporating spray. The unusual egg-shaped balls of calcium carbonate, called **oolites,** also form by precipitation induced by evaporation. Modern examples of oolite formation are found on the Bahama Banks where calcium carbonate precipitates on sand grains or fecal pellets. The rounded shapes result from continual rolling of the oolites in the agitated seawater.

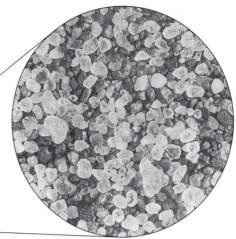

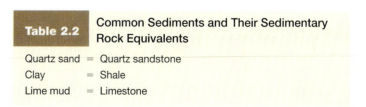

Figure 2.19 Sediment. The sediment consists of a layer of cross-bedded sand between layers of coarser-grained gravel. Rounded grains of quartz sand are prominent in the enlarged section.

Biochemical Deposition The slight changes in acidity found in seawater near certain plants and animals prove that they affect the chemistry of the water. Some sedimentologists believe that most precipitation in seawater is a result of biochemical processes. Sediments composed of shells or skeletons of animals are good examples. Many marine invertebrates remove calcium carbonate from seawater and use it to construct their shells. Some type of algae and a group of single-celled animals take silica out of seawater to build their hard parts (see chapter 8). Excreta of marine invertebrate animals, usually composed of calcium carbonate and impurities held together by a mucous material, are a third example of biochemical processes involved in formation of sediment.

Formation of Rocks from Sediment

Although most sediment deposited on land surface is eroded away, deposits on the sea floor, especially in places where they are buried, survive for many millions of years. Sediment may accumulate to a thickness of hundreds, even thousands, of feet. As this burial takes place, the weight of the overlying sediment compresses and compacts the loose sediment. Eventually, silica, calcium carbonate, or other substances in the water that move through pore spaces in the sediment cement the loose grains together. Some sediments, especially limy muds, recrystallize as they are buried and become hard limestones. In this fashion, sediments gradually change from a loose or soft material into a hard, compact sedimentary rock, a process called **lithification** (table 2.2).

Compaction is usually the first stage in lithification. The weight of the overlying sediment compresses the deeper sediment, forcing water out, eliminating open spaces and pressing soft materials together. Colloids are highly compressible. The volume of clay can be reduced by 40% or more, but sands and other clastic sediments are only slightly compressible.

When mud is present, it may act as a cementing material, but, more commonly, cement is formed when solutions passing through a sediment carry ions. These ions precipitate in pore spaces, cementing the particles of sediment. The most common cementing materials are calcite, quartz, iron carbonate (siderite), and iron oxide (hematite). These may be derived from the sediment itself, or they may be introduced from outside by the water flowing through. Some cements are economically important. For example, the iron oxide that cements some marine sandstones in the Appalachians is rich enough to make the sandstone an iron ore.

Inorganic materials compose most sedimentary rocks, but under special conditions, organic matter may form rock-like material, such as coal. Where the crust is subsiding, great thicknesses of plant matter may accumulate in swamps. If other sediments such as sand or clay accumulate on top of the plant matter, the increasing pressure and rising temperature deep in the accumulation alters the remains of the plants. The organic materials gradually change color and texture as they change into coal.

Table 2.2	Common Sediments and Their Sedimentary Rock Equivalents
Quartz sand	= Quartz sandstone
Clay	= Shale
Lime mud	= Limestone

Classification of Sedimentary Rocks

Sedimentary rocks are usually classified according to their texture and chemical or mineralogical composition, but a few are classed according to their mode of origin. The two major textural divisions, fragmental materials and chemical or biochemical precipitates, are generally easy to recognize. However, some fine-grained fragmental sediments, clay-sized particles for example, are so small that individual fragments are not easily seen. They can, therefore, be confused with chemically precipitated rocks.

Among the fragmental rocks, more precise identification is based on the size and shape of the fragments and on their composition (see appendix C). Some fragmental sediments consist of particles with a relatively narrow range of sizes. Others contain a wide range. The consistency of particle sizes is described in terms of the degree of sorting. A well-sorted sand is one in which the range of sizes is not great; a poorly sorted sand contains a broad range of sediment sizes, most of which fall within the sand-size range. Some fragmental sedimentary rocks are further subdivided according to the compositions of the fragments. The following adjectives indicate specific composition: **siliceous,** containing silica; **carbonaceous,** containing calcium carbonate; and **argillaceous,** containing clay.

Many of the fragmental rocks are the remains of plants or animals; some sandstones are entirely made up of sand-size fragments of the shells of marine invertebrates. Such rocks may be called calcareous sandstones, but most sedimentary rocks that are formed largely of calcium carbonate are called *limestones,* or they are given special names such as chalk, a limestone consisting of the shells of microscopic-sized animals.

Chemical precipitates tend to be much more compact and uniform in texture than fragmental sediments are. Many precipitates have a fine banding or laminated structure, and examination with a microscope may reveal a mosaic of intergrown crystals resembling igneous rock textures. These and other characteristics of sedimentary rock texture and composition enable us to deduce the environment in which the rock formed.

Sediments and Sedimentary Rocks as Environmental Indicators

Each type of sediment and its sedimentary rock equivalents contains information about the conditions under which it was deposited. The composition alone reveals part of this story. Features formed during deposition provide additional details, and the presence of fossils provide especially valuable clues. Although animals that inhabit the land may range through a variety of environments, terrestrial plants and many marine organisms require special conditions for survival. For example, corals prosper in clean, shallow

(a)

(b)

Figure 2.20 Stratification. (a) Stratification is one of the most characteristic features of sedimentary rocks. A sequence of different layers is exposed in these cliffs in Utah. A sandstone layer caps the mountain and is underlain by silts and clay. (b) Interbedded limestone and shale. The limestone layers are more resistant to weathering and erosion than the shale.

water that is warm and rich in microscopic organisms on which the corals feed. Some invertebrates live only in beach swash zones. Others are attached to rocks in water at the sea surface. Still others crawl around on the sea floor but only on the continental shelf. A few are adapted to the great pressure at the bottom of the ocean depths. Most fossils reveal at least some critical information about the environment. Most are restricted to either fresh water, brackish water, or marine water, and many are good indicators of water depth and temperature.

Primary features formed during sedimentation provide additional information about the environment. These features may include the presence or absence of layering in the sediment. This layering, also known as **stratification,** is a common characteristic of sediments (figure 2.20). Some strata extend over great distances. It is easy to imagine how the material settling to the sea floor can create a widespread layer. As long as the type of matter supplied to the sea remains the same, the layer becomes thicker. However, if something happens that changes the character of the sediment supplied to the sea, the composition or texture of the sediment on the sea floor changes, and a new layer starts to form. Today, most of the sediment that accumulates off the coast of Florida is composed of the remains of organisms, and a layer of limestone muds composed of shell fragments is forming. Imagine what would happen if huge clouds of dust blew in from Africa or if volcanic eruptions in the Caribbean spread large quantities of ash over the region. A new layer of clay or volcanic ash would cover the limestone muds. The changes could be much slower as well. In the geologic past, clay sediments derived from rising mountain chains have flooded into the ocean, covering limestone that formed when the water was clear and organic sediment dominated the supply.

The shape of some sedimentary rock bodies reveals how they formed. For example, stream gravel is usually deposited in a stream channel, and the shape of the channel may be preserved. Similarly, the sediments deposited as deltas, as sand dunes, by the wind, or by glaciers have characteristic shapes and internal features. The presence of cross-bedding is one such feature (figure 2.21). This type of bedding forms when grain-sized sediment is carried by wind or water currents (see chapter 17). In deserts, the wind moves sand by rolling or bouncing it along the ground surface. Where the sand accumulates, it moves gradually upslope on the windward side of a dune and then slips down the leeward side. The windward side has low slopes, and the combination of the laminations of sand produces the cross-bedding. Currents in water may produce similar crossbeds on a smaller scale. Several other types of features that form when sedimentation takes place are shown in figure 2.22. These features play an important role in the interpretation of the rock record discussed in chapter 4.

METAMORPHIC ROCKS

Like sediments, most **metamorphic rocks** provide information about the environment in which they form. Most metamorphism occurs in three distinctly different types of environment. In one type, the country rock intruded by magma is transformed by the heat of the intrusion. This is called **contact metamorphism.** The second type of metamorphism occurs as a result of elevated pressure and temperature caused by **burial** of sedimentary rock where the crust subsides. The third type, known as **regional metamorphism,** takes place where rocks formed at shallow depths sink into

Figure 2.21 Cross-bedding. Cross-bedded sandstone forms where currents move granular materials. They are most common in stream channels and wind-blown deposits.

(a)

(b)

Figure 2.22 Features Formed at the Time Sediment Is Deposited Reveal the Environmental Conditions Under Which the Sediment Accumulated. (a) Mud cracks form where fine sediment dries out. They are common features in lake deposits in arid or semiarid regions, on tidal flats, and along streams and small bodies of water, even in humid regions. (b) Cross-bedding is a common feature in streams and wind-blown sand deposits.

a much hotter and higher pressure environment, which may be into the mantle (figure 2.23). Many such rocks are elevated in mountain systems. The rocks formed in all three of the metamorphic environments occur at the surface where the crust is uplifted and the original cover on the metamorphic rocks is removed by erosion. The processes involved in all metamorphism are closely related to four conditions—high heat; high pressure, called **confining pressure,** caused by the weight of overlying rock; localized stress; and directed pressure caused by deformation of the crust, as in mountain building. Metamorphism generally also involves chemical reactions that occur within a rock mass. The chemical reactions are especially sensitive to changes in temperature, pressure, and the amount of water present. Metamorphism may occur without the addition of fluids or the introduction of new elements or compounds. The changes take place by the exchange of component elements among the minerals in the rock or, in some instances, in the

recrystallization and growth of new crystals of the same minerals that were present before metamorphism. However, where metamorphism occurs near bodies of magma, fluids are often introduced and affect the metamorphism.

In each type of metamorphism, the texture and the mineral composition of the rocks involved usually change. These changes occur because the minerals in the rock are chemically or physically unstable in the presence of such high temperature and pressure. These metamorphic processes commonly produce a new rock of different texture and mineralogical composition. New minerals form by recombination of elements present in the original minerals or by recrystallization of the original minerals. As larger crystals grow, new textures develop, and the internal structure and fabric of the rock change.

Minerals in the original rock that are stable under the temperature and pressure that prevails during metamorphism may survive

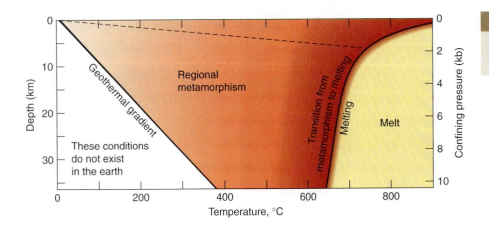

Figure 2.23 Fields of Metamorphism and Melting are Related to Temperature and the Depth of Burial.

Table 2.3 — Common Metamorphic Rocks and Their Unmetamorphosed Equivalents

Original Rock	Metamorphic Equivalent
Quartz sandstone	Quartzite
Limestone	Marble
Limestone and Quartz	Siliceous Marble
Shale	Slate (if metamorphosed under high directed pressure)
	Phyllite (if low-temperature metamorphism)
	Schists (if medium-temperature metamorphism)
	Gneiss (if high-temperature metamorphism)
Volcanic tuff	Same as for shale; if the tuff is rich in magnesium and iron, it may become amphibolite (a rock containing large amounts of hornblende)
Basalt or gabbro	Greenstone (if low-temperature metamorphism)
	Amphibolite (if medium- to high-temperature metamorphism)
Granite or diorite	Granitic gneiss

and even grow to form larger crystals. However, minerals in the original rock that are unstable under metamorphic conditions undergo reactions that tend to establish an assemblage of minerals that are stable under metamorphic conditions. If the metamorphism takes place over a long time and the reactions continue to completion, all of the unstable minerals in the original rock may disappear, and new stable minerals formed from the same elements take their place. However, if metamorphic processes are interrupted or if reactions fail to go to completion because they are too slow, some of the original unstable minerals may remain. Metamorphism clearly demonstrates that under certain conditions, rocks undergo transformation from one rock type to another.

Composition and texture provide the basis for identifying and classifying metamorphic rocks. Metamorphic rocks commonly contain minerals that exhibit their natural crystal form. These perfectly shaped crystals grow during metamorphism. If the rock is being deformed during metamorphism, some of the new minerals may grow so their flat or elongated shape is nearly perpendicular to the directed pressure. This produces an alignment of platy minerals, such as mica, or elongate minerals, such as hornblende. In some metamorphic rocks, these alignments are so well developed that the rock breaks along the planes in which minerals are aligned. Slate, phyllite, and schists all exhibit strong mineral alignments (table 2.3) (see appendix C). Gneisses contain thin bands in which different minerals concentrate. Marble (metamorphosed limestone) and quartzite (metamorphosed quartz sandstone) are generally more homogeneous than other metamorphic rocks. But they may also contain streaks or bands with different mineral compositions formed by impurities in the original rock.

Because all types of igneous, sedimentary, and metamorphic rocks may be metamorphosed, original textures and compositions vary greatly (see table 2.3). The recrystallization and recombination of elements that takes place during metamorphism tend to restrict the range of compositions in metamorphic rocks. If the rock being metamorphosed contains mainly quartz (e.g., quartz sandstone, quartz siltstone, vein quartz, or quartzite), the resulting metamorphic rock will consist mainly of quartz also. All such rocks become quartzite when metamorphosed. Quartz may recrystallize during metamorphism, and in the sandstones, the grains of quartz sand fuse together as a result of high temperature, but the composition remains unchanged. In a similar manner, metamorphism of all limestones, regardless of their original texture, produces marble containing enlarged crystals of calcite ($CaCO_3$). Other rocks that contain a variety of original minerals generally undergo more drastic changes in mineral composition during metamorphism.

The minerals formed during metamorphism from original rocks that contained a variety of mineral components indicate the temperature, water content, and pressure conditions under which metamorphism takes place (figure 2.24). Of course, the minerals form only if the elements in each of these minerals are present in the original rock.

Micaceous minerals, such as chlorite, biotite, and muscovite, are generally present in metamorphic rocks formed by the metamorphism of shale and other rocks containing clay minerals. As the micas crystallize in rocks that are compressed, the micas grow parallel to one another (and generally perpendicular to the direction of maximum stress or differential pressure). This

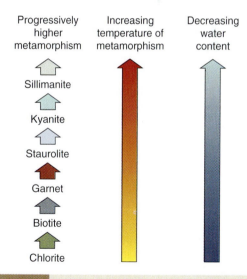

Figure 2.24 Relationship of Temperature and Water Content to Mineral Formation. The sequence of minerals shown at left form in metamorphic rocks at progressively high temperatures; however, the water content of the rock also influences their formation. Thus, the mineral sillimanite is most likely to form at high temperatures in dry rocks, but chlorite is most likely to form at low temperatures where water is present in the rocks.

Figure 2.25 Metamorphic Rocks. (*a*) Large garnets surrounded by hornblende, a member of the amphibole group. (*b*) A banded gneiss. The dark layers contain hornblende; the lighter colored layers contain quartz and feldspar. (*c*) Tight folds in a banded gneiss. (*d*) Slate. (*e*) Marble.

produces rocks with strongly aligned minerals, a texture known as **foliation.** The three metamorphic rocks with the strongest alignment of minerals are phyllite, slate, and schists. If metamorphism continues, recrystallization and recombination of elements may ultimately lead to formation of a rock containing bands in which different minerals become segregated. These banded rocks are gneisses (figure 2.25). At higher temperatures, metamorphism gradually gives way to melting and the transformation of metamorphic rock to magma and to a major transition in the rock cycle.

The rock cycle makes it clear that processes that take place on Earth's surface are closely related to processes that occur deep inside the planet. Many of the surficial processes, such as weathering, transportation and deposition of sediments, and changes in the physical climate system, cause rapid changes in the environment. These surficial changes are especially fast when compared with internal processes such as movement of lithosphere plates. Examination of the largest features of Earth's surface and interior, the Earth model, will clarify relationships between the surface and the deep interior.

SUMMARY POINTS

1. Melting of rocks forms a liquid, called magma, that is rich in silicon and oxygen. Magma that cools inside Earth forms plutonic rocks. These commonly intrude other rocks surrounding the magma. Magma may reach the surface and pour out as lava or erupt as volcanic fragments. Magma that cools beneath the surface crystallizes slowly and forms coarse-grained rocks; magma that cools rapidly at the surface forms fine-grained rocks or glasses.

2. Minerals, naturally occurring elements or inorganic compounds, combine to form three major types of rocks—igneous, sedimentary, and metamorphic rocks.

3. Many of the most common rock-forming minerals, including quartz, feldspar, mica, pyroxenes, amphiboles, and olivine, are silicate minerals. The silicon-oxygen tetrahedron is the basic structure of this group of minerals. Tetrahedra, consisting of four oxygen atoms surrounding one silicon atom, may be present in some minerals as independent tetrahedra, but commonly the tetrahedra combine as pairs, chains, or sheet structures.

4. Magma or lava crystallization and lava freezing to form glass produce igneous rocks.

5. The settling out of fragments or crystallization of solids from water, air, or ice produce sediment and sedimentary rocks. The

term sediment applies if the material is unconsolidated. Loose sand, clay, and mud are examples. If the sediment compacts or if crystals become cemented together forming a solid material, the resulting rock is a sedimentary rock.

6. Metamorphic rocks form through the alteration of igneous, sedimentary, or other metamorphic rocks. This alteration generally takes place deep in Earth through the effects of heat, chemically active fluids, and pressure.

7. Geologists classify lava according to its composition (see figure 2.13). The lava from which rhyolite forms is rich in silica; basaltic lavas contain low silica; andesite lava is intermediate in composition.

8. Temperature, composition, and the amount of gases present affect the viscosity of lava. High temperature, low-silica content, and large quantities of gases tend to produce low-viscosity lava. In general, basaltic lavas (see figure 2.13) have low viscosity; rhyolites have high viscosity.

9. Pumice and scoria are fine-grained or glassy rocks containing holes formed by gases. These are derived from silica-rich (rhyolite) and silica-poor (basalt) lavas (see figure 2.13). They may form as pieces of lava blown into the air or at the top of lava flows when gases escape, leaving the vesicles.

REVIEW QUESTIONS

1. How do minerals differ from rocks?
2. How can you distinguish igneous, sedimentary, and metamorphic rocks from one another on the basis of their texture?
3. What stages in the rock cycle does clay go through as it is transformed into slate?
4. Under what conditions do rocks become metamorphosed?
5. Metamorphism takes place deep inside Earth. How then is it possible for these rocks to be exposed at Earth's surface?
6. What determines the texture (grain size or crystal size) of igneous rocks?
7. What processes can cause rocks of different composition to crystallize from a single magma?
8. How can so many different types of igneous rocks originate from a single basaltic magma?

9. Could as many different types of igneous rocks originate from a granitic magma?

THINKING CRITICALLY

1. Why would you expect to find different types of metamorphic minerals at increasing distances away from the margin of a large igneous intrusion?
2. Make a list of plants and animals that live in restricted environments. Their presence in sediment reveals that environment.

KEY WORDS

amphiboles	mica
andesite	minerals
argillaceous	obsidian
basalt	olivine
burial	oolites
calcite	pluton
carbonaceous	plutonic rock
clastic	porphyry
cleavage	pumice
colloid	pyroxenes
confining pressure	quartz
contact metamorphism	quartz sand
country rock	regional metamorphism
feldspars	rhyolite
foliation	rock
hot spots	scoria
hydrologic cycle	sediment
igneous rock	sedimentary rock
intrusions	silicate minerals
lava	siliceous
lithification	stratification
metamorphic rock	volcanic rock

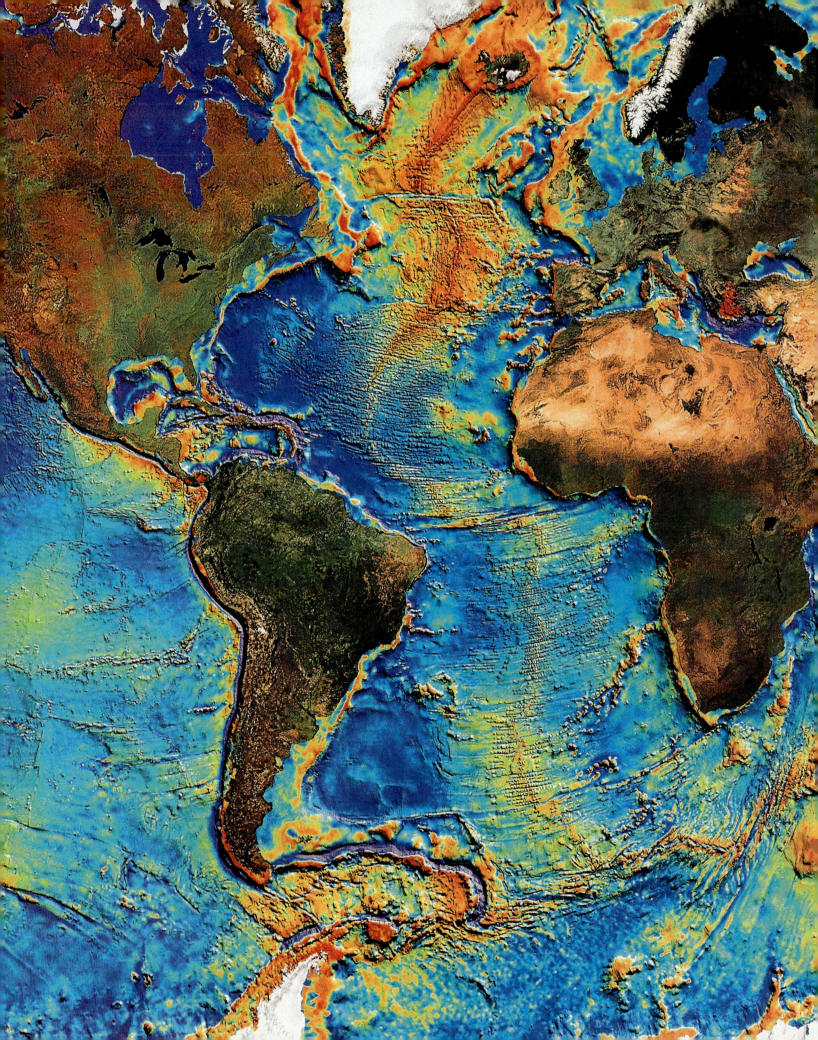

chapter three

Earth Model— Core-Mantle System

The varied landscapes of the continents and ocean basins are products of processes that take place deep inside Earth, as well as those that shape the surface.

Chapter Guide

Processes acting deep inside Earth have profound effects on the environment at the surface. Although Earth's magnetic field originates in the core, it extends far beyond the core, and influences processes ranging from the guidance systems of migratory birds to the aurora. Less obvious is the importance of heat that rises from the core. It makes material in the lower mantle buoyant. As this material rises, it spreads laterally in the upper mantle and causes lithospheric plates to move. These movements generate earthquakes, cause volcanic activity, and lead to the formation of mountains. The gravitational attraction of the mass in Earth's interior exerts a pull on all materials on the surface and causes them to move downslope. This drives the movement of streams and glaciers. Gravity also causes the lithosphere to respond to shifts in mass on the surface which may cause parts of the lithosphere to subside and other parts to rise. Thus, Earth's interior is the place to look for understanding the dynamics of the planet. This chapter examines the character of the major divisions inside Earth and the methods scientists use to study a part of Earth that is far removed from direct observation.

INTRODUCTION

Earth is a dynamic planet, and processes that act deep inside Earth affect its surface. For insights into the deeper parts of Earth, indirect observations and inferences are the keys. Because the deepest wells have not penetrated the crust, most of what we know about Earth's interior comes from studying variations in gravitational force, vibrations caused by earthquakes, and the magnetic field. Studies of gravity help define Earth's shape, mass, density, and the distribution of rock masses of different densities. Earthquakes generate vibrations, making it possible to identify the major divisions of the interior and to determine the character of the processes acting there. Finally, the magnetic field reveals additional information about processes acting near the center of Earth. Based on these indirect observations and an understanding of the origin of Earth and its early history, geophysicists have formulated a model of Earth's interior. Understanding what takes place on the surface depends on knowing the cause-and-effect relationships between Earth's interior and exterior.

EARTH'S SURFACE

Continents and ocean basins are the most prominent physical features of Earth's surface (figure 3.1). If the ocean basins were dry, the difference would be even more dramatic. Although the average elevation of Earth's surface is about 2.5 kilometers (1.6 miles) below sea level, only a small percentage of the total surface area lies near this elevation (figure 3.2). Instead, the average elevation of continents is about 0.48 kilometer (0.3 mile) above sea level, and the average elevation of the ocean basins is nearly 3.7 kilometers (2.3 miles) below sea level. Although ocean basins and continents are both part of Earth's outer layer referred to as the crust, this separation in elevation alone indicates that continents and ocean basins are fundamentally different.

Continents

Although continents differ from one another in shape and size, they share many characteristics. Sedimentary rocks form a relatively shallow cover over parts of all continents. Below the sedimentary cover, the upper part of continents are granitic in composition. The lower parts are closer to gabbro in composition. Parts of every continent have been stable for many hundreds of millions of years. Stable areas largely devoid of volcanic activity, with few earthquakes, and no high mountains are called **cratons** (figure 3.3). The North American craton includes central Canada, the Great Plains, and the interior lowland west of the Appalachian Mountains. Regions where ancient rocks of the cratons lie exposed at the ground surface are called **shields.** These rocks are all that remain of huge masses of igneous rocks and metamorphosed sedimentary and volcanic rocks, most of which formed more than half a billion years ago. Erosion that continued over this vast time span reduced former mountain ranges to nearly flat surfaces. Rocks that were once buried deep in the core of the mountains crop up at the surface.

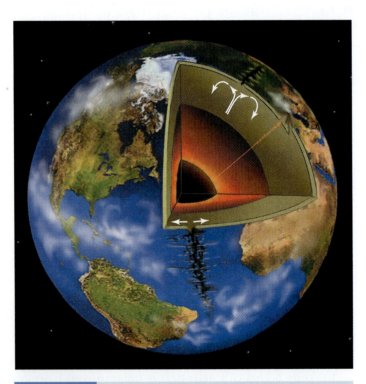

Figure 3.1 Schematic Diagram of Earth Systems Showing the Atmosphere, Oceans, Organic Matter, and Major Internal Features.

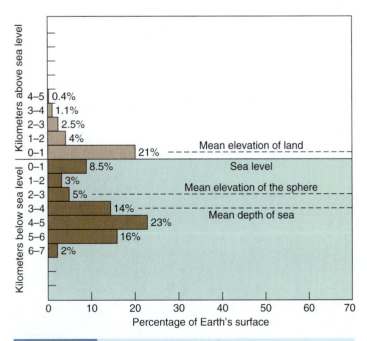

Figure 3.2 Bar Graph Showing the Percentage of the Surface Area in Each Elevation Range.

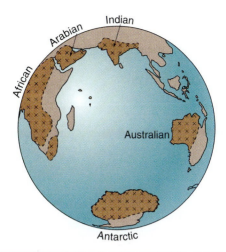

Figure 3.3   Sketch Maps Showing the Approximate Extent of Continental Cratons.

Continental platforms are portions of cratons that are covered by a "veneer" of sedimentary rocks. The North American platform includes the Great Plains and the lowland region located between the front ranges of the Rocky Mountains and the Appalachian Mountains (figure 3.4). Across most of this platform, sediments range in thickness from a few hundred to a few thousand feet thick. Closer to the mountains, the sedimentary rock cover is thicker, and some parts of the platform have subsided, producing large sedimentary basins. Large basins occupy all of the state of Michigan, most of the Dakotas, and the area around Denver, Colorado. A few areas on the platform, such as the Ozark and Adirondack mountains, have remained as high, dome-shaped uplifts. Much more dramatic and young mountainous regions lie along the margins of many cratons, including those in North America, Eurasia, Australia, and South America (figure 3.5). These mountain belts form long and complex systems, such as the Appalachian and Rocky mountains, along their margins. The presence of mountains along so many continental margins led geologists to think that continents grow by the addition of mountains along the edges of cratons.

Mountain Systems

Mountain belts, such as the Appalachians, Rockies, Alps, Himalayas, and Andes, differ from one another in topography, in details of their internal structure, and in the age of the mountain building, but all of them involve strong deformation of the crust. Many of the sedimentary rocks found in mountain belts originally formed along or near continental margins. They include sandstones, shales, and limestones identical to those found on the shallow sea floor adjacent to continents and on the slopes leading into greater water depths. During mountain building, older sedimentary rocks deposited before mountain building begins are compressed as though they are in a great vise. The layers bend, forming folds. As deformation continues, the rocks break and slip along planes called **faults.** Commonly, the layers stack up, one on top of an-

Figure 3.4   The Major Crustal Divisions of North America.

other, forming great piles of deformed rock like that seen today in the Appalachian and other mountain belts (figure 3.6).

Ultimately, during mountain building, rocks deposited below sea level are elevated to great heights. In the Himalayas, rocks

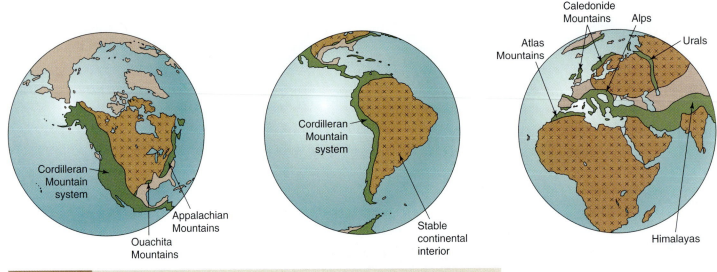

Figure 3.5 Young Mountainous Regions. Young mountain belts extend along the western margins of North and South America and across southern Eurasia. These mountain belts are still actively forming.

containing marine fossils crop out near the top of Mount Everest. In both the Alps and the Rocky Mountains, igneous and metamorphic rocks that were deep in the crust before mountain building began now form the highest peaks. The long systems (or belts) of mountains that extend from Antarctica to Alaska, (the Cordilleran Mountains) and the mountains that extend from North Africa to Indonesia (the Alpine-Himalayan Mountains) are relatively young mountains (see figure 3.5).

Many parts of these systems rose during the past 20 million years, and in the Himalayas, uplift continues today. Eventually, erosion will lower the high mountains, but this takes many millions of years. In general, younger mountains are higher than older, inactive mountains. Mountains such as the Appalachian Mountains and their European counterparts in Scotland and Norway are much older than the Rocky Mountains or the Alps. Most of the uplift of the Appalachian Mountains took place more than 200 million

years ago. Since they initially rose, erosion has beveled the high peaks, exposing the igneous and metamorphic rocks that generally form the deeper parts of the mountains.

Continental Margins

Continental margins are zones of transition from continents to ocean basins. Some margins have broad, flat continental shelves; deep trenches in the sea floor and active volcanism flank others. These marginal zones are categorized as passive or active depending on the amount of earthquake and volcanic activity and the presence or absence of mountain belts less than 500 million years old.

Atlantic-type Continental Margins

Atlantic-type continental margins are those that have little or no volcanic activity, few earthquakes, and no active mountain building. Most of the continental margins around the Atlantic Ocean are of this type (figure 3.7). Similar margins surround Africa, India, Australia, and most of Antarctica and the Arctic Ocean. Along most passive margins, a pronounced change in the slope of the sea floor at a depth of between 100 and 200 meters (330 and 660 feet) defines the edge of the continental shelf (figure 3.8). Beyond this break in slope, the **continental slope** descends to greater depth, where a reduction in sloping defines the base of the continental slope and the beginning of the **continental rise.** Gradually, the continental rise becomes flat. At a depth of about 5 to 6 kilometers (16,000 to 20,000 feet), it merges with the edge of the **abyssal plains,** the expanses of flat sea floor at the bottom of the ocean. These plains are nearly featureless, being disrupted only by a few hills and isolated seamounts, most of which are extinct volcanos.

The continental shelves of both passive and active continental margins are portions of the continents that have been repeatedly

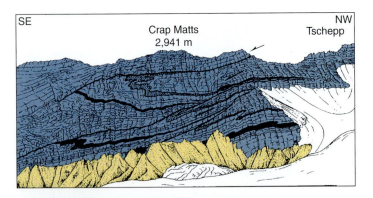

Figure 3.6 Cross Section Depicting the Structure in Part of the Swiss Alps. In many mountain belts, sedimentary rock units are folded and stacked by displacements on faults.

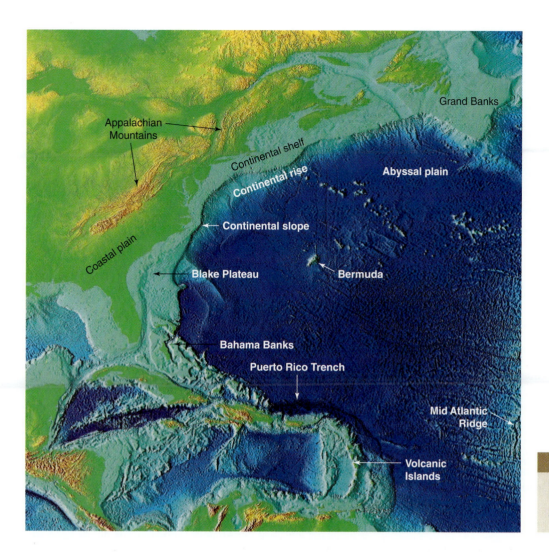

Appalachian Mountains

Continental shelf

Grand Banks

Continental rise

Abyssal plain

Continental slope

Coastal plain

Blake Plateau

Bermuda

Bahama Banks

Puerto Rico Trench

Mid Atlantic Ridge

Volcanic Islands

Figure 3.7 The Eastern Margin of North America Showing the Continental Shelf, Slope, Rise, and Abyssal Plains.

inundated by the sea. During the last 2 million years, sea level has fluctuated from about 75 meters (250 feet) above the present level to 130 meters (425 feet) below the present level. Changes in the amount of water stored as ice on land are responsible for most of these fluctuations. Portions of what were once continental shelves are now above sea level, and parts of what was dry land during glacial advances are now submerged. Even greater fluctuations in the level of the sea took place many millions of years ago when the sea rose above the edge of the continent and invaded the continental interiors. Sediments now exposed in the Atlantic and Gulf Coastal plains (see figure 3.4) were deposited during marine transgressions that took place nearly 70 million years ago. Today, portions of these layers of sediment lie exposed on land, but they continue seaward beneath the continental shelf. The sea floor on most continental shelves is featureless, characterized by flat or low topography formed by the erosive effects of wave action or the deposition of sediment carried into the sea by rivers (see chapter 8).

Active Continental Margins

Most of the active continental margins are sites of volcanic activity, earthquakes, and crustal deformation. Not only do most earth-

quakes take place along active margins, but also the deepest and, with few exceptions, the most powerful of all earthquakes occur along active margins (see chapter 6). Many of the active margins are located along the edge of the Pacific Ocean, which is commonly called the "ring of fire." On the western side of the Pacific, many arcuate chains of active volcanoes, commonly referred to as **island arcs,** lie offshore, separated from nearby continents by seas that are not as deep as the ocean basin (figure 3.9). An almost continuous succession of island arcs lies across the northern and western margins of the Pacific Ocean. Some island arcs, such as the Aleutian Islands, the Indonesian Islands, and the West Indies, form near-perfect arc patterns; others, such as the islands in the southwestern Pacific north of Australia, have more irregular patterns. Similar island arcs are present in the Caribbean, in the Scotia Islands at the southern tip of South America, and in the Aegean Sea. Particularly violent eruptions that often include devastating explosions characterize the volcanoes in the island arcs. Deep-sea trenches, often twice the depth of the ocean basin, separate the volcanic islands from the deep-sea floor.

Mountains are rising along the active margins of some continents. High mountains, for example, the Andes, with active volcanoes and an offshore deep-sea trench, lie along the western margin

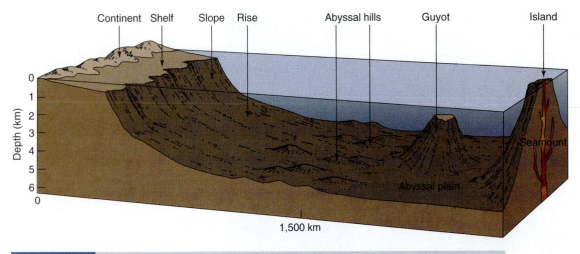

Figure 3.8 Diagram of the Atlantic Margin Showing Major Features of the Sea Floor.

of South America (figure 3.10). Volcanoes, earthquakes, and a trench lie along the coast of Central America, and a highly complex margin is present in North America where a sliver of the western part of the continent is moving northward relative to the rest of the continent along the San Andreas fault. Farther north, volcanoes again lie along the margin, and another fault similar to the San Andreas is present along the Canadian margin. Plate tectonic theory, outlined in chapter 5, provides explanations for the origin of both Atlantic-type and active continental margins.

Ocean Basins

Although oceans cover nearly 70% of Earth's surface, little was known about them until the Second World War. When submarine warfare started, the Allies recognized the need for detailed maps of the sea floor, and a program of mapping began. It continued as a scientific effort after the war, culminating in the production of maps that revealed a much more varied sea floor than was expected. Many scientists had anticipated that most of the sea floor would be flat and featureless plains covered by sediment. Such plains occupy large portions of the ocean basins, but the continuity and length of oceanic ridges and the presence of numerous straight and deep valleys that cross these ridges came as a surprise (see figure 3.10).

Abyssal plains, which lie at depths of 5 to 6 kilometers (3 to 3.5 miles), are the largest flat surfaces on Earth (figure 3.11). They owe their flat surfaces to sediment composed of the fine remains of plants and animals, dust carried over the sea by strong winds, remains of burned-out meteorites, and coarser sediment carried into the abyssal plains by submarine landslides. This blanket of sediment buries the rocks of the oceanic crust. Low hills that rise slightly above the surface and towering volcanic mountains, called **seamounts,** break the monotony of the abyssal plains. Many seamounts form lines and sometimes ridges that rise from the bottom of the ocean. A few active volcanoes, such as those in the Hawaiian Islands, rise from the abyssal depths to the surface of the ocean. These are gigantic piles of lava formed over hot spots in the mantle where magma rises from great depths. Measured from its base on the sea floor, the volcano Mauna Loa on Hawaii is slightly higher than Mount Everest. Once eruptions cease, volcanoes that rise above sea level stop growing, and wave action quickly erodes the edges until the top is gone.

Early ocean explorers recognized the presence of a shallow-water ridge near the middle of the Atlantic Ocean (see figure 3.10). Later, detailed mapping of the sea floor revealed that a long and continuous ridge system exists throughout the ocean basins. The most prominent part of this ridge system, the Mid-Atlantic Ridge, extends the length of the Atlantic. Although it is not as steep, the difference in elevation between the abyssal plains and the volcanoes on the Mid-Atlantic Ridge is comparable to the relief in the Alps (see figure 3.10). All of the islands located along the oceanic ridge are volcanoes, and samples collected from the top of the ridge confirm that it is a massive pile of volcanic rocks. In looking for answers to how such massive amounts of lava come to the surface, why some continental margins are passive but others are so violently active, or what causes mountains to form along the margins of continents, it is necessary to know more about what lies deeper inside Earth.

EXPLORING EARTH'S INTERIOR

Studying Earth's interior is like putting together a giant jigsaw puzzle. As pieces are put in place, a picture of Earth's inner structure, dynamics, and composition emerges. It is the excitement of finding the missing pieces and seeing a picture of some place that is so utterly inaccessible slowly emerge that makes this kind of research so exciting and rewarding. (From William Bassett, *Science*, 1994).

Deep wells, mine shafts, and rocks and lava erupted from volcanoes provide the only opportunities to directly examine rocks

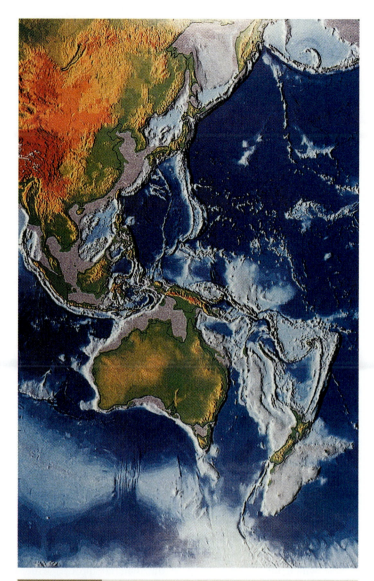

Major Features of the Western Pacific Ocean. The dark blue indicates abyssal plains.

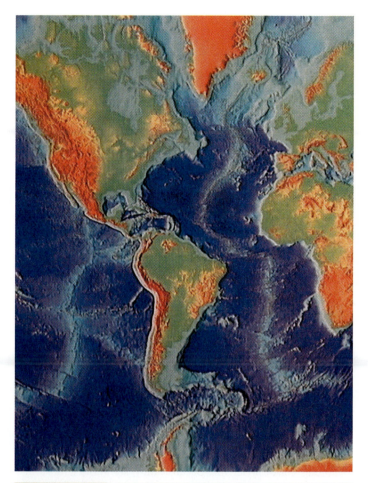

Figure 3.10 Major Features of the Atlantic and Eastern Pacific Oceans. The map includes the Mid-Atlantic Ridge, the East Pacific Rise, and the abyssal plains of the Atlantic and eastern Pacific.

from Earth's interior. Even the deepest mines descend only a few thousand feet, and the deepest core samples come from a depth of about 10 kilometers (six miles). These samples reveal the types of rocks that make up the outer portions of Earth's crust.

USING SEISMIC WAVES TO MODEL EARTH'S INTERIOR

Movements on faults and volcanic eruptions cause the surrounding rocks to vibrate. Because rocks have elastic properties, these vibrations, called **seismic waves,** move away in all directions from the place where they originate, known as the **focus** (figure 3.12). Some seismic waves remain near the surface and within the lithosphere; others pass deep through Earth's interior. The point on the surface directly above the focus is the **epicenter.** If the focus of a major earthquake is shallow, people may actually see waves that resemble waves on a pond move across the surface of the ground. The magnitude of the motion decreases with distance from the focus, but sensitive instruments designed to detect ground motion can measure minute movements of the ground at great distances. By analyzing these waves, seismologists study the character of Earth's interior. Seismic waves that travel paths through Earth are the main source of information about the composition and structure of the deep interior. Chapter 6 provides additional information about the generation, detection, and interpretation of seismic waves.

Seismologists explain the generation and transmission of seismic waves through Earth in terms of the elastic response of rocks. All materials that return to their original shape after being deformed exhibit elastic behavior. The maximum amount of pressure an elastic material can bear without undergoing permanent deformation or breaking is its **elastic limit.** When a hard rock is hit with a hammer, the hammer bounces. Both the rock and the hammer are

Figure 3.11 Profile Across the Abyssal Plain of the Western Coast of France. This profile, based on echo-sounding records, shows the flat and featureless character of the abyssal plains.

Figure 3.12 Using Seismic Waves to Locate Earthquakes. The epicenter is the point on Earth's surface directly above the focus, where the shock originates. The distance from the recording station to the epicenter is the epicentral distance.

exhibiting elastic behavior. Because they are elastic, rocks and liquids transmit seismic waves.

Explosions and displacements along faults cause several types of seismic waves. Some of these waves are due to compression of rocks near the focus; others result from shear (distortion) of the rock. When a volcanic eruption or the detonation of a bomb takes place inside Earth, the rock all around the point of the explosion is compressed. Because the rocks are elastic, this compression travels out in all directions from the focus. Immediately after the passage of the compression, the rock begins to expand elastically, creating a **rarefaction** (figure 3.13).

Thus, in an earthquake, a succession of vibrations in the form of compressions and rarefactions travel out in all directions away from the focus. Because the rock along their path moves back and forth in the same direction the wave is moving, seismologists call these **longitudinal waves** (also called **compressional seismic waves**). These are the fastest of the seismic waves generated during an earthquake, and they arrive before any other types of seismic waves at distant points. For this reason, they are also called **primary,** or **P waves.** Sound waves are also of this type. Thus, compressional seismic waves are also known as P waves, primary waves, and longitudinal waves. They can travel through any compressible material, including liquids, gases, and solids.

Prior to movement, the two parts of the rock mass separated by the fault are fixed in place. The rocks across the fault distort as pressure builds up (figure 3.14). The pressures affecting the rock within the fault zone subject it to what is known as a **shear stress.**

This type of pressure causes the rock on either side of the fault zone to move in opposite directions. The rock behaves somewhat like a spring that is being wound progressively tighter. The distorting rock stores energy until it finally fails. Suddenly, slip occurs along the fault plane, and the rock at the place of failure snaps back like a metal rod that is bent and then released. As rocks along the break snap to a new position, they compress the rocks ahead of them and stretch the rocks behind them. A compressive wave radiates outward in one direction while rock behind it is drawn in toward the focus. On the opposite side of the fault, the directions of compression and stretching are reversed. The elastic rocks vibrate back and forth, causing a succession of compressions and rarefactions to move out in all directions from the focus.

Motions at the focus of earthquakes involve shearing of the rock as well as compression and expansion. The distortion associated with shearing moves outward from the focus by elastic response of the rock as does the compressional wave (P wave), and the shear, or S, wave travels the same path as the P wave, but more slowly. Because S waves always arrive at a seismograph station later than the P waves, they are called **secondary (S) waves.** Shear waves, which involve distortion of the material through which they travel, cannot pass through liquids. The motion involved in a shear wave is much like the wave that travels along a rope tied to the wall at one end and flipped at the other (see figure 3.13).

Although earthquakes occur as deep as 700 kilometers (420 miles), most originate closer to the surface. These shallow earthquakes generate seismic waves that travel within the rocks at and near the surface. One type of **surface wave,** called a **Rayleigh wave,** resembles the up and down, rolling-motion waves that occur when a stone is thrown into a pond. A second type of surface wave, called a **Love (L) wave,** causes the surface to shift back and forth at right angles to the direction the wave is traveling. The combined effect of these two types of motion is devastating for most types of masonry structures and is also responsible for dislodging rock on steep slopes.

Seismic Wave Paths

Earthquakes generate waves that move away from the focus in all directions. The leading edge of the disturbance, the *wave front,* is like an expanding shell (figure 3.15). Lines drawn perpendicular to

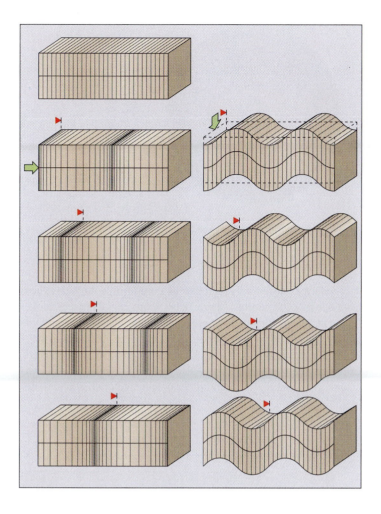

Figure 3.13 Types of Seismic Waves—Longitudinal and Transverse Waves. If the bar shown at the top left is given a sharp blow on its left end, a compressional wave moves toward the right. If the particles move to and fro in the same direction as the wave is traveling, the wave is said to be longitudinal. A downwards blow, as shown on the right, starts a transverse wave like the one in a rope shaken at one end. Earthquakes send both kinds of waves through Earth.

the wave front, called **rays,** illustrate the path the wave takes as it moves from one point to another.

Because rocks of different compositions have different densities and elasticity, seismic waves travel through them at different velocities. If the rock around the focus is homogeneous, the velocity of the wave is uniform in all directions, and the wave front is spherical. If the rock composition varies, the velocity is greater in some directions than others, and the wave front is more irregular in shape.

Seismic waves resemble sound waves. When the wave strikes a boundary between materials that have different physical properties, part of the wave energy reflects from the boundary, and part may bend as it moves across the boundary—a phenomenon known as **refraction** (figure 3.16). Such changes occur where waves pass from one layer of rock into another of different seismic velocity. Depending on the angle at which the waves approach the contact between layers, part of the wave energy reflects and part refracts across the contact. If waves approach the contact at a low angle, all of the wave energy reflects from the contact. If the waves travel a path that is perpendicular to the contact, they continue straight into the next layer. If the waves approach at other angles, they bend as they pass through the contact. Waves bend away from the perpendicular if the seismic velocity of the lower layer is higher than that

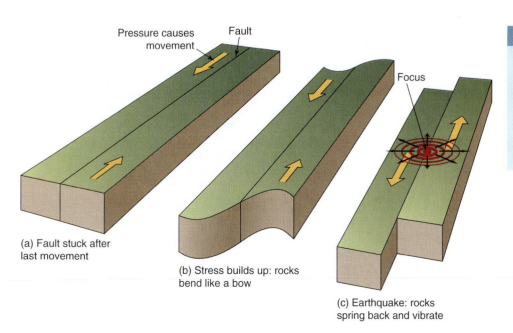

(a) Fault stuck after last movement

(b) Stress builds up: rocks bend like a bow

(c) Earthquake: rocks spring back and vibrate

Figure 3.14 Cause of an Earthquake. (a) Movement along a fault causes the rocks on both sides to become distorted. When the amount of strain reaches the maximum the rock can bear, a sudden displacement occurs. (b) The strain is suddenly released. (c) Strain may continue to accumulate and eventually cause another earthquake.

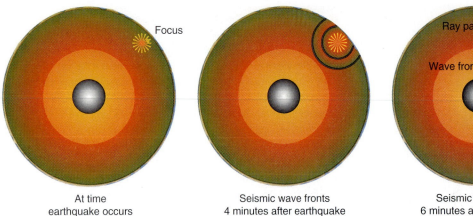
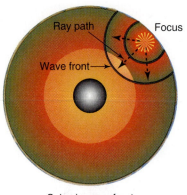

| At time earthquake occurs | Seismic wave fronts 4 minutes after earthquake | Seismic wave fronts 6 minutes after earthquake |

Figure 3.15   Seismic Wave Fronts. This cross-sectional view of Earth shows the propagation of wave fronts and rays emanating from an earthquake focus. When an earthquake occurs, several different types of elastic waves are generated at the same time.

above, and they bend toward the perpendicular if the seismic velocity of the lower layer is lower than that above.

On their journey through Earth's interior, seismic waves reflect or refract where they encounter boundaries between materials of different composition or density. By understanding how reflection and refraction affect the time necessary for a seismic wave to pass from the focus of an earthquake to a station where seismometers detect the waves, it is possible to interpret the structure of Earth's interior. With this information, seismologists can model Earth's interior. The interior may be subdivided on the basis of composition or physical properties (figure 3.17). In terms of physical properties, the inner core is solid, the outer core is liquid, the lower and middle parts of the mantle are solid, the asthenosphere is plastic, and the lithosphere is solid and strong. In terms of composition, the core is mainly iron, the mantle is composed of magnesium and iron-rich silicate minerals, continental crust is granitic, and the oceanic crust is primarily basalt.

THE CRUST

As initially used in the Middle Ages, the term *crust* signified the outer solid, brittle part of Earth, a thin cover over what was thought to be Earth's fiery hot interior. The term *lithosphere,* meaning "rocky sphere," also designated the outer solid portion of Earth. The two terms, crust and lithosphere, were used interchangeably until the early 1900s. In 1909, the Yugoslavian seismologist Andrija Mohorovičić discovered a sharp change in seismic velocities at a depth of about 40 kilometers (25 miles) in Europe. The discontinuity is a sudden increase in seismic velocity of P waves from about 6 to 8 kilometers per second (3.7 to 5 miles per sec). As seismologists began to find this discontinuity, called the *Mohorovičić discontinuity* (commonly shortened to *M* or *Moho*), in many other places, it was used to define the base of the **crust.**

The crust varies considerably in composition and thickness from one place to another (figure 3.18). Continental crust is thicker (about 40 kilometers on average) than oceanic crust (10 kilometers on average). Maximum thickness occurs beneath mountain belts where the crust may reach a thickness between 60 to 80 kilometers. Continental and oceanic crusts also differ in composition. Continental crust contains a much higher percentage of silicon and alu-

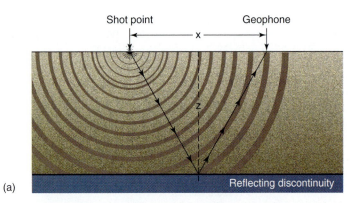

Figure 3.16   Reflection and Refraction. (*a*) Waves are reflected from an interface between two different materials in the crust. The line with arrows depicts the ray path from the shot point to a geophone that can detect the ground motion. (*b*) The ray paths of waves bend as the wave passes through rocks with different physical properties such as density and rigidity. On passing through Earth's layers, the waves are bent and follow a curved path.

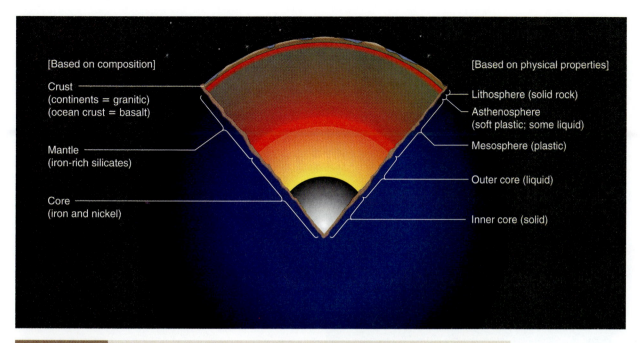

Internal Divisions of Earth. The internal divisions of Earth may be differentiated on the basis of their composition or their physical properties as indicated.

minum than oceanic crust. On continents, granitic rocks and gneisses are at the surface in shield areas. Drilling demonstrates that they also lie beneath the sedimentary rock cover in platform areas. These old igneous and metamorphic rocks also reach the surface in mountain belts where they have been elevated from great depths.

Information about the crust beneath the sea comes from seismic studies, supplemented by cores taken by ships outfitted to drill in deep water. Oceanic crust contains basaltic rocks rich in magnesium- and iron-bearing minerals. A relatively thin layer of sediment covers nearly all of these igneous rocks. In the deep sea, the oceanic crust that lies beneath approximately 5,000 meters (16,400 feet) of water consists of an upper layer of semiconsolidated sediment underlain by a layer of basaltic lava. Most of this lava contains distinctive pillow-shaped masses that form where basalt is extruded under water. Beneath these lavas, narrow, sheet-like bodies of basalt, called **dikes,** lead from the lava flows downward into large masses of basalt and its coarse-grained equivalent gabbro. These intrusions formed as a result of the cooling and crystallization of magmas that rose from the upper part of the mantle.

THE LITHOSPHERE AND ASTHENOSPHERE

Blocks of rock thought to be from the upper mantle are present at the surface in a number of places. These blocks are pieces of a rock called **peridotite,** composed of magnesium-iron-rich minerals such as olivine and pyroxene. This same rock type occurs in lavas that come from great depth in the mantle. It also is found in long

carrot-shaped bodies that resemble pipes. They originate in the mantle and rise toward the surface. In Africa, miners recover diamonds from these pipes. With depths close to 3.2 kilometers (2 miles), such mines are among the deepest on Earth. In some mountain belts, mountain building has forced oceanic crust up

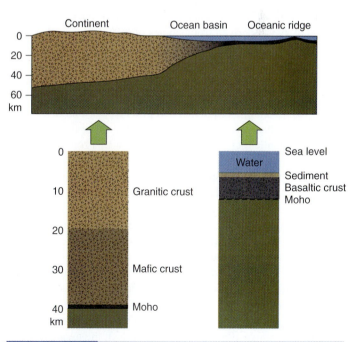

Figure 3.18 Representative Crustal Structure of Oceanic and Continental Crust.

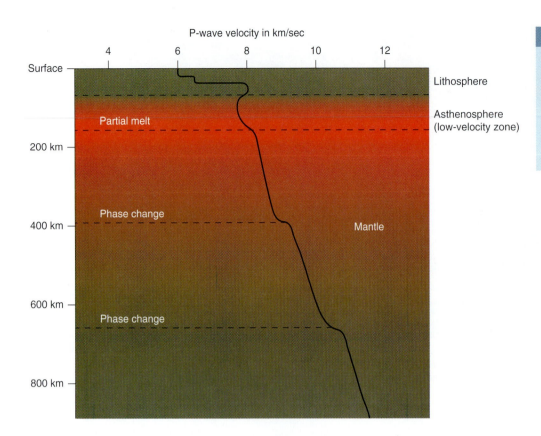

P-wave velocity in km/sec

Figure 3.19 P-Wave Velocity Varies with Depth. Breaks in the slope of the curve indicate changes in composition or physical properties of material in Earth. Such changes indicate the base of the lithosphere, the asthenosphere, and the location of phase changes in the mantle.

onto the edge of the continent. Exposures of this type occur in Newfoundland, in the Appalachian Mountains near Baltimore, Maryland, and in Cyprus. All of these upper-mantle rocks contain the same group of magnesium and iron-bearing minerals, notably olivine. These rocks have seismic velocities and densities that correspond to values derived from studies of seismic waves that pass through the mantle.

A significant change in seismic velocity occurs within the depth range of 100 to 200 kilometers (62 to 124 miles) in the mantle. Beno Gutenberg, an American seismologist, discovered that seismic waves traveling through this part of the upper mantle move at a slower rate than waves traveling above it in the solid, overlying rigid rock. Mantle rock both above and below this zone has higher seismic velocity. After Gutenberg discovered the low-velocity zone, called the **asthenosphere,** the overlying rigid layer came to be known as the **lithosphere** (figure 3.19). Thus, the lithosphere includes both the crust and the uppermost part of the mantle.

Both temperature and pressure determine the melting point of rocks. Measurements in mines and wells indicate that temperature below the surface increases with depth at a rate of about 30°C per kilometer (1°C for every 100 feet). If this rate of increase, known as the **geothermal gradient,** continues downward, at a depth of about 100 kilometers (62 miles), olivine-rich rock should be close to its melting point. This defines the upper boundary of the asthenosphere. Above 100 kilometers (62 miles), the temperature is normally too low for melting to occur in the lithosphere. Below 300 kilometers (186 miles), the pressure due to the weight of the overlying rock is too great for the mantle rock to melt despite in-

creasing temperature. Within the asthenosphere between approximately 100 and 300 kilometers (62 and 186 miles), the combined temperature and pressure make the hot rock plastic, somewhat like asphalt. These conditions may even produce a small amount of liquid melt. The zone of weak material plays a critical role in processes that affect the lithosphere. Most basaltic magmas originate in the asthenosphere, the same zone over which lithospheric plates move. The asthenosphere behaves like a plastic base on which the lithosphere floats.

Earth's Floating Lithosphere

The first evidence that parts of Earth's interior are not solid and rigid, that a liquid or plastic layer must exist beneath the rigid lithosphere, came from an unexpected discovery early in the history of tectonics. In 1850, George Everest, for whom Mount Everest is named, made a survey to determine the length of one degree of latitude in India. His party was working on the same problem—determining the shape of Earth—that Pierre Bouguer had studied earlier in the Andes. In making the survey, Everest used two surveying methods. One involved taking sightings on stars with a telescope. The other involved locating points on the ground surface by surveying. At the end of the survey, he found that the two sets of measurements did not agree (one set differed from the other by 5.25 seconds of arc). Double-checking, he found no error in the methods and reported the results.

The results of Everest's survey became the subject of investigations by John Pratt, the archbishop of Calcutta, and George Airy, a

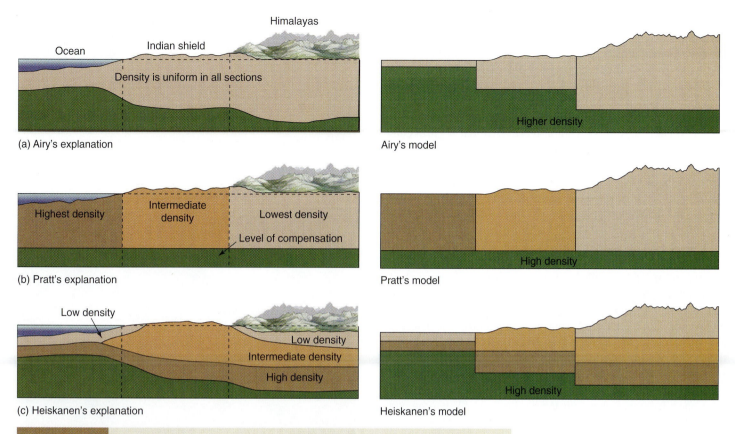

Ocean
Indian shield
Himalayas
Density is uniform in all sections

(a) Airy's explanation

Airy's model

Higher density

Highest density
Intermediate density
Lowest density
Level of compensation

(b) Pratt's explanation

Pratt's model

High density

Low density
Low density
Intermediate density
High density

(c) Heiskanen's explanation

Heiskanen's model

High density

Figure 3.20 Three Models Illustrating How Density Might Be Distributed Within the Lithosphere That Is in Floatational Equilibrium with an Underlying Layer (the asthenosphere). (a) depicts the model suggested by Airy. In this model, the outer part of Earth is of uniform density. Consequently, mountains have roots. (b) illustrates the model suggested by Pratt. In his model, mountains are high because they have lower density than areas of lower elevation. (c) shows a more modern model in which density varies from one part of the lithosphere to another. Model (c) incorporates what is known about the lithosphere and asthenosphere from seismic studies.

scientist in Britain. Neither knew of the other's work, but both concluded that the discrepancy in the survey could be traced to the pull exerted by the mass of the Himalayas on the plumb bob used to level the telescope in the survey. They estimated the mass of the mountains, calculated how much that mass should have thrown off the results, and found that the discrepancy in the measurements made by Everest should have been at least three times greater than it was. In other words, the gravitational attraction of the mountains pulled the plumb bob off vertical, but not as much as Pratt or Airy had expected. Both Pratt and Airy concluded that the mountains contained less mass than they estimated. However, they developed two very different ideas about how to account for the discrepancy (figure 3.20).

Pratt suggested that the rock in and beneath mountains has a lower density than that beneath plains. He concluded that mountains are high because they represent rock that has expanded on heating, just as dough rises. According to this model, rocks near the surface vary in density, but at some depth, density is uniform. On the other hand, Airy thought that the density of crustal rocks was about the same everywhere and that mountains must be high because they have roots that extend down into a dense molten layer.

Pratt and Airy devised their models long before scientists knew much about the structure of Earth's crust and lithosphere. Later seismic studies show that the outer portions of Earth differ from both of their models. As suggested by Pratt, continents differ from ocean basins in average density, but the crust is not uniform in density from top to bottom. As suggested by Airy, the thickness of continental and oceanic crust is different, and the thickest crust lies beneath the highest mountains, but no evidence indicates that roots of mountains extend into an underlying liquid layer. Today, most scientists think the asthenosphere is a plastic, perhaps partially molten layer in the upper mantle. The existence of this layer favors a model more like that of Pratt. In the modern model, the lithosphere "floats" on the asthenosphere. The lithosphere is relatively rigid throughout, but it does bend. By studying its response to shifting of loads, such as the growth or decay of ice sheets and the formation of large volcanoes on Earth's surface, scientists can judge the strength and flexibility of the lithosphere. As scientists accumulated more information about the outer part of Earth, it became clear that different parts of the crust do have different compositions and densities as postulated by Pratt. It also became evident that high mountainous regions extend to greater depths than other areas. Mountains do have "roots" as suggested by Airy. In addition, vertical variations in composition and density are present in different parts of the crust. These discoveries led a geophysicist, W. Heiskanen, to suggest the model shown in figure 3.20c.

The models devised by Airy and Pratt set the stage for C. E. Dutton, one of the early U.S. Geological Survey leaders, to formulate the **theory of isostasy.** Dutton recognized that if Earth consisted of homogeneous material, it would be a perfect spheroid, but if it is heterogeneous and some parts were denser or lighter than others, the lighter matter would tend to form bulges, and the denser matter would form flat areas or depressions. He proposed the term *isostasy* for this tendency.

Applying the Theory of Isostasy

The theory of isostasy tells us that the weight of all columns of Earth materials over the center of Earth, and probably over the asthenosphere, are equal. Anything that upsets that balance sets in motion movements in the asthenosphere and lithosphere that tend to compensate for the imbalance. Isostasy is an extremely valuable theory that may be used to predict how the lithosphere will respond to changes in the distribution of mass brought about by such processes as volcanism, mountain building, and erosion. The lithosphere has enough strength and rigidity to enable it to support loads of rock for short time spans, such as hundreds of years, but over long periods (thousands or millions of years), the lithosphere sags under the weight of heavy loads.

Though 97% of the lithosphere is in a state of isostatic equilibrium, the 3% that is not in equilibrium is unstable and is either rising or subsiding. Erosion, igneous activity, formation and melting of ice sheets, and mountain-building processes shift loads from one place to another on the crust.

Tops of mountains erode rapidly, and the downslope movement of materials, running water, and glaciers carry the eroded material away. The removal of mass causes the column of rock extending through the lithosphere under the mountains to weigh less relative to other parts of the lithosphere. As erosion removes rock, the pressure at the base of the lithosphere decreases, causing the plastic rock in the asthenosphere to flow under the mountains, uplifting them again. Thus, as a result of these isostatic adjustments, mountains tend to rise even as they erode away.

Some parts of Earth's crust are undergoing isostatic adjustments today. In parts of Greenland and Antarctica, the surface of the crust is depressed as much as 250 meters below sea level by the weight of the ice sheets. As recently as 15,000 years ago, central Canada and Scandinavia were sites of thick accumulations of ice (figure 3.21). These areas were depressed when the ice sheets were large, but since the ice began to melt, the land has been rising. Recent level surveys prove that the land is still rising. Marks made by Vikings at dock sites along the coast of Norway are several meters above the seawater level today. Portions of both Canada and Scandinavia are rising at rates of as much as several millimeters each year (figure 3.22). These same regions are places where the pull of gravity is lower than average. This indicates that the

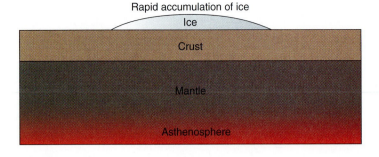
Rapid accumulation of ice

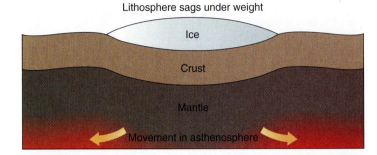
Lithosphere sags under weight

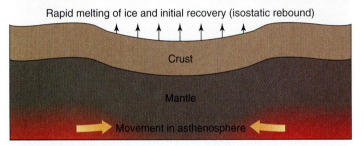
Rapid melting of ice and initial recovery (isostatic rebound)

Figure 3.21 Isostatic Adjustments. This schematic diagram illustrates the isostatic depression of the lithosphere that would result from the placement of a load on the surface. That load might be a large quantity of sediment deposited near an uplifted area, a thrust sheet moved in an active mountain belt, or growth of a continental ice sheet as shown here. If the ice sheet grows rapidly, the lithosphere is strong enough to sustain the weight, but as the ice grows thicker, the lithosphere subsides. Plastic material in the asthenosphere moves out from beneath the ice, and the crust sags. After the ice melts, material in the asthenosphere returns, and the lithosphere rebounds.

lithosphere under these areas is less massive than average. When ice was present, the extra weight caused material in the asthenosphere to flow away from the centers of ice accumulation. Once the ice started to melt, a return flow began, and the present uplift indicates that this flowage continues. As the central part of North America rises, Hudson Bay will grow smaller. Similar adjustments are taking place around the Great Salt Lake in Utah, which was filled to depths of 300 meters (985 feet) after ice in North America melted. As the water evaporates, the region rises because these adjustments are taking place.

By measuring the acceleration caused by gravity, symbolized by g, one can identify places where the value of g is greater than it

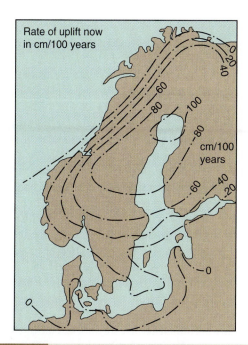

Rate of uplift now in cm/100 years

cm/100 years

Figure 3.22 Crustal Uplift Following Removal of a Load. Isostatic rebound is taking place in the area occupied by an ice sheet in Scandinavia. This sketch shows the rate of uplift for every hundred years.

should be if mass were uniformly distributed around Earth. If the value of g is abnormally high, the measurement, called a *positive gravity anomaly,* indicates a mass surplus underground. A low value, a *negative anomaly,* indicates a mass deficiency. Positive anomalies are present over many volcanoes. If the area around a volcano is not sinking, the crust is locally strong enough to support the weight of the volcano. Positive gravity anomalies might be expected over the Mississippi Delta, where large quantities of sediment are deposited annually on the continental shelf, but measurements in the region reveal no positive anomaly. It appears that compensation is taking place. The lithosphere is subsiding fast enough to maintain floatational equilibrium as the delta grows. The theory of isostasy provides insights to why the lithosphere rises and falls as mass shifts from one place to another on the surface. Later, the theory of plate tectonics demonstrated how large sections of the lithosphere shift horizontally.

THE MANTLE

All mantle rocks are somewhat similar, but seismic studies indicate that they are not identical. A number of abrupt changes in seismic velocities, referred to as **seismic discontinuities,** are present in the mantle (see figure 3.19). From laboratory experiments with olivine, geologists have found that the crystal structure of olivine changes as pressure increases. The pressure causes atoms in the olivine crystal to become more closely packed, creating the min-

eral spinel. At greater depth, spinel changes to another mineral, called perovskite. These changes in internal structure of the minerals result in changes in density and hence in seismic velocity. The step-like velocity increases in the upper mantle occur at depths where the pressure is great enough to cause a change in the crystal structure of olivine. As the crystal structure becomes more compact, the density of olivine increases, and seismic velocity rises.

Convection in the Mantle

Based on evidence from seismology, the interior of Earth consists of layers of matter that differ in density, with the more-dense matter beneath the less-dense matter. Thus, Earth's interior is density stratified. This model supports the idea that Earth was once molten. Because the solid outer portion of Earth provides a good insulation, residual heat from the early stages of Earth's formation is likely responsible for the high temperature inside Earth. The decay of radioactive isotopes in the mantle also generates heat. Regardless of the sources of heat, as the temperature of matter rises, its density decreases, and at some temperature, the solid begins to behave as a plastic.

How is it possible that mantle material can behave as a plastic but be rigid enough to transmit shear waves? Evidence from experiments with materials at high temperature and under great pressure proves that many materials are brittle if deformed rapidly or at low temperature and pressure. Those same materials behave plastically if deformation occurs slowly or at high temperature and pressure. Consequently, the mantle rock behaves as an elastic material in response to the deformation that occurs during the passage of seismic waves that move at miles per second. For such rapid deformations, the mantle rock is rigid, but over long periods, the mantle rock is plastic. The mantle rock acts like "silly putty" that shatters if suddenly hit, but if left undisturbed and under pressure, it flows. The rate of flow in the mantle is extremely slow; however, mantle materials in contact with the outer core may be more fluid, and some spots within the mantle are much hotter and more fluid than the surrounding material.

As heat from the molten core rises into the lower part of the mantle, the mantle rock becomes less dense, and it expands and flows upward through the surrounding material (figure 3.23). This type of overturn in the mantle is referred to as **convection.** It is analogous to the flow in a pan of water heated on a stove or to the rise of clouds on a summer afternoon. In all cases of convection, less-dense matter generally flows upward through higher-density material. Mantle rock flows because it is plastic. The higher-temperature plastic material flows through the lower-temperature, more-dense material.

Although those who have studied the question agree that convection takes place in the mantle, the exact pattern of convection is still debated. Some think two levels of convection are more likely—one in the lower mantle, another in the upper mantle. Others favor the idea that large convection cells operate throughout the mantle from the top of the outer core to the base of the lithosphere (figure 3.24). Initially, most scientists envisioned symmetrical patterns of convection. Today, most favor a much less symmetrical system of convection defined by slight

Figure 3.23 Convection in the Mantle. The rise of the hot plastic material in the mantle probably resembles the rise of clouds formed where hot air rises from Earth's surface.

differences in the seismic velocity of waves traveling in different directions through the mantle. This field of study is called **seismic tomography.**

As the amount of data from earthquakes has increased and high-speed computers enable more rapid and precise analyses of the data, geophysicists have discovered that seismic waves passing through the mantle travel with slightly different velocities in different directions. The elastic properties and temperature of mantle rock determine the velocity of seismic waves. The elastic properties of olivine crystals, one of the main constituents of mantle rock, vary with the orientation of the crystal structure. Because the crystals of olivine are elongated, they align in the direction of flow of the plastic mantle rock. Thus, mantle velocities are related to crystal orientations as well as temperature. This makes it possible to map convection flow patterns in the mantle (figure 3.25). These mappings reveal that some slabs of lithosphere sink all the way to the boundary between the mantle and outer core and that material rises from near this boundary back toward the surface. The cool down-going slabs sink near deep-sea trenches, and the returning hot material rises near the surface at spots beneath volcanic centers, such as the Hawaiian Islands and along oceanic ridges.

THE CORE

Dramatic changes in the composition, physical properties, and state of matter take place at the boundary between the mantle and the core—at a depth of nearly half Earth's radius. Because the crust and mantle have relatively low density, the core must be composed of a high-density material in order to account for the average density of Earth as a whole. Seismic studies indicate that the core is liquid, high density, probably composed of iron magma. This liquid core lies below the solid, silicate rocks in the mantle. Studies of mete-

orites indicate a probable composition for the core. Most meteorites are one of two principal types—**siderites,** composed of iron and nickel, and stony meteorites, called chondrites, composed mainly of olivine. Some planetologists have suggested that meteorites are either fragments of a planet that broke up or pieces of matter that failed to form planets. Thus, meteorites provide the best evidence of the types of materials making up the interior of Earth. The seismic velocity and density of siderites are approximately the same as Earth's core. In composition, chondrites resemble the core and mantle combined. The velocity of seismic waves traversing the outer core suggests that it has a density slightly less than pure liquid iron. This finding has led researchers to think that some lighter elements, possibly silicon, sulfur, or oxygen, are combined with the iron.

Seismologists identified the core by studying the seismic waves that travel through the deep interior. Soon after seismograph stations were set up in many different parts of the world and seismologists began to compare records, they found that most of the stations located more than 103° away from the focus of an earthquake did not receive direct compressional (P) or shear (S) waves (figure 3.26). Something, now identified as the core, blocked the direct path these waves were expected to travel through Earth. The core casts a shadow in much the same way a ball illuminated by a light casts a circular shadow. However, unlike light waves, P waves travel through the core along a curved path and return to the surface. Seismologists recognized the **shadow zone** by studying graphs showing the time required for P waves to travel from an earthquake focus to seismograph stations at various distances. **Travel time** is the time needed for a seismic wave to travel from the focus to any point on the surface. These graphs reveal evidence of layers in Earth's interior.

Careful study of the waves arriving at distances greater than 103° from the epicenter reveals that no seismic waves that travel direct paths appear between 103° and 142°. The travel time for the waves that come in at 142° is consistent with that expected for P waves that travel a path from the focus to the core-mantle boundary where they bend (refract) into the core. They pass through the core and bend again as they cross the boundary between the core and mantle before returning to the surface (figure 3.27). Thus, the shadow zone for direct P waves lies between 103° and 142°. The shadow zone for direct S waves is the entire zone beyond 103° (no direct S waves arrive beyond 103°). Thus, shear waves cannot pass through the core. Because liquids cannot transmit shear, failure of S waves to pass through the core indicates that at least the outer part of the core is liquid. Once seismologists determined the approximate depth to the core-mantle boundary, they were able to look for other seismic evidence of this boundary. They predicted, and later discovered, that some seismic waves reflect off the core surface and return to Earth's surface. These waves, called PcP or ScS waves (the lowercase c indicates a reflection from the core), appear on many seismograms (see figure 3.27). Waves that pass through the core are labeled PKP or PKIKP if they go through the inner core.

If the entire core were liquid and uniform in composition, P waves would pass through it with the same velocity regardless of their path. However, compressional waves travel through liquids at a lower velocity than they would if moving through a solid. This explains why P waves cross the outer core at a reduced velocity. Later,

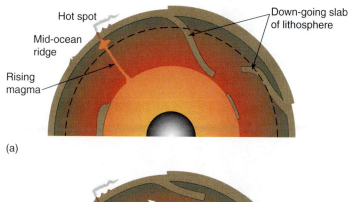

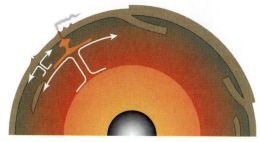

(a)

(b)

Figure 3.24 Models of Mantle Convection. (*a*) In this model, plumes of hot material rise from the outer edge of the core, and cool material from the lithosphere that sinks in subduction zones descends deep into the mantle. (*b*) In this model, heat rises from the core, but the formation of plumes of magmas occurs high in the mantle, and slabs of lithosphere do not sink more than a few hundred kilometers into the mantle.

Figure 3.25 Convection Flow Patterns in the Mantle. Recent geophysical data indicate that portions of the slabs of the lithosphere that sink into the mantle continue to much greater depths than previously thought. The rocks in the slabs are altered as they descend. Some portions of the altered slabs sink to the boundary between the core and mantle. As rock near the core-mantle boundary becomes hotter, its density decreases, and it rises. The rise of this hot material may create the hot spots at the surface.

Inghe Lehman, a Dutch seismologist, recognized that P waves traveling almost directly through the center of Earth travel at a slightly higher velocity than those that go through the outer portions of the core. This indicates that the inner portion of the core is solid.

Both the solid inner core and the liquid outer core rotate, but recently seismologists have found evidence that the inner core rotates more rapidly than the surrounding liquid (figure 3.28). The solid inner core rotates in the same direction as the rest of Earth, but it makes one more complete rotation than the rest of Earth every 400 years. This rotation of the core has significant effects on Earth's magnetic field.

EARTH'S MAGNETIC FIELD

Magnets have poles commonly called the north and south or the positive and negative poles. Scientists arbitrarily define the polarity of the field in terms of the north and south poles of Earth. The end of a magnetized needle that points toward Earth's magnetic north pole is the north-seeking, or north-end, of the needle. If the polarity of Earth's field reversed, as it has many times in the geologic past, the north-seeking end of the needle would point toward the south pole. If opposite poles of two magnets are brought close together, they attract one another; similar poles repel one another.

In a magnetic field, metals such as iron become magnetized. Because of this process, called **magnetic induction,** a paper clip picked up by a bar magnet becomes magnetized and will attract other paper clips.

Although the Chinese observed and described the magnetic effects of certain rocks 4,000 years ago, scientific study of Earth's magnetic field did not start until the sixteenth century. William Gilbert demonstrated that the magnetic field of Earth as a whole resembles that of a great bar magnet in which magnetized needles align themselves in the field. Both the Chinese and Gilbert were familiar with lodestone, a naturally magnetized rock that contains the mineral magnetite. Gilbert thought large masses of lodestone inside Earth created a permanent magnetic field around Earth. Scientists now know that the strength of the magnetic field varies, and they think the field originates in the molten outer core.

Since the time of Gilbert, scientists have learned much more about the character of Earth's magnetic field. For example, Earth's north magnetic pole does not presently lie at or even near the geographic North Pole where the axis of rotation of Earth penetrates the surface. Instead, the magnetic north pole is in the islands of northern Canada at about latitude 75° N. The south magnetic pole is located near Antarctica at latitude 68° S, south of Australia. Thus, if Earth's magnetic field were caused by a bar magnet, the axis of that bar would not pass through the center of Earth, but would miss

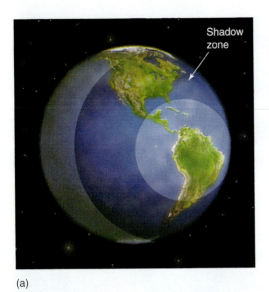

(a)

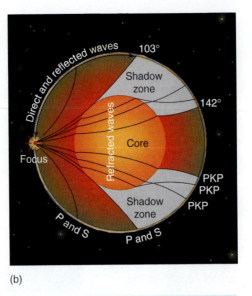

(b)

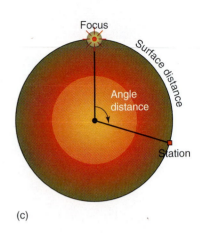

(c)

Figure 3.26 Paths Followed by Seismic Waves in Earth. (a) The shadow zone. (b) When P waves cross the boundary between the mantle and core, they are sharply bent. This creates a shadow zone for P at distances between 103° and 142° from the focus. Note that the inner core is not shown in this diagram; refer to figure 3.27. (c) Distances from earthquake epicenters and seismography stations are usually expressed in terms of the angle between lines connecting the station and focus with a point at Earth's center.

it by almost 1,000 kilometers (600 miles). Even more difficult than the position of the poles to explain, in terms of Gilbert's model, are observations that the positions of the poles have shifted over time; the polarity of the field has changed; and the strength of the field slowly, but continually, varies. These changes prove that some dynamic process acting inside Earth causes the strength and orientation of the main magnetic field to change. Clearly, the field is not that of a permanent magnet, as Gilbert had proposed.

Strength, polarity, and orientation define the magnetic field. Two measurements define the field orientation. These are declination and inclination. **Magnetic declination** is measured in a horizontal plane. It is the angle between the direction to the geographic North Pole (the pole of Earth's rotation), called **true north,** and the direction the compass needle points. Along a north-south line passing from the Great Lakes to Florida, declination is almost zero, but it rises to 15° west of true north in Maine and over 20° east of true north in Washington state. The **inclination** of the field is the angle between a magnetized needle mounted on a horizontal axis and the horizon. (Compass needles do not indicate inclination because a weight is tied on the south end of the needle.) Envision the magnetic field as imaginary lines showing the angle a magnetized needle would assume at any place around Earth. These lines are nearly horizontal near the equator. Closer to the magnetic poles, they are steeply inclined and become nearly vertical at the magnetic poles. In middle latitudes, lines of the magnetic field have an inclination of 50° to 70°.

The strength of a magnetic field is the force of attraction it exerts on a magnet. That force is directly proportional to the strength of the magnets (called pole strength) and inversely proportional to the square of the distance between them. This explains why two bar magnets with opposite poles held close together exert a strong pull that decreases as the magnets are separated. The relationship resembles the law of gravitational attraction. The force of attraction Earth's field exerts on small magnets provides a measure of Earth's field strength. The strength of the field is highest near the poles, but it is not constant.

Variations in Earth's Magnetic Field

Measurements of the magnetic field collected by scientists for over a hundred years indicate that the magnetic field not only varies from place to place, but also over time at the same place. Some of these changes are periodic, but others are irregular. Magnetic field strength varies from night to day and with the season of the year. These changes indicate that the Sun influences the magnetic field. Cycles related to motions of the Moon are also known. Generally, these variations account for a small part of the total field, and they are predictable because they have fixed periods. Sunspot activity causes highly irregular variations in the strength of the field. When sunspots are active, high-energy charged particles from the Sun bombard Earth, causing problems with radio reception. They also cause unusually bright auroral displays. At such times, Earth's magnetic field changes abruptly, showing drastic increases and decreases at different places on Earth's surface. Sunspot activity is cyclic, reaching a maximum about every eleven years, but occasional "magnetic storms" caused by solar flares may occur at any time. These storms may have a detrimental influence on orbiting satellites and interfere with electric transmissions on Earth.

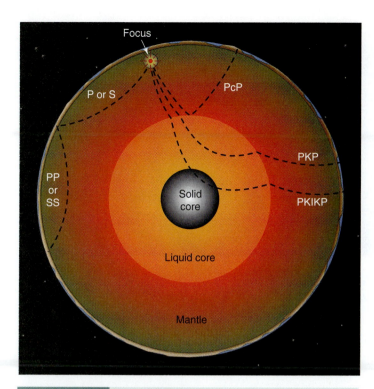

Figure 3.27 Wave Paths of Selected Seismic Waves Generated During an Earthquake at a Single Focus. P = compressional wave path; S = shear wave path; c = reflections from the core; K = path through the outer core; and I = path through the inner core.

Long-term changes in the orientation, strength, and polarity of the magnetic field also take place. (Changes in polarity are discussed in chapter 5.) As early as 1635, Henry Gellibrand, an astronomer in Britain, noted that the position of the north magnetic pole was shifting. Later scientists, who referred to the effect as magnetic drift or polar wandering, suggested that the magnetic pole may be moving about the geographic North Pole. As more data became available, they identified two types of polar wandering. One type involves movement of the magnetic poles around the geographic pole; the second is an apparent motion that results from the movement of lithospheric plates.

Both the orientation and the magnetic field strength continually change everywhere across the face of Earth. These changes show up as small but continuous increases or decreases in the declination, inclination, and field strength. Contour maps showing the rate of change indicate that the field is growing stronger in some places and becoming weaker in others. Convection in the core where the main part of the magnetic field originates is responsible for these changes. Strong temperature differences between the inner core and the mantle could cause convection in the outer core. Crystallization and sinking of minerals in the outer core could also stir up the outer core. If this happens, the inner core is growing larger. In addition to these changes in the magnetic field, the field overall is shifting towards the west and losing strength at a rate of about

1/1,500 of its value each year. These variations prove that the magnetic field does not originate from a permanent magnet, that it must, therefore, originate from a dynamic process.

Origin of Earth's Magnetic Field

Earth's total magnetic field results from the combination of three distinctly different magnetic fields. Charged particles that reach Earth's atmosphere from the Sun cause one of these fields (figure 3.29). The strength of this field varies greatly and reaches a maximum at times of high sunspot activity, but it normally causes only a small part of the total field. Permanently magnetized rock in the crust and upper mantle creates a second magnetic field that remains stable over time. These magnetic rocks account for about 20% of Earth's total field. The remaining field, called the **main magnetic field,** which normally accounts for nearly 80% of the total field, arises from the core. The orientation and strength of this part of the field undergo slow progressive changes.

Electric currents cause magnetic fields. In 1820, Hans Øersted observed that passing a wire carrying an electric current over a compass needle causes the needle to rotate until it points at right angles to the wire. This demonstrates that electric charges not only have electric fields associated with them, but also set up a magnetic field oriented perpendicular to the direction of motion of an electric current. This connection between electric and magnetic fields set the stage for modern ideas about the origin of Earth's magnetic field.

Some of the baffling changes observed in the main field, such as variation in field strength and polarity, are easier to understand if electric currents rather than a magnet cause the magnetic field. The iron composition of the core is favorable for electric currents, and

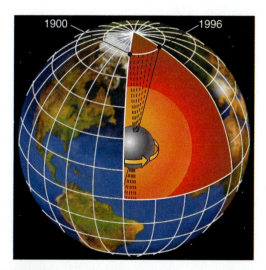

Figure 3.28 Inner and Outer Cores of Earth. In this breakaway view the solid inner core lies at the center of a much larger liquid outer core. The inner core rotates slightly faster than the outer core.

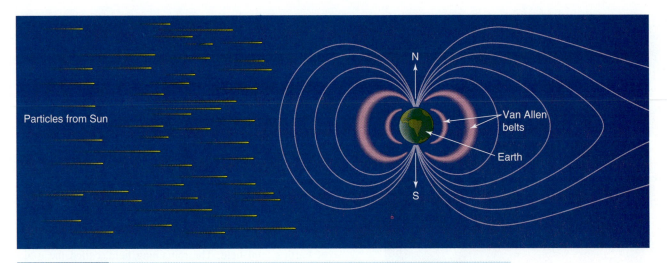

convection in the hot, fluid outer core is probably responsible for sustaining currents once they exist. Electric currents might begin in the core as a result of chemical processes similar to those in a battery. Slow movements within the molten outer core or movements along the boundary between the solid inner core and the liquid outer core may cause such electric currents. Differences in the rate of rotation of the inner core, outer core, and possibly the mantle could also cause electric currents to flow and generate magnetic fields.

The theory described, generally referred to as the **self-exciting dynamo theory,** agrees with what is known about Earth's interior from seismology. This theory also explains why Earth's magnetic field varies. Small, eddy-like convection currents present in the core explain the local variations observed at Earth's surface. The rotation of Earth imparts a general pattern of symmetry to the entire field, causing alignment of the fields arising from convection cells. Thus, the magnetic effects observed at the surface of Earth are the combined effects of all of the fields caused by convection cells, differences in rotation inside Earth, and minor variations due to permanently magnetized rock in the lithosphere and astronomic processes.

SUMMARY POINTS

1. The average elevation of continents is much higher than that of the ocean basins. Density differences arising from the composition of the rocks in these two types of crust account for this difference.

2. Portions of all continents are composed of rocks more than 540 million years old. Most of these occur in cratons where the structure of the rocks indicates that parts of the cratons were once parts of mountain systems. The mountain building took place so long ago that erosion has reduced the mountains to nearly flat regions.

3. Rocks in mountain systems are strongly deformed as though the rocks had been compressed in a vice folding and faulting them.

4. Passive continental margins have little seismic or volcanic activity, and they have been only slightly deformed. In contrast, active continental margins are sites of many volcanoes, most earthquakes, deep-sea trenches, and mountain building.

5. Most information about the deep parts of Earth's interior comes from geophysical studies, especially from analyses of the velocity of seismic waves passing through the interior and from Earth's magnetic field. Experimental work in crystal physics and petrology has provided the basis for comparing geophysical observations with natural materials.

6. Two types of seismic waves pass through Earth's interior. Compressional (P) waves, similar to sound waves, involve the passes of compressions and rarefactions through the interior. Shear (S) waves involve distortion of the rock.

7. Seismic discontinuities enable geophysicists to identify the major divisions of Earth's interior. These divisions, forming concentric shells in the interior, are the crust, lithosphere, mantle (including the upper mantle, asthenosphere, and lower mantle), outer core, and inner core.

8. The Mohorovičić discontinuity (Moho) lies at the base of the crust. A rigid part of the upper mantle lies directly beneath the crust. A plastic zone in the mantle, the asthenosphere, defines the base of the lithosphere. Thus, the lithosphere includes the crust and the solid upper mantle located above the asthenosphere. The shadow zone and the reflections from the core allow seismologists to identify the boundary between the core and mantle. The higher velocities of waves traveling

through the center of Earth make it possible to identify the inner core.

9. The seismic velocity of waves passing through the various divisions of Earth's interior make it possible to identify their compositions. Continental crust has an average chemical composition similar to that of granite (see appendix C). The oceanic crust has an average composition similar to that of basalt. Iron-magnesium silicate minerals, including olivine, pyroxene, and plagioclase feldspar, are the main components. The outer core appears to be molten metal, largely consisting of nickel and iron and similar to metallic meteorites.

10. Although the mantle is rigid enough to transmit elastic waves, it is so hot that the rock undergoes slow plastic flow. Temperature differences between the crust and the core drive convection in the mantle.

11. Geophysicists describe Earth's magnetic field as its strength, polarity, and orientation. Unlike the field of a permanent magnet, Earth's field varies. The magnetic field gradually changes in strength and orientation. Its polarity has also changed in the past.

12. The main magnetic field probably arises from within the outer core of Earth where movements in the liquid iron create electromagnetic effects that generate the magnetic field. Permanently magnetized rock in the crust and mantle cause a small part of the total field. In addition, the field is subject to fluctuations related to sunspot activity.

13. Rocks containing certain minerals, especially magnetite, may become permanently magnetized in Earth's field.

REVIEW QUESTIONS

1. How do the inner core, outer core, and mantle differ from one another in composition and physical properties?
2. What is the distinction between the crust and lithosphere?
3. What evidence supports the idea that Earth has a liquid core?
4. How are gravity measurements used to define the shape of Earth?
5. What causes the magnetic field, and where does the field originate?
6. Why do geologists think some connection exists between convection in the outer core and the magnetic field?
7. How do the various types of seismic waves generated during earthquakes differ from one another?
8. What characteristic do scientists use to describe the magnetic field?
9. How is evidence of the magnetic field that existed in the geologic past preserved in rocks?
10. How is it possible to determine what is deep inside Earth?
11. What can one learn about the interior of Earth by measuring the acceleration due to gravity?
12. Why do rocks deep in the mantle remain solid, while those in the asthenosphere are close to melting?

13. Describe how the angle at which a seismic wave approaches a discontinuity determines whether or not it is reflected or refracted.
14. What changes does olivine undergo with depth in the mantle?
15. How do mountains in the ocean basin differ from those on land?
16. Why are some mountain belts so much higher than others?
17. Why are continents so much higher than ocean basins?
18. How does the average composition of continents differ from that of oceans?
19. How do passive continental margins differ from active continental margins?

THINKING CRITICALLY

1. What properties of the asthenosphere affect surface processes?
2. What processes operating in the core affect life on Earth's surface?
3. How would humans be affected if Earth's magnetic field disappeared?
4. How would humans be affected if the polarity of Earth's magnetic field suddenly changed?

KEY WORDS

abyssal plains
asthenosphere
Atlantic-type continental margins
compressional seismic waves
continental platforms
continental rise
continental slope
convection
cratons
crust
dikes
elastic limit
epicenter
faults
focus
geothermal gradient
inclination
island arcs
lithosphere
longitudinal waves
Love (L) wave
magnetic declination

magnetic induction
main magnetic field
peridotite
primary (P) waves
rarefaction
Rayleigh wave
rays
refraction
seamounts
secondary (S) waves
seismic discontinuities
seismic tomography
seismic waves
self-exciting dynamo theory
shadow zone
shear stress
shields
siderites
surface wave
theory of isostasy
travel time
true north

chapter four

Time and Change in Earth Systems

Chapter Guide

By examining the present, we can unlock the mysteries of the past. Rocks in Earth's crust preserve a vast but incomplete account of Earth history. Many environmental processes began operating in the early years of Earth's evolution, which spans 4.5 billion years. By looking at processes causing changes today, we can understand the changes of the past and the potential for global changes in the future.

The primary goals of this chapter are to identify the time frames within which natural processes function, to learn some ways geologists have deciphered Earth history, and to identify some pivotal events in that history. Some processes bring about change in short time intervals. Others take place so slowly that changes are barely noticeable even over all of human history. Most of what is known about these processes is learned from studying the rock record. How are events from the past recorded in the rock record? How can scientists determine the age of these events? How is knowledge of the present used to understand the past? This chapter explores answers to these questions.

Rock layers exposed in the walls of the Grand Canyon preserve a history spanning hundreds of million years. This view is of Mount Hayden.

INTRODUCTION

Discovering the rates at which natural processes alter the environment is a critical part of understanding the operation of Earth systems. Is mountain building taking place today? How fast are mountains rising? How fast are species becoming extinct? Is the atmosphere becoming warmer? How would a warmer Earth affect the growing seasons? How fast does ice on Earth melt? Is sea level rising? How fast might it rise? How would patterns of rainfall change on a warmer Earth? Such rate-related questions have immediate implications for humans and public policy. One way to answer these questions is through studying Earth history. Humans have observed and recorded only the last few thousand years of this history. Sediments, rocks, and the fossil record contain information about past changes that extend almost to the origin of Earth. This record contains information about how rapidly the climate changed at the onset of the last glacial advance and how the climate changed as ice sheets grew and decayed. The rock record contains answers to a host of other questions such as how long it took for the Grand Canyon to form, how quickly mountains rise, and how fast organisms evolved from relatively simple forms to the diverse complex animals that live today.

Earth is constantly evolving. Some of the processes responsible for this evolution, such as the drifting of plates, affect large areas and take place so slowly that they pass unnoticed. Nevertheless, these imperceptible shifts eventually alter the environment dramatically. Imagine the changes that must have taken place as the plate on which India is located moved from a position adjacent to Antarctica to its present location north of the equator. Thousands or even millions of years may pass before such changes are evident. As India collided with Eurasia, the Himalayan Mountains rose to their present height. While mountain building is imperceptibly slow, earthquakes generated by the collision take place quickly and may have devastating impacts on people who live nearby. Figure 4.1 depicts the time frame and area affected by some important processes that shape the environment. Studying processes that modify the environment over short time spans requires distinctly different methods of measuring time from those used in examining processes that require thousands to millions of years. In the last few decades, it has become clear that Earth's environment is changing. Concern about changes that might affect the balance of the Earth system in such a way that it will not support human life is a strong motivation for monitoring life-support systems, such as protection from ultraviolet radiation and the supply of essential nutrients.

TIME FRAME—HOURS TO DECADES

Violent storms, lightning, and earthquakes can cause drastic local changes in seconds, minutes, or hours. Hurricanes, volcanic eruptions, plant growth cycles, seasonal temperature changes, melting of snow, and growth and decay of sea ice in polar regions occur from days to a year (figure 4.2). Over decades, human activities, such as burning fossil fuels, dumping waste into rivers and oceans, using toxic chemicals, excessive irrigation of agricultural land, and modification of coasts, significantly impact processes that cause change.

Many of these rapid processes directly affect the daily lives of humans and are much more likely to attract attention than those that operate in longer spans. Government agencies conduct most

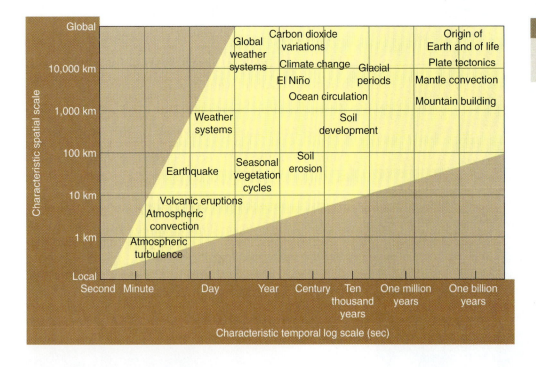

Figure 4.1 Characteristic Spatial and Temporal Scales of Selected Earth System Processes.

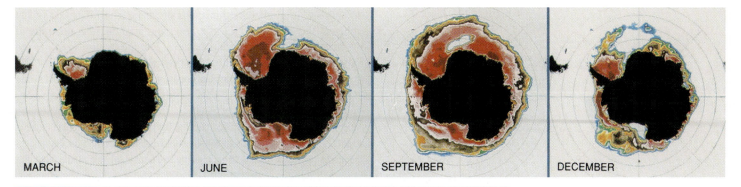

MARCH   JUNE   SEPTEMBER   DECEMBER

Figure 4.2 Satellite Images Depicting Seasonal Changes in the Amount of Sea Ice (shown in color) in the Antarctic. Satellites provide ways of monitoring conditions and changes in many parts of Earth's environment.

environmental monitoring. They are responsible for measuring air and water quality, the quantity of water flowing through streams, the health of animal and plant populations, crop production, shifts in the position of the margins of glaciers, changes in shorelines, and for tracking the weather.

Today, many of the most cost-effective systems used to monitor Earth's environment operate from satellites. Before the advent of satellites, most Earth systems' monitoring occurred on the ground or in the oceans, but new technology has dramatically improved the documentation of short-term changes in the environment. Pictures taken by handheld cameras gave way to air photographs that produced map-like images. Later, color and infrared film that detect heat sources added variety to the types of images. Major advances accompanied the development of satellites capable of circling Earth in a matter of hours. Satellite images are improving human ability to document these changes. For example, infrared

images taken in 1992 and 1997 of part of the Missouri River flood plain show the formation of a new channel in one of the big bends (figure 4.3). In another case, Airborne Topographic Mapper data collected between October 1997 and April 1998 reveal the dramatic retreat of a sea cliff near Pacifica, Oregon, during the six-month period (figure 4.4) when a succession of storms struck that coast.

In a fraction of the time needed to collect data at ground level, satellites provide information from large areas, including those that are remote and inaccessible on the ground. Consequently, satellites provide a practical method for tracking many different types of global and local changes. Images obtained from satellites make it possible to track the movement of icebergs, measure the growth and decay of ice sheets, and follow seasonal changes in plant-life production in the oceans and on land. These images reveal where fires are destroying forests or spot the source of sediments or pollutants that are entering streams and the oceans. Although it is convenient to distinguish short-term from long-term processes, many processes that have significant impact in short time spans are ultimately caused by those that act over much longer periods. For example, continents driven by convection currents in the mantle may move toward one another at rates of a few inches or less each year, but as they collide, the strain that gradually builds up in the crust may release in seconds or repeatedly over a few days.

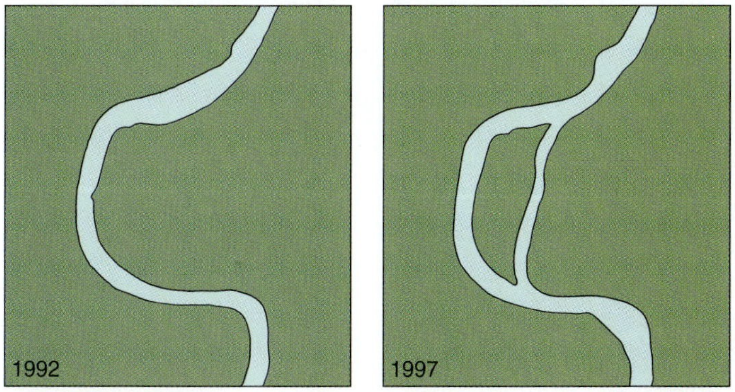

1992   1997

Figure 4.3 Sketches of Part of the Flood Plain Along the Missouri River North of Boonville, Missouri. The sketches are drawn from a Landsat image taken in 1992 and a SPOT image taken in 1997 of an area 7 kilometers (4.2 miles) wide. Note the new channel on the 1997 image.

TIME FRAME—DECADES TO CENTURIES

Many of the natural processes that shape the human environment require decades or even centuries to produce noticeable differences. Changes in climate, soil erosion, the atmosphere's chemical composition, circulation in the deeper layers of the oceans, and terrestrial and marine biological systems take place slowly. Historical records provide accounts of dramatic events that occur within time

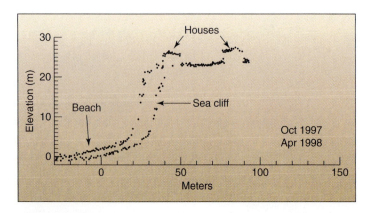

Figure 4.4 Profiles Across the Shoreline Made by Airborne Topographic Mapper in October 1997 and April 1998. Note the retreat of the cliff that occurred over this six-month period.

frames of decades to centuries. For example, changes in climate during the twelfth and thirteenth centuries were of such magnitude that the period is known as the **Little Ice Ages.**

TIME FRAME—THOUSANDS TO MILLIONS OF YEARS— THE GEOLOGIC PERSPECTIVE

Plate movements, mountain building, and isostatic adjustments like those described in chapters 3 and 5 occur at imperceptible rates. Slight changes in the tilt of Earth's axis and in the shape of the path Earth takes in its journey around the Sun affect the climate. Formation of soil and changes in the geographic distribution of

plants and animals take thousands of years. The formation of the planet, the origin of Earth's core and mantle, the opening of oceans, the rise of mountain systems, and the evolution of life are all products of processes that require millions of years.

Although most long-term processes go largely unnoticed, some subtle changes eventually have significant effects on the environment. Changes in the shape of Earth's orbit around the Sun, in the tilt of the axis of rotation, and a slight wobble of the axis all affect the amount of solar radiation that reaches Earth. Since solar radiation is the primary source of heat energy on the surface of the planet, changes in the amount of solar radiation affect the heat balance in the atmosphere and contribute to the growth and decay of ice caps. As the ice melts, sea level rises. About 18,000 years ago, sea level stood nearly 100 meters (330 feet) lower than it is today. Since that time, sea level has risen more than 0.5 centimeter (0.20 inches) per year. Some oceanographers estimate that sea level continues to rise 1 millimeter (0.003 inch) or more each year. At that rate, sea level will be another 1 meter (3.3 feet) higher by the year 3000. Careful studies of the rate at which sea level is rising are taking place, and some scientists think the rate may soon accelerate.

Other slow-acting processes produce dramatic changes over long time spans. Formation of new sea floor at the Mid-Atlantic Ridge is increasing the width of the Atlantic 4 centimeters (1.6 inches) per year. If this spreading rate continues, the Atlantic Ocean will be 40 kilometers (25 miles) wider 1 million years from now. A mountain range that rises a centimeter each year would be as high as the Himalayas in a million years.

These slow processes trigger other processes such as volcanic activity or earthquakes that occur in time frames of minutes to years. These natural events may be catastrophic for humans (figure 4.5). Geologic evidence indicates that life on Earth went through long periods of slow evolutionary change punctuated by rapid change. Learning how these interactions occur and discovering how to

Figure 4.5 Earthquake Damage During a California Earthquake That Caused Freeways to Collapse.

quantify them are among the goals of Earth systems science. The rock record is the only archive of these processes. To understand these geologic processes, one must learn to interpret the rock record and understand the implication of what is called *geologic time*.

THE ROCK RECORD

The accretion of materials composing Earth continued after the oldest known rocks formed. What had formerly been thought to be a large gap in time has recently been closed because geologists found grains of zircon in rocks in Australia that are 4.4 billion years old. This leaves a gap of only 100 million years between the formation of the planet and the crystallization of the oldest rock. These old rocks were part of an early-formed continental crust rich in silicate minerals. Rocks found during exploration on the Moon provide another line of evidence used to infer conditions during the early history of Earth and the solar system.

Like the rocks on the Moon, Earth is about 4.6 billion years old. It is hard to comprehend the implications this vast time span has for the processes acting on the surface of Earth and within its interior. The slow evolution of organisms, and movements of lithospheric plates cause dramatic changes if continued for millions of years.

Changes in the landscape illustrate the importance of long time spans in the operation of geologic processes. For those who accept the idea that Earth is only a few thousand years old, how rivers like the Colorado can cut canyons the size of the Grand Canyon is a perplexing problem (figure 4.6). For people who know how streams erode and transport rock debris, it is not difficult to visualize the deepening of the canyon over several million years. The rocks ex-

posed in the walls of the canyon gradually crumbled, broke away, and fell into the Colorado River, which transported them to the sea before dams blocked their downstream movement. The canyon is being enlarged today in about the same way and perhaps as quickly as it ever has been in the past. The face of Earth has never before looked exactly the way it does today; it will never be the same again. The landscape changes even from one day to the next, but the changes are so slow that people generally fail to perceive them. In the short term, these changes seem insignificant; in the long term, they may convert areas occupied by seas into mountains, and mountains into plains. The cumulative effects of natural processes over time have caused the changing record of Earth's history.

Most of the oldest rocks on Earth are igneous and metamorphosed sedimentary rocks formed long before invertebrate animals like mollusks, corals, and gastropods came into existence (figure 4.7). These crystalline rocks contain a record of events that took place during nearly nine-tenths of Earth's history, but they are difficult to interpret. Only the youngest contain any fossil remains. Many of these rocks have been greatly altered and deformed, and in most places, younger sedimentary rocks cover them. The largest exposures of crystalline rocks occur in continental shields (see chapter 3).

Sedimentary rocks are exposed over three-quarters of Earth's land surface, and sediments cover most of the sea floor. Because this sedimentary cover is hundreds to thousands of feet thick, rarely is it possible to see deeply into this pile of rock. The Grand Canyon of the Colorado River is one of those rare places where it is possible to see so deeply into the crust. Much of the sedimentary rock exposed on continents formed when shallow marine waters covered parts of the continents. Changes in the shape and depth of ocean basins cause oceans to flood over the edges of continents.

Figure 4.6 Photograph of the Grand Canyon. Note that layers that resist erosion form cliffs. Less-resistant layers have lower slopes.

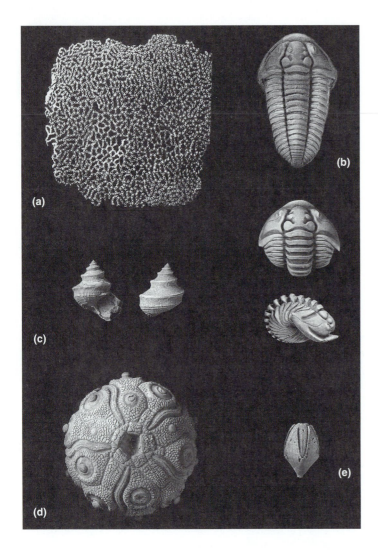

(f)

Figure 4.7 Fossils. (*a*) Coral, (*b*) trilobite, (*c*) gastropod, (*d*) echinoderm, (*e*) blastoid, and (*f*) photograph of algae. Fossils such as these indicate the time during which they lived.

Even today, oceans cover continental shelves—the edges of the continents. The sea has advanced over and retreated from portions of continents many times during the past half billion years, leaving a history of these events in the stratified rocks. Many of these dramatic changes on the continents coincided with the formation of mountain systems.

When interpreting Earth's history, the present is a key to the past. In the mid-eighteenth century, George Füchsel, a physician, first expounded this idea known as the **theory of uniformitarian-**

(a)

(b)

Figure 4.8 Records Left in Rocks. (*a*) The layer that looks like a fence in this cliff in Yellowstone National Park is a buried lava flow. (*b*) Lava flows exposed south of Santa Fe, New Mexico.

ism. In essence, this theory states that the principles that govern the operation of natural processes shaping Earth today have remained unchanged through time. Put another way, the observable is the key to the unobservable. Uniformitarianism does not imply that Earth has remained unchanged or that rates of change are constant; it implies only that the functioning of "natural laws," such as the law of gravitational attraction, has remained the same. The idea did not gain widespread attention until a Scottish physician and scientist named James Hutton (1726–1797) used the idea to show that basaltic rocks exposed in rock exposures far from modern volcanoes formed by volcanic processes (figure 4.8). The theory of uniformitarianism provides the basis for much of the reconstruction of Earth's history.

The theory of uniformitarianism helps us understand the nature of the record left in rocks—the **stratigraphic record.** By studying the way sediments form today, geologists infer how similar sediments formed in the past. For example, in the Bahama Banks, sediment composed largely of calcium carbonate accumulates on a large area of the sea floor. Farther north, sands composed of quartz eroded from rocks in the Appalachian Mountains form beaches along the seashore, and silts and muds also derived from the mountains settle in deeper water on the continental shelf. Today, distinctly different types of sediment are settling to the sea floor at the same time. Each one is a product of the physical and biological environment in which deposition occurs. Animals and plants that eventually become fossils in the sediment are also products of this same physical and biological environment. Some plants and animals live in restrictive environments. For example, corals grow in shallow marine water, and the sandy sediment around the reef is composed of the broken fragments of the reef. Other organisms such as sharks and algae inhabit the open ocean. Their remains occur in many sedimentary environments.

The concept of uniformitarianism has not always been widely accepted. In the nineteenth century, as scientists began to examine and interpret the rock record over large areas, they discovered evidence that prominent changes in the rock record took place about the same time in different places. Some scientists favored the idea that catastrophic events affecting large parts of Earth brought about massive changes. In part, the belief that Earth was only a few thousand years old was responsible for this view. Gradually, **catastrophism** gave way to the notion of slow evolutionary changes, which included many minor catastrophes, like hurricanes and floods. More recently, evidence of catastrophic meteorite impacts on Earth supports the concept of gradual evolution punctuated by events that caused rapid global changes. In trying to determine which view is correct, dating events is critical.

From earliest recorded history, humans have based their concept of time on the apparent movements of the Sun and Moon around Earth. Even prehistoric monuments, such as Stonehenge, in England, indicate that early humans tracked the movement of the Sun and Moon. Today, everyone accepts the idea that a year is the time required for Earth to make one complete revolution around the Sun, and the lunar month is the time between two consecutive full Moons. This system provides a precise way of keeping track of events and the time interval between them, but only the most recent portion of Earth's history has taken place since humans began using these absolute measures of time to keep records and record events.

For the portion of Earth's history before written records were made, geologists date events in two distinctly different ways: (1) by establishing the relative age of events and (2) since the 1950s, by determining their absolute age in years. The first method involves learning to interpret the record preserved in rocks. The second depends largely on decay rates of radioactive isotopes.

Figure 4.9 Disturbed Sequences of Strata. This cliff in the French Alps exposes a large fold in which the layers near the bottom are upside down.

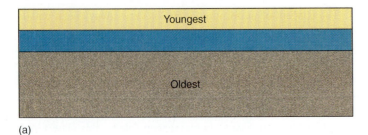

(a)

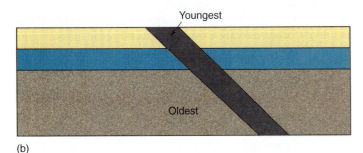

Youngest

(b)

Fault is youngest event

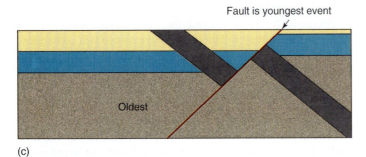

(c)

Figure 4.10 Sequences of Events Are Established by Using the Laws of Superposition and Crosscutting Relationships. (*a*) In layered sedimentary rocks, the oldest layer is on the bottom. (*b*) Sedimentary layers cut by an igneous intrusion are older than the intrusion. (*c*) Layers and the intrusion later cut by a fault are older than the fault.

Relative Ages of Events

Most modern marine sediments form as flat-lying layers that accumulate slowly on the sea floor. Similarly, ancient layers that crop out in the upper parts of the Grand Canyon remain horizontal hundreds of millions of years after they formed the canyon (see figure 4.6). In the seventeenth century, philosopher Nickolas Steno (1631–1687) recognized that most sediments are horizontal at the time of deposition and used this idea to establish the relative ages of the sequence of events recorded in layered sedimentary rocks. He recognized that in piles of undisturbed sedimentary rocks, the layers on top are younger than the layers under them. This concept, known as the **law of superposition,** provides an invaluable tool widely used to reconstruct sequences of events in Earth's history. As he developed his theory, however, Steno realized that this principle does not always apply to badly disturbed sequences of strata. He had come across places where layers of sedimentary rock fold back on themselves in such a way that part of the sequence is upside down. In other places where faults break and displace the sequence, older layers lie on top of younger layers (figure 4.9).

In addition to the law of superposition, which establishes the relative age of layers deposited in sequence, the principle of **crosscutting relationships** is a valuable way to establish relative ages and sequences of events. Near large intrusions of igneous rock, thin, tabular-shaped intrusions, called **dikes,** commonly cut across the rocks surrounding the large intrusions (see chapter 7). The dikes are younger than the rock through which they pass. In general, rock bodies are older than events that modify them in any way (figure 4.10). The principle applies to many similar relationships. For example, a fault is younger than rocks displaced by the fault; a lava flow is younger than the rock over which it flows. The time of metamorphism of a sedimentary rock is younger than the formation of the sedimentary rock, and fragments of a rock

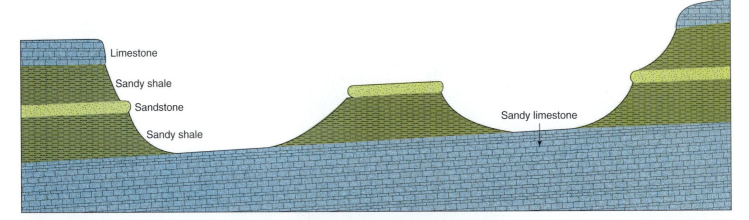

Figure 4.11 Correlation by Similarity of Rock Type and Position in a Sequence. In the region depicted by this cross section, rock types are constant enough to make correlation of layers possible by similarities in texture, in color and composition, and by their position in an established sequence. Correlation is also achieved by comparing fossil fauna and flora.

enclosed in another rock are older than the enclosing rock. Application of such simple logic provides a convenient way to interpret the history of events preserved in layers of rock. Like the pages in a book, each layer of sedimentary rock is a record of what was happening when it formed and often holds clues to what preceded its formation.

Piecing the Rock Record Together

In trying to establish the geologic history of a large area, correlation of the sequence of events that took place simultaneously in different parts of the area is necessary (figure 4.11). Geologists make these correlations by matching fossil content, rock types, and age of layers. Some strata, especially those that resist weathering and erosion, produce distinctive landforms, such as cliffs, or have distinctive colors that show up on aerial photographs and satellite images as continuous features over vast areas (figure 4.12). These can be seen from the rim of the Grand Canyon where distinct, continuous layers extend from horizon to horizon (see figure 4.6). Even where individual layers thin out and disappear, they are usually part of a sequence of layers that recurs throughout a large region. In some places, the sequence may not be exposed, or it may lie buried beneath younger layers. In these places, records from deep wells may be needed to establish its presence.

Parts of the rock record indicate uninterrupted deposition for many millions of years, but breaks in the record are common. These breaks show up in a number of ways. The groups of fossils present in one layer may be quite different from those in the overlying layer. In a few instances, it appears that most of the lifeforms in the older layer died out before the younger layer formed.

The contacts between these strata commonly contain evidence that the top of the lower layer was exposed to weathering and erosion before the upper layers were deposited. Geologists call these breaks in the stratigraphic sequence **unconformities.** The breaks are more obvious if the erosion surface separates igneous or metamorphic rocks from overlying sedimentary layers or if the layers below the break were deformed before the overlying layers were deposited (figure 4.13). By the early part of the nineteenth century, geologists had succeeded in correlating strata by their position relative to unconformities as well as by their fossil content (figure 4.14). This served the important purpose of providing a basis for the modern geologic time scale.

Fossils as Guides to the Rock Record

Some fossils consist of plant or animal remains like a shell or bone, but many fossils consist of casts or molds of the original shells and skeletons or tracks and trails left as impressions in sediment (figure 4.15a). Sometimes a hard and durable material such as silica replaces or fills the remains, as it does in petrified wood (figure 4.15b). Excellent fossils of organisms that lived 500 million years ago are preserved in sedimentary rocks.

Many of the plant and animal groups fossilized in ancient sedimentary rocks have modern, living counterparts. Just as modern plants and animals provide clues to the environments in which they live, fossilized organisms provide information about the environment in which they lived. Some groups, such as oysters and corals, live attached to the sea floor. Others roam around on the bottom of the sea in shallow water. Still others, such as sponges, live in deep water. Fossils of animals that live only in water of a

Figure 4.12 Photograph Showing Rock Units That Can Be Traced over Long Distances.

certain depth indicate the water depth in which the sediment formed. Some animals live only in marine or fresh water; others are restricted to land environment. Consequently, some fossils provide keys not only to the type of environment and the age of the sediment in which they occur, they may even indicate the water temperature and water depth.

Early in the nineteenth century, William Smith, an English engineer, observed that many sedimentary rock layers contain par-

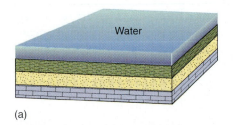

Deposition of sediment in the sea

(a)

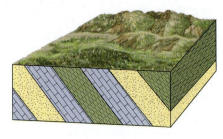

Deformation of sedimentary layers causes tilting of layers and uplift. Once uplifted, erosion removes part of uplifted region.

(b)

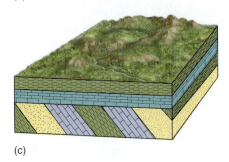

Submergence of region is followed by deposition of sediment on top of tilted and eroded layers.

(c)

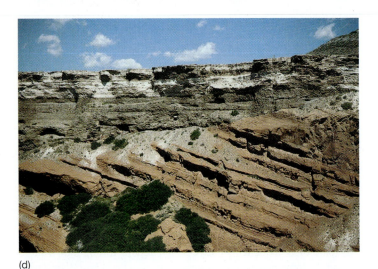

(d)

Figure 4.13 Development of an Angular Unconformity. (a) Sediment is deposited, forming a layered sequence, with the youngest layer on top. (b) Deformation of the crust causes the sedimentary layers to tilt or fold. The tilted layers are shown uplifted and eroded in this drawing. (c) The tilted layers are submerged, and new layers of sediment are deposited on the tilted layers. These layers are shown after uplift and development of a new landscape on the uplifted layers. (d) An angular unconformity exposed near Cody, Wyoming.

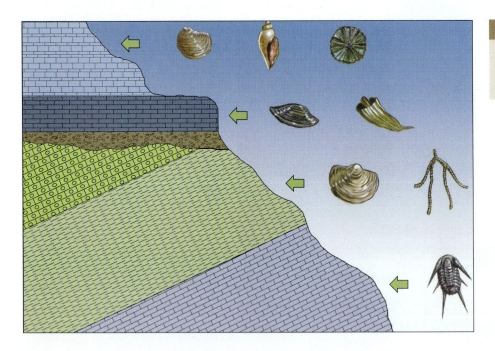

Figure 4.14 The Fossil Content of Layers Below an Angular Unconformity May Be Quite Different from Those Above the Unconformity.

(a)

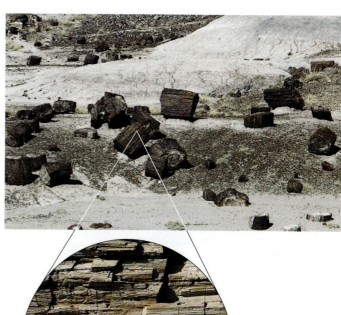

(b)

ticular assemblages of fossils that are unique to the layer or part of the stratigraphic section in which they occur (figure 4.16). Smith discovered that different groups of fossils occur in an orderly sequence, a principle known as the **law of faunal succession.** Smith's observations suggest that plants and animals come into existence and become extinct at different times and that the assemblages of fossils in different rock layers are of different ages. Fossils of some groups of plants and animals prove that they survived over a hundred million years. Other fossils, called **guide,** or **index fossils,** occur only in rocks that originated over a short time, a few million years. Their presence identifies the age of the rocks. Various members of an extinct group of arthropods, called trilobites, are guide fossils for sedimentary rocks deposited in the early part of the Paleozoic (figure 4.17). Knowing the age of different fossils and how to differentiate them from those that are younger or older, it is possible to identify the relative age of the rock by using the fossil assemblage. Smith's observations and deductions formed the basis for development of a geologic time scale. Later, his observations were important in the formulation of the theory of evolution, but Smith did not draw this conclusion.

THE GEOLOGIC TIME SCALE

In the late eighteenth and early nineteenth centuries, British and European geologists first began to construct a geologic time scale, but they had no way of knowing the absolute age of rocks. The scale they put together is based on the laws of superposition and faunal succession, the position of sedimentary rock sequences relative to unconformities, and abrupt changes in the fossil content of sequences.

The contacts between these units of rock with distinctively different characteristics became the basis for time distinction.

A dramatic contact was found in Great Britain between unfossiliferous crystalline rocks and overlying sedimentary rocks that contain recognizable fossil remains. The rocks below this contact were designated **Precambrian** (figure 4.18). In some places, this contact between Precambrian and younger rocks lies within sedimentary sequences. The older rocks may even contain the remains of primitive life-forms such as worms or bacteria. The oldest and most obscure part of the Precambrian, known as the **Hadean eon,** covers the period from the origin of Earth to the age of the oldest known rocks—a period for which no rock record exists. The oldest granites are about 4.4 billion years old. So, more than 100 million years remain almost completely lost, and the record for the early part of the Precambrian is fragmentary. The name **Archean eon** applies to the middle part of the Precambrian. The other large divisions of the time scale are based on the types of fossil remains found in rocks of each age. The suffix *zoic,* which means *life,* is attached to the names of time divisions in which fossils were prominent. The younger part of the Precambrian, known as the **Proterozoic eon,** began about 2.5 billion years ago and contains primitive fossils such as bacteria, algae, and worms. The time

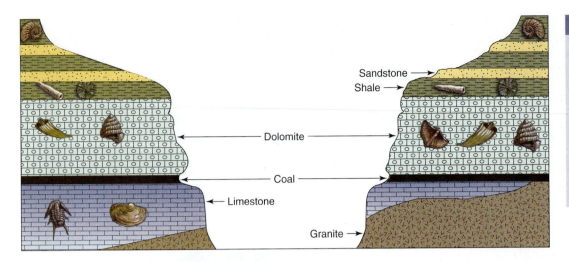

Figure 4.16 Representatives of Some of the Invertebrate Fossil Groups. Fossils of invertebrate animals occur in sedimentary rocks deposited in oceans hundreds of millions of years ago. Similar fossils occur in the same sequence at these two outcrops. In this case, the rock types are also similar.

Labels in figure: Sandstone, Shale, Dolomite, Coal, Limestone, Granite

Figure 4.17 Guide Fossil. Trilobites, such as the ones shown here, are among the most widely used index fossils for rocks of Cambrian age. Trilobites range in size from a few centimeters (less than an inch) to more than 50 centimeters (20 inches) in length.

following the Precambrian, the last half billion years of Earth history, is called the **Phanerozoic eon.**

Because the invertebrate animals that lived during the earliest part of Phanerozoic time, called the Cambrian period, were relatively advanced life-forms, it is clear that life had a long history of development before the Cambrian. Fossils of bacteria and fossils of stromatolites are present in rocks that are 3.5 billion years old. **Stromatolites,** which resemble heads of cabbage preserved in rock, are masses of calcium carbonate or silica that became trapped in mat-like masses of algae and bacteria. Stromatolites preserve a rich variety of cell types. Thin layers of cells that may indicate a rhythmic growth due to seasonal changes characterize their structure. Despite their primitive character, stromatolites live today in Shark Bay, Australia, and at a few other localities. The presence of fossil stromatolites in rocks that are more than 3 billion years old in North America, Africa, and Australia proves that algae spread rapidly around Earth. They undoubtedly played an important role in altering the atmosphere and preparing it for other forms of life. Because algae carry on photosynthesis, they were an important source of oxygen for Earth's atmosphere.

These fossils prove that life originated early in Earth's history. Sedimentary rocks formed during the later parts of the Precambrian contain fossils of algae, a variety of worms, and other soft-bodied animals that resemble arthropods, jellyfish, and mollusks. These primitive animals are known as the **Ediacaran faunas.** The differences in the diversity and complexity between Cambrian and Precambrian fossils took place over a relatively short-time interval of a few tens of millions of years. During this time, animals evolved from soft-bodied organisms to those with hard parts. These hard parts make up most of the fossil record. Gradually, as paleontologists find rocks formed during the latest part of the Precambrian, connections between the Precambrian and Cambrian fossils should become better documented.

Most sedimentary rocks and the incredibly diverse fossil record they contain belong to the Phanerozoic eon. These rocks are also the ones most widely exposed at the surface. It is not surprising that even though they contain a record of only about 12% of Earth history, they are studied in great detail and separated into many

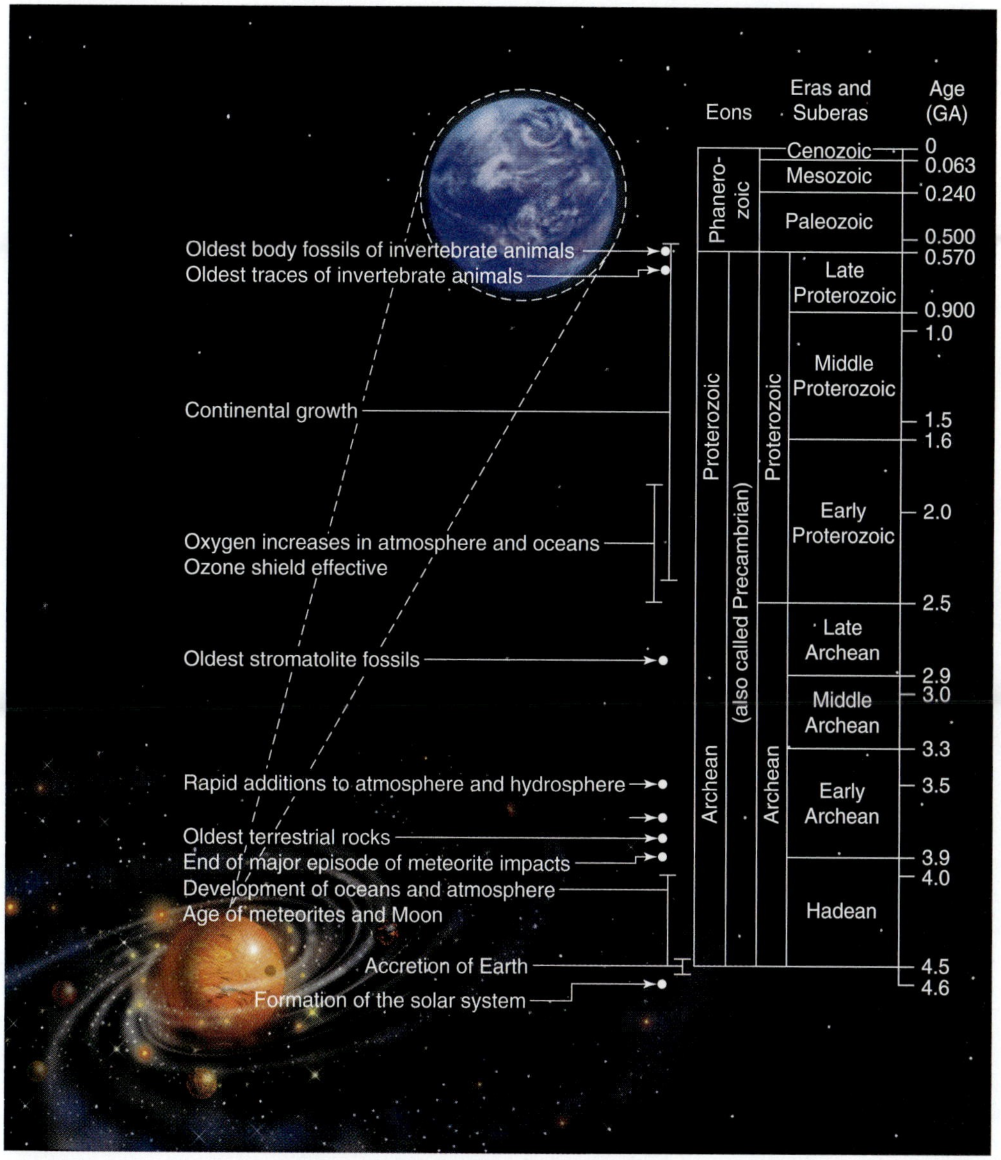
Eons
Eras and Suberas
Age (GA)
Phanerozoic
Cenozoic
Mesozoic
Paleozoic
Proterozoic
Late Proterozoic
Middle Proterozoic
Early Proterozoic
(also called Precambrian)
Archean
Late Archean
Middle Archean
Early Archean
Hadean
0
0.063
0.240
0.500
0.570
0.900
1.0
1.5
1.6
2.0
2.5
2.9
3.0
3.3
3.5
3.9
4.0
4.5
4.6
Oldest body fossils of invertebrate animals
Oldest traces of invertebrate animals
Continental growth
Oxygen increases in atmosphere and oceans
Ozone shield effective
Oldest stromatolite fossils
Rapid additions to atmosphere and hydrosphere
Oldest terrestrial rocks
End of major episode of meteorite impacts
Development of oceans and atmosphere
Age of meteorites and Moon
Accretion of Earth
Formation of the solar system

Figure 4.18 **Important Events in Earth's History.** The geologic time scale is shown at left.

subdivisions. Prominent unconformities and distinctive changes in fossils within the rock units led to the subdivisions of the Phanerozoic eon into **eras, periods, epochs,** and **ages.**

Most fossils in the early part of the **Paleozoic era** (the oldest era in the Phanerozoic eon) are remains of single-celled protozoans and invertebrate animals; shellfish, including brachiopods and clams; corals; bryozoans; sponges; and animals like the extinct trilobites (see figure 4.17). Fossils of fish appear early in the Paleozoic era as do club mosses, horsetail rushes, and ferns. Before the end of the Paleozoic era, amphibians and reptiles roamed a land surface on which pine and ginkgo trees grew. Early in the **Mesozoic era,** small dinosaurs appeared for the first time, but before the end of the era, their kin occupied the sea as well as the land. Some grazed on plants; others dominated the animal kingdom for millions of years. Birds and small, primitive mammals shared the land surface with the reptiles. At the end of the Mesozoic era, dinosaurs suddenly disappeared as did many other groups of land and marine animals. During the **Cenozoic era,** mammals diversified and took the place of dinosaurs as the dominant animals. Flowering plants appeared and expanded. Finally late in the Cenozoic era (modern life), fossils of humans appeared in the rock record (figure 4.19).

The Cenozoic era is a time of dramatic global changes. The rock record for this period is especially rich, and it provides evidence of the progressive changes that lead to the present. The two periods, the Tertiary and Quaternary, comprise the Cenozoic era. The **Quaternary period,** which started about 1.8 million years ago and includes all human history, is the youngest and the shortest of all geologic periods. Its history is known in great detail. The Quaternary includes two epochs, the **Pleistocene epoch,** which is commonly referred to as the Ice Ages, and the **Holocene epoch,** which is also called Recent (figure 4.20).

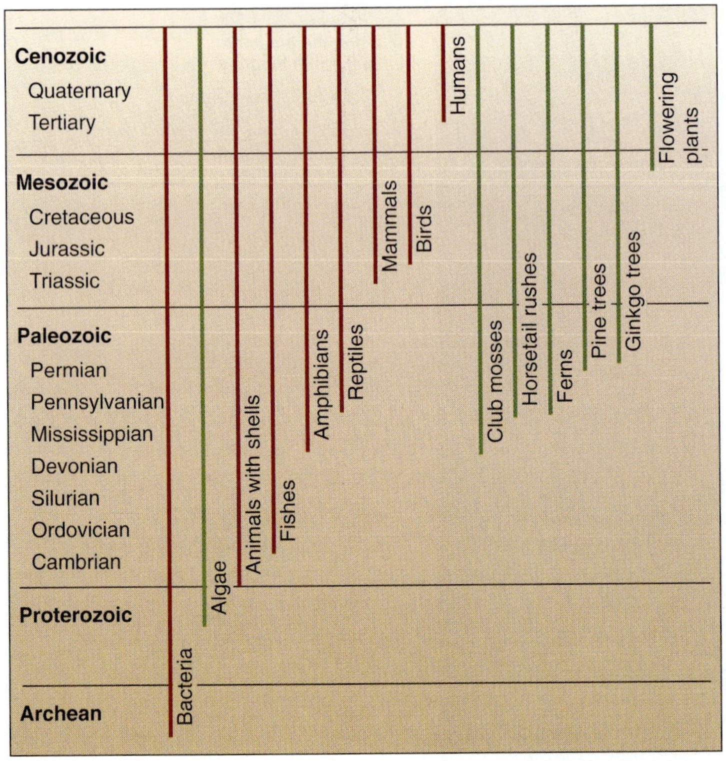

(a)

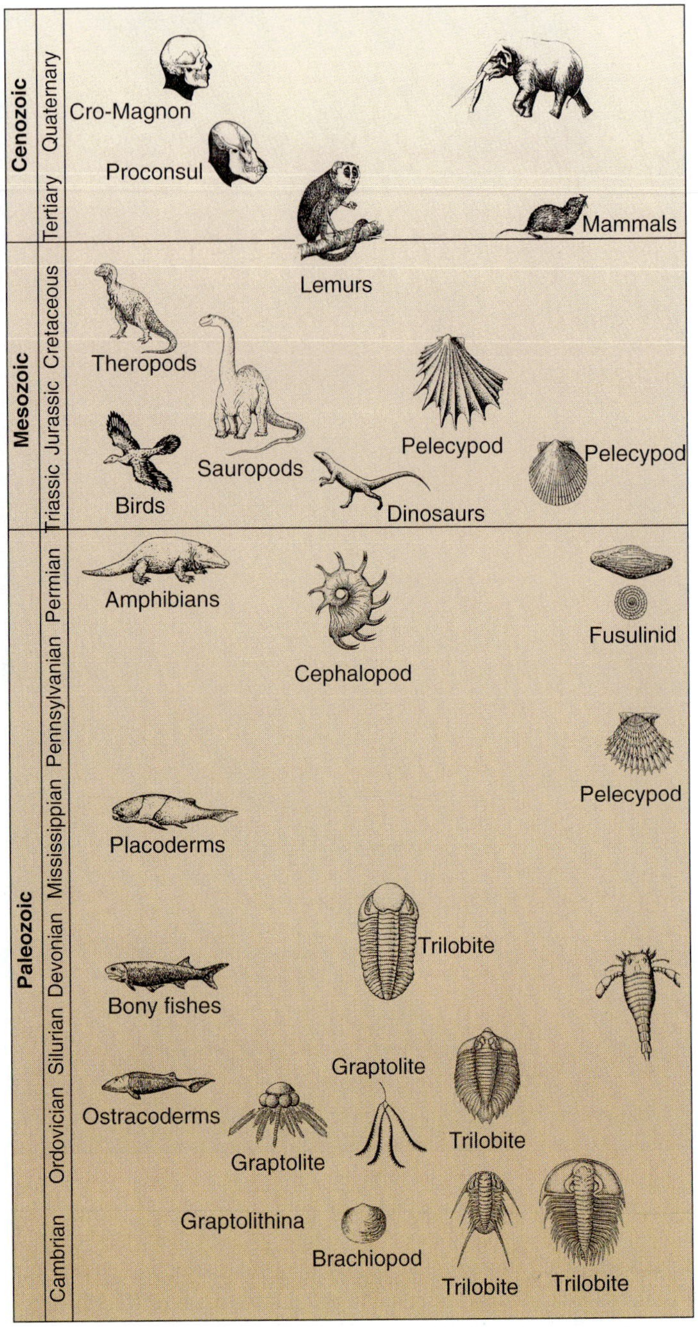

Figure 4.19   Geologic History. (a) The first appearance of groups of fossils is indicated by bold lines. (b) Representative fossils found in the Paleozoic, Mesozoic, and Cenozoic eras.

(b)

The Holocene epoch, which includes the last 10,000 years of Earth history, is the time following the melting of vast sheets of ice that occupied large parts of the Northern Hemisphere including most of Canada and parts of the northern United States. Melting of the ice started about 15,000 to 18,000 years ago. The Holocene may be a time between advances of ice sheets of continental dimensions. Many such advances separated by warmer periods, known as *interglacial intervals,* have taken place during the Pleistocene epoch. Therefore, continental ice sheets may cover much of North America in the future. The Pleistocene epoch started about 1.6 to 1.8 million years ago when Earth cooled enough for continental glaciers to form in the Northern Hemisphere. As ice accumulated on land, it advanced into the middle latitudes, and sea level dropped as water accumulated in ice sheets on land.

The geologic time scale existed long before it was possible to establish absolute ages for events in Earth's history. Some geologic indicators of the passage of time, such as deposition of sediment or erosion of the land surface, showed the order in which events occurred but not how much time passed between them. Sequences of events indicate the relative age of events, but they do not tell absolute time. The evolution of life provides the most valuable record for relative age dating. Evolution has gone on since the first forms of life appeared on Earth, and evidence of the process is present in the fossil record. Although a few species of plants and animals have survived for hundreds of millions of years, such organisms are the exception. Most species survive for much shorter periods. As William Smith discovered, the fossils of plants and animals that lived during any one period of geologic time are a unique assemblage. These assemblages provide a reference system for determining the order of succession of geologic events.

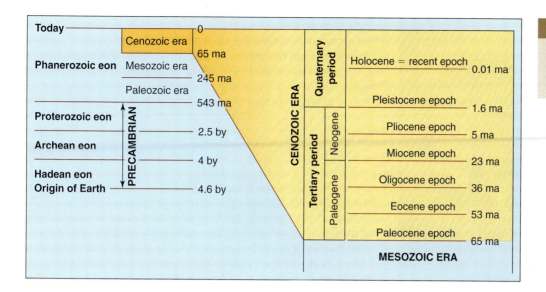

MEASURING ABSOLUTE TIME

Although determining the relative age of events made it possible to construct a time scale, more precise dating of events required scientists to identify processes that, like the movement of the hands of a clock, are unidirectional and take place at known rates. These *geologic clocks* leave a record that can be related to **absolute time,** the time measured in years rather than sequences of events. Unfortunately, some of these natural or geologic clocks are useful for dating only the recent past. For example, **tree rings** result from seasonal variations in the rate of growth of trees (figure 4.21).

Figure 4.21 Tree Rings. The wide, light-colored tree rings mark years of rapid growth. The narrow, dark-colored rings formed during years of slow growth.

Growth is slow during the winter months and more rapid during summer months. Variation in the growth rate results in rings of slightly different color and density. As climatic conditions change, the width of seasonal growth rings also changes so that tree rings provide an indication of climate as well as the age of the tree. By matching the growth rings of many trees, it is possible to establish a chronology of climatic events. In parts of the southwestern United States, bristlecone pine trees have survived nearly 5,000 years. By matching tree rings of living trees with those of older dead trees, it has been possible to construct series of tree rings that extend this chronology over 8,000 years, still relatively recent by geologic standards.

Seasonal changes also occur in lakes that freeze over during winter months. A thick, light-colored layer forms during summer when sediment flows into the lake. A thin, darker layer of fine-grained sediment settles to the lake bottom during the winter when the surface is frozen. Layers that indicate seasonal variations of this type are **varves** (figure 4.22). The number of varves indicates the age of the sediment in years. By correlating varves from one lake to another, geologists have extended the varve count to over 17,000 years in Scandinavia. However, in the long context of geologic time, lakes are temporary features in the landscape. Consequently, this method of dating, like the use of tree rings, is most valuable for the most recent times.

Dating based on the rates at which certain natural processes proceed provides another way of establishing chronology. For example, certain species of lichen appear to grow at constant rates. Consequently, the size of these lichens provides a measure of the length of time the lichen has grown and indirectly indicates the approximate length of time the rock surface to which the lichen is attached has been exposed to the atmosphere. Dating the natural glass, obsidian, involves a somewhat similar process.

Obsidian forms where lava cools so fast that crystals do not have time to form. This commonly happens where lava flows into water. The usefulness of obsidian for dating followed discovery that this glass absorbs water from the atmosphere at a constant rate. The

Figure 4.22 Varved Silt and Clay. Light-colored bands are silt; dark-colored bands are clay.

added water changes the refraction of light rays passing through the glass, a property known as the *index of refraction.* Water diffuses into glass at a steady rate, so the depth of penetration is a measure of the length of time the glass has been exposed to the atmosphere. This method of dating is useful in dating artifacts like chipped obsidian, commonly used to make scraping devices and arrowheads. This method also makes it possible to determine the age of lava flows that cracked open as the lava cooled and contracted. Based on this method, the lavas now exposed at Obsidian Cliff at Yellowstone National Park poured into a lake nearly 200,000 years ago.

Radiometric Age Determination

Other methods of dating, notably those involving the measurement of the proportions of radioactive isotopes remaining in a rock, make it possible to date much older events. Radioactive decay processes are excellent geologic clocks, and the dates obtained in this way are absolute dates. Using this method, a variety of radioactive substances make it possible to determine ages over a long time span, even exceeding the age of Earth. Radioactive-dating techniques can answer a wide variety of questions, such as how long it takes for water to circulate from the surface into the deepest parts of the ocean or how rapidly rock exposed to the atmosphere weathers.

In the 1950s, geochemists discovered how to determine the age of certain minerals that contain radioactive elements. These elements change into other elements that are not radioactive. Because the length of time required for one-half of a radioactive isotope to decay and form another isotope or a stable element is constant, this method of dating provides a means of measuring the

time interval over which decay has taken place. Isotopes are atoms of elements that have the same number of protons in the nucleus but differ from one another in that they contain different numbers of neutrons in the nucleus. Carbon-14, uranium-235 and 238, and potassium-40 are examples of isotopes useful in dating rocks. Some isotopes are stable; others, referred to as *radioactive isotopes,* decay spontaneously by emitting radiant energy, alpha particles (the equivalent of the nucleus of a helium atom), beta particles (electrons), and high-frequency waves known as gamma rays.

Pierre and Marie Curie, who first isolated radioactive elements, produced only 1 gram of radium, a radioactive element, from 5 tons of the mineral pitchblende. Later, chemists discovered that stable isotopes or elements result from the decay process and that the rate of decay for any given radioactive isotope is constant. The **half-life** of an isotope is the amount of time needed for one-half of the atoms of a radioactive isotope to decay (figure 4.23). With the passage of each half-life, half of the amount of the isotope that was present at the start of the half-life decays. Some radioactive isotopes decay to form other radioactive isotopes, but eventually, a stable isotope results. Knowing the quantity of each isotope present in a sample, the exact sequence of steps in the decay process, and the half-life of each step make it possible to determine the length of time the decay process has lasted and when the decay process started.

Two types of radiometric dating provide the ages of material that contains radioactive isotopes. In one type, the radioactive isotope, called the **parent isotope,** breaks down. It may break down to form another radioactive isotope, but eventually the end product is a stable isotope, called the **daughter isotope.** This type is the most widely used in dating inorganic materials. Several steps (changes from one isotope to another) may be involved before the final stable daughter product appears, but the amount of the daughter present at any time is directly related to the amount of the parent that is present and to the length of the decay process. If radioactive elements are present in the minerals of an igneous rock when it solidifies, the length of the decay process measures the time that has elapsed since the cooling and crystallization of the magma took place.

Two assumptions are involved in radiometric-dating techniques. One is that the system has remained closed so that neither parent nor daughter atoms moved into or out of the material being dated. In most rocks and minerals, the system closes when the mineral forms. The second assumption is that the daughter product was not present in the system at the time it became closed. Because these assumptions are not always valid, the natural system should be examined to determine whether or not these assumptions are justified.

Only a few radioactive isotopes are suitable for use in radiometric age determination. Isotopes of uranium, potassium, rubidium, and samarium are especially useful (table 4.1). Some of these isotopes, such as uranium, have such long half-lives that they can be used to date the oldest rocks, but others decay in a matter of a few years. Uranium, mica, feldspar, and materials like wood, charcoal, and shells that contain radioactive carbon are among the

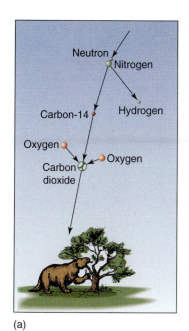

(a)

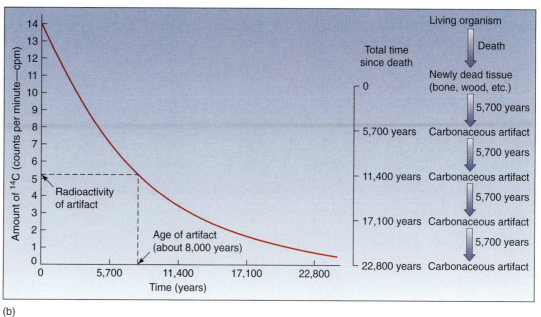

(b)

Figure 4.23 Carbon-14 and Radioactive Dating. (*a*) Atoms of carbon-14 form as neutrons coming in from space strike nitrogen atoms in the atmosphere. As a result of the collision, the nitrogen becomes radioactive and quickly decays to form carbon-14. The carbon-14 combines with oxygen, forming carbon dioxide which is taken up by plants. The plants may be eaten, transferring some of the carbon-14 into animals. The remains of the plants or animals may later be dated by use of the carbon-14. (*b*) This plot of the amount of carbon-14 left over time shows how little of the original carbon-14 remains after 30,000 years. The amount of carbon-14 is indicated by the number of atoms that decay (counts per minute). By measuring the number of counts per minute from a specimen, the age can be determined from this plot.

most important materials that can be dated by radiometric age determination.

Using Radiation from Space to Study Earth Systems

The second type of radiometric dating assumes that the rate at which the radioactive isotope is produced and enters a natural system remains constant until something happens to stop the introduction of the isotope into the system. For example, a tree takes in carbon dioxide continually during its life, but when the tree dies, the process stops, and the quantity of carbon in the tree is fixed. When a system closes, any radioactive isotope in that system begins to decay at a fixed rate. The quantity of the isotope remaining in the system at any time after the system closes depends on the amount of time that has passed since the system closed.

Cosmogenic isotopes are produced by radiation from the Sun and from deeper space. High-sensitivity mass spectrometers detect and measure minute quantities of these isotopes. Isotopes produced by the interaction of cosmic radiation with matter at Earth's surface are used for dating. Aluminum-26, beryllium-10, carbon-

14, helium-3, neon-21, thorium-230, and tritium are among the cosmogenic isotopes used to study Earth's environments.

Carbon-14 Carbon, including radioactive **carbon-14** (C-14), is present in all living materials and is preserved in wood, bones, shells, charcoal, water, and grains of carbonate minerals. Its presence in artifacts makes it especially valuable in anthropological and archeological studies. C-14 originates in the upper atmosphere where cosmic rays bombard nitrogen atoms in the atmosphere (see figure 4.23). Neutrons from the cosmic rays may combine with nitrogen-14 to form unstable nitrogen-15, which breaks down in seconds to form radioactive carbon-14. This radioactive carbon immediately begins to decay, with a half-life of 5,730 years, to again form nitrogen-14, a stable isotope. As the quantity of radioactive carbon decreases, it is more difficult to measure the quantity accurately. For this reason, materials that are more than about 50,000 years old contain too little C-14 to date by this method.

While it is in the atmosphere, C-14 becomes involved in processes such as photosynthesis and respiration. Plants absorb carbon dioxide that contains radioactive C-14 and nonradioactive

Table 4.1 Isotopes Commonly Used in Radiometric Age Determination

Ga = billion years; Ma = million years

Cosmogenic Isotopes		
Isotope	Half-life	Uses
Beryllium-10	1.6 Ma	Erosion rates; exposure at surface
Carbon-14	5,730 years	Used to date wood, shells, other carbon-bearing materials
Thorium-230	75,400 years	Used as a tracer in oceans
Tritium	12 years	Used as a tracer in water

Methods Using Parent/Daughter Ratios			
Parent	Half-life	Daughter	Used to Date
Uranium-234	244,500 years	Protactinium-231	Carbonate rocks
Uranium-235	0.7 Ga	Lead-207	Zircon, uraninite, pitchblende
Uranium-238	4.5 Ga	Lead-206	Zircon, uraninite, pitchblende
Potassium-40	1.3 Ga	Argon-40	Mica, hornblende, whole rock
Rubidium-87	47 Ma	Strontium-87	Mica, biotite, feldspar, whole rock
Samarium-147	106 Ga	Neodymium-143	Whole rock

carbon, and the carbon becomes part of the plant matter. If an animal eats the plant, both the carbon-14 and the stable form of carbon become part of its living tissue. When the plant or animal dies, it stops taking in carbon, but the C-14 in its system continues to decay. Gradually, the ratio of C-14 to stable carbon decreases.

C-14 also enters natural systems when carbon dioxide gets into the surface layers of the oceans where animals use the carbon to build shells composed of calcium carbonate ($CaCO_3$). Because atmospheric gases are present in ice and snow, radioactive carbon also may become trapped in glacier ice. Ice cores provide one of the best clues to past climatic conditions. The gases trapped in the ice tell us the composition of the atmosphere, and C-14 allows us to establish the age of the ice. The results of these studies are discussed more fully in chapter 13. Unfortunately, the short half-life of C-14 limits its usefulness to relatively recent times.

Radiometric age determination makes it possible to assign absolute dates to the divisions of the geologic time scale. Dating makes it possible to determine when many of the most important events took place in the early parts of Earth history.

PIVOTAL EVENTS IN EARTH'S HISTORY

The origin of Earth, the solar system, and the universe were pivotal events in Earth's history. These are discussed in chapter 21. The following sections on pages 98 to 100 concern events that took place after Earth had become a distinct planet. Most of the history of these events exists in the rock record.

Origin of the Core, Mantle, and Crust

The progressive increase in density toward the center of Earth supports the idea that the internal divisions of Earth originated as a result of the gravitational separation of materials in a liquid Earth (see figure I.9). All modern scientific hypotheses concerning the origin of Earth postulate that Earth was molten at an early stage in its history. In such a molten mass, if cooling is slow, the denser matter sinks toward the center and lighter materials rise toward the surface. This type of separation caused by differences in density is especially effective if some of the liquid components behave like oil and water and will not mix. As minerals begin to crystallize, those with a density greater than the density of the melt sink; less dense minerals float. Gravitational separation is responsible for the high-density iron in the core, less-dense minerals like olivine, pyroxene, and garnet in the mantle, and even less dense minerals like quartz and feldspar in the continental crust. Most of this separation of materials according to density was complete early in Earth's history. The outer portions cooled and crystallized first, forming an insulating blanket—the lithosphere—over the deeper molten materials in the outer core. Although a hot core and somewhat cooler mantle are shown as concentric shells in most drawings of Earth (see figure I.9), evidence derived from studying seismic waves and Earth's gravity field indicate that Earth is not perfectly spherical nor are its shells homogeneous.

Origin and Evolution of the Atmosphere

Earlier, scientists thought the atmosphere was a residue of gases that enveloped Earth when it formed. However, the composition of the present atmosphere (see figure 1.3d) is quite different from what one might expect if it were a residue of a primitive atmosphere. Today, most scientists think the atmosphere formed from gases that escaped from Earth's interior through volcanic activity. Some scientists suggest that comets may have brought much of the atmosphere and water in the ocean to Earth. In either case, over time, the composition has changed. Chemical reactions between gases in the atmosphere and the solid earth began as soon as the solid crust formed. Reactions also occurred between the atmosphere and solar radiation. Later, as organic materials evolved, biochemical processes caused further changes.

Opponents of the residual atmosphere hypothesis further point to the low concentration of inert gases that do not enter chemical reactions in Earth's present atmosphere compared with their concentrations on the Sun. If Earth's atmosphere started out with a composition similar to that of the Sun, the atmosphere should have significant amounts of the heavy inert gases that occur on the Sun. Light gases such as hydrogen and helium were lost because Earth is not sufficiently massive to hold them, but Earth should have held the heavy gases such as krypton and xenon. Instead, these two gases are thousands of times less abundant on Earth than are some of the nongaseous elements close to them in weight. This is exactly what one might expect if Earth formed by the accretion, the clumping together, of bodies that were too small to hold these heavy inert gases. As the bodies of rock accumulated to form Earth, heat generated by impacts and decay of radioactive elements caused rock to melt. Gases rose from the molten rocks, and, with the exception of hydrogen and helium, these gases are held in place by gravity.

According to the **degassing** hypothesis, the atmosphere evolved from gases emitted by volcanoes. Measurements around modern volcanoes indicate that the main gases produced are water vapor, nitrogen, sulfur dioxide, carbon monoxide, and carbon dioxide. The first three of these are major constituents of Earth's atmosphere. Oxygen is notably absent from this list, and the oldest rocks formed at Earth's surface show no sign of oxidation, indicating that oxygen was not initially present in the atmosphere. Oxygen, one of the most abundant and important gases on Earth, must derive from other sources.

The atmosphere is a product of an evolutionary process that continues today. If it began as volcanic gases from Earth's interior, it changed dramatically by the addition of oxygen and by reactions between rocks, the atmosphere, and oceans. Later, as plants evolved, photosynthesis increased the amount of oxygen in the atmosphere.

Today, Earth's atmosphere contains the following gaseous constituents (percentage by volume): nitrogen, 78%; oxygen, 20.9%; argon, 0.9%; carbon dioxide, 0.04% (see figure 1.3). The atmosphere also contains water vapor, but unlike the dry gases, the quantity of water vapor varies greatly and may constitute as much as 4% of the total volume. Even though nitrogen is not the most abundant volcanic gas, its concentration builds up in the atmosphere because it does not become involved in chemical reactions with minerals and rocks. The first free oxygen in the atmosphere probably formed in the upper layers of the atmosphere when ultraviolet radiation split water molecules emitted from volcanoes. This process, called **photochemical dissociation,** is taking place today at a rate that, if continued throughout geologic time, would provide several times the amount of oxygen presently found in Earth's atmosphere. Unlike nitrogen, much of the oxygen produced in the past has been incorporated in oxygen compounds in the minerals of Earth's crust. Oxidation of iron-bearing minerals is the most significant of these chemical reactions. Through oxidation, oxygen in the atmosphere becomes incorporated in iron-bearing minerals like hematite and limonite—the mineral names for rust.

For at least 2 billion years, photosynthesis by green plants has been important in maintaining the high oxygen content of the atmosphere, but oxygen was an important gas in the atmosphere long before plants became abundant on Earth. Oxygen-producing cyanobacteria (also known as blue-green algae) became significant contributors to Earth's atmosphere about 3.5 billion years ago. By 2 billion years ago, the quantity of oxygen in the atmosphere was sufficient to cause oxidation of iron compounds. Before that time, the atmosphere contained so little oxygen it was a reducing atmosphere (one that lacks sufficient oxygen to cause oxidation). As oxygen and other atmospheric gases approached their present concentrations, reactions between the atmosphere, oceans, biosphere, and land, similar to those occurring today, began to take place. The reactions between plants and the atmosphere are an example. Plants remove carbon dioxide from the atmosphere and use it to build cellulose. As many invertebrate animals consume plants, they too take up some carbon and secrete calcium-carbonate shells. The action of bacteria, high temperature, and increased pressure contribute to the transformation of the remains of these organisms into coal or petroleum, or they are disseminated as carbon in sedimentary rocks.

Many chemical reactions and mass transfers take place between the atmosphere and Earth's crust. These interactions are part of the rock cycle described in chapter 2. Important among these interchanges is a group of processes known as *chemical weathering.* Chemical reactions take place between gases in the atmosphere and certain minerals. These reactions tend to remove certain constituents from the atmosphere as they combine with the minerals to form new compounds. One example is the oxidation of iron-bearing minerals. Another is a complex series of reactions between carbon dioxide, water, and the most abundant mineral in igneous rocks, feldspars. Some of the liberated constituents of the igneous rocks, especially calcium, recombine with carbonate ions in water to form limestone. The net effect of these reactions has been to remove carbon dioxide from the atmosphere and to store it in sediments on the bottom of the ocean.

Origin of the Oceans

During early stages of Earth history, the atmosphere was too hot for water to exist in its liquid state, but as temperatures near the surface dropped, water vapor in the atmosphere condensed and fell to the surface where it accumulated to form the oceans. The composition of isotopes in the oldest known rocks suggests that water was already present on Earth 4.4 billion years ago.

The origin of the oceans involves two quite different questions. One is how the basins in which seawater resides came into existence. The other is how such a tremendous volume of water came to Earth's surface. The answer to both of these questions hinges on events that took place during the earliest part of Earth's history, and both are debated. The first oceans may have formed while Earth was undergoing transformation from a totally molten body to one with a solid external shell. When minerals first began to crystallize in the molten Earth, low-density minerals like quartz and feldspar floated to the surface and formed large "rafts" of less-dense rock. These rafts became the first continental crust. The ocean basins were depressions separating the continents. Others think it is more likely that initially a thin layer of less-dense rocks covered the entire Earth. According to this view, this layer broke apart, and movements caused portions of it to become thicker. The thicker parts of this low-density crust rose and formed continents, leaving other parts of Earth's surface lower in elevation. Formation and evolution of ocean basins are discussed in greater detail in chapter 8.

The water on Earth's surface may have come, as it does today, from volcanoes. If this is the case, water was present in the atmosphere early in the history of the planet. It began to condense as the temperature of the surface cooled enough for water to exist (see figure 1.15). As rains began to fall, the face of Earth bore both the scars of impact craters and stream channels, a combination now seen on Mars (figure 4.24). Gradually, the action of water and plate tectonics erased the impact scars. If the theory that the atmosphere evolved from the degassing of Earth's interior is correct, the oceans accumulated gradually over millions of years as volcanic eruptions gradually but steadily added water to the atmosphere. Unfortunately, no rocks that formed during these first days on Earth remain to reveal what happened during the earliest part of Earth history.

Large quantities of steam continue to reach the surface of Earth through volcanoes today; water vapor comprises about 98% of the total gas content of volcanic emanations. Most modern volcanoes lie near the margins of the oceans or within the ocean basins. Part of the steam that comes from them must be water that is being recycled. However, even if no more than 0.8% of the water brought out of volcanoes today is derived from magma, over geologic time this would be enough to account for the amount of water present in the oceans.

Appearance of Life on Earth

The origin of life on Earth has long been interesting to scientists. This interest intensified after the theory of evolution first gained a following. However, many questions must still be answered before anything resembling a consensus can emerge. Nevertheless, much more is known today, and scientists are better able to frame the questions that need to be answered than they were just a few years ago. Evidence of simple life-forms is preserved as fossils in some rocks more than 3 billion years old in North America, Australia, Greenland, and Africa. In one of the Australian localities, J. William Schopf, head of the University of California, Los Angeles center for the study of the evolution and origin of life, found fossil strings of primitive cells that resemble blue-green algae. Lava flows above and below these fossils provide radiometric ages of 3.46 and 3.47 billion years. How much earlier life existed on Earth remains unknown. Most scientists agree that living cells would not have been able to survive during the earliest part of Earth's history when huge meteoroids bombarded the surface, raising the temperature, vaporizing any accumulations of water, breaking through the thin crust, and causing widespread volcanic activity. That stage of Earth history ended about 4 billion years ago. By whatever means, life may have originated much earlier and much faster than most scientists had suspected.

In 1909, Charles Walcott, the secretary of the Smithsonian Institution, discovered a remarkable fossil locality in British Columbia. These rocks, known as the Burgess Shale, formed about 540 million years ago. They contain the remains of a large fauna of invertebrate animals. These fossils were preserved in an oxygen-free environment that protected them from oxidation. Bacteria and scavengers reveal how diverse and advanced living organisms were at the beginning of the Paleozoic era, about 540 million years ago.

Figure 4.24 Topography of Mars. This photograph of the surface of Mars shows a type of topography that might well have existed on Earth early in its history, when the surface was marked by both impact craters and stream channels. Some impact craters are older than the channels; others are younger. Any water remaining on Mars is now frozen and remains beneath the surface as ice and as polar ice caps.

In investigating the origin of life, most scientists approach the question as a problem in the chemistry of evolution—how and under what conditions is it possible to form a cell that is capable of reproducing itself? Despite the vast numbers and complexity of organic molecules, organisms consist of a relatively small number of elements, mainly carbon, hydrogen, and oxygen with smaller quantities of nitrogen, phosphorus, and sulfur. All of these elements exist on Earth as inorganic substances. Something happened early in Earth's history that enabled these elements to combine to form organic molecules. In 1953, this problem prompted Harold C. Urey and Stanley L. Miller of the University of Chicago to undertake a now-famous series of experiments. They placed ammonia and methane, gases that they expected to exist in Earth's primitive atmosphere, in closed bottles and subjected them to electrical discharges meant to simulate lightning. After a few weeks, the mixture had changed color, and in it, they found amino acids, the primary organic molecules of which proteins in cells are made. Although ideas about the composition of the atmosphere have changed, Urey and Miller demonstrated that organic molecules could come from inorganic materials in a simple process that could have taken place on the early Earth. The results of these experiments stimulated others to seek answers to the question of how more complex organic molecules may have formed. Exactly how remains a lively topic of investigation.

Some scientists think that organic molecules originated in space. They point to the identification of amino acid molecules in meteorites. Others, like Urey and Miller, favor the idea that organic molecules initially formed in the atmosphere as a result of reactions among atmospheric constituents facilitated by lightning discharges, ultraviolet radiation, radioactivity, or cosmic radiation. In either case, organic molecules rained down in the oceans where organic molecules combined to form cells. The oceans are favored because ocean water provides protection from the breakdown of organic molecules by ultraviolet radiation.

Living organisms have managed to survive on Earth for more than 3 billion years. The fossil record indicates a progressive increase in diversity and complexity of living forms over that time, but this record is marked by several dramatic breaks when many life-forms disappeared.

Close Calls in the History of Life

Life on Earth has had several narrow escapes from extinction. One such widespread die-off occurred near the end of the Paleozoic era, about 250 million years ago. Toward the end of the Paleozoic era, continental collisions caused the rise of long, high mountain belts along the eastern margin of North America and Scandinavia. Other mountain belts rose in Australia and Russia, and a great continent known as Pangaea formed (see chapter 5). The elevation of the continents that took place at this time caused widespread unconformities in the rock record. During this break in the rock record, a dramatic change took place in life on Earth. Some estimate that 95% of all species disappeared from Earth over a period of less than a few hundred thousand years and possibly as short as a few days. The reason for this rapid extinction is uncertain, but two explanations seem plausible. One is that an asteroid hit Earth and created a cloud of dust that drastically reduced incoming solar radiation. A second is that extensive volcanic eruptions, known to have occurred at that time, threw Earth into darkness, which killed most plants. In either case, changes in the environment were unfavorable for living organisms.

Another dramatic change in fauna took place about 65 million years ago at the end of the Mesozoic era. In 1980, Luis and Walter Alvarez, scientists at the University of California at Berkley, suggested that the impact of an asteroid on Earth at the end of the Mesozoic era caused mass extinctions. The dinosaurs were among the 70% of all species living at that time that were victims of this event (figure 4.25). While studying the sedimentary layers at the contact

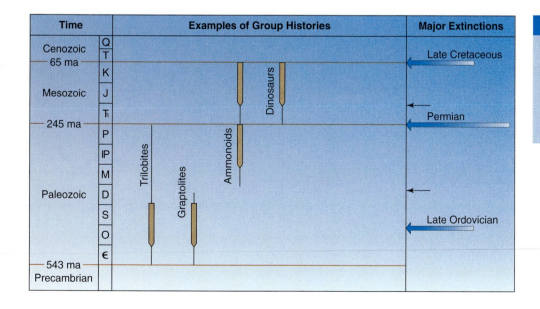

Figure 4.25 Geologic Time Scale Showing the Rise and Fall of Various Groups of Animals. Note the dramatic extinctions that took place at the end of the Paleozoic era and for the dinosaurs at the end of the Mesozoic era.

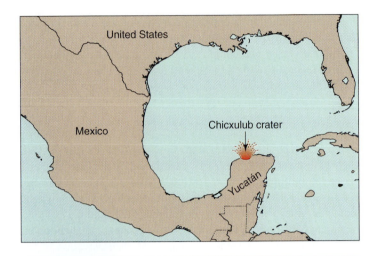

Figure 4.26 Location of Chicxulub Crater. A huge impact crater that formed at the end of the Mesozoic era is located along the northern border of the Yucatán Peninsula. Many scientists think this impact caused dinosaurs to become extinct.

iridium—an element that is extremely rare in most rocks but common in most meteorites. The Alvarezes concluded that a large asteroid or comet hit Earth and created a huge cloud of dust that spread over much of Earth. The dust cloud blocked sunlight and changed the climate so drastically that many forms of plant and animal life died out on the land surface and in the upper layers of the oceans.

Scientists have now examined the contact between Mesozoic- and Cenozoic-era rocks with great care in many places. A thin layer of iridium-bearing clay marks this contact throughout the world. In some places, this clay layer also contains a variety of quartz, which forms only under extreme pressure like that found in nuclear explosions or meteoroid impacts. The search was on for the impact crater. Now, many agree that a large impact crater lies on the edge of the Yucatán Peninsula in Mexico (figure 4.26). This huge impact crater at Yucatán is buried under a kilometer-thick layer of limestone. Petroleum geologists with Petroleos Mexicanos discovered the crater while looking for oil and gas. Drill samples from the impact site, called the Chicxulub structure, found quartz that had been shocked by high pressure and partially melted. Surrounding it at distances of several hundred kilometers, other scientists found breccia and other debris blown out of the crater. These rocks formed at the end of the Mesozoic era at the same time the mass extinctions took place. This coincidence in time and similarity of materials in the crater with those found around the world at the extinction level are widely accepted as evidence that impacts were responsible for the extinctions at that time. Another somewhat smaller and younger crater is located near the mouth of the Chesapeake Bay (figure 4.27). Neither Yucatán Peninsula nor Chesapeake Bay craters are obvious because both are buried beneath sediments.

point between rocks of Mesozoic and Cenozoic eras in Italy, they found richly fossiliferous marine limestones below the contact, a thin layer of clay at the contact, and limestones that contained almost no fossils above the contact. Reasoning that the clay formed during the event that marked the end of the Mesozoic era, they began to study this clay in detail. Chemical analyses of the clay revealed that it contained unusually large amounts of the element

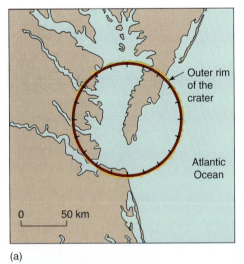

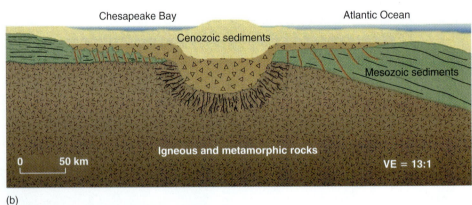

(a)   (b)

Figure 4.27 Chesapeake Bay Crater. (a) A large impact crater lies beneath undeformed sediment near the mouth of the Chesapeake Bay. This impact took place about 36 million years ago in the Tertiary Period. (b) Cross section of impact crater based on geophysical information.

SUMMARY POINTS

1. Earth's environmental processes operate within time frames that vary from a few seconds to millions of years. Most of these processes can be categorized in one of five temporal bands: millions to billions of years, thousands of years, decades to centuries, days to years, and seconds to hours. Most processes do not operate independently of other processes. In many instances, processes that operate over exceedingly long periods (e.g., movement of plates and mountain building) directly affect other processes that operate within much shorter time frames (e.g., earthquakes or volcanic eruptions).

2. Scientists study by direct observation those processes that operate within the relatively short time frames of hours, days, years, and even decades. Most scientific investigations in physics, chemistry, biology, and physical geology involve processes that take place in short time frames. It is much more difficult to follow the development of processes that function within time frames that range from hundreds to billions of years in duration. Most such processes are examined in the fields of historical geology and astronomy.

3. When scientists first began considering processes that function in long time frames, they had no way of establishing accurate dates for their observations. They devised a geologic time scale based on the relative age of events. Using the order in which layers are deposited, crosscutting relationships, succession of fossil groups, and unconformities, a workable time scale was established early in the nineteenth century. This scale did not provide absolute ages of events and left important questions, such as the age of Earth, unanswered.

4. Tree rings and varved sediments provide indications of the passage of time in terms of years—the standard measure of time. The ages determined using such indicators provide measurements of absolute time. In the 1950s, geochemists found ways to use the breakdown of radioactive isotopes to measure absolute ages when they discovered that radioactive isotopes break down at a constant rate. The half-life of a radioactive isotope is the length of time required for one half of the atoms of a parent isotope, such as uranium, to break down to form a stable daughter product, such as lead. Radioactive isotopes of different elements vary in the length of their half-lives and in their mode of occurrence in natural systems.

5. Because C-14 moves through so many environmental systems (water, ice, wood, charcoal, bones, shell, etc.), carbon dioxide is especially valuable for radiometric dating. The short half-life of C-14 limits its application to events that took place within the last 50,000 years. In contrast, uranium, a component of minerals found in igneous rocks, has a long half-life, which makes it possible to unravel geologic events that occurred during early stages in the formation of Earth.

6. Both the relative time scale (the geologic time scale) and absolute age determinations are widely used. However, absolute age determination is restricted to certain materials.

REVIEW QUESTIONS

1. Briefly identify the types of fossils found in sedimentary rocks of the Precambrian, Paleozoic, Mesozoic, and Cenozoic ages.
2. What is stratigraphic correlation?
3. What is the distinction between relative age determination and absolute age determination?
4. What methods of absolute age determination are useful for dating materials that formed during the Holocene (Recent) epoch?
5. What types of natural materials can be dated by radiometric age determination?
6. What was used to identify the major divisions in the geologic time scale?
7. What types of catastrophic events occur almost every year?
8. What types of catastrophic events have global effects?
9. How was it possible to correlate strata that are separated by great distances before the use of radioactive age determination became widespread?
10. How are the major divisions (eras, periods, epochs) of the geologic time scale distinguished from one another?
11. How do scientists measure the passage of time for periods before humans kept records?
12. How does the use of cosmogenic isotopes differ from the use of parent-daughter isotopes?
13. Which of the dating methods might you use to determine the age of a basaltic dike?

THINKING CRITICALLY

1. Why should people be concerned about imperceptibly slow processes?
2. How does an understanding of Earth history help in prediction of its future?
3. What causes differences to occur in the sedimentary rocks that are forming today? Will all of them contain the same types of animal remains? Explain why or why not.

KEY WORDS

absolute time
ages
Archean eon
carbon-14
catastrophism
Cenozoic era
cosmogenic isotopes
crosscutting relationships
daughter isotope
degassing
dikes

Ediacaran fauna
epochs
eras
guide fossils (also index fossils)
Hadean eon
half-life
Holocene epoch
law of faunal succession
law of superposition
Little Ice Ages

Mesozoic era
Paleozoic era
parent isotope
periods
Phanerozoic eon

photochemical
   dissociation
Pleistocene epoch
Precambrian

Proterozoic eon
Quaternary period
stratigraphic record
stromatolites

theory of uniformitarianism
tree rings
unconformities
varves

unit two The Plate Tectonic System

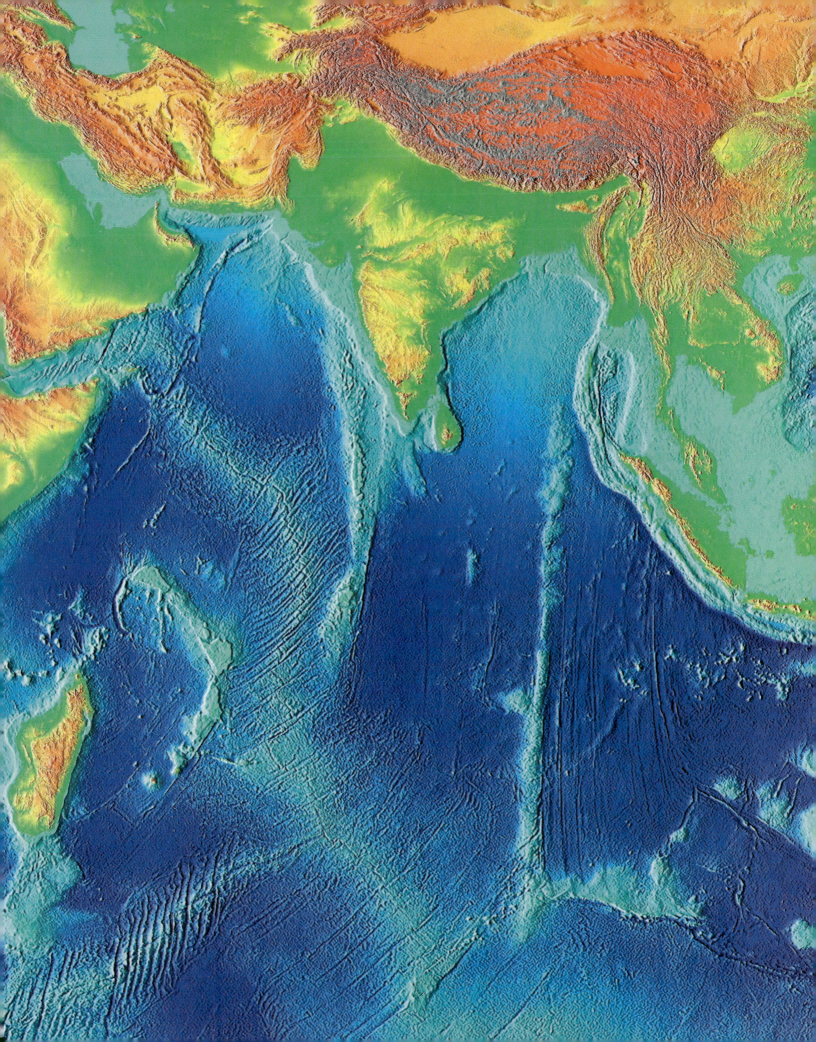

chapter five

Plate Tectonics and Mountain Building

INTRODUCTION

THE ROLE OF PLATE TECTONICS
IN ENVIRONMENTAL SYSTEMS

EVOLUTION OF PLATE TECTONIC
THEORY

TECTONIC CYCLES

MOUNTAIN BUILDING

The Himalayan Mountain system formed as the Indian subcontinent moved northward, closed a former ocean, and collided with Eurasia. Sea-floor spreading in the Indian Ocean is clearly marked by the oceanic ridge and the faults that cut across it.

Chapter Guide

The idea of continental drift surely began when the first maps of the Atlantic were published showing the similarity in the coastlines of South America and Africa. The idea was widely rejected until the 1950s when new discoveries in the ocean basin convinced scientists that the idea had merit. What causes continents to move and the consequences of these movements is called plate tectonic theory. In its current form, the theory of plate tectonics provides the basis for the synthesis of much that is known about earthquakes, volcanoes, mountain building, and the evolution of the physical world for hundreds of millions of years.

This chapter extends the discussion of the lithosphere that began in chapter 3 and traces the important developments that led to modern interpretations of the lithosphere, crustal structure, and the history of divisions of the lithosphere. Some of these divisions, such as high, geologically young mountains like the Alps and Rockies, have attracted the attention of generations of geologists. Careful study of other divisions, such as the deep sea floor, began only a few decades ago. How these major features of Earth formed and how they are related to one another is the subject matter of **plate tectonics.**

INTRODUCTION

The lithosphere is the major division of Earth's interior that most directly affects Earth's surface environment. The lithosphere is not static; dynamic processes that directly affect human lives take place here. Movements of lithospheric plates cause earthquakes, volcanic eruptions, and indirectly cause all of the side effects related to these—landslides, tsunami, and the climatic effects of volcanic gases. Although these dramatic effects happen abruptly, they are the products of processes that act over a much longer time frame and on a much grander scale—processes that ultimately lead to the movement of continents and cause mountains to rise from the sea.

After World War II, an explosion of new information about Earth became available, which led to a revolution in thinking about global tectonics in the 1960s. New geophysical techniques used for studying Earth's gravity and magnetic fields emerged after World War II. Use of seismic methods, the application of radiometric dating techniques and studies of remanent magnetism to geologic problems, and mapping of the ocean floor also began after the war. More recently, establishment of a worldwide network of seismograph stations, development of high-speed computers, and use of satellites to study Earth's surface have helped confirm the earlier discoveries.

Almost all geologists now agree that continents move, that new sea floor and oceans continue to form, and that the lithosphere is composed of a number of thin plates bounded below by the asthenosphere and separated from one another by zones in which the crust is strongly deformed (figure 5.1). Most plate boundaries are one of four types: (1) **Divergent plate boundaries** are boundaries along which plates are moving away from one another. Most of these lie along oceanic ridges and are places where new oceanic crust is forming. (2) **Convergent plate boundaries** are those where plates move toward one another. Along these boundaries, plates collide and mountain belts form, as is now taking place in the Himalayas. (3) **Subduction zones** are other convergent boundaries where one plate sinks under the other, as is happening south of the Aleutian Islands in Alaska. (4) **Transform faults** are shear zones where plates may move laterally by one another. These faults cut through the lithosphere and make it possible for whole plates to shift position. The San Andreas fault in California, described in chapter 6, is one of the best known of these faults.

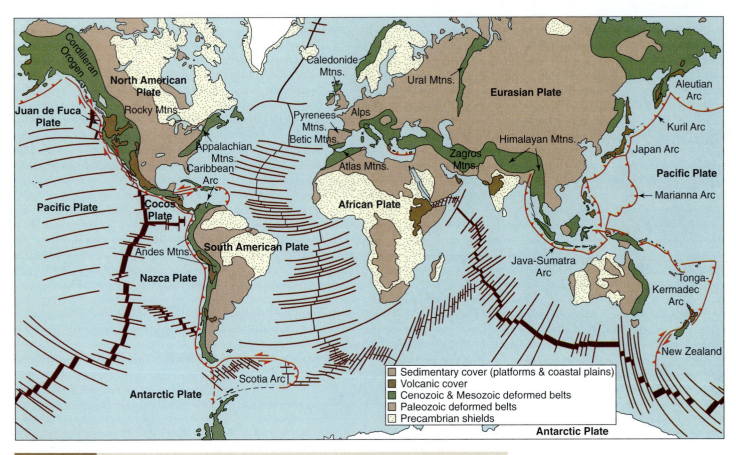

Figure 5.1 **Major Structural Divisions of Earth's Crust.** Shield areas have a light pattern; lines with barbs indicate subduction zones; the darker shading indicates young mountain belts; the width of the dark lines that mark spreading plate junctions indicate the rate of spreading.

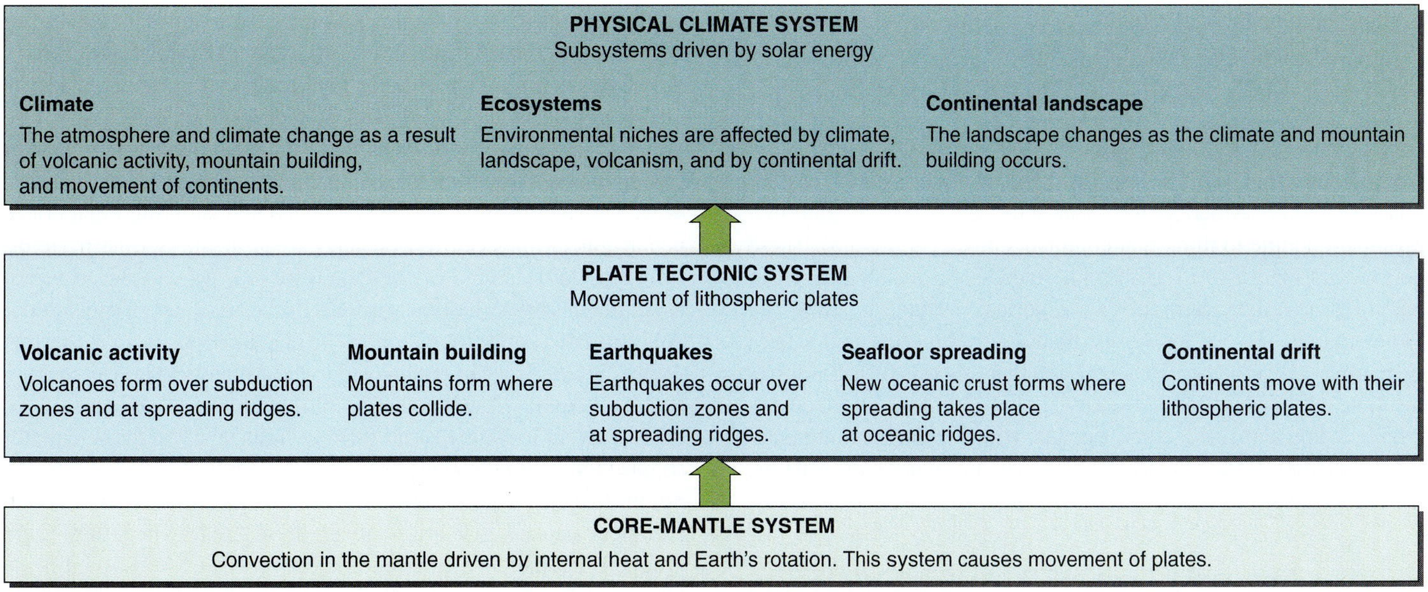

PHYSICAL CLIMATE SYSTEM
Subsystems driven by solar energy

Climate
The atmosphere and climate change as a result of volcanic activity, mountain building, and movement of continents.

Ecosystems
Environmental niches are affected by climate, landscape, volcanism, and by continental drift.

Continental landscape
The landscape changes as the climate and mountain building occurs.

PLATE TECTONIC SYSTEM
Movement of lithospheric plates

Volcanic activity
Volcanoes form over subduction zones and at spreading ridges.

Mountain building
Mountains form where plates collide.

Earthquakes
Earthquakes occur over subduction zones and at spreading ridges.

Seafloor spreading
New oceanic crust forms where spreading takes place at oceanic ridges.

Continental drift
Continents move with their lithospheric plates.

CORE-MANTLE SYSTEM
Convection in the mantle driven by internal heat and Earth's rotation. This system causes movement of plates.

Figure 5.2 **Plate Tectonic System.** The plate tectonic system has great impact on surficial processes, ranging from the formation of new sea floor to volcanism, earthquakes, mountain building, and drift of continents. Each of these, in turn, has impacts on the environment. For example, continental drift, mountain building, and volcanism all cause changes in the landscape and affect the physical climate system.

Plate movements affect most aspects of Earth's physical environment. For this reason, plate tectonics not only makes it possible to synthesize much of what scientists know about the lithosphere, it also explains important connections between the internal and surficial processes that change Earth (figure 5.2).

THE ROLE OF PLATE TECTONICS IN ENVIRONMENTAL SYSTEMS

Although plate tectonics takes place inside Earth and plate movements occur so slowly that they pass unnoticed by humans, plate movements play a major role in shaping Earth's surface. Plate movements have clearly changed the face of Earth for at least the last half-billion years, and some evidence indicates that plate movements were taking place in the Precambrian era 2.7 billion years ago, soon after continents first formed. Crustal structures, such as ocean basins and mountain systems, evolve slowly over hundreds of millions of years, but plate movements affect the surface environment in many ways. The most obvious effects are those caused by volcanic activity and earthquakes. Volcanic activity impacts the atmosphere. So, plate tectonics influences the atmosphere and the physical climate system. These, in turn, greatly impact biogeochemical processes that affect the surface and influence the human environment. Even the slow movement of continents eventually causes continents to move from tropical to frigid parts of the globe. Indeed, the development of plate tectonic theory has changed our view of the world and made us aware of some of the environmental changes that may lie ahead.

EVOLUTION OF PLATE TECTONIC THEORY

Plate tectonics permits a unified description of Earth's crustal features and the processes that affect them. It incorporates the earlier idea of continental drift with a more recent understanding of how new sea floor forms, how portions of the lithosphere sink into the mantle, and how mountains develop. Elements of plate tectonic theory go back to observations made soon after the first accurate maps of the Atlantic Ocean appeared around the turn of the twentieth century. The idea that new sea floor forms along oceanic ridges originated in 1961 and was soon followed by the concept of **subduction,** the idea that parts of the lithosphere sink and melt into the mantle. Once this critical information about the lithosphere was confirmed, the theory evolved so fast that it is considered revolutionary.

Over a period of about a decade, most American geologists changed their conception of the way Earth processes involving the lithosphere operate. They abandoned the idea that continents and ocean basins have been static in position and permanent features since Precambrian time. They abandoned the notion that compression in the lithosphere causes great down-buckles to form, and that mountain belts develop when a down-buckle is squeezed causing sedimentary rocks caught in the down-buckle to rise. They gave up the idea that many continental margins are moving out over the edge of the ocean basins. They adopted the ideas that continents move, that continents have changed position repeatedly over geologic time, and that most volcanoes are associated with the formation of oceanic crust or subduction of oceanic crust. The first hints of plate tectonic theory resulted from a long debate about the possibility that continents may shift position.

Continental Drift

Compare the shapes of the western coast of Africa and the eastern coast of South America and you will see evidence that first prompted people to think that continents move (figure 5.3). These continental margins fit together like pieces of a jigsaw puzzle. In an article on the origin of the Americas written in 1858, an American, A. Snider, proposed that South America and Africa were originally together. He suggested that this was the situation at the time of Noah's flood. He envisioned a violent event during which a volcano-lined crack opened within a large continental landmass and a large section, including the island of Atlantis, drifted westward to become the Americas.

Half a century later, continental drift became the focal point of a protracted scientific discussion following the publication of an article by an American, F. B. Taylor, in 1910 and a book titled *The Origin of Continents and Oceans* by German meteorologist Alfred Wegener in 1915. Taylor described how movements of continents ending in collisions could account for the origin of mountains. Wegener knew about geologic evidence that suggested that continental glaciation occurred on land areas in the Southern Hemisphere toward the end of the Paleozoic era. An unusual group of fossil plants called the **Glossopteris flora,** all of which grew under cold climatic conditions, appear in glacial deposits in Australia, Africa, South America, India, and in the Antarctic. He was seeking a way to explain how this distribution might be possible and especially how an ice sheet could form in India so close to the equator (figure 5.4). He reasoned that India must have been close to the South Pole when the ice sheet formed and that a single continental ice sheet caused glaciation of all these areas when they were grouped together. Wegener and others proceeded to fit continents back together like pieces of a jigsaw puzzle. He favored a fit in which all land areas in the world originally formed one huge continent called **Pangaea** (figure 5.5). Other scientists favored two supercontinents, a southern landmass called **Gondwanaland** and a northern continent called **Laurasia,** separated by a body of water known as the Tethys Sea.

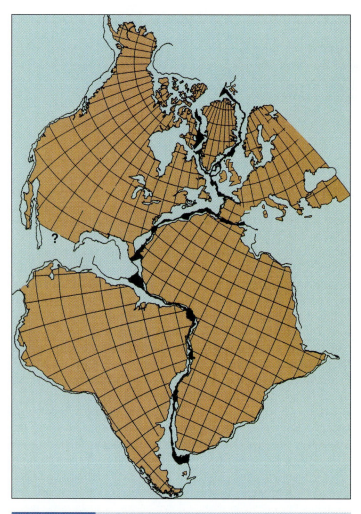

Figure 5.3 Pre-drift Reconstruction. At the depth of 2,000 meters (6,000 feet), the western margin of Africa fits almost perfectly against the eastern margin of South America. This reconstruction of continents to their predrift position is based on a computer program developed by Sir Edward Bullard.

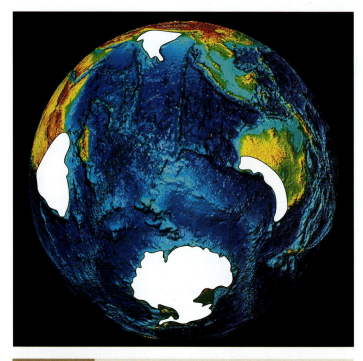

Figure 5.4 Late Paleozoic Glaciation. Shaded areas were covered by ice during the late Paleozoic continental glaciation. Deposits of rocks transported by glaciers and grooves in bedrock, formed as the glaciers moved, provide evidence of this glaciation. The southern part of South America was also glaciated in the late Paleozoic.

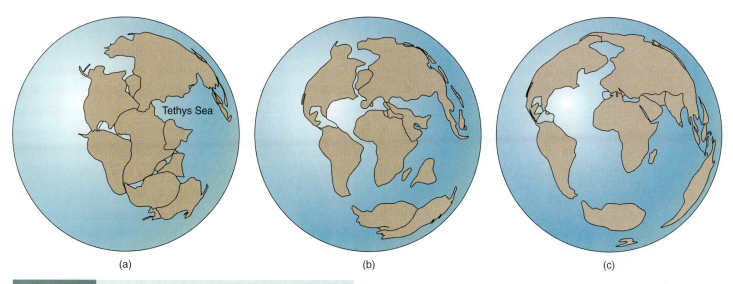

(a) (b) (c)

Figure 5.5 Pangaea. (a) This reconstruction of the continents indicates their position relative to one another about 200 million years ago, near the end of the Paleozoic era. The Atlantic, Indian, and Antarctic Oceans had not formed. A broad sea, known as the Tethys Sea, occupied a position between Eurasia, Arabia, and India. (b) At the stage shown here, the continents had split apart and were moving. The Atlantic Ocean had begun to form, and the Tethys Sea was closing. (c) This sketch shows the position of the continents today, as seen on this projection.

Geologists immediately attacked the idea of continental drift. They found much to criticize because Wegener put continents and continental fragments together without regard for the geology of each piece. Geologists pointed out many instances in which rocks of the same age or type on different fragments did not fit together. Wegener's background in meteorology made him less aware of, and less concerned about, the details of regional geology than scientists with more geologic training. This led the geologists to ignore his suggestions. No one would seriously suggest that the east coast of North America resembles the west coast of Europe. Furthermore, geophysicists showed that the gravitational drag proposed by Wegener as the force that moved continents was totally inadequate to move continental masses through oceanic crust. Opponents of continental drift ignored Wegener's suggestion of seafloor spreading.

Interest in continental drift fell to a low point when geophysicists applied seismic methods to study the continental margins and ocean basins. These studies revealed a much thicker continental crust than the crust in the oceans and proved that both were solid. Wegener had compared continents to great ships moving across the oceans. The geophysicists asked, "How could a continent 40 kilometers (25 miles) thick move through a sea of solid basalt?" Until the 1960s, no answer was forthcoming.

Despite the lack of a mechanism to explain continental drift, interest in this concept continued among African, Australian, and European geologists who worked in the Southern Hemisphere and who were seeking to understand the distribution of late Paleozoic

glaciation. They were unable to find a better explanation for the geographic distribution of late Paleozoic glaciation and the Glossopteris flora than continental drift. In 1958, Warren Carey, an Australian geologist, made the first careful comparison of the fit between Africa and South America. He found that the continental margins of Africa and South America fit together almost perfectly at a depth of 2,000 meters (6,400 feet).

To resolve the question of matching the geology of Africa and South America along their continental margins, an international group of geologists, led by Patrick Hurley of Harvard University, studied the geology of the margins of these continents. Most began thinking that the study would prove the two margins could not have been together. They examined the bedrock geology and mapped the ages of the rocks, rock types, and the trends of faults, folds, metamorphic belts, and bodies of igneous rocks, all of which exist in ancient rocks formed long before the time postulated for the breakup of Pangaea. By the end of the study, the results demonstrated conclusively that both the general regional geology and details fit together nearly perfectly (figure 5.6). Other evidence supporting this type of fit came from totally different types of investigations. Notable among these were studies of Earth's magnetic field and the preservation of magnetic effects in some rocks.

Apparent Polar Wandering

The needle in a compass lines up in the direction of Earth's magnetic field. Extensions of lines defined by the orientation of the magnetic field should cross somewhere near the pole of rotation. If the continents and ocean basins had been static since Precambrian time as many geologists believed until the 1950s, the position of the poles and the direction to the poles from any place on Earth should have been static also.

Although scientists discovered remanent magnetic effects in bricks, pottery, and rocks in the early 1900s, these magnetic effects were not used to locate and trace movements of Earth's magnetic poles until the 1950s. Scientists knew that rocks such as basalt

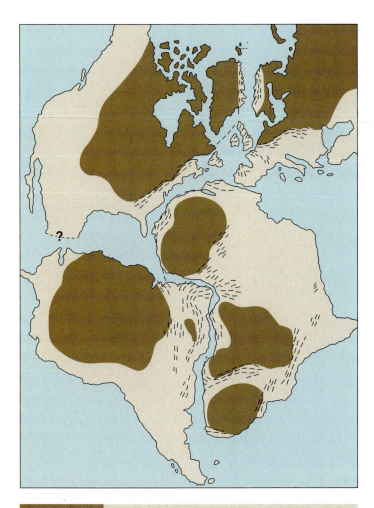

Pre-drift Configuration of Ancient Mountain Belts and Cratons. This is the position postulated for the continents in the late Paleozoic. The position of the continental cratons and younger orogenic belts form a coherent pattern when the continents are placed in the configuration shown.

contain iron-bearing minerals that retain characteristics of the magnetic field in which they solidify. Using the orientation of the magnetic field preserved in lavas as they cooled, it is possible to determine the approximate direction of the magnetic field at the time of cooling. The direction is expressed in terms of the inclination of the field, which varies with latitude, from zero near the equator to near vertical at the poles, and its direction relative to true north is known as *declination*. If the orientation of the magnetic field in rocks of the same age is known at widely separated places, those lines should also coincide near the pole. However, when geophysicists started determining the orientation of the magnetic field in ancient rocks from different parts of the world, they found that the direction to the magnetic north pole was quite different from today's magnetic north pole. Moreover, the direction to the north pole for rocks of different ages, even from the same continent, did not match. Instead, the direction to the pole was found to shift progressively as

shown in figure 5.7. From the perspective of North America, the north pole shifted from a position in the Pacific Ocean during Precambrian time to a position in Siberia by the Triassic period. This phenomenon was given the name **polar wandering.** At first, some thought that changes in the orientation of the magnetic field deep inside Earth caused polar wandering. This idea was short lived.

When polar-wandering paths determined from rocks on different continents failed to coincide, geologists who thought that continents and ocean basins remained fixed in position were surprised. Even more surprising was the discovery that if the continents are shifted into positions like those used to assemble Pangaea, the paths for polar wandering based on different continents almost coincide. Although polar wandering did not provide a mechanism for continental drift, this discovery was significant in making skeptical scientists reconsider the possibility that continents move relative to one another.

Seafloor Spreading

Mapping of the sea floor, paleomagnetic studies, and clarification of crustal structure set the stage for the major breakthroughs that changed thinking about global tectonics. The next major step came in 1961, when Harry Hess, a geologist at Princeton University, suggested that new sea floor is being created along the oceanic

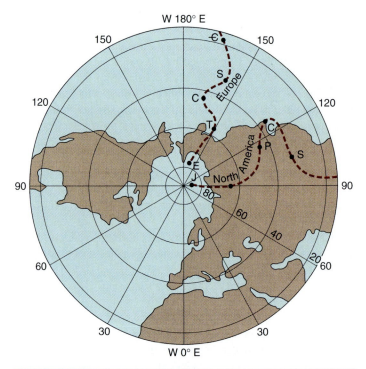

Polar Wandering. The two heavy lines show the position of the magnetic north pole as determined in Europe and in North America at various times in the geologic past. These apparent polar-wandering paths are not identical because North America and Europe have drifted apart.

ridge system. Hess proposed that oceanic ridges are located over rising portions of mantle convection cells. He envisioned that the upwelling in the mantle elevates the crust, causing it to split. As the convecting mantle comes to the surface, it diverges away from the ridge crest, breaks the overlying crust, and drags the sea floor apart. Measurements of the flow of heat through the sea floor show unusually high heat-flow along ridges over the places where convection currents are thought to rise. Low heat-flow values are present over deep-sea trenches where convection currents sink.

In this model, ridges are high because hot convection currents rise under them, elevating the crust along the ridge. A central valley forms along ridge crests where the ridge pulls apart, as hot, plastic mantle beneath the ridge moves in opposite directions away from the ridge crest. Heat associated with rising convection currents generates magma beneath oceanic ridges. As the cooler upper crust cracks and breaks open, magma invades the open spaces, forming dikes. Over a long period of time, the continual addition of new dikes and lava flows at a spreading ridge crest creates new sea floor.

This theory, called *seafloor spreading,* quickly became the focus of studies designed to determine whether or not new oceanic crust forms along oceanic ridges. One of the most important of these studies analyzed remanent magnetic effects. These effects, combined with knowledge about reversals in the polarity of the magnetic field, provided proof of seafloor spreading.

Using Magnetic Anomalies to Trace the Movement of Plates

Systematic examination of the strength of the magnetic field off the western coast of North America was conducted from Scripps Institute of Oceanography in the 1950s. Maps prepared as part of this study revealed a strange-looking pattern, characterized by long stripes of alternating abnormally high and low values of magnetic-field strength, referred to as *magnetic anomaly stripes* (figure 5.8). It seemed obvious that the magnetic properties of the rock in the oceanic crust or upper mantle caused this pattern, but what caused the alternating stripes was unclear. In places, the long, linear stripes were offset. Because some of them coincide with cliffs in the sea floor, scientists quickly identified them as faults. One of these cliffs in the sea floor connects with the San Andreas fault.

The origin of the magnetic stripes remained a mystery until the early 1960s when Fred Vine from Princeton University and D. H. Matthews of Cambridge University, England, worked with the U.S. Navy to study magnetic-field strength over the Mid-Atlantic Ridge just south of Iceland (figure 5.9). They found long stripes of abnormally high and low values of magnetic-field strength like those off the west coast of North America. South of Iceland, the stripes are parallel to the crest of the ridge, and the patterns on opposite sides of the ridge are symmetrical. These observations suggest that the origin of the stripes is related to the formation of the ridge. Vine and Matthews concluded that the magnetic field currently produced in Earth's core is uniform in strength across the area and that the observed differences, which give rise to the stripes, arise from remanent magnetic effects caused by the rocks in the ridge. They reasoned that the dikes and

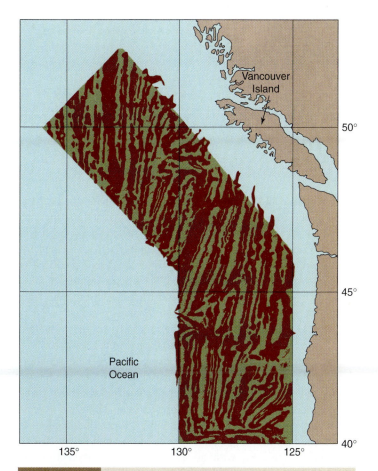

Figure 5.8 Magnetic Anomaly Stripes. By coloring anomalies in the strength of the magnetic field, this pattern was obtained for the field off the western coast of North America. All red areas have unusually high magnetic-field strengths. The green areas have unusually low magnetic-field strengths. This study was one of the first in which this type of alternating banded anomalies was discovered.

lava flows found over the crest of the ridge take on the polarity and direction of the existing magnetic field as they cool and solidify (figure 5.10). The iron-bearing minerals in basalts preserve the magnetic field in which they crystallize when the temperature of the rock reaches about 573°C (1,000°F). Thus, the high magnetic-field strength found over the ridge crest is there because the magnetic effects caused by the new crust that forms along the ridge crest reinforces the strength of Earth's main field. The low values occur where portions of the oceanic crust formed during a time of reversed polarity. The remanent magnetic effects of these rocks are negative and reduce Earth's present field. Thus, each stripe of high field strength coincides with crust formed during a time when the magnetic field had the same polarity as the present field, and each low stripe coincides with a time of reversed polarity. As the sea floor spreads, the crust near the crest of oceanic ridges splits and carries its record of polarity reversals in the remanent magnetism of the rocks in both directions from the ridge.

Figure 5.9   Correlating Magnetic Anomaly Stripes with Age of Crust.
Sea floor at the ridge crest is polarized in the same direction as the present (normal) magnetic field. The flanking negative anomalies are caused by reversed polarity of the remanent magnetization of the sea floor that solidified at the ridge crest when the field was reversed. Red parts of the scale represent times when the magnetic field had the same polarity it has today. During time represented by white areas, the field was reversed.

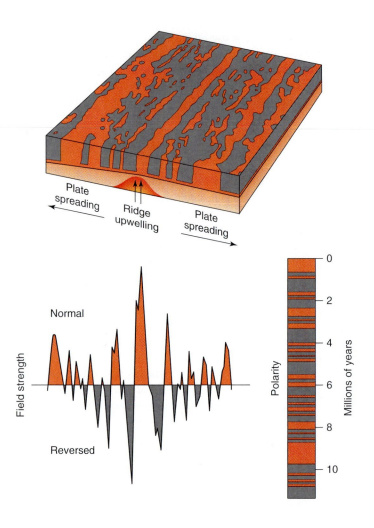

If samples could be collected from the basalts that lie beneath the sea floor, it would be possible to determine the age of the stripes and the rate at which sea floor spreads. Some samples have been collected and dated, but the cost of drilling in the deep sea makes this an impractical solution. Instead, the stripes are dated by collecting samples from basalts exposed on land. The age and polarity of these land samples make it possible to construct a **geomagnetic time scale** that shows the age of reversals in magnetic-field polarity (figure 5.11). Thus, by counting reversals out away from modern ridge crests, it is possible to determine the age of the sea floor.

The magnetic stripes provide strong proof of the seafloor-spreading hypothesis. They make it possible to determine the age of the sea floor, and, by measuring the distance of any given stripe (of known age) from ridge crests, scientists can determine the rate at which the ridge spreads (see figure 5.9). Rates of spreading in the Atlantic Ocean range from about 1 to 6 centimeters (0.4 to 2.4 inches) per year, with an average of about 4 centimeters (1.6 inches) per year. Other ridges, such as the East Pacific Rise, spread at rates in excess of 10 centimeters (4 inches) per year.

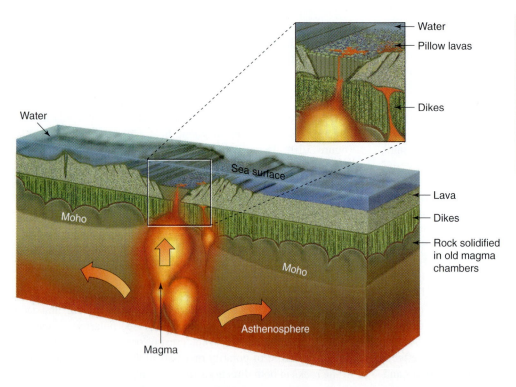

Figure 5.10   Oceanic Ridge Cross Section.
Oceanic ridges are places where new sea floor forms. Beneath the ridges, the oceanic crust is thin. The median valley forms by extension across the ridge. Rising bodies of magma beneath the ridge crest and the dikes and lavas form as the magma makes its way to the surface.

Seafloor spreading has provided an explanation for an observation that had long troubled oceanographers. Scattered seismic studies of the sea floor revealed much thinner layers of sediment than expected. In the deep sea, sediment is rarely more than a kilometer thick. Based on current rates of sediment accumulation, sediment in the sea floor should be much thicker if the oceans are as old as the continents. If the oceans are relatively young, the sediment should be thin, and if new sea floor continuously forms along ridges, sediment should be thinnest on ridges and thicker away from the ridges. Seismic profiles and drilling from ships designed for this purpose soon established that sediment is thinner over ridges and thicker away from ridges. These studies demonstrated that sediment thickness is greatest on older oceanic crust and that the oceanic crust is young relative to the continents. The oldest oceanic crust is less than 200 million years old.

Using Transform Faults to Trace Plate Movements

When oceanographers first recognized the straight, steep valleys that cut across and commonly offset the oceanic ridges, they assumed that faults caused the valleys. At that time, geologists thought that offsets of features, such as the ridges across faults, involved horizontal displacements like those illustrated in figure 5.12. These type of faults are called **strike-slip faults**—essentially vertical faults across which the two sides move laterally by one another as shown in figure 5.12. However, the Canadian geophysicist J. Tuzo Wilson recognized that the concept of a mobile sea floor opens the way for new possibilities of movement patterns. He realized that, because seafloor spreading involves movements in opposite directions across the ridge crest, the present movement along the faults across ridge crests may be the opposite of what scientists expected (see figure 5.12). Wilson coined a new term, trans-

form faults, for special types of strike-slip faults like the ones that cross spreading ridges. Later studies confirmed the movement he suggested. Transform faults also occur in places where part of a plate descends into a subduction zone.

Traces Left by Hot Spots

Many volcanic centers, such as the ones at Hawaii and at Yellowstone National Park, occur far from plate boundaries. Seismologists have tracked the movement of magma from deep in the mantle to the surface. Some of these hot spots (see chapter 7) rise beneath oceanic crust. Others like the one at Yellowstone Park rise beneath continental crust.

Wilson recognized the hot-spot phenomenon and went on to point out its use in tracking the movement of plates. His reasoning was that if the hot spot arises from a fixed position in the mantle, the volcanic activity on the surface would appear to shift its position as the plate moves over the hot spot. The Hawaiian Islands provide one of the best examples. The only active volcanoes along the Hawaiian Ridge lie on the island of Hawaii. A new volcanic center, a seamount known as Loihi, is rising from the sea floor on the eastern side of Hawaii. The "big island" is situated at the end of a long chain of inactive submarine volcanic centers that make up the Hawaiian Ridge (figure 5.13). Cores taken from these submerged volcanic centers revealed that the volcanic activity is increasingly older toward the west. As hypothesized by Wilson, these old volcanic centers are tracking the movement of the Pacific Plate.

The northwest trending Hawaiian Ridge ends at Midway Island, and the submarine ridge turns sharply to the north and forms the Emperor Seamount chain. This ridge is a continuation of the Hawaiian hot-spot track, but the plate changed direction when Midway occupied Hawaii's present position over the hot spot. The

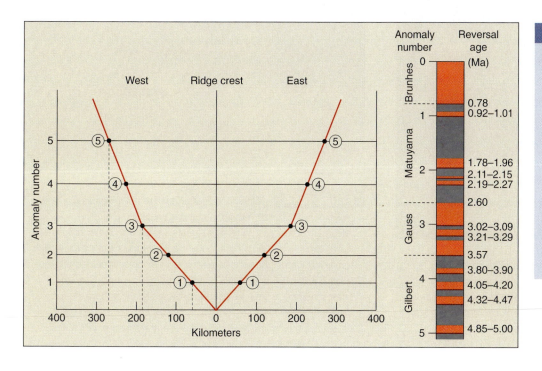

Figure 5.11 Matching Magnetic Anomaly Stripes with the Geomagnetic Time Scale. Each clearly recognizable band of magnetic anomalies on the flank of an oceanic ridge is assigned a number. These numbers correspond to a particular time in the geologic past. Therefore, a plot of the anomaly number against distance of the anomaly from the ridge crest allows us to determine spreading rates. In the example shown here, spreading between anomalies 3 and 5 was much slower than that between anomalies 1 and 3.

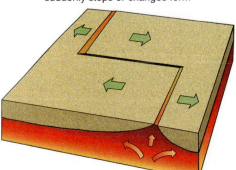

Strike-slip fault
Steeply inclined faults with lateral displacement

Transform fault
Strike-slip fault along which the displacement suddenly stops or changes form

Figure 5.12 Comparison of the Geometry and Movement Associated with a Conventional Strike-slip Fault and Transform Fault.

tracks left by hot spots provide a valuable clue to the direction of plate movement. As the plate moves over the stationary hot spot, volcanoes spring up on the plate above the hot spot. As the plate continues to move, the older volcanoes die as they lose their heat and magma source, and new ones form. The resulting chain of volcanoes is a physical topographic ridge that reveals the direction the plate moved over the hot spot.

Measuring Plate Movements

By determining the age of the extinct volcanoes along the Hawaiian Ridge, it is possible to calculate the rate at which the Pacific Plate is moving. That rate is about 10 centimeters per year, a value close to that obtained by measuring how far magnetic anomaly stripes are away from the crest of the East Pacific Rise. Detailed surveying provides another means for measuring plate movements (figure 5.14). By carefully resurveying the distances and position of points located on different plates, it is possible to document the movements of plates relative to one another. The rates of plate movement (in the range of 4 to 10 centimeters per year) are so low that conventional methods of surveying lack the precision needed to detect the movements over the long distances involved. In the late 1980s, United States government agencies provided public access to a precise satellite navigation system called the *Global Positioning System* (GPS) (figure 5.15). This system makes it possible to determine the

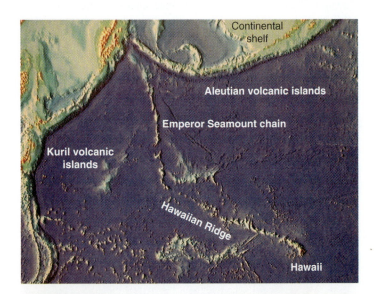

Figure 5.13 Map of the Sea Floor in the Vicinity of the Hawaiian Islands. Note that the islands lie along a prominent, largely submerged ridge. Volcanoes on Hawaii at the eastern end of this ridge are the only ones that are active. Rocks from other islands and seamounts along the ridge are progressively older to the west.

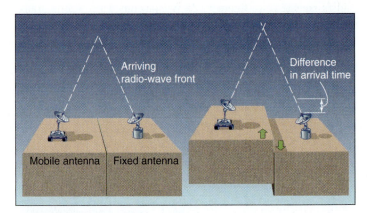

Figure 5.14 Measuring Plate Movements. Measuring movements on Earth with an interstellar yardstick, a mobile-radio telescope homes in on the faint signal generated by a star billions of light-years distant. An atomic clock measures the slight difference in the arrival time of the radio-wave fronts at the telescope and at a second unit located across a plate boundary. The time delay and antenna aiming angles are used geometrically to compute the distance between the stations; both factors change when crustal deformation shifts the relative positions of the stations.

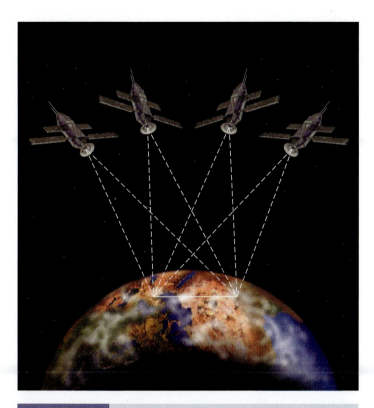

Satellites of the Global Positioning System provide a highly accurate frame of reference for the location of points on Earth's surface.

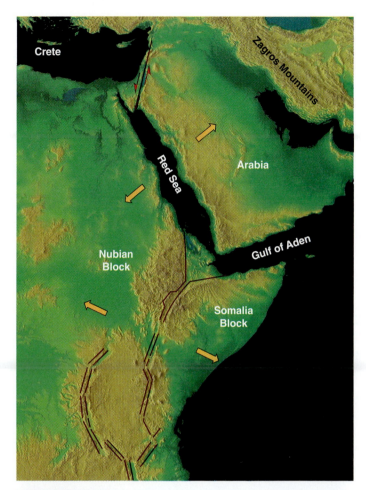

Figure 5.16 Movement of African and Arabian Plates. The Red Sea, Gulf of Aden, and the extensive system of fault blocks in eastern Africa are all caused by extension of the crust in this region. The arrows indicate the local relative directions of extension.

position of points on the surface with great accuracy. Measurements made with this system confirm earlier predictions of plate movement and reveal even more details than had previously been possible. For example, these measurements indicate that the African and Arabian plates are moving northward toward Eurasia. The Arabian Plate is moving faster to the northeast where a collision has been in progress between the Arabian Plate and the Eurasian Plate (figure 5.16). A young mountain belt, the Zagros Mountains, has formed where this collision took place. As Africa approaches Eurasia, the oceanic crust under the Mediterranean Sea is sinking back into the mantle. This subduction is taking place just south of the island of Crete (see figure 5.16).

Destruction of Oceanic Crust in Subduction Zones

If new oceanic crust forms everywhere along the global encircling system of oceanic ridges (see inside cover) and if the sea floor spreads away from these ridges, then either Earth is getting bigger or some process is destroying portions of the lithosphere. Most scientists think the evidence supports the idea that the oceanic lithosphere sinks back into the mantle in subduction zones.

Long before the advent of plate theory, Hugo Benioff, a seismologist at California Institute of Technology, recognized that earthquakes in island arcs most commonly occur in distinct zones. These zones start near the surface under deep-sea trenches and slope under the volcanic arcs, extending to depths of as much as 700 kilometers (435 miles). Later, as earth scientists began to look for places where the oceanic lithosphere disappears, these zones, now called Benioff zones, seemed likely places (figure 5.17). To test this idea, a group of geophysicists from Columbia University set up a network of seismograph stations in the southwest Pacific. They monitored the passage of seismic waves across and within the Benioff zone beneath the Tonga and Kermadec islands just north of New Zealand. The data obtained in the study of the Tonga-Kermadec earthquakes showed that waves traveling just beneath this Benioff zone behaved like waves transmitted through cool, rigid lithosphere, while those waves that traveled deeper paths through the asthenosphere lost energy rapidly, as expected in less-rigid rock. The presence of rigid rock deep beneath the islands supports the idea that a flap of lithosphere is under the Benioff zone (figure 5.18). This is convincing evidence that subduction is real, and that it is presently taking

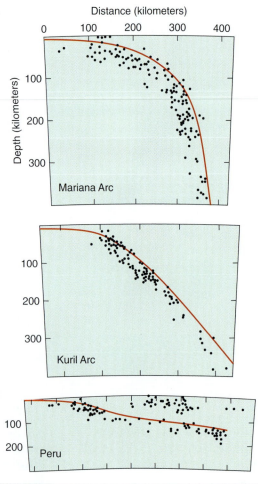

Distance (kilometers)

Depth (kilometers)

Mariana Arc

Kuril Arc

Peru

Figure 5.17 **Benioff Zones.** Hugo Benioff first recognized the zones of concentrated seismic activity located beneath island arcs. He plotted earthquake foci in sections across island arcs and found the well-defined zones of earthquake foci. These earthquakes are now interpreted as originating in the down-going slab of lithosphere in subduction zones.

place in Benioff zones like those beneath the island arcs and along the coast of South America.

TECTONIC CYCLES

Plate tectonics provides mechanisms by which rocks in the lithosphere can be recycled. Materials from continents are carried into the oceans where they reside on top of oceanic crust formed at spreading ridges. As oceanic crust moves into subduction zones, the materials in the down-going slab of oceanic lithosphere pass into the mantle where they may be melted and eventually brought back to the surface through volcanic activity or where plates collide. Plate theory also explains how areas on the globe that were once continents may become ocean basins. This process begins with the splitting of a continent.

Breakup of Continental Crust

The continental crust shows signs of being pulled apart in many places. The most obvious evidence of stretching is normal faulting and development of **grabens** (figure 5.19 and see figure 5.28). The crustal extension in all of these places results from the pulling apart of the lithosphere over divergent plate boundaries. In the eastern United States, grabens formed when Pangaea broke up at the beginning of the Mesozoic era. Sediments that eroded from adjacent highlands quickly filled the depressions. Younger grabens formed as the Red Sea opened (see figure 5.16). This extension created the East African Rift valleys, the floors of which form great plains, such as the Serengeti plain where herds of wild animals roam today. Magmas have risen along the steeply inclined faults at the edge of these grabens forming volcanoes like Kilimanjaro. In east Africa, stretching of the crust stopped short of pulling the continental crust completely apart, but in the Gulf of Aden, the Red Sea, and the Gulf of California, this type of stretching has been a prelude to the formation of new oceans.

The Formation of Ocean Basins and Oceanic Crust

Following the initial rupture of continental crust, grabens form as the crust pulls apart and some blocks of crust sink. Further extension pulls the fragmented continental crust apart, allowing new sea floor to form. Today, this is taking place in the Red Sea and the Gulf of California (figure 5.20 and see figure 5.16). The oceanic ridge system, complete with high-heat flow, transform faults, and magnetic anomaly patterns, extends into these seas where shores were adjacent to one another only a few million years ago. First, the crust swelled over rising mantle plumes or convection cells in the mantle, the continental crust cracked, and grabens developed in the crust. Finally, the two sides of the Red Sea and the Gulf of Aden opened, allowing new oceanic crust to form. Today, crustal extension continues in the Red Sea and the Gulf of Aden.

New sea floor is forming in the developing oceans. Based on the age of the magnetic anomalies in the Gulf of California, scientists estimate that the Gulf is only about 5 million years old. Baja California and a strip of California west of the San Andreas fault are part of the Pacific Plate that is moving northwest relative to the remainder of the continent. If movement along this boundary continues, a sliver of land including Baja and the West Coast will eventually move out into the Pacific and become an island.

Volcanoes Formed over Subduction Zones

Island arcs in the western Pacific from the Aleutians to New Zealand mark places where the oceanic lithosphere is sinking into the mantle (figures 5.21 and 5.22). Almost all of these chains of active volcanoes have a deep-sea trench on the oceanward side; a Benioff zone of high seismic activity beneath them; and lavas, many of which are andesitic in composition (see figure 2.14). Trenches are topographic expressions of the subsidence of the sea floor caused by the bending and sinking of the oceanic lithosphere. As the down-going slab sinks, sediment and upper parts of the slab split off and pile up beneath the inner wall of the trenches as shown

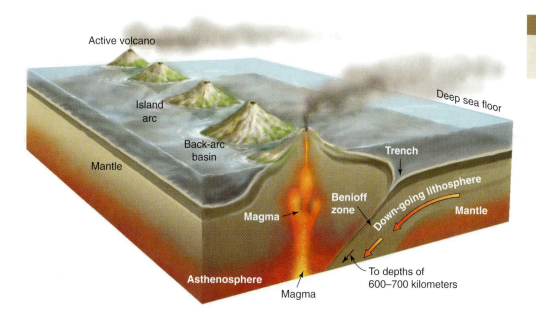

Active volcano

Island arc

Back-arc basin

Mantle

Magma

Asthenosphere

Magma

Deep sea floor

Trench

Benioff zone

Down-going lithosphere

Mantle

To depths of 600–700 kilometers

Figure 5.18 Block Diagram of a Subduction Zone and the Overlying Volcanic Islands.

in figure 5.23. Seismic reflection and refraction surveys between the trench and the volcanic islands reveal slices of oceanic crust between the trench and the active volcanoes.

Volcanoes form over down-going slabs of the lithosphere because upper portions of the oceanic crust contain mixtures of basaltic lavas and water-saturated sediment that begin to melt when they reach a depth of about 100 kilometers (62 miles). Although the oceanic crustal rock is initially cool, as the slab sinks, its temperature rises, and the rocks that were stable near the surface become unstable and undergo metamorphism. Eventually, parts of it melt. The presence of water in the mixture lowers the melting point of the basalt in the oceanic crust and the peridotite that is present in the overlying mantle. The resulting magma contains elements that were in both the down-going slab and the overlying asthenosphere. The composition of this magma is intermediate between that of rhyolite and basalt. For this reason, most of the volcanoes over subduction zones erupt a distinctive type of lava, known as **andesite** (see chapter 2).

Signs of volcanic activity also occur behind island arcs in areas referred to as **back arc basins** (figure 5.24). These basins are shallow seas that separate many volcanic arcs from adjacent continents. Heat-flow measurements indicate that these are abnormally hot areas. It appears likely that they are hot because the sinking oceanic lithosphere in front of the arc pulls the arc forward, causing

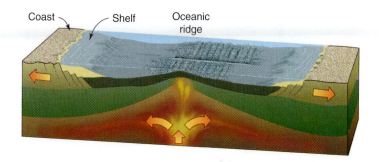

Coast   Shelf   Oceanic ridge

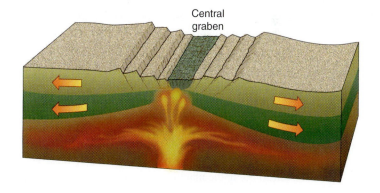

Central graben

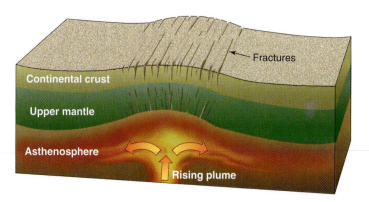

Fractures

Continental crust

Upper mantle

Asthenosphere

Rising plume

Figure 5.19 Steps in the Evolution of a New Ocean. Fractures form in the continental crust as hot mantle rises beneath the continent. As extension continues, a graben forms along the central part of the spreading zone, and more normal faults develop along the flanks of the separating continental fragments. Finally, the continents are far apart. The older sea floor is covered by sediment, but new sea floor continues to form along the oceanic ridge.

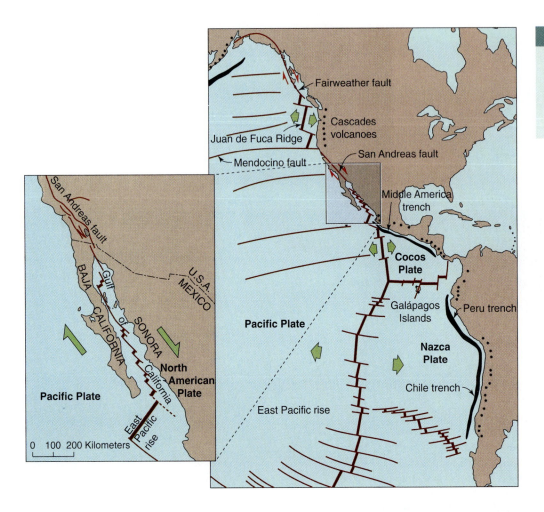

Figure 5.20 Major Features of the Eastern Pacific Region. The Gulf of California started to form about 5 million years ago as the region west of the San Andreas fault began to slip northwest.

extension and spreading to occur behind the volcanic arc. Heat then rises as the hot underlying mantle rock rises.

The Causes of Subduction

Three factors facilitate the movement of oceanic crust into subduction zones. Initially, many attributed the movement of plates almost entirely to convection in the mantle. Most agree that mantle material involved in convection rises under oceanic ridges and returns downward and deeper into the mantle in subduction zones. This convective overturning in the mantle exerts drag on the more-rigid rocks of the lithosphere. Secondly, the hot, partially molten, plastic asthenosphere forms a zone in which the lithosphere can slip over the underlying mantle. As parts of a slab of lithosphere move away from the oceanic ridge where they form, the rocks cool, and the density of the cooler rock increases. When the cool edge of the slab converges toward a continent or a less-mobile part of the oceanic crust, it sinks. Gravity pulls the lithosphere down the slope away from oceanic ridges where oceanic crust forms and toward subduction zones. The third condition that contributes to subduction results from changes that occur in the down-going slab as it moves deeper into the asthenosphere. Some of the minerals in mafic rocks, magnesium, and iron-rich rocks such as peridotite undergo metamorphic changes as temperature and pressure increases, causing the atoms to pack together more tightly in higher-density forms. Thus, deep in the sinking slab, density increases and pulls the slab down.

MOUNTAIN BUILDING

Mountain ranges, such as the Sierra Nevada in California, Blue Ridge in Virginia, and the Colorado Front Range, are small elements of much larger tectonic units, referred to as **mountain belts,** or **orogens.** Some of these mountain belts are hundreds of kilometers wide and thousands of kilometers long. All of these larger mountain belts are products of interactions between plates along plate boundaries. Although many older mountain belts exist on Earth, two relatively young belts, the Alpine-Himalayan and the Cordilleran, are especially prominent (see inside cover). The Cordilleran mountain belt, located along the western margin of North and South America, includes the Rocky Mountains, the deformed region west of the Rocky Mountains in North America, and the Andes in South America. The Alpine-Himalayan belt, which lies between Eurasia on the north and Africa, Arabia, and India on the south, includes all of the mountains between North Africa and southern Spain to southeast Asia. Uplift of these mountain belts started millions of years ago, and both of them continue to grow today.

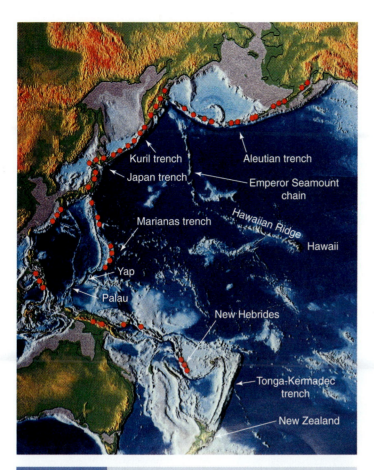

Figure 5.21 Island Arcs in the Western Pacific. Heavy red dots indicate volcanic island arcs.

the type exposed in the shields, may shear off and force their way into the overlying pile of sediment and sedimentary rocks. In a few places, even pieces of ancient oceanic crust are forced up on the continent. This has happened at the Long Range Mountains of Newfoundland. If the temperature and pressure are great enough, the deep crust may fold as it has in the Alps (see figure 4.9).

Plates converge at rates of a few centimeters (1 to 2 inches) per year, so mountain building is a slow process. As the mountains rise, weathering and erosion accelerate on the uplifted crust, and streams carry the erosion products into low-lying regions nearby. As loads shift from one part of the crust to another, isostatic adjustments in the elevation of the crust take place because the lithosphere floats on the deeper interior, as discussed in chapter 3.

The Appalachian Mountains—Collision with Africa

The Appalachian Mountains are but one part of a mountain system that extends from the Marathon Mountains in West Texas and includes the Ouachita Mountains in Arkansas and Oklahoma (see figure 3.4). At the time the mountains formed, the system included the Caledonian Mountains of Ireland, Scotland, and Norway. The Appalachian-Caledonian mountain system formed along the margin of an ocean that separated Africa and Eurasia on the east from North and South America and Greenland on the west. At various times during the Paleozoic era, volcanic island arcs stood along this continental margin and became part of the present North American continent as subduction along the continental margin plastered the arcs against the continent.

Most mountain building results from the collision of two continents, a continent and an island arc, or fragments of continents with continents or island arcs. Because of its higher density and lower elevation, oceanic crust generally sinks and subducts under continental crust where plates converge. However, a single plate may contain both oceanic and continental crust. As convergence and subduction continue, eventually a continent or continental fragment on the subducting plate approaches the subduction zone. When this continental crust reaches the trench, it collides with the inner wall of the trench, but because of its lower density, it cannot go down the subduction zone. As the plates are driven together, they compress the sediments caught along the margins of a colliding continent and an island arc or between colliding continents (figures 5.25 and 5.26). Caught as though between the jaws of a gigantic vise, the sediment and rock along the edges of the two colliding elements shorten by folding (figure 5.27) and faulting (figure 5.28) and uplift. In front of the colliding plates, layers of sedimentary rock commonly slip along bedding planes and form great piles as layers stack one on top of another (figures 5.29 and 5.30). Slowly, the mountains rise, lifting the sediments deposited in the sea higher and higher. Eventually, large pieces of the deeper continental crust,

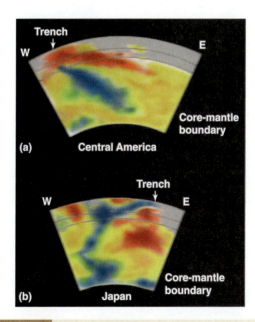

Figure 5.22 Oceanic Lithosphere Sinking into the Mantle. Vertical slices through the crust across the western margin of South America (a) and across an island arc (b) indicate colder material (blue) descending from the trench area into the mantle.

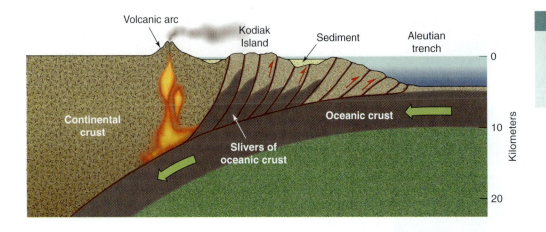

Volcanic arc   Kodiak   Sediment   Aleutian
                Island   trench

Continental crust

Oceanic crust

Slivers of oceanic crust

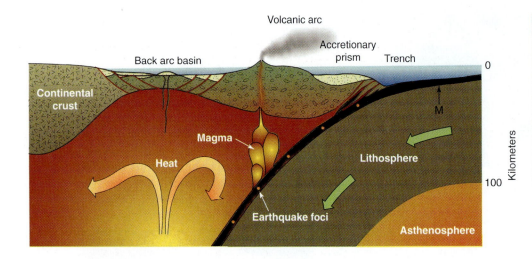

Volcanic arc

Back arc basin   Accretionary prism   Trench

Continental crust   M

Magma

Heat   Lithosphere

Earthquake foci

Asthenosphere

Figure 5.24 Cross Section of an Island Arc. This generalized cross section across an island arc shows the trench, slivers of sediment scraped off the down-going lithosphere, magmas formed below the volcanoes in the arc, the back arc basin, and adjacent continental crust. Earthquake foci occur within and all along the top edge of the down-going lithosphere.

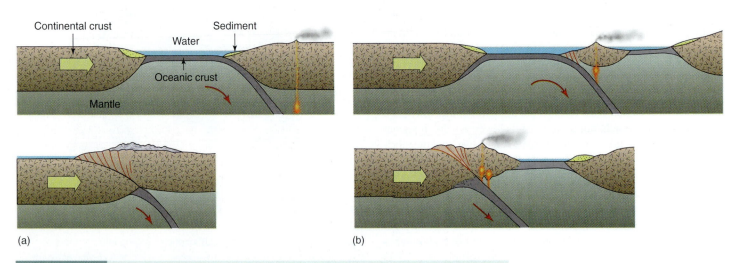

Continental crust   Sediment
        Water
Oceanic crust
Mantle

(a)   (b)

Figure 5.25 Continental Collision. (a) Two continents collide as subduction takes place along the continental margin. (b) Collision of an island arc with a continent.

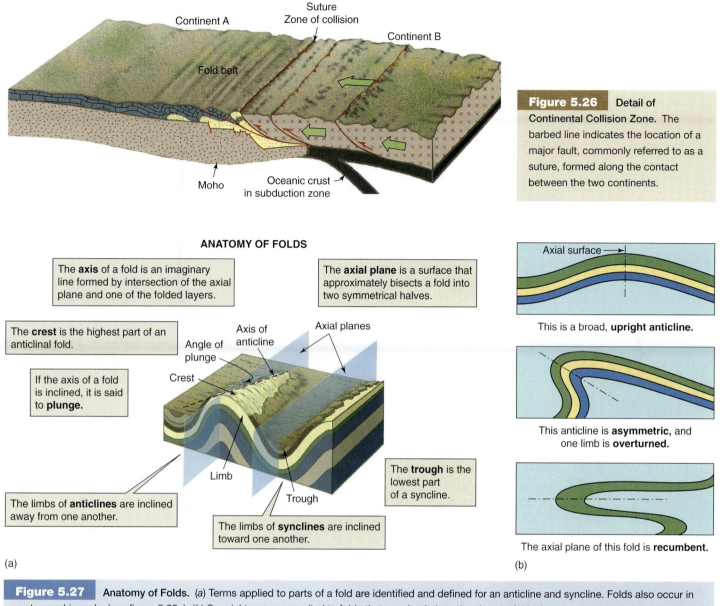

Continent A

Suture
Zone of collision

Continent B

Fold belt

Moho

Oceanic crust
in subduction zone

Figure 5.26 Detail of
Continental Collision Zone. The
barbed line indicates the location of a
major fault, commonly referred to as a
suture, formed along the contact
between the two continents.

ANATOMY OF FOLDS

The **axis** of a fold is an imaginary
line formed by intersection of the axial
plane and one of the folded layers.

The **axial plane** is a surface that
approximately bisects a fold into
two symmetrical halves.

The **crest** is the highest part of an
anticlinal fold.

Axis of
anticline

Axial planes

Angle of
plunge

If the axis of a fold
is inclined, it is said
to **plunge.**

Crest

The limbs of **anticlines** are inclined
away from one another.

Limb

Trough

The **trough** is the
lowest part
of a syncline.

The limbs of **synclines** are inclined
toward one another.

(a)

Axial surface

This is a broad, **upright anticline.**

This anticline is **asymmetric,** and
one limb is **overturned.**

The axial plane of this fold is **recumbent.**

(b)

Figure 5.27 **Anatomy of Folds.** (*a*) Terms applied to parts of a fold are identified and defined for an anticline and syncline. Folds also occur in metamorphic rocks (see figure 2.25c). (*b*) Special terms are applied to folds that are simple broad arches, to folds that are asymmetric, and to folds in which the axial plane is horizontal or nearly horizontal. One limb of the asymmetric fold shown here is overturned. The beds on the overturned limb are upside down.

At times during their long histories, these mountains may well have been as high as the modern Alps or Himalayas, but mountain building stopped over 200 million years ago before opening and growth of the Atlantic Ocean started. As the Atlantic Ocean formed, the mountain belt split apart. Erosion that started 200 million years ago continues to reduce the height of the mountains. After nearly 100 million years, the ocean invaded the edge of North America, and sediments covered the eastern part of the Appalachian Mountains. These sediments and sedimentary rocks form the modern coastal plains along the Atlantic and Gulf coasts. They separate the Appalachian, Ouachita, and Marathon mountains from one another (see figure 3.4). Nevertheless, enough of

the old, deeply eroded mountain system remains to piece together its structure and history in considerable detail.

The culmination of Appalachian mountain building took place near the end of the Paleozoic era when Africa collided with North America, finally closing an ocean, known as the Iapetus Ocean, that had separated these continents for half a billion years. During this collision, large slices of rock from deep in the continental crust sheared off and forced their way upward into the overlying sedimentary rocks that had formed along the edge of the continent (see figure 5.26). As the collision continued, mountains rose, and the rising pile of rock forced its way northwestward toward the continental interior. In the internal portions of this mountain

ANATOMY OF FAULTS

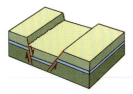

The structure formed where a block drops down between two normal faults is called a **graben.**

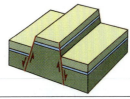

The feature formed where a block between two normal faults is up is called a **horst.**

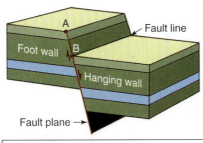

The **footwall** of a fault is the block beneath the fault plane.

A **fault line** is the trace of the fault along the ground.

The **hanging wall** of a fault is the block above the fault plane.

The hanging wall has moved down relative to the footwall in **normal faults.**

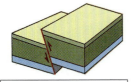

The hanging wall has moved up relative to the footwall in **reverse faults.**

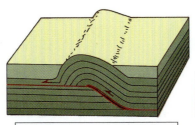

The fault zone is inclined (dips) at a very low angle on **thrust faults.**

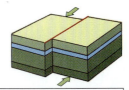

One block moves laterally relative to the adjacent block in **strike-slip faults.**

(a)

Figure 5.28 Anatomy of Faults. (a) Faults separate adjacent crustal blocks. Classification of faults is based on the relative movement of these blocks. The most common types of displacements of the blocks are illustrated in this figure. (b) Two examples of fault zones are illustrated in these photographs. The fault in the right photograph is a thrust fault. The top (hanging wall) side of the fault has moved to the right, relative to the lower (footwall) side of the fault. Note the drag fold formed in the hanging wall. The left photograph illustrates a normal fault. In this case, rocks on the left side of the faults have moved down, relative to those on the right side. Note the features indicating drag in the fault zone.

(b)

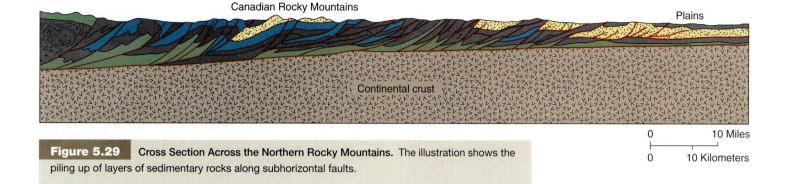

Canadian Rocky Mountains

Plains

Continental crust

0   10 Miles
0   10 Kilometers

Figure 5.29 Cross Section Across the Northern Rocky Mountains. The illustration shows the piling up of layers of sedimentary rocks along subhorizontal faults.

belt, Precambrian-aged rocks that were in the lower part of the continental crust were uplifted. They are exposed in the Blue Ridge (see figure 5.30). In the external portions of the uplifted area, pressure arising from the collision folded and pushed sedimentary rocks deposited on the continental margin to the northwest (see figure 5.30). Huge layers of these sedimentary rocks responded by sliding over one another as thrust sheets. Today, these folds and thrust sheets form the Valley and Ridge Province. Structural features similar to those found in the Valley and Ridge are present in the front ranges of the Canadian Rockies.

Assembly of the Western Margin of North America

The Pacific and Atlantic margins of North America have distinctly different modern plate tectonic settings. Today the eastern margin of North America is a passive margin, part of the North American Plate that extends east to the Mid-Atlantic Ridge. As the Atlantic Ocean grows, North America moves westward. This westward drift moves North America over parts of a spreading ridge. This spreading ridge is still evident as the East Pacific Rise and the Juan de Fuca Ridge (figure 5.31). Numerous transform faults cut across the East Pacific Rise as it passes into the Gulf of California. The East Pacific Rise ends where the San Andreas fault cuts across and displaces it (figure 5.32). The San Andreas fault is a transform fault connecting the East Pacific Rise with another transform fault at Cape Mendocino. This second fault cuts across the Juan de Fuca Ridge. Deformation of the region during the Cenozoic era involves subduction of the Juan de Fuca Plate along the coast of Washington and Oregon and northwestward slip of a sliver of North America including Baja and the part of California west

of the San Andreas fault. About 5 million years ago, slip along the San Andreas fault caused the Gulf of California to open, but deformation along the West Coast started millions of years earlier, and its effects extend eastward into Utah. The crust in parts of this region which includes all of Nevada and western Utah, known as the Basin and Range Province, is being pulled apart as the western margin of the continent moves northwest. This modern deformation is quite different from the earlier tectonic history of the western margin of North America.

Early in the Paleozoic era, long before mountain building started in western North America, the edge of the continent was located approximately where the Rocky Mountains are located in Canada and along the western edge of the Rocky Mountains in Utah (see figure 5.32). Since that time, the western margin of the continent has grown by the accretion of island arcs and fragments of continental crust. These fragments came from the west as oceanic crust moved east and subducted along the western North American margin. A number of major episodes of mountain building took place. One that occurred near the end of the Mesozoic led to the uplift and deformation of the Rocky Mountains. The best-preserved folded and thrust faulted portions of this mountain belt, called the Sevier Belt, occur in Canada and along the western border of Wyoming. Pieces of this fold belt continue southward through Arizona. After this fold belt developed, large blocks of continental crust rose during a phase of mountain building known as the Laramide Orogeny and formed the Middle and Southern Rocky Mountains. The largest of these uplifts is a huge crustal block that includes portions of Colorado, New Mexico, Arizona, and Utah. This area, called the Colorado Plateau, is well known for the spectacular formations that occur in the National Parks and Monuments of that region, such as the

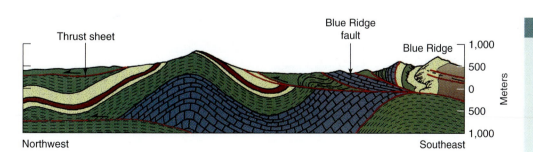

Thrust sheet

Blue Ridge fault

Blue Ridge

1,000
500
0
500
1,000
Meters

Northwest

Southeast

Figure 5.30 Detailed Cross Section Across the Contact Between the Blue Ridge and the Valley and Ridge in Central Virginia. A cross section across the southern or central Appalachian Mountains would resemble the section across the northern Rocky Mountains shown in figure 5.29.

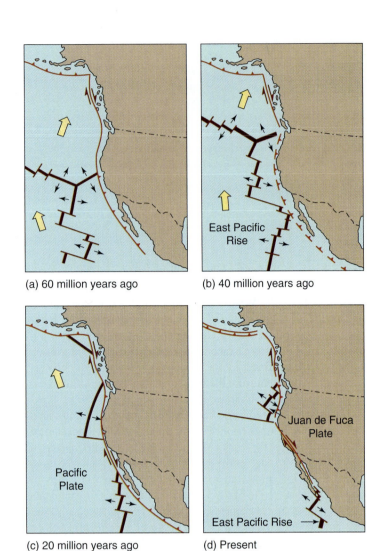

(a) 60 million years ago

(b) 40 million years ago

East Pacific Rise

(c) 20 million years ago

Pacific Plate

(d) Present

Juan de Fuca Plate

East Pacific Rise

Figure 5.31 Evolution of Fault Systems, Subduction Zones, and Spreading Ridges Along the Western Margin of North America over the last 60 Million Years. (*a*) A triple-spreading junction is shown. The junction, (*b*) and (*c*), migrates into the Aleutian subduction zone. At the same time, North America is moving westward, and a subduction zone off the southwestern margin of the continent is slowly transformed, (*d*) into a strike-slip fault as the North American Plate moves over the spreading ridge.

Grand Canyon, Zion, Bryce Canyon, Canyonlands, and Arches. The folding in the Laramide orogenic belt and uplift of the block-like mountains resulted from strong forces that compressed the western margin of North America. For large parts of the western United States, that compression ended about 20 million years ago and was replaced by extensional deformation that began to pull the crust from Oregon to Arizona apart. That extension continues to stretch the crust in that region today.

Recent earthquakes in California, Washington, and in the Basin and Range Province bear witness that deformation in the Cordilleran mountain belt continues. Chapter 6 describes some of these earthquakes.

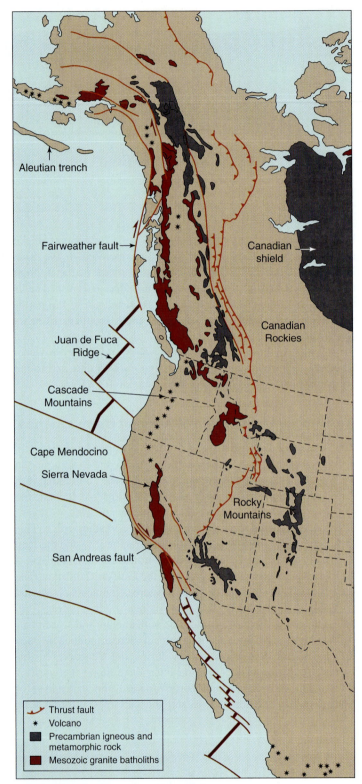

Aleutian trench

Fairweather fault →

Juan de Fuca Ridge →

Cascade Mountains →

Cape Mendocino

Sierra Nevada →

San Andreas fault →

Canadian shield →

Canadian Rockies

Rocky Mountains →

⌒ Thrust fault
* Volcano
■ Precambrian igneous and metamorphic rock
■ Mesozoic granite batholiths

Figure 5.32 Major Tectonic Elements of the North American Cordillera. The dark-shaded areas indicate where Precambrian rocks crop out. The barbed lines indicate a major thrust fault. Folds are associated with these faults. The asterisks mark positions of modern volcanoes.

SUMMARY POINTS

1. Plate tectonic theory makes it possible to explain so many different kinds of observations in a unified theory. It ties together ideas and observations that once seemed unrelated and difficult to understand.

2. The first suggestion of continental drift resulted from fitting continental margins, especially Africa and South America, together. The discovery that ice sheets covered portions of Africa, South America, Antarctica, India, and Australia near the end of the Paleozoic strengthened the theory, as did discovery of fossils of the same land plants and amphibians on these now far-removed continents.

3. The idea that continents are moving was revitalized by the discovery that polar wandering paths determined from different continents coincide if the continents are joined in the pattern postulated from Pangaea.

4. The idea that continents ride passively on lithospheric plates and that the sea floor moves apart as new crust forms at oceanic ridges followed extensive research in the ocean basins during the 1950s. Dikes emplaced near the ridge crest as it pulls apart and lava flows extruded on the sea floor comprise most of the new lithosphere formed at oceanic ridges.

5. Demonstration that magnetic anomaly stripes form a symmetrical pattern across the Mid-Atlantic Ridge provided the initial confirmation of the theory of seafloor spreading. These stripes correspond to basalt extrusions and intrusions emplaced in the crust at a spreading ridge crest during periods of normal and reversed polarity of Earth's magnetic field.

6. Drilling and seismic refraction studies confirm that sediment thickness increases away from ridge crests.

7. The oceanic crust is older away from ridge crests. By matching the magnetic anomaly belts over any part of the sea floor with corresponding anomalies on the geomagnetic time scale, geologists can estimate the age of the sea floor.

8. The trace of volcanic centers located over hot spots in the mantle indicates the direction of plate movement. Hawaii is a modern hot spot; the ridge to the northwest of Hawaii reveals the direction of movement of the Pacific Plate.

9. Transform faults cut across oceanic ridges and may connect the ridges with subduction zones or other crustal elements. These connections allow a change in the type of movement taking place across the fault.

10. Oceanic lithosphere is subducted in island arcs and along some continental margins, such as the western margin of South America. As the lithosphere sinks, drag and bending of the lithosphere causes earthquakes. At depths of 100 to 200 kilometers, the sinking oceanic crust dehydrates, releasing water that rises into and partially melts the overlying upper mantle. Some of the sinking crust also partially melts, forming magma of andesitic composition that rises and forms volcanic island chains.

11. Several processes probably contribute to the movement of plates. Convection currents exert drag on the base of the plates in the asthenosphere. Gravity causes plates to move toward trenches because oceanic plates are inclined away from ridges toward trenches. Cooling of the slab and pressure-induced changes in the down-going lithosphere cause its density to increase, pulling the slab downward.

REVIEW QUESTIONS

1. What led early geologists and Alfred Wegener to suspect that continents move relative to one another?
2. Why did geophysicists initially oppose the idea of continental drift?
3. What methods are used to trace the direction of plate movements?
4. What causes the magnetic anomaly stripes observed in the ocean basins?
5. How are the magnetic anomaly stripes identified?
6. What causes continents to break up and drift apart?
7. How are volcanic eruptions, earthquakes, and mountain building caused by plate movements?
8. What is a lithospheric plate, and what forms the boundaries of plates?

THINKING CRITICALLY

1. Based on what is known about modern plate movements, what can you predict about the probable future changes in global geography?
2. How are plate movements and past climatic conditions on Earth related?
3. If convection in Earth's mantle stopped, what processes that act on the surface would be most directly affected?

KEY WORDS

andesite
back arc basins
convergent plate boundaries
divergent plate boundaries
geomagnetic time scale
Glossopteris flora
Gondwanaland
grabens
Laurasia

orogens (also **mountain belts**)
Pangaea
plate tectonics
polar wandering
strike-slip faults
subduction
subduction zones
transform faults

chapter six

Earthquakes

Chapter Guide

Every year, several areas from around the world are destroyed by the effects of earthquakes, landslides, or tsunamis triggered by the sudden movement of the crust. These stories provide convincing evidence about the close connection between processes taking place deep inside Earth and the human environment.

In chapter 3, we saw how seismic waves produced during earthquakes provide the most direct evidence about the nature of Earth's interior and how they result from plate movements. The drift of plates is imperceptibly slow, but the release of the stored energy resulting from that movement in earthquakes takes place over extremely short time spans, commonly seconds to minutes. What causes earthquakes? What happens during an earthquake? How is energy stored and then suddenly released? Is it possible to predict where and when earthquakes will occur? How can people compare earthquakes? Do they differ in the amount of energy expended and the effects they have on Earth's surface? Can people protect themselves and their property from earthquake damage? These are among the questions discussed in this chapter.

The western edge of California is shifting to the northwest relative to the rest of the continent along the San Andreas fault.

INTRODUCTION

Much of the most advanced research in the study of earthquakes takes place in Japan, and the Japanese have taken more precautions to reduce damage from earthquakes than any other people. Nevertheless, when a major earthquake struck the city of Kobe in January 1995, over 5,000 people perished, and over 50,000 buildings were destroyed. Many wooden buildings burned as fires started in some sections of the city, elevated highways collapsed as pillars under them broke, roads twisted, trains derailed, and buildings as tall as five stories crumbled. Property damage exceeded $30 billion (figure 6.1). Despite the destruction, many of the precautions the Japanese had taken substantially prevented greater damage. Electric trains that travel at 160 miles per hour in Japan are tied into a seismic warning system that cuts power when major earthquakes strike, thus avoiding derailments. More stringent building codes enacted in 1981 saved many buildings constructed after that date.

Earthquakes and severe storms cause more loss of life and property than any other natural disasters. In 1976, an earthquake striking during the night destroyed much of Tangshan, China. Estimates of death were between one-quarter and three-quarters of a million people. Earth movements originated only 11 kilometers (6 miles) directly beneath the city. As blocks of the crust moved along a fault, the grinding of rock produced rumbling sounds, a prelude to vibrations that leveled or severely damaged 95% of the residences in the city and its suburbs. Nearly half of the steel-reinforced concrete structures toppled; railroad tracks buckled; fissures opened in the ground and along roadways; and, where saturated with water, the ground flowed like a liquid.

Many cities are located in unstable crustal zones. As more people move into these population centers, losses caused by earthquakes will almost certainly increase. Although earthquake hazards are widespread, the areas most subject to earthquakes are known. Nevertheless, exactly where and when an earthquake will occur is unpredictable. Some of the methods currently used to predict earthquakes are described later in this chapter under the section entitled "Earthquake Prediction," page 145. The frequency of earthquakes in California is well known, but based on the history of seismic events, other areas of the United States, such as the Mississippi River Valley, New England, the St. Lawrence River Valley, the Pacific Northwest, and portions of the Southeast, are also vulnerable. A series of earthquakes, including the most intense earthquake in U.S. history, occurred at New Madrid, Missouri, in late 1811 and early 1812 before the Midwest became heavily populated. If a similar earthquake happened in that area today, property damage would probably exceed $60 billion, and the death toll would approach 5,000 with nearly half-a-million people left homeless.

This chapter begins with a discussion of the causes of earthquakes and what happens at the **focus,** the place where earthquakes originate (see chapter 3). Seismologists use the seismic waves generated at earthquake foci to study the deep interior parts of Earth. Records of these waves collected from widely separated seismograph stations reveal where earthquakes occur and how deep they

Figure 6.1   Earthquake Damage in 1995 at Kobe, Japan.

are. Compilations of these data reveal striking patterns indicating that most seismic activity takes place at plate boundaries (see figure I.10). Case histories of major earthquakes describe what happens during earthquakes and how individuals and communities can reduce the damage. The final section of this chapter considers the methods being used to predict where earthquakes are expected to occur, when they may occur, and how severe they could be.

CAUSES OF EARTHQUAKES

Abrupt movements along faults and volcanic eruptions cause most earthquakes. In both of these, rocks fracture, setting up vibrations in the brittle and highly elastic rocks of the crust and upper mantle. A few additional earthquakes originate deeper in the mantle where slabs of lithosphere sink into the more-ductile mantle rock.

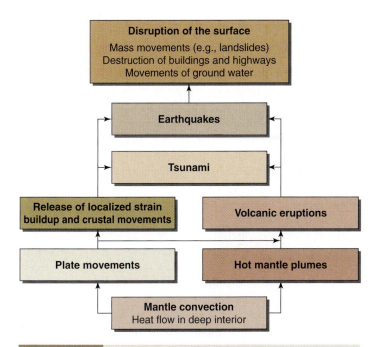

Disruption of the surface
Mass movements (e.g., landslides)
Destruction of buildings and highways
Movements of ground water

Earthquakes

Tsunami

Release of localized strain buildup and crustal movements

Volcanic eruptions

Plate movements

Hot mantle plumes

Mantle convection
Heat flow in deep interior

Figure 6.2 Flow Diagram Showing the Relationship Among Various Processes Related to Earthquakes. Convection deep inside Earth causes heat to rise toward the surface at hot spots and along oceanic ridges. These may lead to volcanic eruptions that cause earthquakes along divergent plate boundaries. Other plate movements cause stress to build up in the lithosphere where plates collide. Earthquakes along these convergent plate boundaries are products of mountain building or subduction of oceanic lithosphere. The abrupt movements in the oceanic crust may cause large waves (tsunami) to form in the ocean, as well as earthquakes.

Ultimately, the source of movements along faults and volcanic eruptions is traced to convection deep in the mantle (figure 6.2).

Heat-flow in the mantle creates hot, rising columns of mantle material and drives plate motions. Plate movements result in slip along plate margins. Some of these margins, such as the zone along the west coast of North America, are major fault zones where plate movements distort the rock until the rocks fracture and slip. Distortions of materials caused by applied pressure (stress) are called *strains*. Along divergent plate boundaries, crustal rocks break as plates move apart. Where plates converge, collisions between plates or subduction of one plate beneath another produces the strain buildup that eventually causes brittle failure of the rocks and consequent earthquakes.

Earthquake Mechanisms

Minor tremors may accompany large mass movements such as landslides and volcanic activity, but abrupt movements along faults in bedrock cause most near-surface earthquakes. Most shallow earthquakes originate within tens of kilometers (a few miles) of the surface where rocks of the crust and lithosphere are brittle

and elastic. However, some earthquakes originate several hundred kilometers deep, and in a few places, they occur between 600 and 700 kilometers (350 and 400 miles) deep. At these depths, the temperature and pressure are so great that rocks are not brittle; they are too plastic to break. The mechanism of these deeper earthquakes remains uncertain. They may result from sudden changes in the crystal structure of olivine. Seismologists explain most shallow, fault-related earthquakes based on the San Andreas fault model in California (figure 6.3).

One of the most devastating earthquakes in the United States happened in 1906 in San Francisco, California. It resulted from movement along the San Andreas fault, part of a complex system of faults that extends from the Gulf of California to Cape Mendocino in northern California. The region on the west side of the fault moves northwest relative to the rest of the continent (see figure 6.3). In the years immediately before the 1906 earthquake, H. F. Reid, a seismologist, had repeatedly made precise surveys measuring the changes in location of markers on opposite sides of the fault. Between 1850 and 1905, these surveys revealed that the region west of the fault had moved slightly northward relative to the region east of the fault, but earthquakes did not accompany these movements. Reid theorized that the rock within a zone about 1.6 kilometers (1 mile) wide along the fault had distorted without actually breaking. He concluded that the rock within the fault zone behaves elastically. According to Reid's idea, called the **elastic rebound theory,** strain in the fault zone builds up until it eventually

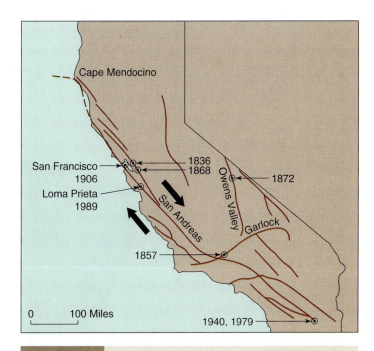

Figure 6.3 The San Andreas Fault System and Other Large Faults in California. The illustration shows locations of some of the major earthquakes that have taken place in California. Most are located along the San Andreas and related faults along the west coast.

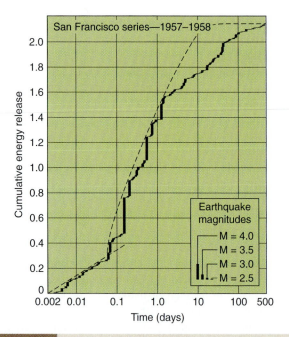

Figure 6.4 Energy (vertical scale) Released in a Series of Earthquakes Near San Francisco in 1957–1958. Note that the energy (indicated by earthquake magnitude, M) was released in a large number of closely spaced earthquakes rather than in a single event.

becomes so great that the rock ruptures (figure 6.4). With this sudden release of the strain, the severed rock on either side of the fault displacement snaps back to its original shape. During this abrupt displacement, the rock vibrates. These vibrations move through Earth as wave motion. Some of these waves resemble sound waves (also called compressional, primary, or P waves), but others are more like the vibrations in a piano wire, which are shear waves. These are called transverse, secondary, or S waves.

In elastic rebound, three processes combine to cause the shaking associated with earthquakes: (1) fracturing of the strained region accompanied by sudden release of strain, (2) grating of rock along the fault surface due to frictional drag as rock on one side of the fault moves past rock on the other side, and (3) additional abrupt breaks and displacements along the fault. As a break and the resulting movement take place at one point, relieving the strain at that point, the stress increases sharply farther along the fault, eventually forcing another break. A long series of smaller earthquakes, called **aftershocks,** commonly follow a major earthquake.

Since the 1906 earthquake, seismologists have studied earthquake mechanisms carefully. They hope to learn what happens before and during an earthquake. Successes will help them predict when and where future earthquakes may occur. More recent studies indicate that conditions along a major fault, such as the San Andreas, vary from place to place. Some sections of the fault undergo almost continuous movements, called **fault creep.** Microearthquakes occur along these sections of the fault, and small movements release strain as quickly as it builds up. Along other sections, the crust across the fault remains locked for long periods, and strain builds up. As strain increases along these sections, they become sites of potentially dangerous earthquakes.

The San Andreas is one of many interconnected faults (figure 6.5 and see figure 6.3), all of which are caused by movement between the Pacific and North American plates. When movement or a break occurs on any one of these faults, it affects not only the stresses at other places on that fault, but also the stress and strain on other nearby faults.

The elastic rebound theory of earthquake generation applies to earthquakes that occur in the crust. The mechanism of deeper

Figure 6.5 Major Faults in the San Francisco Bay Area. All these faults belong to the same system of faults.

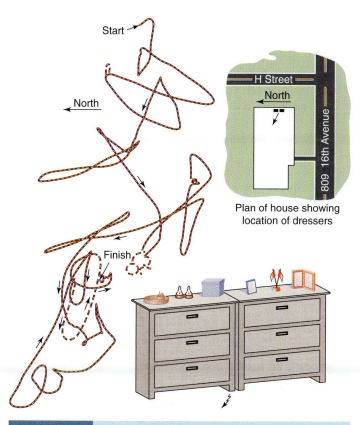

Figure 6.6 The Trace on Asphalt-Tile Flooring Left by a Dresser Leg During the 1964 Anchorage, Alaska, Earthquake.

Start →

← North

Finish

H Street

North

809 16th Avenue

Plan of house showing location of dressers

EFFECTS OF EARTHQUAKES

Most of the damage and loss of life that occur during and immediately after earthquakes results from the effects the vibrations have on the materials at the ground surface (figure 6.6). Surface waves generated by shallow-focus earthquakes cause the ground to roll in a vertical plane and to move sideways with a back-and-forth movement (figure 6.7). These waves shake structures so severely that most unreinforced masonry structures—stone, brick, cinder block, stucco, and adobe—disintegrate. Because elevated freeways have so much mass, once they begin to move back and forth, the pressure on the columns holding them up often exceeds the shear strength of the material, even where that material is steel-reinforced concrete. Engineering studies have repeatedly underestimated the strength needed to resist this stress.

Soil and poorly consolidated sediment offer little resistance to the stresses created when waves form near the ground surface. As surface roll takes place, soil and sediment stretch and the ground cracks open. The magnitude of ground motion increases in unconsolidated materials like soil, fill, sand, or mud. When this happens, buried water mains, gas lines, and drainage pipes rupture. Fires may start, igniting gas and other flammable materials. Broken water mains make fire fighting difficult even if the roads remain open so fire trucks can respond to the blazes.

earthquakes is more difficult to explain. These earthquakes occur at depths where both the great pressure of the overlying rock and the high temperatures cause the rock to flow like a hot plastic rather than rupture like a brittle, elastic solid. Experimental studies of rock behavior at high temperature and great pressure confirm that rock at these depths should yield by plastic deformation before the strain can accumulate enough to cause brittle failure in which rock cracks or shatters like a piece of glass. Despite the plastic behavior of rocks in and below the asthenosphere, deep earthquakes that originate there may still involve abrupt movements along fault zones, but the movement differs from that near the surface. This deeper movement most likely involves a phenomenon called **stick-slip.** Adhesion across potential planes of movement causes them to stick together. Eventually they yield, moving by a series of sudden slips before once again sticking. Although most major earthquakes take place along fault zones in the lithosphere and subducted lithosphere, some earthquakes have other causes.

Several processes that occur during volcanic activity, including violent explosions and collapse of slopes, may cause earthquakes. Movement of magma toward the surface often causes tremors. After an eruption, collapse of the central part of the volcano may generate earthquakes. The extent of damage and the intensity of wave motion generated during earthquakes associated with volcanoes tend to be more localized than those associated with major fault systems, but in many regions, particularly around the perimeter of the Pacific Ocean, volcanic activity and faulting are closely associated.

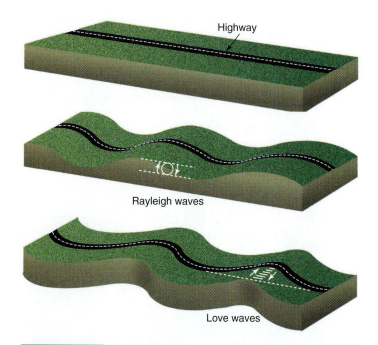

Highway

Rayleigh waves

Love waves

Figure 6.7 Surface Waves. Two distinctly different types of waves are depicted on the ground surface. Rarely is the amplitude of the ground motion great enough to distort the surface as shown here, but movements of these types are associated with most shallow-focus earthquakes.

Figure 6.8 The Cracks in the Ground Surface Formed During an Earthquake in Montana in 1959.

If displacements along a fault reach the ground surface, as they did in the earthquake near Hebgen Lake in Montana, 1959 (figure 6.8), and in Anchorage Alaska, 1964 (figure 6.9), the movements open fractures in the ground and destroy structures built on the fault. For this reason, dams, hospitals, schools, power stations, nuclear reactors, and other structures that are likely to contain or affect large population concentrations should not be built astride faults. Where the ground contains water in soil and sediment, seismic surface waves may cause the soil or sediment to turn into mud. Water and mixtures of water and sand may spurt into the air through cracks in the ground, or the liquified layers may flow like a mass of wet mud. In Anchorage, much of the damage resulted from shifting of a layer of soft, wet clay (see figure 6.9).

Unusually intense earthquakes in oceanic crust or under large lakes can cause large water waves, called **tsunami.** Although some people call them "tidal waves," tsunami are unrelated to tides. Tsunami result from the sudden displacement of water caused by an offset in the sea floor, a volcanic eruption, or a submarine landslide. Tsunami have caused heavy loss of life in villages located along shorelines of Pacific islands where most tsunami occur.

Earthquakes occurring in areas with steep slopes commonly trigger large downslope movements of soil and rock. In 1959, an earthquake west of Yellowstone National Park caused a massive landslide that dammed the Madison River (see figure 16.11). Slope failures take many forms, ranging from the collapse of cliffs composed of solid rock to the downslope flowage of wet soils.

DETECTION OF SEISMIC WAVES

The **epicenter** is the place on the ground surface directly above the earthquake focus. Near the epicenter, ground motion resulting from shallow focus earthquakes may be violent enough to turn buildings or shift them off of their foundations (see figure 6.9) or throw objects into the air. At greater distances from the epicenter, the amplitude of the ground motion decreases rapidly because the wave energy spreads over an ever-increasing surface area. Deep focus earthquakes also produce low-amplitude surface wave motion. Rarely do people located more than a few hundred kilometers from the epicenter of an earthquake feel the ground movement. Suspended objects such as chandeliers may move even though no one feels the shock wave. At first, a suspended chandelier appears to move because the building around it is moving. Actually, the chandelier remains relatively still because of its **inertia,** the tendency of mass to resist changes in its state of motion. As vibration continues, the chandelier begins to swing like a pendulum. Its motion is still largely independent of the motion of the building and ground on

Figure 6.9 Wreckage of a School in Anchorage, Alaska, Following the 1964 Earthquake.

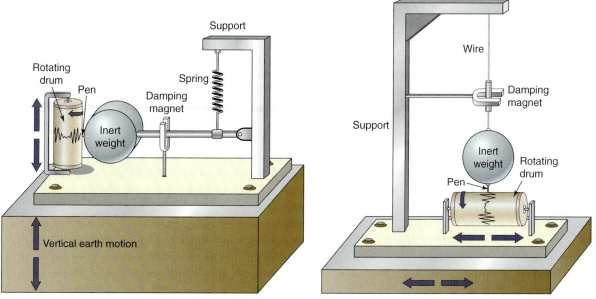

Support

Rotating
drum

Pen

Spring

Damping
magnet

Inert
weight

Wire

Damping
magnet

Support

Inert
weight

Rotating
drum

Pen

Vertical earth motion

Horizontal earth motion

Figure 6.10 **Schematic Models of Mechanical Seismographs.** Seismograph stations generally have one instrument designed to measure vertical movements and two (set at right angles) to measure horizontal movements. In their simplest configurations, a vertical seismograph is a weight suspended from a spring, and a horizontal seismograph operates like a pendulum. In both cases, an inertial weight has a pen attached that traces the motion of the ground relative to the mass. The pen traces this movement on a rotating drum.

which it stands. Instruments used to record ground motion, called **seismographs** (figure 6.10), also operate on the principle that masses have inertia and tend to remain still if they are isolated from the ground motion. Many seismographs resemble pendulums.

Seismographs detect the rate of change of ground motion (acceleration) and the direction of ground movements. Most seismographs detect and record the relative motion between the ground and the center of some type of pendulum. The basic element in the design of most seismographs is a mass suspended by springs from a frame anchored to bedrock. The mass tends to remain still when the frame moves. Modern seismographs contain a heavy magnet for the mass and a coil of wire that is fixed to the frame and surrounds, but does not touch, the magnet (figure 6.11). As the ground moves, the coil moves through the field of the magnet. This generates an electric current in the wire that is proportional to the movement of the ground. The magnitude of the current depends on how fast (the acceleration) and over what distance the ground (and therefore the coil) moves. An amplifier increases the current and feeds it into a recording device that translates variations in the electric current into a digital recording or into the motion of a pen. The pen draws a line on a piece of paper, called a **seismogram** (figure 6.12). This line, called a **trace,** shows no deflection unless the ground moves. The amount and direction of the trace deflection measure the amount and direction of ground movement. Seismographs of this kind usually operate with a magnification of 500 to 20,000, which is sufficient to produce good records of ground motion. Magnifications exceeding 100,000 are

possible, but at such high magnifications, background noise due to the wind blowing through trees or traffic makes the record difficult to read. Time marks imprinted at one-minute intervals on the seismogram make it possible to determine the time of arrival of any particular ground motion at the seismograph station.

Because one instrument can detect motion in only a single direction, a seismograph station normally consists of at least three instruments. One of these detects vertical motion; the other two record movement in east-west and north-south directions. Other specialized seismographs detect waves that have different **periods,** the time between passage of adjacent waves.

Locating Earthquake Epicenters

If an earthquake disrupts lines of communication in remote areas, people outside the area must determine both where and when the earthquake took place. The arrival of seismic waves is the first indication of an earthquake at a seismograph station some distance away from the epicenter. P and S waves travel the same path but at different speeds. Thus, the distance to the epicenter from any seismograph station is a function of the time lag between the arrivals of these two waves. This distance does not indicate the exact location. The epicenter could be anywhere on a circle that is drawn around the station with a radius equal to the distance from the station to the epicenter. However, seismologists can determine the precise location of the epicenter from three seismograph stations.

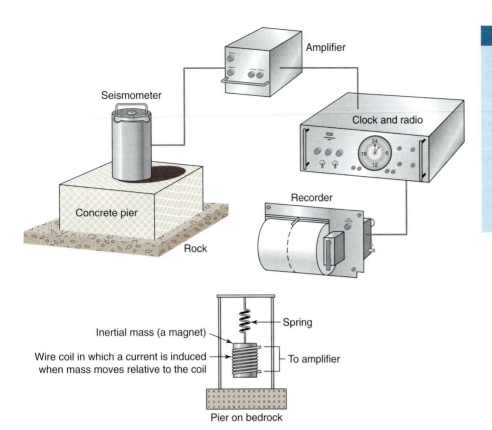

Amplifier

Clock and radio

Seismometer

Recorder

Concrete pier

Rock

Inertial mass (a magnet)

Spring

Wire coil in which a current is induced
when mass moves relative to the coil

To amplifier

Pier on bedrock

Figure 6.11 Modern Seismographs. Most modern seismographs utilize the principle of electromagnetic induction (an electric current is generated in a wire as it moves through a magnetic field) to measure ground movement. Using electronic amplification, it is possible to amplify ground movements over 100,000 times. Thus, it is possible for seismographs in North America to detect earthquakes in China. A radio with a clock is also used in these installations so the exact time of arrival of a seismic wave can be recorded.

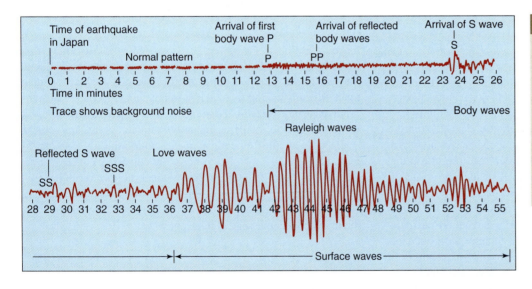

Time of earthquake in Japan

Arrival of first body wave P

Arrival of reflected body waves

Arrival of S wave

Normal pattern

P

PP

S

0 1 2 3 4 5 6 7 8 9 10 11 12 13 14 15 16 17 18 19 20 21 22 23 24 25 26
Time in minutes

Trace shows background noise

Body waves

Rayleigh waves

Reflected S wave

Love waves

SSS

SS

28 29 30 31 32 33 34 35 36 37 38 39 40 41 42 43 44 45 46 47 48 49 50 51 52 53 54 55

Surface waves

Figure 6.12 A Seismogram. This seismogram of an earthquake originating in Japan was recorded at the University of Tasmania. The time marks below the seismogram trace indicate time elapsed after the earthquake. The first body wave (P) arrived in Tasmania about 13 minutes after the earthquake began. S waves arrived at 23 minutes, and surface waves began to arrive at 36 minutes.

They do this by drawing a circle with the radius equal to the distance to the epicenter around each station. The point at which these three circles intersect is the epicenter (figure 6.13).

Determining Focal Depth

Seismologists determine the depth at which an earthquake occurs by comparing the time required for compressional (P) waves to travel two different paths from the focus to a seismograph station (see figure 3.12). One of these waves travels the most direct path from the focus to the station. The other wave travels from the focus

to Earth's surface where it reflects off the surface and then travels to the station. The difference in the travel time of these two waves is a measure of the depth of the focus.

The Distribution of Earthquakes

For several decades, a network of more than 100 seismograph stations has been operating throughout the world. Compilations of earthquake epicenters located using these stations provide a picture of the distribution of earthquakes in Earth's lithosphere (figure 6.14). Plots showing the distribution of earthquakes that

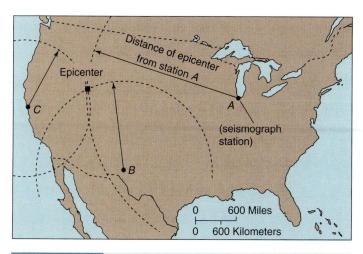

Figure 6.13 Location of an Epicenter. Once the distance to the epicenter from each of three seismograph stations is known, it becomes possible to locate the epicenter by triangulation, as shown.

occur within different ranges of focal depths provide evidence of Benioff zones (see figure 5.17) and demonstrate the shape of down-going slabs of lithosphere in subduction zones.

Most earthquakes occur within long and relatively narrow belts separated by large areas where earthquakes are rare. The zones of highest seismic activity coincide with and help identify plate boundaries. Shallow-focus earthquakes, those with focal depths of 25 kilometers (15.5 miles) or less, occur along all plate boundaries. Only shallow-focus earthquakes occur along oceanic ridge systems where the crust pulls apart as new sea floor forms and along transform faults that cut across and offset oceanic ridges (see figures 5.1 and 5.12).

Shallow- and intermediate-depth earthquakes occur in mountain belts where plate convergence takes place. Many of these earthquakes result from one plate overriding the margin of another along zones of thrust faulting. Major faults, such as the San Andreas, that cut through the lithosphere commonly produce shallow- and intermediate-focus earthquakes as one plate slides by another.

Shallow-, intermediate-, and deep-focus earthquakes are closely associated with subduction zones where cool, brittle lithospheric slabs sink into the mantle. Most subduction zones experience numerous earthquakes with focal depths ranging from near surface to 400 kilometers (250 miles). In addition, most of the deepest earthquakes with focal depths of 600 to 700 kilometers (350 to 400 miles) also happen in subduction zones.

The most stable regions—those with the fewest earthquakes—are deep ocean basins and portions of continents outside young mountain belts. Nevertheless, even the stable parts of Earth's crust are not completely exempt from earthquake activity. In January 2001, one of the deadliest earthquakes on record occurred in the Bhuj-Anjar-Bhachau region of western India. More than 50,000 people perished. In 1886, an earthquake severely damaged Charleston, South Carolina, in a region that rarely experiences major earthquakes. This earthquake is described in more detail later in this chapter on page 137.

MEASURING EARTHQUAKE INTENSITY AND MAGNITUDE

Historians and scientists described the effects of earthquakes for many years before the development of seismographs. In 1902, Italian seismologist G. Mercalli compiled a set of observations to create a scale that indicates progressively greater earthquake effects. He selected observations that most people could report objectively and referred to the scale as a measure of intensity. This scale, shown in table 6.1, classified earthquakes from I to XII.

Seismologists still use a modified version of Mercalli's intensity scale. Despite its value as an indicator of the effects of earthquakes, the level of intensity does not provide a measure of the focal depth or the type of material under the observer, both of which affect intensity. For earthquakes that release the same amount of energy, deep earthquakes have lower intensities than shallow earthquakes. Because intensity depends only on surface effects, it is not a reliable measure of the amount of energy released at the focus of an earthquake. Unlike other scales, intensities can be interpreted from historical records written long before seismometers were available. A more precise method of measuring energy came later after the invention of seismometers.

With the development of the sophisticated seismometer, designed to measure the amount and acceleration of ground motion, seismologists began to estimate the relative amounts of energy released in earthquakes. In 1935, Charles Richter, a seismologist at the California Institute of Technology, devised a scale that represents the magnitude of an earthquake. Richter found that the amplitude of the ground motion as recorded on a certain type of seismograph makes it possible to estimate the amount of energy released during an earthquake (figure 6.15). This scale is logarithmic; so a unit increase in magnitude number indicates a tenfold increase in the amplitude of ground motion. Because the amplitude of surface waves provides the basis for this scale, it is called a *magnitude scale*. Many refer to it simply as the *Richter magnitude*. Although it is a measure of ground motion, it is not a direct measure of the amount of energy released at the focus.

An increase in magnitude of 1 unit represents an increase in the amount of seismic energy released by a factor of about thirty. The amount of energy released during a magnitude-6 earthquake is about ten times that released during the atomic explosion at Bikini Island, an atoll in the Pacific Ocean where the United States tested atomic bombs. The magnitude-8 earthquake that struck San Francisco in 1906 released energy equivalent to nearly 10,000 times that of the hydrogen bomb exploded at the Bikini atoll in a test.

Although the Richter magnitude scale is open-ended, few earthquakes register above 8 on the scale. Of the thousands of earthquakes recorded each year worldwide, only about 150 have magnitudes of 6 to 6.9; only one or two dozen have magnitudes of 7 to 7.9; and on average, only one has a magnitude of 8 or more (table 6.2).

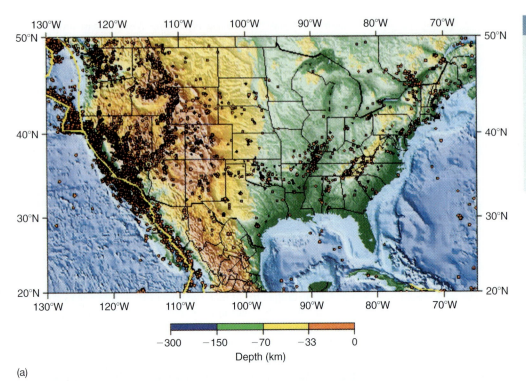

Distribution of Earthquakes. (a) Location of earthquake epicenters in the contiguous United States over a twenty-year period. (b) The dots mark the location of earthquakes. Notice how closely these coincide with the margins of crustal plates in figure 5.1. The color-coded dots indicate the depth and location of earthquakes.

(a)

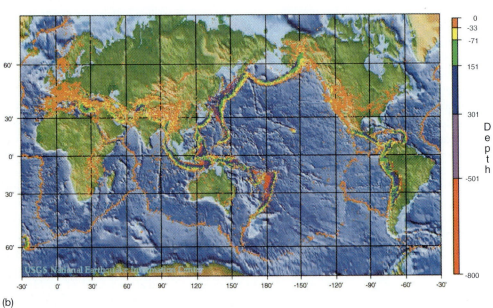

(b)

EARTHQUAKES IN STABLE AREAS

Faults associated with plate boundaries cause most earthquakes, but occasionally, large earthquakes affect areas far removed from such boundary zones. Some areas of the world rarely experience earthquakes, and they are essentially aseismic. However, most of these areas occasionally experience low-magnitude earthquakes, and some have been sites of major earthquakes. Although few magnitude-5 or greater earthquakes occur in New England, in 1638, the Plymouth Pilgrims experienced an earthquake soon after they arrived. Chimneys fell during a 1727 earthquake centered near Newbury, Massachusetts, and three magnitude-5 earthquakes have affected New England since 1982. The Southeast and the midcontinent, where major earthquakes are rare, were sites of two of the biggest earthquakes on record in North America. One destroyed much of Charleston, South Carolina, in 1886, and a second struck New Madrid, Missouri, in 1811–1812. This was probably the highest-magnitude earthquake recorded in North America.

Table 6.1 The Modified Mercalli Scale

I. Not felt, except by a few people under especially favorable conditions.

II. Felt by only a few people at rest. Delicately suspended objects may swing.

III. Felt quite noticeably under favorable circumstances. Parked automobiles may rock slightly.

IV. Felt by many or most people; some awakened. Dishes, windows, doors disturbed; walls cracked. Sensation like that of a heavy truck striking a building.

V. Some dishes, windows, and so on, broken; a few instances of cracked plaster; unstable objects moved.

VI. Felt by all; many frightened and run outdoors. Some heavy furniture moved; a few instances of fallen plaster or damaged chimneys.

VII. Everyone runs outdoors. Damage negligible in buildings of good design and construction; slight to moderate in well-built ordinary structures; some chimneys broken. Noticed by people driving cars.

VIII. Panel walls thrown out of frame structures. Fall of chimneys, factory stacks, columns, monuments, walls. Heavy furniture overturned. Sand and mud ejected in small amounts.

IX. Buildings shifted off foundations; ground cracked conspicuously. Underground pipes broken.

X. Ground badly cracked; rails bent; landslides considerable from riverbanks and steep slopes. Water splashed over banks.

XI. Few masonry structures remain standing. Bridges destroyed. Broad fissures in ground. Earth slumps and land slips in soft ground.

XII. Damage total. Waves seen on ground surface. Lines of sight and level distorted. Objects thrown into the air.

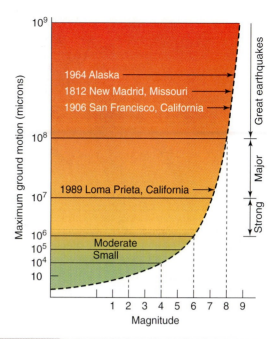

Figure 6.15 Richter Magnitude Scale. The Richter magnitude scale was originally designed to be logarithmic. Thus, a magnitude 5 earthquake has a tenfold greater amplitude in the ground motion than a magnitude 4 earthquake.

New Madrid, Missouri The number and intensity of earthquakes that struck New Madrid make it clear that not all earthquakes strike in major seismic belts (figure 6.16). Earthquakes that destroyed New Madrid affected a large elongated area extending from southeastern Missouri into Arkansas. Recent studies in the Mississippi Valley indicate that a major zone of crustal movement lies beneath the sediment that fills the valley. Steeply inclined faults bound a region that dropped down and created a graben, known as the Reelfoot graben, or rift (figure 6.17). Sediment filled the graben long ago when the Gulf of Mexico extended into the region. More recently, deposits from the Mississippi River have covered the older marine sediments. It was these unconsolidated sediments that became so unstable during the 1811 and 1812 earthquakes.

Based on eyewitness accounts, the course of the Mississippi River changed and may have even reversed course for a few hours. During the earthquakes, huge masses of land slipped into the river, and Reelfoot Lake formed behind a scarp pushed up in western Tennessee. No evidence exists of how many people perished during the New Madrid earthquakes. A few people were known to have died when the riverbanks collapsed. A few probably fell overboard from boats, and empty canoes were reported on the river downstream. However, because the population around New Madrid and in northeastern Arkansas was small, the loss of life was minimal for such a major event. If comparable earthquakes occurred in the Mississippi Valley today, loss of life and property would be huge. A major earthquake anywhere in the area would cause serious damage in Memphis and St. Louis, as well as in hundreds of small towns.

Charleston, South Carolina In 1886, an earthquake severely damaged Charleston, South Carolina. This earthquake occurred before the invention of modern seismic instruments, but based on accounts of the damage, seismologists can estimate the intensity of the Charleston earthquake. The city fell within a zone of intensity IX on the modified Mercalli scale (figure 6.18 and see table 6.1). About sixty people died. People reported ground movements throughout most of the southeastern states. These reports suggest an intensity of V as far away as Chicago, Illinois.

Table 6.2	Richter Scores and Death Tolls for Some of the Highest-Magnitude Earthquakes on Record		
Year	Place of Earthquake	Richter	Loss of Life
1906	San Francisco	8.3	700
1960	Chile	8.5	5,700
1964	Alaska	8.6	131
1985	Mexico	7.9	9,500
1999	Turkey	7.6	30,000+

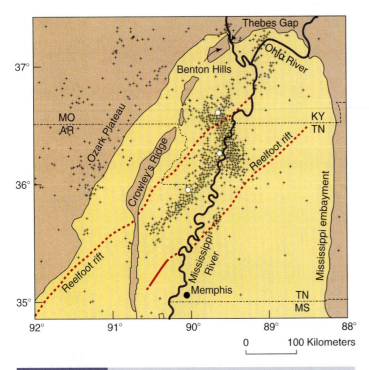

igneous and metamorphic rocks buried 900 meters (2,950 feet)
below sediments of the Atlantic Coastal Plain. Apparently, a
number of steeply inclined faults have broken the crystalline basement layer into big blocks that are shifting. The faults may be
ancient breaks formed as a result of continental rifting when North
America and Africa began to separate. How long they will continue to be active remains unknown.

EARTHQUAKES IN SEISMICALLY ACTIVE AREAS

Most seismic activity is concentrated along plate boundaries (see
figure I.10). The most dangerous earthquakes generally occur along
transform faults and in subduction zones.

Seismic Activity on Great Strike-Slip Faults

Turkey, 1999 At 3:02 A.M. on August 17, 1999, the latest in a
series of eleven devastating earthquakes (magnitudes exceeding
6.7) along the North Anatolian fault struck near the town of Izmit,
Turkey. This major strike-slip fault extends from the Zagros fold-
and-thrust belt in Iran into the Aegean Sea and forms the boundary
between the Anatolian Plate (most of Turkey) and Eurasia. Turkey
is moving westward relative to Eurasia (figure 6.20). The Anatolian
Plate is a relatively small plate located in a highly complex zone
between the African and Eurasian plates.

The latest sequence of earthquakes along the North Anatolian
fault began in 1939. At that time, a 360-kilometer- (225-mile)
long segment of the fault ruptured with displacements of as much
as 7 meters. The August 17, 1999 earthquake happened in two
phases. The first, located at the western end of the active part of the
fault that caused a 90-kilometer- (56-mile) long rupture in the
ground, lasted 12 seconds and had a magnitude of 7.4. After an
18-second pause, a second major earthquake ruptured a zone
30 kilometers (18 miles) long farther east. This caused an offset of
5 meters (16 feet) across a small country road.

Seismic activity in the region continues today at a level higher
than it was prior to 1886. A number of recent earthquakes lie along
a line that trends northwest from Charleston, South Carolina, into
eastern Tennessee, as shown in figure 6.19. It seems probable that
earthquakes originated as a result of movements on faults in

Figure 6.17 **Generalized Cross Section Across the Reelfoot Graben.** The Precambrian rocks
that lie at depth under the Mississippi embayment crop out in the Ozark Mountains and in the Blue
Ridge. Paleozoic rocks, shown in yellow (the lower layer) and blue, crop out around the Ozark Mountains and in the Appalachian Plateau region in Tennessee. Mesozoic and Tertiary sediments crop out in
the Mississippi embayment, and stream deposits underlie the Mississippi River Valley.

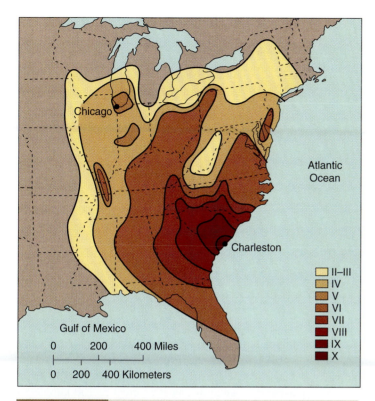

Figure 6.18 Intensity Map for the 1886 Charleston Earthquake. Map of the eastern United States, contoured to show the broad regional patterns of the reported intensities for the 1886 Charleston earthquake. Contoured intensity levels are color-coded according to the Mercalli scale.

The tragic loss of more than 30,000 lives and well over $6 billion in property resulted largely from the collapse of poorly constructed apartment buildings (figure 6.21). As in many other places, block walls between concrete floors failed, and the floors collapsed into a pile. At Izmit, the fault runs near the shore of the Marmara Sea where it crosses several deltas. Displacements of the unconsolidated sediment in the deltas caused massive landslides that extended landward and resulted in sediment under the town slipping as much as 100 meters (328 feet) toward the sea. Part of the town, including a hotel, shops, and restaurants, slid under water. People observed several unusual events over the Marmara Sea during the earthquakes. A large ball of flame erupted from the water, and people heard a loud explosion. Large balloon-shaped masses of light shot out of the water. Methane gas and possibly radon released from the unconsolidated sediment probably caused these events.

California California is known for the high level of seismic activity located along the west coast where a strip of land between Baja California and Cape Mendocino is moving northwest with the Pacific Plate. The San Andreas fault is the best known of a system of interconnected faults that define the boundary between the Pacific and North American plates (see figure I.10). Earthquakes along these faults plague the residents of California and pose some

of the greatest seismic hazards in the United States. Although numerous smaller earthquake events punctuate the history of this region, the great earthquake of 1906 at San Francisco and the more recent earthquakes at Loma Prieta in 1989 and at Northridge in 1994 reveal how serious the threat is for residents of this region.

San Francisco Movement on the San Andreas fault caused the major earthquake that devastated San Francisco in 1906. In the San Francisco area, the resulting lateral displacements exceeded 4.5 to 6 meters (15 to 20 feet). These displacements were almost entirely horizontal with little or no vertical components. The entire region west of the fault moved north relative to the east side. Damage occurred over a large part of California, but the greatest losses were in San Francisco and nearby towns where between 500 and 700 people perished and approximately 20% of the city was destroyed (figure 6.22). Fires that swept the city after the shock caused much of the damage. Water mains had broken during the earthquake, making it nearly impossible to get water to extinguish the fires, some of which raged out of control for days. In a last-ditch effort to stop the spreading fire, firefighters resorted to dynamiting buildings in the flame's path. Unfortunately, this action only spread the fire.

The San Francisco earthquake had an estimated magnitude of 8.3 on the Richter scale. Its intensity, however, varied greatly from place to place depending on the type of soil or bedrock in the area. Intensities exceeded IX and X in construction areas over landfills in the Bay Area. People in areas underlain by solid bedrock reported much lower intensities. Similar effects were observed during the 1971 San Fernando earthquake in southern California. Buildings located on weak materials suffered extensive damage. Some buildings tipped over and collapsed; reservoir dams were damaged; and overpasses fell. In San Fernando and Los Angeles, this moderate-sized (Richter magnitude 6.6) earthquake resulted in the death of sixty-four people and caused approximately half-a-billion dollars in property damage.

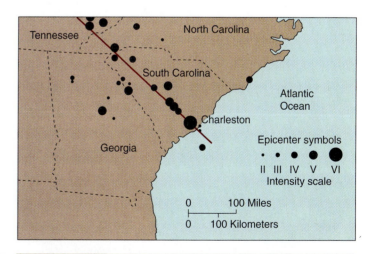

Figure 6.19 Earthquake Epicenters in the Charleston Area, for Shocks from 1961–1975.

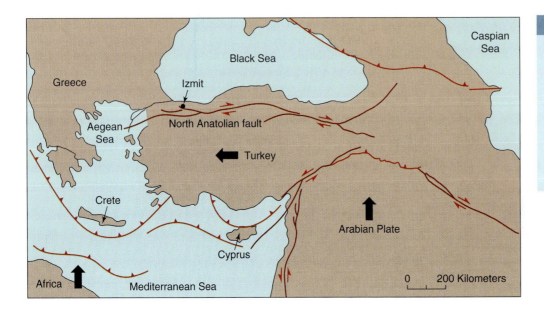

The North Anatolian fault is similar to the San Andreas fault in California, but they differ in their tectonic settings. The San Andreas fault connects two parts of a divergent plate boundary. The North Anatolian fault connects the north end of the Zagros fold belt in Iran with the orogenic belt in Greece.

Loma Prieta, California In 1989, a major earthquake struck Loma Prieta, a town near Santa Cruz, California. This earthquake occurred along a subsidiary branch of the San Andreas fault. Because no movement had taken place between Loma Prieta and downtown San Francisco since 1906, seismologists had previously identified this section of the San Andreas fault system as a "locked section" and the probable site of a future earthquake (figures 6.23 and 6.24). The movement in 1989 occurred on a previously unidentified fault. One side of the fault moved up almost as much as it moved laterally. In contrast, the 1906 movement on the San Andreas fault was purely horizontal. At that time, the region west of the fault shifted northward relative to the rest of the continent, offsetting roads and streams as much as three to seven meters (10 to 23 feet).

The magnitude of the Loma Prieta earthquake was 7.1, a much less severe earthquake than the 1906 earthquake. Nevertheless, property damage reached nearly $30 billion. The shock waves caused extensive damage at distances as great as 100 kilometers away from the epicenter. Many of the frame and stucco buildings in the older sections of Santa Cruz collapsed; frame buildings in the marina section of downtown San Francisco buckled; and fires destroyed more property. A section of the Bay Bridge slipped out of place, dumping vehicles into the waters below. However, most deaths resulted from the collapse of a multistory apartment building, and many deaths occurred on the Oakland freeway. The columns holding the upper part of the Oakland freeway snapped, dropping the top level of the freeway onto cars on the lower level.

Figure 6.21 Photographs Showing Damage in the Streets in Adapazari Following the Kocaeli, Turkey, Earthquake.

Figure 6.22 Photograph of San Francisco After the 1906 Earthquake. Uncontrolled fires caused much of the damage shown here. Having lost all water lines, the fire department attempted to put out the fire by blowing up and burning down buildings.

Most of the hardest-hit areas were places where the foundation materials were unconsolidated sediment or artificial landfill (figure 6.25).

The Loma Prieta earthquake was no surprise. Its approximate location and magnitude had been predicted more than a year earlier, but the predictions lacked the critical information about the exact time and place. In 1996, seismologists issued a similar but even more vague warning for the Pacific Northwest.

Subduction-Related Earthquakes

The Japanese Islands are part of the long chain of island arcs that rim the western edge of the Pacific Ocean. Offshore along the northern islands, a deep-sea trench marks the place where the Pacific Plate sinks beneath the Japanese Islands. Japan has a long history of seismic activity arising from this subduction zone. In 1923, one of the most devastating earthquakes on record killed 140,000 people in Tokyo. Fires caused much of the damage at that time. Ten years later, the largest known earthquake, a magnitude 8.9, that hit a less populated portion of the island, caused much less loss of property and life.

Figure 6.23 Location of the Main Earthquake and Aftershocks of the Loma Prieta Earthquake, 1989. The cross section A–A′ is drawn along the fault zone to show the focal depths of the tremors. The cross section B–B′ depicts the focal points projected into a plane at right angles to the trend of the fault zone. Section A–A′ shows that most earthquakes occurred within 20 kilometers of the surface. Section B–B′ indicates that the fault is steeply inclined. Symbols are assigned according to focal depth and scaled by magnitude.

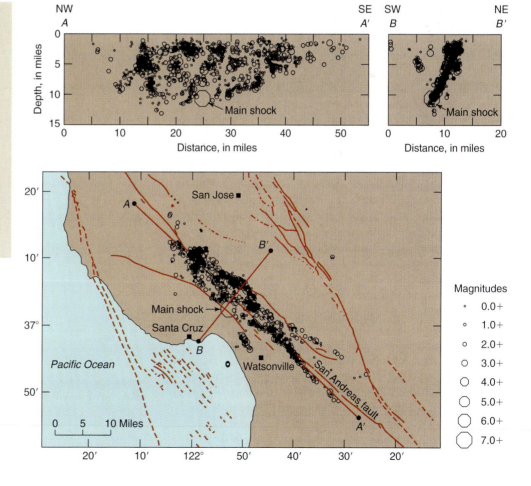

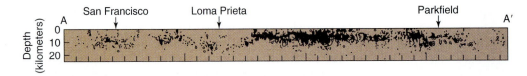

San Francisco   Loma Prieta   Parkfield

Figure 6.24   Cross section of Seismicity of the San Andreas Fault from 1969–1989, Before the Loma Prieta Earthquake. Note the small number of earthquakes in the region near Loma Prieta.

In 1995, a major earthquake destroyed large parts of the city of Kobe, Japan. Much of the damage in Kobe resulted from construction on stream deposits, poorly consolidated sediment, and landfills built into the wetlands along the coast. Kobe is the sixth largest port in the world, so it is not surprising that high land values around the harbor encouraged the construction of large areas on fill that is notoriously unstable during earthquakes.

Although they are part of the same circum-Pacific belt, most earthquakes in Alaska and the Pacific Northwest are quite different from those along the California coast and from north of Vancouver Island to the Gulf of Alaska. From Cape Mendocino in northern California to Vancouver Island in British Columbia and all along the Aleutian Island chain, earthquakes originate in subduction zones (see figure 5.21). The Aleutian deep-sea trench marks the place where the Pacific Plate sinks beneath the Aleutian Islands. A smaller plate named for the Juan de Fuca Ridge subducts along the coast of Washington and Oregon. New sea floor that forms along the Juan de Fuca Ridge moves east and slides under the edge of the continent, giving rise to the volcanoes of the Cascade Mountains (figure 6.26). This plate, which has a low slope, extends as far east as Montana. The absence of a deep-sea trench and the relatively low level of seismicity along the Washington and Oregon coasts distinguish this from other subduction zones. A trench is not present here because sediment carried into the ocean by the Columbia River fills in any depression that forms along this subduction zone.

Because most of the largest earthquakes occur in subduction zones, efforts are being made to discover why such a small number of earthquakes occur in this particular subduction zone. Although it is possible that the faults associated with the subduction zone are locked, it seems more probable that strain dissipates quickly. Because the distance between the Juan de Fuca Ridge and subduction zone is short, the down-going slab is relatively young and warm. This should make the rock less brittle and promote creep rather than strain buildup. Evidence from the rock record is more foreboding.

Sediment samples collected from wells offshore contain evidence that coastal swamps, which existed along the coast of Washington state, sank simultaneously about 300 years ago. A layer of sand appears to have washed over the swamp deposits as if a large wave, a tsunami, suddenly struck the coast. To verify this sequence of events, model experiments have been conducted to see whether an offshore earthquake of magnitude 8 or even 9 could have generated such a tsunami. The answer appears to be yes. More disturbing is evidence from the other side of the Pacific

that a 2-meter-(6.5-foot) high tsunami swept along the coast of Japan on the night of January 27, 1700. Indians of the Pacific Northwest relate stories about an earthquake that caused flooding along the coast of the Northwest many years ago. This evidence that a high-magnitude earthquake struck the Northwest coast about 300 years ago might be taken as a warning of the earthquake that struck south of Seattle in 2001 and that others may occur in this area in the future. Just how big any future event may be depends on the extent to which faults in the subduction zone are locked and the length and depth of the fault that breaks when movement happens. Some seismologists estimate that a break along most of the length of the fault could generate an earthquake of about magnitude 8.

Washington Nisqualty Earthquake, 2001 Subduction of the Juan de Fuca Plate along the continental margin caused the magnitude-6.8 earthquake that struck south of Seattle on February

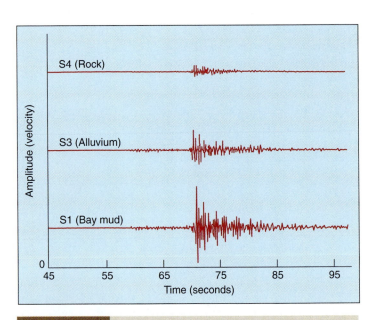

Figure 6.25   Seismograms Recorded for a Magnitude 4.1 Aftershock of the Loma Prieta Earthquake. Station S1 was located on mud near the collapsed freeway, S3 on alluvium near the uncollapsed freeway, and S4 on Franciscan rock in the Oakland Hills. All three traces are plotted at the same scale.

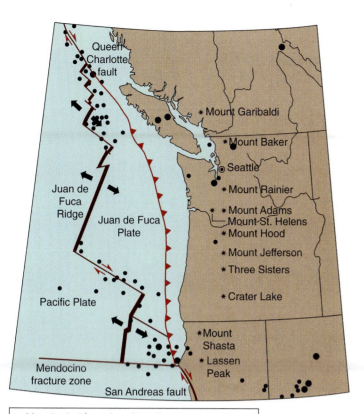

Magnitude 6⁺ earthquake epicenters
(larger dots indicate high magnitude earthquakes)
* Volcanoes

Figure 6.26 Seismicity and Plate Tectonic Setting of the Pacific Northwest. The heavy barbed line shows the approximate location of the east-dipping subduction zone that marks the edge of the Juan de Fuca Plate. Except for the sediment deposited along this coast, a deep-sea trench would be located here. Large dots mark locations of earthquakes of magnitude 7 or more. Smaller dots mark locations of earthquakes of magnitude 6 to 6.9. Spreading is taking place along the Juan de Fuca Ridge.

28, 2001. The epicenter was located near the epicenters of earthquakes that occurred in 1949 and in 1965. Studies of the mechanism at the Nisqualty focal point indicate that bending of the down-going slab probably caused the earthquake. As a down-going slab bends, tension causes the top of the slab to extend, and normal faults form as the extension occurs. The earthquakes at this location all happened at depths of 50 to 70 kilometers (30 to 43 miles). At these depths, most moderate earthquakes cause low-amplitude surface waves. For this reason, damage in the nearby urban areas of Seattle and Olympia was not great. In addition, improved building codes were effective in reducing damage.

Anchorage, Alaska, Earthquake One of the most powerful earthquakes in the history of North America hit Anchorage, Alaska, in 1964. The Richter magnitude was 8.3 to 8.7. Land-slides, fracturing, lateral shifting of the ground, and large surface waves caused much of the damage during this earthquake (see figure 6.9). The vibrations that accompanied this earthquake continued for between one and seven minutes and were particularly destructive. The focus was shallow, and the epicenter was located in the high mountainous region near the head of Prince William Sound, about 128 kilometers (80 miles) east-southeast of Anchorage (figure 6.27). A long series of aftershocks were distributed over a region of about 260,000 square kilometers. Aftershocks affected the continental margin of the Aleutian trench from Prince William Sound to Kodiak Island. Twenty-eight aftershocks of magnitude 6 or higher happened during the first day, and about 12,000 shocks of magnitude 3.5 or higher followed in the seventy-day period after the main shock.

The Anchorage earthquake was one of many that occur within the most seismically active area in the world, the nearly continuous belt of earthquakes that circles the Pacific Ocean (see figure 6.27). Most earthquakes in this zone result from subduction of oceanic crust beneath chains of volcanic centers like the one that extends from the Alaskan peninsula to Kamchatka, the long peninsula at the western end of the Aleutian Islands (see figure 5.21).

During the Alaskan earthquake of March 27, 1964, the level of the land changed over a vast area from Prince William Sound to Kodiak Island. A large block of Earth's crust tilted (figure 6.28), elevating the area southeast of Kodiak Island, while the region to the northwest subsided. Maximum uplifts of 8 meters (26 feet) took place on Montague Island, and portions of the Kenai Mountains sank by nearly 2 meters (6.5 feet). The movement involved tilting of blocks of the crust along a fault that comes to the surface along the Aleutian trench and slopes under the island arc (see figure 5.23).

PLANNING TO AVOID DISASTER

With the possible exception of hurricanes, earthquakes and earthquake-induced phenomena have killed more people than all other natural processes combined. Over the past thousand years, more than 100,000 people have died in each of seven earthquakes. An estimated 800,000 people died in the 1556 Shensi, China, earthquake, the most disastrous earthquake on record. An estimated 650,000 people perished in the 1976 earthquake in China. Over 30,000 died in Turkey in 2000, and more than 50,000 died in India in 2001.

The potential loss of life and property is so great it might seem unlikely that people would choose to live on land in areas of proven high earthquake risk. They do so because the areas of high risk are extremely large and the exact location of future earthquakes is highly uncertain. Earthquake hazards exist in many of the world's most heavily populated areas. Most of the populations of California and Japan live in seismically active areas; millions more live in dangerous areas in the Middle East, in China, and South and Central America. Although earthquake prediction is highly uncertain, individuals and government agencies can take steps to greatly reduce much of the potential damage.

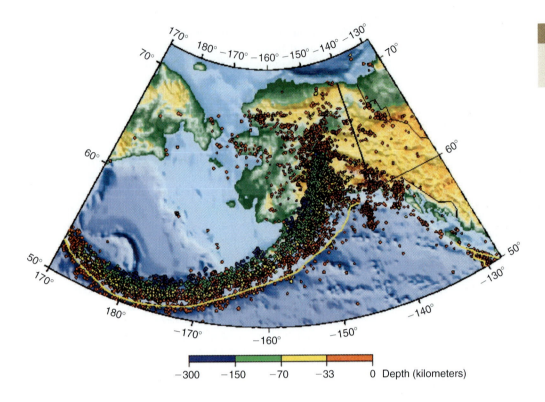

−300	−150	−70	−33	0 Depth (kilometers)

Reducing Earthquake Damage

Individuals can reduce their exposure to seismic hazards by being well informed about the risk and taking appropriate action to eliminate or minimize that exposure.

- Where are fault traces located and how close are they to your home or place of work? Federal and state agencies, such as the U.S. Geological Survey, have this type of information (figure 6.29).
- What information about seismic risks has been compiled for your area?
- Is your house constructed out of stone, adobe, or brick? If you live in a structure that is vulnerable, consider strengthening part of the building with reinforced concrete or heavier framing. In an earthquake, doorways and bathrooms tend to be safer than other parts of a structure. Stay away from large glass windows during earthquakes.
- Does a reservoir or other body of water exist that might endanger your home in the event of a dam failure? If so, find out what measures have been taken to inform and evacuate residents in an emergency.
- What type of foundation material lies beneath your building? Bedrock provides the greatest safety; landfills on former wetlands, unconsolidated lake bed, and alluvial deposits are the most dangerous.
- Are steep slopes located either above or below your house? Have they failed in the past? If your house is cantilevered on a steep slope, it may be especially vulnerable to ground motion.

Local government agencies may follow a variety of actions to mitigate seismic hazards by

1. Providing the public with information about known natural hazards.
2. Reducing risks through planning and enforcement of zoning ordinances and building codes.

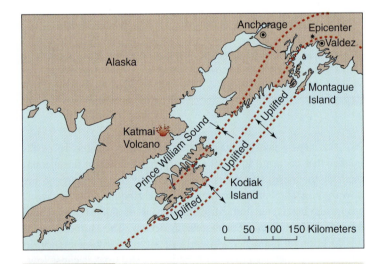

Figure 6.28 The Alaskan Earthquake of 1964. The epicenter of the 1964 Anchorage, Alaska, earthquake was located east of Anchorage. A large block of the crust tilted during this earthquake. The southeast side of Kodiak Island went up relative to the areas to the northwest. Some areas rose more than a meter. The Aleutian trench lies southeast of the fault line.

Aerial View Along the San Andreas Fault. This aerial view shows housing subdivisions constructed near and directly over one of the most active faults in North America.

3. Purchasing especially high-risk areas (e.g., areas along active faults) for use as parks and green space.
4. Providing rescue and relief work after a disaster.

Earthquake Prediction

From a quick review of the devastation and loss of life and property caused by earthquakes, it is not surprising that studies to discover ways of predicting earthquakes have been given top priority among research projects funded by government agencies. In recent years, efforts to predict and mitigate earthquake damage have cost about $100 million each year.

Efforts to predict earthquakes go back at least to early Greek civilization. Some of the earliest known manuscripts show that the early Greeks observed the fluctuation of the water level in wells before and during earthquakes. These fluctuations were apparently used to predict or monitor earthquake activity. For many years, the Chinese and Japanese have monitored animal behavior in hopes of identifying changes related to impending earthquakes. Unusual animal behavior has been frequently reported, such as that of squirrels fleeing New Madrid, Missouri, before the devastating earthquakes of 1812 and birds flying off Hebgen Reservoir near Yellowstone National Park shortly before an earthquake tilted the reservoir.

Individuals who take different approaches from those of seismologists have issued dire warnings of impending earthquakes. A consultant on weather prediction became convinced that earthquakes result from Earth tides, and in 1990, he predicted that the New Madrid area would be affected by a major earthquake during an unusual astronomical event. During this event, the Sun, Moon, and several planets became aligned on December 3, 1990, in what is known as the Jupiter effect. His predictions were widely advertised and resulted in such great concern that schools were dismissed on that date in northeastern Arkansas. Later he reversed his prediction, but that received little notice in the press. The prediction turned out to be wrong. The good news from this nonevent was that many people learned that they live in a seismically active part of the world. For once, emergency drills took on a sense of urgency.

Most efforts to predict earthquakes are concentrated on changes in natural systems that may be precursors of seismic activity. The close connection between earthquakes and movements on faults is well established, as is understanding of earthquake mechanisms and how activity on one part of a fault affects other parts of the system. Fault displacement has been identified as the primary cause of earthquakes; therefore, looking at past seismic history is useful for determining where future events may occur. Because minor earthquakes frequently precede major earthquakes, areas of persistent small earthquakes receive careful attention. In recent years, persistent small earthquakes have caused concern about the Mississippi Valley region, near Charleston, and along the Pacific coast. In the San Francisco Bay area, intermediate-sized earthquakes have generally preceded major events. The frequency of earthquakes with magnitudes of 5 to 6.2 increased in the years immediately before the three largest seismic events in California's history, including the 1906 earthquake. Foreshocks precede about a third of all southern California earthquakes of magnitude 5 or more. On this basis, seismic activity in this region creates concern for the future of both the San Francisco Bay and Los Angeles areas.

In the 1980s, seismologists selected Parkfield, California, a small community about halfway between San Francisco and Los Angeles, for an intense study of earthquake prediction. Since 1857, six moderately strong earthquakes have occurred there. Each was separated from the next earthquake by about twenty-two years. If this interval were the amount of time needed for strain to reach the critical level, another event might have occured in 1988. It did not. Since selecting Parkfield as the site for the most comprehensive experiment in earthquake prediction, scientists have installed a vast array of monitoring equipment in hopes of identifying some consistent, reliable, and measurable phenomena that precede earthquakes. Even before selecting Parkfield, they knew from geodetic measurements that the fault under Parkfield is locked and that the amount of strain across the fault is equal to or greater than that before the last major Parkfield earthquake in 1966. The U.S. Geological Survey and thirteen other institutions are monitoring the Parkfield area as they search for precursors. They have networks of equipment monitoring a great variety of phenomena, such as (1) the number, magnitude, and focal depth of all seismic events, (2) the direction and amount of displacement of points on opposite sides of the fault trace (figure 6.30), (3) the rate of movement of points across the fault,

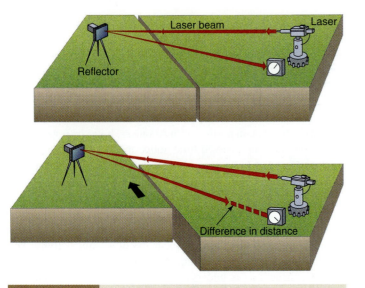

Figure 6.30 **Measuring Displacement Across Faults.** Modern technology provides very precise ways of measuring displacements across faults. Any change in the position of the reflector relative to the laser gun causes a change in the travel time of the beam. These devices help detect the buildup of strain across a fault.

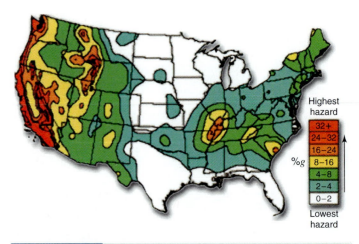

Figure 6.32 Seismic Hazard Map for the Lower United States. The color bars indicate the degree of hazard. Shaking is expressed as a percentage of the acceleration of a falling object due to gravity (g), which is 32 feet/sec^2.

(4) ground-water level in wells, (5) the strength of the magnetic field, (6) the tilt of the ground surface (figure 6.31), (7) the amount of radon in ground water, (8) the amount of hydrogen in the soil, (9) the resistance of the rock to the passage of electric currents across the fault, (10) the acoustic emissions, and (11) the temperature in wells. Having missed the 1988 prediction, seismologists have waited anxiously ever since for the big event. In 1992, a first-level alert was issued following a 4.7-magnitude earthquake, but, much to almost everyone's surprise, nothing happened.

Seismologists still differ about how best to look for precursors of major earthquakes. Many think the approach used at Parkfield is the best hope. Others point out that events far removed from epicenters sometimes forecast major earthquakes. For example, in northern California, the interval between eruptions at geysers changed several days before the three largest earthquakes, including the Loma Prieta event. Unfortunately, this type of information fails to pinpoint the location of a future break and does not give precise information about timing either.

Modern studies of seismic activity have identified the most dangerous seismic zones (figure 6.32). Geodetic measurements enable seismologists to isolate locked portions of fault traces, and the shape of fault traces and past history of movements help identify areas most subject to near-term earthquakes. Despite all of the efforts to find precursors of major earthquakes, identifying both the time and place of future earthquakes remains elusive.

Although the regions were recognized as places where major earthquakes might occur, predictions have not been successful. A variety of ideas have been advanced about what should be done next. Some seismologists advocate concentrating attention on reducing hazards rather than trying to predict earthquakes. Yet, in the November 1995 issue of *Science,* an article titled "Quake Prediction Tool Gains Ground" discusses an idea developed in Greece. The technique monitors electrical signals in the ground. During the last nine years, this group claims to have successfully predicted 10 out of

Figure 6.31 **The Tiltmeter.** Tiltmeters similar to those used to monitor volcanoes indicate changes in the level of the land across fault zones. The tiltmeter operates like a carpenter's level except that it is much longer and thus sensitive to very slight changes in level.

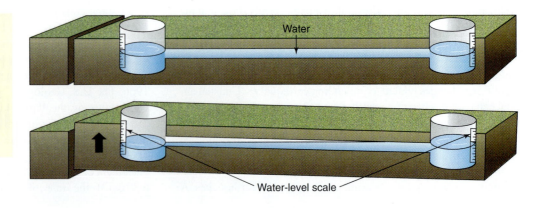

11 earthquakes of magnitude 5.8 or more that affected the area being monitored. The results have met with a barrage of criticism, but just before the Loma Prieta earthquake in 1989, geophysicists at Stanford detected a strange burst of noise. Even this promising prediction tool fails to give the exact time and place of impending events. For the present, the increased levels of seismic activity in a number of heavily populated areas provide warning that those communities should give earthquake preparedness a high priority.

SUMMARY POINTS

1. Most shallow-focus earthquakes result from movements along faults, many of which are located at plate boundaries. Volcanic activity and mass movements are responsible for other earthquakes. According to the elastic rebound theory, strain gradually builds up in the rock in the fault zone until the rock breaks. The elastic response of the rock causes earthquakes and generates elastic shock waves that move through and across the surface of Earth.

2. Slippage at one point along the fault increases stress elsewhere, often causing a cluster of aftershocks as stress is released in a series of lesser breaks.

3. Deeper earthquakes are also thought to involve movement along faults, but because the temperature and pressure at depth give the rock greater plasticity, the movement is stick-slip rather than elastic rebound.

4. Compressional (P) and shear (S) waves are transmitted through the interior of Earth. Other waves travel mainly through the lithosphere. These surface waves cause most earthquake damage.

5. Seismologists describe the position of an earthquake based on its focus, the subsurface point of origin, and the location on the surface just above the focus, the epicenter. The time lag between arrival of the P and the S waves provides a measure of the distance of a seismograph station from an earthquake epicenter. By using the distance of the epicenter from three stations, seismologists can locate epicenters.

6. Most earthquake activity occurs in narrow belts that coincide with areas of volcanic activity and with the boundaries of lithospheric plates.

7. Seismologists use two scales to express the size of earthquakes. The Mercalli intensity scale reflects observed surface effects of the tremor. The Richter magnitude scale derives from the amplitude of surface waves as recorded on seismographs.

8. Earthquakes pose serious threats to people who live in earthquake-prone areas or near sea level around the edge of the Pacific where most tsunami occur. In areas subject to earthquakes, location of housing developments, selection of building materials, and layout of highways and water mains deserve careful planning to minimize damage.

9. Most earthquakes result from plate movements. Only shallow-focus earthquakes arise at divergent plate junctions; most of the intermediate and deepest earthquakes lie within subduction zones.

REVIEW QUESTIONS

1. How do the magnitude and intensity scales used to describe earthquakes differ?

2. In what plate tectonic setting(s) do shallow-focus (and deep-focus) earthquakes occur?

3. Cite examples of earthquakes that have affected areas outside the plate tectonic settings described in question 2.

4. What measurements must be made on seismograms to determine the distance from a seismograph station to an earthquake epicenter?

5. If you live in an area affected by earthquakes, what should you do as an individual to protect yourself?

6. What natural processes cause earthquakes?

7. What techniques of earthquake prediction seem most promising?

8. What areas in the United States have the highest frequency of earthquakes?

9. What secondary effects of earthquakes account for most property damage?

10. What methods are used to predict the time and place of future earthquakes?

THINKING CRITICALLY

1. Should seismologists issue public warnings when they think an earthquake might affect a major city?

2. What is the seismic risk where you attend college? What factors are important in assessing seismic risk?

3. Could the sudden displacements of the ground associated with earthquakes affect ecosystems? If so, how?

4. Examine the map in figure 6.14, which shows the location of earthquake epicenters. Approximately what percent of earthquake epicenters are not located close to plate boundaries?

5. If you lived in a community with high seismic risks, what precautions would you advocate community government take to minimize the damage that might result from an earthquake? Consider the location of utilities, transportation routes, hospitals, fire stations, and schools.

KEY WORDS

aftershocks	periods
elastic rebound theory	seismogram
epicenter	seismograph
fault creep	stick-slip
focus	trace
inertia	tsunami

Volcanic Activity

Molten lava pours out of Pu'u O'o vent in Hawaii Volcanoes National Park.

Chapter Guide

Mount Vesuvius, Tambora, Krakatau, and Mount St. Helens are legendary in human history. Mount Vesuvius destroyed the Italian city of Pompeii in A.D. 79. The volcano Tambora destroyed an entire island in 1815. In 1883, Krakatau erupted so violently that it spewed dust, gases, and debris into the atmosphere around the world, which did not dissipate for two years. The explosion at Mount St. Helens in 1980 was so powerful it could be heard 217 miles away. Volcanoes do not just cause destruction, however; they are the probable sources of the gases in the atmosphere and the water in the oceans. As an important source of carbon dioxide in the atmosphere, volcanic eruptions may also be important contributors to changes in the climate on Earth.

Volcanism is a product of igneous activity that starts below the crust and rises to the surface. As discussed in chapters 2 and 5, plate tectonic theory explains where eruptions occur. This chapter explores the relationships between deep-seated plutonic activity and volcanism and why some eruptions are violent explosions, but others are more passive outpourings of lava. It examines the many forms igneous intrusions and volcanic activity may take and their significant impact on the environment.

INTRODUCTION

When thinking about a volcano's destructive power, many think only of explosions or rivers of molten lava consuming everything in their path. Few think of the deadly mudflows a volcano can instigate. Nevado del Ruiz is a 5,386-meter- (17,666-foot) high active volcano in the Andes Mountains of Colombia. On the afternoon of November 13, 1985, explosive eruptions began in the crater. Hot fragmental materials ejected from the volcano, and heavy rains caused by the emission of huge quantities of water vapor melted snow and ice on the mountain. Lava did not cause the disaster that accompanied the eruptions. Mudflows, called **lahars** (figure 7.1), did. The water combined with ash to form a mudflow that swept down the steep slopes along narrow valleys leading to the town of Armero on the banks of the Rio Lagunillas. The mudflow arrived so quickly that it caught the population unprepared, and few escaped. The mud spread across the valley and poured over the town, killing 23,000 people. Thousands more were injured or left homeless. It was the worst volcanic disaster in the history of Colombia and the worst in the world since the eruption of Pelée on the island of Martinique in 1902.

In part, the tragedy of the Nevado del Ruiz eruption was because the people of Armero were not adequately warned of the danger. Volcanic activity started a year earlier. Scientists working in the region noted earthquakes, minor explosions, and increased activity around holes emitting fumes in the crater. Efforts were made through the United Nations disaster relief program to increase surveillance, and seismic monitoring began months before the major eruption. People in the region became more concerned in September when an eruption caused ash to fall over a large area. By mid-October, volcanologists had prepared a map showing the types of hazards that might be expected from a major eruption, and this map was being discussed with officials in nearby cities. When the eruptions began shortly after noon on the thirteenth, an alert was sent out to communities around the volcano. By 4:00 P.M., evacuation was urged for many communities, but it is not clear whether the people of Armero received the order. The major eruption happened at night, and heavy rains deterred evacuation, as did calls by prominent local citizens for people to remain calm. The experience at Nevado del Ruiz demonstrates the serious hazards that active volcanoes present and the need for both an understanding of the hazards and an effective, unambiguous way of warning people when dangers are present.

GEOGRAPHY OF MODERN VOLCANISM

The historically active volcanoes lie in clearly defined zones (figure 7.2). Plate tectonics provides explanations of both the location and the composition of the lavas extruded from volcanoes. Basaltic lavas pour from submerged volcanoes along the crest of the oceanic ridges, where seafloor spreading is taking place, and along some transform faults in the oceanic crust. These magmas originate in the upper mantle by the melting of rocks, such as peridotite, that contain iron and magnesium-rich silicate minerals, like olivine and pyroxene, and calcium-rich silicate minerals of the feldspar group. Most volcanoes that erupt andesite magma, which is intermediate in composition between basalt and granite, form over subduction zones. Magmas generated in subduction zones contain a mixture of elements derived from the mantle and from oceanic crust. The Cascade Mountains in Washington, Oregon, and southern British Columbia are examples of volcanoes that formed over subduction zones. Their lavas are similar in composition to that of volcanoes on

(a)

(b)

Figure 7.1 Destruction by Lahars. (a) View of Nevado del Ruiz in Colombia. A mudflow is moving down the stream. (b) Aerial view of the remains of the town of Armero after the mudflow destroyed most houses near the river.

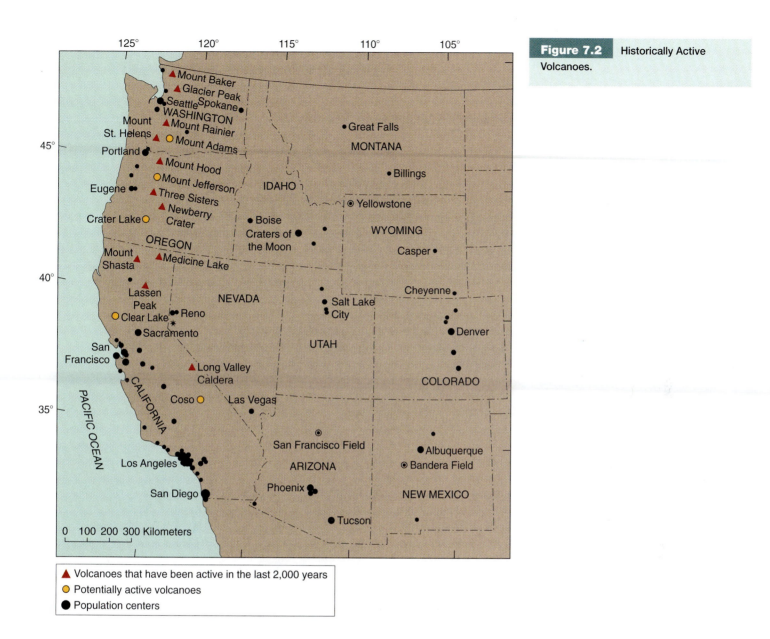

Figure 7.2 Historically Active Volcanoes.

▲ Volcanoes that have been active in the last 2,000 years
◯ Potentially active volcanoes
● Population centers

the western side of the Pacific and along portions of the coasts of Central and South America. These volcanoes form a more-or-less continuous ring around the Pacific called the *ring of fire* (see inside cover). Volcanoes associated with subduction are also found in Italy, the Greek Islands, the East Indies, and the South Sandwich Islands between South America and Antarctica.

Other volcanic centers, including the famous ones in Hawaii and Yellowstone National Park, lie above hot spots, where heat and magma rise from the mantle through the crust. Studies of the deep interior suggest that the heat and possibly some magma at hot spots may rise from near the bottom of the mantle (see figure 3.24). Lavas of basaltic composition erupt where hot spots lie beneath oceanic crust, but lavas rich in silica or intermediate in composition and mafic lavas rise to the surface from hot spots beneath continental crust. These lavas cool to form rhyolite, andesite, and basalt.

BENEATH VOLCANOES

Although the magma chambers beneath volcanoes are inaccessible, plutons that form thousands of feet below the surface are exposed in the continental shields and in mountain belts. Erosion has stripped away the rock intruded by the plutons, providing three-dimensional views of these masses. Like roots growing out of the magma, offshoots and feeder systems leading from the plutons toward the volcanoes that once stood above them are plainly visible. The tops of some large plutons in the Sierra Nevada are still in place, and the deeper interior portions are also exposed, making it possible to study the once-buried magma chambers and their long-since frozen contents.

As magma moves closer to the surface, it follows the course of least resistance. This course may be cracks in the country rock around the intrusion or along contacts between different types of

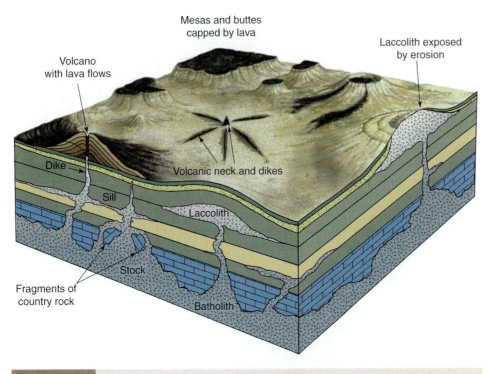

Figure 7.3 depicts labels: Mesas and buttes capped by lava; Volcano with lava flows; Laccolith exposed by erosion; Dike; Volcanic neck and dikes; Sill; Laccolith; Stock; Fragments of country rock; Batholith

Figure 7.3 Block Diagram of Various Types of Igneous and Volcanic Rock Bodies. The diagram includes dikes, sills, laccoliths, a stock, and a volcanic neck.

rocks. Figure 7.3 depicts the shapes of the most common intrusives. Some develop within the structure of a volcano; others occur at great depths, and many of these may develop in places where the magma does not break through to the surface and form a volcano. In addition, intrusions range widely in sizes and form. The largest plutons, called **batholiths,** exceed 100 square kilometers (40 square miles) in surface area. The Sierra Nevada in California consist largely of batholiths, as pictured in figure 7.4. Smaller bodies called **stocks** commonly protrude from the sides or tops of batholiths. Maps and aerial images of batholiths depict oval bodies that pushed surrounding rocks aside as they moved into place (see figure 7.4). The magma also invades the country rock by melting it and by displacing large blocks that sink into the magma.

What happens at the bottom of batholiths is unclear. Nowhere is the bottom of a batholith exposed. Most of them appear to increase in size at the bottoms of deep valleys that cut across them, but seismic studies indicate that batholiths taper deep in the crust. Many extend deep into Earth's crust and may have the shape of an inverted water drop, a shape assumed by the magma as it rose toward the surface.

PRODUCTS OF VOLCANIC ACTIVITY

Three types of materials come out of volcanoes: gases, lava, and fragmental materials called **pyroclastics,** which include rocks blasted out of the volcano and bits of lava that cool in the air.

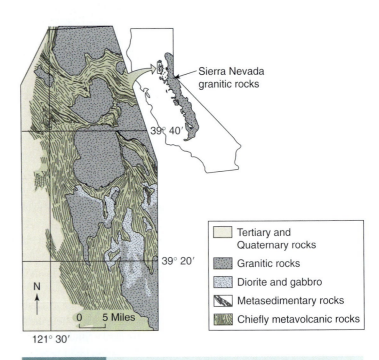

Legend:
- Tertiary and Quaternary rocks
- Granitic rocks
- Diorite and gabbro
- Metasedimentary rocks
- Chiefly metavolcanic rocks

Sierra Nevada granitic rocks

39° 40′
39° 20′
121° 30′
0   5 Miles
N

Figure 7.4 Map of Several of the Many Batholiths Exposed in the Sierra Nevada of California. The Sierra Nevada batholith is comprised of many small plutons generally separated by thin walls of metamorphosed volcanic rock that were pushed aside as the plutons were forced into position.

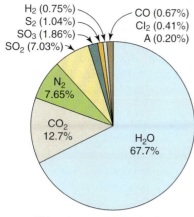

(a) Volcanic gases at Hawaii (Kilauea)

(b)

(c)

Figure 7.5 **Volcanic Gases.** (a) The gases produced by the volcanoes in Hawaii contain the constituents shown here. Note that water vapor and carbon dioxide make up a large portion of the total. (b) Cloud of gases escaping from Kilauea Volcano on the Island of Hawaii. Reactions between sulfur dioxide, water vapor, and oxygen produce this smog. (c) Huge blocks of volcanic rock mixed with fragments of many other sizes smolder at the site of this active volcano.

Gases

During most eruptions, gases rise from the magma to the surface (figure 7.5), and they may continue to escape long after the eruption of lava and small pyroclastic fragments, such as dust and cinders, stops. Gases may continue to rise even after lava solidifies in the feeder system. These gases emerge through cracks in the solid surface layer of lava flows and from holes formed in the crust of lava lakes. These small vents are aptly named **fumaroles,** after the fumes that come out of them. Gases may also break through to the surface of cone-shaped piles of cinders after making their way through the ash and cinder that fill the passageways. In 1912, a particularly violent explosion at Katmai volcano in Alaska produced an area called the Valley of Ten Thousand Smokes. The ground around fumaroles is usually yellow from the crystallization of

sulfur and sulfur-bearing minerals, such as pyrite, that form encrustations in and around the vent.

Water vapor is the most abundant gas coming from volcanoes (see figure 7.5). In fact, most of the water on Earth probably came from degassing Earth's interior through volcanic eruptions early in Earth's history. Carbon dioxide, hydrogen sulfide, and sulfuric and hydrochloric acids are also commonly present.

Lava

Because most volcanoes are unpredictable and the temperature is so hot, collecting samples is hazardous. A few volcanologists have paid for their curiosity with their lives. Fortunately, some volcanoes, especially those in the ocean basins, are less prone to violent eruption and are more accessible for study. Volcanologists collect samples of lavas like those in Hawaii even while the lava is fluid, but because of the danger involved at more explosive volcanoes, they generally collect samples of solidified lava. Based on their chemical composition, most lavas that are nonexplosive fall into one of the three major types described in chapter 2. These differ from one another especially in the amount of silica they contain—basaltic lava with low silica content, andesitic lava with intermediate silica content, and rhyolitic lava with high silica content (see figure 2.14).

Crystals grow as atoms attach to the surface of tiny crystal nuclei. Generally, this growth is a slow process. Consequently, the faster lava cools, the less time crystals have to grow. Because lava cools rapidly at the surface or near the surface, the crystals in most rocks formed from lava, such as basalt, andesite and rhyolite, are visible only with magnification. When lava cools fast, as it does when it flows into water, not even tiny crystals have time to form. The resulting solid lacks an internal orderly arrangement of its atoms. It is a natural glass called obsidian.

The chemical composition, temperature, and amount of water and gas present determine the fluidity of lava. **Viscosity,** a measure of the resistance a liquid offers to flow, is used to describe how fluid lava is. Liquids like water that flow easily have low viscosity, and those that are sluggish like tar have high viscosity. A lava is less viscous when it is hot, when large amounts of water and gases are present, and when the lava has a low silica content. The color of molten lava indicates its temperature. Orange, golden yellow, and white incandescence signify increasing temperature from 900° to 1,200°C. At these temperatures, basaltic lavas like those in Hawaii flow like streams following valleys and form lava lakes in depressions. However, when the temperature of lava decreases, it becomes sticky like cold molasses. Below about 750°C, even Hawaiian lavas barely move.

Lava Flows When low-viscosity lava emerges from a fissure or single vent, it spreads over the land surface until it flows into a valley. Such lavas may become ponded, forming lava lakes. If a lake forms in the crater of a volcano, pressure builds up in the feeder system and may either force lava to break through the flank of the volcano or hold in gases until an explosion clears the vent.

As it solidifies, the lava of most basaltic flows usually assumes a rope-like form called **pahoehoe** (pronounced pa-hoi-hoi) or a blocky form known as **aa** (pronounced ah-ah) (see figure 7.6). (Both terms are Hawaiian in origin.) As the outer surface of a thin sheet of low-viscosity lava cools and begins to crystallize, the semisolid surface wrinkles and slowly twists into the ropy form. Aa forms when a thick crust develops over the surface of a slowly moving or more viscous lava flow. The crust, usually composed of the porous rock scoria, breaks up as the liquid interior of the flow continues to move. The flow buries blocks of the hard crust as they fall forward off the steep front of the flow.

If it is extruded into or under water, basaltic lava assumes rounded or pillow-shaped masses. The surface of the pillows freezes almost instantaneously when the lava enters water, leaving a glassy rim, but the interior remains fluid. As the lava moves forward through surface cracks, it squeezes through the openings like blobs of toothpaste.

Because of its high silica content, its low gas and water content, and its relative low temperature, rhyolitic lava is much more viscous than basaltic lava. As a result of its viscosity, rhyolite tends to form bulbous masses over the vent or fissure from which it issues. The resulting forms have steep slopes and may assume flow patterns like those seen in warm asphalt. The Glass Mountains of California and some flows in Yellowstone National Park are this type. Lava domes of this sort are relatively small because they

(a)

(b)

Figure 7.6 Varieties of Lava. (a) Lava that flows while the surface is still very hot commonly develops a ropy surface and is called pahoehoe. (b) Lava that moves after the surface cools commonly forms aa, or blocky lava masses.

solidify so quickly. The lava plugs the feeder system before much magma comes to the surface. With gases trapped beneath the plug, pressure builds beneath the volcano. For this reason, volcanoes that erupt rhyolitic lavas are more explosive than those that erupt basaltic lava.

Pyroclastic Materials

Most of the material erupted from some volcanoes consists of fragments (figure 7.7). Coarser fragments fall near the vent; fine particles may remain suspended in the air for long periods. The smallest fragments, volcanic dust, may travel hundreds, even thousands, of miles. In 1980, winds carried particles of ash from Mount St. Helens in Washington state almost completely across the country. High-altitude winds carry dust from volcanoes in southern Italy into North Africa. During particularly violent eruptions, such as the 1991 eruption at Pinatubo in the Philippine Islands, dust and sulfate aerosols reach the upper atmosphere and circulate around Earth, causing brilliant sunsets and blocking solar radiation enough to affect Earth's climate.

Gases held under pressure while magma is underground expand as pressure drops. When small bits of lava enter the air, gases in the lava expand, forming bubbles called **vesicles** (see figure 2.12). Pumice, a light-colored, vesicular, glassy rock, forms when bits of viscous rhyolitic lava containing large quantities of gas suddenly cools in the air. Basaltic lava may form scoria, a similar but dark-colored porous rock. In A.D. 79, the volcano Vesuvius erupted and buried Pompeii, Italy, under pumice and ash. The deposits accumulated so fast that few people escaped.

Consolidated pyroclastic rock called **tuff** forms when airborne volcanic ash and dust settle out of the air or from airborne particles that form blanket-like deposits. If the temperature in the cloud of ash and dust is sufficiently high, the molten or partially molten particles stick together as they fall, forming a **welded tuff.** When the thickness of the ash and dust increases, the weight of the overlying material compresses material lower in the pile.

ANATOMY OF VOLCANOES

A volcano is any vent at the surface through which magma, gas, fragmental materials, or all of these erupt. Beneath many volcanoes, the magma follows planes of weakness, such as faults, fractures, or bedding planes, as it rises through the surrounding rock. The conduits through which magma moves to the surface form the feeder system for the volcano. Magma may break through to the surface along faults or fractures causing **fissure eruptions** (figure 7.8), but in many volcanoes, material erupts through a single pipe-like opening called a **central vent** (figure 7.9). The shape of the volcano depends on the amount and type of material erupted and on less obvious factors, such as lava viscosity, the amount of gases present, the shape of the land before the eruption, the type of eruption, and the character of the feeder system.

Volcanoes with central vents are generally more or less round when viewed from above. If fragmental materials are ejected, the volcano becomes a **cinder cone,** similar to the one shown in figure 7.9. The slopes of cinder cones are usually about 35°, the angle at which loose fragments can stand in piles. If the slope becomes steeper, fragments roll and slide downslope, restoring the 35° angle. The vent itself commonly resembles the inside of a funnel. Because of the steep slopes and the loose fragments, cinder cones erode quickly. These features are beautifully exposed at Capulin volcano in New Mexico and Sunset Crater near Flagstaff, Arizona (see figure 7.9).

Mixtures of pyroclastics and lava from central-vent volcanoes build **composite cones,** also called **stratovolcanoes** (figure 7.10). Fragmental materials build the cone higher; **ash flows** and lavas, if they are fluid, extend the outer margins of the cone. Low-viscosity lava flows may continue far from the base of the cone along stream valleys.

A third type of volcanic form results when larger quantities of highly fluid (low-viscosity) lava erupt. Because such volcanoes resemble a shield in profile, geologists call them **shield volcanoes** (figure 7.11). They are broad and have low slopes of 10° or less. Although the slopes are low, such volcanoes may be huge, reaching great heights. Mauna Loa, the highest volcano in Hawaii and the largest shield volcano in the world, stands over 4,400 meters (14,400 feet) above sea level. Measured from the adjacent abyssal plains, Mauna Loa is a pile of lava higher than Mount Everest. Like many shield volcanoes, Mauna Loa formed over a mantle hot spot.

Magma eventually solidifies in the feeder system through which it has moved toward the surface. As time passes, pyroclastic materials erode, quickly exposing rock that solidified in the feeder system. At Shiprock, New Mexico, erosion has stripped away the pyroclastic materials that once formed a large volcano, exposing the feeder system. The central pipe-like feeder composed of basalt rises above the surrounding plain forming a **volcanic neck** (figure 7.12). A system of radial dikes may radiate away from the pipe as they do at Shiprock, New Mexico, and in the Spanish Peaks of Colorado.

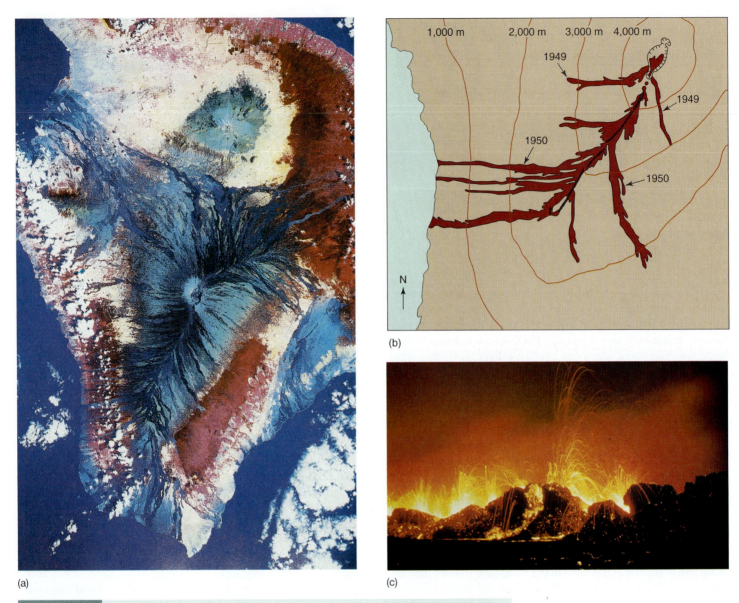

(a)

(b)

(c)

Figure 7.8 Fissure Eruptions. (*a*) Satellite image of Mauna Loa in Hawaii. Note the fissures from which huge amounts of basaltic lava have poured out over the land surface. (*b*) Sketch map of one fissure. Dates indicate the time of the most recent flows seen on the image. (*c*) Photograph of lava spurting from a fissure in Hawaii.

Calderas

The eruption of large quantities of lava, pyroclastic debris, or gases or withdrawal of lava in the vent and feeder system may leave an empty space under the volcano. Voids may form as gas pressure drives lava to the surface during eruptions or as the magma remaining inside the chamber contracts while it cools and crystallizes. As pressure in the magma chamber lessens, the loose volcanic debris and lava making up a volcano may slowly subside or violently collapse into the empty space, creating an enclosed or partially enclosed depression called a **caldera.** Several calderas are aligned

near the top of Mauna Loa (figure 7.13). Explosions during which the top of a volcano blows away may also create calderas.

Crater Lake, Oregon, occupies a caldera that formed following the collapse of a large volcanic cone. Sketches by volcanologist Howell Williams depict the sequence of events thought to have led to the present form of Crater Lake (figure 7.14). Formation of the lake in the caldera and development of a small cinder cone, Wizard Island, are the most recent events in the long history of this volcano.

Volcanic activity continues within the caldera in Kilauea volcano on Hawaii. A deep pit-like opening lies within a large area

(a)

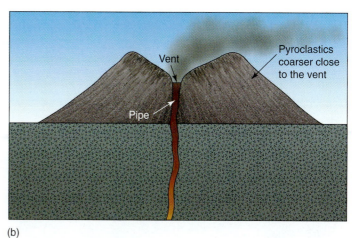

(b)

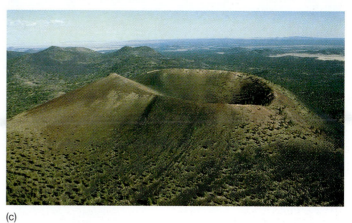

(c)

Figure 7.9 **Central-Vent Eruptions.** (*a*) View of volcano Paricutin erupting in Mexico. (*b*) Sketch of a cinder cone. (*c*) View of Sunset Crater cinder cone located near Flagstaff, Arizona.

(a)

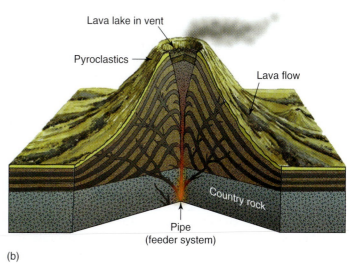

(b)

Figure 7.10 **Composite Cones, or Stratovolcanoes.** (*a*) Aerial view of Mount St. Helens in eruption in 1980. (*b*) Block diagram showing the internal structure of a typical composite cone.

that has subsided. This pit has been the site of numerous lava lakes. High-angle faults ring the caldera, forming the cliffs around it. These cliffs form a system of concentric escarpments (sudden breaks in the level of the ground surface) along which the central sections have dropped downward. An even larger caldera lies at the top of Mauna Loa (see figure 7.13).

ERUPTIVE STYLES

As rock melts and expands, it causes tremendous pressures to build up in magma chambers. If the pressure over the magma is less than that in the magma, the expanding body of magma rises. The liquids move along any existing fractures and faults, or they push the overlying layers of rock upward into domes, causing new cracks through which the liquids may eventually travel.

Where large quantities of gas under high pressure reach the surface, an explosion accompanied by little or no lava may occur. If small quantities of gas are present or if they escape as the magma rises, the eruption may consist mainly of the extrusion of lava flows. When the magma contains large quantities of gases that cannot escape, the pressure beneath the volcano may rise until it is great enough to blow away the top or side of the volcano. Such violent eruptions are associated with composite volcanoes formed of andesitic magmas. These silica-rich magmas have high viscosity, and they retain gases. Gases escape more readily from basaltic magmas, releasing pressure and making these volcanoes less explosive.

Particularly violent and potentially lethal eruptions of hot gases, red-hot ash, and exceedingly small drops of lava so hot they glow are called **nuée ardents.** An eruption of this type destroyed the town of St. Pierre on the island of Martinique. The lava involved in this type of eruption is andesitic. The gas in these clouds lubricates the solid, enabling them to flow rapidly. Volcanologists estimate that some flows of this type travel at rates up to 100 kilometers per hour

(60 miles per hour). These intensely hot clouds may spread great distances over the surrounding land. As they cool, the gases condense into extremely hot liquids and ash that fall, forming a deposit in which the ash and particles of glass may weld together into blanket-like deposits. Not all blanket-like deposits are welded. Clouds of hot gases carry materials heavier than air in suspension at or near ground level. Depending on the material that flows in these clouds, they may be called debris, ash, or, more generally, **pyroclastic flows** (figure 7.15). These flows are held close to the ground by their weight. They consist of a hot mixture of volcanic gases and ash. In some instances, they are produced by the explosive disintegration of lavas in a volcanic crater. Others result from the explosive emission of gases and ash that are so hot the ash particles weld together. The solids included in a typical ash flow may consist of pumice, scoria, dust, or even large blocks. The cloud of gases and pyroclastic material sweeps down the flanks of the volcano and across the ground surface, destroying everything in its path. Huge clouds of fine ash and gases rise from the mass as it moves.

Most volcanoes have a characteristic style of eruption that is repeated in successive eruptions. These characteristics include the type of material erupted, the physical properties, especially the viscosity of the lavas, and the violence of the eruptions. These characteristics determine the shape of the volcano. Most volcanoes exhibit characteristics that resemble those of famous or well-known volcanoes. These include explosive volcanoes like Tambora, Krakatau, Peleé, and Mount St. Helens; the relatively passive outpouring of lavas as seen in Hawaii and in some vast lava plateaus; and persistent eruptions like those at Stromboli in Italy.

Explosive Eruptions

Tambora The most violent explosion in recorded history destroyed a small island in the East Indies in 1815, when the volcano Tambora blew about 200 cubic kilometers (48 cubic miles) of

(a)

(b)

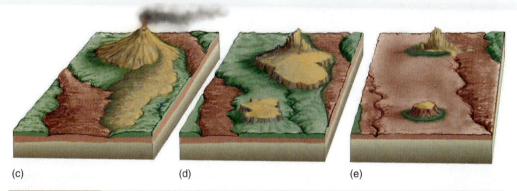

(c)   (d)   (e)

Figure 7.12 Volcanic Necks. (a) Devils Tower, Wyoming. The central high mountain is all that remains of a much larger volcano. (b) Other necks, like this one in Utah and Shiprock, New Mexico, are composed of lava that solidified in the pipe. The surrounding ash and flows have eroded away. (c–e) Block diagrams show the evolution of a volcanic neck.

Earth's crust into the air. Debris created by this tremendous explosion covered islands for hundreds of miles around. Explosions of this sort are rare, and most of them involve stratovolcanoes that have been dormant. The eruption of another volcano, Krakatau, located in the same region is much better known.

Krakatau One Sunday afternoon in August of 1883, a few mild explosions rocked the volcanic island of Krakatau, located in the East Indies. The next morning, an explosion ripped the cone of the volcano apart, sending more than 4 cubic kilometers (1.5 cubic miles) of rock into the air. A cloud of dust, gases, and debris rose 27 kilometers (17 miles) into the atmosphere. The heavier debris fell back to Earth nearby, but gases and dust caught in upper-air currents circled Earth, coloring sunsets around the world for the next two years.

The explosion blew away most of Krakatau (figure 7.16). A depth sounding taken after the explosion over the site of the former peak, which had stood 792 meters (2,600 feet) above sea level, was 304 meters (1,000 feet) below sea level. Although few people lived on Krakatau, a gigantic tsunami generated by the awesome eruption killed thousands of people in the lowlands along the coast of Java and nearby islands.

The explosion of Krakatau in 1883 was but the latest in a long series of explosions at that site. The outlines of older volcanoes destroyed long ago are still visible. A fringe of small islands marks the rim of a former volcano that collapsed or exploded before recorded history (see figure 7.16). A second rim of islands defines the perimeter of what was Krakatau, and now a new volcano (Anak Krakatau) is growing near the site of the 1883 eruption.

Figure 7.13 Calderas. (*a*) An aerial view of the calderas on the crest of Mauna Loa in Hawaii. The calderas are aligned along one of the major fissures. (*b*) Geologic sketch map of Kilauea caldera showing the circular faults formed as the central part of the caldera collapsed. The cliffs mark the down-thrown side of the faults. The ages of lava flows in the crater are indicated by color and date. These faults are also shown in cross section.

(a)

0   1 Kilometer

1919

Kilauea 1 Km crater

Kilauea crater

1954

1959

1961

1885–1894

1921   1877

Kilauea caldera

Halemaumau

1919   1894   1921

1961

(b)

Cascade Mountains These prominent mountains include a string of modern volcanoes located near the coast, extending from northern California to southern British Columbia (figure 7.17). The magmas beneath these volcanoes originate from melting caused by subduction of the Juan de Fuca Plate (see figure 6.26). Several million years ago, the region at the western edge of the Columbia Plateau rose, forming a long, broad arch. Some of the oldest lava flows folded at that time. Younger flows and ashfalls then covered this arch. The large composite cones of Mount Hood, Mount Lassen, Mount Adams, and Mount Rainier (see figure 6.26), formed at this time. The nearly perfect shape of these easily eroded cones indicates their youth. Many of them have been active in historic times. Mount Lassen was active in 1913–1917; Mount Baker heated up in 1978; both Mount Rainier and Mount Hood have shown signs of warming, probably caused by rising magma; and Mount St. Helens erupted in 1980.

Mount St. Helens, Washington Long before the eruption at Mount St. Helens started (see figure 6.26), studies of the lava flows and ash deposits around the volcano helped geologists identify it as the volcano in the Cascades most likely to erupt. Activity at Mount St. Helens in April 1857 was the last of a long series of eruptions at 100- to 150-year intervals that extends over the past few thousand years. Despite the danger, and the close monitoring by the U.S. Geological Survey, sixty-five people died in the blast of the 1980 eruption.

Seismic activity beneath the volcano began in March with an earthquake 5 kilometers (3 miles) below the surface. Increasing seismic activity continued for a week following the initial shock. By March 24, low-magnitude earthquakes were occurring every

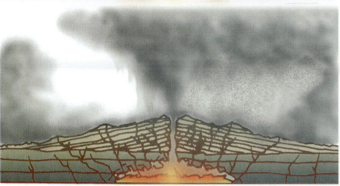

Figure 7.14 Steps in the Evolution of Crater Lake. A large composite cone rose over a long period of time as a series of eruptions added to its mass. Eventually, a major eruption spread pumice over a vast area around the volcano. As much as 70 cubic kilometers (17 cubic miles) of lava erupted. As this eruption came to an end, the cone collapsed into the space below the summit area.

few minutes. Several days later, eruptive activity began with an explosion audible from 15 kilometers (9 miles) away. A cloud of steam and ash rose from the summit. When the air cleared, a new crater about 70 meters (230 feet) in diameter appeared inside the large summit crater (see figure 7.10). The next day, a second explosion sent a cloud of dust and steam 2 kilometers (1.2 miles) high, and an avalanche of ash flowed down the northwest flank. Observers found

two arc-shaped fissures, 1.5 and 5 kilometers (0.9 and 3 miles) long, running east-west across the summit. By dawn, a dense ash cloud rose 3 kilometers (1.8 miles) above the top of the volcano, and an explosive outburst blew blocks from the crater. As meltwater from glaciers near the summit mixed with ash, small mudflows formed and flowed down the northeast flank. Engineers lowered water levels in Swift Reservoir 24 kilometers (15 miles) southeast of Mount St. Helens to accommodate possible meltwaters.

In early April, ash and steam eruptions continued about once every three hours. Two new craters merged to form a larger one about 350 meters (1,150 feet) across. Minor eruptions of ash and steam continued but with diminishing intensity. Sometimes, large blocks of ice that had fallen into the crater blew back out. Tiltmeters installed on the sides of the mountain showed erratic ground movements. Generally, the area rose before the minor eruptions. Later in April, explosive activity ceased, but steam continued rising from the new crater until early May. Comparison of a 1952 topographic map with new maps reveals that the upper part of the volcano rose as much as 100 meters (330 feet). Careful topographic surveys revealed that one area rose 6 meters (20 feet) between April 24 and 29. The two fissures identified earlier defined a graben across the summit area, and the north flank of the volcano rose and moved northward.

Park rangers closed the Spirit Lake area north of the volcano because major ice avalanches were falling from glaciers near the summit that had been severely broken up by the uplift of the north side. The bulge on the north flank continued to grow in the next weeks, and steam came out of the vent, but otherwise the volcano remained dormant until May 18.

On the morning of May 18, a bulge on the north flank of the volcano began to grow much more rapidly. An earthquake occurred as the oversteepened slopes slumped down the side of the volcano. This slumping released the pressure on gases below the volcano and triggered an explosion that blew away much of this northern side of the summit. People heard this explosion 350 kilometers (217 miles) away. It sent a cloud of ash 15 kilometers (9 miles) into the atmosphere. The eruption had three distinct phases: (1) a directed blast threw hot ash out of the north side of the cone and blew down trees up to 24 kilometers (15 miles) away from the summit. (2) A pyroclastic flow (see figure 7.15) and a landslide carried material from the north flank of the volcano across the lower slopes and down the Toutle River for 28 kilometers (17 miles), burying the riverbed as deep as 60 meters (197 feet) in places. (3) Flows of pumice, containing some fresh magma, erupted through the break in the north flank. Prevailing winds carried ash from this May 18 eruption east where 10 to 12 centimeters (4 to 5 inches) of ash fell 140 kilometers (87 miles) downwind. In Spokane, Washington, 500 kilometers (310 miles) northeast of the summit, visibility dropped to a few feet.

After the eruption, the former summit area of Mount St. Helens was 100 meters (328 feet) lower (figure 7.18). A lava dome formed in the crater, and most subsequent eruptions occurred around this dome. Because lava domes like this act like plugs, bottling up lava and gases below the volcano, geologists continue to monitor it closely. Future activity here remains uncertain.

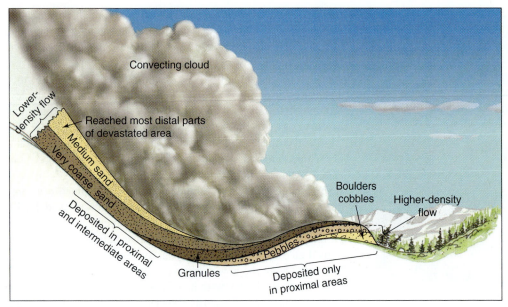

Convecting cloud

Lower-density flow

Reached most distal parts of devastated area

Medium sand

Very coarse sand

Deposited in proximal and intermediate areas

Granules

Pebbles

Deposited only in proximal areas

Boulders cobbles

Higher-density flow

(a)

(b)

Figure 7.15 Pyroclastic Flows. (a) Ash flows form when a cloud of ash and coarser pyroclastic debris sweep downslope from a volcanic eruption. These flows may carry fine clastic materials many miles away from the volcano, destroying everything in their path. (b) This pyroclastic flow moved down the slopes of Mt. St. Helens in 1980. These flows may move at speeds over 100 km/h (60 mph) and reach temperatures in excess of 400°C (750°F).

Incandescent Cloud Eruptions

Mount Pelée This volcano, situated on the island of Martinique in the West Indies, was the site of a devastating eruption in 1902. This eruption consisted mainly of the emission of a cloud of extremely hot, heavier-than-air gases and hot ash particles. The name **Peléan** refers to this type of eruption. Such eruptions are common in island arcs. Residents of St. Pierre, located along the coast below the mountain, knew that Mount Pelée had previously been active, but eruptions were rare, and none had caused serious damage. So residents were not seriously alarmed when in April 1902, activity started with formation of fumaroles in the upper valley of the volcano. A few weeks later, ash came out of the crater, and residents smelled sulfur gases. Late in April, explosions began, and increasing amounts of ash fell on St. Pierre (figure 7.19). The population of the island became uneasy, but officials reassured them.

By the morning of May 8, activity had subsided, and only a thin cloud of smoke was rising from the crater. Suddenly, at 8:00 A.M., the volcano exploded in four great bursts. A huge black cloud, a nuée ardent, spread out from the volcano. Because the cloud contained fine drops of lava and hot ash, it was heavier than air and consequently flowed down the slope of the 1,340-meter- (4,400-foot) high mountain, engulfing the city and port in two minutes. The hot gases killed almost the entire population of the city, estimated to be about 30,000, in a matter of minutes. The only survivors were people on ships in the harbor, a prisoner confined in a dungeon, and a cobbler in his shop.

Mount Pinatubo, Philippines Before its climactic eruption in June of 1991, Mount Pinatubo was one of many dormant stratovolcanoes located around the Pacific. It had last erupted over 400 years ago. When the recent eruption came, over 300 people died, mainly

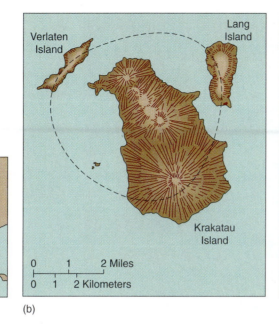

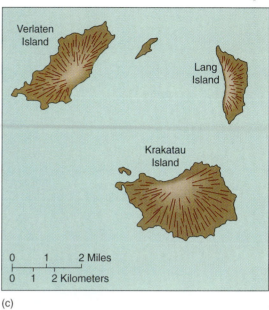

(a)

(b)

(c)

Figure 7.16 **Krakatau Eruption.** (a) Locator map showing Krakatau. (b) The islands before the eruption of 1883. The dashed line indicates the area affected by the explosion. (c) The present configuration of the islands at Krakatau.

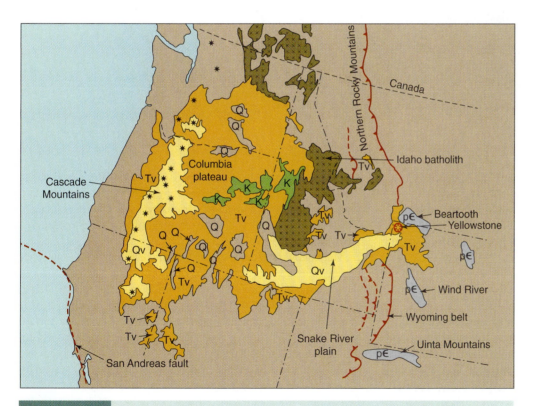

Figure 7.17 **Distribution of the Volcanic and Plutonic Rocks of the Northwestern States.** The dark pattern identifies large batholithic intrusions of Mesozoic age. Orange and yellow patterns define Tertiary and Cenozoic lava fields. The asterisks indicate the location of Quaternary volcanoes.

(a)

Figure 7.18 The Mount St. Helens Eruption. (a) During the eruption of Mount St. Helens, a major explosion blew away the top and much of one side of the volcano. The blast blew down trees in the foreground and covered much of the flat area in the middle ground with pyroclastic materials that came down the slope of the mountain.
(b) Profiles across Mount St. Helens show the changes following the eruption in May 1980.

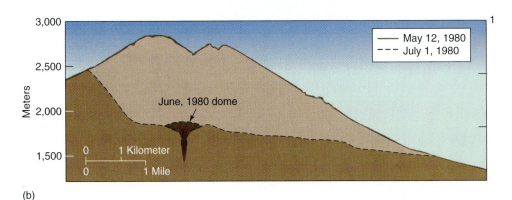

(b)

from the collapse of ash-laden roofs (figure 7.20), despite early warning and evacuation of nearly 60,000 people. Geologic surveys made long before Mount Pinatubo's latest eruption revealed that the cone is surrounded by large fans of young (600 to 8,000-year-old) pyroclastic flows and volcanic mudflows that formed when rains fell on the ash during earlier eruptions. The modern volcano is located within the caldera of an older and deeply eroded stratovolcano. As in the case of Krakatau, volcanism persists over long periods.

In 1991, activity began with a series of steam-induced explosions coming from a line of small vents along one face of the volcano. The next day, a line of new vents opened along an extension of the first vents. A portable seismograph installed near Pinatubo recorded about 200 small earthquakes during its first day of operation. Evacuation of people located within a 10-kilometer (6-mile) radius of the volcano started. A team of U.S. Geological Survey volcanologists called in to help evaluate the situation installed a more extensive network of seismographs and tiltmeters (see fig-

ure 6.31) designed to detect small changes in the slope of the ground. Over the next two months, they prepared a volcanic hazard map showing areas subject to various types of danger and established a system to alert local government officials and residents of areas thought to be in danger.

Maps showing the location of earthquakes displayed a cluster of activity about 5 kilometers (3 miles) north of the summit. Most of these earthquakes had focal depths of 2 to 6 kilometers (1.2 to 3.7 miles). About the middle of May, the quantity of sulfur dioxide increased tenfold, indicating that new magma was rising toward the surface. In late May, the location of the earthquake cluster moved toward the active steam vents. During this time, harmonic tremors associated with movement of magma in a feeder system indicated that an eruption was imminent. Then in early June, the east flank of the volcano began to tilt just before a lava dome extruded near one of the steam vents. At this time, earthquake activity stopped, but the dome continued to grow, indicating the presence of a magmatic system at a shallow depth. Fearing an imminent

eruption, authorities increased the size of the area being evacuated, and on June 10, the U.S. military evacuated personnel at Clark Air Base. On June 12, an explosive eruption blew a column of ash 19 kilometers (12 miles) into the stratosphere. In the following days, a series of strong explosions shook the area; huge pyroclastic flows engulfed Clark Air Force base. Satellite images showed the plume extending 35 to 40 kilometers (22 to 25 miles) into the air.

Pyroclastic flows spread down the volcano's slope into valleys. In one valley, the flow traveled 16 kilometers (10 miles) (see figure 7.20). A typhoon passed 50 kilometers (31 miles) north of Pinatubo. This storm carried coarse ash far from the eruption and dumped huge quantities of rainfall on the deposits, remobilizing the pyroclastic debris and debris flows. Until the end of August, ash continued to pour through the vent system. At times, it rose over 300 meters (1,000 feet) and spread a blanket over the surrounding region. Because these eruptions coincided with the yearly wet season, extensive mudflows buried some lowland towns and washed out all bridges within 30 kilometers (19 miles) of Pinatubo.

By the time volcanic activity stopped, over 100,000 people had moved into refugee camps, and over 700 had died because of either the eruption or health problems in the camps. Without the early warnings, thousands more would have perished.

Persistent (Strombolian) Eruptions

Stromboli, located at the northern end of the Aeolian Islands in the Mediterranean Sea, typifies the persistent, moderate, and uniform type of volcanic activity called **Strombolian.** Aristotle described

its activity much as volcanologists do today. Because a red glow follows each explosion on Stromboli, some call it the "lighthouse of the Mediterranean."

Stromboli has a near-perfect cone rising about 3,500 meters (11,500 feet) from the sea floor and standing 915 meters (3,000 feet) above sea level. Although the crater has steep-sided walls around most of its perimeter, one side is lower than the others. A prehistoric eruption probably blew that side away. When current eruptions happen, lava and ash flow down a steep slope on this breached side of the crater. Large blocks that have fallen from the surrounding cliffs protrude from the crater floor. The crater itself has three vents in which lava rises and falls by as much as 6 meters (20 feet). During a typical eruption, large bubbles form in the rising lava. When the bubbles burst, clouds of steam and glowing fragments rise into the air; then the lava subsides. At other times, lava flows almost continuously, cascading down the slope toward the sea. This type of relatively passive activity prevails as long as the vents remain open, but when the vents become temporarily blocked by solidified lava, more violent explosions follow, breaking up the material in the vent and producing ash-laden clouds.

Eruptions of Basaltic Lava

Compared with Krakatau, volcanic activity in Hawaii is quiet and passive. These two types of activity are as different as a dynamite explosion and the oozing of hot tar out of a hole in a tank. In Hawaii, pressure in the magma is sometimes sufficient to spurt lava upward in fountains several hundred feet high, but generally, pressure does not build up enough to cause violent explosions. Most Hawaiian eruptions involve the extrusion of highly fluid basaltic lava that rises and flows out of either a central vent or fissures located on the flanks of the broad volcano. Lava often emerges from numerous low fountains along a fissure, creating a curtain of red-hot lava (see figure 7.8). At other times, it rises within a pit-like crater, a caldera, forming a lava lake (see figure 2.15). The frequent eruptions in Hawaii have been so passive that the U.S. Geological Survey maintains a volcano observatory at Kilauea crater on the Island of Hawaii (figure 7.21). This research laboratory, situated on the rim of the huge pit-like caldera formed on the flank of this massive shield volcano, provides a ringside seat for many dramatic eruptions like the 1959 eruption at Kilauea.

Lava Plateaus

Low-viscosity lava once covered vast areas in India, Antarctica, and the northwestern United States. Most of these flows are basaltic in composition, and some are thousands of feet thick. The lava poured from long fissures rather than from central vents.

The Columbia Plateau This region of a quarter-of-a-million square kilometers in the northwestern United States and southern British Columbia is a vast area built up by the extrusion of a succession of lava flows (see figure 7.17). Although individual

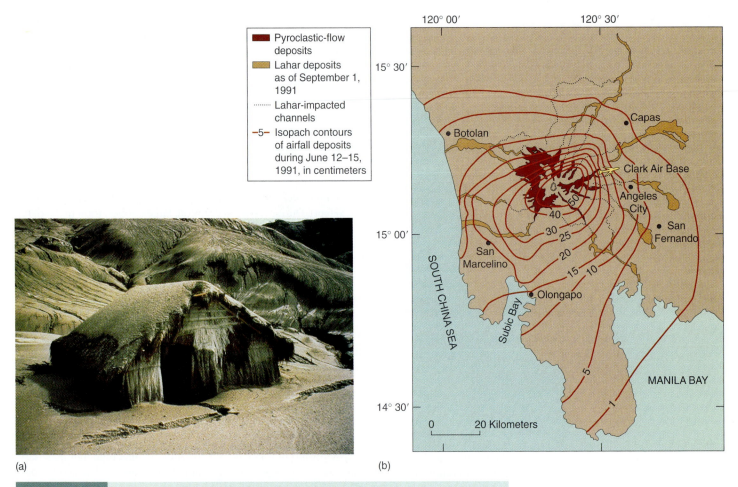

(a)

(b)

Figure 7.20 **Mount Pinatubo Eruption.** (*a*) Many who died during the Mount Pinatubo eruption were killed when the roofs of their homes collapsed under the weight of ash and mud. (*b*) This map illustrates the distribution of different thicknesses of Pinatubo's 1991 pyroclastic deposits in west central Luzon and the extent of valley-filling pyroclastic flows around Pinatubo. Isopachs are lines that connect points where the thickness of the ash is the same.

flows are only a meter (a few feet) thick, the accumulation of flow atop flow has resulted in a pile of flows more than 1.6 kilometers (a mile) thick. The most recent activity in the region occurred along the eastern margin of the Cascade Range. Volcanism began 38 to 53 million years ago in the Eocene epoch, but younger flows, 7 to 27 million years old (Miocene age), cover most of the plateau. Even younger volcanics extruded over the last 2 million years during the Pleistocene cover most of the Cascade region and extend eastward along the Snake River and Yellowstone areas. Because individual flows are so extensive, some covering over 160-square-kilometer (60-square mile) areas, and uniform in thickness and composition, it is believed that these flows originated as fissure eruptions. As flow after flow of these fluid basaltic lavas poured out of the fissures, they filled low areas and moved toward the edge of the large structural basin where the Columbia Plateau resides.

Snake River and Yellowstone Yellowstone National Park (figure 7.22) lies east of the Columbia Plateau lava fields and at the eastern end of a strip of volcanic rock lying along the Snake River in Idaho (figure 7.23). The volcanic activity along this strip marks the track of the crust over a mantle hot spot. Because the magmas extruded are rich in elements concentrated in continental crust, heat and solutions rising from the mantle must have melted part of the continental crust. Unlike the basaltic lavas of the Columbia Plateau area, andesitic and rhyolitic flows are prominent components of the volcanics in Yellowstone Park.

The record of volcanic activity in the Yellowstone area indicates that it has been continuous over many millions of years. Eruptive centers of 36 to 53 million years old (Eocene age) are present in the high mountains both east and west of Yellowstone National Park. At that time, basalt and andesite, ejected as angular fragments called **volcanic breccias,** spread across the land in

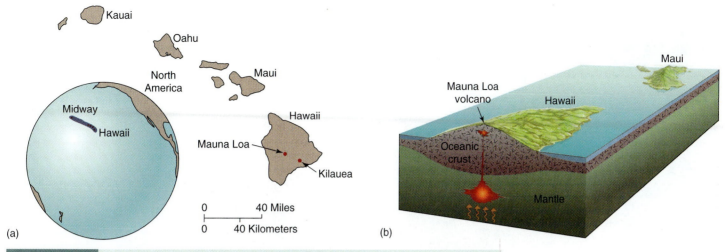

(a)

(b)

Figure 7.21 **Volcanic Activity in Hawaii.** (*a*) Only the big island of Hawaii has active volcanoes. Geologists think the islands formed over a mantle hot spot as the Pacific Plate moved northwest over the hot spot. (*b*) Schematic representation of the Hawaiian hot spot. The magma rises from great depth. It has been tracked with seismographs from depths of at least 60 kilometers (37 miles). As the Pacific Plate moved to the northwest over the Hawaiian hot spot, a ridge of volcanic material formed on the sea floor, building the Hawaiian ridge.

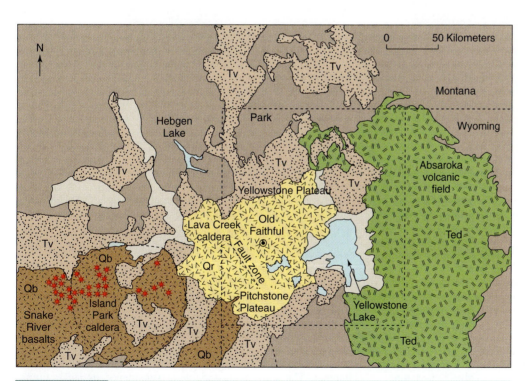

Figure 7.22 Sketch Map of the Yellowstone National Park and Surrounding Areas Showing **Volcanic Rocks.** Tv is Tertiary volcanics; Qb is Quaternary basalt flows; Ted is Tertiary volcanic breccia deposits; Qr is Quaternary rhyolite flows; asterisks indicate small volcanoes.

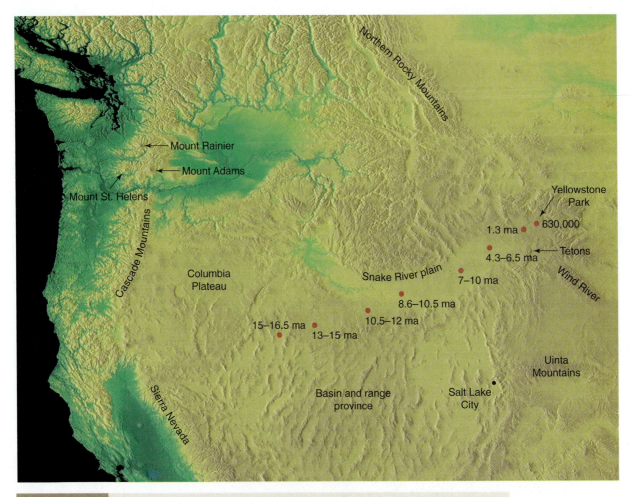

Mount Rainier
Mount Adams
Mount St. Helens
Cascade Mountains
Northern Rocky Mountains
Yellowstone Park
1.3 ma • 630,000
4.3–6.5 ma
Tetons
Columbia Plateau
Snake River plain
7–10 ma
Wind River
8.6–10.5 ma
10.5–12 ma
15–16.5 ma
13–15 ma
Uinta Mountains
Sierra Nevada
Basin and range province
Salt Lake City

Figure 7.23 Hot Spots in the Snake River and Yellowstone Area. The large dots indicate the position of the hot spot that the United States has moved over. The ages beside the dots indicate the approximate ages of the volcanic activity at each spot.

huge debris flows. Other more violent eruptions produced welded tuffs of rhyolite, which also blanketed the countryside. During the last few million years, faulting in the Rocky Mountains has up-lifted large blocks of the crust and displaced the older volcanic deposits.

Volcanic activity in Yellowstone (see figure 7.22) differs from that in the Cascades (see figure 7.17). Stratovolcanoes, of the type formed over the subduction zone in the Cascades, are absent in the Yellowstone area. Both regions have been sites of violent volcanic eruptions, but individual eruptions in Yellowstone covered larger areas and involved lavas of higher silica content than those in the Cascades. Several welded tuffs formed during particularly violent eruptions. At other times, vast quantities of highly viscous rhyo-lite poured from fissure zones along the crest of the Yellowstone Plateau. These flows extend almost 50 kilometers (31 miles) along former river valleys and contain about 400 cubic kilometers (96 cubic miles) of lava. Distinctive concentric ridges mark the surfaces of these flows. Their leading edges, still well preserved and moderately steep, indicate that the lava was highly viscous.

Some of these flows, located on the Pitchstone Plateau, are obsidian.

Most geologists think that volcanic activity in Yellowstone could become active again. The rocks in the park record three cycles of volcanism over the past 2 million years. The most recent cycle started 1.2 million years ago with the formation of two ring-shaped surface fracture zones above and around deep magma chambers. About 600,000 years ago, an explosive eruption of rhy-olite was followed by the collapse of the surface, and the formation of two large calderas. In 1990, leveling studies in the park revealed that part of the Yellowstone caldera was rising at a rate of about 14 millimeters (0.5 inch) per year. This rapid rate of uplift was probably caused by the rise of magma in the underlying magma chamber. The same area underwent subsidence in 1987. Future movements are uncertain, but it seems probable that volcanic activity will eventually affect the Yellowstone area again. When such activity will resume is impossible to predict. Hot magma in the chamber beneath Yellowstone National Park is responsible for the current high temperature in the ground—over a hundred times normal in places—and for the hot springs and geyser activity for which the park is famous.

The preceding case histories, from the explosive eruptions at Krakatau to the outpourings of lava in the Columbia River Plateau, demonstrate the wide range of styles exhibited by volcanoes and the impacts eruptions have on Earth environments. Volcanism originates in Earth's interior, and it affects all Earth systems.

THE ROLE OF VOLCANIC ACTIVITY IN EARTH SYSTEMS

Volcanic activity is the most direct connection between Earth's interior and systems that operate at the surface (figure 7.24). Volcanic activity results from processes caused by movement of plates and to convection that extends deep into the mantle.

Volcanism in the Plate Tectonic System

Unlike the volcanic activity during the early stages of Earth history when magma was close to the surface everywhere, most modern volcanoes are closely related to plate boundaries. Magma rises to the surface above subducting slabs of oceanic crust, above the rising limbs of convection cells, or as plumes ascending from deep-seated hot spots in the mantle. Volcanic activity is a critical part of the plate tectonic system. At spreading ridges, rising magma lifts the oceanic crust while cooler oceanic crust subsides. The combined effect creates the slope away from ridge crests. This slope facilitates the movement of the oceanic crust away from ridges. In subduction zones, volcanic activity drives part of the rock cycle that recycles materials in the lithosphere.

Volcanic Effects on the Physical Climate System

Although the character of igneous processes has remained largely unchanged since the early stages of Earth history, the importance of molten matter in Earth systems has changed dramatically. Most scientists think that during those early times, before any of the rocks exposed at the surface today formed, molten materials covered Earth's surface. Volcanic processes dominated Earth systems. The crust consisted of a thin layer of solidified lava and rafts of low-density minerals crystallized at depth in magma that rose to the surface. Lava flows filled the depressions formed where large asteroids hit the surface. Volcanic gases streamed from fractures in the crust into the early atmosphere, providing the primary source of water vapor and other gases that compose the atmosphere. By comparison, modern volcanism plays a much subdued, but still important, role in the environment.

About 100 volcanoes have erupted in recent years (see inside cover). The fact that many more could become active at any time provides ample evidence that generation of magma is taking place inside Earth. Most of this magma never reaches the surface, but volcanic activity is occurring someplace on Earth almost every day. It not only influences the lives of people who live nearby, it also changes the

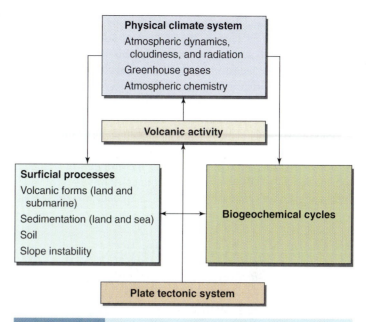

Figure 7.24 Schematic Diagram Illustrating the Place of Volcanism in Earth's Environmental Systems.

landscape and the local environmental systems. Gases and dust from eruptions change local weather patterns, bury or reroute streams, and destroy the habitat of indigenous plants and animals. The effects of this process are far-reaching. Volcanic ash and dust may be blown high into the atmosphere and transported great distances. Eruptions under water may generate tsunami, and massive eruptions alter weather patterns over large portions of Earth's surface.

Dust and ash ejected into the atmosphere reduce the amount of solar radiation reaching Earth's surface. Some of the gases, especially sulfur gases, become fine droplets or aerosols that absorb radiation. This warms the upper part of the atmosphere but cools the lower part of the atmosphere. Carbon dioxide is another volcanic gas that has critical effects on the climate by absorbing radiation from Earth's surface and warming the atmosphere. These gases are held within the magma below the ground, but once a conduit to the surface forms, the gases escape.

Sulfur content seems to be a critical ingredient in these weather-altering clouds, and volcanoes differ tremendously in the amount of sulfur they emit. NASA estimated that the cloud emitted by the Mexican volcano El Chichon in 1982 contained over 3 million tons of sulfur dioxide. By comparison, Mount St. Helens produced only a tenth of that amount. During violent eruptions, sulfur dioxide, dust, and sulfur may rise into the stratosphere. These dust and gas clouds slowly circulate around Earth at that altitude, far above the clouds and rains that normally cleanse dust from the lower atmosphere. Because little mixing of air occurs between the higher stratosphere and the lower atmosphere, heat contained in the stratosphere has little effect on the temperature of Earth's surface. However, sulfur dioxide in the stratosphere changes to sulfuric acid, which absorbs solar radiation. Thus, the stratosphere traps solar radiation high above ground level, which reduces the amount of solar radiation that reaches the lower part of the atmosphere and the ground. This causes cooling of the lower levels of the atmosphere.

Torrential rains that accompany many volcanic eruptions provide dramatic evidence of the connection between eruption and the weather near the volcano. These weather effects can extend far beyond the volcanic center. Frigid temperatures in North America and Europe followed the eruptions of both Tambora in 1815 and Krakatau in 1883. The year following the eruption of Tambora became known as "the year without a summer." It was a year of widespread crop failures caused by an unseasonably cool growing season.

VOLCANIC HAZARDS

Volcanoes constitute a serious and widespread hazard to human life, especially around the perimeter of the Pacific and in the Caribbean and Mediterranean (table 7.1). The blast effects of volcanic explosions, falling debris, and hot ash and gases pose the most immediate hazard, but the hazards associated with a volcanic eruption are far-reaching. Deposits of welded tuffs cover most of the central portion of the North Island of New Zealand. Similar deposits of rhyolite cover large portions of several states in the southwestern United States. These deposits indicate that large volcanic eruptions affected these areas during the Quaternary period. If such an eruption, covering several states with incandescent gases and ash, took place today, the loss of life and property would be immense. Unfortunately, similar events may occur at some time in the future.

The great clouds of water vapor usually present over erupting volcanoes pose a different type of hazard. Water vapor condenses as it rises, creating torrential rainfalls over volcanoes. The rainwater mixed with the ash on the flanks of the volcano produces debris slides and mudflows that extend far beyond the volcano.

Ashfalls and lava flows may temporarily dam streams. When water flows over these dams, they collapse or erode quickly. The river washes the ash away, and a wall of water floods the river valley below. People living near an erupting volcano have to cope not only with the ash and lava at the volcanic center, but also with landslides, floods, torrential rains, blocked roads, and failure of communication systems.

Table 7.1	Hazards Associated with Volcanic Eruptions

Falling materials—falling pyroclastics and ballistic projectiles
Flowage processes
    Pyroclastic flows
    Laterally directed blasts
    Debris avalanches
    Floods
    Lava flows
Explosions
Volcanic gases and acid rains
Indirect hazards
    Earthquakes and ground movements
    Tsunami
    Posteruption erosion and sedimentation problems
    Posteruption famine and disease

Source: (After Tilling, 1989)

Explosive eruptions or collapse of volcanoes in the ocean can create tsunami. These enormous waves travel rapidly out from their source area and can cross the Pacific Ocean in a day, causing destruction as they reach shores around the ocean. In extreme cases, tsunami destroy trees and structures tens of feet above sea level.

Volcanoes warrant careful attention for the hazards they present. To be forewarned, to know what type of eruption is likely, and to know what areas around a volcano are most vulnerable to the different side effects of volcanic activity are the best defenses against eruptions. Several modern techniques make it possible to monitor volcanoes and predict activity. These techniques include seismographs to record seismic tremors related to the movement of magma. Tiltmeters allow volcanologists to detect changes in slope due to swelling of the summit as magma moves toward the surface. Satellite images recording in infrared bands make it possible to identify heat sources on volcanoes. Changes in the chemistry of the gases escaping from volcanoes also provide clues about impending activity. Scientists are hopeful that data collected from these types of monitoring will prevent tragedies like those that destroyed Pompeii, the city near the famous volcano Vesuvius in Italy, and St. Pierre, which is located on the island Martinique.

SUMMARY POINTS

1. Volcanoes form where gases or liquids that compose magma break through Earth's surface. Gases escape into the atmosphere, but lava and all but the finest pyroclastics remain near the vent and build volcanoes.
2. Geologists classify lava according to its composition. The lava from which rhyolite forms is rich in silica; basaltic lavas contain low silica; andesite is intermediate in composition.
3. Temperature, composition, and the amount of gases present affect the viscosity of lava. High temperature, low silica content, and large quantities of gases tend to produce low-viscosity lava. In general, basaltic lavas have low viscosity; rhyolites have high viscosity.
4. The fragmental materials erupted from volcanoes consist of pieces of the cone, feeder system, and pieces of lava that cool

and solidify in the air. Because they cool so fast, these pieces are fine-grained or even glasses. Pyroclastic particles such as dust, ash, and blocks differ from one another in size. Masses of angular fragments form volcanic breccias. The term cinder is a general term applied to small fragmental materials. Tuffs are rocks formed by the consolidation and compaction of ash. If the pieces of ash are sufficiently hot, they may fuse together, forming a welded tuff.

5. Pumice and scoria are fine-grained or glassy rocks containing holes formed by gases. These are derived from silica-rich (rhyolite) and silica-poor (basalt) lavas. They may form as pieces of lava blown into the air or at the top of lava flows as gases escape, leaving the vesicles.

6. The type of eruption depends on the quantity of gases present in the magma and whether or not they can escape easily. When gas pressure builds up, violent explosions are possible. In general, volcanoes that erupt rhyolite and andesite are more likely to have violent explosions than those that erupt basaltic lava.

7. Geologists commonly describe volcanoes based on the resemblance of their activity to that of Krakatau (explosive), Pelée (nuée ardente or incandescent clouds), Stromboli (persistent moderate explosive), or Hawaii (passive lava flows).

8. The shape of both lava flows and volcanic cones is generally related to how viscous the lava is. Rhyolite is highly viscous and forms bulbous extrusions with steep slopes. Andesite volcanoes often form composite cones. Basalts tend to form shield volcanoes or lava plateaus like the Columbia Plateau. Cinder cones form when pyroclastics are the primary eruptive material. Composite cones form when volcanoes erupt both pyroclastics and lava.

9. The shape of a volcano depends on the way magma rises to the surface and the composition of the lava. Most eruptions occur over a central vent or along a long crack (a fissure eruption). After volcanic activity ceases, erosion quickly strips away the loose pyroclastics, exposing plutons, such as the dikes that formerly fed fissure eruptions or the magma frozen in feeder pipes like that at Shiprock, New Mexico.

10. Pit-like depressions called calderas form where the top of a volcano collapses into the empty chamber left underground after magma has erupted or where a large crater is left after a violent explosion blows away the top of the volcano.

11. Modern volcanic activity occurs in three main areas: along oceanic ridges where plates are moving apart, over subduction zones in island arcs and young mountain belts where plates are moving toward one another, and over hot spots in Earth's interior.

12. The onset of volcanic activity is commonly preceded by earthquakes and changes in ground level and slope near the volcano. The composition of the lavas from earlier eruptions and the location or tectonic setting of a volcano provide information needed to predict the type of eruption.

13. Dust and gases carried into the stratosphere during violent eruptions affect the weather. These gases absorb radiation and can cause abnormal cooling of Earth's surface.

REVIEW QUESTIONS

1. How do eruptions of volcanoes at Hawaii and other locations on oceanic crust differ from eruptions of volcanoes in island arcs like Pinatubo?
2. Why are heavy rains commonly associated with volcanic eruptions?
3. How do the shapes of volcanoes that erupt mainly pyroclastic materials differ from those that erupt rhyolite and basaltic lavas?
4. How can volcanic eruptions affect global climate?
5. What dangers does a volcanic eruption of the type found in island arcs pose to people who live within 25 kilometers (15 miles) of the volcano?
6. In what plate tectonic settings are the most active volcanoes located?
7. How do calderas form?
8. Why do some volcanoes explode, while others emit only gaseous clouds?
9. How do the lavas erupted by volcanoes in ocean basins differ from those in islands around the border of the Pacific?
10. How can so many different types of igneous rocks originate from a single basaltic magma?
11. Could as many different types of igneous rocks originate from a granitic magma?
12. Why do volcanoes differ in shape and the type of material they erupt?

THINKING CRITICALLY

1. What steps should the city of Seattle take to protect its citizens from a potential eruption of Mount Rainier?
2. What hazards other than lava flows are associated with volcanic eruptions?
3. How might you recognize the type of eruptions associated with a dormant volcano?

KEY WORDS

aa	pyroclastic flow
ash flow	pyroclastics
batholiths	shield volcano
caldera	stocks
central vent	stratovolcano
cinder cone	Strombolian
composite cone	tuff
fissure eruption	vesicles
fumaroles	viscosity
lahars	volcanic breccias
nuée ardents	volcanic neck
pahoehoe	welded tuff
Peléan	

unit three **Earth's Physical Climate System**

The Sea Floor and Marine Environments

Starfish occupy a tidal pool at Big Sur, California.

Chapter Guide

Water covers almost an entire hemisphere of Earth's surface. But, ironically, as vast as oceans are, they are one of the least explored parts of Earth. Satellite images of Mars and Venus provide more precise information about the surface of those planets than is available for the floor of the oceans. Only in fairly recent years has equipment been designed that enables scientists to study the ocean's floor. Studies of the ocean basins have provided most of the information on which the theory of plate tectonics is based. The oceans have long provided a significant amount of the world's food supply, and like those in the Caribbean and at the Great Barrier Reef in Australia, they contain some of the most exotic and beautiful environments known anywhere. The greatest mass movements that occur on Earth take place in the oceans in the form of landslides and their resulting turbidity currents. Chapters 3 and 5 introduced the sea floor; however, this chapter provides greater insights about the landscapes on the sea floor and the sources of the sediment that covers the ocean bottoms. It shows the close relationship between the geology and ecology of the oceans and touches on some of the resources held in the oceans.

INTRODUCTION

Marine environments occupy over 70% of the surface area of Earth—far more than all terrestrial and transitional environments combined. The size of the oceans is particularly impressive to astronauts as they encircle Earth in space (figure 8.1). Viewed from space high over the central Pacific Ocean, it is clear that water covers almost an entire hemisphere. This great reservoir of water distinguishes Earth from all other planets in the Solar System. Through interaction with the atmosphere, the water in the oceans plays a critical role in governing Earth's climate, and the oceans provide the habitat for thousands of species of plants and animals. Gradually, scientists are uncovering the mysteries of this vast realm and its highly varied environments.

The oceans have held many surprises for those who have explored and studied them. In 1938, a South African fishing trawler hauled in a large **coelacanth,** a fish that can reach 2 meters in length. Until that event, no one had ever reported seeing a live coelacanth. Even fossils of coelacanths were unknown in rocks younger than 60 million years old. Geologists had assumed that coelacanths became extinct long ago. In 2000, live coelacanths were filmed off the coast of South Africa. Other surprises awaited the scientists who began intensive exploration of the oceans following the Second World War. Until the 1950s, most scientists thought the sea floor would be a featureless plain. Instead, echo soundings reflected off the sea floor gradually revealed a highly varied seascape. Prior to the 1960s, most scientists thought the ocean basins had formed during the early stages of Earth's history and had remained essentially unchanged since then. As more data accumulated, it became clear that the cratons are ancient, but the sea floor is an active crustal element and little, if any, of it is older than 165 million years. In 1977, scientists diving in a submersible vessel off the Galápagos Islands in the Pacific were amazed to find biological communities of long tube worms, white clams, and eyeless shrimp living around hydrothermal vents from which clouds of hot, sulfurous water emanate (figure 8.2). No one expected such abundant and diverse life on the sea floor in deep, oxygen-poor water. Undoubtedly, because less than 1% of the sea floor has been studied in detail, more surprises about marine environments await future explorers.

Scientists first studied the deep parts of the ocean using nets and dredges towed behind ships. They succeeded in collecting a variety of marine organisms, but the deeper parts of the ocean remained a mystery. Then, after World War II, serious efforts to map submarine topography using echo-sounding devices began. At the same time, seismic techniques used to detect atomic explosions also provided insights into the oceanic crust, and piston-coring devices made it possible to bring up sections of core tens of meters long from the sea floor. Using these cores, oceanographers started to reconstruct the history of the oceans based on the record left in sediments.

Diving suits and submarines make journeys for people into the deep possible. Scuba divers can descend more than 33 meters (100 feet). Early submarines could descend much deeper than divers, but they were poorly suited for scientific study. The *bathysphere* was the first vessel designed specifically for study of the

Figure 8.1 **The Blue Planet.** From space the vast expanses of oceans are one of Earth's most distinctive features.

deep sea. It descended to a depth of 1 kilometer but was not maneuverable. Finally, the Swiss engineer Auguste Piccard designed and built the first successful research diving vessel called a *bathyscaphe.* It had both maneuverability and the ability to withstand the great pressure in the deep parts of the ocean. In 1960, the bathyscaphe *Trieste* successfully descended to the bottom of the deepest place in the oceans—the Challenger Deep in the Mariana Trench, about 11 kilometers (7 miles) below the surface of the ocean. The bathyscaphe and newer remotely operated vehicles called R.O.V.s., such as those used to explore the remains of the *Titanic,* are equipped with cameras, arms that can grab samples, and other sophisticated water- and sediment-sampling devices (figure 8.3). These devices allow scientists to study the sea floor, and satellites provide new ways to collect information about the surface of the sea floor. One satellite uses radar measurements that detect slight changes in the elevation of the sea surface caused by submarine topography. Changes in the shape of the sea floor cause slight variations in the force of gravity, and these show up as changes in the height of the sea surface (figure 8.4).

THE SEA FLOOR— OCEANIC RIDGES

Although the presence of shallow water in the middle of the Atlantic had been known for many years, the great length and continuity of the oceanic ridge system was discovered in the 1950s

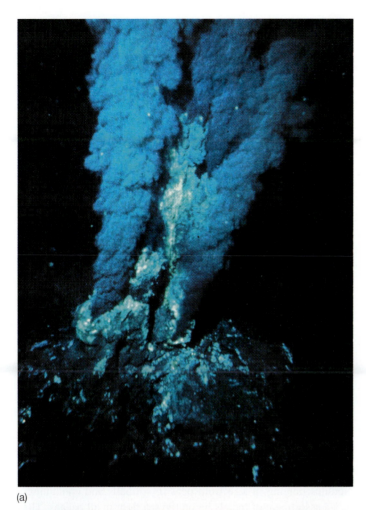

(a)

(b)

Figure 8.2 **Hydrothermal Vents.** (*a*) Three-dimensional image of a cluster of hydrothermal vents located on the Juan de Fuca Ridge in the Pacific Ocean. (*b*) White crabs resting on basaltic rock near a thicket of tube worms at a hydrotherm vent.

(see figures 3.10 and 5.1). This remarkable ridge system is the largest feature in Earth's crust. Along much of its length, waters that are over 5 kilometers (about 3 miles) deep lie on either side of it, and it breaks the surface in a number of volcanic islands, such as Iceland, the Azores, and the Galápagos. Exploration of the ridge system has confirmed that volcanic rocks of basaltic composition form the ridge everywhere and that these lava flows are exposed on the sea floor near the crest of the ridge. As the theory of plate tectonics evolved, the ridges were recognized as places where new oceanic crust is forming and the sites of divergent junctions between plates. Grabens and high-angle faults have shaped the topography of the ridge crest (see figure 5.19). Faults expressed by sharp ridges with cliffs on one side form as a result of extension along the ridge crest. These cliffs remain as topographic features on the flanks of the ridge crest as the sea floor spreads. Some of these escarpments persist until the accumulation of sediment buries them. Escarpments located along transform faults are prominent topographic features in the sea floor. Some of these faults show up as steep valleys that run across the ridge crest. They appear as heavy lines on many maps of the ocean basin as in figures 5.1 and 5.20.

THE SEA FLOOR— ABYSSAL PLAINS

A blanket of sediment covers the flat expanses of sea floor known as the *abyssal plains.* These surfaces are present in every ocean at depths of 5 to 6 kilometers (15,000 to 18,000 feet). In some parts of the ocean, this sediment originates in submarine landslides. In other areas, the sediments are the remains of organisms. Before the advent of plate tectonics, geologists thought that the oceans were permanent features that had formed early in Earth's history. If the ocean basins were primitive features, even at extremely slow rates of sedimentation, layers of sediment many thousands of feet thick should have built up over the several billion years since the first oceans formed. Seismic profiles taken of the sea floor in the abyssal plains revealed a surprisingly thin layer of sediment less than a few thousand feet thick over large areas of sea floor. Drilling confirmed the evidence from the seismic profiles, and it soon became clear that sediment thickness varies from oceanic ridges, where sediment is thin or absent, to passive continental margins, where several miles of sediment have accumulated beneath some continental slopes. Fossils in the sediment make it clear that the oldest sediment in the oceans is no older than the early Mesozoic, about 165 million years. This ancient sediment is buried along passive continental margins like those around the North Atlantic Ocean.

Cores that penetrate completely through the sediment in many abyssal plains are now available. The *Glomar Challenger* obtained many of the first cores. New ships, such as the *JOIDES Resolution,* continue to obtain deep cores. To remain over the drill hole during coring, these ships are especially designed with multiple propellers. Drilling is often passed over in favor of a less-expensive technique called *piston coring* (figure 8.5). These cores, in combination with

Figure 8.3 The Submersible *Alvin.* Submersibles like *Alvin* make it possible to explore the depths of the oceans. *Alvin* carries three people. It is equipped with cameras and can collect specimens. It can explore in water less than 3 kilometers (10,000 feet) depth.

Seamounts and Guyots in the Abyssal Plains

High mountains and submarine ridges rise from the abyssal plains in many places. All of these are volcanic in origin, and some of them have been active recently enough to stand as islands. Some of these volcanoes, such as those on Hawaii, have been active in historic times (see inside cover). Others, like Bermuda, are capped by coral reefs. Some of these volcanoes are located over active mantle hot spots. In the case of the Hawaiian and Emperor seamount ridges, the active volcanic center has shifted as the oceanic crust moved over the hot spot (see discussion in chapter 5). Other volcanic centers formed along transform faults that extend laterally away from ocean ridges.

The tops of many seamounts, especially in the abyssal plains of the Pacific, lie far below the surface of the ocean. The shapes of these seamounts suggest that they are also of volcanic origin, and volcanic rocks have been dredged from some of them, but many of them have flat tops. Sediments collected from the tops of these flat-topped seamounts, called **guyots** (see figure 3.8), indicate that at one time they were at sea level. Their peaks were leveled by wave action, and they were later submerged as the sea floor subsided more than a mile. Some had been atolls when they were at the surface. Others may have subsided too rapidly for corals to maintain the reef top near the surface, and still others occur at high latitudes and are in waters that are now too cold for corals to survive.

Riches on the Deep-Sea Floor

Nodules ranging in size from peas to softballs and containing a broad assortment of metals lie on the sea floor in all oceans. They are especially numerous south of Hawaii in the Pacific and in the southern Atlantic. These bodies, called **manganese nodules,** have long been a curiosity, but as the quantity and types of metals present have become clear, their economic value has attracted attention. Although manganese is a major constituent, these nodules also contain nickel, copper, cobalt, iron, and small quantities of a number of other metals. Billions of tons of metals are present in these nodules, and the quantity of nickel and copper exceeds the amount of these metals known in deposits on land. The metals are present in concentric layers that have grown over time through processes of chemical precipitation. The gases and fluids that rise into the sea from fumerole-like stacks of **hydrothermal vents,** located along oceanic ridges and submarine volcanoes, are likely sources of the metals (see figure 8.2*a*). The distribution of the nodules far removed from ridges indicates that the metals remain in solution in seawater as it slowly circulates across the bottom of the oceans. The nodules are best developed in areas where sediment deposition is extremely slow. Otherwise the sediment would cover the deposits before they could form nodules. Radiometric techniques indicate that the nodules grow at rates of tenths of an inch every million years. Because these rates are even slower than the low rates of sedimentation, it seems likely that the nodules remain on the sea floor as a result of currents or some other process that elevates them.

photographs and dredge samples, make it possible to map the distribution of modern sediments.

Layers of clay, known as **red clay,** cover large areas of the abyssal plains on the Pacific sea floor. The fine sizes of sediment in this clay come from minute pieces of dust, ash, and clay carried by currents. An unusually high content of iron and nickel in red clay suggests that it may contain the remains of burned-out meteorites. Hydrothermal vents (discussed later in this chapter on page 190) from which gases escape on oceanic ridges are also potential sources of much of this metal. Although this clay is present in most parts of the ocean, it forms a distinct layer only in areas far removed from land and in parts of the ocean where the production of organic materials is low. Of all sediments, red clay accumulates at the slowest rate—between 0.04 and 1 millimeter per 1,000 years.

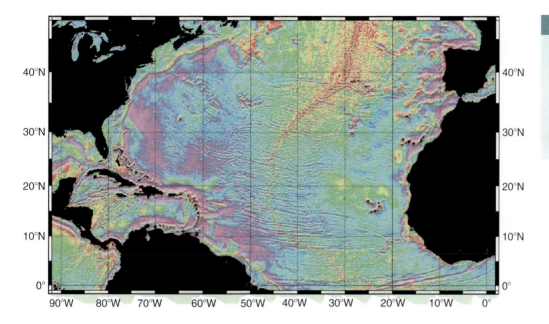

THE SEA FLOOR—PASSIVE CONTINENTAL MARGINS

Shallow, nearly flat surfaces called *continental shelves,* which are generally about 100 meters (330 feet) deep at their outer edges, are present along most continental margins (figure 8.6, and see figures 3.7 and 3.8) (see discussion in chapter 3). At the edge of most shelves, the slope of the sea floor is steeper, commonly about 6° on the continental slope. At depths of about 1 to 4 kilometers (3,000 to 4,000 feet), the slope of the sea floor generally decreases to form continental rises that pass into the flat surface of the abyssal plains at depths over 5,000 meters (16,400 feet).

Interactions between the dynamics of the lithosphere and the effects of wave erosion and deposition clearly shape continental margins. Erosion and deposition have greater impact along passive continental margins where recent crustal deformation is absent or minimal. Continental margins around most of the Atlantic and Indian oceans are of this type. These margins began to form about 165 million years ago when these oceans started to open. At that time, what is now the western margin of Africa fit snugly against the eastern margin of South America and parts of North America (see figure 5.3). The margin of Europe was in contact with the northern part of North America. As the landmass, known as **Gondwanaland,** began to break apart, step faults and grabens defined the edge of the ocean just as they do today along the margins of the Red Sea, which is in an early stage of opening.

The Mid-Atlantic Ridge (see chapter 3 opening image) began to form as the Atlantic opened. Volcanoes and lava flows grew over the rising magma chambers. From that time onward, the continents drifted farther away from the oceanic ridge, and little crustal deformation or volcanism took place along the margins of the Atlantic. The exceptions occurred in the Caribbean and at the southern tip of South America. Both of these places are located between large continents that rotated and moved independently of one another as the Atlantic grew. Along the rest of the Atlantic margin, wave erosion cut into the edge of the continent, and streams flowing from inland carried sediment into the ocean as they do today. The streams drop sediment to the sea floor soon after they enter the ocean. Some of this sediment takes the form of deltas like those of the Mississippi River in the Gulf of Mexico and the Ganges River in the Indian Ocean. Where currents carry the sediment into deeper water, it moves down the

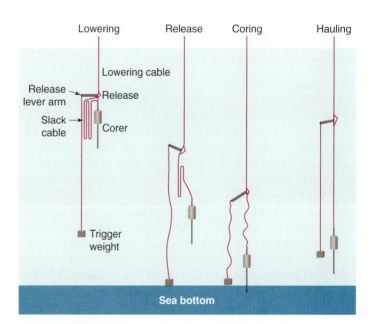

Image of the Continental Margin off the California Coast. Note the flat surface of the continental shelf.

Sediments of this type are common in the Gulf of Maine where continental ice sheets scraped rocks from the continental interior and dropped them out in deep water.

The offshore sediment in tropical regions is dramatically different from that found in polar regions. Glacial deposits and cold-water fauna present at high latitudes give way to remains of warm-water fauna in temperate and tropical regions. In the tropics, sediments composed of fragments of corals may be present. Another distinct environment exists in the Bahama Banks where the strong currents that flow across the ocean from east to west carry warm water with high concentrations of calcium carbonate. The water evaporates as waves break across the shallow platform in the Bahamas, and calcium carbonate precipitates from the saturated water. The deposits build up on pellets excreted by marine organisms and fragments of shells. The resulting sediment, oolitic limestone, is a distinctive sediment characterized by concentric shells of calcium carbonate. Most oolites are spherical and about 1 to 2 millimeters (less than 0.1 inch) in diameter.

Landslides and Turbidity Current Deposits on Passive Margins

Sediments that accumulate at the edge of the continental shelf are not always stable. The importance of this instability became apparent in the 1950s after an investigation designed to find out why cables from North America to Great Britain that had been laid on the sea floor east of Newfoundland and south of the Grand Banks suddenly failed (figure 8.8). Those closest to the shelf broke first and were followed in sequence by those in increasingly deeper water. A small earthquake happened just before the first break occurred, and many assumed that movement on a fault stretched and

continent slopes, and deep currents carry it along the margins where it is banked up to form great wedge-shaped masses that thin both toward the continental shelf and the abyssal plains (see figure 3.11). As the Atlantic opened, the Appalachian Mountains were young. Large quantities of sediment eroded off the high, steep slopes of the mountains were carried into the ocean where the sediment now lies buried beneath the continental shelf, slope, and rise (figure 8.7).

Sediments on Passive Margins

The physical and biological environments vary greatly along continental shelves. The sediments on the shelves reflect the conditions that prevail today and through the Pleistocene. During the ice ages, sea level has varied about 150 meters (500 feet). During low stands of the ocean, streams eroded across the continental shelves and left canyons that are now submerged. These canyons are present on the continental shelf offshore from most major rivers that were located south of the ice sheets during the glacial advances. Areas located farther inland and not submerged bear the same marks of erosion seen above sea level today.

In high latitudes, the ice sheets gouged out large valleys as they advanced across the shelves and deposited materials as they retreated. Continental ice sheets such as the modern glaciers in Antarctica carry debris frozen into the ice. During the last glacial advances, these sheets extended far out into the ocean where the edge of the ice eventually broke off, forming huge icebergs. Once freed from an ice sheet, the icebergs move with the currents until the ice melts. As the ice melts, rocks and other debris in the ice sink to the sea floor. Through this process, large blocks of rock may drop and become incorporated in other types of marine sediments.

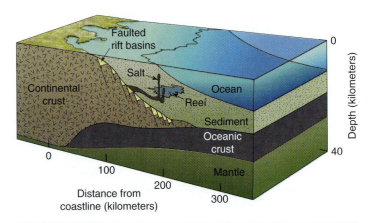

Figure 8.7 Schematic Cross Section Across the Atlantic Margin of North America. Normal faults, formed when the continent broke away from Africa, lie buried beneath the thick accumulation of sediment. During the Mesozoic, salt deposits accumulated. Some of these have risen as carrot-shaped bodies into the sediment. Later in the Mesozoic, reefs formed along the continental border, but eventually they died out and sediment covered them.

broke the cables. A number of cores of sediment were taken near the failures. Cores taken from hills on the sea floor far offshore contained the types of deep-water sediment expected, but sediments originally deposited in relatively shallow water on the edge of the continental shelf filled the valley-like depressions between these hills. These shallow-water sediments were present even in cores collected in much deeper water. Scientists concluded that a large landslide occurred at the edge of the shelf and caused the earthquake. The shallow-water sediment then moved down the continental slope, abrading the cables and eventually settling out on the sea floor. Since this study, detailed seismic profiles have revealed numerous huge masses of material that have slumped and slid down continental slopes. Sediment carried in turbidity currents is spread widely in deep waters of the continental rise and into the abyssal plains.

Most such landslides occur along the edge of the continental shelf or in submarine canyons where deposits of sediment derived from land become unstable (figure 8.9). Many of these canyons

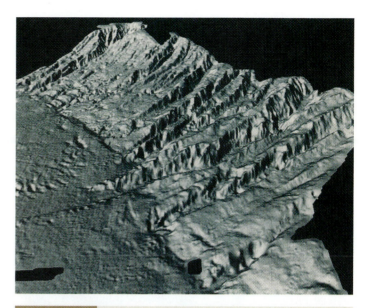

Figure 8.9 Image of the Continental Margin off the Coast of Delaware. Turbidity currents have cut deep canyons into the edge of the continental shelf and the continental slope. Some of these currents extend into the abyssal plains. The smooth plain at the base of the slope indicates where the sediments settled.

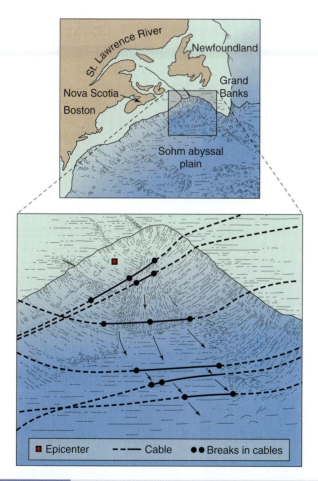

Figure 8.8 Failure of Cables Caused by Turbidity Currents. Cables lay across the continental shelf and the Grand Banks. A turbidity current formed as sediment at the edge of the continental shelf broke loose and flowed down the continental slope and rise, abrading the cables as it moved toward the abyssal plains.

formed when sea level was lower during the ice ages. At that time, most of the continental shelf was above sea level, and streams cut channels across the shelf area. Sediments were deposited where the streams entered deep water, and this unconsolidated sediment breaks free and moves downslope. Parts of these heavy masses of water-saturated, unconsolidated sediment turn into a slurry of water and sediment. This mixture is heavier than water, and it remains close to the sea floor and flows as a turbid mass downslope. These fluid masses carry large quantities of sediment into the abyssal plains or into trenches. These submarine landslides and resulting turbidity currents are the largest of all mass movements on Earth.

Density or turbidity current deposits are easy to identify because they contain shallow-water sediments in areas where deep-water sediment is expected. The deposits contain shallow-water fauna and primary sedimentary features, such as cross-bedding and ripple marks that indicate the presence of a current. Graded bedding in which the grain sizes decrease from bottom to top of a layer is a characteristic feature of turbidity current deposits. One can simulate the formation of graded bedding by placing sediment containing a variety of grain sizes in a bottle, shaking the bottle vigorously, and letting the sediment settle. The large grains settle first, followed by progressively smaller-sized grains.

Gulf of Mexico and Salt Deposits on Passive Margins

Although passive, the margins around the Gulf of Mexico (figure 8.10) are dramatically different from the eastern margin of North America. Both have been sites of accumulation of sediments since the opening of the Atlantic Ocean. Both contain thick wedge-shaped masses of sediment eroded from the continental interior, but as the Atlantic opened, it was a narrow, shallow body of water, and the region that is now the Gulf of Mexico remained partially enclosed. This region was also arid, evaporation rates were high, and extraordinary amounts of salt precipitated on the sea floor.

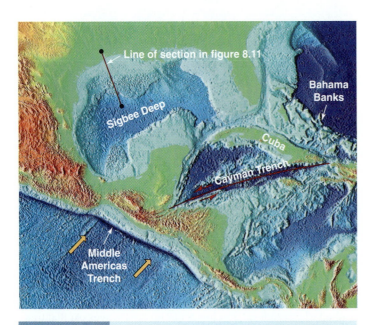

Figure 8.10 The Margins Around the Gulf of Mexico. The North American continental crust lies north and west of the Gulf of Mexico. Cuba and the southern part of Mexico are fragments of continental crust that became separated when North and South America drifted apart during the Mesozoic era. The Cayman trench is a major fault zone. The northern side has been moving west relative to the southern side. The Middle Americas trench is an active deep-sea trench. Major earthquakes related to subduction in this trench have cause devastating damage in Mexico and Central America. A line of volcanoes lies along this thin connection between North and South America.

Today, thick layers of that salt lie deeply buried beneath other sediments along the margins of the Gulf. These salt deposits, some of which are thousands of feet thick, lie beneath the southern part of the Gulf coastal plain and extend hundreds of miles off the coast south of Louisiana. Salt has a lower density than most sediment. Under the weight of the sediment, the salt has flowed. In some places, pillow-shaped ridges or wall-like masses of salt rise into the overlying sediment. Most distinctive of all are the carrot-shaped domes (figures 8.11 and 8.12). Movements of the salt have created a highly irregular sea floor on the continental slope and rise in the northern part of the Gulf and west of the Yucatán peninsula (see figure 8.10). The sea floor contains huge, enclosed depressions that resemble sinkholes. Some of these salt domes are so close to the surface that springs yielding salt water have formed. Deep shafts into some salt domes open the way for extensive mines from which comes much of the salt used in America.

Thick deposits of salt formed in several basins far inland from the Gulf (figure 8.13), and it was in one of these basins in East Texas that major deposits of oil and gas were discovered.

Oil and Gas on Continental Margins

Many of the major oil and gas fields in North America are located in sediments that accumulated along the continental margin in the Gulf of Mexico. Most of these fields are associated with salt features in the Gulf coastal plain and continental shelf. Decay of organic matter in marine environments leads to the formation of oil. Even the oil and gas accumulations now found far inland formed when those areas were submerged. As sediments accumulate and compact, the oil migrates out of the sediments in which it forms and moves into beds of sand that have porosity and permeability. Because these sands are deposited in a marine environment, salt water fills most of the pore spaces. Once oil moves into the sands, it moves upward through the salt water until it becomes trapped. Many of these traps formed where the sediments are folded, forming anticlines or domes (figure 8.14). In the Gulf, many of the domes form where the upward movement of salt pushes sediment into dome-shaped features. Oil also accumulates against faults that juxtapose porous and permeable

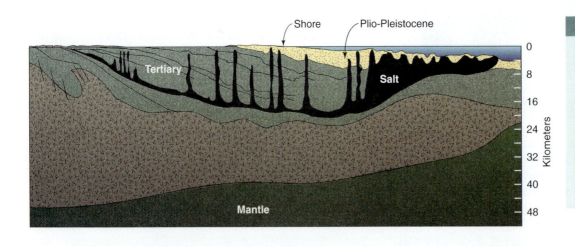

Figure 8.11 Cross Section of the Northern Margin of the Gulf of Mexico. Salt deposits shown in black lie at depth along the northern and southern margins of the Gulf of Mexico. Salt has invaded the sediments and rises close to the surface. Thick accumulations of Cenozoic age sediment lie along the margin.

layers with impervious rocks, in places where porous and permeable layers thin out, and where porous layers butt into impervious rocks like salt.

Exploration for oil and gas has led to major discoveries in the North Sea, off the coasts of Newfoundland, western Africa, and in the Gulf of Mexico and the Persian Gulf. The quantity of oil in the crust is limited, and demand is increasing rapidly. In recent years, production has exceeded 20 billion barrels per year. The increase in demand for 2001 was 1.8 million barrels per day. Many, perhaps

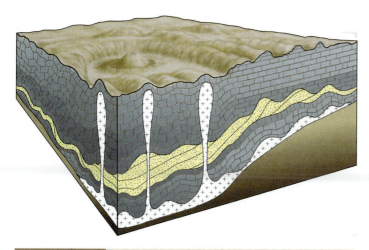

Figure 8.12 Salt Penetrates Overlying Sediment. Salt is unstable under the weight of the overlying sediment and rises as domes, ridges, or elongate carrot-shaped bodies referred to as salt domes.

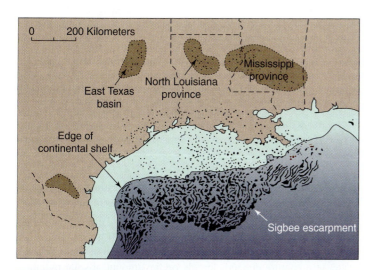

Figure 8.13 Salt Domes Along Gulf Coast. Places where salt has risen above the layer in which it originates are shown in black. Most of the salt lies beneath the continental slope and rise, but salt is also present in basins located far inland.

most, of the major oil accumulations on land in North America are known and being used. Consequently, more attention is given to exploration for offshore accumulations. This is accompanied by increasing concerns about the potential for pollution arising from oil spills and the problems they create in the ocean.

THE SEA FLOOR—ACTIVE CONTINENTAL MARGINS

Along active continental margins where subduction is taking place, the continental slope either steepens and passes into a deep-sea trench, as it does off the Andes in South America, or it flattens out to form a marginal basin before becoming shallow again as it approaches a volcanic island arc and trench system. Margins of this type are common in the western Pacific Ocean (see figure 3.9).

Deep-Sea Trenches on Active Margins

All deep-sea trenches are located over subduction zones, and most are associated with volcanic island arcs near continents. But the Peru-Chile trench (see chapter 3 opening image) lies along the western margin of South America (figure 8.15), and the Mariana trench is located in the open ocean. Trenches differ in size, depth, and shape. Those like the Mariana trench located far from land are deeper and show the shape of the subsiding sea floor much better than those closer to land. Generally the side of the trench closest to the volcanic arc has a steeper slope than the side that opens toward the ocean and the oceanic lithosphere that is moving into the subduction zone. In the Japan trench, a seamount is moving toward the trench and will eventually be crushed against the inside wall of the trench as the lithosphere sinks.

All trenches are natural sediment traps; so sediment carried by rivers that flow into the ocean near a trench passes into the trench. No sediment from continents reaches the Mariana trench, but the eastern end of the Aleutian trench is partially filled with sediment, and sediment from the Columbia River has completely filled the trench over the subduction zone located off the coasts of Washington and Oregon. Although most trenches lie beneath waters that contain large quantities of organisms that have shells, the trenches contain little or no shells composed of carbonate minerals. Calcium carbonate becomes more soluble in cold water like that in the deep trenches, so, shells dissolve as they sink toward the bottom of the trench. Only skeletal remains composed of silica collect as sediment in the bottom of deep trenches.

Volcanic Ash and Desert Dust

Ash and dust are important sources of deep-sea sediment. Ash generated by violent eruptions along the western rim of the Pacific Ocean rises high into the atmosphere where winds blowing from west to east carry it over the trenches and spread it great distances across the Pacific Ocean. Winds that blow across North Africa from east to west pick up dust from the Sahara desert and

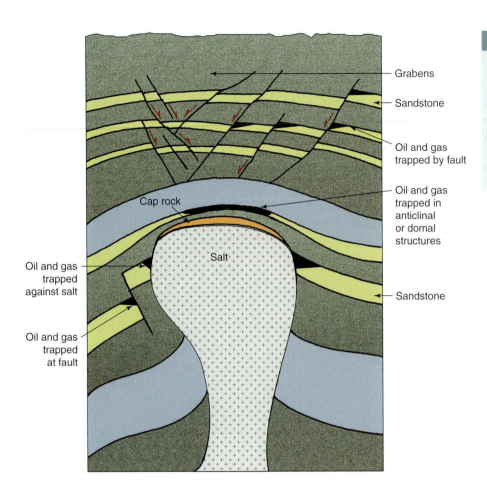

Grabens

Sandstone

Oil and gas trapped by fault

Cap rock

Oil and gas trapped in anticlinal or domal structures

Salt

Oil and gas trapped against salt

Sandstone

Oil and gas trapped at fault

Figure 8.14 Oil and Gas Accumulates Around Salt Domes. Major oil and gas accumulations occur over and around salt domes. Often oil, gas, and water fill the spaces between grains of sand in layers of sandstone. They separate according to their density. The oil floats on the water and rises until it stops where a barrier stops further migration. The barriers may be the salt of a salt dome or layers of clay-rich sediment.

distribute it across the Atlantic Ocean (see Chapter 17). Much of this ash and dust falls into the oceans and settles to the sea floor where it spreads as sheet-like deposits. These deposits are generally thinner downwind, but they may cover vast areas of the sea floor and become part of the sediment in a great variety of marine environments.

MARINE ENVIRONMENTS

The critical physical components of marine environments are their geologic framework, the composition of seawater, water temperature, and solar radiation. The shape of the ocean basins forms a framework in which marine environments develop. This framework consists of the major physiographic provinces, such as continental shelves, continental slopes and rises, abyssal plains, oceanic ridges, deep-sea trenches, and seamounts (see figures 3.9 and 3.10). Each of these physiographic provinces formed over millions of years through geologic processes such as volcanic and tectonic activity, erosion, and deposition of sediment.

Geologic processes also directly affect the chemistry of seawater. Volcanic activity along spreading ridges brings chemicals from the mantle and deep in the crust into seawater. Activity along these ridges may even cause slight alterations in water temperature. Weathering and erosion on land also release elements and

compounds from rocks that streams carry, along with large quantities of water, into the marine environment. The continuation of this transfer over geologic time has caused seawater to become salty. Biochemical processes and sedimentation remove some of these chemical constituents from seawater. For example, most marine organisms build skeletons out of elements removed from seawater. When the organisms die, the hard parts of marine plants and animals, constructed from elements in seawater, settle to the ocean bottom and become part of the sediment on the sea floor. Thick accumulations of these remains cover vast areas of the sea floor. Inorganic sediments may also contain some of the elements, such as sodium and chloride present in seawater. If seawater becomes cut off from the open ocean and evaporation is high, deposits of salt may form through chemical precipitation.

The temperature of seawater is critical for the survival and reproduction of marine organisms. The rapid decrease in the number of cod in waters off Iceland during the Little Ice Age (thirteenth to nineteenth centuries) demonstrated how sensitive cod are to water temperature. Water temperature also limits the growth of modern reefs. Water temperature depends largely on the amount of solar radiation reaching Earth's surface. Maximum heating due to solar radiation occurs where incoming solar radiation strikes the surface most directly in equatorial latitudes, and where it is neither intercepted by cloud cover nor reflected by snow and ice. For these reasons, surface temperature in the oceans is highest in equatorial latitudes and lower

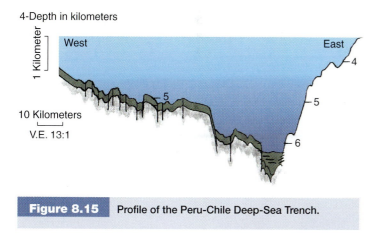

4-Depth in kilometers

1 Kilometer

West   East

10 Kilometers
V.E. 13:1

Figure 8.15   Profile of the Peru-Chile Deep-Sea Trench.

at higher latitudes. Circulation in the oceans also has a major effect on water temperature and climate (see chapter 13).

Solar radiation shapes marine environments through its effect on the process of photosynthesis and its influence on water temperature. The maximum depth of penetration of sunlight, approximately a hundred meters in the open ocean, varies depending on the clarity of the water. The sunlit zone, called the **photic zone,** is one of the most significant subdivisions of the marine environment. In this zone, marine plants, notably the unicellular diatoms—a type of algae—carry on photosynthesis. These plants make up the base of the food chain in the ocean.

Stratified Marine Environments

Two major communities of life inhabit the ocean. One includes animals and plants that live attached to or confined to the sea floor, the **benthic community** (figure 8.16). The other inhabits the water above the bottom, the **pelagic community** (figure 8.17). The sea floor itself is a unique part of the ocean, called the *benthic zone.* The special environments, such as estuaries, deltas, and tidal flats, described in chapter 10, are parts of the benthic zone. The rest of the open ocean, between the bottom and the surface, is the *pelagic zone.* Pressure increases with depth in a uniform manner throughout the ocean, but temperature, salinity, and nutrient supply, which play major roles in shaping the environments in the open ocean, vary from one part of the ocean to another. Fish, turtles, and marine mammals—such as whales, dolphins, and sea otters—and huge numbers of smaller plants and animals inhabit the open ocean. Some of these live mainly in the sunlit zone, while others move freely into deeper waters in zones oceanographers define in terms of the water depth.

BIOLOGICAL COMMUNITIES AND THEIR SEDIMENTARY RECORD

Far from land, sediments composed mainly of the remains of organisms cover vast areas of the floor of the open ocean (figure 8.18). Closer to land, even though organisms are more numerous, their remains are a small fraction of the huge volumes of inorganic sediments that move into the sea from land areas. In the transition zone between open ocean and beaches, marine sediments are commonly mixtures of organic and inorganic materials. The organic materials belong to one of three distinct biological communities: (1) **plankton,** the plants, animals, and bacteria that float and cannot move independently of the currents; (2) **nekton,** the fish and marine mammals that are strong enough to swim freely against the current and also live in the pelagic zones; and (3) **benthos,** the plants and animals that dwell on or close to the bottom. Attention here is focused on the organic communities that are significant contributors to marine sediment. Most of these are minute in size.

Planktonic Communities

Although the sunlit surface layer of the oceans, the photic zone, teems with microscopic-sized plants and animals (figure 8.19), only a few plants, such as the sargassum seaweed, are large. Sargassum initially grows attached to the sea floor in shallow water, but waves break it off and it floats. Floating sargassum gathers in a large, quiet area of the central North Atlantic Ocean, known as the Sargasso Sea. Sailing ships were often becalmed in this area.

All types of plankton share an inability to direct their own movement. A few are weak swimmers, but they, too, drift with the prevailing current. Plant plankton, known as **phytoplankton,** include vast numbers of algae.

Bacteria (see figure 1.21), which are abundant in the ocean, are among the simplest of all living organisms. These extremely primitive organisms adapt to more diverse living conditions than any other organisms and are present around the hot fumerole vents on the oceanic ridges. They are present in soil and water, on the surfaces of plants and animals, and even in the digestive tracts of animals. They survive extreme cold in the Antarctic, high pressure in the deep sea, low pH in springs, even boiling temperatures for short periods. Most require some outside source of food, but a few species make their own food by oxidizing sulfur and iron compounds, nitrates, and hydrogen. Some bacteria are photosynthetic, but many survive in the absence of sunlight, even in deep parts of the ocean.

Diatoms are simple, minute plants that hold chlorophyll. They live in the surface layers throughout the oceans but are most abundant in the cold water around the polar ice caps. Like other plants, diatoms absorb carbon dioxide and minerals dissolved in seawater. They use these materials to manufacture carbohydrates. Diatoms absorb silica among other elements. They use silica to make a hard framework that supports their soft parts. The solid framework survives the plant and sinks to the sea floor where it becomes part of the sediment. Although all of the approximately 20,000 species of diatoms have solid parts, each species has a distinctive shape. Some look like little drums or discs, and others resemble highly ornate hats (see figure 8.17). All diatoms reproduce by cell division. Under favorable conditions, a single diatom may multiply several times a day. When this happens, diatoms may become so numerous that they change the color of the water. The diatoms and other

Figure 8.16 **The Benthic Community.** (*a*) A few of the inhabitants of sea cliffs. (*b*) Some of the animals commonly found on seashores composed of sand or soft sediment.

Periwinkle

Common acorn barnacle

High tide

Crab

Low tide

Small acorn barnacle

Seaweed

Sea urchin

Clam

Snail

(*Brachiopod*)

(a)

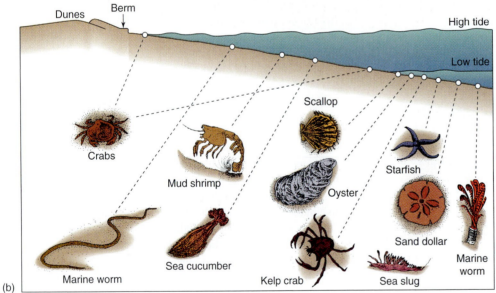

Dunes

Berm

High tide

Low tide

Scallop

Crabs

Mud shrimp

Oyster

Starfish

Sand dollar

Marine worm

Sea cucumber

Kelp crab

Sea slug

Marine worm

(b)

algae are said to "bloom" when they multiply this rapidly. In the North Sea, algae blooms sometimes make the water slimy and change the color to brown.

Flagellates are phytoplankton that have one or two whip-like tails (see figure 8.17). Flagellates live in all parts of the oceans, but they are most abundant and diverse in warm water. Like other plants, they prosper in the photic zone where they carry on photosynthesis. They are so numerous that a liter of water may contain millions of flagellates. Like the diatoms, they occur in a tremendous number of shapes, and they rival the diatoms in abundance. Unlike diatoms, however, they have no hard parts that become part of marine sediment. Some species of dinoflagellates multiply in such great

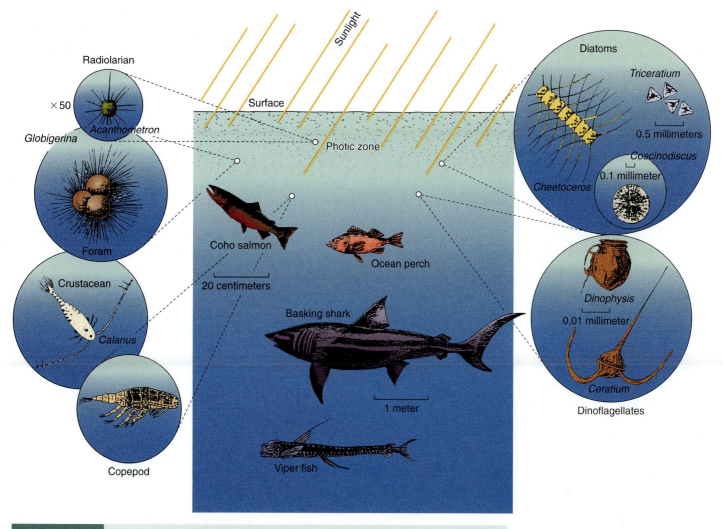

Radiolarian

×50

Acanthometron

Globigerina

Foram

Crustacean

Calanus

Copepod

Sunlight

Surface

Photic zone

Coho salmon

20 centimeters

Ocean perch

Basking shark

1 meter

Viper fish

Diatoms

Triceratium

0.5 millimeters

Coscinodiscus

0.1 millimeter

Cheetoceros

Dinophysis

0.01 millimeter

Ceratium

Dinoflagellates

Figure 8.17 **The Pelagic Community.** Sketches illustrate the most abundant plankton in the photic zone and a few representatives of the nekton.

numbers that they cause a reddish brown discoloration of the water, known as *red tide.* At the time of these blooms, toxins in the cells of flagellates pose serious problems for other marine organisms. The toxins may be deadly to many fish, shellfish, and pose danger even to marine mammals like dolphins and whales. Consumption of contaminated seafood can also be hazardous to humans.

Coccoliths are another type of minute and highly abundant phytoplankton. Coccoliths are most abundant in tropical and semitropical waters, especially near the coasts, but they differ from diatoms and flagellates in that they produce plates composed of calcium carbonate that may become part of the sediment on the sea floor. If compacted, this sediment turns into chalk. Coccoliths also live in deep water far below the photic zone. Oceanographers have discovered specimens in water over 4 kilometers (2.5 miles) deep. At such great depths, coccoliths do not depend on photosynthesis for food. Instead, they assimilate dissolved organic matter from seawater.

Most **zooplankton** are unicellular, planktonic animals that belong to the kingdom known as protista. Many zooplankton remain free-floating and suspended in water throughout their lives, but others are larvae of benthic organisms, such as corals that travel with the currents for part of their lives until they settle and attach themselves to the sea floor.

Of the single-celled animals called **foraminifera,** one group, the **globigerina,** constructs its shells out of calcium and carbonate from seawater (see figure 8.17). These shells or skeletons take many forms; some are coiled like gastropod shells. Because foraminifera do not carry on photosynthesis, they are not restricted to the photic zone. Instead, they capture their food, most of which consists of phytoplankton. Like other single-celled organisms, they are exceedingly small. A teaspoon of chalk such as that forming the chalk cliffs at Dover, England, may contain half a million individual foraminiferas.

Another abundant zooplankton group, **radiolarians,** have a distinctive hard framework with radiating forms composed of silica

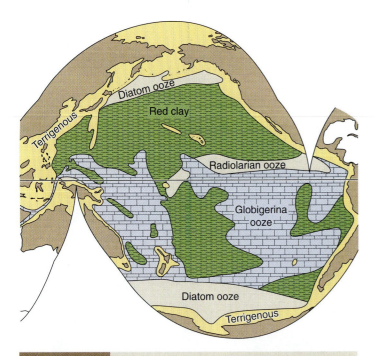

Figure 8.18 Sediment Distribution on the Sea Floor of the Pacific Ocean. Deposits of sediment carried into the ocean by ice sheets surround Antarctica. Deposits from streams are present on continental shelves and slopes where major streams enter the ocean, but large reefs are present on the shelf in the tropics. Siliceous oozes composed of the remains of diatoms occur at high latitudes, and oozes containing abundant radiolarian remains lie along an equatorial belt. Red clay covers much of the northern Pacific and a large area in the south. Manganese nodules are common where red clay is present. Calcareous oozes composed of globigerina shells cover vast areas in the south Pacific.

(see figure 8.17). Silica does not dissolve as readily as calcium carbonate does in deep water where the pressure is great. Consequently, remains of radiolarians survive in the sediment in deep-sea trenches where globigerina shells are rare.

Crustaceans include such familiar marine animals as lobsters, crabs, and shrimp. The smaller crustaceans are zooplankton, but larger, more-familiar crustaceans are benthic. All crustaceans are multicellular animals that, like other arthropods, have jointed legs and a hard external covering. Crustaceans eat tremendous quantities of plants, such as diatoms, and turn that plant food into animal tissue. They, in turn, become a prime source of food for fish and whales. **Copepods** (see figure 8.17), shrimp-like crustaceans, about the size of a grain of rice, are the most numerous of the multicellular crustaceans. The number of copepods may exceed all other multicellular animals combined. **Krill,** another shrimp-like crustacean, about 5 centimeters (2 inches) long, are especially abundant in the cold waters around Antarctica where they are the prime food source for some whales. A large blue whale consumes about 450,000 kilograms (a million pounds) of krill each year.

Nektonic Communities

All members of nektonic communities are strong enough to swim against most of the currents they encounter in the oceans. Most nekton are fish (see figure 8.17), but marine mammals, including whales, dolphins, seals, and walruses and some reptiles, including turtles, snakes, and the marine iguana, are also nekton. Although people are most familiar with these creatures that provide human food and are most visible along the seashore, their remains make up a small part of marine sediment. Nevertheless, their fossil remains are often important indicators of water depth in which ancient sediments formed. Fish remains first appeared in the fossil record in the early part of the Paleozoic era, nearly half-a-billion years ago. Since then, fish have evolved in ways that allow some species to adapt to almost all marine environments.

Benthic Communities

Many plants, such as seaweed, and animals, such as oysters, corals, and sea lilies (crinoids), live firmly fixed in sediment or attached to rocks on the sea floor (see figure 8.16). Being confined to a fixed position on the ocean bottom imposes serious constraints on the environments in which these plants and animals can survive. The invertebrates that live attached to the bottom require special conditions. Most of them live in shallow water penetrated by sunlight. Approximately forty-five species of flowering plants live in the shallow waters of the sea. Their roots and leaves provide shelter and food for more than 400 species of animals, including such invertebrates as urchins and also clams, crabs, sea horses, shrimp, many species of fish, turtles, and manatees. They must have sources of nourishment at hand, and they cannot escape pollution, so they survive and multiply best in clean water. The corals are among the best known of these animals. Other members of the benthic community are free to move and live in less-restricted environments than those like the corals that are attached to the bottom. Worms and gastropods bore into the bottom and survive on organic matter in the sediment. Still others, such as lobsters and crabs, roam the sea floor in search of food.

Many of the organisms in the benthic community survive in very specific environments. For some, it is water depth; for others, it is the water temperature or salinity. For this reason, fossils of these specialized organisms are a key tool for those who want to reconstruct ancient environments, and they are widely used as indicators of paleoenvironmental conditions.

Threats to Marine Biodiversity

When social and natural scientists first began to express concerns that Earth's resource base might be insufficient to support the projected future population, many looked to the ocean as the ultimate source of an "unlimited" food supply. Increasing harvests of fish have revealed that the potential size of these marine food resources is much more limited than they appeared to be earlier. Marine organisms are also subject to a number of threats to both the size of marine populations and to marine biodiversity. These threats take

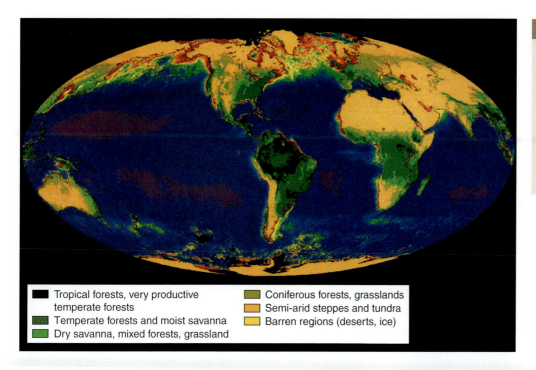

Figure 8.19 Global View of the Earth's Biosphere. These satellite images show the concentration of phytoplankton in the surface layers of the ocean. Purple indicates low concentrations; yellow, orange, and red indicate high concentrations in the oceans. On land, the vegetation index is indicated by yellows (low) to dark green (high).

Tropical forests, very productive temperate forests

Temperate forests and moist savanna

Dry savanna, mixed forests, grassland

Coniferous forests, grasslands

Semi-arid steppes and tundra

Barren regions (deserts, ice)

many forms, such as the destruction of habitat and pollution of the oceans as well as excessive harvesting of fish and other marine species.

Estuaries, bays, marine swamplands, and deltas are important breeding grounds for many fish. These fish spend the early part of their lives in relatively protected waters. Human activities in coastal areas are damaging the habitat in many of these near-shore marine environments, along the rivers that feed into marine waters and the ocean itself. The damage comes from many, often unintended, sources. A large part of the pollution of near-shore environments comes from land. Streams flowing off land areas carry fertilizers, animal waste, oil, and residues of oil from parking lots and residue of emissions from automobiles and factories. Pollution also comes from human activities in the oceans. A number of major oil spills have occurred when tankers have run aground. The wreck of the *Torrey Canyon* tanker in 1967 was among the first of these. This ship released over 700,000 barrels of oil off the coast of Britain. Much of the oil washed ashore in France and southern England. In 1989, the *Exxon Valdez* ran aground in Prince William Sound, Alaska, releasing 11 million gallons of oil that washed ashore along several thousand miles of shoreline. Other substances also pose serious hazards to the ocean environment. During a hurricane, a freighter spilled thousands of tons of sulfuric acid in a Mexican port. Exotic species of plants are spread through discharge of ballast water in foreign ports. Motorboats have devastating effects on the manatees in Florida and on the scallop populations near Java in the southwestern Pacific. Coastal developments also take a toll on the oceans. It is estimated that nearly half of the world's mangrove forests have been destroyed to make way for various types of coastal developments, and landfills in shallow waters have destroyed the breeding grounds for many marine organisms.

Land reclamation projects have filled in large areas along coasts. Generally, these filled areas become sites for housing or industrial developments that increase water pollution caused by sedimentation of clay and silt derived from construction sites. These fine sediments make water murky and reduce the penetration of sunlight. The wetlands are also damaged by the use of chemical toxins and fertilizers. The phosphates and nitrates in fertilizers promote the growth of algae in the wetlands, and these may choke out the supply of oxygen to animals and carbon dioxide to plants. Boston's Back Bay area, wetlands in coastal New Jersey, and areas around San Francisco Bay are all examples of wetlands lost in this manner.

Disposal of waste products in marine waters continues to damage the marine environment. An estimated 50 million tons of waste is dumped in the oceans each year. Much of this waste (80%) consists of materials dredged out of streams and other bodies of water on land. Ten percent of these wastes are industrial and about 9% are sewage. Until the 1990s, a number of major cities, including New York City, disposed of their garbage by dumping it at sea. These waste materials commonly carry high concentrations of some metals, and some wastes contain other toxic substances. The ocean-bottom habitat is always altered at long-term dump sites. Some organisms disappear entirely from these sites; others carry infections or exhibit fin rot.

Highly toxic substances, even radioactive wastes, have been dumped in the ocean. Serious consideration has been given to using the oceans, especially the deep-sea trenches, as repositories for high-level radioactive wastes. In theory, sediment would bury watertight canisters containing the waste. Eventually, as subduction continues, the canisters would move deep into the crust. Uncertainties about how long the waste might reside on the sea floor and what might happen if the canisters failed has made this alternative for waste disposal unappealing.

In many parts of the world, overharvesting of marine species for food has led to conflicts among nations. Technical advances have made it possible to find and trace schools of fish much more efficiently than was previously possible. Federal restrictions limit fishing of some threatened species along the coast of the United States. International agreements bind many nations to agreements that regulate harvesting of some species, particularly marine mammals. But for many species, limits on fishing are not enough. Some nations refuse to limit their fishing. In the North Atlantic, excessive fishing has depleted the population of cod to such an extent that other species may take over its niche. In the Pacific Northwest, salmon populations are rapidly declining. Dams on the rivers in the northwest prevent salmon from returning to their upstream breeding grounds to spawn. Pollution, which destroys salmon habitat, and overfishing also contribute to the population decline.

Threats to marine biodiversity also arise from changes in the atmosphere. Over the past 18,000 years, changes in the global climate have caused both the temperature of seawater and sea level to rise. In recent years, changes in climate appear to be causing an acceleration in warming. Increases in carbon dioxide in the atmosphere appear to be an important factor in this change. Carbon dioxide and methane in the atmosphere absorb radiation from the ground surface and cause temperature to rise. This process is an important cause of an ongoing increase in global temperature, which is discussed in more detail in chapter 11. Rising water temperature has had an impact on the fish populations along the California coast where cold-water species are decreasing in numbers due to increases in local water temperatures. Increasing water temperature is also a threat to the survival of coral reefs.

SPECIAL OCEANIC ENVIRONMENTS

Hydrothermal Vents

Studies of the oceanic ridges have revealed that hot solutions rising from volcanic rocks on these ridges carry a number of metals. In 1977, scientists aboard submersibles began to study the oceanic ridge crest south of the Galápagos Islands in the Pacific Ocean. At water depths over a mile, they came across one of the least-expected discoveries of recent decades. They found chimney-like mounds composed, in part, of metals and extending tens of meters above the ocean floor (see figure 8.2). The metals include many of those found in manganese nodules. Black clouds of hot, sulfur-rich water—hydrothermal solutions—were emanating from these chimneys at temperatures of 350°C or more. About 5 to 10 centimeters (a few inches) away from the vent, water temperature is close to 2°C. When this hot water encounters the cold, deep water of the ocean, metals precipitate, building chimneys called *black smokers*. Most of these chimneys are only a few meters high, but the highest, named Godzilla, located off the coast of Oregon, is over 50 meters (165 feet) high. These chimneys, composed, in part, of metals, indicate that either large quantities of metals rise

from the magmas in Earth's upper mantle or that the metals are leached out of parts of oceanic crust by hot fluids that pass through the crust into the water and the sediment near ridge crests. Previously unknown fauna and an unusual type of bacteria that can use the chemicals coming from the hydrothermal solutions live in these deep waters around the hydrothermal vents. Black mats, several centimeters thick, composed of these bacteria surround some of the smokers, and at times, the bacteria appear to form clouds in the hot water.

In general, the biological communities living in the deep sea have low population density, high species diversity, and they produce small quantities of biological matter. In contrast, the number of species found around hydrothermal vents is low, but the population density and biomass are extraordinarily high. Large populations of clams, mussels, and tube worms attached to the lava flows on the sea floor make up these biological communities. Grazing animals, such as white crabs and pink "vent fish," remain close to the vents.

It is surprising to find such thriving communities of animals at such great depth in the ocean because the plants that carry on photosynthesis in the photic zone are not present in deep water. Unlike phytoplankton, the bacteria around these vents do not need sunlight. The bacteria apparently provide the nutrients at the bottom of the food chain for the vent communities. They fulfill the role of the phytoplankton in the photic zone. They are capable of living on hydrogen sulfide, a substance that is highly toxic to most lifeforms. The heat coming out of the vents may provide energy for the bacteria much as sunlight does for plants that are photosynthetic. In turn, the bacteria break down the chemicals in the environment into food used by the animals in this community.

The discovery that bacteria provide food in this deep-water environment may help solve another problem bearing on the early evolution of life. Bacteria, the oldest known fossils, may have provided a source of food for the other primitive forms of life that were on Earth before photosynthesis became an important process.

Reefs

The animals and plants that make up reefs flourish in near-surface waters. Corals are responsible for most modern reefs, but many other invertebrate animals share this habitat and contribute to the reef mass. Many, like corals, favor warm water that is relatively free of suspended dirt. Because corals flourish in warm water, most modern coral reefs lie within a belt that extends about 20° north and south of the equator. In this belt, ocean currents that flow from east to west carry the coral larvae toward the western side of each ocean. For this reason, major reefs in the Pacific occur near Australia and in Indonesia. Similarly, most reefs in the Atlantic are in the Caribbean. The coral larvae attach to solid materials on the sea floor, and the colony begins to grow. They are not free to move about to gather food. They prosper in clean, agitated water where currents bring food to them. Because corals cannot survive in muddy water, gaps in reefs are usually present where muddy waters from streams enter the ocean.

Figure 8.20 Reef Flats. (a) The reef flat lies close to sea level. (b) Several different types of coral colonies on the reef flat near Green Island on the Great Barrier Reef.

(a)

(b)

Reefs located on the seaward side of islands, where the open sea brings in a constant supply of water rich in food and other nutrients, contain an abundant and varied fauna. The slope of the sea floor, the direction of prevailing waves, and the direction and strength of currents determine which organisms live along the seaward margin of the reef. Strong grooves form on the seaward side of reefs where water returns to the sea after waves break on the edge of the reef. Red algae thrive in the strong, steady surf. They cover and tend to flatten the ridges between the grooves.

The top surface of most reefs consists of a thin layer of living coral, called the **reef flat** (figure 8.20). Most corals live close to mean sea level. During low tide, the water level may drop below the top of the reef flat and expose the corals to air for a short time each day. Some species of coral can survive this exposure, but most corals must remain below water. From the reef flat, water depth increases rapidly on the seaward side. Fragments of the reef that break off along the edge of the reef during storms accumulate and form these steep slopes.

In recent years, it has become apparent that many problems beset coral reefs. In the southwest Pacific especially, the use of blasting and cyanide to harvest fish are damaging reefs. In many other areas, such as the Great Barrier Reef of Australia, the Caribbean, and along the eastern coast of Africa, the cause of reef destruction is less obvious. Several diseases are killing coral. The latest of these is a condition called *rapid-wasting disease*. The diameter of the affected area on the coral can spread as much as 7 centimeters (3 inches) per day. Not only does this kill the living coral, it also destroys the calcium carbonate skeleton. Some scientists think that increases in water temperature are making corals more susceptible to disease. In any case, many are concerned that the problems affecting corals are symptomatic of greater problems involving the oceans.

Fringing and Barrier Reefs Most modern coral reefs grow as **fringing reefs** within a few hundred meters of shore or as **barrier reefs** offshore, commonly near the edge of the continental shelf (figure 8.21). Many modern fringing reefs lie around volcanic islands in the Pacific and around many nonvolcanic islands near continents.

Usually, a shallow lagoon separates the main part of the fringing reef from adjacent land. Although they are not as densely populated as reef flats, quiet waters in the lagoon provide an ideal place for delicate forms of coral and other invertebrates to live protected from the effects of breaking waves. These lagoons, only a few meters to a few tens of meters deep, contain the remains of algae and fine sands derived from the seaward side of the reef.

The **Great Barrier Reef** of Australia, the largest in the world, covers nearly 200,000 square kilometers (76,000 square miles) and extends 1,900 kilometers (1,180 miles) along the east coast of Australia (see figure 8.21). This complex of scattered reefs, separated by channels, lies on the continental shelf. Although one reef is about 960 kilometers (600 miles) long, the Great Barrier Reef is not a single, continuous reef mass. Long reefs aligned parallel to the coast form an outer barrier near the edge of the continental shelf. A relatively open body of water called the Barrier Reef Channel lies between the outer barrier reefs and the mainland. The Barrier Reef Channel is 240 kilometers (150 miles) wide near its southern end and contains over 600 islands. Many of these are small islands composed of coral sand and reef fragments; others consist of rock masses like those on the mainland. Fringing reefs surround most of these islands.

During the 1960s, students of the reef noticed an unusually large population of a giant starfish, the Crown of Thorns, which eat coral. Over a few years, this population expanded rapidly and destroyed large areas of the reef. Ecologists immediately undertook studies to determine the reason for this explosive population growth. These studies focused on the natural predators of the starfish, such as the coral shrimp, which eats starfish larvae. Coral shrimp are highly sensitive to the pesticide DDT, which had been used extensively in Australia. Some ecologists think runoff from land areas sprayed with DDT may have gone into the breeding grounds of the shrimp and killed them. The decrease in the shrimp population would have allowed the population of starfish to explode and eventually would have led to the destruction of large areas of reef. However, the number of starfish found as fossils in layers of sediment in deep wells suggests that fluctuations in starfish populations may follow a boom-bust cycle that extends much farther back in time than the use of DDT.

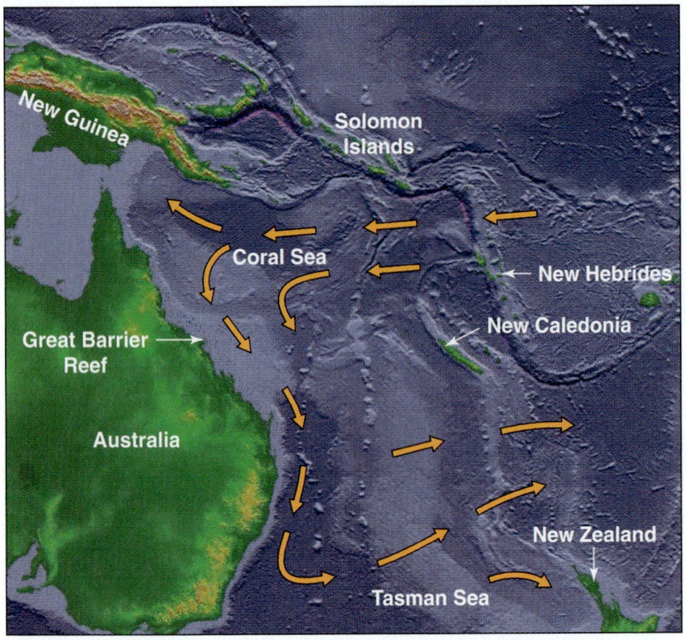

Water wells drilled on some islands on the reef complex penetrate hundreds of feet of coral debris. The thick coralline rock under the modern live reef supports the idea that the reef grew upward as sea level rose following the last glacial advance. Upward growth of coral reefs also leads to the formation of atolls.

Atolls are ring-shaped coral reefs that enclose lagoons (figure 8.22). Charles Darwin suggested that atolls form when a volcanic island with a fringing reef subsides. As the sea floor subsides, the reef continues to grow upward to maintain the living organisms at the necessary water depth. Eventually, water covers the island, leaving the circular atoll around a water-covered depression, or lagoon. Drilling at Bikini and Eniwetok confirms Darwin's prediction that volcanic rocks lie beneath atolls.

Bikini Atoll, an oval ring of islands in the southwestern Pacific Ocean, is about 42 kilometers (26 miles) long and 24 kilometers (15 miles) wide (figure 8.23). It is probably best known as the site of early American tests of atomic bombs after World War II. To evaluate the effects of atomic explosions on marine life, government surveys conducted extensive studies of Bikini Atoll before and after the atomic explosions.

Marginal reefs up to about 1-mile wide are almost continuous around the atoll. The central lagoon, which has an average depth of 48 meters (157 feet), covers approximately 648 square kilometers (259 square miles). The islands that surround the lagoon stand at a maximum height of 8 meters (26 feet) above sea level. The submarine topography around Bikini shows that the atoll occupies the top of a flat-topped mountain. The shape of this mountain suggested that it is the base of a volcano, and drilling on Bikini confirmed that lava flows lie beneath the reef limestone. Although

Figure 8.21 The Great Barrier Reef. Warm currents that move across the Pacific in equatorial latitudes bring warm water to the coast of Australia where the Great Barrier Reef has formed. Fringing reefs and atolls surround most of the island in the area north of New Caledonia. The water is too cold for reefs along the southeastern coast of Australia and in New Zealand.

Types of coral reefs

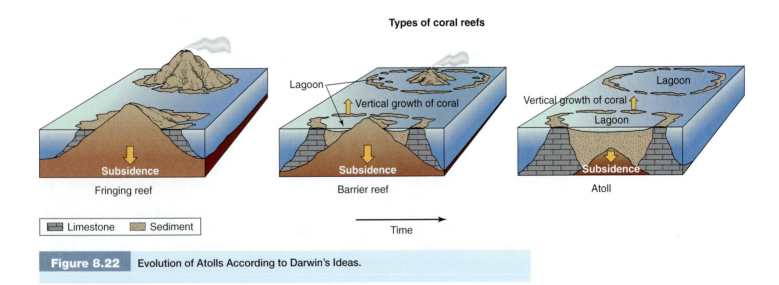

Figure 8.22 Evolution of Atolls According to Darwin's Ideas.

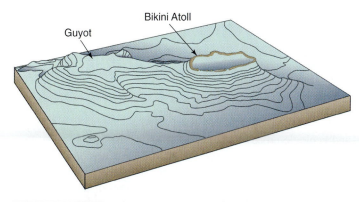

Figure 8.23 **Bikini Atoll.** The submerged portions of Bikini Atoll have the shape expected for a volcano.

scientists expected to find that the atoll caps a volcano, the thickness of the coral beneath the living part of the reef came as a surprise. One well penetrated more than 700 meters (2,300 feet) of coralline limestone, much too great a thickness to result from sea level changes caused by variations in the water-ice balance. A possible explanation of this much reef limestone is that subsidence of the sea floor must have taken place to lower the tops of many guyots to depths of 100 to 1,000 meters (330 to 3,300 feet). At least part of this subsidence is due to the increase in density of oceanic crust that occurs as it cools and moves away from ridge crests. Warping of the sea floor may also contribute to the subsidence.

THE ROLE OF THE OCEANS IN BIOGEOCHEMICAL CYCLES

All of the elements involved in the biogeochemical cycles, described in chapter 1, move through marine environments. Additional elements, such as silica, move mainly between the water in the oceans and the land, sea floor, and mantle where plate tectonics and surficial processes are most important in the exchanges. Carbon moves mainly between plants, animals, the surface layer of the oceans and the atmosphere.

The buildup of carbon dioxide in the atmosphere and the prospect of significant global warming in coming decades make the carbon cycle especially significant (figure 8.24 and see figure 1.23). Although the carbon dioxide content of the atmosphere has increased steadily over recent decades, the amount of carbon dioxide now in the atmosphere is lower than some scientists' studies predicted. Other natural processes called *sinks* must remove carbon dioxide from the atmosphere. Plants on land and plankton in the photic zone of the oceans are the most likely sinks. Much of the carbon moving through the biosphere exists in the photic zones of the oceans. Thus, marine biological processes play an important role in controlling the concentration of CO_2 in the atmosphere.

At the surface of the ocean, carbon dioxide in the atmosphere readily enters seawater, where algae and other marine plants take it up and use it in photosynthetic processes. The plants either die and sink or become part of the food chain for marine animals. The animals produce more carbon dioxide through respiration, and this carbon dioxide passes into the water. Eventually, when the animals die, their shells sink. As the dead plant and animal matter sinks, it decomposes through the processes of oxidation and bacterial action. The carbon dioxide produced as a by-product of these processes also helps maintain the concentration of CO_2 in the upper part of the ocean. When the concentration of carbon dioxide is not the same in the water and air, it moves from the fluid containing the greater concentration to the one with lower concentration. When the concentration of carbon dioxide in the water equals that in the air, no net transfer takes place. However, natural processes and human-induced changes may upset this balance.

Human activities, mainly the burning of fossil fuels, has introduced large quantities of carbon dioxide into the environment. Part of this carbon dioxide moves from the atmosphere into the oceans where plants take it up. Plants convert some of the carbon to organic molecules which, in turn, are consumed by animals. Both plants and animals may use part of the carbon to make shells or other hard parts which sink to the sea floor as sediment when the plants and animals die. Deep circulation in the oceans also removes carbon dioxide from near the surface. That carbon dioxide remains with the water as it slowly makes its way through the deep parts of the ocean. Carbon dioxide also reacts with calcium and sinks to the sea floor as sediment, where it may remain many millions of years. Given enough time, these natural processes remove significant quantities of the carbon dioxide from the atmosphere. Once removed from the surface layer, more carbon dioxide moves from the atmosphere into the ocean. However, the rate at which these processes move carbon through these cycles is unknown.

The carbon cycle in the ocean is complex. The supply of nutrients in the ocean determines the abundance of phytoplankton in the surface layers of the sea, and the number of plants determines the amount of carbon dioxide they will take out of the water. In recent decades, streams carrying excess fertilizer applied to farmland have increased the nutrient supply in some parts of the ocean. These nutrients increase the amount of plant life in the oceans. As the plants die and sink to the bottom, they will carry carbon with them and thus reduce the amount of carbon near the surface.

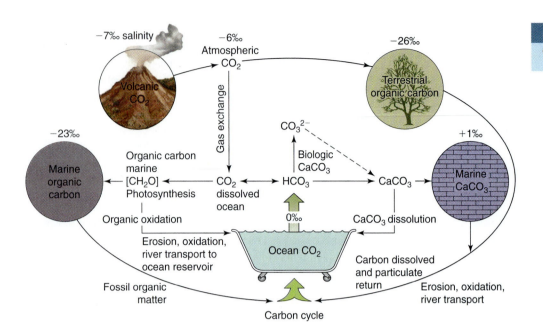

Figure 8.24 Movement of Carbon Through the Environment.

Ozone levels also affect the carbon cycle indirectly. Because ozone reduces the amount of ultraviolet radiation reaching the surface, ozone affects organisms in the upper layer of the ocean. Ultraviolet radiation is harmful to living organisms and could reduce the productivity in the photic zone. Any change in the productivity of the photic zone affects the exchange of carbon between plants, animals, and the water.

Scientists have achieved a good understanding of how biologic and geologic processes in the ocean affect the carbon cycle. Preliminary predictions of the rate of movement of carbon dioxide between the oceans and the atmosphere have been made, but much work remains to be done before all movements of carbon through biogeochemical cycles are understood.

SUMMARY POINTS

1. Ridges composed of basaltic rocks rise from the depths of the ocean and can be traced through all oceans. These basalts have come from the mantle along divergent plate boundaries.
2. Large numbers of shallow-focus earthquakes and active volcanoes lie along the oceanic ridges and the transform faults that cut across the ridges.
3. Guyots are flat-topped seamounts that are submerged. Although the tops were flattened by wave erosion, guyots are now submerged at greater depths than any changes in sea level. They establish that the sea floor has subsided.
4. Great accumulations of sediment lie beneath the continental shelves and slopes of passive continental margins.
5. Deposits of salt that formed early in the history of the opening of the oceans are among the sediments that underlie passive margins along the eastern margin of North America and the western margin of Africa. Oil and gas accumulations are associated with many of the salt structures formed under these margins.
6. Deep-sea trenches mark many of the places where subduction is taking place. However, sediments carried into trenches close to land have filled some of these features, including the margins off the shores of Washington and Oregon states.

7. Sediments buried beneath continental margins are important sources of oil and gas.
8. Manganese nodules found on the deep-sea floor contain large quantities of copper, cobalt, and other metals.
9. Major factors in shaping marine environments include water temperature, currents, sunlight, and water depth.
10. The surface layer of the ocean that is penetrated by sunlight teems with life. The plants and animals that float with currents and are collectively called plankton form the base of the food chain. Nekton, mainly swimming fish and marine mammals, inhabit all depths and feed on plankton and other nekton. The bottom community, referred to as benthos, includes some organisms such as oysters, corals, and sea lilies that are attached to the bottom. Others crawl on the bottom or burrow into the soft sediment.
11. Diatoms, a form of algae, are extremely numerous in the cold waters at high latitude. They have a siliceous framework that sinks to the sea floor upon death and becomes an important part of the sediment in parts of the ocean.
12. Two groups of single-celled animals are especially important contributors to the formation of marine sediments. One of these, a species of foraminifera known as globigerinas, has

calcite shells that resemble snail shells. The other group, known as radiolarians, has hard parts composed of silica. Many of the siliceous and calcareous oozes on the sea floor are accumulations of the remains of these two groups.

13. Crustaceans, many of which are shrimp-like organisms, are especially important in the food chain of whales and fish.

14. Unusual communities of organisms surround hydrothermal vents. The chimney-like formations consist of metals precipitated from clouds of black smoke that pour out of magmas beneath the oceanic ridges. Remarkably, bacteria that thrive in the heat of the hydrothermal vents make up the base of the food chain for this community, which includes clams, mussels, tube worms, crabs, and fish. These communities thrive far below the penetration of sunlight and do not depend on the plankton that support most life at higher levels in the ocean.

15. Reefs are among the most diverse of all Earth environments. Corals build the calcium carbonate structure of reefs, but many plants, notably red algae, and a vast array of invertebrate animals and fish inhabit the environment.

16. The larvae of corals float freely and attach themselves to rocks. Many reefs start as the larvae attach to rocks along the shorelines of volcanoes in the Pacific. As the reef grows, it forms a fringe around the island. Where the sea floor subsides slowly, the reefs grow upward at the same time the island sinks. Eventually all that is left is a circular reef that surrounds a lagoon. Deep drilling indicates that the atolls were originally fringing reefs around volcanoes.

17. Corals prosper in clean, warm water. They depend on algae for a source of oxygen, so most reefs are in shallow water penetrated by sunlight.

18. Sediments cover most of the sea floor. Close to continents, these sediments consist of materials derived from the land, such as sand, silt, clay, glacial sediment, and dust. The ash and dust from active volcanoes is spread extensively across the oceans. In the open ocean, most of the sediment consists of the hard parts of diatoms and foraminifera or red clay, an extremely fine clay derived from a variety of sources, including meteoritic dust. Red clay builds up extremely slowly.

19. Submarine landslides carry sediment from the edge of the continental shelf into the abyssal plains. These landslides carry sediment that originated in shallow water into the abyssal plains.

REVIEW QUESTIONS

1. What properties of marine environments distinguish them from one another?
2. What is a food chain? Cite an example from the marine environment.
3. What are major threats to marine biodiversity?
4. What is so unusual about the fauna that lives around the hydrothermal vents?
5. Briefly describe Darwin's theory about the origin of atolls.

6. What has drilling on guyots revealed about the origin of atolls?
7. What are the main types of inorganic sediment on the sea floor?
8. How would you recognize a turbidity current deposit in a drill core?
9. Where are organic sediments most widespread in the oceans?
10. How does carbon move through the marine environment?
11. Compare the submarine topography of the continental margins of North America.
12. What factors cause the sea floor in the abyssal plains to be so flat?
13. Describe the sea floor you would expect to find on the flanks of oceanic ridges.
14. In what way does the sea floor topography off the Atlantic coast differ from that you would expect to find near an island arc in the western Pacific?

THINKING CRITICALLY

1. Discuss how discoveries in the marine environment have confirmed or called into question theories about Earth processes such as sea-level rise, plate tectonics, and evolution.
2. Discuss the potential impacts, both positive and negative, of tourism on coral reefs.
3. What terrestrial systems impact marine environments? How do they do this? What role do humans have in affecting marine environments?
4. Generally speaking, humans are most familiar with nektonic communities in marine environments. Consider the role of planktonic and benthic communities in supporting nektonic communities. How do they interact? How might nektonic communities be affected by changes in salinity or increased amounts of chemicals in the oceans?

KEY WORDS

atoll	Gondwanaland
bacteria	Great Barrier Reef
barrier reefs	guyots
benthic community	hydrothermal vents
benthos	krill
Bikini Atoll	manganese nodules
coccoliths	nekton
coelacanth	pelagic community
copepods	photic zone
crustaceans	phytoplankton
diatoms	plankton
flagellates	radiolarians
foraminifera	red clay
fringing reef	reef flat
globigerina	zooplankton

Ocean Dynamics

A breaking wave forms a beautiful curl as the wave moves toward shore.

Chapter Guide

Water covers nearly 70% of the planet's surface; so, it is not surprising that the properties of seawater have a great impact on the planet's environment and other Earth systems. The salts in seawater came largely from land environments, and important interchanges continue between the oceans, atmosphere, and biosphere. The characteristics of seawater are important determinants of the ecology of transitional and marine environments, and they control the deeper circulation of water through the ocean basins. Scientists first pieced together the global pattern of near-surface circulation and its relation to prevailing winds. The movement of water through the deeper parts of the oceans remained unknown until oceanographers discovered the connection between the properties of seawater and its movement. Although oceanographers have learned how to trace these movements, the oceans are vast, and many questions regarding the pattern of deep circulation, and the relation between oceanic circulation and climate, remain unresolved. Nevertheless, it is clear that air and seawater are part of the same large-scale system, and that understanding the dynamics of the sea is a critical foundation for learning about Earth's climate.

INTRODUCTION

The Greek historian Herodotus compiled a map of the world as it was known in 450 B.C. His map shows that the Greeks had a good understanding of the geography of the Mediterranean area. Early sailors followed the coast around southern Africa into the Red Sea and as far as the Indus River, but their notion of the shape of the South Atlantic and Indian oceans was vague. The exploits of another group of early voyagers, the Vikings, are known through stories passed from one generation to the next. According to one of these stories, Bjarni Herjolfsson, a merchant sailor, sailed west from Iceland around A.D. 986. He sighted land but turned around without landing and returned to Iceland to find that his father had sailed off with Eric the Red to a western land Eric the Red referred to as "green land." Ruins of early Viking settlements remain on Greenland. Eric the Red described bare, rocky coasts with deep valleys leading into green hills capped by mountains of ice. He settled there with his family, including his oldest son Leif Eriksson. Bjarni sailed off to the west looking for his father, but sailed too far south and missed the land with mountains of ice. Instead, he found a low coast with flat, forested lands. He had landed somewhere along the coast of Labrador or Newfoundland and was probably the first European to discover North America.

These early adventurers sailed through waters that even modern sailors consider perilous. The early sailors must have learned to take advantage of the complex systems of currents that flow in the northern part of the Atlantic Ocean (figure 9.1). The East Greenland current helped carry Bjarni, Eric, and later Leif on their voyages from Iceland to Greenland and eventually to North America. They probably used the North Atlantic current to return home. Without these currents, the history of exploration during the days of sailing ships would have been quite different. Columbus was fortunate enough to discover the strong easterly winds in the trade-wind belt that drive the North equatorial current across the Atlantic (see figure 9.1). Modern maps of this current help explain why Columbus landed in the Caribbean even though he departed to the west from Spain. Currents, and the physical properties of seawater that govern them, have numerous other impacts on humans and environmental systems.

The vast expanses of seawater—70% of the surface area of Earth—and the capacity of water to absorb and store heat energy make the oceans one of the most important elements in Earth's climatic system. The oceans play a critical role in heat balance on Earth by exerting a moderating influence on climate. Seawater absorbs heat energy in the daytime, especially in summer, and at low latitudes where the amount of incoming solar radiation is greatest. The water loses heat energy at night and in winter, and eventually, water heated near the equator moves in currents to other parts of the ocean where it helps warm the air at high latitudes. In addition to this interchange of heat energy between the oceans and air, other substances—water vapor, oxygen, and carbon dioxide—also move between the surface of the sea and the air. Through these interactions between the surface of the sea and the overlying atmosphere, air masses acquire their temperature characteristics and moisture content. Thus, climate is affected by changes in the temperature of water near the surface of the oceans, which depends in large part on oceanic circulation. Surface oceanic circulation is largely influenced by wind, but deeper circulation depends on the temperature and chemical composition of the water. So, climate depends largely on the ocean, and the ocean, in turn, depends somewhat on climate.

COMPOSITION AND PROPERTIES OF SEAWATER

Dissolved Gases in Seawater

The surface of the sea is a contact zone between atmospheric gases and a saltwater solution. Gases, such as water vapor, oxygen, nitrogen, and carbon dioxide, move across the surface. Wave action facilitates this interchange, but the exchange takes place even if the air-water interface is perfectly still. The interchange of gases continues until the gas pressure in the atmosphere equals that of gases in the upper layer of seawater. Under this condition, the movement of gases from the air into the water is equal to the movement of gases from the water into the air—an equilibrium condition. Temperature and atmospheric pressure govern this equilibrium, but because pressure at sea level varies so slightly, temperature is more important as a cause of the variations in the gas content in seawater. As the temperature rises, gases move more freely between water and air.

Ocean waters are so well mixed that the proportions of salts and inert gases, such as nitrogen and argon, to one another remain relatively constant (table 9.1). In contrast, the concentrations of materials involved in biological processes, such as oxygen, carbon dioxide, and nutrients (e.g., nitrates, phosphates, and silicates), vary considerably. These vary because they enter into processes like respiration and photosynthesis, and they may combine to form minerals such as calcite (calcium carbonate), a major constituent of sediment on the sea floor.

In the upper levels of the ocean, huge numbers of microscopic plants live in water that sunlight penetrates, the **photic zone** (figure 9.2). Here, through the process of photosynthesis, plants use the energy of sunlight to transform carbon dioxide and water into carbohydrates and oxygen. Animals in the photic zone use the carbohydrates as their primary source of food and the oxygen in their respiratory processes. When the animals die and their remains sink, decay processes consume the oxygen, which remains low at depth.

Below the photic zone, where sunlight does not reach and thus oxygen-producing photosynthesis cannot occur, concentrations of oxygen decrease. It is also here that oxygen reacts with dead organic matter and is used in respiration. At a depth of several hundred meters, oxidation causes a rapid decrease in oxygen (see figure 9.2). Circulation of seawater carries oxygen to depths below its origination point, but if oceanic circulation is poor, anaerobic (oxygen-poor) conditions may prevail. Such conditions are present in some deep basins, such as the Black Sea, where water is stagnant and renewal of oxygen takes place slowly. Circulation of

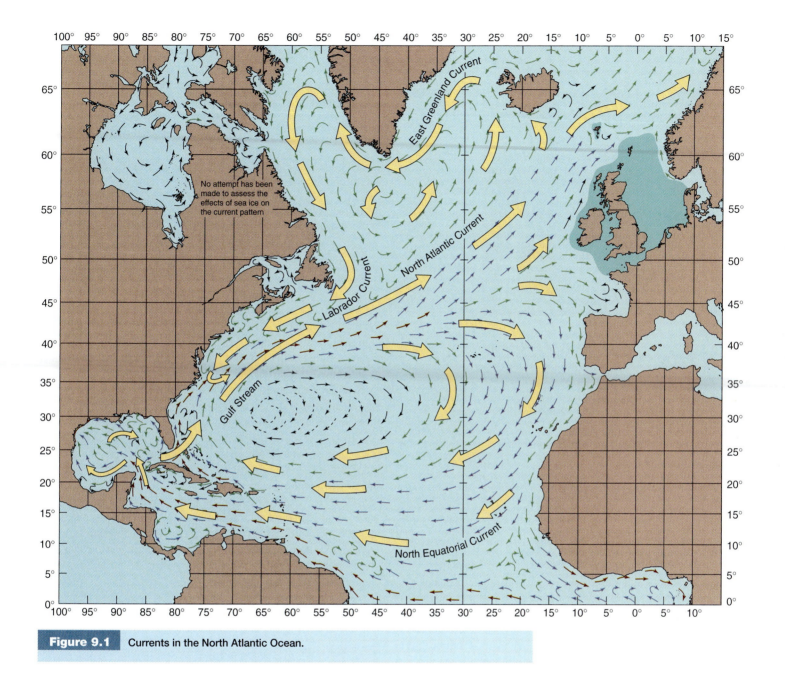

Figure 9.1 Currents in the North Atlantic Ocean.

water at depth in the oceans depends primarily on the physical properties of the water, and the salt content of the water is a primary determinant of these properties.

Salts in Seawater

Ocean water contains a variety of salts that occur as ions and gases mixed with the water. Although most of the known elements occur in trace quantities in seawater, only a few elements make up most of the salts. The salt content of an average sample of seawater weighing 1,000 grams (2.2 pounds) contains the elements shown in table 9.1.

The composition of seawater is somewhat similar to that of human body fluids, an interesting observation that supports the idea that humans evolved from marine vertebrates. The salts in seawater and their abundance in crustal rocks are about the same (with the exception of chlorine, bromium, borate, sulfate). The salt content of the oceans comes, in large part, from elements released from crustal rocks through weathering processes and carried into the oceans by streams. Chlorine, the most abundant ion in seawater (see table 9.1), and sulfate come mainly from volcanic eruptions.

Wave action and oceanic circulation mix the constituents in seawater so thoroughly that the proportions of the various salts in seawater are nearly constant. Because the proportions of the salts

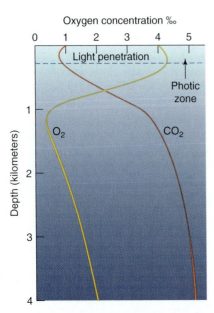

Oxygen concentration ‰

Figure 9.2 **Photic Zone.** This profile shows representative concentrations of oxygen and carbon dioxide in the upper 4 kilometers of the open ocean. Oxygen enters the surface of the ocean where an interchange takes place between the atmosphere and the upper layer of water. Algae, heavily concentrated in the upper part of the ocean where photosynthesis occurs, also produce oxygen. Oxygen is consumed as organic matter from the surface layer sinks. This process causes an oxygen minimum to form at a depth of about 1 kilometer. In this diagram, oxygen concentrations are expressed in milliliters of oxygen per liter of water.

from adjacent land areas. Rainfall, melting ice, and the addition of fresh water where streams enter the sea dilute seawater and reduce its salinity. Evaporation and freezing of seawater increase salinity. Salinity is important in oceanic circulation mainly because of its role in determining seawater density.

Physical Properties of Seawater

Density depends especially on temperature, salinity, and to a lesser extent on pressure. The density of seawater increases as salinity or pressure increases or as temperature decreases. This temperature effect changes, and water density decreases as water temperature drops below 4°C. Because water is almost incompressible, the effects of pressure, which increases at a rate of 0.1 bar of pressure per meter of depth (1.47 pounds per 3.28 feet), are slight. Some of the least-dense water at the surface of the ocean (density 1.018 grams per cubic centimeter) has a salinity of 30‰ and a temperature of 30°C. Some of the highest-density seawater occurs at a depth of 8 kilometers (5 miles), has a density of 1.063 grams per cubic centimeter, a salinity of 35‰, and 0°C temperature. Even these seemingly slight density variations cause more dense water to sink and drive the deep circulation in the ocean basins.

If nothing disturbed the water in the oceans, a stable density-stratification would exist with the lowest-density water at the surface and water of increasing density at greater depth. This condition does not exist on Earth for many reasons: (1) surface temperatures vary (more or less with latitude), reaching a minimum in the polar regions (figure 9.4); (2) Earth rotates; (3) surface rates of evaporation vary; (4) fresh water enters the sea in most coastal areas; (5) distribution of plants and animals in the sea varies; and (6) some salts removed from seawater by biological and chemical processes contribute to sediment formation. All of the above cause temperature and salinity and, therefore, water density to vary with depth and from one part of the ocean to another (figures 9.5 and 9.6).

remain constant, the quantity of any single constituent provides a measure of the total salt content. Oceanographers refer to the total number of grams of salts per unit per kilogram of water as **salinity.** Because it is abundant and for measuring ease, geochemists generally use chlorine to determine salinity. They express salinity as parts of salt per thousand parts of water. This is symbolized by ‰. Salinity determines some of the physical properties of seawater, such as the density and ability of the water to transmit electrical current. For this reason, it is possible to determine the salinity of seawater by measuring these physical properties.

Salinity ranges from about 33‰ to 37‰ in the open ocean, with average values of 34.9‰ in the Atlantic and 34.62‰ in the Pacific (figure 9.3). Salinity is lower in coastal waters where streams carry fresh water into the ocean, and it tends to be high in arid or semi-arid regions where much evaporation occurs and where streams carry small quantities of fresh water into the sea. The Mediterranean (salinity of 38‰) and Red Sea (salinity of 40‰) have the highest salinity of all oceans. Both of these seas lie within arid and semiarid regions, and neither receives a large inflow of fresh water

Table 9.1	Average Salt Content of Seawater Expressed in Grams of Salt per Kilogram of Water, or ‰ (see figure 1.3)
Chlorine	18.97 grams
Sodium	10.47 grams
Sulfate	2.65 grams
Magnesium	1.28 grams
Calcium	0.41 grams
Potassium	0.38 grams
Bicarbonate	0.14 grams
Bromine	0.06 grams
Borate	0.02 grams
Strontium	0.01 grams

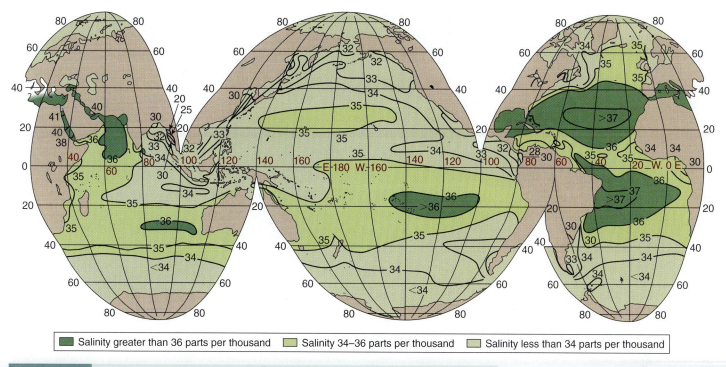

Salinity greater than 36 parts per thousand   Salinity 34–36 parts per thousand   Salinity less than 34 parts per thousand

Figure 9.3   Distribution of Seawater Salinity Near the Surface of the Ocean. Salinity is expressed as the number of parts of salt per thousand parts of water. The contours connect points of equal salinity. Note the high salinity of seawater between latitude 20–40° north and south of the equator. Heavy rainfall near the equator dilutes the seawater there.

Variations in Temperature and Salinity in the Ocean

Near-surface temperatures in the oceans range from 35°C in equatorial latitudes to freezing in polar latitudes (see figure 9.4). Because of its salt content, seawater does not freeze until the water cools to −2°C. Although the range of water temperature in the ocean is great, the bulk of seawater has a temperature less than 6°C and nearly half is under 2°C. Temperature in the oceans varies with depth and with latitude. Strong seasonal variations in temperature occur near the surface in the zone that sunlight penetrates. This upper 100 to 200 meters (330 to 660 feet) of the ocean is called the **surface layer** (see figure 9.5). In polar regions, water temperature does not vary much with depth, but toward the equator, warming in the spring and summer produces a shallow warm layer. As winter approaches, the temperature drops, and storms cause mixing of the water to a greater depth. Below the surface layer and down to about 800 meters (2,600 feet), temperature decreases rapidly with depth, in a zone called the **permanent thermocline** (see figure 9.5). The thermocline is a zone of transition between lower-density, warmer water above and cold and more-dense water below. Water in the surface layer above the thermocline moves in a near-surface circulation.

NEAR-SURFACE CIRCULATION IN THE OCEANS

Study of ocean currents began soon after the first attempts to navigate the seas. Because currents have dramatic effects on the speed of sailing vessels, sea captains took special notice of them. By the middle of the nineteenth century, Admiral Matthew F. Maury, who started naval research in the United States, had compiled remarkably accurate maps showing the general pattern of near-surface currents. Since then, oceanographers have progressively defined current patterns and identified the underlying forces that give rise to them.

Tracing Near-Surface Circulation

The large-scale pattern of near-surface circulation forms a relatively simple picture (figure 9.7). Near the equator, water moves from east to west in the **equatorial currents.** In the middle latitudes of both hemispheres, water returns from west to east, the **west-wind currents.** Between the prevailing westerly currents and the equatorial easterly currents, near-surface water flows in large circular patterns called *gyres.* The movement in these gyres is

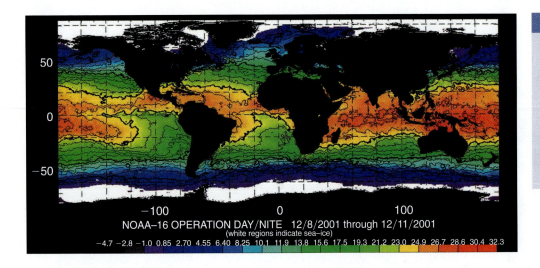

Figure 9.4 Map Showing the General Distribution of Temperatures in the Oceans in August. As expected, the warmest water lies near the equator. Currents carry warm water farther north and south on the western side of both the Atlantic and Pacific oceans. In January and February, the warm water shifts south of the equator.

clockwise in the Northern Hemisphere and counterclockwise in the Southern Hemisphere. Land areas in both hemispheres break up and deflect the equatorial and westerly currents.

In the Southern Hemisphere, the strongest, simplest of all currents flows around Antarctica where the prevailing westerly winds cause the current to follow a simple, circular pattern around Antarctica from west to east. The narrow passage between South America and Antarctica, the Drake Passage, constricts the currents and causes the velocity of the circum-Antarctic current to increase, making this the most treacherous stretch of water in the ocean. Otherwise, the flow is unobstructed. This circum-Antarctic current brings about mixing of waters in all of the oceans in the Southern Hemisphere.

In the North Pacific, the Aleutian Islands block large-scale movement of water between the Arctic and Pacific oceans. Consequently, a large clockwise gyre in the North Pacific dominates the

current pattern. On the western side of the Pacific, the equatorial currents swing northward along the coast of Japan, becoming the **North Pacific current.** This current continues from west to east as it flows past the Aleutian Islands (see figure 9.7) and impinges on the coast of Washington where it splits. One part deflects northward along the coast of British Columbia and southern Alaska, forming the **Alaska current.** The other component turns to the south and becomes the **California current,** which eventually rejoins the equatorial drift. The California current brings cold waters and marine life down the West Coast that are usually found much farther north.

In the Atlantic, the Northern equatorial current strikes the northern coast of South America and deflects to the north, where it flows through the Caribbean and into the Gulf of Mexico. Here, the **Gulf Stream** (figure 9.8) forms and flows from the Gulf of Mexico through the strait between Florida and Cuba (see figure 9.7). This

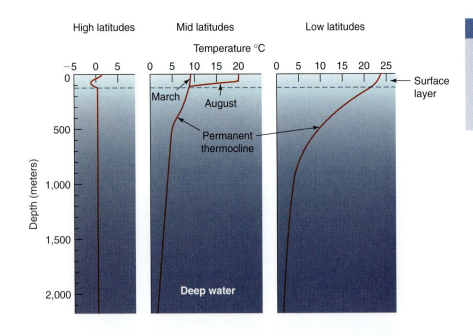

Figure 9.5 Temperature Structure of the Ocean at Different Latitudes. Seasonal variations in temperature are marked in middle and low latitudes, but only slight warming occurs at high latitudes. Seasonal variations are most prominent in middle latitudes.

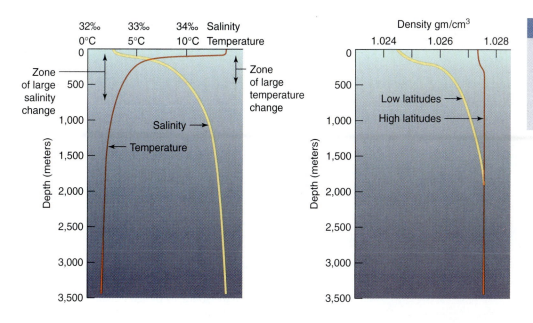

32‰   33‰   34‰ Salinity
0°C   5°C   10°C Temperature

Zone of large salinity change

Zone of large temperature change

Salinity →

← Temperature

Depth (meters)

Density gm/cm³

1.024   1.026   1.028

Low latitudes →

High latitudes →

Depth (meters)

Figure 9.6 Representative Profiles of the Ocean Showing Variations in Salinity, Temperature, and Density with Depth. Note the rapid changes that occur in the thermocline.

constriction increases the velocity of the current to nearly 5 kilometers per hour (3 miles per hour). This segment, known as the **Florida current,** flows northward along the southeastern coast of the United States, where it is joined by other currents deflected to the northern side of the Caribbean Islands. This stream of warm, northward moving, equatorial water is as much as 80 kilometers (50 miles) wide and over 1 kilometer (3,300 feet) deep. At times, it follows a meandering course and occasionally breaks up into several threads flowing in a braided pattern with disconnected swirling loops called *eddies* that lie along the edge of the main flow. North of Cape Hatteras, the Gulf Stream moves farther from the coast. It becomes

the North Atlantic current and flows eastward almost all the way across the Atlantic. Off the western coast of Europe, it splits into two parts; one moves southward back toward the equator, and the other moves northward along the coast of Norway.

Water moves freely through the passages between the Arctic and North Atlantic oceans. In the North Sea, currents move northward into the Arctic along the coast of Norway. In a return flow, a strong southerly current, the **Labrador current** (see figure 9.1), flows southward from the Arctic Ocean between Greenland and Labrador. South of Newfoundland, some of the water in the Labrador current mixes with the Gulf Stream, but the balance of the water in the

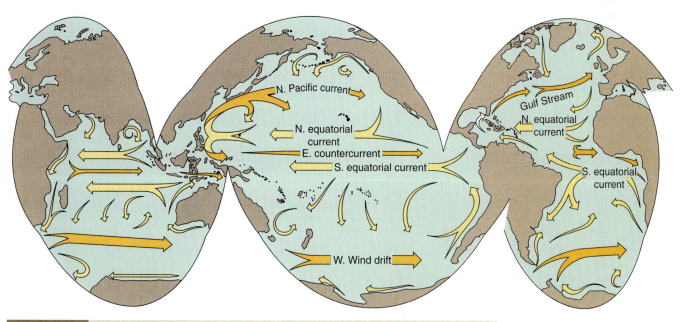

Figure 9.7 Schematic Map Showing Circulation of Water Close to the Surface of the Oceans.

current continues south along the east coast of the United States as far as Cape Hatteras.

Causes of Near-Surface Circulation

Several mechanisms drive currents that move near the surface (figure 9.9). The most important of these mechanisms are

- Prevailing winds that cause wind-driven currents
- Coriolis and Ekman effects caused by Earth's rotation
- The slope of the ocean surface
- Gravitational attraction of the Moon and Sun, which cause tidal currents

Wind-Driven Currents

A seasonal reversal of currents in the northern part of the Indian Ocean afforded one of the first indications that winds drive ocean currents (figure 9.10). In the southern Indian Ocean, the west wind drives a persistent west to east movement, but north of the equator, the currents follow the direction of the prevailing winds and reverse direction with the seasons. During the summer season (May–September), prevailing winds, called *monsoons*, blow from the southwest toward India and Asia. They bring drenching rains of moisture evaporated from the Indian Ocean. When the southwest monsoons cease and the winds begin to blow out of the northeast, as they do in the winter, the currents in the ocean also change.

Those who observe the ocean for a long time find that strong, persistent winds generate waves and surface currents. The mechanisms by which wind energy transfers to the surface of the sea are less obvious. Usually, enough turbulence is present in the wind to create eddies, minor updrafts, and irregularities in the flow pattern of air over the water surface. Turbulence and updrafts initiate small depressions and rises in the surface of the water. Once these irregularities in the sea surface form, even a slight wind will continue the movement. If the wind blows consistently from one direction, it exerts a more or less continuous pressure on one side of the waves. As the wind passes the crest of the waves, it exerts a pull, a suction, on the other side. In this manner, currents develop

Figure 9.8 Satellite Image of the Gulf Stream. Note the prominent eddy currents associated with this current. At times, it follows a meandering course.

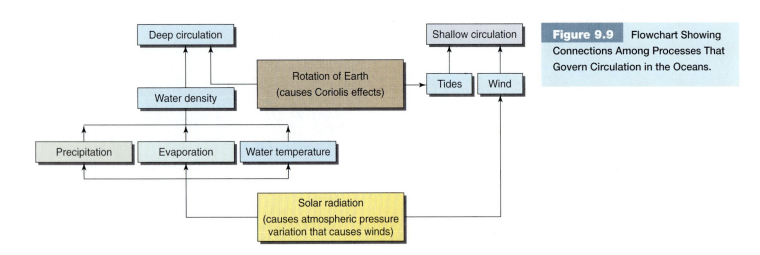

Figure 9.9 Flowchart Showing Connections Among Processes That Govern Circulation in the Oceans.

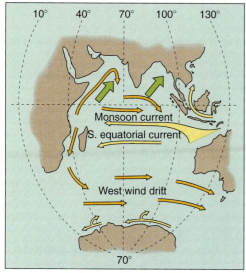

Current pattern in the Indian Ocean during May to September when monsoon winds flow from the southwest, indicated by solid arrows.

Currents from May to September

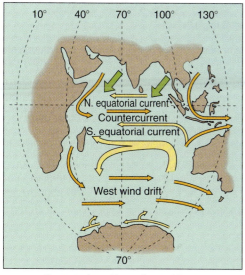

Generalized current pattern in the Indian Ocean during November to March when winds blow southwest in the northern part of the ocean.

Currents from November to March

Figure 9.10 **Wind-Driven Currents Cause Monsoons.** Prominent seasonal changes occur in near-surface circulation in the Indian Ocean. Changes in the wind direction (green arrows) associated with the monsoons cause these circulation changes.

where the wind blows consistently from one direction. In the tropics, trade winds generate strong currents of this type on both sides of the equator. In middle latitudes, westerly winds set up currents that flow from west to east.

Coriolis and Ekman Effects

G. G. Coriolis, a French engineer for Napoleon, demonstrated that objects in motion on a rotating sphere tend to follow curved paths. This effect takes place because stationary objects at the equator, such as an island, travel at a higher velocity than objects located

closer to the poles. Consequently, something like water or air that moves from a high latitude toward the equator passes over parts of the surface that are moving west to east at a higher velocity than the water or air. In the Northern Hemisphere, the current appears deflected to the right. Similarly, water that moves north or south from the equator passes over an Earth surface that is rotating more slowly, and the path followed by the current bends to the right in the Northern Hemisphere and to the left in the Southern Hemisphere. These deflections are possible because the frictional drag between Earth and the atmosphere and oceans is so small.

The Coriolis effect varies with latitude. It is minimal along the equator and increases toward the poles. Because of the Coriolis effect, surface-water currents in the Northern Hemisphere move along paths to the right of the prevailing wind direction, and currents in the Southern Hemisphere move to the left of prevailing winds. The importance of the Coriolis effect in the ocean became evident in the late nineteenth century when the Norwegian scientist, artist, and Arctic explorer Fridtjof Nansen noticed that the direction Arctic pack ice drifts is 20 to 40° to the right of the prevailing wind direction. In 1902, a Scandinavian physicist, V. W. Ekman, showed that in a similar fashion, Coriolis effects cause wind-driven currents at the surface of the ocean to be directed 45° to the right of the wind. Ekman suggested that this deflection should extend to successively greater depths in the ocean. The water behaves as a succession of layers in which each layer exerts a drag on the layer below. The effect of each layer is like that of the wind on the uppermost layer. The angle of deflection increases, and the velocity of the wind-driven current decreases with depth, as shown in figure 9.11 (the length of the

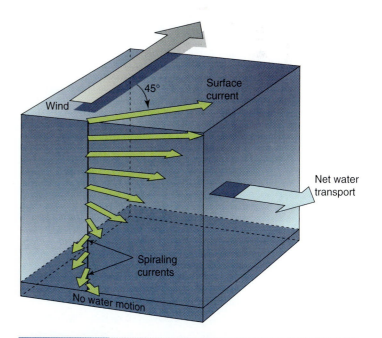

Figure 9.11 **Ekman Spiral in the Northern Hemisphere.** The Ekman spiral represents the direction and strength of water movement resulting from the wind blowing across the surface as shown.

arrows corresponds to the velocity). The Ekman effect mainly involves water in the upper 100 meters (330 feet) of the oceans. According to this analysis, currents should flow in different directions at different depths. The flow should turn progressively to the right until at some depth, the flow should be in the opposite direction to the surface currents. Later, oceanographers confirmed that parts of the water beneath the Gulf Stream actually do flow in the opposite direction from the surface flow.

As a result of the Ekman effect, the net movement of water in the surface layer is almost at right angles to the direction of the wind everywhere in the Northern Hemisphere. Thus, if the wind blows from the north along a north-south trending coast in the Northern Hemisphere, wind-driven currents flow out to sea along the coast (figure 9.12). As surface water moves away from the coast, deeper water rises toward the surface in a process known as **upwelling.** Where this movement takes place, the deep water, which is rich in nutrients, supports a large fauna and flora near the surface. If the wind direction reverses, near-surface water moves toward the shore and tends to pile up along shore. The buildup of water along the shore blocks the rise of water from depth.

Slope of the Sea Surface as a Cause of Near-Surface Circulation

Movement of water by wind-driven currents causes and maintains variations in the level of the sea surface that contribute to surface circulation. Some of the differences in the level of the sea surface are persistent. Across the Florida Peninsula, the water on the Gulf of Mexico side is nearly 18 centimeters (7 inches) higher than the water on the Atlantic side. Because of differences in level, locks are necessary in the Panama Canal to lower ships from the Atlantic to the Pacific Ocean and raise those going the other direction. A

strong fan blowing over water in a pan causes this type of effect. The difference in level results from the piling up of water against the shore. This piling up is most evident on the western side of the oceans in the trade-wind belts.

As a result of the Coriolis effect, water in the surface layer in the Northern Hemisphere moves to the right of the prevailing wind direction and piles up in a broad dome in the North Atlantic. Less-dense water piles up in the center of each of these rotating currents, called gyres, raising the level of the ocean. As the water piles up, water moves off the rise, tending to restore the level of the sea, but the surface of the sea never becomes perfectly level. Instead, the rate of movement of water down the surface of the elevated areas approximately equals the rate of movement deeper in the surface layer toward them. As soon as the water starts to flow downslope, Coriolis effects cause it to veer to the right. The combination of downslope flow and Coriolis effect produces currents, called **geostrophic currents** (figures 9.13 and 9.14), that move around the dome.

Influence of the Shape of the Ocean Basins on Near-Surface Circulation

The shape of the ocean basin confines near-surface ocean currents and deflects them when they approach continental margins. For this reason, surface currents in the oceans of the Northern Hemisphere, which contains four-fifths of all land area on Earth, are more complex than the currents in the Southern Hemisphere, most of which is covered by water.

If no land areas were present, equatorial currents would flow all the way around Earth from east to west, but South America obstructs the currents in the Atlantic; Australia and southeast Asia obstruct currents at equatorial latitudes in the Pacific; and Africa interferes with equatorial currents in the Indian Ocean. In all three

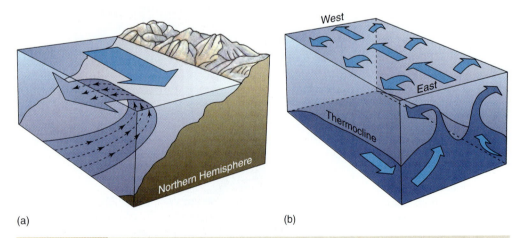

(a)   (b)

Figure 9.12 Ekman Effects. Upwelling of deep water occurs when the Ekman effects cause water to move offshore. (a) Upwelling along a coast in the Northern Hemisphere. The Ekman effect would cause water to move toward shore if the surface winds blew as shown in this diagram. (b) Upwelling takes place along the equator because water in the northern equatorial current moves north while that south of the equator moves south. This divergence causes upwelling along the equator.

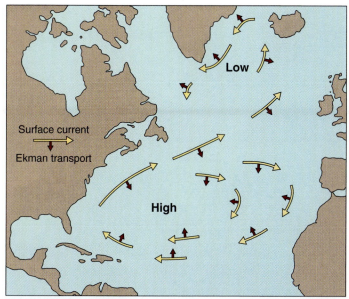

Surface current

Ekman transport

Low

High

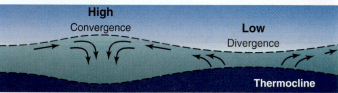

High
Convergence

Low
Divergence

Thermocline

Figure 9.13 Geostrophic Currents. In the Northern Hemisphere, geostrophic currents originate as a result of the piling up of water in the center of the large gyres. Water flows away from the center of the gyre formed by counterclockwise currents. This results in a depression in water level. Where the gyre is caused by clockwise currents, the water piles up. As water near the surface flows off of the pile, Coriolis effects cause it to follow a path around the high water.

oceans, the currents split as they approach land and become part of the large gyres (see figures 9.7 and 9.14). Close to shore, the configuration of the coast causes many currents to split and flow around obstacles in their paths. Although they vary and form complex patterns in many places, much more is known about these near-surface currents than about the currents that flow deep in the oceans.

DEEP-SEA CIRCULATION

Before oceanographers had ways of studying the deep sea, many assumed that deep-sea currents were too weak to transport sediments on the bottom of the ocean. As cameras capable of photographing even the deepest parts of the sea floor began to provide pictures of the sediment on the ocean bottom, it became obvious that oceanographers were wrong. Photographs revealed ripple marks similar to those found in shallow streams or along beaches. Currents flow-

ing over sediments, even in the abyssal plains in water depths of 5 to 6 kilometers (3 to 4 miles), caused these deep ripples. Only currents strong enough to move sand grains can cause such ripples. The rush of submarine landslides causes currents that can produce ripples on the ocean floor in deep water. But ripples also develop in places where these currents are unlikely. Currents associated with the Gulf Stream extend to depths of over 1 kilometer (0.62 mile), and the presence of ripple marks on almost all seamounts indicates that relatively strong deep-sea currents must be present in many parts of the ocean.

Tracing the Movement of Deep Currents

While sampling water off the coast of Brazil, one of the pioneers of oceanic circulation exploration, Alexander Humbolt, discovered cold water at shallow depth. Because this cold water could not have formed in these equatorial waters, Humbolt concluded that it originated in polar regions and flowed toward the equator at depth. He suggested that much of the near-surface circulation is essentially a return flow of currents that originate at high latitudes where surface seawater cools and sinks into the ocean basin.

Variations in temperature and salinity that affect water density are the primary causes of deep-water circulation. However, most of the processes responsible for variations in temperature and salinity occur at the surface of the sea where large water masses acquire their physical characteristics. At or near the surface, cooling and freezing of seawater causes seawater density to increase. Thus, the cold temperatures in polar regions increase the density of seawater. High temperature and large quantities of rainfall that lower salinity in the tropics reduce the density of seawater. Volcanic activity on the sea floor may also cause the temperature of deep water to increase locally, but it is not clear that this affects deep circulation. Evaporation and cooling of seawater increase its density, while dilution resulting from precipitation, influx of fresh water from streams, and melting of ice all reduce salinity and therefore density.

At high latitudes, oceans radiate heat and give up an amount of heat that is about equal to the total amount of solar radiation reaching Earth's surface at those latitudes. Thus, the oceans moderate the air temperature in these regions. As the seawater releases heat at high latitudes, its temperature decreases. The resulting cold, high-density water sinks, sliding downslope into the deep sea, pushing the bottom water ahead of it. The shape of the sea floor directs the path of the sinking waters.

Because deep currents are generally extremely slow, it is impractical to determine movement of deep water with devices fixed to the sea floor. The vast stretches of the deep sea and the cost and difficulty of performing studies there make it difficult to measure deep circulation. For decades, oceanographers have traced the movement of water masses by measuring their physical properties (temperature and salinity). These studies revealed that vast quantities of cold water (about 2.5°C with salinity of 35‰) sink in the northern part of the Atlantic Ocean, forming a mass of water

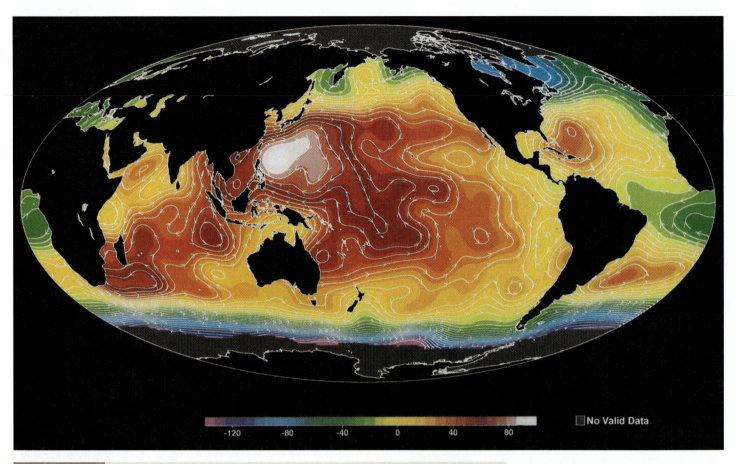

No Valid Data

-120   -80   -40   0   40   80

Figure 9.14   Map of the Ocean Surface Showing Sea-Surface Height Data Taken by Radar
Altimeter. Arrows indicate the direction of flow of geostrophic currents that move along contours in a
clockwise direction in the Northern Hemisphere and counterclockwise in the Southern Hemisphere.

known as the **North Atlantic deep water** (figure 9.15). This water
sinks and moves south through the Atlantic. The flow carries
nearly 15 times that of the flows of all the rivers in the world.
Studies of the distribution of tritium, a radioactive isotope of hydrogen, in seawater confirm that vast areas of the North Atlantic
contain North Atlantic deep water. Tritium is an extremely rare
component of seawater, but during the atmospheric testing of
atomic bombs in the 1960s, large quantities of tritium were introduced into the atmosphere and oceans. Because water that contains
tritium must have been near the surface at the time of the atomic
tests, years later it is possible to trace the movement of water that
contains tritium (figure 9.16). As water in the North Atlantic deep
water moves south, it encounters a mass of cold (about 0°C), high-
salinity water that originates near Antarctica and moves in a strong
current that extends to great depth around Antarctica. This
extremely cold, high-density water sinks from the continental shelf
around Antarctica, particularly in the Weddell Sea, which is
located south of the southern tip of South America.

The water mass formed in the Weddell Sea, called *Antarctic
bottom water* (see figure 9.15), has the highest density of any water

in the oceans. This water mass has a temperature of −1°C and
contains about 34.6‰ salt. As this water mass sinks, it moves
about 2 to 3 cubic kilometers (70 to 100 million cubic feet) of
water per second into the southern oceans. This water spreads
northward across the sea floor. Because of its high density, Antarctic bottom water sinks to the bottom of the ocean and lies beneath
the North Atlantic deep water, which eventually rises in the South
Atlantic. Oceanographers defined the deep circulation in the
Atlantic Ocean before they had much information about the other
oceans. In recent years, more data from other parts of the world
have led to the development of a model of circulation commonly
referred to as the conveyor belt.

The Conveyor Belt

The **conveyor belt model** of oceanic circulation defines and inte-
grates the movement of water at all depths throughout the oceans.
According to this new model, the mass of cold water that originates in the northern Atlantic moves southward along the western
side of the Atlantic until it encounters and becomes part of the

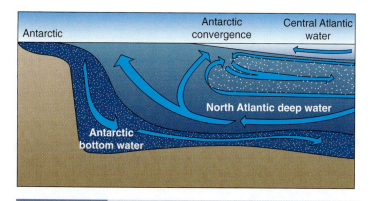

Figure 9.15 Schematic Cross Section from Antarctica North in the Atlantic Ocean Showing the Major Water Masses. The highest density water is in the Antarctic bottom water. As this bottom water moves north, it encounters huge quantities of North Atlantic deep water that flows over the bottom water. Because the central water mass originates near the surface in equatorial latitudes, its high temperature and low salinity give it low density. Consequently, it remains near the surface.

circum-Antarctica flow. This strong, deep current carries the water into the Indian and southern Pacific oceans. As the deep water in these currents mixes with warmer water, it rises to a higher level in both the Indian and Pacific oceans and begins a return flow toward the west in equatorial latitudes (figure 9.17). The return flow eventually carries water up along the eastern side of the Atlantic back to the area in which it originates. Studies of radioactive tracers indicate that it takes about 600 years for water to complete this journey.

OCEAN-ATMOSPHERE INTERACTIONS

Many of the largest air masses acquire their temperature and moisture content while over the ocean. That air then follows the patterns of atmospheric circulation driven by Earth's rotation and other processes described in chapter 12. Most ocean-atmosphere interactions occur at the surface of the sea where gases can move between the two fluids. The temperature of the air located over the ocean gradually approaches that of the water. Hot air that moves over the sea from hot desert regions cools, and cold air from polar regions warms as it moves across the surface of the ocean. Similarly, dry air picks up moisture as water evaporates from the ocean, and air heavily laden with moisture loses part of its moisture content by precipitation. In general, the ocean exerts a moderating influence on the atmosphere, and this affects the climate of the regions over which the air circulates. Thus, changes in the temperature or moisture content of the atmosphere alter the temperature and salinity of the surface waters. Conversely, changes in the

temperature of the near-surface water in the oceans affect air masses and the climate of regions over which the air passes.

Changes in the circulation of seawater played an important role in the climatic changes that occurred during the ice ages. Two processes increase the density of the water in polar regions. The cold temperature at high latitudes is a primary cause of increased density. The migration of salt out of freezing water increases the salinity and the density of water near the ice. As the dense water sinks, it pushes deeper water ahead of it and drives the deep circulation. A decrease in the density of water at high latitudes decreases the rate at which cold water sinks. Dilution of seawater at high latitudes caused by increased precipitation or by warming decreases the density of surface waters and slows their sinking. Oceanographers have developed computer models to illustrate the effects global warming can have on the circulation of the ocean. All of these models indicate that a warmer atmosphere would slow circulation in the oceans. Increased quantities of carbon dioxide in the atmosphere is one potential cause of global warming. One

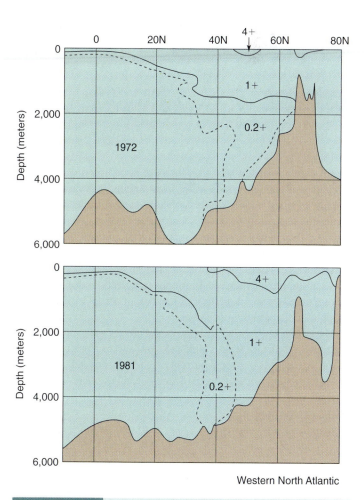

Western North Atlantic

Figure 9.16 Cross Sections Drawn from North to South Through the Western Part of the North Atlantic Showing the Concentration of Tritium in Water Samples Collected in 1972 and in 1981.

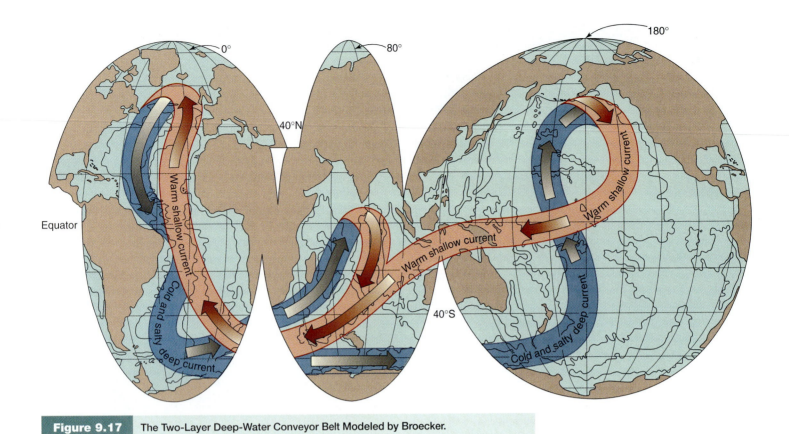

Figure 9.17 The Two-Layer Deep-Water Conveyor Belt Modeled by Broecker.

model indicates that a four-fold increase in carbon dioxide would warm the air enough to stop water from sinking in polar regions. If sinking of cold surface water stopped, deep currents would become slower or stop altogether, and surface waters would become warmer. This model predicts that the sinking of cold, heavy water that drives deep currents and causes water in the deep sea to return to the surface would die out within 200 or 300 years. This would cause a significant increase in global temperature that would be accompanied by rapid melting of ice, rise in sea level, and drastic shifts in the geography of biomes.

Mechanical processes and buoyancy effects may influence the rate at which surface waters sink and cause an overturning of water in the oceans. The strong currents that flow around Antarctica move huge quantities of water. Some oceanographers think these currents may pull water in from the bottom of the Atlantic, facilitating overturning in the North Atlantic. Another phenomenon known as El Niño shows how changes in one part of the ocean or atmosphere may have far-reaching effects on another.

El Niño

The name El Niño, literally, "the child," was given to a phenomenon first recognized along the coast of Peru, an area known for the cold waters rich in nutrients and teeming with fish. In some years, about Christmastime, a warm current called El Niño moves in from the west along this coast. Occasionally, the current is unusu-

ally strong and warm. The warm, moist air that accompanies this warm current also causes heavy rains to fall inland in an area that is normally dry. As described by R. C. Murphy (1926),

> . . . the sea is full of wonders, the land even more so. First of all the desert becomes a garden . . . The soil is soaked by the heavy downpour, and within a few weeks the whole country is covered by abundant pasture. The natural increase of flocks is practically doubled and cotton can be grown in places where in other years vegetation seems impossible.

During these warm spells, changes in the life in the sea are equally dramatic. The normal fish and marine life ordinarily present disappear, and long, yellow and black sea snakes from the rain forest farther north appear in the ocean.

Not until the 1960s did scientists begin to realize that this coastal phenomenon in Peru was related to changes that affect the oceanic circulation over large parts of the Pacific or that the El Niño effects might be connected through the atmosphere to failure of the monsoons in India. Sir Gilbert Walker, director of observatories in India in the early part of the twentieth century, recognized that major changes in the patterns of rainfall and wind take place over the tropical Pacific and Indian oceans on an irregular basis. He called these the *southern oscillation*. The end members of the oscillation are referred to as El Niño and La Niña. In recent decades, these oscillations have become the object of careful studies. The most strongly developed El Niño on record affected the central Pacific in 1997 and 1998.

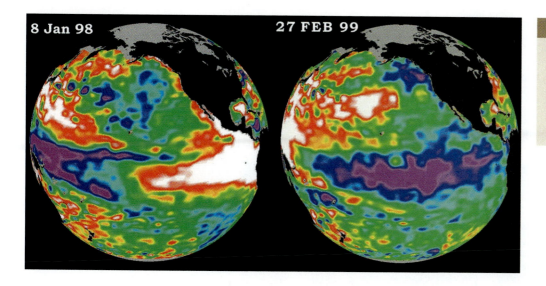

8 Jan 98 **27 FEB 99**

Figure 9.18 Development of El Niño Effects in the Central Pacific from January, 1998, to February, 1999. These satellite data images depict water temperature in the Pacific Ocean. Red and white colors indicate warm water.

Processes involved in El Niño demonstrate the close connection between oceans and the atmosphere. Normally in the Pacific and Indian oceans, currents located near the equator flow from east to west (figure 9.18). These currents located in the trade-wind belt carry surface waters along the coast of South America westward. As these surface waters move westward, cold, nutrient-rich water wells up from the deep sea along that coast. These deep waters are rich in the nutrients needed to support the plants that feed the large fish population that abounds along this coast when currents move offshore. As the water moves westward in the trade-wind belt, it gradually becomes warmer. The overlying air picks up moisture as the current moves across the Pacific. When the warm water and moist air reach the western side of the Pacific, the air rises in convection cells over islands in that region and heavy rains fall. The name La Niña is applied to this type of circulation and the accompanying weather when the currents are unusually strong.

When the oscillation swings in the opposite direction, El Niño trade winds weaken and may even reverse direction. A countercurrent arises along the equator, carrying water to the east. This warm water eventually reaches the coast of South America where it blocks the upwelling of deep, cold water. The warm water produces heavy rains along the coast of South America. These rains in South America coincide with reduced precipitation in Indonesia where the supply of moist air decreases. In 1997, extremely dry conditions in Indonesia made it impossible to control fires set by farmers trying to clear land for agricultural use and promote the growth of grasses for grazing. The fires raged through tropical rain forests. The effects of El Niño extend far beyond the central Pacific. Because the temperature of ocean waters strongly influences atmospheric circulation and the quantity of moisture in the atmosphere, El Niños cause drought and flooding in areas far removed from the central Pacific (table 9.2).

The southern oscillation involves the thermocline. During normal conditions, the thermocline slopes from South America, where it is near the surface, toward Indonesia (figure 9.19). During El Niño conditions, the thermocline sinks to greater depth along the coast of South America. As scientists continue their analysis of El Niño phenomena, it appears that the disturbances associated with the southern oscillation have effects that extend far beyond the tropics of the Pacific and Indian oceans. Shifts in tropical rainfall and wind patterns take place over much of Earth. When tropical Pacific waters are unusually warm, the rapidly moving

Table 9.2 Damage and Loss of Lives Connected to the 1982–1983 El Niño

Mountain and Pacific states	Storms	45 dead	$1.1 billion
Gulf states	Flooding	50 dead	$1.1 billion
Northeastern U.S.	Storms	66 dead	—
Ecuador, northern Peru	Flooding	600 dead	$650 million
Southern Peru, western Bolivia	Drought		$240 million
Southern Brazil, northern Argentina, eastern Paraguay	Flooding	170 dead, 600,000 evacuated	$3 billion
Australia	Drought, fires	71 dead	$2.5 billion
Indonesia	Drought	340 dead	$500 million
Southern India, Sri Lanka	Drought	unknown	$150 million
Southern Africa	Drought	disease, starvation	$1 billion
Western Europe	Flooding	25 dead	$200 million

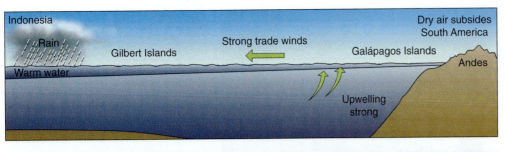

The Oscillation of Weather Conditions in the Equatorial Pacific Are Related to the Strength of the Trade Winds. When the trade winds are strong and persistent, the surface layer of the ocean moves from east to west, causing water to pile up on the western side of the Pacific. At these times, air over Indonesia is hot and rises, producing abundant rainfall. At the same time, upwelling is strong along the coast of South America, and dry air promotes arid conditions on the western slope of the Andes. During El Niño events, the trade winds become weak, and warm water in the equatorial countercurrent moves eastward. This causes precipitation to shift eastward. As warm water enters the coastal regions of South America, upwelling is blocked, and rains fall along the coast.

high-altitude wind currents become stronger and carry more moist air to the western part of North America. Consequently, unusually wet winter conditions arise along the West Coast. From the effects of El Niño, the close connections between the oceans and the atmosphere become clear. What initiates these events that ripple through the system is unclear. They may originate in the atmosphere or in the ocean. The way changes in one part of this system translate through the system altering weather at great distances from the origin of the disturbance reveals how sensitive the climatic system is to change.

TIDES

Even in ancient times, people recognized the relationship between the periodic rise and fall of sea level, called **tides,** and the phases of the Moon. They measured the time between successive high tides, called the **tidal period,** variations in the level of the water surface between high and low tide, known as the **tidal range,** and variations in the period and range from day to day and from month to month. As the quantity of data mounted, it became clear that tides are complex. Tide charts (figure 9.20) show that tides differ from place to place, even along the same coast. Some places have **diurnal tides,** consisting of one high tide and one low tide each day. Other localities have **semidiurnal tides,** two high and two

low tides each day. Generally, the range of these two are different. Still other localities have **mixed tides** (see figure 9.20), and a few localities may have no tides at all. It became clear that the effects of gravitational attraction between the water in the ocean and the Moon and Sun can explain many, but not all, observations of tidal phenomena. The astronomical theory of tide generation explains these observations.

Astronomical Theory of Tides

The Sun and Moon are the only extraterrestrial bodies that exert gravitational force on the oceans sufficient to affect the level of the water. Although it is 150 million kilometers (93 million miles) away, the Sun is such a massive body that it exerts nearly as much gravitational attractive force on Earth as does the Moon. However, because the difference in the attraction exerted by the Sun on water on opposite sides of Earth is so slight, tidal phenomena are more closely related to the Moon than the Sun.

Tide-Generating Forces

The gravitational attraction between masses (in this case, the Sun or Moon and water in the ocean) is proportional to the product of the masses and inversely proportional to the square of the distance

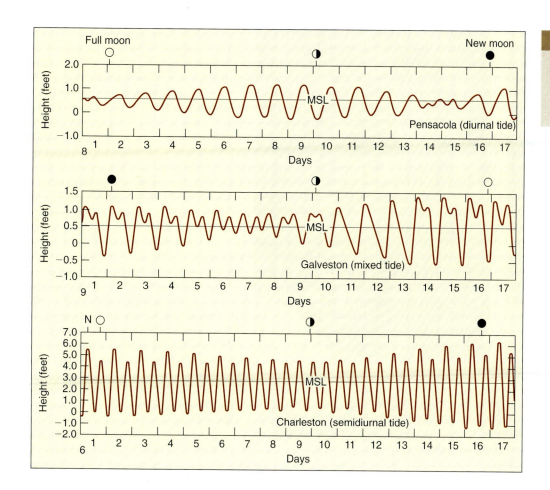

between the masses. Water on the side of Earth facing the Sun or Moon is 12,756 kilometers (7,921 miles) closer to those bodies than is water on the opposite side. Thus, the force of attraction between the Sun or the Moon and the water on Earth is different on the two sides of Earth. The mass of the Sun is so great and the distance between the Sun and Earth is so far that the difference in distance from the Sun to water on opposite sides of Earth has little effect on the gravitational attraction exerted on the water. However, the distance from the Moon to water on opposite sides of Earth makes a significant difference in tidal effects. Because of the difference in distance, water on the side of Earth facing the Moon experiences a much greater gravitational attraction toward the Moon than does water on the opposite side.

To understand how the gravitational attraction of the Moon affects Earth's tides, assume that a spherical Earth covered with water of uniform depth rotates on its axis and revolves around the Sun. Because the Moon and Earth revolve around a common center of gravity, some additional force must counterbalance the forces of attraction between the two bodies. Otherwise, the Moon would crash into Earth. Centrifugal effects, directed away from the center of rotation of the Earth-Moon system, account for this counterbalancing effect. Two tidal bulges occur on this model of Earth. One forms on the side toward the Moon, where centrifugal effects reinforce the Moon's gravitational force of attraction. The second

forms on the opposite side (facing away from the Moon), where the centrifugal effect is greater than the force of attraction. These two effects are close but not equal in strength, giving rise to a slightly higher bulge on the side toward the Moon.

The gravitational effect of the Sun is similar in character, but slightly less than half as strong as that of the Moon. It is the net effect of the gravitational forces of both the Sun and the Moon that determines tidal phenomena. When the Sun and Moon are in line with Earth (figure 9.21) the gravitational attraction of the two bodies on Earth is cumulative. Unusually high and low tides, called **spring tides** (unrelated to the season), result. When lines connecting the Sun and Moon to Earth are at right angles to Earth, the tide-producing forces of the Sun and Moon tend to cancel each other out. This gives rise to unusually low tidal ranges, called **neap tides.** These conditions occur twice each month, once when the Moon is at first quarter and again when the Moon is at third quarter. Most of the time, the gravitational effects of the Sun and Moon are out of phase. The size and shape of the two bulges gradually change as the water responds to the slow changes in the position of the Sun and Moon relative to Earth. These close connections between tidal ranges and the positions of the Sun and Moon validate the astronomical theory of tide generation.

As a result of the gravitational attraction exerted by the Moon and the centrifugal effects caused by Earth's rotation, one would

(a)   (b)

Figure 9.21   Conditions for High and Low Tides. (a) Conditions giving rise to the lowest monthly tides. (b) Conditions giving rise to the highest monthly tides. Two tidal bulges form in response to the gravitational attraction of the masses of the Sun and Moon for water on Earth and to centrifugal effects caused by the rotation of the Moon around Earth.

expect that two high tides and two low tides should occur each day. These would occur as Earth rotates on its axis under the tidal bulges formed by tide-generating forces on opposite sides of Earth. Most coasts do experience two high tides each day, and the interval between these high tides is 12 hours and 25 minutes. Because the Moon revolves around Earth, this interval between high tides is more than the 12 hours one would expect if the Moon remained in one place over Earth. The Moon revolves around Earth once every 28 days. Because of this revolution, the Moon goes through phases and is fully illuminated once every 28 days. The progressive movement of the Moon around Earth also causes the highest high tide to occur about 50 minutes later each day.

The gravitational attraction of the Moon is primarily responsible for generating tides, but high tides do not normally coincide exactly with the passage of the Moon overhead. Frictional drag causes the tide to arrive slightly after the passage of the Moon overhead, and the combined effects of the Sun and Moon may cause the high tide to arrive slightly before or after the passage of the Moon overhead. The interval varies from place to place.

Unfortunately, the astronomical theory does not explain all tidal observations. On the Pacific coast of North America, for example, high tides sweep from south to north. High tides normally come to Cape Flattery in northern Washington three hours later than they do to San Diego in southern California. A similar effect occurs along the coasts of Labrador and Newfoundland. If the tides were generated solely by astronomical effects, high and low tides should arrive at the same time all along these coasts. Furthermore, fifteen places in the ocean (seven in the Pacific and four each in the Atlantic and Indian oceans) do not have tides. As more data about tides have become available, it has become clear that the astronomical theory cannot provide an explanation for many observations. After studying the tides around the North Sea, oceanographers have concluded that the shape of the ocean basins must be a factor in tidal phenomena.

Effects of the Shape of the Ocean Basins on Tides

If Earth were a perfect sphere covered by water to a uniform depth, the astronomical theory of tides should explain all tidal phenomena. Instead, continents and oceans of varying depths interfere with the movement of tidal bulges around the globe. In many parts of the ocean, the tide resembles a **standing wave,** like the wave formed when coffee swirls around in a cup. As Earth rotates, the crests of these standing waves sweep around a central nodal point where tides, if observed at all, have small ranges (figure 9.22).

One of these nodal points is located near the center of the Gulf of St. Lawrence. Tides sweep around the shores of the Gulf of St. Lawrence. Typical tidal ranges along the coast of the United States are 2 to 7 meters (7 to 23 feet), but in some locations, such

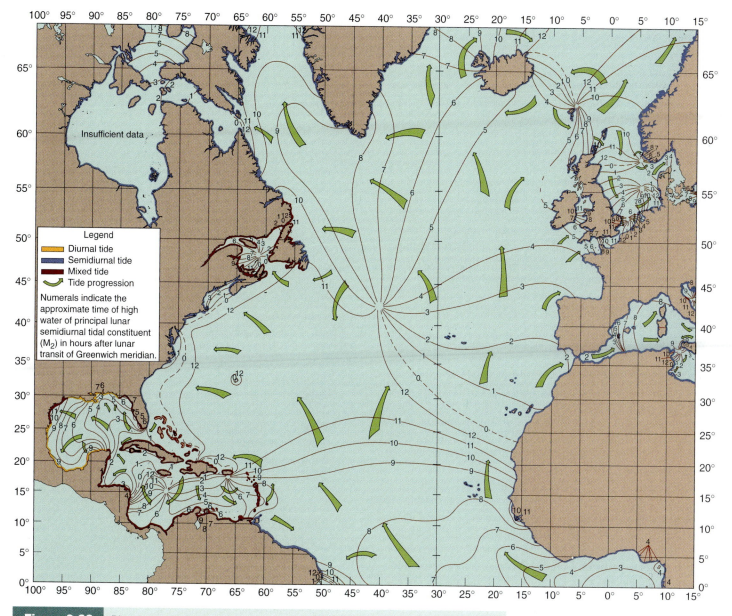

Figure 9.22 Diagram Depicting the Movement of Tidal Standing Waves Around Ocean Basins. Lines connect points that experience high tides at the same time. Other lines connect points that have the same tidal range.

as in the Bay of Fundy in Nova Scotia, the tidal range may reach 17 meters (56 feet). In the open ocean, changes in sea level caused by tides have little effect on currents, but closer to land, especially where islands are close together, tides generate strong currents. As the level of the ocean rises or falls on one side of a string of closely spaced islands, water flows through the straits between the islands, sometimes with great velocity. As the tide approaches the coast, it creates a horizontal current, called a **flood tide,** which moves into bays and rivers and may continue upstream for great distances. When the tide reverses direction and flows back out to sea, the tide is said to be **ebbing.** Tides are exceptionally long waves that under

certain circumstances create currents. Tides affect the Hudson River 208 kilometers (129 miles) upstream at Troy, New York, where the tidal range is 1.5 meters (5 feet). In rivers where tidal ranges are particularly high, the incoming tide forms a turbulent wave, called a **tidal bore,** that moves upstream at high velocity. An unusually high tidal bore forms on the Amazon River. This bore is 4.8 meters (16 feet) high and moves at a velocity of 22 kilometers per hour (14 miles per hour). These tidal currents scour the channel, keeping fine sediment in suspension and preventing its deposition. Tides also affect the sediment on the continental shelf, especially near shore where the effects are similar to those of other waves.

SUMMARY POINTS

1. Seawater contains some constituents, such as sodium, chlorine, sulfate, and magnesium, that are well mixed in the water and remain in relatively constant proportions to one another. Oceanographers use these to define and measure salinity. Other constituents, such as oxygen, carbon dioxide, and various nutrients, are involved in biological processes and are not constant in proportion to one another.

2. Salinity, temperature, and pressure determine the density of seawater, the most important factor in deep-water circulation.

3. Ocean water temperature is highest in equatorial latitudes and lowest near the poles. The colder, high-density water sinks, initiating circulation in the deeper portions of the ocean basins.

4. In a vertical section through the ocean, the upper surface layer where photosynthesis takes place is warmer than lower layers, and its temperature changes much more seasonally than it does in lower layers. Beneath the surface layer in the "permanent thermocline," temperature drops. Deep water below the thermocline is cold, about 2°C on average.

5. Wind and Coriolis effects are the most important factors that determine the currents in the surface layer. Near the equator, strong currents flow from east to west in all oceans. North and south of the equator, currents move in large gyres. Those in the Northern Hemisphere turn clockwise; those in the Southern Hemisphere turn counterclockwise. A strong current flows around Antarctica where land offers few obstructions. In all oceans, islands and the irregular shape of continents deflect and break up currents.

6. Circulation in the deep parts of the ocean basins is much harder to measure than surface circulation. Deep water moves slowly in response to slight changes in the density of overlying water. A relatively strong current circles Antarctica, but most other deep currents are slow and sluggish. Models of deep circulation depict currents that sink from the North Atlantic, move south through the Atlantic, flow south of Africa into the Indian and Pacific oceans where the water rises, and flows back to the Atlantic at shallower depths.

7. Gravitational attraction between the waters on Earth and the Moon and Sun provide tidal bulges in the ocean. These bulges move as Earth rotates, and they change position as a result of the revolution of the Moon around Earth. As the tidal bulges move around Earth, the shape of the ocean basin modifies the shape of the bulges and sets up large-standing waves that rotate around points that experience little or no tides.

8. Interactions between the atmosphere and the ocean are responsible for equatorial currents that reverse direction from time to time giving rise to El Niño. Generally, this equatorial current flows from east to west, but in some years, about Christmas-time, the current reverses direction. These reversals have great effects on the weather and fishing in equatorial latitudes, and their effects extend far beyond the tropics and influence weather around much of the globe.

REVIEW QUESTIONS

1. How do gases get into seawater?
2. What conditions cause the salinity of seawater to increase?
3. What is the effect of temperature on seawater density?
4. Describe the near-surface current gyres in the North and South Atlantic Ocean.
5. What conditions cause upwelling?
6. Why isn't the sea surface level?
7. How do oceanographers trace the movement of deep water?
8. What causes monsoons?
9. Describe the thickness of the surface layer in the oceans, and explain how it differs from one part of the ocean to another.
10. What problems are encountered in trying to determine the pattern of oceanic circulation referred to as the conveyor belt?
11. What positions of the Sun and Moon relative to Earth produce the highest tidal ranges?
12. What observations suggest that tides are caused by gravitational attraction between water on Earth and the Moon?
13. What features of tides cannot be explained by the astronomical theory of tide generation?
14. How does a diurnal tide differ from a semidiurnal tide?
15. Why does tidal range vary from one part of the coast to another?
16. What causes heavy rains to fall in Peru during El Niño events?
17. What interactions occur between the surface layers of the ocean and the atmosphere?
18. How is it possible for El Niño events to affect weather in mid-latitudes? What evidence exists that this happens?

THINKING CRITICALLY

1. What would be the effect on weather in North America if the Arctic Ocean were ice-free?
2. How does circulation of water in the deep ocean affect the environment of coastal areas? Give examples of how coasts in different parts of the world might be affected if deep water ceased to move.
3. What could cause the salinity of water in the North Atlantic to decrease?
4. What would happen to the deep circulation in the oceans if the salinity of the water in the North Atlantic decreased?
5. Long Beach, California, and Long Beach, North Carolina, are located at the same latitude; both are next to a major ocean. Why do they have different characteristics of weather and climate?

KEY WORDS

Alaska current	Florida current	North Pacific current	surface layer of the ocean
California current	geostrophic currents	permanent thermocline	tidal bore
conveyor belt model	Gulf Stream	photic zone	tidal period
diurnal tides	Labrador current	salinity	tidal range
ebbing tide	mixed tides	semidiurnal tides	tides
equatorial currents	neap tides	spring tides	upwelling
flood tide	North Atlantic deep water	standing wave	west-wind currents

chapter ten

Coasts and Coastal Environments

View of part of the western coast of Scotland. Coarse pebbles form the beach in the foreground, but sandy beaches are present on the far side of the river that enters this bay. Note the way the waves bend as they move into the bay.

Chapter Guide

Evidence of marine life can be found far inland away from the ocean's edge. How can this be? At times during the past 2 million years, sea level was nearly 70 meters (230 feet) higher than it is today. At other times, it dropped 130 meters (426 feet) lower than it is today. Changes in sea level occur because of changes in the amount of ice stored on land and because the plate tectonic system causes the lithosphere to move. On the coastline where continents and oceans meet, fresh water and salt water mix, creating a rich, diverse assortment of ecosystems. Changes in sea level cause these ecosystems to shift position as sea level rises and falls. Streams, glaciers, and wind on land combine with wave action, currents, and marine organisms in the oceans to contribute to the abundant, varied ecosystems in the transition zones between land and sea. This combination of marine and terrestrial influences account for transitional environments, such as sea cliffs, beaches, swamps, tidal flats, estuaries, and deltas. Could global warming cause sea level to dramatically rise once again? This chapter begins with a description of the processes that affect sea level. Then it examines the main processes that shape the shore—waves and nearshore currents. The chapter concludes with a brief review of some of the most distinctive transitional environments.

INTRODUCTION

The **shoreline,** the line where sea and land meet, is the boundary separating the coast from the continental shelf. Changes in climate and the shape of ocean basins and especially the balance between ice and water cause shorelines to shift back and forth across the continental shelf and low-lying coastal regions.

The parts of this zone that are influenced by shallow marine water, marine wetlands, and adjacent lands contain the greatest diversity of ecological niches on Earth. Huge populations of marine, brackish, and freshwater organisms inhabit the shallow water along the coast, and many of the fish in the deep sea spend the early part of their lives in these relatively protected waters. The ecological niches along the shore are as varied as the organisms that populate them. The transitional zone between land and ocean includes the ice marginal shores of the Arctic and Antarctic at one extreme, and at the other, it includes equatorial reefs. All the agents of change that affect the land surface (streams, glaciers, wind, ground water, and mass movement) affect the coast in one place or another. In addition, waves, currents, and marine organisms shape the shore and contribute to the great diversity of environments (figures 10.1 and 10.2).

The land is rising about 1 centimeter (0.4 inch) per year along the coast of Norway and Sweden. Uplift also affects central Canada and extends down the east coast of North America as far south as Cape Cod. This slow uplift is due to rebound of the lithosphere after great masses of ice that covered these regions 10,000 years ago melted. Land along the coast of Louisiana is subsiding under the combined effects of the weight and compaction of the mass of sediment deposited by the Mississippi River. Many millions of years ago, oceans occupied much of the interior of North America. Changes in the shape of the ocean basins caused by plate movements were probably responsible for these inundations of the continents.

When sea level rises, waves, currents, and marine organisms come in contact with land previously shaped in terrestrial environments. Where mountains are located along the edge of the continent, the continental shelf is usually narrow, and waves quickly cut into the high topography to produce cliffs. Where the sea advances across coastal plains with low relief, features develop such as offshore islands, beaches, swamps, and estuaries like those along the Atlantic coast. If sea level drops, the shoreline recedes progressively toward lower parts of the continental shelf. Because most continental shelves are relatively flat, the newly developing coasts resemble those present along modern coastal plains. Geomorphic agents that work on the land play an important role in shaping some coasts. Where volcanoes or crustal deformation are active, they may dominate the development of coastal features. Organisms dominate a few coasts where reefs or mangroves prosper. All of these coasts are subject to change as the level of the sea varies.

SEA LEVEL

At any point along the shore, **mean sea level** is the average of the position of the sea surface over a long period. Mean sea level provides a reference used in surveying both submarine and land

Figure 10.1 Seashores Are Sites of Rapid Change. Here along the shore in California, waves surge ashore. During storms the waves hit the low straight cliff.

topography. The contour lines on topographic maps indicate elevation relative to mean sea level as measured with tide gauges at a particular place on the coast. Although one might think of mean sea level as if the surface of the ocean had the characteristics of a spherical surface with the same elevation everywhere, this is not true. Currents and variations in the gravitational attraction between the water and the crust cause the level of the sea to vary in different parts of the ocean.

Sea-Level Changes

The level of the surface of the sea changes continually. Variations in atmospheric pressure and storm waves cause short-term changes. Changes in the volume of ice stored on land, water-volume changes caused by heating and cooling of the oceans, uplift and down-warping of the floor of the ocean, the addition of sediment or extrusion of volcanic rocks on the sea floor, and seafloor spreading all cause long-term variations in sea level (figure 10.3). Evidence of these long-term shifts in the level of the shoreline comes from a variety of sources. Beaches, notches cut in cliffs at sea level, and terraces cut by wave action at or just below mean sea level are among the features most frequently used to identify former sea levels. Fossils also provide useful information. The remains of trees, such as mangroves that grow only in shallow marine water, and shells of invertebrate animals that live attached to the sea floor in water of a certain depth may indicate former sea-level positions.

Effects of Glaciation on Sea Level During the ice ages, sea level repeatedly stayed at one level long enough to leave a clear record of former sea-level positions before shifting rapidly to another level. Records of these former "stands" of the sea appear on both the submerged inner part of some modern continental shelves and on the emerged part of the continental margins. Glacial advances and retreats account for most of these changes in sea level.

Worldwide changes in sea level accompanied each of the major advances of continental ice sheets, described in chapters 13 and 20. During the advance of ice, sea level stood more than 100 meters (330 feet) below its present level, and during interglacial periods when virtually all ice melted, sea level was 70 to 100 meters (230 to 330 feet) above its present level (figure 10.4). Wave-cut cliffs and other sea-level indicators found at the same elevation in many parts of the world provide evidence that changes in level were global. Carbon-14 dates of these sea-level indicators confirm that many of these changes took place simultaneously.

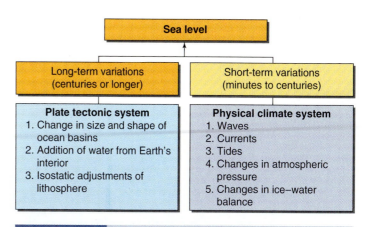

Figure 10.3 Sea-Level Variations. The level of the sea varies over both short and long time intervals. The plate tectonic system causes most long-term changes. The physical climate system is responsible for most short-term changes.

Antarctica contains the largest single accumulation of ice in the world. Whether its ice mass is increasing or decreasing is a critical question in the debates concerning global warming and its effects on sea level. One study based on a comparison between snow accumulation in Antarctica's continental interior and ice loss around the ice sheet's margins concluded that the ice mass is growing. However, satellite imagery along the fronts of the ice shelves indicates that they are retreating. Any large changes in the mass of ice should alter the gravity field over the ice. Studies of gravity anomalies indicate little change in the ice mass. If this conclusion is correct, changes in the quantity of ice in Antarctica are not responsible for the most recent changes in sea level. In contrast, the thickness of sea ice in the Arctic has decreased in the last few decades, and most glaciers in mountains are rapidly retreating.

Sea level also varies in response to other processes. Many oceanographers attribute the most recent rise in sea level to warming and

SYSTEMS THAT SHAPE COASTAL ENVIRONMENTS

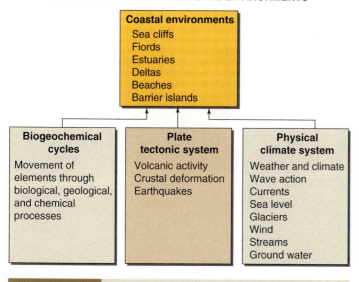

Figure 10.2 Systems That Shape Coastal Environments. The schematic diagram shows how processes that are part of the physical climate, plate tectonic, and biogeochemical systems combine to create transitional environments.

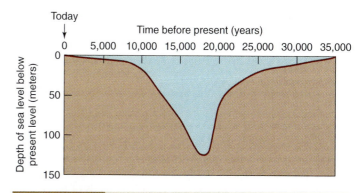

Figure 10.4 Rate of Sea-Level Change over Time. Sea level reached a low point about 18,000 years ago at the peak of the last major advance of ice sheets. Sea level rose rapidly as the ice melted until about 10,000 years ago. Since then the rate of rise has been slow.

expansion of near-surface water. The movement of sediment from land areas into ocean basins also causes sea level to rise. The addition of water to the oceans through volcanic activity would also raise sea level. However, neither addition of water nor sediment is thought to account for significant variations in sea level. Far more significant than either of these are movements of the lithosphere related to plate tectonics.

Tectonic Effects on Sea Level According to plate tectonic theory, both the size and shape of the ocean basins change, slowly but continually. The Atlantic Ocean is increasing in size at the expense of the Pacific Ocean. Upward movements near spreading ridges, downward movements at subduction zones, deformation associated with plate collision, and growth of new volcanic centers at hot spots all affect the level of the sea. Some plate movements change the level of the sea throughout the world. Growth of the Atlantic Ocean during the Mesozoic almost certainly affected the level of the oceans, and tectonic activity was probably responsible for changes in sea level in the Paleozoic when marine waters repeatedly invaded the continental interior of North America.

In more modern times, the best documented instances of tectonic effects on sea level are localized along sections of the coast and continental shelf where crustal movements involve uplift and down-warping of the coast. Some reefs in Indonesia, formed during the ice ages, stand nearly 320 meters (1,050 feet) out of water despite the recent rise in sea level. Along the west coast of the United States, terraces formed during earlier stands of the sea are now high above the sea surface (figure 10.5). In other places, subsidence and compaction of sediments cause sea level to change. The northern margin of the Gulf of Mexico is currently subsiding as sediment under the continental shelf and slope adds weight to the crust and causes the underlying sediment to compact. This compaction is accompanied by gradual slip of the sediment toward the deeper part of the Gulf.

Recent Sea-Level Trends

Sea level started to drop on a global scale about 35,000 years ago and reached a low point during the last maximum glacial advance, between 17,000 and 21,000 years ago. Thereafter, sea level rose rapidly at first and then much more slowly during the last

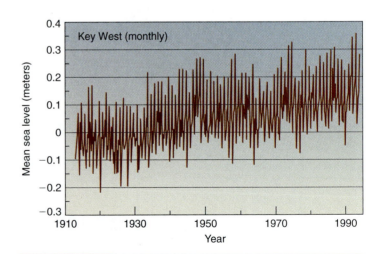

Figure 10.6 Recent Sea-Level Trends. Data obtained by NOAA at Key West, Florida, shows short-term fluctuations in sea level that occur on a scale of months to years. The long-term rise is about 2 millimeters per year.

Figure 10.7 Condemned Houses Await the Next Storm.
Severe erosion of beach front property has destroyed many homes
along the Outer Banks of North Carolina.

6,000 years (see figure 10.4). Recent measurements indicate that
sea level is again rising rapidly. Since 1940, the rate of rise has
been nearly 2.5 millimeters (0.1 inch) per year, that is, three times
the rate of rise between 1890 and 1940. This rise appears to be due
mainly to melting ice and to expansion of seawater as the temperature of the upper levels of the ocean rises. Transfer of water
drawn from ground-water reservoirs to the ocean also contributes
to the volume of water in the ocean. Sea level is rising today
(figure 10.6), causing many problems along the heavily populated
portion of the east coast of the United States where intensive development of waterfront property is taking place (figure 10.7). If
the recent rise in sea level continues, long sections of the shore will
be threatened by inundation, and heavily populated areas will require protection. The Gulf and Atlantic coasts of the United States
are especially vulnerable.

PROCESSES THAT SHAPE THE SHORE

Waves and Currents

As sea level rises, seawater inundates low wetlands along the shore,
and waves and currents begin to modify the shoreline. Most natural
waves in the ocean differ from the regular, simple waveforms familiar from geometry (figure 10.8). One type of natural wave, called
swell, resembles these ideal waves in shape, but most natural waves
have shapes that resemble small, elongated humps or pointed peaks
separated by troughs. These irregular shapes form where several
waves coming from different directions interfere with, and are
superimposed on, one another. Wind generates most of these waves.

Most wind-generated waves are **progressive waves,** similar to
the waves formed when a rock hits the surface of a pond. The wave-

form moves across the surface of the sea, but the water does not
travel with it. If the water did advance with the waveform, seas
would be impassable to most surface vessels. Instead, the water
moves in a nearly circular path, like a bobbing cork. The size of the
circular path decreases with depth until the motion is negligible at
a depth equal to about half the wavelength (figure 10.9). Below this
depth, waves have no effect either on the water itself or on the
ocean bottom. Storm waves may have wavelengths over a thousand
feet; as a result, they cause motion in water hundreds of feet below
the surface. As a wave enters water that has a depth of less than half
the wavelength, the bottom interferes with the motion of water in
the wave. This interference reduces the velocity of the wave motion. The amount of interference varies with the depth of the water.
Thus, wave velocity depends on water depth. Because water depth
is important in determining wave velocity, oceanographers commonly classify waves as shallow-water or deepwater waves. In
deep water, the velocity of a wave is a function of the wavelength.
But when the wave moves into water that is shallow relative to its
wavelength, wave velocity is a function of water depth.

Wave Reflection and Refraction Where waves encounter the
shore, part of the wave energy reflects off the shore. If the wave
encounters a solid, rocky cliff, a high percentage of the energy
reflects. If the wave moves across a wide, shallow offshore area,
interference of wave motion with the bottom dissipates much of
the wave energy before the wave reaches the shoreline.

Where waves approach the shore obliquely, the wave crest
bends as a result of a process called **wave refraction** (figure 10.10).
While the part of the wave in shallow water slows down, parts of
the wave moving in deeper water travel with higher velocity. This
causes the wave crest to bend. It swings around so the wave
approaches shore more nearly perpendicular to the shoreline.

Wave refraction spreads wave energy along recessed parts of
the shore and in bays. Refraction also focuses wave energy on
headlands, those parts of the shore that protrude into the ocean
(figure 10.11). As a result, erosion by wave action is much more
concentrated on headlands than it is in embayments. This often results in the shorelines being smoothed as the headlands are eroded
and the embayments are filled with sediment.

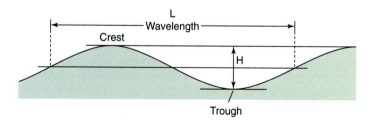

Figure 10.8 A Theoretical Waveform Seen in Cross Section.
Terms commonly used to describe waves include crest (the top of a
wave), wavelength (the distance between two successive crests),
trough (the bottom of a wave), wave height (the vertical distance
between a trough and a crest), and amplitude (one half of the height).

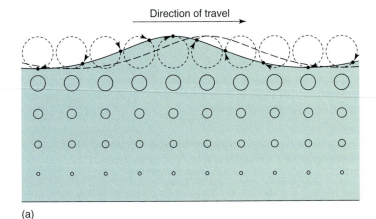

Direction of travel

(a)

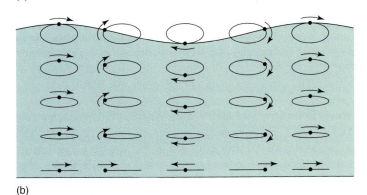

(b)

Figure 10.9 Relationship of Water Movement to Water Depth. (a) Water movement associated with waves in deep water dies out gradually with depth. (b) In shallow water, water movement extends to the bottom where drag interferes with the movement and causes the water to move back and forth along the bottom.

Despite refraction, waves rarely arrive exactly perpendicular to the shore. As the waves break obliquely on the shore, they cause a current to move parallel to the shore. This current, called the **longshore current,** is strong enough to move fine sediment along the shore. These currents generally lie between the shore and sandbars commonly found offshore. Movement of this sediment is called *longshore drift* and may amount to several hundred meters per day.

Breakers As waves enter shallow water, the depth of the water restricts the movement accompanying the wave motion. The nearly circular path of water involved in deepwater wave motion flattens into an elliptical path. Near the bottom, movements decrease to a back and forth motion. This restriction on water movement causes the wave height to increase and wave velocity to decrease. These changes occur rapidly when the water depth is between one-tenth and one-twentieth of the wavelength. As the wave slows down, the velocity of the top part of the wave begins to exceed the forward velocity of the wave as a whole, and the wave breaks. Water at the top of the wave spills over in front of the lower part of the wave, forming a **breaker** (figure 10.12). As the wave breaks, its regular

geometric form breaks down, and the water moves onto the beach as a rushing turbulent mass. Beyond the zone of breakers, little more than a slight shifting of sediment back and forth by wave movement takes place, but within the zone of breakers, the breakers churn up sediment on the bottom and bring it into suspension. A slight depression, or trough, a few meters deep may form within this zone as sediment moves from the depression and forms a submerged bar on the landward side of the depression. In some instances, this bar may eventually become a beach.

The water depth near shore and the wave height and length determine the size of breakers. Breaking waves are generally higher along shores where deep water lies near shore than they are where shallow water extends far offshore. The high breakers make these areas more appealing to surfers, who prefer breakers that have large curling forms.

Mechanics of Wave Erosion Where water near shore is deep, waves break directly against the rock and sediment exposed at the shoreline. Storm waves crash against the shore with enough force to drive water into cracks in the rock, open fractures, and dislodge blocks of rock. A moderate-sized wave, 1.3 to 4 meters (4 to 13 feet) high, can exert pressures from 50,000 to 100,000 pounds per square yard against the rocks exposed where it breaks. Waves pick up the dislodged blocks and fragments and hurl them against cliffs, in effect using them as tools to facilitate erosion. Waves can erode the shore rapidly. In 1957, a new volcano appeared at the edge of Fayal, one of a group of islands known as the Azores, located in the central Atlantic. The volcano grew to a height of 300 meters (1,000 feet) before volcanic activity ceased. Within a few weeks after the eruption subsided, wave action completely destroyed the cone, leaving only a shallow platform.

Waves also erode shorelines where sand and silt lie beneath the breaking waves. As water along the front edge of a breaker plunges forward, it digs into and churns up several centimeters of bottom

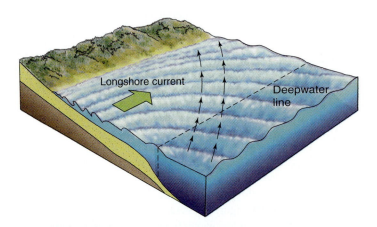

Longshore current

Deepwater line

Figure 10.10 Wave Refraction. Because wave velocity decreases in shallow water, waves bend as they approach shore. The wave crest swings around, becoming more nearly perpendicular to the shore as it moves into shallow water. This movement produces a current that moves along parallel to the shoreline.

sediment. Once suspended, currents can easily move this fine sediment. Because currents are so effective in moving suspended clay and silt-sized sediment, these sizes rarely occur on beaches. Fine-sized sediment, especially clay, may remain suspended until it settles in deep water where the bottom lies below the level of wave action.

Tsunami

Tsunami is the Japanese name for the unusually large waves that sweep onto the margins of the Pacific following large earthquakes with epicenters in water-covered areas. Sudden displacements of the sea surface usually caused by an abrupt movement on a fault in the sea floor, a submarine volcanic eruption, or a submarine landslide cause these waves. Tsunami may also be formed by the impact of a large meteorite in the ocean. Once formed, tsunami move out away from the place of origin as a succession of progressive waves. In the open ocean, tsunami have wavelengths that are long relative to their height, which is only 10 to 20 centimeters (4 to 8 inches) in the open ocean, so people on ships in the open ocean do not notice them as they pass. As the waves approach shore and the sea floor begins to interfere with the circular water motion associated with waves, the wave height increases. The wave height continues to build up until the wave breaks and the water crashes against the shore.

In shallow water and along the shore, tsunami are capable of doing tremendous damage, overturning ships, washing away buildings, and eroding beaches. **Runup** is the term applied to the elevation above mean sea level reached by a tsunami (see discussion of

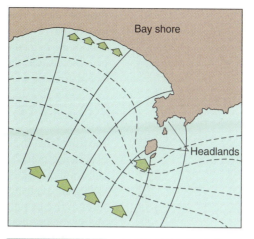

Figure 10.11 Wave Refraction Affects Headlands. Wave refraction concentrates wave energy on headlands and spreads the wave energy out along embayments.

tsunami in chapter 7). In rare instances, runup may reach 20 meters (65 feet) above mean sea level. A runup of nearly 25 meters (80 feet) in one bay accompanied the 1964 earthquake in Anchorage, Alaska. The size of the wave and the shape of the sea floor near the shore determine how high a wave will runup. The highest runups occur where deep water is close to shore and where embayments confine the approaching wave. The eruption of Krakatau in 1883 created a 30-meter- (100-foot) high wave that resulted in the loss of an estimated 36,000 lives in Indonesia.

Most tsunami arrive as a series of high waves that gradually diminish in height. Withdrawal of water from the shore often signals the initial arrival of the wave train. Following the earthquake that destroyed much of Lisbon, Portugal, in 1755, water in the estuary at Lisbon withdrew, and many people fleeing fires that were raging through the city went out on piers where they perished as the first high wave reached shore.

Most tsunami happen in the Pacific Ocean, where major earthquakes are associated with subduction zones in island arcs. Because of their close association with earthquakes, tsunami are also called **seismic sea waves.** Nations around the Pacific Rim have established a system designed to warn people who live near shore of an approaching tsunami (figure 10.13). The system goes into action when seismologists detect an earthquake of magnitude 6.5 or more with its epicenter in a water-covered area. Following detection of such an earthquake, coastal localities near the epicenter monitor tide gauges to see whether water levels are abnormally high or

Figure 10.12 Breakers. Where waves break, the top of the wave spills or plunges into the surf. Sediment on the bottom churned up by this movement travels with the water in the breaking wave onto the beach.

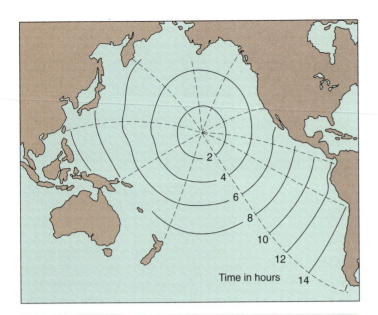

Figure 10.13 Diagram Showing the Time Required for a Tsunami to Cross the Pacific.

Time in hours

rising rapidly. If it appears that a tsunami has formed, a warning is sent from the headquarters in Hawaii to all nations with Pacific coastlines. It is then the obligation of the country to notify people in coastal areas and prepare for the arrival of the waves.

Displacements of the sea floor along faults or submarine landslides that occur near shore may generate exceptionally large waves as they did during the 1964 earthquake at Anchorage, Alaska. Unlike tsunami that have crossed the ocean, however, these waves reach shore too rapidly for the seismic sea-wave warning system to

provide alerts. The only protection people in these areas have is knowledge of potential unstable submarine slopes or active faults near shore. Armed with this knowledge, a community can take steps to avoid disaster through zoning or other land-use decisions.

COASTAL ENVIRONMENTS

Environments in the transition zone between the ocean and land experience both rapid and slow changes. Many slow changes result from movements associated with the plate tectonic system or with gradual changes in the physical climate system that cause sea level to rise or fall. Rapid changes occur during powerful storms, altering many transitional zone environments. Equally dramatic are the impacts people have as they try to control the shoreline to protect property or to claim land for development. Most habitats in transitional environments are fragile. Water draining from the land surface in streams ultimately arrives along the coast. Where that water contains sewage, fertilizer, or toxic substances, longshore currents distribute it along the shore. Oil spilled from ships and drilling platforms washes ashore as does much of the floating debris dumped overboard in the open ocean. Comparison of major environments in the transition zone, such as sea cliffs, beaches, mudflats, swamps, estuaries, and deltas, reveals how natural processes operate in each of the environments and how they respond to these natural and human-produced impacts.

Coasts with Sea Cliffs

Sea cliffs form where waves approach and break against steeply sloping land. If the slope of the sea floor is gentle, waves lose their energy before they reach the shoreline. Sea cliffs (figure 10.14)

Figure 10.14 A Sea Cliff Along the Pacific Coast. Note the small sand beach that lies within this small bay.

Figure 10.15 Wave-Cut Terraces in Tasmania, Australia. The lower terrace is frequently washed by breaking waves that undercut the cliff capped by an older terrace.

characterize the landscape of more of the world's coasts than any other coastal landforms. The long-term rise in sea level has brought the seashore to rest against landforms initially created by processes like stream action that operate on land rather than underwater. Cliffs form where the rising sea has drowned former stream or glacial valleys, such as the **fiords,** described in chapter 20. Others develop where young volcanoes grow from the sea floor or where tectonic uplift has elevated the land. Cliffs do not form where the sea inundates low-lying coastal plains, such as those along the south Atlantic and Gulf coasts of the United States.

Most sea cliffs are steep, and many are hundreds of meters high. Typically, a sudden break in slope occurs at the foot of the cliff, separating it from a nearly flat bedrock surface covered with a thin sediment veneer that slopes gently seaward (figure 10.15). As waves break against the foot of the cliff, it is undermined and gradually retreats, leaving a **wave-cut terrace.** Rocks dislodged from the cliff may lie on this terrace, or a sand beach may occupy it part of the time. During storms, these rocks become the tools used by the wave to cut the base of the cliff. In this way, the waves cut a **notch** near the base of sea cliffs, even in massive, erosion-resistant rock types. Eventually, the cliff face breaks away and falls into the sea.

Because breaking waves are not effective agents of erosion under water, cliffs cut by wave action stop abruptly just below water level. Water depth limits the width of wave-cut terraces. As the terrace becomes wider, an increasing proportion of the wave energy dissipates before the wave reaches the base of the cliff and reduces the ability of the waves to erode the cliff. Normally, terraces are narrow, but if sea level rises slowly over an extended period, the terrace width increases. The part of the terrace formed first gradually submerges, and waves can continue to move across the terrace and break near the shoreline.

Wave erosion is rarely uniform along the shore. Consequently, in the short-term, irregularities in the coast develop as wave erosion progresses. Retreat of a wave-cut cliff may leave columns of rock standing isolated as islands just offshore. These columns of rock, called **stacks** (figure 10.16), resemble smokestacks; however, they are not necessarily small or round. Most stacks are remnants of hard rock that remain offshore. In time, erosion by waves will also remove them.

Sea caves form where wave action hollows out a portion of rocks at the base of a cliff, forming a shallow cave-like chamber. Caves open to the ocean may also form where wave action cuts into the side of an existing cave, allowing the sea to move into a cavity formed by solution of bedrock by ground water. Long caves also form where waves expose lava tubes. **Arches** may form where stratified rocks of varying hardness make up the cliff (see figure 10.16). Wave erosion quickly undercuts the layer of weaker rock, sometimes leaving the resistant capping layers as a natural bridge arching over them. Such features, like stacks and beaches that connect stacks with the mainland, eventually disappear, although new ones may form as erosion continues.

Beaches of sand, pebbles, or cobbles frequently lie along portions of coasts characterized by cliffs, stacks, and other forms created by wave erosion (figure 10.17). Such beaches may be temporary features of the shoreline that form during summer when waves are smaller, less effective agents of erosion than they are in winter. When storms set in, these beaches quickly shrink and may disappear altogether. This commonly occurs along the coast of northern California, Oregon, and Washington.

Cliff Habitats

The upper parts of high sea-cliffs are normally dry habitats. Sea spray may bathe the cliffs in salt water during storms, but the cliffs drain quickly. The water table is usually low along a cliff, so little ground water reaches the higher parts of the cliff face. Consequently, a restricted assortment of plants grows on most sea cliffs.

strong shells and can withstand the continual battering they take from breaking waves. Many of these organisms occupy zones determined by the water level. Some, such as periwinkles, lichens, and blue-green algae, live at a level where they receive only the water thrown up on the rocks as waves break. Barnacles, mussels, limpets, red algae, and some worms and corals can survive exposure to the air during low tides, but they must remain underwater part of each day. Other species, such as kelp and sea squirts, may live at the base of cliffs, where they can stay continually underwater.

Beaches

Beaches almost always form where large quantities of sediment reach the shore, and they are most common where low topography lies inland from the sea. In addition, patches of sand may accumulate in embayments along even the most rugged shores. A source of sediment is of prime importance. Usually, the sediment comes from streams entering the ocean or from wave erosion and disintegration of rocks exposed along the shore. Sediment may also come from volcanoes, melting glaciers, or windblown deposits. In all cases, wave action and longshore drift spread the sediment along the shore, forming beaches. The sediment shifts onshore and offshore as the character of the waves change.

Sand Supply The supply of sediment that nourishes the beaches on the eastern shore of North America comes mainly from streams that drain the continent. After reaching the coast, the finest sediment moves out to sea, but the longshore current moves the sand-sized fraction along the shore. Off the New England coast, sediment derived from glacial outwash deposits, like those exposed on Cape Cod and Long Island, provides a ready source of beach material. Farther south, water from the Hudson River, the Delaware River, the Chesapeake Bay, and smaller rivers along the

Figure 10.16 Stacks. Pieces of the cliffs that formerly lined the coast may remain offshore as stacks. Wave erosion has cut the cliffs back toward the land. In time, the stacks located along this part of the California coast will erode away. A small arch is located in the stack at the bottom of this photograph.

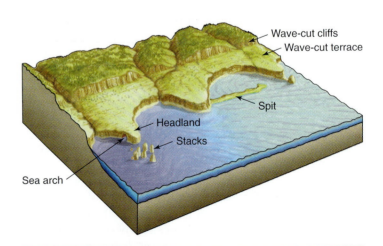

Figure 10.17 Features Commonly Found Along Shorelines That Have Cliffs. Features shown include a curved deposit of sediment (spits), stacks, and headland.

These include some grasses and succulent plants that either require little moisture or have the capacity to store water in their stems and leaves. Although the steep slopes and limited plant growth provide habitats for relatively few animals, cliffs are favorable places for birds to live. Because they afford protection from most predators, sea cliffs are favorite breeding grounds for seabirds. Gulls, auks, and other seabirds usually nest and lay their eggs close to other members of their species. By being close together, the group can more easily defend eggs against predators. Thus, one section of a cliff may contain many breeding birds, while other sections are vacant.

Low cliffs and rocky shores unprotected by beach deposits experience especially strong wave and current action. Many of the organisms, such as mussels, that inhabit this environment have

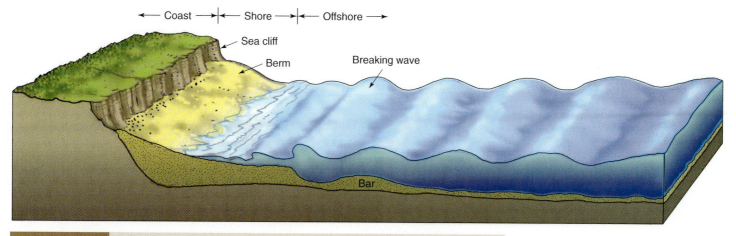

Coast ——→|←—— Shore ——→|←—— Offshore ——→

Sea cliff

Berm

Breaking wave

Bar

Figure 10.18 Profile of a Beach. The shore is the zone of contact between water and land. The actual contact between land and water, the shoreline, moves back and forth as tides and waves cause the water to rise and fall. The coast lies inland from the shore. Offshore is the region on the seaward of the shoreline during the lowest tide.

coast supply sand that was produced by weathering and erosion of the Appalachian Mountains and the Atlantic and Gulf coasts plains.

The beach sand along the coast of Florida is markedly different from that farther north. The proportion of shell fragments increases and quartz sand gradually disappears. This change clearly signifies a change in the source of the sand. Shells are the primary source of sand in eastern Florida where no streams carrying quartz sand enter the Atlantic Ocean. However, on Florida's western coast, both quartz and shells are important sediment sources in these warm, highly productive waters.

Beaches have many different profiles. In some places, a sand beach rests at the foot of a cliff. Other beaches are far removed from high land. However, beaches have three distinct parts: offshore, foreshore, and backshore (figure 10.18). **Offshore** refers to the area located seaward from the position of low tide. The **foreshore** is the zone exposed to air between low and high tide. The **backshore** is that part of the shore that extends inland from the line marking high tide. In figure 10.18, a cliff forms the start of the backshore.

Beach Dynamics

Beaches are transient features of the shoreline. The sandy beach that appears permanent in the summertime may disappear or change to a narrow strip covered by coarse pebbles in the winter. Wave and current action may remove all sediment, exposing the underlying bedrock. Over long periods, beaches move, growing in some places and retreating in others, as the balance between the quantity of sediment supplied to them and the amount removed from them changes.

The profile of a beach changes from hour to hour and from one season to another. Variations in winds and tides quickly change the size of waves and movement of sediment. An increase or decrease in the size of waves or the amount of sediment that reaches a beach

rapidly alters the beach profile. Seasonal variations arise from differences in the amount of storm activity, the balance between sediment supplied to the shore, and rates at which sediment moves. Seasonal variations in the profile of a beach at Carmel, California (figure 10.19), demonstrate the dramatic seasonal changes in the shape of the beach. During the summer, small waves gradually move sand toward the beach from offshore bars, filling in irregularities on the sea floor and producing a smooth bottom and a much wider beach. The larger waves that strike the beach in winter months churn up the sand and move it into deeper water where bars form.

Breaking waves and currents that flow close to shore easily move the sand-sized sediment that accumulates on most beaches. The breaking waves roll and toss pebbles and grains of sand obliquely onto the beach. When the water returns, the sand and pebbles move directly down the slope of the beach. Thus, as each new wave approaches the beach at an oblique angle, the sediment moves along the beach, following a zigzag path. This movement (figure 10.20), called **beach drifting,** or **longshore drifting,** is the main mechanism by which sediment moves along beaches. Both sediment movement and longshore currents result from the oblique approach of waves to the shore. Continued over a long time, beach drifting extends beaches in the direction of the longshore drift. Over time, the drift of sand may close inlets or even build new beaches completely across the mouths of bays. However, in a matter of hours, storms can destroy the work of years of beach drifting. On Cape Cod, a single storm in February 1978 cut a 1,500-meter- (5,000-foot) wide opening through the beach and created a passage between the open ocean and the bay behind the beach.

Storms are not the only mechanism responsible for movement and destruction of beaches. Any upset in the supply of sand to the beach or in the currents along the beach influences beach growth or depletion. Attempts to control erosion and deposition along the shore usually involve construction of **jetties,** piers, or

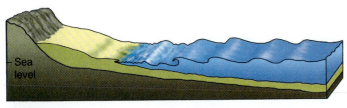

(a) April

(b) May

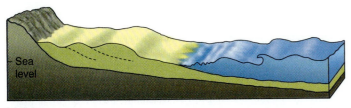

(c) July

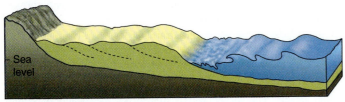

(d) September

Figure 10.19 Seasonal Variation in a Profile Across the Beach at Carmel, California. During the winter, large waves erode the beach and move sand offshore. During the spring and summer, the more gentle surf gradually rebuilds the beach.

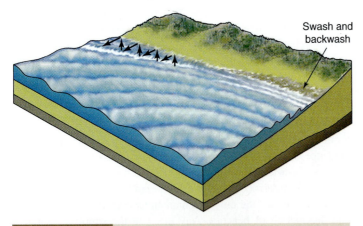

Swash and backwash

Figure 10.20 Sediment Movement Along Beaches. Arrows show the path followed by sand during beach drifting.

breakwaters. Such structures introduce side effects that may cause more damage than they prevent. Building such obstructions may protect and build up one segment of the shoreline while cutting off the normal supply of sand and subjecting another segment of the shoreline to increased erosion (figure 10.21).

Where an obstruction close to shore interferes with the longshore current, sand must pass around the end of the obstruction. If the water depth increases rapidly off the beach, sand slips into deeper water and may not return to the beach. This happens in southern California where deep basins lie near shore along the coast. Sand moving in the longshore drift slips away from the shore and flows into these basins where it is permanently lost from shore processes. This rarely happens on the East Coast, where water depth increases slowly. Consequently, beaches are more or less continuous along the shoreline of the United States south of New England.

Beach Growth

Beaches often develop at the heads of bays or inlets. **Bay-head beaches** form in protected portions of bays where wave action and longshore currents are weak. Headlands on either side of the bay prevent the rapid removal of beach materials. Bars and beaches may even form across the mouth of bays. Initially, a submerged bar grows from the tip of the headland beach. With the addition of more sand, this bar emerges and projects from the land in the direction of the longshore current and beach drifting. The ends of these projections, called **spits,** or **hooks,** generally curve back into the bay (see figure 10.17). If the supply of sediment is adequate, the spit may grow until it extends almost completely across the opening to the sea, leaving only a short channel through which water from streams flowing into the bay can reach the sea. On the Outer Banks of North Carolina, a bridge was constructed across an inlet cut by a hurricane. Today, sand that has drifted down the shore is filling in the opening (figure 10.22).

Sometimes, sand deposits connect islands to one another or islands to the mainland. When the rate of wave erosion decreases or the supply of sand increases, silt accumulates in the small strait between the island and the land. The island shields the area immediately behind it from wave action. As more sand enters this protected area, the submerged bar rises above sea level.

Beach Habitat

Wave action continually disturbs the foreshore portions of sand beaches. Crabs and other burrowing animals find this a suitable environment, but because the sand is in a continual state of movement, plants are unable to put down roots. On some beaches, algae and other microscopic-sized plants abound in the water that washes over the beach with each advancing wave.

Water drains quickly out of dune sand, so only those plants that can survive in a dry environment populate sand dunes. Grasses and some succulent plants—those that store water in their tissues—can grow in sand dunes in backshore areas.

A great variety of seabirds, including gulls, sandpipers, terns, osprey, and cormorants, spend much of their time near the beach.

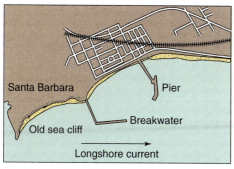

(a) Construction 1928

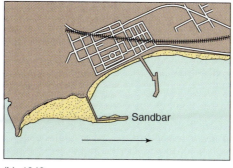

(b) 1948

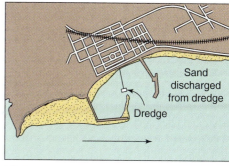

(c) 1965

Figure 10.21 Sketch Maps Showing the Changes That Took Place Along the Coast at Santa Barbara, California, After Construction of a Breakwater in 1928. The shoreline is shown as it looked in (a) 1928, (b) 1948, and (c) 1965. The harbor is now kept open by dredging; sand dredged where it enters the harbor is discharged farther down the shore.

Some of these birds nest in the backshore areas. Others, like the osprey, make nests in out-of-the-way spots such as buoy markers. They feed in the shallow water on small fish and invertebrate animals that live in the sand or wash ashore. Many small creatures, including crabs, sand fleas, and burrowing insects, dwell in the sand. Turtles also use beaches as nesting ground. Many marine mammals, including walrus, seals, and sea lions, mate and raise their pups on beaches. Marine mammals such as whales, dolphin, and porpoises beach themselves where they may die. No one knows why they do this. And, of course, humans use beaches for recreation, relaxation, and often for housing, condominiums, and resort developments.

Barrier Islands

The sand beaches that lie along much of the eastern shore of North America from Cape Cod to Mexico comprise some of the longest beaches in the world. The Hudson, Delaware, Savannah, James, and other rivers break the continuity of the beaches. Breaks also occur at the southern tip of Florida where great mangrove swamps in the Everglades lie along the coast and in Louisiana where the deltas of the Mississippi extend into the Gulf of Mexico. Elsewhere along the Atlantic and Gulf coasts, sand beaches are present. Along some sections of the shore, the beach lies along the shore of the mainland; at other places, the beach stretches as a

Figure 10.22 Aerial View of Oregon Inlet at Outer Banks, North Carolina.

long, thin ribbon of sand separated from the mainland by a bay or marshland. These offshore sand deposits form **barrier islands** (figure 10.23 and see figure 10.22).

The rise in sea level over the past few thousand years has influenced, and is probably largely responsible for, the formation of barrier islands. Many scientists think the islands formed primarily as a result of the gradual buildup of sand from the sea floor, first as a bar and later as an island. According to this idea, wave action initiates island formation, and beach drifting and the extension of spits later modify the islands. The process begins with breakers churning up sediment that settles out as a bar. Storm waves build up this bar deposit until a beach begins to emerge. Once the beach is present, longshore drifting of sand nourishes the beach.

The barrier islands along the Gulf and Atlantic coasts are rarely more than a mile wide. They lie on broad continental shelves, that slope gently toward the deep sea. Bays, swamps, and flatland lie on the landward side of barrier islands, which are slightly asymmetric in profile (figure 10.24). The oceanward side of the island is a "high-energy" environment continually attacked by waves (see figure 10.12). Behind the beach, sand dunes composed of sand

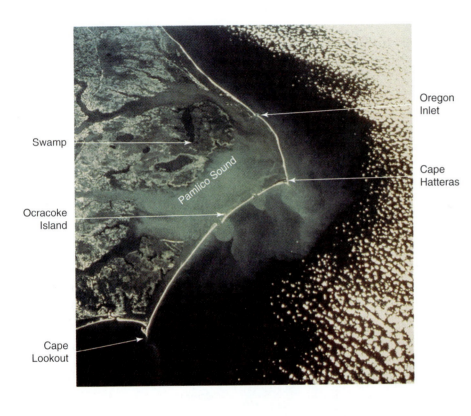

Figure 10.23 Satellite Image of the Outer Banks Region of North Carolina. Sediment transported to the south is visible in the water off Cape Hatteras. Sediment plumes also show in the inlets between the islands. Swamps lie along the edges of the bays. The distance across the area shown is about 100 kilometers.

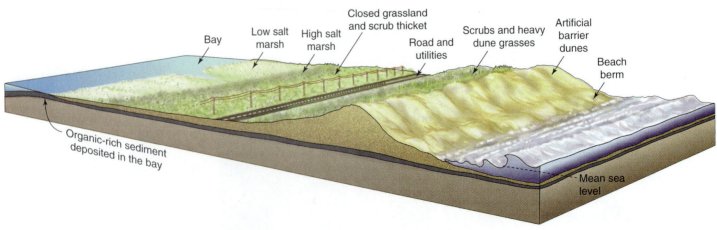

Figure 10.24 Cross Section of a Typical Barrier Island Showing the Different Environments Generally Found Along the East Coast of North America.

blown from the beachfront build the ground higher. Behind the dunes, the land slopes gently into a shallow bay or lagoon.

Grasses and small trees protect the landward side of the islands (figure 10.25). Although water in bays rises and falls daily with the tides, the bay is generally quiet. Strong tidal currents carry sand in and out through inlets that cross the barrier islands, but most sediment in the bays comes from the mainland. Muds that wash into the bay from the mainland quickly become rich in organic material provided by flora and fauna in the bay. The bay environment is one of the most productive of all marine environments. The fauna is diverse, and the bay is densely populated. Many ecosystems exist in bays. Open water covers the central deeper portion of bays, but aquatic grasses and tree-covered marine swamps cover much of the bay margin. Mudflats form where streams laden with heavy loads of clay and silt enter the bays.

Because sea level is rising, most barrier islands are slowly migrating toward the bays that separate them from the mainland (figure 10.25). This migration became clear when scientists recognized that many of the shells eroding out of the sand on the beach side were shells of invertebrate animals that lived in the bays. Layers of peat and even forests that once stood on the bay side (figure 10.26) are exhumed on the beach side of the islands. Several processes cause landward migration of some islands. Perhaps the most important is the removal of sand from the beach and deposition of it in the bay during storms. Storm waves that wash over the beach (figure 10.27), strong winds that blow sand into the bay, and tidal currents flowing through inlets (see figure 10.22) also accomplish part of the transfer.

The development of housing and tourist facilities on many barrier islands has stimulated great interest in preventing any further landward migration of the islands. Property owners on barrier islands face a constant threat. The history of destruction associated with coasts illustrates the extensive damage and high costs associated with human habitation near the shore. At Galveston Island,

Texas, in 1900, 6,000 people drowned when a hurricane struck the beach. A storm in 1962 severely damaged property between Long Island and Cape Lookout, North Carolina. Since 1977, Miami has spent millions of dollars per mile of beachfront to restore its beaches to their former width. In 1993, Hurricane Andrew hit southern Florida with such force that it has taken years for the region to recover from the loss of billions of dollars worth of

Figure 10.26 Soil in the Bay Behind a Barrier Island Is Exposed During Low Tide. This soil is rich in organic matter.

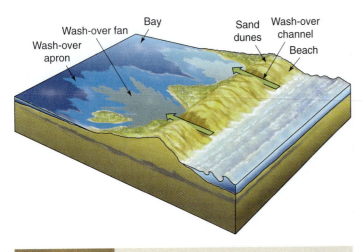

Wash-over apron Wash-over fan Bay Sand dunes Wash-over channel Beach

Figure 10.27   Block Diagram of a Barrier Island Showing the Effects of Storm Waves That Wash over the Islands.

property. Destructive storms strike somewhere along this coast almost every year. New inlets have formed across the barrier island beaches in numerous places where hurricanes have crossed them, wiping real estate off the map. A constant struggle goes on at many places to maintain the beach and save the homes, hotels, and towns built at the water's edge.

These efforts usually involve replacing sand eroded from the beach face or constructing solid structures along the beach to stop erosion. Some communities import sand from inland; others dredge sand from offshore. Jetties built perpendicular to the shoreline interfere with the longshore transport of sand. Offshore breakwaters built parallel to the beach tend to slow longshore transport of sand and help rebuild beaches, but to the extent that these structures hold sand in one place, they deplete the sand supply farther down the beach.

Tidal Flats

Tidal flats are low-lying lands near sea level that are protected by the shore from wave erosion and strong currents. At low tide, the surface of the tidal flat stands slightly above water level, but the sea advances over them at high tide. Tidal flats form near an abundant source of sediment. Some tidal flats originate almost entirely from deposition of suspended matter transported to the sea in streams, but others appear to be swamp or glacial deposits reworked by currents acting in the tidal zone. Organic-rich mud is the main sediment on tidal flats. Although many organisms are present, some invertebrates cannot survive in this environment because the clay and silt clog their filtration and breathing structures. Anoxic conditions just below the sediment surface also make it impossible for many burrowing organisms to survive. However, some gastropods, annelid worms, and crustaceans, notably crabs, thrive on the abundant supply of organic matter. The remains of these invertebrates are an important part of the sediment. Fine silt, clay, and some sand mixed with varying amounts

of shell fragments, sea urchin spines, and fine plant matter produce the soft, water-soaked mud of the tidal flats. Excrement of worms and other mud-ingesting invertebrates also composes a significant part of the sediment.

The amount of sediment and the rates of sedimentation on modern tidal flats are surprisingly high. In a three-year period, more than 7 meters (23 feet) of sediment settled in the harbor entrance to Wilhelmshaven, Germany. Continuous dredging is necessary to keep the channel open for shipping.

Where waves build up tidal flats close to sea level or slightly above it, people may drain the deposits and use them for farmland. The Dutch have constructed extensive dikes and elaborate drainage facilities for this purpose. The land reclaimed from the sea makes excellent farmland. It is flat, rich in nutrients, and wet. However, the low elevation of these lands makes them vulnerable to flooding, and, if dikes fail, to inundation by the ocean. As sea level rises, these threats are increasingly serious. Many marine swamps are also vulnerable to a rise of sea level.

Marine Swamps

Swamps, bogs, and **morasses** are low, spongy lands generally saturated with moisture. Abundant rainfall or some other water supply and an impermeable or water-filled substratum that prevents drainage are necessary for the continuance of a swamp. Land of this type is frequently present on flat ground near sea level, on the flood plain of major rivers, in glaciated terrain where glacial deposits have choked drainage, and even in some mountain valleys. The area of land covered by swamps worldwide exceeds 2.6 million square kilometers (1 million square miles).

Marine swamps form along low-lying portions of coasts where brackish water and fresh water merge. These swamps are so close to sea level that a slight rise of the sea inundates the swamp. Where inundation happens, marine deposits cover and protect the swamp deposits from rapid decay. If the deposits were exposed to the atmosphere, oxidation and bacterial action would destroy them.

Large marine swamps lie inland from the seashore along the Atlantic and Gulf coasts. These swamps form behind offshore bars or islands where sediments from the continent and plant matter fill lagoons and bays. Portions of the coast with low tidal ranges and without strong tidal currents favor development of marine swamps and the growth of swamp plants. Because most freshwater plants die in salt water, marine swamps contain a distinctly different flora. In tropical climates, like that in southern Florida, mangrove trees prosper in marine swamps (figures 10.28 and 10.29). Mosses prosper in many marine swamps, and most of these swamps contain an abundance of grasses and reeds that tolerate salt water. As plants die, they fall into the water and begin to decay, forming a mat of dark, organic-rich material called *peat*. People have dried and used peat as fuel in many parts of the world. It is also an excellent preservative. Peat bogs have yielded nearly perfectly preserved human artifacts as well as human and animal remains.

Sedimentary rocks composed of swamp deposits record evidence that this environment was widespread at many times in the geologic past. Late in the Paleozoic, conditions favorable for the

Figure 10.28 A Marine Swamp. Mangroves hold sediment in place and provide a protected environment for small fish and other animals.

ters (1,300 feet) of plant matter. The plant matter was covered by sands and shale as the subsidence continued. Eventually, the plant matter was compressed and altered to form coal.

Swamps provide a habitat for a great variety of plants and animals. In the tropics, swamps rival rain forests in their diversity. Swamps, such as the Everglades in Florida and the Dismal Swamp in Virginia and North Carolina, provide cover and food for huge flocks of migratory birds. In tropical swamps, the warm water and patches of wet ground provide homes for alligators, snakes, invertebrates, turtles, fish, amphibians, raccoons, and numerous rodents.

Early settlers rarely inhabited swamps because wet land is not suitable for agriculture. Later, people began to convert swamps by draining them for use as farm or pastureland. Large swamps on the flood plain of the Mississippi River were drained and transformed into some of the richest farmland in North America. In some places, the wet land has been converted to rice production, which requires long periods of inundation. In urban and suburban areas, people have commonly filled the wet land and used it for housing and other developments. The resulting loss of wetlands has destroyed the habitat of many animals and plants and has led to a major controversy regarding the future of remaining wetlands, discussed in chapter 14.

development of swamps became extensive throughout the world. Carbon-bearing sediments are so abundant in the rocks from this period that geologists named the time from about 350 to 270 million years ago the Carboniferous period.

Organic matter in swamps may eventually turn into coal. For this to happen, the organic material must not decay. Burial is the first step in this process. If burial is not rapid, oxidation and bacterial action break down the organic material. Burial occurs most rapidly where the region is subsiding. Only thin accumulations of swamp deposits form if the level of the swamp remains stable. Because water is shallow in most swamps, slow subsidence of the swamp relative to the sea is necessary for the accumulation of more than a few meters of plant matter. Approximately 38 meters (125 feet) of plant matter will form 1 meter (3.28 feet) of coal. A slowly subsiding area is most favorable for preservation of such sediment. Some of the coal seams in Pennsylvania and West Virginia are several meters thick. A 9-meter- (30-foot) thick coal seam in Russia represents the accumulation of more than 400 me-

Estuaries

The rise in sea level during the past 18,000 years has flooded the lower reaches and former valleys of most streams that flow into the ocean. Where these streams flow on low slopes and especially where tidal range is great, tides affect the lower part of the stream system. **Estuaries** are the seaward end of stream valleys where salt water and fresh water meet. The shape of the landscape that is flooded as the sea inundates the coast determines the shape of an estuary. Some estuaries, such as the lower Hudson River and the lower Mississippi River, retain their river-like channel shape. Others, such as San Francisco Bay, Delaware Bay, and Chesapeake Bay, are so large that they no longer resemble the rivers that formerly flowed through them.

As high tide approaches shore, the sea surface rises and a wedge-shaped mass of seawater moves into the estuary. As the tide recedes,

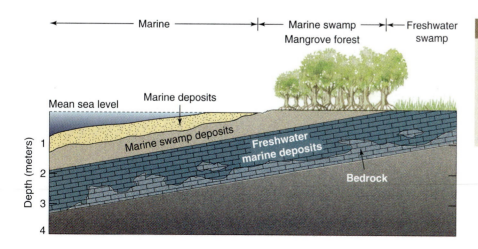

Figure 10.29 Cross Section of the Swamp Environment Along the Coast of Southern Florida. Mangrove forests grow at the edge of the shore in salt water. Farther inland, the swamps contain brackish water and fresh water. Distinctively different plants inhabit each of these environments.

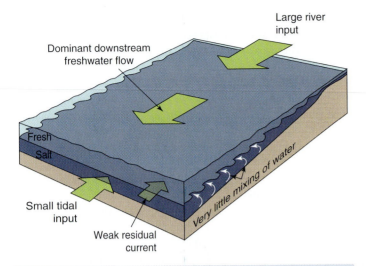

Large river input

Dominant downstream freshwater flow

Fresh
Salt

Small tidal input

Weak residual current

Very little mixing of water

Figure 10.30 Diagram of an Estuary. In some estuaries, the salt water and fresh water mix, but generally, the fresh water lies above the more dense salt water. Estuaries show varying degrees of density stratification depending largely on the ratio of salt water to fresh water. Commonly, the salt water moves as a wedge-shaped mass upstream in the estuary.

the wedge retreats ahead of the fresh water that continues to enter the upper end of the estuary. Where fresh water comes into contact with salt water, the fresh water floats on the heavier salt water, but as the tides come and go, the water mixes (figure 10.30). The ratio of fresh water to salt water in an estuary usually indicates how well mixed the waters are. Scientists use the ratio of fresh water to salt water in an estuary to distinguish one estuary from another. The ratio in Delaware Bay is 1 part fresh water to 100 parts salt water, whereas in the Mississippi estuary, the ratio is 1:1. Although the shape and the ratio of fresh water to salt water in these estuaries vary greatly, both are highly important to these ecosystems.

Most estuaries are important breeding grounds. Extremely high levels of productivity and low species diversity characterize estuarian fauna and flora. Which plants and animals inhabit an estuary depends on water temperature and other factors in their physical environment. Fish, oysters, clams, and other invertebrate animals, as well as birds, snakes, and other reptiles populate many American estuaries.

Estuaries were attractive places for the early settlers to build ports and towns. Some of these, such as New York, Baltimore, and San Francisco, sprang up at the coast; others, such as Philadelphia, Washington, D.C., and Houston, however, developed closer to the head of navigation on the rivers flowing into the estuaries. These ports reduced reliance on land transportation, which was much more costly than ship transport. As the population grew around the estuaries and in the drainage basins of the streams that flow into them, environmental problems multiplied.

Modification of the shore around many estuaries has destroyed habitats and polluted the water. Construction of harbor facilities, dredging of channels, filling of wetlands for housing develop-

ments, and stripping of land cover for all types of construction has increased the amount of sediment pouring into estuarine waters. In addition to the direct destruction of wetlands, sediment carried out into the estuary settles to the bottom and changes the environment of the plant and animal communities that inhabit the bottom. Sediment- and pollutant-charged water shows up in color satellite images of most estuaries.

Pollution reaches estuaries from a variety of sources. Before dumping was outlawed in American ports, ships commonly discarded all types of materials, including hydrocarbon-laced water, into estuaries. Sewage still gets into estuaries from ships and shoreline communities. Runoff carrying excess fertilizer from cultivated land also remains a serious problem. The fertilizer and sewage cause algae to grow. Oxidation burns up the dead algae but leaves the water oxygen-deficient and incapable of supporting many living organisms. Many of the environmental problems that beset estuaries also affect deltas.

Deltas

Deltas are collections of sediment deposited where rivers enter other bodies of water. Most big deltas form where rivers enter the sea, but small deltas form where streams enter lakes, or more rarely, another larger and slower stream. Where a river enters a quiet body of water, the velocity drops, and the river loses its ability to transport sediment. Even clay colloids form clumps and sink where streams enter bodies of salt water. The heavier sediment settles out first; finer sediment stays suspended long enough to move farther away from shore. As sediment accumulates, the delta grows.

Figure 10.31 Satellite Image of the Yukon Delta in Alaska. Note the exceptionally well developed distributaries. This is a fluvial-dominated delta.

Wave action, tides, and the amount of sediment brought to the delta determine its shape. Wave action dominates and shapes the front of some deltas; the rise and fall of water caused by tides controls the shape of others, and streams are the dominant factor in still others. Compare the shape of the deltas of the Yukon and Mississippi Rivers (figures 10.31 and 10.32), which are shaped by deposition of sediment from the streams, with that of the Nile River (figure 10.33), which has been eroded by wave action.

A characteristic feature of fluvial-dominated deltas is the way the river breaks up into a number of smaller streams, called **distributaries,** when it reaches the delta. These streams distribute the water and sediment along the front edge of the delta. This is clearly seen on the "birdfoot" delta of the Mississippi River. The Mississippi discharges over a million cubic feet of water per second onto the delta. This water carries nearly a tenth of a cubic mile of solid matter and about 200 million tons of dissolved matter onto the delta each year. Most clay remains suspended and moves into the Gulf, but the bed load, consisting mainly of sand and silt in the distributary channels, settles on the delta or near the delta front. During floods, natural levees form along the channels, building the channels slightly higher than the surrounding topography. In the case of the Mississippi River, the Corp of Engineers has also built artificial levees to maintain shipping channels across the delta.

Bays and swamplands that stand at or close to sea level fill the areas between distributary channels. Where levees along the distributaries break, sediment-charged water floods out into these bays and swamps, leaving the stream's sediment load in the bay or swamps as the floodwaters recede. Under normal conditions, the swamps are wet grasslands that surround and partially separate bodies of shallow, open water.

The waters on deltas contain a rich fauna and flora. Many fish spend the early stages of their lives in the shallow, open water on the delta, and worms, gastropods, and clams inhabit the shallower water in great numbers. This abundant food source attracts larger animals, such as birds, some reptiles, and small mammals, to the delta.

The Mississippi River has flowed into the Gulf for many millions of years, and older deltas preceded the more recent ones. The modern Mississippi Delta is one of sixteen different lobe-shaped masses of sediment deposited as deltas in the Gulf over the past 6,000 years (figure 10.34). The land is so flat in southern Louisiana that the course of the Mississippi changes with ease. Floodwaters from the Mississippi once flowed into the Gulf through the Atchafalaya River. The two rivers are now very close together at one point, and the Atchafalaya River is steeper than the Mississippi River. So, to maintain the Mississippi in its present course, the U.S. Army Corps of Engineers had to build a system of enormous structures, called the Old River Control System, that allows 30% of the river water to flow down the Atchafalaya River and keeps the other 70% flowing down the Mississippi River past Baton Rouge and New Orleans.

Changes in the river's course took place as sea level fell and rose in response to the amount of ice on land. Before the Pleistocene, sea

Figure 10.32 Satellite Image of the Mississippi River Delta.
Deposition of sediment transported to the delta by the river has determined its shape. New Orleans is located along the river in the upper-left part of this image.

Figure 10.33 Satellite Image of the Delta of the Nile River.
The sharply defined edge of the delta indicates that it is shaped by wave action.

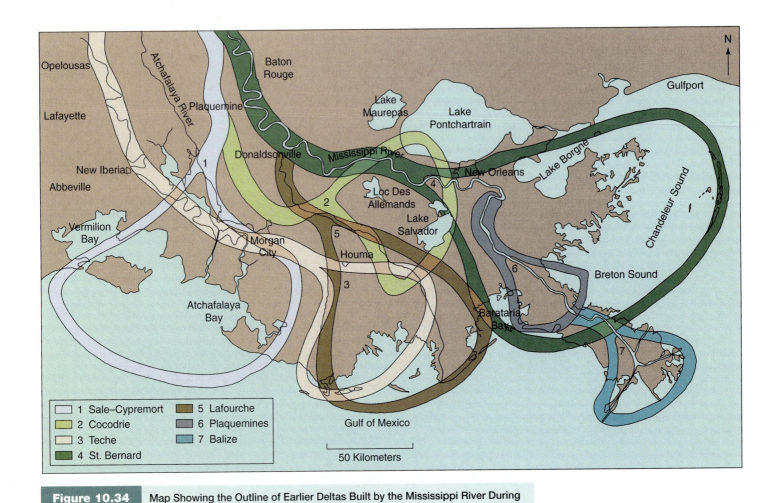

Opelousas
Baton Rouge
Atchafalaya River
Plaquemine
Lafayette
New Iberia
Abbeville
Lake Maurepas
Lake Pontchartrain
Gulfport
Donaldsonville
Mississippi River
New Orleans
Lake Borgne
Chandeleur Sound
Loc Des Allemands
Lake Salvador
Vermilion Bay
Morgan City
Houma
Breton Sound
Atchafalaya Bay
Barataria Bay
Gulf of Mexico

☐	1 Sale–Cypremort	▨	5 Lafourche	
☐	2 Cocodrie	▨	6 Plaquemines	
☐	3 Teche	▨	7 Balize	
▨	4 St. Bernard			

50 Kilometers

Figure 10.34 Map Showing the Outline of Earlier Deltas Built by the Mississippi River During the Ice Ages.

level was much higher, and the Mississippi entered the Gulf north of Memphis. At the peak of the last glacial advance, when ice covered the northern part of North America and lowered sea level, the Mississippi entered the Gulf far out on the continental shelf. At that time, the river flowed in a deeper valley. As sea level rose, sediment filled that valley. Ever since humans first settled on the delta, they have been caught up in the changes that are a natural part of this environment. These changes, plus problems resulting from human developments, pose serious threats to the ecosystems on the delta.

Dams on the Missouri and Arkansas rivers trap sediment that would have reached the delta. In addition, engineering projects cause millions of tons of sediment to build up where the Mississippi and Atchafalaya come close together. Since these projects first began, dams and diversions have reduced the sediment reaching the delta by about 50%. The rate of replenishment of sediment to the delta is not sufficient to compensate for the combined effects of dams that trap sediment in their reservoirs, a global rise in sea level, subsidence of the delta surface as underlying sediments in the delta compact, and regional subsidence caused by the shifting of coastal plain sediments toward the Gulf of Mexico. As a result, land loss along the lower Mississippi Delta has accelerated throughout this century. If the present trend continues, some estimate that the entire surface of the delta will lie underwater within a few decades.

SUMMARY POINTS

1. The level of water in the oceans is constantly changing—over the short term as the result of waves, currents, and tides. Over the long term, changes in the ice–water balance and tectonic activity affect sea level.

2. Many geologic features, such as reefs, deltas, sea cliffs, and beaches, form at sea level. Where sea level drops, these features remain in the landscape until erosion removes them. If sea level rises, these features lie on the sea floor until sediment buries them. Because many features survive for long periods, they provide a record of past changes in sea level.

3. The changing ice–water balance that accompanied the advance and retreat of major glaciations produced sea-level changes of as much as 100 meters above and below present sea level. Tectonic activity has caused even greater differences in the elevation of continents and the sea.

4. Since the last major advance of glaciers about 18,000 years ago, sea level has been rising in most places. As the ice sheets began to melt, the rise in sea level was especially rapid. The rise has continued but at a much slower pace during the last few thousand years.

5. Wave action shapes many shorelines. Waves erode headlands at a faster rate than they do bays and thus help to even out the coastline. The turbulent water of a breaker and the sand and pebbles it carries are effective agents of shoreline erosion, forming sea cliffs and wave-cut terraces. As the sea cliff retreats, stacks, sea caves, and arches may remain offshore as remnants of earlier stands of rock.

6. Ocean waves and longshore currents shift the products of shoreline erosion and stream sediment that reach the shore. This sediment lies temporarily in beaches, spits, and bars where it is subject to erosion and redeposition. Where these deposits form barrier islands, they protect bays from wave action and create one of the most productive marine environments.

7. Sudden displacements of the sea floor caused by displacements on faults or volcanic eruptions may create waves of exceptionally long wavelength, called tsunami. The wave height of tsunami increases dramatically as they enter shallow water. Consequently, they may cause great damage to shorelines and human developments that are high above normal sea level.

8. A great variety of environments are present along shorelines. Climate, wave and current action, and the geologic setting of the coasts determine the character of these environments.

9. Beaches frequently form in protected bays or along low-lying shores. From Cape Cod to Mexico, many islands with sand beaches have formed offshore along the coast. These barrier islands slowly shift position as sea level changes. They are especially subject to storm damage during hurricanes.

10. Tidal flats may form where the coastal and shallow water topography has low relief, is near the mouths of major rivers, and is protected from wave erosion.

11. Large marine swamps are present in southern Florida and along the coast of parts of Virginia and Georgia. These swamps are habitat for a great variety of distinctive plants and animals. Mats of organic matter form in these quiet bodies of water. This organic matter may later become peat and finally coal.

12. The rise in sea level has inundated the mouths of many rivers, forming estuaries. Fresh water from streams mixes with sea water in estuaries, creating an unusual environment that is especially valuable as a breeding ground for fish.

13. Deltas form where large streams enter the sea. Streams drop their sediment load as they reach base level, and if wave and current action do not remove the sediment, it builds up as a delta. Distributary channels, levees, bays, swamps, and the delta front each have distinctive environments and habitat for different animals and plants.

REVIEW QUESTIONS

1. Name two ways an increase in global temperature affects sea level.
2. Describe the trend of global changes in sea level over the past 25,000 years.
3. How is it possible for sea level to drop relative to the shore in some places while sea level is rising relative to the shore in most other places?
4. In what ways do the environments of deltas differ from those of tidal flats?
5. What are the primary threats to the ecology of estuaries? Why should people care?

THINKING CRITICALLY

1. How do coastal processes and human intervention in these processes affect beachfront real estate?
2. Discuss the interactions among land environments, marine environments, and the coastal environments between them. What Earth systems affect the interactions between environments, and how?
3. How would the character of the coast change if sea level rose 10 to 20 meters? Discuss the changes that would affect a portion of the coast with which you are most familiar.
4. Discuss the effects of natural disasters, such as earthquakes, volcanic eruptions, and tsunami, on coastal environments and their inhabitants (both human and nonhuman).
5. Discuss the advantages and disadvantages of draining wetlands in coastal areas.

KEY WORDS

arches	mean sea level
backshore	morass
barrier island	notch
bay-head beaches	offshore
beach drifting	progressive waves
bog	runup
breaker	sea caves
breakwaters	seismic sea wave
coast	shoreline
delta	spit
distributary	stacks
estuary	swamp
fiord	swell
foreshore	tidal flat
hook	tsunami
jetties	wave-cut terrace
longshore current	wave refraction
longshore drifting	

chapter eleven

Earth's Atmosphere

Chapter Guide

Concern about the consequences of global warming, the destruction of rain forests, and the melting of polar ice caps has sparked great interest in environmental issues. Popular magazines now publish articles with titles like "The Great Climate Flip Flop," "The World's Hottest Year in More Than a Century," or "Too Much Hot Air" that discuss evidence obtained from ice cores that indicates drastic past changes in climate or the relationship between atmospheric chemistry and climate. Such topics were formerly discussed only in technical journals. People are better informed than ever, and many are aware of the urgent debates about global warming, the ozone hole, and climate change. This chapter and chapters 12 and 13 provide the background needed to understand these debates. This chapter explains the composition and structure of the atmosphere. Chapters 12 and 13 examine atmospheric circulation, weather systems and storms, the close relationship between oceanic circulation and climate, modern climates on Earth, and the history of climatic change that has led to present atmospheric conditions.

INTRODUCTION

COMPOSITION AND STRUCTURE
OF THE ATMOSPHERE

HEAT BALANCE IN THE
ATMOSPHERE

THE ATMOSPHERE—PROBLEMS AND ISSUES

The atmosphere is greatly influenced by the oceans, continents, and solar radiation.

INTRODUCTION

Weather is the state of the atmosphere at any given moment (figure 11.1). It normally changes from hour to hour and season to season. Long-term averages of weather conditions define **climate.** Large areas such as the Arctic or the tropics are well-known because of the climatic conditions that prevail there most of the time. However, like weather, climatic conditions change. Today, an ice sheet covers most of Greenland, and sea ice has closed the entire Arctic Ocean to shipping. About 18,000 years ago, ice sheets covered all of Canada and extended as far south as the Ohio and Missouri rivers. What caused the climate to change so drastically that the ice sheets melted? How rapidly do such changes occur? These are obviously significant questions. But even less-severe changes in climate have had great impact on human history. In the 1930s, failure of crops because of several consecutive years of drought caused many thousands of people to move from Oklahoma and Texas to California. Today, people are concerned about changes in the atmosphere's chemistry that result from burning of hydrocarbons and the subsequent release of carbon dioxide. These problems have stimulated great interest in how Earth's climate system works.

The climate system directly involves the structure, chemical composition (table 11.1 and see figure 1.3*d*), and circulation of the atmosphere. However, the atmosphere is not isolated from other Earth systems or from human actions. Gases are freely exchanged among the atmosphere, the surface of the ocean, soils on land, and the plants and animals that live both on land and in the sea. Gases also enter the atmosphere because of volcanic activity.

COMPOSITION AND STRUCTURE OF THE ATMOSPHERE

Initially, meteorologists measured the composition and physical conditions of the atmosphere including temperature, moisture content (humidity), and wind direction at ground level. Later, measurements obtained from balloons carrying radio transmitters and by aircraft supplemented ground-level data. More recently, rockets and satellites have provided new opportunities for investigating the character of the atmosphere. The first soundings of the atmosphere provided information about variations in temperature with altitude and made distinguishing major structural divisions of the atmosphere possible (figure 11.2). New, more sophisticated instruments make it possible to measure the composition and density of the atmosphere as well as its temperature. Temperature normally decreases at higher elevations. Some mountain tops, even some like Kilimanjaro in Africa located near the equator, are so cold that ice remains year round. The rate of decrease in temperature is 6.5°C per kilometer (3.5°F per 1,000 feet) rise in elevation. The temperature decreases to about −60°C (−108°F) at an altitude of 10 kilometers (6 miles), only slightly higher than Mount Everest. Temperature remains stable between 10 and 20 kilometers (6 and 12 miles); then it begins to rise.

Based on changes in temperature, meteorologists subdivided the atmosphere into the **troposphere, stratosphere, mesosphere,** and **thermosphere** (see figure 11.2*a*). Zones in which temperature changes direction separate these divisions. Temperature drops with elevation in the troposphere and stabilizes at an elevation of 10 kilometers (6.2 miles) in the **tropopause** before reversing

Table 11.1	Composition of the Atmosphere
Dry gases	
Nitrogen	78.1%
Oxygen	20.9%
Argon	0.9%
Carbon dioxide, neon, helium, methane, krypton, nitrous oxides, hydrogen, and ozone	0.1%
Pollutants of national concern	
Sulfur dioxide	
Nitrogen oxides	
Carbon monoxide	
Hydrocarbons	
Ozone	
Particulates	
Radioactive particles	
Dust, ash, cement	
Fumes with metallic oxides	
Liquid mist	
Greenhouse gases	
Carbon dioxide	
Methane	
Water vapor	
Nitrous oxides	
Chlorofluorocarbons	
Ozone	

direction in the stratosphere. Similar gradual transitions separate each of the divisions in the atmosphere. The second temperature reversal, the **stratopause,** occurs at an altitude of about 50 kilometers (30 miles), and a third change, the **mesopause,** takes place at an altitude of about 90 kilometers (56 miles). The Sun is the only significant source of heat energy for the atmosphere, but because the composition and concentration of certain atmospheric gases vary with altitude, some levels in the atmosphere are warmer than others. The ground surface absorbs solar radiation that heats the lower part of the atmosphere. A very small part of this heat is transferred to the air by conduction. In addition, the ground absorbs and reradiates energy that is absorbed by water vapor and other gases in the atmosphere. The gases that absorb energy and become warm, called *greenhouse gases,* are less abundant toward the top of the troposphere. Consequently, temperature drops at higher altitudes in the troposphere and reaches its lowest level at the tropopause. The stratosphere contains ozone. This gas also absorbs solar radiation and causes the temperature to rise in the middle of the stratosphere. Temperature resumes its decline above the ozone concentration and reaches a minimum at the mesopause. Above the mesopause, in the thermosphere, the atmosphere contains so few molecules that an astronaut walking in space at that altitude would feel cold. However, solar radiation causes the molecules at high altitude to move rapidly, and it is that movement that is used as a measure of temperature. Consequently, the upper levels of the atmosphere have a high temperature.

Composition of the Lower Atmosphere

In the troposphere and lower part of the stratosphere, nitrogen and oxygen are by far the most abundant gases. Smaller quantities of water vapor, liquid mists, and solid particles, as well as a number of gases are present (see table 11.1 and figure 11.2b). Water vapor enters the atmosphere primarily through evaporation of surface waters, volcanic eruptions, and transpiration by plants. Solid particles get high into the atmosphere from many sources, including dust blown from the ground surface, salts from the ocean, dust from meteorites, pollen, microorganisms, and volcanic dust. Humans add to this load by introducing industrial waste, soot, and, in the past, products of nuclear explosions.

In the lower part of the atmosphere, the rise of hot air and circulation mixes the dry gases (nitrogen, oxygen, argon, and others), creating a homogeneous mixture called the **homosphere.** This relatively uniform mixture of dry gases extends from ground level to an altitude of about 80 kilometers (50 miles). The upper parts of this well-mixed layer do not include water vapor, particulate matter, or most pollutants that exist in large concentrations closer to ground level, and the upper parts vary greatly in composition at different altitudes. Above the homosphere, turbulence is greatly reduced, and concentrations of the various atmospheric constituents (e.g., oxygen and nitrogen) form thick, poorly mixed layers. This part of the atmosphere, referred to as the **heterosphere,** contains concentrations of electrons and ionized particles of oxygen and nitrogen. The highest part of the atmosphere is extremely rarified and contains hydrogen and helium that are escaping from Earth and incoming cosmic rays, meteorites, and particles coming from the Sun (see figure 11.2b).

Water Vapor

Water is an exceptionally important component of the atmosphere even though it is present in relatively small quantities (0 to 4%) by volume in the homosphere. Approximately 425 centimeters (14 feet) of water evaporates from Earth's surface each year. At any given time, the atmosphere holds an amount of water vapor equivalent to the water contained in the upper 2.54 centimeters (1 inch) of the oceans. This water plays a critical role in many processes in the atmosphere, and, more than any other component of the atmosphere, it determines weather and climate.

Humidity, the quantity of water held in a given volume of air, is expressed in two ways. **Absolute humidity** is the quantity of water vapor contained per unit of volume of air or the number of grams of water vapor contained in a cubic meter of air. **Relative humidity** is the ratio of the amount of water vapor in a volume of air to the amount of water vapor that could be contained in the air at a given temperature without having some of it change state and become liquid. Air is **saturated** with water vapor when the quantity of water vapor is sufficient for some of it to condense to a liquid. Saturation represents an equilibrium condition between evaporation and condensation. The maximum quantity of water vapor that any volume of air can contain depends on air temperature. Air can contain more water vapor at higher temperatures. For this reason, air in the summer may contain large amounts of water and feel "muggy," while air in the winter is usually much "drier" and feels crisp.

Condensation happens more readily if some object is present to act as a nucleus on which condensation can form. Microscopic particles serve this purpose in the air. If the air temperature is above freezing, water vapor in saturated air condenses as liquid droplets. If the temperature is below freezing, water vapor may crystallize, forming snowflakes (figure 11.3), or freezes, forming ice (figure 11.4). Initially, these droplets of water, snowflakes, or ice may be so small that they remain suspended in the air. Clouds consist in large part of droplets or ice crystals that are too small to fall. Droplets that form near ground level create ground-level clouds known as **fog.** As the droplets or ice crystals grow or hit one another and coalesce, they become heavier and fall. **Precipitation**

Figure 11.2 Composition and Structure of the Atmosphere.

(a) Temperature structure and major divisions of the atmosphere. The solid curve depicts the temperature at different altitudes in the atmosphere. (b) Compositional structure of the atmosphere. The lower part, called the homosphere, is well mixed. At a higher level, layers of charged particles are present. At still greater altitude, in a region known as the magnetosphere, the atmosphere consists primarily of charged particles approaching the Earth from the Sun. Ozone is an important constituent in the stratosphere. It protects humans from ultraviolet radiation. However, near the surface, ozone, pollutants and greenhouse gases are hazardous to humans.

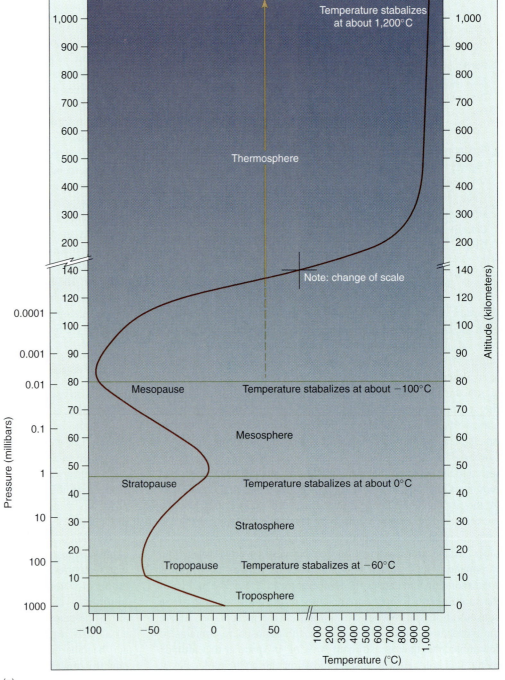

(a)

consists of all types of falling water or ice (see figure 11.4). Most precipitation falls from particular types of clouds.

Types of Clouds

Clouds are classified based on their shape, composition, and altitude. Most clouds fall within a few categories (figure 11.5). **Stratus clouds** are layered and generally occur below altitudes of 2,000 meters (6,560 feet). Little or no sunlight illuminates the ground surface, and rain or drizzle usually falls from these dark, gray clouds. Stratus clouds at high altitude are called *altostratus clouds*.

Cirrus clouds have a distinct wisp shape (see figure 11.5*b*). They consist of ice crystals that are blown by winds at high altitudes, often over 6 kilometers (20,000 feet).

Cumulus clouds (see chapter-opening photograph) are puffy and resemble balls of cotton. Generally, they have a relatively flat base that is often dark because the lower part of the cloud is shaded

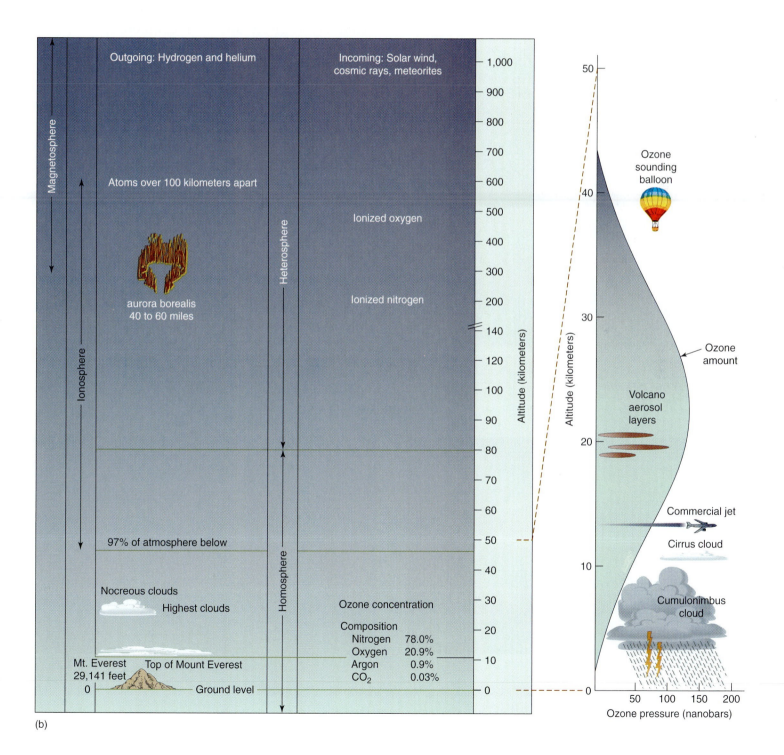

(b)

Figure 11.3 Crystallized Water Vapor. Snowflakes are hexagonal-shaped crystals that exhibit a great variety of fine details.

Figure 11.4 Frost Forms When Air Reaches the Dew-Point Temperature That Is Below Freezing.

by the upper portions. Cumulus clouds that have high tower-like forms are called *towering cumulus*. Others called *cumulus congestus clouds* cover large areas and may have high vertical forms with tops that resemble cauliflower. **Cumulonimbus clouds** (see figures 11.5 and 12.16) are huge and may have an anvil shape produced when the rising air reaches the warmer air in the stratosphere. As a result of losing its buoyancy, the cloud spreads laterally and assumes the form of an anvil. These clouds rise high into the atmosphere and produce thunderstorms.

Composition of the Upper Atmosphere

Layers of different composition characterize the upper parts of the atmosphere. Ultraviolet radiation breaks down oxygen (O_2) and nitrogen molecules (N_2), forming oxygen and nitrogen atoms (O and N). The oxygen and nitrogen atoms absorb ultraviolet radiation, lose electrons in the process, and become ionized. Concentrations of oxygen ions are present in the upper parts of the atmosphere—the thermosphere—and nitrogen ions predominate in the lower part of the thermosphere. Electrons freed in the development of these ions form layers at certain altitude levels in the ionosphere (see figure 11.2*b*). They reflect low-frequency radio waves and absorb high-frequency radio waves.

Ozone

Ozone is a highly reactive gas formed by the combination of an oxygen molecule with an oxygen atom ($O_2 + O \rightleftharpoons O_3$) (figure 11.6). This process occurs most frequently in the stratosphere where ozone produced in this way concentrates in a zone located between 15 and 30 kilometers (9 and 18 miles). Ozone is continually created and destroyed by natural processes. As ozone reacts with ultraviolet radiation, it breaks down to form oxygen atoms (O) and molecules (O_2). Ozone shields the lower parts of the

(a)

(b)

Figure 11.5 Types of Clouds. (*a*) Cumulonimbus clouds rise toward the top of the troposphere, where the rise of the clouds will stop and spread. (*b*) Cirrus clouds are wisps of ice crystals.

atmosphere and Earth's surface from large amounts of ultraviolet radiation. Depletion of ozone in the stratosphere, discussed later in this chapter, is a matter of great concern.

Outer Reaches of the Atmosphere

The atmosphere does not have a distinct outer edge. At higher levels in the thermosphere, the concentration of ionized oxygen decreases. Above 500 kilometers (300 miles), oxygen atoms are as much as 167 kilometers (100 miles) apart. At that level, the atmosphere has less gas in it than the best vacuums created at ground level. Near the outer edge of the atmosphere, hydrogen and helium are present, but both gases are slowly escaping from Earth's gravity field. Finally, above 3,000 kilometers (1,800 miles), the only things detected in the "atmosphere" are emissions from gigantic explosions on the Sun's surface that are coming toward Earth. These particles are electrons and hydrogen ions (H^+), part of the **solar wind** ejected during eruptions, which resemble gigantic nuclear explosions. These eruptions appear as spots on the surface of the Sun. As these charged particles reach Earth, they distort Earth's magnetic field (figure 11.7). Some of the particles trapped in the magnetic field move down into the atmosphere toward the polar regions. When the charged particles enter the atmosphere, they strike the nitrogen and oxygen in the upper atmosphere and cause them to emit radiation that produces colors in the sky. These colorful atmospheric displays known as the *northern* (aurora borealis) or *southern* (aurora australis) *lights* are **auroras** (figure 11.8).

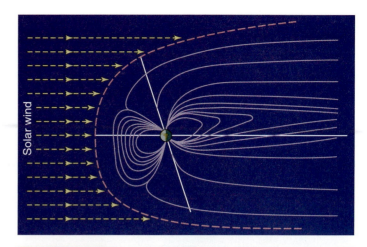

Figure 11.7   Schematic View of the Solar Wind and Its Effects on the Lines of Magnetic Force Surrounding the Earth. Some particles in the solar wind penetrate the bow shock and approach Earth where the magnetic field traps incoming charged particles from the Sun and directs them toward the polar regions.

HEAT BALANCE IN THE ATMOSPHERE

Although small quantities of heat reach the atmosphere through volcanic activity and by conduction of heat from the interior, nearly all heat on and near Earth's surface comes from the Sun. Based on measurements made over decades, the amount of solar radiation intercepted from the Sun does not vary greatly. It appears to remain almost constant at about two calories per square centimeter per minute, a value referred to as the **solar constant.** The solar constant = $1,367$ W/m^2. Most of the energy in this solar radiation lies in the visible, ultraviolet, and shorter infrared parts of the electromagnetic spectrum (figure 11.9). If all of this energy remained on Earth, the atmosphere would become intolerably hot. Under normal conditions, about as much heat is lost as is gained.

The Concept of Heat Balance

A balanced condition exists when the amount of heat introduced into the atmosphere is equal to the amount of heat lost. Earth gains heat when the surface and gases in the atmosphere absorb solar radiation and become warm. Earth loses heat when it radiates energy out into space. Heat is also transferred from water on Earth's surface to water vapor in the atmosphere as a result of evaporation of water, which requires heat energy. Heat energy absorbed during evaporation remains available as latent heat (see chapter 1), which is released when condensation occurs.

Movement of heat energy through evaporation and condensation is one mechanism by which heat is transferred from one place on Earth to another. Large-scale movements of heat energy happen through circulation of water in the oceans (see chapter 9) and of air

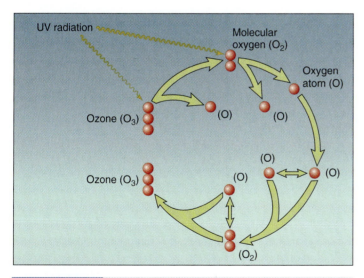

Figure 11.6   Formation and Breakdown of Ozone. Ozone (O_3) in the stratosphere is continually breaking down and reforming. Ultraviolet radiation breaks the ozone down into its components, oxygen molecules and oxygen atoms. Radiation also breaks oxygen molecules into oxygen atoms that can recombine as oxygen molecules and then ozone. Other reactions between the oxygen atoms or molecules and other chemicals introduced into the stratosphere disrupt this cycle and cause a decrease in the amount of ozone.

Figure 11.8 Photograph of an Aurora. Auroras form as the charged particles enter Earth's gaseous atmosphere and react with nitrogen and oxygen ions in the ionosphere.

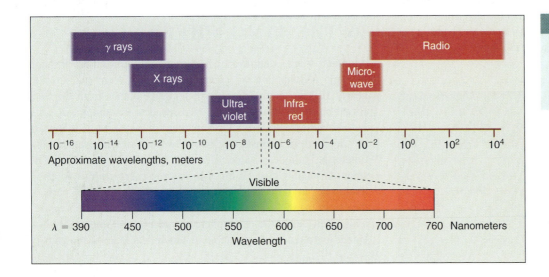

Figure 11.9 The Electromagnetic Spectrum. Radiation from the Sun lies in all parts of the electromagnetic spectrum. Only the central part of the spectrum is visible.

in the atmosphere (see chapter 12). These movements of heat moderate the temperature of both equatorial and polar regions and are a critical factor determining climate.

Over a long time, imbalances in the heat budget of the atmosphere cause changes in climate. The geologic record contains a long and impressive history of climatic change. Major variations in climate appear to result from the changing position of continents caused by continental drift and in slight changes in the amount of solar radiation reaching Earth. Obviously, human activities did not cause climatic changes that took place thousands or millions of years ago, but modification in atmospheric composition since the Industrial Revolution may be sufficient to cause climatic changes.

Before examining these changes, it is important to understand the operation of the natural system.

Earth's Path Around the Sun and Incident Solar Radiation

The nearly circular path Earth follows around the Sun lies in a plane called the **ecliptic.** Earth is slightly closer to the Sun during the winter months in the Northern Hemisphere, but this difference in the amount of solar radiation intercepted by Earth has little effect on Earth's seasonal variations. Seasons change mainly because Earth's axis of rotation is inclined to the ecliptic at an

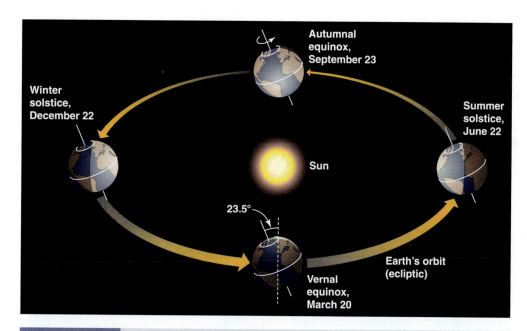

Earth's Ecliptic and Seasonal Changes. Earth's equator is inclined 23.5° to the plane in which Earth revolves around the Sun. Because the direction of inclination remains constant, the Northern Hemisphere receives more direct rays from the Sun in June, July, and August; the Southern Hemisphere receives more direct rays from the Sun in January, February, and March.

angle of 23.5°, and the orientation of the axis relative to the ecliptic does not change through the year (figure 11.10).

Not all parts of the nearly spherical Earth receive the same amounts of solar radiation. The angle of incidence of solar radiation ranges from 0 to 90°. The maximum amount of radiation per unit of surface area occurs where radiation comes in perpendicular to the surface. The least radiation per unit of surface area occurs where radiation strikes Earth's surface at a small angle (figure 11.11). At any moment, the amount of incident solar radiation varies across the hemisphere illuminated by the Sun. As the Earth rotates on its axis, radiation spreads around the globe. Because Earth's axis is tilted, heating of Earth does not coincide exactly with the latitude (see figure 11.11). Greater amounts of solar radiation strike the surface when the sun is located directly overhead. Because the axis is tilted, solar radiation strikes the surface from directly overhead between latitude 23.5° north, called the **Tropic of Cancer,** and 23.5° south, called the **Tropic of Capricorn.** Earth's axis remains tilted with the same orientation as it revolves around the Sun. In June, July, and August, the Northern Hemisphere receives more solar radiation than does the Southern Hemisphere. At the time of the autumnal (September 23) and vernal (March 20) equinoxes, when the sun is located directly over the equator, solar radiation is equally spread over the Northern and Southern hemispheres. In January, February, and March, more radiation reaches the Southern Hemisphere (see figure 11.10). Nevertheless, the portion of Earth on either side of the equator, between the tropics of Capricorn and Cancer, receives much

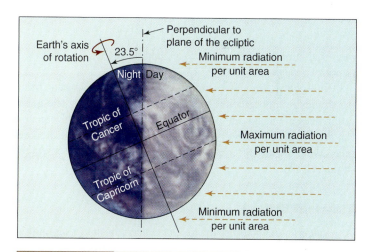

Incidence of Solar Radiation. The equatorial latitudes on Earth receive much more solar radiation than high latitudes. Beams of identical size striking the Earth near the equator are concentrated in small areas, but those near the poles spread over areas of much larger size. Greater amounts of solar radiation strike the Earth between the equator and latitude 30°N and S than are reradiated back to space, leading to a buildup of heat. At higher latitudes, the amount of radiation leaving Earth exceeds the amount reaching the Earth. Consequently, the polar regions are cool. Atmospheric and oceanic circulation spreads the heat surplus from the equatorial regions to higher latitudes.

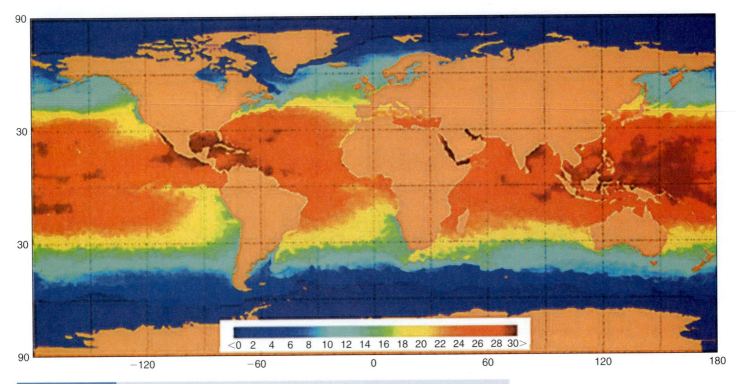

90
30
30
90
−120   −60   0   60   120   180

<0 2 4 6 8 10 12 14 16 18 20 22 24 26 28 30>

Figure 11.12 Mean Surface Temperature of the Earth for June 1988 as Detected by Instruments on Meteorological Satellites. High temperatures are dark red and low temperatures are blue.

more radiation throughout the year than do higher latitudes (see figure 11.11).

Atmospheric circulation and ocean currents move heat from the tropics to higher latitudes. Without these processes, the great differences in the amount of incident solar radiation in the equatorial regions as compared with that in polar regions would cause much more drastic differences in temperature than those observed (figure 11.12).

The Heat Budget

Atmospheric scientists use the budget concept in analyzing Earth's heat balance. As with any budget, this is an effort to trace the income and outgo—in this case, of heat (figure 11.13). Most of the heat reaching Earth as solar radiation is either absorbed or reflected. The surface of Earth absorbs 51%; clouds reflect 20%; gases in the atmosphere absorb 19%; and scattering by the atmosphere sends 6% back into space. The surface of Earth reflects 4% of the solar radiation that strikes it. High percentages of radiation are reflected from ice and desert regions.

About 30% of all solar radiation returns to space as losses resulting from reflection and scattering. Scattering results when small particles in the atmosphere, mainly gas molecules, redirect the path of radiation that hits or comes close to them. These gases and the solids also absorb much of the radiation that hits them. Materials vary greatly in their absorption characteristics. Each atmospheric gas absorbs specific wavelengths of radiation (figure 11.14). Ozone and oxygen absorb ultraviolet radiation, but water vapor absorbs longer wavelengths of radiation.

Gases and particulate matter, such as carbon soot in the atmosphere, water, or solids on Earth, that absorb radiation, warm and give off radiation, but the emitted radiation is in longer wavelengths than the energy they absorb. Part of the radiant energy absorbed by clouds is reradiated back into space, and part of the energy reradiated from clouds reaches Earth's surface where materials on the surface absorb it. These materials reradiate part of the energy. Clouds absorb part of that energy, and part of it leaves Earth and its atmosphere. In addition, some of the heat energy that reaches the Earth is used in processes that transform the heat. These processes are responsible for the formation of clouds and for maintaining Earth's heat balance. For example, heat absorbed by Earth heats and causes hot air to rise through **convection.** Many of the clouds visible on hot summer afternoons form as convection lifts moist air. Some of the heat at the surface evaporates water and contributes to plant growth. Much of the remaining energy is reradiated from the surface and is absorbed by clouds before it returns to space as reradiated energy. A large part of this reradiated energy causes what is known as the *greenhouse effect.*

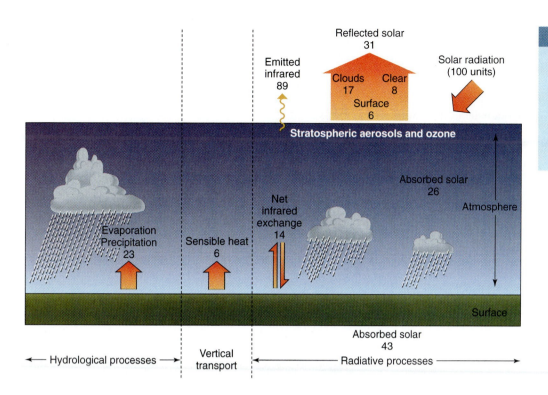

Emitted infrared
89

Reflected solar
31

Solar radiation
(100 units)

Clouds
17

Clear
8

Surface
6

Stratospheric aerosols and ozone

Absorbed solar
26

Atmosphere

Net infrared exchange
14

Evaporation Precipitation
23

Sensible heat
6

Surface

Absorbed solar
43

← Hydrological processes → | Vertical transport | ← Radiative processes →

Figure 11.13 Schematic Representation of the Heat Budget of Earth. Note that an equal number of units of solar radiation comes into and leaves Earth. Of the radiation leaving Earth, 30% reflects off the surface and 70% returns to space as infrared radiation.

The Greenhouse Effect and Global Warming

Most greenhouses have glass walls and a glass ceiling. Ultraviolet and visible radiation pass through the glass and strike plants and other materials in the greenhouse causing them to become warm. The heated plants and other materials reradiate energy in long wavelengths that cannot penetrate the glass. As a result, heat energy remains trapped inside the greenhouse. In some respects, the atmosphere resembles a greenhouse in that solar radiation heats the ground surface, which reradiates long-wavelength energy. Some atmospheric gases, called greenhouse gases, absorb the long-wavelength radiation from the Earth, increasing atmospheric temperature (table 11.2). The greenhouse effect is a normal condition on Earth and is responsible for making much of Earth's surface habitable. Water vapor and carbon dioxide (CO_2) have long been important in maintaining surface temperatures. Without these gases, the temperature at Earth's surface would be below freezing. Over the last decade, rising global temperature and concerns about the consequences of this rise have focused attention on the gases humans introduce into the atmosphere. Because carbon dioxide is such a major contributor to warming, it

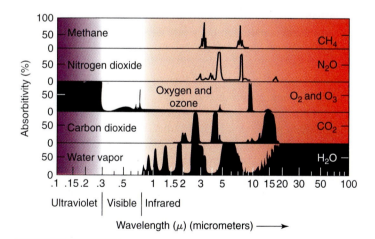

Methane — CH_4

Nitrogen dioxide — N_2O

Oxygen and ozone — O_2 and O_3

Carbon dioxide — CO_2

Water vapor — H_2O

Absorbitivity (%)

.1 .15 .2 .3   .5   1 1.5 2   3   5   10 15 20 30 50   100

Ultraviolet | Visible | Infrared

Wavelength (μ) (micrometers) →

Figure 11.14 Absorbitivity of Solar Radiation by Atmospheric Gases. Each atmospheric gas absorbs radiation from particular parts of the electromagnetic spectrum. Oxygen molecules and ozone absorb ultraviolet radiation. Water vapor and carbon dioxide absorb infrared radiation, most of which comes from Earth's surface. The shaded section indicates visible radiation.

Table 11.2	Greenhouse Gases*	
Gas	**Concentration 1989**	**Annual Increase**
Carbon dioxide	354.00 ppm	0.4%
Methane	1.70 ppm	0.7%
Nitrous oxides	308.00 ppb	3.4%
Chlorofluorocarbon	0.75 ppb	Banned in much of the world by an international agreement in 1987, but CFCs have a long residence time in the atmosphere.

*Note that water vapor is also an important greenhouse gas.

ppm = parts per million

ppb = parts per billion

has been the focus of the public debate on the greenhouse effect. However, water vapor, methane, ozone, and other greenhouse gases such as nitrogen oxides and CFCs also affect atmospheric temperature (figure 11.15).

Concern about global warming began with publication of observations of the concentration of carbon dioxide over Hawaii (figure 11.16). These observations documented a continual rise in carbon dioxide levels since 1958. Projections of future increases in carbon dioxide concentrations in the atmosphere depend on the rate of emissions. At the current rate of emissions, atmospheric carbon dioxide concentration could double in about 180 years. If the rate of emissions increases by 4%, the concentration could double in 50 to 60 years. Seasonal variation caused by the absorption of carbon dioxide by plants during the growing season gives the steady upward trend in carbon dioxide concentration a step-like pattern. In part, this steady rise results from the burning of coal, oil, wood, and other compounds that contain carbon and hydrogen. Because the amount of carbon dioxide produced by the

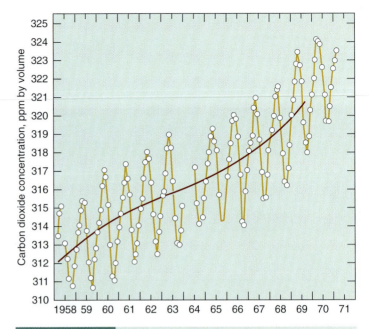

Figure 11.16 Record of the Quantity of Carbon Dioxide in the Atmosphere as Measured at an Observatory in Hawaii. Seasonal variations are responsible for the sharp ups and downs. The level of carbon dioxide drops in summer when plants take up large quantities of carbon dioxide.

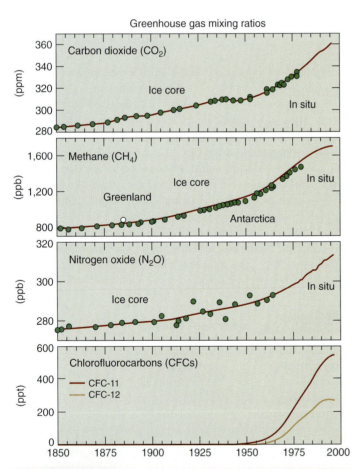

Figure 11.15 Concentrations of Greenhouse Gases in the Industrial Era. The concentrations of carbon dioxide, methane, nitrogen oxides, and chlorofluorocarbons have increased dramatically since the Industrial Revolution. These rises are thought to result from human activities.

burning of wood and gas is small compared with the amount produced during the combustion of coal and oil, current atmospheric concentrations of carbon dioxide result mainly from the burning of coal and oil. Large-scale burning of hydrocarbons began with the onset of the Industrial Revolution. The rapid increase in the concentration of carbon dioxide and the resulting alteration in the composition of the atmosphere is surprising (see figure 11.15).

The concentration of carbon dioxide in the atmosphere during prehistoric times is measured by taking samples of gases trapped as air bubbles in ice. Changes in the carbon dioxide content of the atmosphere in prehistoric times resulted in part from variations in the amount of gases emitted from volcanoes. Measurements show that carbon dioxide has increased from about 280 ppm in 1880 to the present levels over 350 ppm. What happens in the future depends on the rate at which carbon dioxide is introduced into the atmosphere and on the rate at which carbon dioxide becomes involved in other processes that remove it from the atmosphere. These processes include sinks, such as plant matter, and formation of limestone from marine plankton discussed earlier. Many scientists are studying the effect of increased levels of carbon dioxide on the natural processes in which it is involved. Only after the processes, collectively known as the carbon cycle, are fully documented will it be possible to obtain an accurate determination of how much of the increased carbon dioxide will remain in the atmosphere and how it will affect atmospheric temperature.

In 1997, the Framework Convention on Climate Change, an international committee operating under the auspices of the United Nations, recommended that countries limit the amount of carbon dioxide emissions. These recommendations, discussed at a meeting at Kyoto, Japan, are known as the *Kyoto Protocol*.

The Carbon Cycle Revisited

Because carbon dioxide absorbs infrared radiation and becomes warm, it is a major factor in making the lower levels of the atmosphere warm enough for human habitation. The carbon cycle, described in chapter 1 (see figure 1.23), involves complex movement of carbon in a variety of forms, such as carbon dioxide and calcium carbonate, through Earth's environment. The buildup of carbon dioxide in the atmosphere drives other processes involving the hydrosphere, biosphere, and lithosphere. Some of these processes may remove carbon dioxide and reduce greenhouse-warming effects.

The lithosphere contains large quantities of carbon in coal, oil, gas, and limestone ($CaCO_3$). Plants, living tissue, and the shells of invertebrate animals also contain carbon. The atmosphere is a major reservoir of carbon in the form of carbon dioxide, much of which volcanoes contributed over many millions of years. Carbon dioxide continually moves into and out of these reservoirs. For example, plants take in carbon dioxide released during respiration by organisms. Some carbon dioxide moves through the surface of the sea into the atmosphere (an estimated 99 billion tons per year). Carbon dioxide also gets into the atmosphere as a result of decay processes sometimes called "soil respiration" (an estimated 80 billion tons per year), and by burning fossil fuels (estimated 5.5 billion tons per year). Plants remove unknown amounts of carbon dioxide from the atmosphere. The carbon becomes fixed in the plant, and when the plant dies, the carbon may either be oxidized in the decay process, in which case it returns to the atmosphere as carbon dioxide, or it may remain in the plant matter. If the plant matter is buried, the carbon becomes part of the soil. Carbon dioxide also moves from the atmosphere into the upper layers of the ocean, where algae use some of it for photosynthesis. Invertebrate animals use carbon to construct shells composed of calcium carbonate. Thus, carbon is removed from this cycle and stored indefinitely when seashells sink to the bottom of the ocean and become incorporated into marine sediment. The remains of the soft parts of organisms contain carbon that may eventually be transformed into hydrocarbons such as oil and gas. Carbon is also removed from the cycle when plants in swamps die, fall to the floor of the swamp, and eventually become buried and transformed into peat and coal. The carbon in oil, gas, and coal remains in storage unless humans extract and burn it or it is exposed to the atmosphere naturally.

Although the processes by which carbon moves through the carbon cycle are known, only general estimates of the quantities and rates at which carbon moves from one of these reservoirs to another are available. Nevertheless, a few general conclusions are clear: (1) Humans are affecting the carbon cycle. Carbon dioxide produced by human activities such as burning of hydrocarbons and

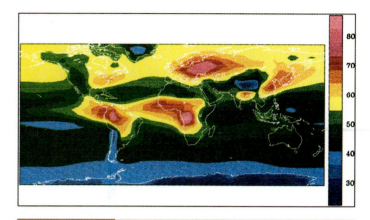

Figure 11.17 Computer Model Depicting the Global Distribution of Carbon Monoxide in the Atmosphere. Most of this gas rises into the air from industry or from the burning of forests.

forests (figure 11.17) has increased the atmospheric concentration of carbon dioxide and carbon monoxide. (2) The movement of carbon dioxide into other reservoirs has not kept pace with additions to the atmosphere. (3) Because plants are an important "sink" for atmospheric carbon dioxide, removal of plant cover reduces the amount of carbon dioxide that can move out of the atmosphere through plant photosynthesis. It now appears that plants may remove carbon dioxide from the atmosphere more quickly than any other mechanism. This is one of the reasons for concern about the destruction of tropical rain forests.

THE ATMOSPHERE—PROBLEMS AND ISSUES

Depletion of Ozone in the Stratosphere

Ozone in the stratosphere shields Earth's surface from excess ultraviolet (UV) radiation and protects living organisms. Intense UV radiation disrupts and breaks apart some complex organic molecules. As a consequence, overexposure to UV radiation not only causes sunburn, but also ages the skin, is the leading cause of cataracts, and may cause skin cancer. Ultraviolet radiation also damages plants.

Concern about stratospheric ozone became a public issue in 1970 during a debate about the development of supersonic transport planes. These planes emit nitrogen oxide that reacts with ozone to produce nitrogen dioxide and oxygen molecules. This reaction leads to a reduction in the ozone concentration in the stratosphere. Because of this discovery, the United States Congress halted the development of the supersonic transport.

In 1974, chemists discovered reactions involving chlorine and ozone that provided valid reasons for even more concern about the depletion of stratospheric ozone. In these reactions, chlorine reacts with ozone to produce chlorine oxide and oxygen molecules. The chlorine oxide then reacts with oxygen atoms to produce more chlorine and more oxygen molecules. Thus, a single molecule of chlorine can continue to break down ozone and reduce the amount of it in the atmosphere indefinitely until it is involved in some other chemical reaction.

Chlorine is a constituent of a group of chemicals known as the chlorofluorocarbons, also known as CFCs (e.g., Freon), which are used in air conditioners, refrigeration, aerosols, foams, and solvents. Large quantities of these CFCs have been and continue to be released into the atmosphere. The gases rise in the atmosphere until ultraviolet radiation breaks them down in a reaction that forms chlorine. In 1979, the United States, Canada, Norway, and Sweden banned the use of CFCs in aerosols (e.g., propellants in spray cans). In the United States, Freon was phased out and replaced with non-CFC refrigerants in the mid 1990s, but the production of CFCs continues in other parts of the world.

Measurement of stratospheric ozone concentrations began in 1957 during the International Geophysical Year—a year during which countries around the world cooperated for the first time in collecting data about the Earth and its environment. By 1985, drastic reductions in ozone concentrations had occurred over the Antarctic, and the resulting ring-shaped area became known as the **ozone hole** (figure 11.18). Since that time, ozone depletion has continued and spread. Some are concerned because large quantities of CFCs are already in the atmosphere, their use con-

tinues in many parts of the world, and the phaseout will be slow. Because chlorine can continue to destroy ozone for a long time, ozone depletion will continue for decades in spite of the steps already taken.

Atmospheric Pollution

The flora and fauna, including humans that inhabited the Earth at the beginning of the Industrial Revolution, lived in an environment that humans altered mainly through agricultural practices. Most flora and fauna were well adapted to that environment. Those that were not became extinct or moved from an unfavorable environmental niche to one for which they were better suited. With industrial development, humans introduced into the atmosphere and surface waters large quantities of toxic materials that can damage or destroy plants and animals. For many decades, the problems arising from the manufacture and handling of toxic materials affected small areas and generally arose from lack of recognition of the hazards they presented. The initial quantities of toxic substances introduced into the environment were so small that the natural system quickly diluted them to safe levels of concentration. Over the years though, both the quantities of pollutants and the population have increased, exposing more people to health risks. Vast quantities of pollutants are currently produced that affect large areas and many people. Table 11.3 lists some of the most important pollutants, their effects, and sources.

Ozone at Ground Level

Ozone is present both at ground level and high in the stratosphere. In the stratosphere, ozone protects the organic matter at the surface of Earth from excess ultraviolet radiation. In this respect, it is a "good" gas, but at ground level, ozone is extremely damaging to plants and animals. It is one of the principal constituents of **smog,** and it is a highly reactive gas. Ozone forms near the surface of the Earth by the same chemical reaction that occurs in the stratosphere—the combination of oxygen molecules with oxygen atoms to form ozone. Most of the oxygen formed at ground level comes from the breakdown of nitrogen oxides and volatile organic compounds in ultraviolet radiation rather than from the breakdown of oxygen molecules. Table 11.4 chronicles the events in the history of ozone depletion.

Reducing Air Pollution

The public has gradually become aware of the severity of air-pollution problems. In December 1952, an unusually heavy fog engulfed London, England. The air became stagnant, and pollutants accumulated, creating smog. By the end of a week of these conditions, approximately 4,000 people had died of pulmonary problems caused by this pollution. The numbers of deaths each day closely paralleled the concentration of sulfur dioxide in the smog on that date. In Athens, Greece, deaths increase dramatically on heavily polluted days. Today, the American Lung Association estimates that air pollution causes more than a hundred thousand deaths worldwide each year. The cost of air pollution has prompted

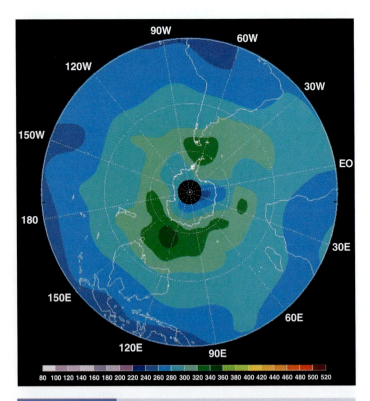

Figure 11.18 View of the South Polar Region of the Earth. This image was taken by satellite in October 2001. The colors depict the concentration of ozone. The most severe depletion is located over Antarctica, but depletion at this time was not nearly as severe as it was earlier.

| Table 11.3 | Effects of Major Atmospheric Pollutants on Humans |

Pollutants from Automobiles

Carbon monoxide: Interferes with blood's ability to absorb oxygen, impairs perception and thinking, slows reflexes, causes drowsiness, and can cause death; if inhaled by pregnant women, may threaten growth and mental development of fetus.

Nitrogen oxides: Can increase susceptibility to viral infections such as influenza; can also irritate the lungs and cause bronchitis and pneumonia.

Ozone in the lower atmosphere: Irritates mucous membranes of respiratory system; causes coughing, choking, and impaired lung function; reduces resistance to colds and pneumonia; can aggravate chronic heart disease, asthma, bronchitis, and emphysema.

Volatile organic compounds (VOC): Benzene is a known carcinogen. Others—toluene, xylene, ethylene dibromide—are suspected of causing cancer, reproductive problems, and birth defects.

Lead (formerly an additive to gasoline and present in some paints): Affects circulatory, reproductive, nervous, and kidney systems; suspected of causing hyperactivity and lowered learning ability in children; hazardous even after exposure ends.

Pollutants from Coal-Fired Power Plants

Sulfur dioxide: Raises the incidence of respiratory diseases including coughs, colds, asthma, bronchitis, pneumonia, and emphysema.

Particulate emissions: Carbon soot is responsible for respiratory diseases. Particulate matter can also carry toxic metals into the lungs.

Nitrogen oxides: See "Pollutants from automobiles."

Pollutants from Failure of Nuclear Reactors

Radioactive particles: Cause cancer, radiation sickness, and severe birth defects.

---

the government to take steps to reduce the problem. Whether public actions are adequate to achieve an acceptable level of air quality remains a topic of public and political debate.

Acid Rain

In the atmosphere, carbon dioxide and water vapor react to produce a weak acid, carbonic acid (H_2CO_3). As a result, rainwater has a pH of about 5.5 (distilled water is neutral with a pH of 7). All forms of precipitation—rain, snow, ice, and fog—can be acidic. Other acids, notably those formed from sulfur dioxide (SO_2) and nitrogen oxides (NO_x), form in the atmosphere as a result of volcanic activity and the burning of sulfur-bearing coal in power plants, industrial boilers, and ore smelters. These compounds are important causes of **acid rain.** Vehicles are also a major source of nitrogen oxides. As the concentration of these contaminants increases over time, rainwater becomes more acidic. Some lakes in the Adirondack Mountains have a pH of about 4. The rainwater near some industrial centers has reached pH levels between 1 and 2, about the same acidity as lemon juice. The highly industrialized areas of Europe, the United States, and China have the most severe problems with acid rain.

In 1963, the United States Congress took an important step toward controlling acid rain and air pollution by passing the Clean Air Act. This act charges the federal Environmental Protection Agency with gathering information, conducting research, and planning ways to control air pollution. The act also establishes control mechanisms that state governments implement. It sets emission standards for automobiles and limits the amounts and types of emissions permitted from stationary sources such as smokestacks.

The effects of acid rain at ground level depend largely on where the rain falls. If limestone is present, it neutralizes the acid in reactions, referred to as **buffering.** However, where the soil and bedrock do not contain buffering agents, the pH of surface streams and lakes drops, and the acid disrupts the biological processes of many plants and organisms. Some cannot survive after the pH reaches certain critical levels. For many species of fish, these levels have been reached in some lakes in the Adirondacks and in Scandinavia where the bedrock is granite and no buffering takes place. The lakes in these locations will no longer support fish life. Acid precipitation also adversely affects forests. An estimated 75% of Europe's commercial forests suffer from damaging levels of sulfur and nitrogen deposition. The World Resources Institute estimates that the cost of pollution damage to European forests is about $30 billion per year. In addition to the loss of timber, these damaged forests cannot absorb carbon dioxide as they should, and this contributes to greenhouse gas problems.

Overview

Largely as a result of population increases and a desire for a higher standard of living, atmospheric pollution has increased and become a serious problem. Energy consumption has grown because air conditioning, central heating, automobiles, electricity, running water, modern highways and buildings, and production of most manufactured products require the expenditure of energy. Most

Table 11.4	Chronicle of Events in the History of Ozone Depletion
1840	Ozone discovered
1928	CFCs invented
1950s	Reaction between water vapor and ozone discovered
1970	Supersonic Transport debate and discovery of the relationship of nitrogen to ozone
1974	Chlorine-ozone chemistry discovered
1979	CFC aerosol ban imposed
1985	Ozone hole discovered
1987	Montreal Protocol to reduce CFC production signed
1988	Ozone depletion in Arctic discovered
1991	Size of Antarctic ozone hole increased from 2 (1985) to 5 million square miles
1992	Ozone losses in middle latitudes of 1 to 3% reported
1993	Ozone levels reach a new low
2000	The Antarctic ozone hole was three times the size of the United States and as deep as Mount Everest

current energy supply comes from fossil fuels. The burning of these fuels increases carbon dioxide levels in the atmosphere. Some manufacturing processes used to produce common household products also create pollutants, some of which are toxic. The highest standards of living exist in North America and Western Europe. If other developing countries succeed in achieving Western standards, the amount of pollution in the atmosphere and waters of the Earth will increase dramatically. Society faces a serious dilemma. How, and whether or not, society will resolve it remains to be seen.

SUMMARY POINTS

1. Today, the atmosphere contains large quantities of nitrogen and oxygen and much smaller quantities of argon and carbon dioxide. Many other gases are present in small amounts, including some that are dangerous to human health.
2. Meteorologists identify divisions in the atmosphere based on changes in temperature and composition. The lower part of the atmosphere is well mixed. Upper parts are compositionally stratified with concentrations of ozone, oxygen, nitrogen, and charged particles forming layers.
3. Ozone forms by the combination of an oxygen molecule with an oxygen atom. If other ions that react with oxygen are present in the stratosphere, they slow the rate of formation of ozone, leading to a reduction in ozone concentration. Several gases, including chlorofluorocarbons (Freon) released by humans, rise into the stratosphere and reduce the amount of ozone. This reduction is most noticeable over the Antarctic where a circular-shaped region known as the ozone hole exists.
4. Water vapor gets into the atmosphere as a result of evaporation of water from the oceans, lakes, streams, and soil. The amount of water vapor that air can contain depends on temperature. When the concentration of water vapor reaches a maximum, water condenses and releases heat.
5. Radiant energy from the Sun accounts for almost all of the heat in the atmosphere. The total amount of heat reaching Earth is approximately equal to the heat energy leaving the Earth through reflection and radiation in long wavelengths from Earth into space. A near balance exists between incoming and outgoing energy.
6. The atmosphere behaves somewhat like a greenhouse. Most incoming solar radiation is in the visible and ultraviolet part of the spectrum. Part of this energy is absorbed and reradiated in longer wavelengths from the surface. These longer wavelengths of radiation are absorbed by water vapor, carbon dioxide, and other greenhouse gases. When this happens, these gases become warmer, leading to heating of the atmosphere. Most scientists think that as a result of burning hydrocarbons (coal, oil, and gas), humans have increased the amount of greenhouse gases in the atmosphere, leading to a concern that the atmosphere will become progressively warmer over time, causing major climatic changes.
7. Carbon dioxide is involved in a number of natural processes including photosynthesis and respiration. Although the movement of carbon through the Earth system is well known, the rates of addition and removal of carbon by natural processes are not known with sufficient precision to make it possible to predict the rate of greenhouse warming, but much evidence indicates that the atmosphere is becoming warmer.
8. Pollution of the atmosphere by the emission of hydrocarbons and other gases, including some that are highly toxic, has prompted the national government and the international community to take initial steps toward reducing emissions of certain gases, including Freon, sulfur dioxide, and nitrogen oxides. Because trees remove carbon dioxide from the atmosphere, the rapid cutting of tropical rain forests is a concern.

REVIEW QUESTIONS

1. Why does Earth have an atmosphere?
2. In what ways do the lower and upper parts of the atmosphere differ?
3. How does ozone in the stratosphere protect people?
4. What is a greenhouse gas? How do these gases lead to warming of the atmosphere?
5. What causes the equatorial latitudes to be warm all year long while the higher latitudes experience seasonal temperature changes?
6. What are the natural sources of carbon dioxide in the atmosphere, and how is carbon permanently removed from the atmosphere through natural processes?
7. What causes normal rainwater to be slightly acidic?
8. What are the bad effects of acid rain?
9. What was the probable source of the first oxygen in the atmosphere?
10. What are the most common natural and human sources of air pollution?
11. What evidence suggests that the amount of heat energy reaching Earth from the Sun has not always been balanced by loss of heat from Earth?
12. How is heat in the atmosphere spread around Earth?

THINKING CRITICALLY

1. Should the government take steps to reduce the use of fossil fuels such as coal, oil, and gas? If so, what should these steps be, and what effects would each have on air quality?
2. What can be done to reduce the amount of carbon dioxide in the atmosphere?
3. What human activities are partially responsible for the rise in global temperature?
4. How would the world be different if no radiation were reflected from its surface?

KEY WORDS

absolute humidity
acid rain
auroras
buffering
cirrus clouds
climate
convection
cumulonimbus clouds

cumulus clouds
ecliptic
fog
heterosphere
homosphere
mesopause
mesosphere
ozone hole

precipitation
relative humidity
saturated
smog
solar constant
solar wind
stratopause
stratosphere

stratus clouds
thermosphere
Tropic of Cancer
Tropic of Capricorn
tropopause
troposphere
weather

chapter twelve

The Atmosphere in Motion

Electrical discharges between the lower part of huge clouds and earth generates lightning.

Chapter Guide

Agnes, Andrew, Camille, and Hugo are names that will long be remembered because of the great damage and loss of life they inflicted. Hurricane Agnes alone caused damage in excess of $3 billion. Even a nameless storm that crossed Galveston, Texas, in 1900, killing 8,000 people is still discussed today. Strong windstorms such as hurricanes and tornadoes are among the most destructive of all natural phenomena. Hurricanes can destroy most buildings and flood long stretches of coasts. Tornadoes are more destructive locally. Hurricanes especially have caused long-term changes in the shape of low-lying coastal areas. As important as they are, these short-term impacts of wind are less important determinants of Earth systems than are the less dramatic movements in the atmosphere referred to as atmospheric circulation. The movement of air masses in conjunction with oceanic circulation plays the critically important role of maintaining heat balance on Earth. When changes occur in either of these great circulation systems, weather and even climate can shift quickly. The causes of air movement, the generation of storms, and the global circulation system in the atmosphere are subjects of this chapter.

INTRODUCTION

Tornadoes may have wind velocities exceeding 500 kilometers per hour (300 miles per hour) and develop winds that can destroy most buildings. Fortunately, tornado paths are less than a mile wide and only a few miles long. Despite their intensity, tornadoes cause less damage than hurricanes do. These huge storms affect areas tens of miles wide and often involve flooding and beach erosion as well as wind damage. For example, most of the damage from hurricane Mitch, one of the most devastating storms on record, which struck Central America in October 1998, resulted from about 50 inches of rain that caused widespread flooding and mud slides in Nicaragua and Honduras. Thousands of people died, whole towns were swept away, the economies of those countries were shattered, and disease spread throughout the region after the storm.

Hurricanes and tornadoes result from extreme differences in atmospheric pressure.

FACTORS THAT GOVERN THE MOVEMENT OF AIR

At any given level in the atmosphere, the weight of the air above that level presses down on the air below and causes atmospheric pressure. Consequently, atmospheric pressure at any given location depends almost entirely on the altitude. Nevertheless, small pressure variations between points at the same altitude are common. Air responds to variations in pressure by moving from areas of high pressure to areas of low pressure, and its flow can be smooth and streamlined or turbulent and irregular.

The uneven heating of Earth's surface, discussed in chapter 11 and figures 11.10 and 11.11, causes most variations in pressure. High temperature causes air molecules to move more rapidly and to expand the volume of the air, resulting in lower pressure. Low temperature results in higher pressure. The air responds to these variations by moving from areas of higher pressure to areas of lower pressure. As air moves, the rotation of the Earth and differences in frictional drag, caused by the variety of surface features on Earth, become important factors governing air circulation (figure 12.1).

Because the Sun's rays strike the Earth nearly perpendicular to its surface in equatorial latitudes throughout the year, much more solar radiation reaches the Earth in equatorial latitudes than elsewhere (see figures 11.10 and 11.11). The high temperature in these latitudes causes the surface layers of the oceans to be warmer, increases the rate of evaporation of surface water, and raises the temperature of the land surface. The buildup of heat would be intolerable for people and many animals were it not for the movements that occur in the atmosphere and ocean. Atmospheric and oceanic circulations disperse much of this heat and transfer it to higher latitudes. These circulations consist of both vertical and lateral movements. Close to Earth's surface, the vertical movement of air is apparent in the rise of the tops of clouds—a phenomenon that is visible on many hot summer days. The movements of clouds and the prevailing wind directions provide

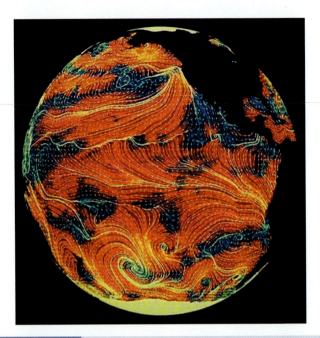

Figure 12.1 Patterns of Atmospheric Circulation. Air moves in response to variations in atmospheric pressure. Satellites make it possible to monitor the circulation. The huge swirling masses of air and the lines along which air from different latitudes converge are prominent features of this image.

the best indication of the patterns of lateral air movements referred to as *atmospheric circulation*. Satellites continuously record these movements, and they are prominent features of television weather reports. The pattern of movement is complex and arises from the effects of Earth's rotation on large masses of air that form over land and water-covered areas. The pressure in these air masses is closely related to their temperature and density, and these characteristics are acquired from the Earth's surface where air masses form.

Atmospheric Pressure

Gravitational attraction holds atmospheric gases to Earth's surface. The atmospheric pressure at a point on the surface is equal to the weight of the overlying gases. The units of pressure are atmospheres, bars, millibars, pascals, and pounds per inch squared. One atmosphere, the pressure at sea level, $= 1.013$ bar $= 1013.25$ millibars, 10^5 pascals, and 14.7 pounds per inch squared. Because pressure variations in the atmosphere are so small, atmospheric pressure is generally expressed as millibars. Weather maps frequently show lines connecting points of equal pressure. These are called **isobars** (figure 12.2).

In static air, air that is not moving, pressure at any given point is equal in all directions, but air pressure does change both vertically and laterally. Changes are noticeable between areas of high and low pressure in both air and water and most noticeable when one climbs a mountain, goes to greater depth under water, or takes off in an airplane. When air is nearly static, the air density decreases

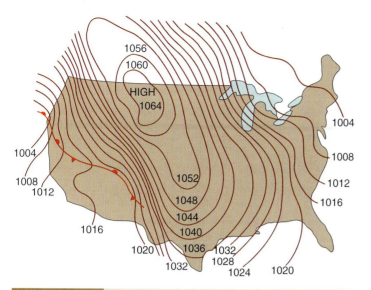

from the surface upward. The air is stable and "density stratified." Air near the surface becomes unstable when the density of the air near the ground becomes less than that above it. This condition occurs when air at a low altitude is heated. The hot air expands, and air pressure decreases. As expansion occurs, the air becomes buoyant and rises. As the warm air rises, denser air moves laterally to the area of low pressure. Conversely, at high altitude, air becomes more dense as a result of cooling and sinks.

Coriolis Effect

Rotation of the Earth causes air masses that move across the surface of the Earth to follow curved paths. Coriolis effects in the atmosphere are similar to those in the oceans (see the discussion in chapter 9) (figure 12.3). Coriolis effects are significant because frictional effects between air and the solid earth are not great. A cloud located near the equator moves around the Earth with the velocity of rotation of all objects on the equator. Because a point on the equator makes a trip of about 40,000 kilometers (24,800 miles) in 24 hours, objects at the equator are moving nearly 1,600 kilometers (1,000 miles) per hour. As this cloud travels northward, it continues to move toward the east at 1,600 kilometers per hour, but points closer to the pole on Earth's surface move at progressively lower velocities. The path of the cloud traced on Earth curves in a clockwise direction—toward the right. As with water currents, all masses of air moving in the Northern Hemisphere tend to follow curved paths with a deflection to the right. In the Southern Hemisphere, paths of moving masses of air or water are deflected in the opposite direction (counterclockwise or to the left).

Movement of Air Around Areas of High and Low Pressure

Differences in heating of air near the ground surface cause most lateral pressure changes. Some areas on the ground absorb more radiation and become warmer than others. Cities with large areas covered by black asphalt are commonly warmer than rural areas where forests are present. Similarly, ice- or snow-covered areas are cooler than water- or rock-covered surfaces. Variations in ground cover and texture also account for other differences in heating.

On a much larger scale, continents have different parcel properties (i.e., heat capacity, reflection of radiation) than oceans do because the temperature of the ground surface changes much more rapidly than does the temperature of the ocean surface. Much more heat from solar radiation is required to raise the temperature of water than to increase the temperature of soil by the same amount. Similarly, water temperature drops much more slowly than soil or rock temperature. For this reason, areas covered by water tend to be warmer than adjacent land areas in winter and cooler in summer (figure 12.4). Warm areas become the sites of low-pressure systems, and areas that are cold become sites of high-pressure systems.

Once an area of high or low pressure forms, air begins to move from areas of higher to areas of lower pressure. Consider a circular area where high temperature has reduced pressure at ground

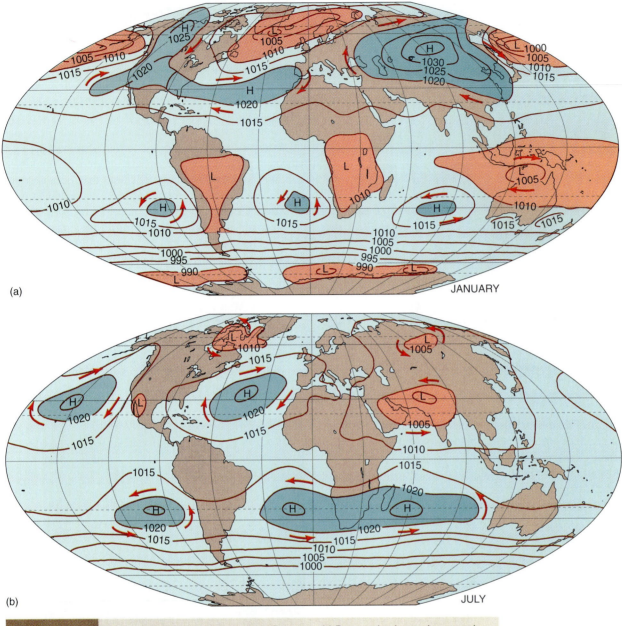

(a)

JANUARY

(b)

JULY

Figure 12.4 **Seasonal Changes of Atmospheric Pressure.** (*a*) Because land areas lose or gain heat more rapidly than water-covered areas, large landmasses become cooler in winter (January in the Northern Hemisphere) than adjacent water-covered areas. The cool air is denser than the warmer air and so has a higher pressure. (*b*) During summer in the Northern Hemisphere, high pressure systems lie over the oceans.

level. The warm air is buoyant and begins to rise. Cooler, higher-pressure air from surrounding areas moves toward the low-pressure center. In the Northern Hemisphere, as air moves, it is deflected to the right and swirls in toward the center of the low in a counterclockwise flow (see figure 12.4). As long as the air near the low-pressure center is warm and buoyant, it continues to rise, and a counterclockwise circulation is set up around the low (figure 12.5).

If, in the Northern Hemisphere, warm air surrounds air that is colder and has a high pressure, air from the high will move outward

from the center of the high and move to the right. A clockwise spiral of air circulation develops. In the Southern Hemisphere, Coriolis effects cause deflection to the left instead of to the right, and the direction of air movements around highs and lows is reversed (see figure 12.4).

Radar now provides an excellent way of tracking air movement across Earth's surface. Conventional radar sends out microwave pulses that scatter when they hit objects such as raindrops. Antennas pick up some of the scattered energy, which makes it possible to

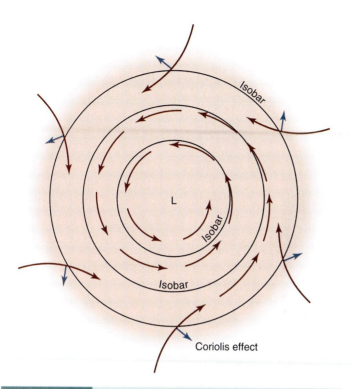

Figure 12.5 Schematic Showing Air Circulating Around a Circular Area of Low Pressure. Air streams toward the low from surrounding higher-pressure areas. As air moves toward the low, the Coriolis effect causes the air to travel in a spiraling path. Air near the center of the low rises to higher altitudes because it is warm and buoyant. At high altitude, the air spreads away from the rising column.

Air in a parcel has different physical properties (temperature and pressure) than air outside the parcel. The behavior of parcel air is somewhat analogous to that of a balloon that contains hot air. As the parcel rises, it passes through air that is steadily becoming less dense. Consequently, the air in the parcel expands, and temperature within the parcel drops because energy is used in the process of expansion. If little or no transfer of energy takes place between air inside and outside a parcel as it expands or contracts, the process is called an **adiabatic process.** Expanding dry air cools 10°C per kilometer (5.5°F per 1,000 feet) as it rises. The temperature of the air outside the parcel generally drops about 6.5°C per kilometer (3.5°F per 1,000 feet) rise in altitude. This is known as the **environmental lapse rate,** or **normal lapse rate.** Air in the parcel cools faster than air outside the parcel. As the parcel rises, the temperature inside the parcel eventually reaches the same temperature as that outside the parcel. At that altitude, the parcel lacks buoyant energy and stops rising.

If the parcel contains water vapor, the relative humidity of the air in the parcel increases as air temperature drops. Because cooler air cannot contain as much water vapor as warmer air, droplets of water condense in the parcel when the temperature reaches the **dew point,** the temperature at which condensation occurs for a particular humidity. The dew point is low for air that has a low water-vapor content and high for more-moist air. At some elevation, depending on the water-vapor content of the air in the parcel, a cloud begins to form, and heat is released during this condensation. Consequently, condensation in the parcel slows the cooling of the air and makes it more buoyant. (The decrease in temperature in a parcel in which water is condensing is about 6°C per kilometer [3.2°F per 1,000 feet], known as the wet adiabatic rate.

determine the position of the object that causes the scattering. Images obtained from radar installations are used on television weather reports. Successive images reveal the direction and speed of storm movements. New radar systems use Doppler effects. These effects are similar to the change in pitch of the sound of a whistle or siren that moves toward or away from the observer. Doppler radars reveal the direction and speed of the movement of rain and may also reveal the details of wind motions. Analysis of Doppler radar provides especially valuable ways to analyze air movements in tornadoes and other storms.

THE VERTICAL MOVEMENT OF AIR

The Rise of Hot Air

As long as air is density stratified with higher-density air beneath warmer, lower-density air, a stable condition exists, and vertical movement does not occur. Once air at a lower level becomes less dense, usually because of heating, an unstable condition develops, and **parcels** of warm air, called **thermals,** rise. Parcels commonly form in the summer, and they are usually easy to recognize because clouds often form in them (figure 12.6).

Figure 12.6 Clouds Forming Within Parcels of Warm Air. Towering cumulus clouds rise because the air associated with them is buoyant. The cloud forms when the temperature in the air is at the dew point—the temperature at which the water vapor begins to condense. Because heat is liberated during condensation, the cloud continues to rise as water condenses.

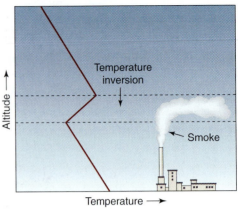

Figure 12.7 Photograph and Diagram Depicting a Temperature Inversion. Temperature inversions commonly form when cool air sinks into valleys or other topographic depressions. This phenomenon commonly occurs on cool, clear nights when the Earth loses heat by long wavelength radiation. The layer of warmer air above the inversion may trap pollutants in the lower layer of air. The graph at right depicts change of temperature with altitude.

The rate depends on the amount of moisture present and ranges between 5° and 9°C per kilometer [9° and 16°F]). Condensation makes the parcel, which is now visible as a cloud, unstable, and the air continues to rise until all the moisture has condensed. When all of the water vapor in the parcel has condensed, a source of heat for the parcel stops, but the dry air may continue to rise until the temperature in the parcel is equal to that outside the parcel. This rising dry air is a source of clear air turbulence.

Vertical Mixing and Air Pollution

The stability or instability of air affects its quality near the ground surface. If a stable condition exists near ground level, pollutants introduced into the air remain near ground level, and they can build to levels that are dangerous to humans. If air at ground level is cold and dense, it moves slowly. For this reason, pollution levels rise early in the day following a cold night. The cold air over a city is stable in the morning when people go to work. Emissions from cars and industrial plants rise into the sluggish air. Pollution levels increase until ground-level temperature begins to rise and parcels start to carry the polluted air higher in the atmosphere and dilute the air near the surface.

Temperature Inversions

A **temperature inversion** exists when air at ground level is colder than overlying air. Inversions may occur in almost any geographic location, but they are most common in mountainous areas where,

in the absence of strong horizontal movement, cold, dense air settles into valleys at night. Once the inversion forms, the boundary between the colder and warmer air becomes a barrier to the upward movement of warm air. Hot emissions from a smokestack or other source introduced at ground level rise through the colder air into which they are introduced, but when the emissions reach the inversion, they encounter air much closer to their own temperature and lose their buoyancy. When this occurs, smoke ceases to rise and spreads laterally along the inversion (figure 12.7). The emissions from cars and other sources are affected in the same way. Temperature inversions pose serious air-pollution problems in many parts of the world. Most inversions dissipate by the middle of the day as the ground is warmed, but occasionally they persist for days.

THE INFLUENCE OF CONTINENTS AND OCEANS ON AIR MASSES

Over large areas of water, land, or ice, uniform temperature conditions prevail for months at a time. Where the ground is warm, the air next to it becomes warm and buoyant, and the pressure drops. Conversely, over bodies of cold ground, water, or ice, air becomes cold and dense, resulting in high pressure. As a result, some large areas tend to remain sites of high, or low, pressure for months. In the winter, the air over the land in midlatitudes and high latitudes is cooled by the cold ground surface. Consequently, in the winter, continents tend to be sites of large masses of high-pressure air referred to as high-pressure cells (see figure 12.4). During winter, water-covered areas at the same latitudes are relatively warm because water changes temperature much more slowly than land does. Large low-pressure cells form over these areas of the ocean. Residents of North America are familiar with the cold, high-pressure air masses that originate over Canada in the winter. These are responsible for long periods of frigid air in central North America. At the same time of year, large low-pressure systems, which form over the Gulf of Alaska, spawn storms that sweep across North America.

Air masses are designated as polar (P) or tropical (T) and maritime (m) or continental (c). Of the air masses that are especially important causes of weather in North America, the cold air mass that forms over Canada is designated cP (see figure 12.2); the warm, moist air mass that lies over the Gulf of Mexico is designated mT; and the cold, moist air mass over the Gulf of Alaska is an mP. Some of these air masses, such as those in the Arctic, the polar maritime, the tropical maritime, and over the equator, persist throughout the year. In contrast, the tropical continental (cT) air masses are most prominent in the summer, and polar continental (cP) air masses are most prominent in the winter. Note the seasonal changes that take place in location of high- and low-pressure systems over continents and oceans shown in figures 12.4*a* and 12.4*b*.

When air is cooled over a continent during winter, the dense air settles and spreads laterally. As it moves out away from the center of high pressure, Coriolis effects deflect it to the right in the Northern Hemisphere in a clockwise direction around high-pressure cells. When cold Arctic air masses are present over Canada, the clockwise circulation around the high-pressure cell moves cold Arctic air into the central United States and accounts for many periods of extremely cold weather.

In the summer, the waters off North America in the Atlantic Ocean are cool relative to the land surface, and a major high-pressure cell develops over the ocean. This air mass, called the *Bermuda High,* dominates weather in the eastern states. The clockwise motion of air around this high frequently brings hot, humid air formed over the Gulf of Mexico into the Mississippi Valley and spreads it across the southeastern part of the country.

GLOBAL PATTERNS OF CIRCULATION

Ground-Level Circulation

Because the prevailing wind direction and its seasonal variability are readily observed at the surface, people have known the general pattern of ground-level circulation for several hundred years (figure 12.8). Observations have been recorded since the early days of sailing. Modern records of prevailing wind directions obtained from satellites (figure 12.9) confirm many aspects of the classical model of ground wind circulation, but the higher-altitude circulation is quite different from that proposed in the classical

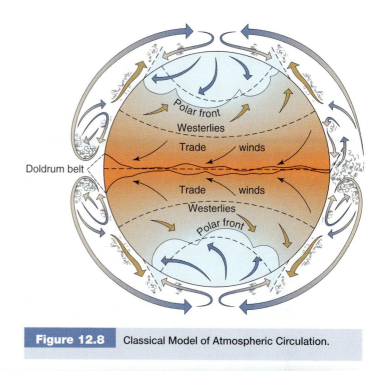

Figure 12.8 Classical Model of Atmospheric Circulation.

model. Movements of high-level clouds provided the first indication of circulation at high altitude. Later, these movements could be tracked by following the movement of balloons, and since 1960, it has been possible to define circulation in the upper parts of the troposphere with satellite data.

Circulation at Low Altitude

In a greatly simplified view, prevailing winds in the lower part of the atmosphere flow in broad zones, which roughly follow parallels of latitude (see figures 12.8 and 12.9). In the polar regions, extremely cold, dense air settles to ground level and moves toward the equator. As it moves, Coriolis effects deflect it, and it becomes an east to west flow known as the **polar easterlies.** Along the equator where temperatures are high, hot, moist air rises to high altitudes. Near the surface, this zone, called the **intertropical convergence zone (ITCZ),** or **doldrums,** is calm and characterized by little lateral movement of air. At about 10 to 15 degrees of latitude north and south of the doldrums lies a belt of strong easterly (east to west) air flow known as the **trade winds.** Early sailing ships took advantage of this belt in their journeys from Europe to North America.

Near latitude 30° north and south, the trade-wind zones give way to yet another belt characterized by calm air, known as the **subtropical high zone,** or **horse latitudes.** These are zones where cool, dry air from high altitude descends to the surface and creates high pressure. In these belts of calm air, where sailing ships often drifted aimlessly, cool air settles back to Earth from high altitudes. Because this air is settling from high altitudes, it is dry. Consequently, many desert regions occur in this belt. The air that descends in these latitudes moves both north and south near ground level and is deflected by Coriolis effects. Air that moves toward the

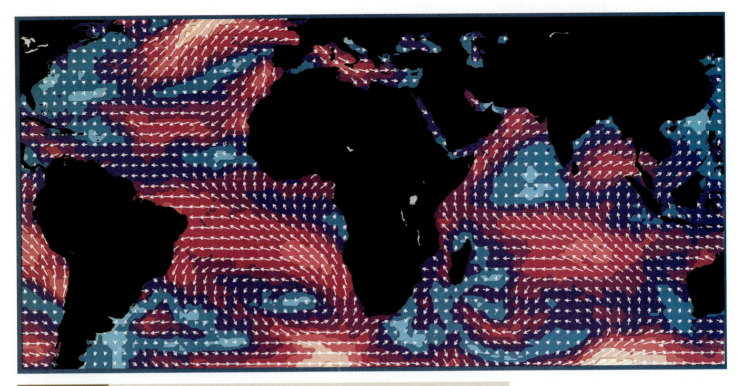

Figure 12.9 Wind Patterns over the Atlantic and Indian Oceans Recorded from Space in September 1979.

equator bends toward the east and becomes part of the trade winds. Air that moves toward the poles is deflected to become part of a belt of strong westerly winds. The **westerly winds,** winds that blow from west to east, are responsible for much of the weather in the United States and the southern part of Canada. It is here that the cold, dense air from high-pressure systems of polar regions encounters the warmer air from middle latitudes.

In winter, the boundary between these masses of air, known as the **polar front,** moves forcefully from the north into the middle latitudes. When this polar air with its easterly component of movement encounters the strong westerly winds in the middle latitudes, air along the front is twisted in a counterclockwise direction in the Northern Hemisphere (clockwise in the Southern Hemisphere), producing low-pressure systems especially associated with winter weather.

Circulation at High Altitude

For a globe that is heated at the equator and cooled at the poles, a simple model of air circulation might consist of hot air rising at the equator, spreading toward the poles, and settling back to the ground in polar regions (figure 12.10). Obviously, circulation of air on Earth is more complex. At one time, many people thought that high-altitude circulation was dominated by single north-south circulation, but when data on high-altitude winds became available, it revealed that the dominant flow of air is west to east and parallel to lines of latitude in the middle latitudes. In addition, meteorologists

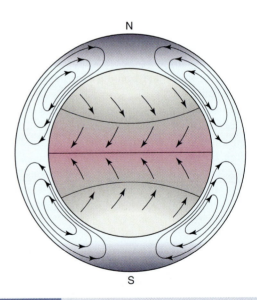

Figure 12.10 An Early Model of Atmospheric Circulation. This early model shows single cells in each hemisphere in which warm air rises at the equator and spreads toward the poles.

became increasingly aware of the important roles sinking of air from high altitudes in the horse latitudes and movement of air from polar regions toward the equator play in high-altitude circulation. A more recent interpretation of the north-south circulation is shown in

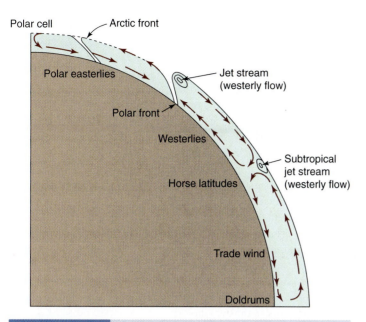

Figure 12.11 **A Modern Representation of High-Altitude North-South Circulation.** A zone of strong currents flow from west to east in mid latitudes. Jet streams are zones of high-velocity west to east flow that occur close to the boundary between the troposphere and stratosphere. They vary in altitude and velocity with the season. The flow has a higher velocity and tends to swing closer to the equator during the winter months.

figure 12.11. Coriolis effects are involved in high as well as surficial circulation. Consequently, currents that move from the equator northward follow paths that bend to the right, and those that move southward from the equator bend to the left.

Wind is actually the movement of the air relative to the surface of the Earth. If no wind existed, air over the Earth would move with the ground surface as the Earth rotates. The velocity of this rotation varies from the equator to the poles. At the equator, the rate of movement is close to 1,600 kilometers per hour (1,000 miles per hour); at the pole, movement is zero. However, the air over Earth is not static. It moves both laterally and vertically. The air that rises into the troposphere in equatorial latitudes is moving from west to east at approximately the same velocity as the Earth beneath it. But when the air spreads toward the poles, it begins to curve to the east because of Coriolis effects, and its west to east velocity is greater than that of the ground below. Consequently, high-speed winds blow from west to east at high altitudes in the middle latitudes. In fact, this is the predominant flow of air in the upper part of the atmosphere.

Jet Streams

Zones of rapidly moving air at high altitudes, known as **jet streams,** can sometimes be detected by watching the movement of high clouds such as thin, wisp-shaped cirrus clouds (see figure 11.5b). At first, the only evidence of this high-velocity move-

ment came from observations of clouds moving rapidly from west to east. After World War II, images of cloud movements made from satellites defined the jet streams, and aircraft could make direct measurements of wind speeds. Now, high-flying jet aircraft use the high-speed streams of air to reduce flying time and fuel consumption. Jet streams are generally broad meandering zones several hundred kilometers wide, located high in the troposphere or low in the stratosphere (figure 12.12). In jet streams, air flows at velocities of 160 to 320 kilometers (100 to 200 miles) per hour. Air movement in jet streams is basically a westerly flow, but their position and strength varies with the seasons. They are stronger and follow more southerly courses in winter months. Because they follow a sinuous course, the direction of flow varies. As jet streams meander, they also move heat from equatorial latitudes toward polar latitudes in summer months and cold air from polar regions toward the equator in winter.

Jet streams occur in two positions. One is along the polar front, the line of contact between polar air masses and middle-latitude air. The second is situated in the horse latitudes on the edge of zones of high pressure that tend to persist in regions north and south of the equator (see figure 12.11). These jets are known, respectively, as the **polar front jet** and the **subtropical jet.** Both jets form along boundaries between air masses formed at different latitudes that have greatly different temperature and pressure.

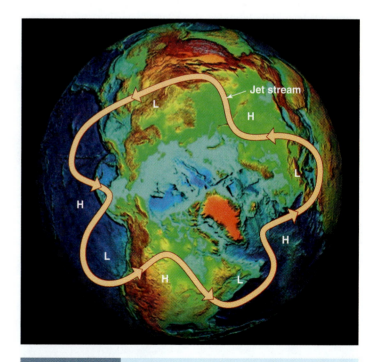

Figure 12.12 **The Jet Stream.** High-altitude air circulation moves around high- and low-pressure air masses that lie over continents and oceans. This produces a flow pattern that can develop large loops. In summer months, the jet streams tend to remain at high latitudes, but in winter months, large meanders develop that bring the cold air much closer to the equator.

Despite their wave-shaped paths, jet streams remain close to the edge of pressure systems. Because the temperature contrasts between the air masses is greatest in winter, the jet streams have higher velocity and shift to lower latitudes in winter.

As the Earth rotates, the ground surface rotates from west to east as does the air high in the troposphere. This pattern of westerly movement is the most prominent feature of the high-altitude wind currents, especially in the middle latitudes. Closer to ground level, air masses developed over continents and oceans are more prominent features in atmospheric circulation.

WAVE-CYCLONES AND MIDDLE-LATITUDE STORMS

Wave-Cyclones

Throughout much of the year, the passage of large, wave-like systems of low pressure, called **wave-cyclones** (figure 12.13), strongly influence weather in middle latitudes. These cyclones are associated with sharp boundaries called **fronts** that separate air of different temperatures and pressures. Four types of fronts—cold, warm, stationary, and occluded—may occur with a wave-cyclone. A front is called a **cold front** or a **warm front** depending on the relative temperature of the air at a given locality and on the temperature of air that is approaching that locality on the other side of the front. In middle latitudes, these large, swirling systems of air

transport cold air from the north toward the equator and warm air from the south toward the pole (see figure 12.13). Because cold air is heavier, it lies beneath the warmer air along the boundary, and the front always slopes up in the direction of the cold air (figure 12.14). Both cold and warm fronts have low angles of slope, but cold fronts are much steeper than warm fronts. They advance more rapidly and usually have more significant weather. Typically, the boundary of a cold front rises 1 meter vertically for every 100 meters (328 feet) laterally (a slope of 1:100). A typical warm front has a slope of 1:200. Most observed slopes are from 1:50 to 1:250. Because the slopes are so low, passage of fronts may take hours. Once a front has passed, weather conditions change.

When a front does not move or is very slow, it is referred to as a **stationary front.** Along these fronts, winds may blow parallel to the front and in opposite directions. Because cold fronts generally move more rapidly than warm fronts, the cold front may move under the portion of the warm front ahead of it, forming what is known as an **occluded front.** This happens near the central part of the low pressure in the cyclone.

Many middle-latitude wave-cyclones form along the polar front. Because wind is blowing in opposite directions along this boundary, a wind shear is established that tends to twist the air in a counterclockwise direction in the Northern Hemisphere. A low pressure center forms at the middle of this swirling air. Low-pressure systems may also form away from the polar front where strong temperature gradients are present. As the low-pressure system develops, the cyclone may develop or die out. Waves of cer-

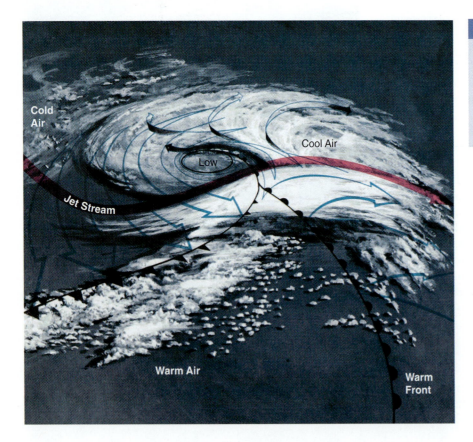

Figure 12.13 Schematic Representation of the Clouds, Air Temperatures, and Fronts Commonly Associated with the Movement of Low-Pressure Systems Known as Wave Cyclones. Cross sections across a warm front and a cold front are shown in figure 12.14. Note that colder air generally lies beneath warmer air.

tain sizes, generally those with wavelengths of 600 to 3,000 kilometers (370 to 1,800 miles), tend to grow and intensify. Smaller waves tend to die out because of friction at ground level.

The weather conditions at a locality on the ground change dramatically as a wave-cyclone passes (see figure 12.14). Weather differs depending on which side, and how close, the locality is to the center of the low pressure. Dramatic changes may occur on the south side of the low as the cold front passes. Ahead of the cold front, air temperature is warm, or moderate patches of cumulus clouds dot the sky, and air pressure is moderate. As the cold front approaches, the cloud cover becomes thicker. The warm (and generally moist) air rises along the edge of the front. As more moisture

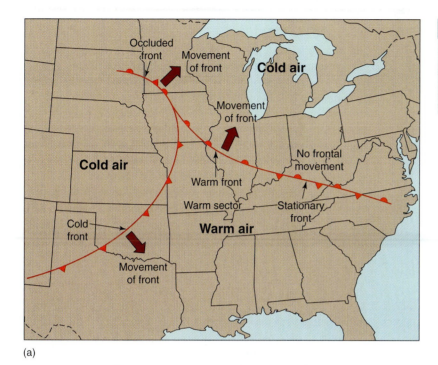

(a)

Figure 12.14 Frontal Systems. (*a*) Typical representation on a weather map that shows a cold front and a warm front, part of which is stationary. (*b*) Cold fronts have much steeper slopes than warm fronts, and the types of clouds formed along both fronts vary with altitude.

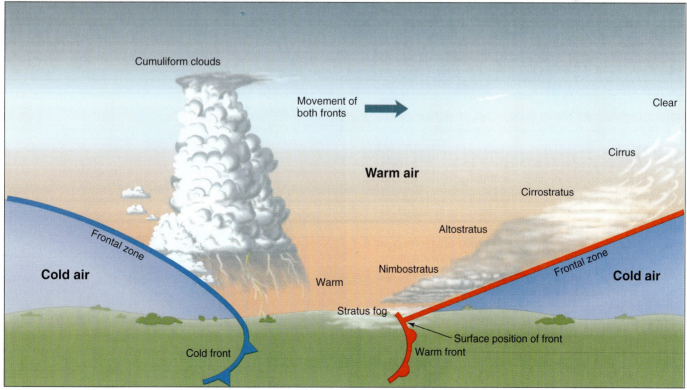

(b)

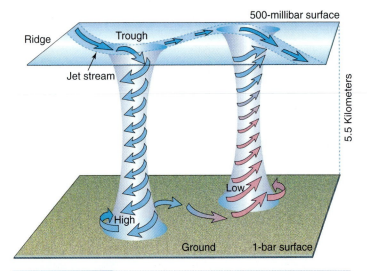

500-millibar surface
Ridge
Trough
Jet stream
5.5 Kilometers
Low
High
Ground   1-bar surface

Figure 12.15 **Relationships Between the Jet Stream and Ground-Level High and Low Pressure Systems.** Warm air rises from a ground-level low-pressure center and sinks from high altitude at ground-level high-pressure centers. The rising air often is pulled into a jet stream near the top of the troposphere.

condenses, rain or snow occurs along the front. If the air on the warm side of the front is moist and hot, thunderstorms and sometimes tornadoes may break out along the front. The heat liberated during condensation fuels their rise. When the cold air behind the front arrives, temperature drops, and in winter, the rain may change to snow or freezing rain. On the warm side of the front, air was flowing from southwest to northeast. After the front passes, airflow changes direction and begins to blow from the northwest. After the front has passed, the air becomes clear and colder.

The sequence of weather conditions on the north side of the center of the low are quite different from those to the south. The north side is affected by the intrusion of warm, moist air into a region of colder air. The warm air rises along the warm front, riding up over the colder ground-level air. As it rises, the dew point is reached, and thick layers of stratus clouds form. Moisture condenses in these clouds, and rain falls until the air dries out. If the ground-level temperatures are sufficiently low, the rain may change to freezing rain, sleet, or snow. After the front has passed, the air temperature becomes warmer, and cumulus clouds frequently occur.

The jet stream may pass directly over the center of low pressure along a cold front, and this center tends to move along with the jet stream (figure 12.15). When the jet stream is meandering and forms loops that extend far south, the center of the low generally follows the same path, bringing unusually stormy weather.

Thunderstorms

Thunderstorms develop from buoyant masses of warm air that contain large amounts of water vapor. Most thunderstorms occur on summer afternoons when the ground is hot. It is estimated that more

than 40,000 thunderstorms occur each day during summer months. In the United States, many of these storms originate in the southern states where hot, moist air from the Gulf of Mexico moves inland. Florida has more thunderstorms than any other state. Thunderstorms are also common in the Rocky Mountains and across the Great Plains. Parcels form and rise. Condensation occurs when the parcel reaches the dew-point temperature, and clouds begin to form. The cloud expands and builds as condensation takes place (figures 12.16 and 12.17). Eventually, enough moisture has condensed for raindrops to develop. Droplets grow, coalesce, and become heavy enough to fall as precipitation. High in the cloud, the temperature is below freezing, and ice or snow forms. As the frozen precipitation falls, it melts. It may reach the ground as pellets of ice or sleet, but more often it falls as rain. However, if rising hot air causes strong updrafts in the cloud, the ice or rain may be carried high into the cloud. If this occurs repeatedly, the size of the masses of ice grow as layers of ice build on one another and form large rounded masses, called **hail.** When the hail becomes too heavy to be lifted in the updrafts it encounters, it falls to the ground. Hailstones can be fractions of a centimeter to more than eight centimeters in diameter. Larger hail can damage or destroy crops, break windows, and even dent the tops of cars or strip limbs from trees.

As rain, sleet, or hailstones fall out of thunderstorms, they drag air downward to ground level. Strong gusts of wind, often capable of blowing down trees, spread out in front of approaching thunderstorms. Strong winds are often responsible for much of the damage done by these storms. Wind gusts descend from high altitudes and are cold, so on a hot summer afternoon, a strong, cold wind often precedes the downpouring of rain and the increase in atmospheric pressure that characterizes the storm passage. The life cycle of a thunderstorm begins with the rise of hot, moist air. Initially, cumulus clouds form, but once condensation starts, the air is unstable, and upward convection increases. The storm clouds rise and may reach the troposphere. A single storm cloud contains many convection

Figure 12.16 **Thunderstorm Clouds.** A huge anvil-shaped cloud of the cumulonimbus type, with which thunderstorms are most commonly associated.

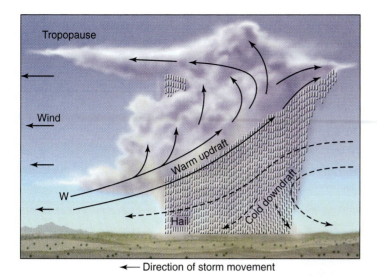

Figure 12.17 Cross Section of a Thunderstorm That Has Risen to the Top of the Troposphere Where a Temperature (parcel) Inversion Inhibits Further Rise. Rain, sleet, and hail commonly fall from such clouds, bringing cold air down with the precipitation.

cells. Heavy rainfall characterizes the mature stage in the storm's existence. As the supply of moisture is depleted, the cloud begins to decrease in size, and the intensity of the storm wanes.

During the rise and fall of precipitation within cumulonimbus clouds, electrostatic charges separate. Hail-sized ice pellets carry negative charges toward the lower part of the cloud, and small ice particles move positive charges high in the cloud. Usually, the separation of negative and positive charges in the clouds reach a maximum in the afternoon and early evening. As charge separation rises, great potential differences estimated at hundreds of millions of volts are created. Eventually, an electric discharge, **lightning,** occurs within the cloud or between the base of the cloud and the ground (figure 12.18). Lightning may involve thirty to forty strokes, each lasting one or two thousandths of a second. The strokes may carry 20,000 amperes of electric current. A lightning strike between the base of a cloud and the ground (figure 12.19) is often preceded by a discharge in the cloud that concentrates negative charges at the base of the cloud. When the potential difference reaches a critical level, a negative charge, called the pilot leader, surges down toward the surface. The surge is repeated until the path almost touches the ground. Then, sharp points on the surface become focal points for formation of a stream of positive charges that advance toward the leader. When they touch, positive electric charges move up the channel followed by a return stroke of negative charges that move to the ground.

Air along the path through which the electric discharge shoots is heated to temperatures of about 10,000°C (18,000°F). The air expands so quickly that it creates a surge of pressure in the atmosphere that we hear as thunder. Sound travels more slowly than light, so lightning appears long before the sound of thunder. The time lag between the two is a measure of the distance of the lightning strike.

Sound travels at approximately 330 meters per second (1,000 feet per second), so the distance to the source of lightning is the number of seconds of lag-time multiplied by 330 meters (1,000 feet).

Electrical discharges also occur above thunderstorms, within the clouds, and between the clouds and the ground. Although rarely seen from the ground, these high-altitude discharges are recorded from satellites, and they take a variety of intriguing forms. Those in the ionosphere often resemble an expanding disc. Below the disc, aurora-like discharges hang down from the ionosphere into the stratosphere where other discharges appear as blue jets shooting up from thunderstorms. These phenomena demonstrate that the upper parts of the atmosphere are more active than previously expected.

Tornadoes

Tornadoes are narrow, funnel-shaped clouds that reach the ground and contain high-velocity winds that may exceed 400 kilometers per hour (250 miles per hour) surrounding an intense low-pressure center where the drop in pressure commonly exceeds 25 millibars (figure 12.20). The funnel develops from the base of a large thunderstorm in which the clouds may rotate. Bag-shaped clouds often appear to hang down below the cloud base before a funnel emerges. Intense damage occurs where the funnel reaches ground level. Typically, the funnel at ground level is between a few and several hundred meters wide and may touch the ground for 5 to 10 kilometers (3 to 6 miles). Along the tornado's path, the high winds are devastating. The wind levels trees, destroys houses, and may even

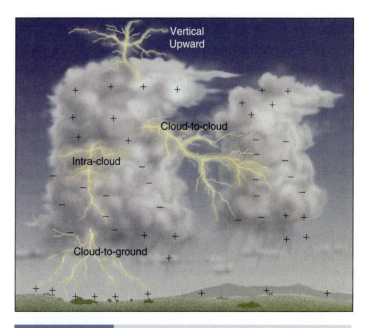

Figure 12.18 Charge Separations in Clouds, and Between Clouds and the Ground. Lightning can join any two centers of opposite charge within a cloud, between two clouds, or between the ground and a cloud. Discharges may also occur vertically between a cloud and the ionosphere.

move trains off their tracks. The effects of tornadoes at ground level are often disastrous. Most damage caused by tornadoes results from the pressure generated by the high winds. The wind may cause pressure amounting to 800 pounds per square foot against buildings.

Tornadoes may form beneath isolated thunderstorms, but more often they occur in thunderstorms that are imbedded along cold fronts. As the front moves forward, cold air pushes under the warm air ahead of it. The rapidly rising hot, moist air is sucked up into cumulonimbus clouds that develop along the front. Release of heat fuels the instability of the air as moisture condenses in the cloud. Often, the jet stream lies above portions of these cold fronts and above the thunderstorms. The jet stream's rapidly moving air is important in creating conditions favorable for the development of tornadoes at lower levels in the atmosphere. The jet stream removes air that rises in the underlying funnel, effectively sucking more air up the funnel.

The areas most frequently affected by tornadoes are areas of flat topography located on the east side of high mountains such as the Andes in South America and mountains in Asia. In North America, most tornadoes occur in the U.S. Great Plains and central southern states during the early summer months. At this time of year, the temperature variations between the northern and southern part of the country reach a peak. Warm, moist air from the Gulf of Mexico moves into the mid continent where it is in the path of wave-cyclones that move cold air eastward across the plains. Occasionally, tornadoes occur in areas far removed from the Great Plains, but most are associated with cold fronts that are moving into areas of hot, moist air.

Tropical Cyclones

Tropical cyclones are large storms with high-velocity winds that develop around low-pressure centers in the tropics (figure 12.21).

Figure 12.20 The Funnel of a Mature Tornado Touches the Ground Near Union City, Oklahoma.

Figure 12.21 Satellite Image of Hurricane Andrew. This image was obtained shortly before the center of the storm moved ashore in the Gulf of Mexico in 1992.

They originate along the boundary between the doldrums and the trade-wind belts (see figure 12.8). They evolve from sinuous wave-like patterns of circulation, called easterly waves, that form in the streaming flow of air in the trade-wind belt. Along this wave, warm low-pressure air from the doldrums rises, and if a low-pressure center intensifies, it is called a *tropical depression.* Air begins to swirl inward and up through the center. As the low-pressure center becomes better defined, wind speeds increase, and the clouds become organized in a circular pattern reflecting the flow of air around the center of the storm. If the winds reach 46 to 118 kilometers per hour (29 to 74 miles per hour), the system is called a *tropical storm;* if they increase above 118 kilometers per hour (74 miles per hour), the storm is called a **hurricane.** These storms are called typhoons in the Pacific Ocean and cyclones in the Indian Ocean. As the wind speeds increase, the clouds near the center of the storm become increasingly organized around a well-defined cylinder of calm air, known as the eye of the storm. Cool air from near the top of the system sinks down through this chimney-like space (figure 12.22). The eye is a relatively small area of calm air that is free from the centrifugal forces that propel the air around the eye. Air flowing at high velocity becomes concentrated in circular bands around the eye. Within these bands, air flows toward the center and rises rapidly.

Hurricanes form in the tropics of both hemispheres, but the storms that do the most damage pass through the Caribbean and Gulf of Mexico or are located in the southwestern Pacific Ocean. They form over water in the tropics, and during their early stages, most tend to move toward the west in the trade-wind belt. The path curves as a result of Coriolis effects, and the storm begins to move into higher latitudes. Once this deflection starts, the storm center moves more quickly. Most storms follow a broadly curved path as they move from the tropics into the westerlies.

As hurricanes travel over warm water, they pick up a large quantity of water vapor that rises as it is sucked in toward the center of the system. When the water vapor cools at higher altitudes, condensation occurs, releasing heat that makes the air less stable and causes it to rise. As long as the hurricane moves over water, this process continues, and the storm increases in size and intensity. If the storm moves over land, the water supply decreases, and the storm begins to lose intensity. Most hurricanes die out as they pass over land. In eastern North America, storm tracks commonly pass through the Caribbean. The supply of water to the storm may be reduced as the storm passes over Florida, Cuba, or Haiti and the Dominican Republic, but if the path takes the storm into the Gulf of Mexico, it may be regenerated. When the hurricane crosses the coast into the continental interior, the storm loses force soon after it crosses the coast. If the storm track passes along the East Coast, the storm may reach high latitudes before being caught in the western flow of air and carried back over the Atlantic (figure 12.23). Here again, the storm loses intensity because its water supply drops as it moves over cooler water.

The amount of damage done by a hurricane can be enormous if it passes over populated areas. The storm system may have a diameter of 500 kilometers (300 miles) with hurricane-force winds that extend 80 to 100 kilometers (50 to 62 miles) from the center, and wind speeds may exceed 100 miles per hour (rarely 150 miles per hour) in the tight zone of high-velocity air located around the center of the storm. Even at a distance of many miles from the center, winds are strong enough to uproot trees, blow houses down, and cause extensive beach erosion. The results can be especially disastrous if the passage of the storm center over the coast coincides with high tide. This happened in 1990 with Hurricane Hugo, which came across the coast near Charleston, South Carolina, during high tide.

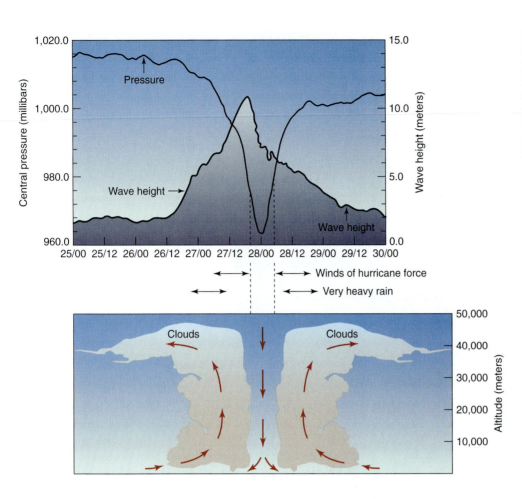

Figure 12.22 Atmospheric Structure and Weather Associated with a Hurricane. Note the extreme drop in pressure at the center of the hurricane and the great increase in the height of waves that is directly related to the pressure.

It is difficult to predict the precise path and the rate of movement hurricanes will follow, but they have a consistent pattern of wind velocity and direction of airflow around the center. Remembering that the flow of air around a low-pressure center is counterclockwise in the Northern Hemisphere, an observer can determine the approximate location of the storm center from observations of the wind direction at a locality. For example, if the wind at the observer's location is blowing from the east, the storm center must be south of that locality; if the wind is from the west, the storm center must be north of the locality (see figure 12.23).

A scale known as the Saffir-Simpson Hurricane Scale indicates the severity of destruction caused by hurricanes as the wind speeds increase. Category-1 hurricanes, with wind speeds of 74 to 94 miles per hour, cause storm surges 1.2 to 1.5 meters (4 to 5 feet) above normal and cause damage to unanchored mobile homes and trees. A category-3 hurricane, with wind speeds of 110 to 155 miles per hour, causes storm surges 3 to 4 meters (10 to 13 feet) above normal and damage to roofs, doors, windows, piers, and structural damage to small buildings. Mobile homes are destroyed, and land lower than 1.5 meters (5 feet) above sea level may flood up to 9.5 kilometers (6 miles) inland. Category-5 hurricanes, with wind speeds above 155 miles per hour, may cause storm surges about 6 meters (20 feet) above normal. Some buildings fail, and the lower floors of all structures located less than 5 meters (about 16 feet) above sea level and within 500 meters (about 1,600 feet) of the shoreline suffer major damage.

Hurricane Devastation in Eastern North America

Hurricane damage in North America is concentrated on islands in the Caribbean, along the Gulf of Mexico, and along the east coast of the United States. The broad, flat land of the coastal plain of the Atlantic and Gulf of Mexico are particularly susceptible to damage from hurricanes. All the way from Mexico to Long Island, a broad zone of flat land extends inland. Low-lying barrier islands make the coast especially susceptible. This was most evident in the September 7, 1900, hurricane that destroyed Galveston, Texas, much of which was built on an island that rose only 1.5 meters above sea level. Luxurious homes lined much of the beachfront. Sand dunes along the beach were removed to improve views of the water. No hurricanes had hit the region in many years, so when the local meteorologist gave warning of an approaching storm, people were not especially impressed. Large numbers went to the shore to see the huge waves that were rolling in under a relatively clear sky. Before most people realized how serious the waves had become, the connection between the island and the mainland had been cut off. They were trapped as the tide rose 2.4 meters (8 feet) accompanied by a 6-meter- (20-foot-) high storm surge that washed across the island

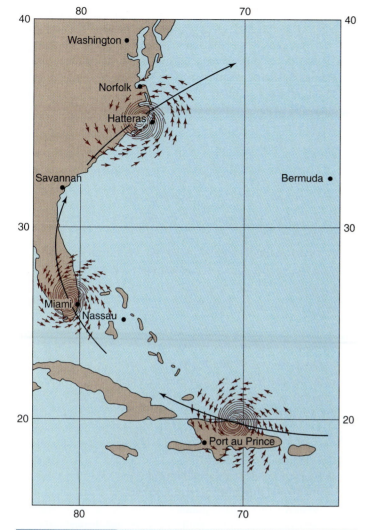

Figure 12.23 Typical Track and Direction of Air Rotation Around a Hurricane That Forms in the Caribbean and Moves Near the East Coast of North America. If the track takes the hurricane over land, the water supplied to the storm decreases, and the disturbance generally begins to die out.

waves washed over islands, destroyed buildings near the beaches, eroded the beaches and sand dunes, and cut dozens of new tidal inlets across offshore islands.

Hurricane Andrew, which struck Miami in August 1992, was certainly one of the most destructive hurricanes to hit North America. Meteorologists had tracked the hurricane by satellite and flew through the storm, so, Andrew's arrival came as no surprise. But the intensity of the storm, with wind speeds over 145 miles per hour, did catch people off guard. Because no major storm had hit this area for several decades, people were very lax about building houses that could withstand hurricane-force winds. The storm passed over substandard housing and large trailer courts, leaving a quarter of a million people homeless. The storm center passed across the Florida peninsula and into the Gulf of Mexico where it regained strength before veering northward into the coast of Louisiana where it did additional damage in a large area of low population.

Devastating hurricanes have opened inlets along the Outer Banks of North Carolina (see figures 10.22 and 10.23). Hurricane Hugo destroyed long stretches of land along the coast of South Carolina. In 1969, Hurricane Camille crossed the northern coast of the Gulf and was ultimately responsible for extensive flooding in Virginia. Severely damaging hurricanes moving along the Atlantic coasts have even reached as far north as Long Island.

Floods Caused by Hurricanes

On August 20, 1969, Hurricane Camille caused rain that set records for the James River basin in Virginia, causing flood levels expected less than once a century. Three years later, in June 1972, Hurricane Agnes caused a second flood to strike the James River basin. Camille came ashore in the Gulf of Mexico near New Orleans and followed a path over the Mississippi Valley where heavy rains fell. Then, it turned east and crossed Kentucky and Virginia. Over the mountains of Virginia, Camille encountered moist air carried inland by a low-pressure system along the Atlantic coast. There the storm intensified, pulling more moist air inland from the Atlantic. Rainfall exceeded 28 inches in an eight-hour period in one county—an amount of rain meteorologists expect to fall about once in a thousand years.

High amounts of precipitation fell all along the storm track, but the heaviest rain fell on the slopes of the Blue Ridge Mountains in the drainage basins of the James River. The mountain soils became saturated and slid down the slopes, leaving thousands of long scars on the sides of the mountains. The resulting mud and debris slides buried homes and added great quantities of sediment to the rising streams. This storm killed more than 150 people.

In the upstream portion of the drainage basin, the amount of water in streams, called **discharge,** increased rapidly in the hours during and immediately after the main storm. Downstream, discharge increased later because of the time required for the flow to reach downstream stations. The discharge became higher downstream because more tributaries contributed their discharge to the main streams. As movement of water down the stream system slowed, flood peaks became broader, and the rivers remained in

destroying large brick buildings and leaving 8,000 people dead. The wind gauge broke showing a wind of over 160 kilometers per hour (100 miles per hour).

Although most hurricanes strike the coast of the Gulf of Mexico or the southeastern states, hurricanes have devastated coastal areas much farther north. In March 1962, a hurricane known as the Ash Wednesday storm struck the coast of New England causing severe damage along a thousand kilometers of the coast. The storm took place during a period of spring tides when the moon was closest to Earth. It lasted long enough for five tidal cycles to occur. This made it possible for waves that were 10 meters (33 feet) high in the open ocean to advance repeatedly across the shoreline. The

flood stage for a longer time. Landslides, inundation of towns and fields by mudflows, removal of topsoil from large areas of the flood plains in the upper part of the basin, scouring of the channels, and inundation of the flood plains throughout the basin caused extensive property damage.

Hurricane Agnes affected a much larger area than Camille did. The total rainfall on land from Agnes set a new record and followed an unusually wet spring. Water saturated the ground near the surface, and runoff was excessive even on low slopes. Hundred-year-plus flows affected a twelve-state area. Because of high-density development on flood plains in the affected region, this was one of the most destructive floods in the country's history. Damage exceeded $3 billion.

SUMMARY POINTS

1. Differences in temperature cause air to move vertically. Hot air has lower density and therefore rises through cooler, higher-density air. Most horizontal movements of air also result from differences in density and pressure. High-density (high-pressure) air tends to settle and move laterally toward air that has lower pressure.

2. Once air begins to move laterally across Earth's surface, it travels in a curved path as a result of the Coriolis effect. In the Northern Hemisphere, these effects cause air to veer to the right. In the Southern Hemisphere, Coriolis effects cause air to veer to the left.

3. The combined effect of pressure gradients and Coriolis effect cause air moving toward areas of low pressure to swirl into the low-pressure center in a counterclockwise direction in the Northern Hemisphere and in a clockwise direction in the Southern Hemisphere. Air moves out from high-pressure centers in the opposite direction from that of the lows.

4. As hot air rises, it expands, and the energy expended in this expansion causes the air to become cooler. Eventually, the rising air has the same temperature and pressure as surrounding air. At this point, the air stops rising.

5. If hot rising air contains water vapor, the air eventually cools to the dew point, and water vapor begins to condense, releasing heat energy. The released heat energy makes the air unstable, and it continues to rise until all of the water vapor has condensed.

6. Air pollution becomes severe where air stagnates. In the winter, stagnation commonly occurs where cold, high-density air fills low places in the landscape. These conditions create temperature (parcel) inversions in which cold air lies beneath warm air. Pollutants introduced into this air remain there until the air moves and disperses them. Movement begins late in the morning when the ground begins to warm and parcels start to form.

7. At ground level, global circulation appears to involve mainly airflow that is nearly parallel to lines of latitude. A belt of hot, calm air, called the doldrums, lies along the equator. Belts of high-velocity east to west winds, referred to as the trade winds, flank the doldrums. Another belt of relatively calm air occurs at about 30° latitude, known as the subtropical high zone, or horse latitudes. Strong westerly winds characterize the adjacent belts. Closer to the poles, belts of easterly flow prevail.

8. At higher altitude, air rises over the doldrums belt and sinks over the horse latitudes and at the poles. These vertical movements are superimposed on a large zone of high-velocity air that flows from west to east over the middle latitudes. The highest velocity in this westerly flow occurs in two streams referred to as jet streams.

9. Air masses and the movement of air are strongly influenced by the character of Earth's surface. Air over continents becomes colder in winter and hotter in summer than that over adjacent oceans. The colder air sinks, spreads laterally, and creates high-pressure air masses. These high-pressure and, in summer, low-pressure areas affect the air circulation that is parallel to latitude. The interaction causes the jet streams to flow in meandering paths.

10. Winter storms most frequently form along cold fronts that result where the meandering latitude-parallel flow causes cold air to move toward the equator. Low-pressure cells form along these fronts and move from west to east beneath the jet streams.

11. Most tornadoes form beneath large thunderstorms or along cold fronts where cold air moves under hot, moist air. These storms cause severe damage along narrow paths (usually about 100 meters [300 feet] wide). In contrast, hurricanes that form because of lateral wind shear near the border of the doldrums belt in the subtropics involve huge volumes of air and are tens to hundreds of kilometers wide.

REVIEW QUESTIONS

1. Why is air pressure higher at sea level than it is on a mountain?
2. If you flew due south in an airplane that took off from Greenland, would you cross the equator at a point due south of Greenland, one west of Greenland's longitude, or one east of Greenland's longitude? Explain.
3. What causes air to flow toward a low-pressure center? Why does it increase in velocity toward the center?
4. Why is air pollution gradually more severe over cities early in the morning than at mid afternoon?

5. How can you estimate the distance to thunderstorms?
6. What causes temperature (parcel) inversions?
7. Describe the major ground-level wind pattern.
8. Where does air from high altitude settle back to ground level? What effect does this air have on the climate of these regions?
9. Describe the change you would expect during the passage of a cold front.
10. What role do continents and oceans play in determining atmospheric circulation?
11. What influences do jet streams have on ground-level weather?
12. Explain why most thunderstorms and tornadoes occur in the mid-to-late afternoon.

THINKING CRITICALLY

1. What steps could a community that has air-pollution problems resulting from temperature (parcel) inversions take to reduce air pollution?
2. If the winds associated with a hurricane are blowing out of the north, where is the center of the hurricane relative to your position?

3. How would a slowing of atmospheric circulation affect surface temperatures on Earth?

KEY WORDS

adiabatic process
cold front
dew point
discharge
doldrums
environmental lapse rate
front
hail
horse latitudes
hurricane
intertropical convergence zone (ITCZ)
isobars
jet streams
lightning
normal lapse rate

occluded front
parcels
polar easterlies
polar front
polar front jet
stationary front
subtropical high zone
subtropical jet
temperature inversion
thermals
tornado
trade winds
tropical cyclones
warm front
wave-cyclone
westerly winds

chapter thirteen

Climate—Past, Present, and Future

Monument Valley. As climatic conditions change, areas that once supported abundant plants may become deserts.

Chapter Guide

The possibility that human actions are changing Earth's climate will be one of the most pressing concerns of the twenty-first century. Present understanding of the climate system is not yet adequate to allow predictions of what climatic changes lie ahead, but the implications of climate changes could be far reaching. The level of the sea, patterns of precipitation, productivity of the soil, variability of weather, and the frequency of storms all depend on climate. Evidence of climate change abounds in the geologic record, and it demonstrates that Earth's climate has been undergoing especially dramatic changes for the past 2 million years. Studying past climates offers the best hope for understanding what causes climate change and what the consequences of changes are likely to be. This chapter begins by defining the present-day climate and identifying the conditions responsible for the modern variations. Even over the relatively short period of recorded history, climatic conditions have changed dramatically in the middle and high latitudes, and the record clearly indicates that changes are in progress today. A record of past climatic conditions is preserved in ice cores, marine and lake-bed sediments, and in rock and fossil records. Studies of these records have generated hypotheses about some of the factors that cause climate to change.

INTRODUCTION

Climate encompasses the long-term averages of weather patterns that tend to repeat themselves. Climatologists characterize the climate of an area primarily by temperature and the amount and type of precipitation. Averages for a year provide a general impression, but averages for each season give a more precise definition. Other characteristics of the weather such as the average hours of sunshine, prevailing wind direction, number of days with extreme temperatures, or frequency of fogs add even more detail.

Keeping systematic records of weather started in the seventeenth century. Daily meteorological observations started at the Academmia del Cimento in Florence, Italy, around 1654. Soon afterward, the Royal Society of London began compiling a record of weather. Record-keeping quickly spread across Europe and even to North America. These measurements documented the high degree of variability in weather from one year to the next. Frequent departures from the "normal" climate for an area became apparent. The high degree of variability in temperature and precipitation from year to year continues today and creates a background noise that tends to mask long-term climatic changes.

Climate is a critical component of the environment. It is one of the most important factors that define various ecological niches (figure 13.1). Despite the exceptional ability of humans to adapt to changing conditions, climate has influenced both human history and day-to-day lives. Climatic conditions determine the limits of human settlement and the cultivation of crops. By controlling the type of crops and the length of the growing season, climate influences the overall size of the population a region can support. In the past, climate influenced human migration and, later, transportation and trade routes, as well as where peoples settled. Some studies suggest that early European societies exerted some control on their population by insisting that no one should marry until they could support a family. That support depended largely on what they could produce on farmland, and agricultural production decreases as average temperature falls. Early in the period from 1560 to 1820, the average age at marriage was 27 years, but as the mean annual temperature dropped, the average age rose to 30 years before dropping to 23 years as temperatures rose.

(a)

(b)

(c)

Figure 13.1 Climate Determines the Types of Plants and Animals That Inhabit a Region. These photographs depict drastically different climatic regions. (*a*) Extremely dry desert conditions seen at Death Valley. (*b*) Vast stretches of snow cover characterize polar regions. (*c*) Lush vegetation is present in rain forests.

EARTH'S MODERN CLIMATE

Global climate refers to the annual or seasonal temperature or precipitation averaged over the Earth as a whole. Change in global climate is a serious concern because small changes in global average temperatures usually indicate much more dramatic changes in the temperature and precipitation of smaller regions.

Most regions contain areas with localized climatic conditions, and climate changes from one side of a mountain to another depending on the wind direction and the amount of solar radiation reaching the slopes. Because human life span is short relative to the rate of global climate change, differences in climate from one region to another are more apparent than are the long-term effects of worldwide changes. Any change in global climate ultimately affects regional climates. Because these regional changes directly affect agricultural productivity and land use, they are of greatest concern to most people.

Factors affecting climate include temperature extremes (highest summer temperature and lowest winter temperature), number of days of temperature above or below some specified level, the frequency of cloud cover, the number of storms of various types, and mean annual and seasonal temperature and precipitation. Using these criteria, climatologists classify the world's climatic regions. Figure 13.2 illustrates the system devised by W. Köppen in 1931. In this system, areas of similar geographic position having essentially the same climatic characteristics define a climatic type. Table 13.1 lists and describes modern climate types.

Modern Climates of North America

As might be expected on a continent that extends from within the Arctic Circle to the tropics, climate across North America is diverse (figure 13.3). North America provides excellent examples of how the location of major high- and low-pressure systems, the jet streams, and mountains affect climate.

A large ice sheet still covers most of Greenland, and although the ice is rapidly thinning, the Arctic Ocean is frozen throughout the year. Polar tundra and permafrost distinguish the northern parts of Canada and Alaska from the remainder of the continent. A broad band that extends from the central part of Alaska across Canada, almost as far south as Winnipeg, is characterized by subpolar, moist climates with severe winters. Ice or snow covers large parts of these areas much of the year, and the average monthly temperature rarely rises above 10°C (50°F). Huge masses of cold air and

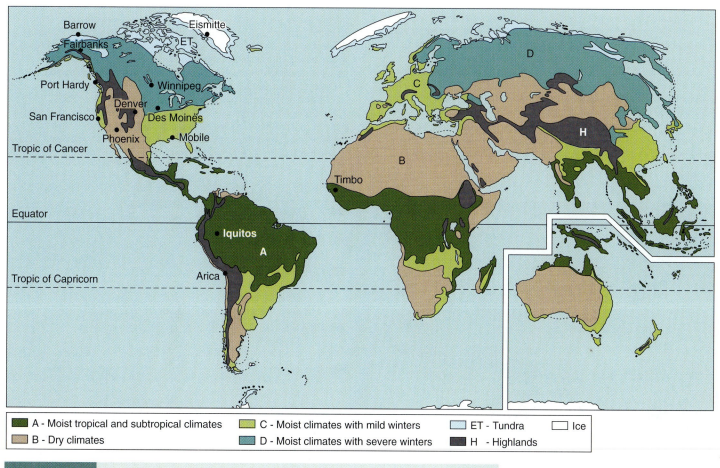

■ A - Moist tropical and subtropical climates	■ C - Moist climates with mild winters	□ ET - Tundra □ Ice
■ B - Dry climates	■ D - Moist climates with severe winters	■ H - Highlands

Figure 13.2 Climatic Regions of the World as Defined by Köppen.

Table 13.1   Data Bank: Modern Climate Types

Wet Equatorial Climate   The wet equatorial climate coincides with the doldrums belt at the equator and is about 10° of latitude wide. Except in mountainous areas, which may be cool throughout the year, all months in wet equatorial climates are warm and moist. Temperature averages about 21°C (70°F), and pressure is low. The Sun is always high in these latitudes, but solar radiation reaches a peak at the equinoxes, and heavy rains may fall almost every day during these peaks. One hundred fifty centimeters (60 inches) or more of precipitation may fall each year, much of it during thunderstorms. The wind is calm to slight except during tropical cyclones.

Trade-Wind Coastal (Littoral) Climate   The trade-wind coastal climate appears along the eastern coastal margins of continents in the trade-wind belts. Most of these regions are long from north to south and relatively narrow from east to west. North-south trending mountains in these areas cause eastward-moving air masses to rise, and this uplift increases precipitation. Trade winds blow steadily in autumn and winter. The mean annual temperature is above 21°C (70°F) with lower averages at high elevations. Normally, more than 89 centimeters (35 inches) of rain falls, and it is distributed fairly evenly throughout the year. The total amount of rainfall varies considerably (Hawaii sometimes has 500 centimeters [200 inches]). The maximum rainfall comes in autumn. Monsoons in these latitudes may interrupt the trade winds and cause a dry season and a wet season, and cyclones are common.

Wet and Dry Tropical Climates   Wet and dry tropical climates are common in areas adjacent to the wet equatorial climate. In these areas, solar heating produces one long, warm period. Most of the rain falls in one period when the doldrums belt expands into the region. During the remainder of the year, trade winds influence the weather and little rain falls. Cold ocean currents that extend close to the equator along the west coasts of South America and Africa also influence these climates. Average annual temperature is above 21°C (70°F) except in regions influenced by the monsoons. Highest temperatures are reached just before the summer rains. Rainfall varies with at least an 89-centimeter (35-inch) average and sometimes several hundred inches in areas that experience monsoons. Winters are always dry with little or no rainfall.

Semiarid Tropical Climate   Semiarid tropical climates are present at higher latitudes than the wet and dry tropical climates. Most of these are in continental interiors, especially on plateaus and next to deserts.

In the summer, most semiarid tropical climates lie on the outer edges of the doldrums belt. In the winter, trade winds or the calm of the horse latitudes influence these belts. Cold ocean currents or the rain-shadow effect of adjacent mountains cause low rainfall in some areas. Temperatures average above 21°C (70°F), often with extended periods of high temperature. Rainfall is low—between 25 and 89 centimeters (10 and 35 inches), and most rain falls in the summer months.

Tropical Desert Climate   Most tropical desert climates occur in the horse-latitude belts of high pressure, but some occupy continental interiors or leeward coasts in the trade-wind belts. Annual temperature averages 21°C (70°F) or more, but the daily temperature range is great, and heat is lost rapidly at night through radiation. During the day, temperatures may reach 50°C (120°F) or more. Rainfall is low, averaging less than 25 centimeters (10 inches) per year, and most of this rain falls during winter months.

Dry Summer Subtropical Climate   The dry summer subtropical climate, known as the Mediterranean type, has average monthly temperatures above 7°C (45°F) in the coldest month and below 24°C (75°F) in the warmest. Generally, temperatures near the coast are lower than those in the continental interior. Rainfall is normally less than 89 centimeters (35 inches) per year, but its range is great, 25 to 230 centimeters (10 to 90 inches), depending on the amount of topographic relief and the position of an area in relation to the coast. Summers are dry. The winter wet season is generally mild, but freezing temperatures may accompany cold fronts. Tropical cyclones may reach into these regions.

Humid Subtropical Climate   Climates of the humid subtropical type appear on the eastern margins of continents in middle latitudes. High pressure over the oceans in summer promotes movement of air inland, as in the eastern section of the United States. Strong cyclonic storms in the winter, with a rapid succession of warm and cold fronts, characterize these areas of the United States, but cold Arctic air dominates the winter weather along the eastern coast of China, which has the same type of climate. The mean temperature in the coldest month is above 4°C (40°F), and the warmest month averages from above 20° to 26°C (70° to 80°F). Precipitation ranges from 89 to 152 centimeters (35 to 60 inches) per year, or sometimes more, with no unusually dry months. Most tropical cyclones come in autumn, and tornado activity reaches a peak during the spring in this climatic region of the eastern United States.

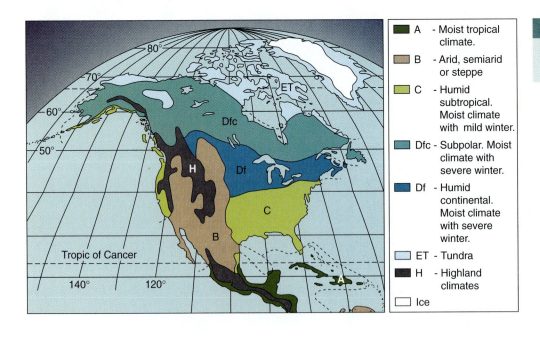

Figure 13.3   Major Climatic Regions of North America as defined by Köppen.

A - Moist tropical climate.

B - Arid, semiarid or steppe

C - Humid subtropical. Moist climate with mild winter.

Dfc - Subpolar. Moist climate with severe winter.

Df - Humid continental. Moist climate with severe winter.

ET - Tundra

H - Highland climates

Ice

Cool Marine Climate The western margins of continents in the regions of the prevailing westerlies have the cool marine type of climate. Intense cyclonic activity in winter interrupts the predominant influence of the westerly wind. Warm ocean currents in these latitudes tend to keep the temperatures in winter higher than is usual for these latitudes. The average temperature of the coldest month is between −1° to 7°C (30° to 45°F), and the warmest month averages about 15°C (60°F). Precipitation is heavy with averages of 90 to 152 centimeters (35 to 60 inches), much of which may fall as snow because the maximum precipitation takes place in winter. Extremes of annual rainfall are close to 500 centimeters (about 200 inches).

Cool Littoral Climate The cool littoral climate occurs in the high middle latitudes, where both marine and continental influences affect the weather. The mean temperature in the coldest month is under 4°C (40°F), and averages in the warmest month may reach 15° to 20°C (60° to 70°F). Precipitation amounts to 64 to 152 centimeters (25 to 60 inches) per year. Rain falls throughout the year in Europe. But some regions of the cool littoral climate have maximum precipitation in winter, as in New Zealand; others experience maximum rainfall in summer, as in Japan. Typhoons influence areas of Japan, bringing excessive rainfall in brief periods of time.

Humid Continental Climate Many areas in the interiors of continents in middle latitudes of the Northern Hemisphere have the humid continental type of climate. The temperature range is large. The coldest month often averages below freezing and sometimes even below −18°C (0°F), but at least three months have averages above 15°C (60°F). In spite of the humidity that characterizes this climate, the precipitation, in the range of 40 to 127 centimeters (16 to 50 inches) per year, is not excessive. Cyclones, with accompanying cold and warm fronts, influence the climate in winter, and thunderstorms are common in summer.

Semiarid Continental Climate The interior of continents in middle latitudes where a mountain barrier cuts off moisture supplies experiences the semiarid continental type of climate. Like the humid continental climate, the semiarid continental climate has a large range of temperatures, but annual precipitation rarely exceeds 40 centimeters (16 inches). Although most of this falls in the summer, the pattern of distribution is erratic; some areas of semiarid continental climate may

not receive any rain in some years. Prolonged periods of drought contribute to dust storms, which often come in the winter. Some climatologists recognize an intermediate type of desert climate that has less than 15 to 30 centimeters (6 to 12 inches) of rain each year.

Marine Subpolar Climate The marine subpolar climate is present in the northern middle latitudes along coasts with high mountains nearby. Ocean waters tend to moderate the temperatures, producing a somewhat smaller range than might be expected for the latitude. Average temperatures are under −1°C (30°F) for several months, and at least one month has an average temperature over 10°C (50°F). Precipitation varies greatly. Cyclonic activity in fall and winter frequently affect these regions, and fog is common.

Moderate Subpolar Climate The moderate subpolar climate is usually a mountain climate currently present in the Rocky Mountains, the Alps, and the Sierra Nevada, where winters are not quite as cold as in the continental interiors in the subpolar latitudes. Temperatures resemble those in the marine subpolar climate, but precipitation falls throughout the year. The average total is in the range of 64 to 100 centimeters (25 to 40 inches) per year.

Extreme Subpolar Climate Temperatures average below −1°C (30°F) for five or more months in the subpolar continental interior regions. Most of these regions are located between the Arctic Circle and about 48° N in North America and 30° N in Asia, where the high altitudes contribute to the low temperatures. Summer temperatures may reach an average of 10°C (50°F) or more for one month, and extreme temperature variation is a common characteristic. Most precipitation falls in the summer months, and the annual average is usually below 25 centimeters (10 inches). Cold air masses so dominate the climate that the air holds little moisture.

Polar Climate The polar circles—latitude 66.5° North and South—approximately define the boundaries of the polar climatic zone, which is a direct result of the low level of incoming solar radiation. Winters are dark, and the sun remains low in the sky even in summer. Cold and dry high-altitude air descends in the polar regions and spreads toward the equator. Precipitation is low because the cold air contains little moisture. Average temperatures never rise above 10°C (50°F) for any month—too low to permit growth of forest or agricultural crops. Ice caps and tundra characterize these regions.

accompanying high-pressure systems develop over this central part of Canada in winter months. At times, the cold air breaks out of the region and brings frigid conditions to the central and eastern United States.

The west coast of the United States and Canada has a cool marine climate. In winter, ocean currents keep the temperatures high for these latitudes. Average temperature for the coldest month is between −1° to 7°C (30° to 45°F), and the warmest month average is about 15°C (60°F). Precipitation averages between 90 to 150 centimeters (35 to 60 inches) per year. Cyclones and frontal systems, many originating over the Gulf of Alaska, cross the coastal belt and influence winter weather across North America. As these storms move east, they bring large quantities of moisture across the coast, where heavy rains and snows fall on the slopes of the coastal mountains in British Columbia, and in the Cascade and Rocky mountains. Large accumulations of snow cover the high mountains located near the west coast. Glaciers continue to move through the valleys of these mountain belts in Alaska and British

Columbia and down the slopes of the Cascades. Very little ice remains year-round in the Sierra Nevada mountains; farther south, little snow falls in the high mountains. In summer months, thunderstorms occur frequently over the mountains.

Although the western slopes of the Sierra Nevada receive rain as a result of the uplift of moist air that moves in from the Pacific, little moisture is left for the region farther inland. As a result, large areas in the western states have a semiarid continental climate. These dry areas are present in eastern Oregon and Washington, and they continue southward into Mexico. Annual precipitation rarely exceeds 40 centimeters (16 inches), and most of that falls in summer months. Some years have no rainfall. Dry winters commonly bring dust storms.

The states occupied by the Rocky mountains have a moderate subpolar climate. These mountains are located much farther inland than the coastal mountains. The high elevations ensure cool temperatures. Average temperature normally exceeds 10°C (50°F) for at least one month. During one or more months, average

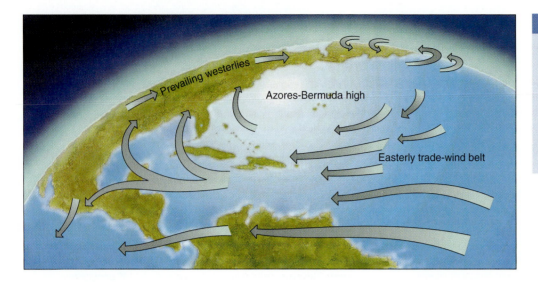

Figure 13.4 A Large High-Pressure Center Called the Bermuda High Develops over the Central Atlantic Ocean Each Summer. Most hurricanes form in the easterly trade-wind belt during late summer when the water is hot. This circulation takes moist air from the Gulf of Mexico into the southern states.

temperature falls below freezing. Some rapidly receding glaciers are present in the Canadian Rocky mountains, but only a few remnants of glaciers remain high in the mountains of Montana. Farther south, winter snows melt during summer months when temperatures rise above freezing. Precipitation in the Rockies ranges from 64 to 100 centimeters (25 to 40 inches) per year and falls throughout the year, but much of that precipitation is snow in winter months. The Rocky mountains block much of the precipitation that might otherwise spread across the Great Plains. The southern plains are left with little rainfall and have a semiarid climate.

The northern midwest has a humid, continental climate typical of continental interiors in middle latitudes. The temperature range is great, with the average temperature in the coldest month significantly below freezing. Precipitation ranges from 127 to 295 centimeters (50 to 116 inches) per year.

The southeastern states have a humid, subtropical climate. A large area of high pressure, known as the Bermuda high, develops over the central Atlantic Ocean in summer (figure 13.4). The clockwise circulation around this high carries warm, moist air inland from the Gulf of Mexico. This makes the region hot and humid during summer months. In the winter, a rapid succession of warm and cold fronts bring strong low-pressure systems, with their cyclonic circulation, across the region. Mean temperature in the coldest month is about 4°C (40°F), and the warmest month averages between 20° to 26°C (70° to 80°F). Precipitation ranges from 150 to 340 centimeters (60 to 135 inches) per year, and no months are consistently dry. Tropical storms, most of which come in the autumn, bring heavy rains and winds to the coastal regions of the Gulf and Atlantic. In the spring, tornadoes like those described in chapter 12, form along cold fronts that encounter the warm, moist air from the Gulf.

The regional climates are interrelated components of Earth's physical climate system. Regional climates have varied considerably over recorded history. The geologic record documents far greater changes in the past. To understand what causes change, it is necessary to know more about the history of ancient climates and how the climate system works.

EARTH'S PHYSICAL CLIMATE SYSTEM

Analysis of global heat balance (see chapter 11) is one of the keys to understanding the operation of Earth's physical climate system. Interactions among the oceans, atmosphere, biosphere, and land make changes in the physical climate system extremely complex and difficult to predict. Figure 13.5 illustrates some of these relationships. The ocean and atmosphere affect one another through such processes as evaporation, precipitation, heat exchange, effects of the wind on the sea surface, and exchanges of gases. An example is the interaction between sea ice, marine waters, and the atmosphere. As the area of the ocean covered by sea ice grows or contracts, the physical properties of seawater change. If sea ice melts, the seawater becomes less dense. If the area covered by ice and snow increases, more solar radiation reflects from the surface, the amount of radiation absorbed by the sea drops, and the temperature of the water decreases. Any changes in the temperature of the sea surface affect the rate of evaporation, and the water content of the atmosphere changes both the cloudiness and the amount of precipitation that fall in the ocean and on adjacent land areas.

Other examples of interactions between the oceans, atmosphere, land, and the biosphere occur in the exchange of gases (especially carbon dioxide and oxygen) between plants and the atmosphere and the exchange of nutrients between plants and the soil. In turn, soil properties affect the amount and types of vegetation that grow, and plants influence the composition of the atmosphere and the development of soil. Even processes acting deep inside the Earth affect the atmosphere. Volcanic eruptions cause rapid changes in both local atmospheric composition and in the amount of solar radiation that reaches the Earth's surface. Most of all, the amount of solar radiation reaching Earth determines the heat balance. At any given time, Earth's climate is a product of all of these factors. Changes in any of them can cause adjustments throughout the system. In recent years, most serious concerns have centered on the effects of changes in the composition of the atmosphere caused by human activity. Discussions of changes involving ozone and greenhouse gases appear in chapter 11.

Measuring the Temperature of the Atmosphere

Determining the average temperature of the Earth's atmosphere at any given time is a much more difficult task than it might appear to be at first. Although thousands of thermometers at meteorological stations record temperature continuously, these instruments are not the same types of instruments nor are they evenly distributed over the Earth. Western Europe, North America, Japan, and other population centers house most recording stations. Large areas of the oceans, Africa, Asia, the Arctic, and Antarctica contain few and widely spaced recording stations. Daily and seasonal temperature variations complicate the problem even more. To solve these problems, measurements of water temperature in the oceans are a better indication of global temperature than is air temperature. Because of its high heat capacity, seawater changes temperature much more slowly than the air. Despite the sluggish behavior of

the oceans, no simple way of determining the average temperature of the oceans is available. Global temperature also influences the advance and retreat of glaciers as well as the amount of sea ice. Measurements of all three of these—air and water temperature and ice volume—indicate that the atmosphere is becoming warmer.

On the basis of measurements made on the land surface, most climatologists conclude that the temperature of Earth's atmosphere has been slowly rising over the last hundred years (figure 13.6). This figure illustrates a compilation of millions of observations made after the mid 1940s. However, much of these data come from western Europe and North America. Old data for the tropical Pacific and the polar regions is scarce. Nevertheless, these data strongly suggest that global temperature has been rising as predicted from the increase in atmospheric carbon dioxide, methane, and other greenhouse gases. Based on the data assembled by the International Panel on Climate Change, the average global temperature rose about 0.56°C (1°F) between 1861 and 1997.

In recent years, measuring surface temperature from satellites has become possible. This technique provides a way of measuring

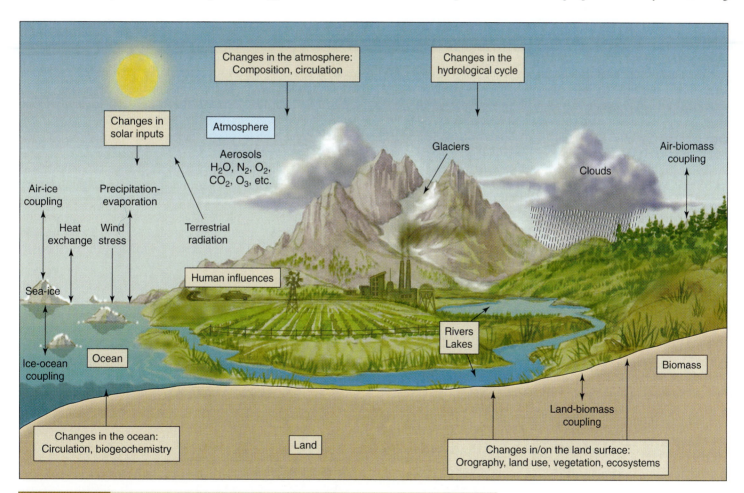

Figure 13.5 **Earth's Physical Climate System.** The climate system involves interactions among incoming solar radiation, clouds and other atmospheric gases, the surface of the ocean, soil, rock, ice, and both plants and animals. The arrows indicate interchanges between different elements in the system.

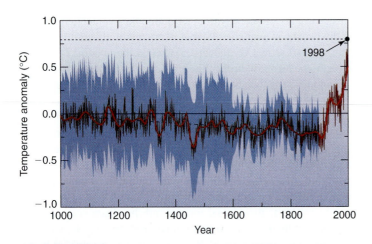

Figure 13.6 Temperature of Earth's Atmosphere over a 1,000-Year-Period. The dark blue indicates the range of variation of global temperature from 1000 A.D. until 1900 A.D. The red line depicts the average values for global temperature. Note the sharp rise that has taken place in global averages since 1900 A.D.

average temperature over large segments of the Earth's surface in short time intervals. The instrument used for this purpose measures the thermal emission of radiation by atmospheric oxygen. Early analyses of the first ten years of data collected between 1979 and 1988 reveal large temperature variability on time scales of weeks to several years, but no obvious trend for the ten-year period emerges. Most of the departures from average temperature are associated with El Niño events. Because calibration and verification of the results of the satellite data are difficult and because measurements must be made when the sky is clear, some climatologists think conclusions about global temperature trends based on these data are premature. Although most climatologists think the trend is toward an increase in global temperature, it is clear that major volcanic eruptions can reduce temperature for periods from months to a year or more.

Modern Climatic Effects of Volcanic Eruptions

After the eruptions in the East Indies of the volcanoes Tambora in 1815 and Krakatau in 1883, it became apparent that volcanic eruptions influence weather. Both eruptions were huge and violent, propelling gases and dust into the stratosphere where they remained suspended while circulating around the Earth. Tiny suspended solid particles or liquid droplets, called **aerosols**, scattered sunlight, creating bright, colorful sunsets around the world for over a year following each eruption. Mean global temperature dropped after both of these eruptions as it did after other major eruptions in the last century (table 13.2). A 44 B.C. eruption of Mount Etna in Italy dimmed the Sun and resulted in cooling, causing crop failures that produced famine in Rome and in Egypt. Temperature drops of about half a degree Celsius after the Tambora eruption produced what is historically known as the "year without a summer." A similar drop in temperature followed the more recent eruption of Pinatubo, a large volcano in the Philippines described in chapter 7.

The dusts and gases blown into the atmosphere during an eruption affect global temperatures. The gases consist mainly of sulfuric acid mixed with water, forming aerosols. The aerosols absorb infrared radiation from Earth's surface and solar radiation. This increases temperature, but the gases, and dust formed by condensation on these nuclei, also scatter sunlight and reflect solar radiation, causing temperatures to decline. The haze formed by aerosols and dust in the stratosphere reflects enough sunlight to cool the troposphere. The net climatic effect of eruptions depends on the quantity and how high gases reach in the atmosphere, and on the latitude of the volcano. For tropical eruptions, the heating effects are great and can displace the jet streams, causing increases in temperature in the Northern Hemisphere. For eruptions at higher latitudes, the reduction in solar radiation caused by aerosols and dust results in cooling effects.

The effects of volcanic activity on surface temperature are well established. Some have suggested that extensive volcanism might cause protracted cooling, even the growth of ice sheets, but the geological records of volcanism and glaciation do not support this conclusion. Today, attention is focused more on what may be causing global warming. The evidence of a warming trend is especially strong in the Polar Regions.

Rapid Changes in Polar Regions

A worldwide retreat of glaciers has been underway for decades. The retreat of glaciers that occupy valleys in places as widely separated as Switzerland, New Zealand, and Alaska is well documented. More recently, rapid changes are being detected in the polar regions. In 1998, images from Antarctica depicted an iceberg over

Table 13.2	Temperature Change Caused by Aerosols		
Volcano	**Date**	**Aerosol Amount**	**Temperature Change**
Tambora	1815	150 million tons	−0.4° to 0.7°C (0.7° to 1.26°F)
Krakatau	1883	55 million tons	−0.3°C (−0.54°F)
Mount St. Helens	1980	Minor	−0.1°C (−0.18°F)
El Chichón	1982	12 million tons	−0.2°C (−0.36°F)
Pinatubo	1991	13.30 million tons	−0.5°C (−0.9°F)

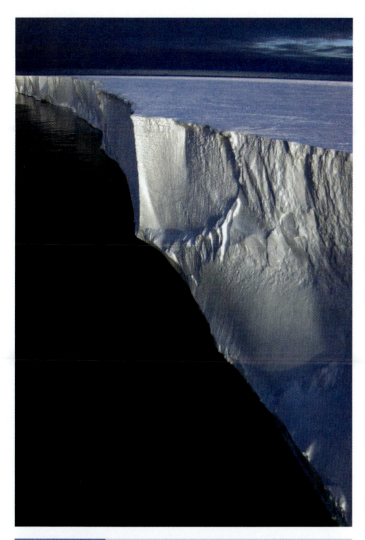

Figure 13.7 Breakup of Antarctic Shelf Ice. This oblique aerial photograph shows one of the largest icebergs to break off of the Ross Ice Shelf in Antarctica. It separated from the shelf ice in March 2000. This iceberg, designated B-15, is 11,000 square kilometers (4,250 square miles) in area and is one of the largest known icebergs.

the continental ice sheet that formed on land in the eastern part of Antarctica expanded westward into the sea and eventually displaced the water from that area. Today, the edge of the West Antarctic ice sheet rests on the former sea floor. Drilling through the ice into the sediment beneath the West Antarctic ice sheet reveals large quantities of rock and sediment dropped from the ice during an earlier breakup. Fossils preserved in this mud are types that live in open water. They sank to the bottom and became incorporated in the mud sometime during an interglacial stage of the Pleistocene when the area now occupied by the thick ice sheet was open water. Clearly, this vast ice sheet may not have been a permanent feature of the Pleistocene landscape.

Studies in the Arctic provide additional evidence of polar change. A research vessel was frozen into and drifted with the Arctic ice pack during 1998. Researchers found more open water in the Arctic than at any time in the last two decades, and the ice was about a meter thinner than it had been twenty years earlier. Surface water temperatures are rising, and the depth of fresh water produced by the melting of sea ice was much thicker in 1998 than it was in the 1970s. The combination of melting below the ice and combined melting and sublimation on the surface exceeded the snowfall by about 35 centimeters (14 inches) during the year. The evidence from studies made in 1998 during a year-long study of the Arctic environment by project SHEBA (Surface Heat Budget of the Arctic Ocean) confirms that the ice is thinning. This is changing the salinity of the seawater and could eventually affect oceanic circulation.

The rapid changes taking place in the polar regions and the evidence of global warming are not unique aspects of modern climate. Change is a theme of Earth's history. Examination of that history provides hope of discovering cause and effect in climate change.

EARTH'S ANCIENT CLIMATES

Records of Past Climates

Before records of meteorological data were kept, references to the severity of winters, the length of the growing season, descriptions of unusual weather events such as floods, the location of vineyards, and other information found in historical documents provided a record of what the climate was like. Some climatologists have prepared detailed accounts of the climate of Europe on the basis of such historical accounts.

Chapter 9 provides a brief account of some of the effects of the "Little Ice Age" on northern Europe, Iceland, and Greenland, and chapter 17 cites evidence of the Dust Bowl conditions that affected parts of North America in the early 1900s.

Prehistoric Records

For prehistoric times, other data must be used to infer climate. The width of tree rings provides a record of the rate of growth of trees. Wide annual rings indicate favorable growing conditions; narrow

160 kilometers (100 miles) long by 50 kilometers (30 miles) wide breaking away from the Ronne ice shelf located in the Weddell Sea. In 1999, an even larger piece, 293 kilometers long by 37 kilometers wide (183 miles long by 23 miles wide), about the size of Connecticut, broke off of the Ross ice shelf (figure 13.7 and see figure 20.2). Cracks outlining other large masses are evident. Ice moves from the continental ice sheet out into these ice shelves at rates of up to half a mile each year. Recent breakup of shelf ice appears related to warming of the sea and loss of ice mass in the shelf ice. Increased frequency and violence of storms may also be involved.

The bed of most of the West Antarctic ice sheet is below sea level. If all ice melted, only the eastern part of Antarctica would stand above sea level. As ice accumulated during the late Cenozoic, rings signify low moisture and adverse conditions. Because many plants survive only within a narrow range of environmental conditions, fossils of their remains indicate limits of certain types of climates. The amount of sediment deposited in lakes is an indication of the amount of rainfall and the discharge of streams flowing into the lakes. Geologic evidence even provides insight into climates that existed millions of years ago (figure 13.8). Information about paleoclimate comes from the following sources:

- Fossil assemblages
- Changes in sediment
- Oxygen isotopes
- Trapped atmospheric gases
- Glacial history
- Changes in sea level

For these various types of information to be of value, climatologists must determine the age of the climate indicators. Methods for dating prehistoric evidence include the various radiometric techniques described in chapter 4. Some of these, notably carbon-14, are highly precise, making it possible to determine ages within a few decades. Unfortunately, the short half-life of carbon-14 limits its usefulness to the last 50,000 years, which covers only the most recent parts of the last major glaciation. Less-precise dates for events are determined with the geomagnetic time scale, sedimentation rates, the types of deposits that form under and around ice sheets, and pollen, which vary as the temperature and precipitation change.

The history of glaciation begins with evidence from rocks in Canada that geologists think are of glacial origin. These rocks dated at about 2 billion years include varved lake-bed deposits and boulders that are flattened. They bear the scratches present today on rocks that have been frozen into the base of glaciers and dragged along over bedrock beneath the ice. More extensive glacial deposits are present in rocks of late Precambrian age. These glacial deposits, which range in age from about 800 to 600 million years, are present in the United States, Alaska, Scandinavia, and Africa.

Ice sheets covered Antarctica and large areas of the southern parts of Africa, South America, Australia, and India near the end of the Paleozoic era when these now widely separated areas were part of a single large continent centered near the South Pole (see figures 5.4 and 5.5). Glacial deposits, striations, and a large body of fossil evidence remain from this glaciation. Nearly 250 million years separate the Paleozoic glaciation from the modern Ice Ages.

The climatic shift into the Pleistocene, the modern Ice Ages, appears to have been gradual. Some evidence indicates that temperatures began to drop nearly 55 million years ago in the Eocene epoch. Microfossils indicate that the waters around Antarctica were already cool enough for sea ice to form by the end of the Eocene and that small glaciers formed in Antarctica over 30 million years ago. The Antarctic ice sheet was growing, and mountain glaciers appeared in the Northern Hemisphere nearly 10 million years ago. During the Pliocene, large sheets of ice started to form and expand in the Northern Hemisphere. As Earth entered the Pleistocene about 1.8 million years ago, the temperature of the oceans fell, and ice accumulated to form huge sheets on North America, Scandinavia, and

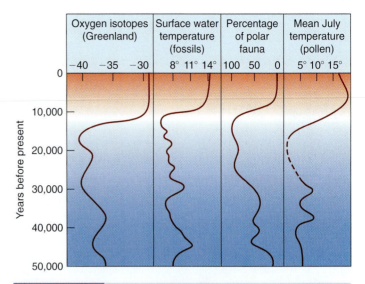

Figure 13.8 Graphs Representing Four Different Methods for Estimating Climate. The oxygen isotopic data was obtained from ice cores taken in Greenland. The surface water temperatures are estimated on the types of fossils present in cores of sediment obtained in the ocean basins. The percentage of fauna that inhabit polar seas is also based on cores obtained in the ocean basins. Mean July temperatures are estimated on the basis of plant pollen found in sediment cores. All four of these methods used to estimate paleotemperatures show the warming trends that started about 15,000 to 18,000 years ago.

Russia. As water from the oceans changed to ice, sea level dropped, reaching levels l00 to 130 meters (330 to 425 feet) below modern levels. During the Pleistocene, climatic conditions fluctuated from cold glacial episodes to warm interglacial periods. Based on studies of the deposits left by glaciers, which are described in chapter 20, four major periods of glacial advance have been recognized in North America.

As more precise methods of studying glaciation by examining the fossil record in marine sediments came into use, it became clear that each major period of glaciation contained many intervals of glacial advance and retreat. The last interglacial period started about 130,000 years ago and ended about 116,000 years ago. The last major glacial advance, known as the Wisconsin glaciation in North America, reached a peak about 35,000 years ago. A rapid increase of about 9°C (16°F) in temperature is recorded in 15,000-year-old ice from Greenland. This rise took place over a few decades and was extraordinarily rapid. It was an early sign of the end of the Wisconsin glaciation. Rapid retreat of the glaciers was in progress by 10,000 years ago. These increases in global temperature and the accompanying rise in sea level mark the start of the Holocene (Recent) epoch of geologic time. The history of advances and retreats of glaciers prompts many to consider the Holocene as an interglacial interval that will likely be followed by another glacial advance. Records of the history of the Pleistocene are preserved in fossils, ice cores, pollen, tree rings, sediments, landforms, and in the fossil record.

Fossils as Paleoclimate Indicators

Fossil assemblages were among the first features in the sedimentary rock record used by geologists to identify ancient climates. Modern plants and animals live within a restricted range of environmental conditions. This use of fossils was a simple step, but one with highly significant implications. Recognition of a group of plant fossils known as the *Glossopteris flora* in late Paleozoic rocks located in India was among the most dramatic discoveries growing out of the use of fossils as climate indicators. Fossils of these plants that ordinarily occurred in sediments deposited at high latitudes in cold climates were discovered near glacial deposits of the same age, but in equatorial latitudes. Ultimately, this discovery became one of the main lines of evidence supporting continental drift and suggesting that parts of India had been glaciated before breaking away from other continents located near the South Pole and drifting to its present location.

Foraminifera are one of the best fossil indicators of climate. These shells are so small that cores of marine sediment may contain hundreds of whole shells. Certain species of foraminifera only live in cold water; others inhabit only warm waters. When the animal dies, its shell sinks to the sea floor where it becomes part of the marine sediment and a general indicator of whether the water was warm or cold at the time it lived. By dating the age of the layers in a sediment core and determining whether the water was warm or cold, it is possible to construct a record of climatic changes that altered the temperature of seawater. Analyses of isotopes of oxygen in shells has made it possible to obtain much more precise information about ancient climates.

Using Oxygen Isotopes to Determine Ancient Temperature

Oxygen trapped in ice and shells provides one of the best indicators of the temperature of the air (in the case of ice) and of seawater (in the case of shells) at the time the oxygen was trapped. Oxygen has three isotopes—^{16}O, ^{17}O, and ^{18}O. Most of the oxygen in seawater is ^{16}O, but a small amount (about 2,000 parts per million) of ^{18}O is also present. The ratio of the heavy oxygen (^{18}O) to ordinary oxygen (^{16}O) in ice or water is a measure of temperature. This ratio changes because the lighter oxygen is easier to evaporate and the heavier oxygen condenses and falls as precipitation slightly more readily than the ordinary oxygen. Most of this rain falls back into the ocean, leaving the air depleted in the heavier oxygen. If atmospheric temperature is sufficiently low, water vapor depleted in ^{18}O condenses as ice or snow and falls to Earth where it may be preserved in glacier ice. As glaciers grow larger, the ratio of heavy oxygen to ordinary oxygen in seawater increases. At the same time, the $^{18}O/^{16}O$ ratio in snow declines. When an animal uses oxygen from seawater to build a shell, the ratio of the two oxygen isotopes in the shell is the same as that in the seawater. Thus, the ratio of these two oxygen isotopes in shells provides evidence of the water temperature and indirectly of air temperature at the time the shells formed.

The ratio of the two oxygen isotopes in snow is the same as the ratio of the two in the air from which the snow crystallizes. The relationship between oxygen-isotope ratio and temperature has been calibrated by comparing these ratios in ice of known age with air-temperature records from Western Europe and Great Britain (figure 13.9). The discovery of this relationship between oxygen isotope ratios and temperature encouraged drilling of several deep cores into ice sheets in Greenland and Antarctica.

Radiometric techniques make it possible to date the age of the ice at various levels in the cores. Gases trapped within bubbles in the ice provide a record of atmospheric composition, and the oxygen-isotope ratios obtained from the ice indicate the approximate temperature of the atmosphere at the time the snow fell. One of the best known of these ice-core records is the Vostok core from central Antarctica. This core provides evidence of atmospheric temperature and composition for the last 150,000 years (figure 13.10). Over this period, temperatures fluctuated by more than 8°C (14°F). Only once before, in all this time, were temperatures as high as they are today. These high temperatures correspond in time to high carbon dioxide and methane content in the atmosphere and to high stands of sea level. This correlation adds to present concerns about the rising levels of carbon dioxide in the atmosphere.

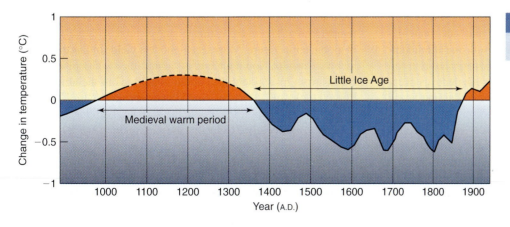

Figure 13.9 Temperature variations over the past thousand years.

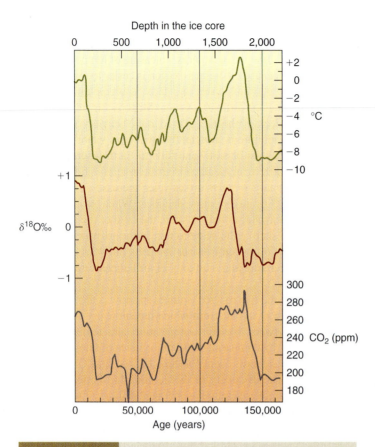

Figure 13.10 Analyses of the Vostok Ice Core. The graph shows how much the carbon dioxide (*bottom*) and oxygen-18 content (*middle*) in the oceans have varied over the past 160,000 years. The top curve shows how much the temperature has varied over the past 160,000 years. Note the close connection between temperature and carbon dioxide in the atmosphere.

Sediment and Sedimentary Rocks as Paleoclimate Indicators

Some types of sediments form only under certain environmental conditions. For example, salt deposits form as a result of high rates of evaporation. Thick salt deposits suggest high rates of evaporation favored by high temperature, low humidity, and drying winds. Peat deposits form under warm, moist conditions, and sand-dune deposits indicate strong winds. Laterite, a type of soil, and aluminum-rich rocks called bauxite (see chapter 15) form where clay-rich soil is deeply leached by warm water. Today, most such deposits occur in the tropics or subtropical regions. Lake beds containing layers that vary in thickness and color with the season form where the lake surface is frozen over part of each year. These types of features made it possible to reconstruct the general climatic conditions that prevailed during each of the periods of geologic time.

More recently, oceanographers have learned how to infer atmospheric circulation patterns from the dust and sand deposited in marine sediment. By studying satellite images, it is possible to delineate the paths followed by modern dust storms. As a dust storm

moves, larger-sized particles settle out and fall to the sea surface. The size of the grains at any distance from the source depends on the strength of the winds. Beyond 1,000 kilometers (620 miles) from the source, the remaining dust is almost uniform in size. The amount of dust moved is a function of the climate of the source region—the more dust, the drier the source region. Figure 13.11 illustrates the general relationship between rainfall and the number of dust storms.

Comparison of observations of modern storm paths, grain size, and thickness of deposits with similar layers of dust and silt in deep-sea cores suggests how arid the continents were and the strength and direction of winds that blew at the time the layers formed. Analysis of the dust that settled forming layers in the cores provides information about the intensity and size of atmospheric circulation systems. These analyses provide a record of dramatic changes that occurred in atmospheric circulation during the Ice Ages. The most impressive of these changes during the Cenozoic took place at the same time continental glaciation started in the Northern Hemisphere. At that time, the amount of dust carried into the northern Atlantic increased by five to tenfold.

To understand how the climate system has changed during the Pleistocene and Recent epochs and how it may change in the future, climatologists are working with computers to model regional and global climates.

MODELING EARTH'S PHYSICAL CLIMATE SYSTEM

Using mathematical models to study Earth's environmental systems began in 1950 when a numerical model for forecasting weather was developed. Since that time, modeling has become a highly sophisticated technique involving the largest of all available computers. The demands on computer memory and calculations are so great that modelers sometimes break problems down into parts that can be run on several computers at one time. Modeling of the climate began in 1975 with the development of a program at the government-operated Geophysical Fluid Dynamics Laboratory. Measurements over Hawaii indicated a rapid increase in the concentration of atmospheric carbon dioxide. Modelers quickly became interested in designing model experiments to estimate the effects doubling carbon dioxide might have on global temperatures. The initial estimate suggested a 3°C (5°F) increase in global temperature. More recent models yield similar though not identical results. Some uncertainty about the results of modeling comes from insufficient data about conditions in the atmosphere. For example, the effects of cloud cover are uncertain. Clouds intercept solar radiation that cools the surface, but clouds also absorb radiation coming from the surface, making the atmosphere warmer.

By 1980, NASA, the Department of Energy, the National Science Foundation, and the National Oceanographic and Atmospheric Administrations had major mathematical-modeling groups at work. By this time, growing concern about the possible consequences of climate change encouraged the government to expand

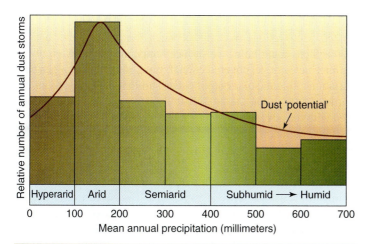

Figure 13.11 The General Relationship Between Rainfall and the Number of Storms Strong Enough to Carry Dust.

funding of this research. To provide controls on results, these four centers used somewhat different approaches to their modeling, and they reached different conclusions. Predictions of the effects of doubling atmospheric carbon dioxide on global temperature ranged from increases of 1° to 5°C (1.8° to 9°F). All models agreed that global temperature would rise, but the differences in magnitude are significant. Many politicians feel that better models are needed if the results are to provide the basis for formulation of policy decisions about energy use and pollution control. Unfortunately, by the time the models are perfected, it may be too late for policy decisions to avert major environmental changes.

Model Construction

Most models are mathematical representations of physical processes. Collecting observations is the first step in making a model. Meteorologists use data about temperature, pressure, cloud cover, and other variables to model the weather on any given day. Before starting to model climate, it is necessary to identify and measure all the variables that are likely to influence the state of the system. Next, it is necessary to define how all of the processes known to affect the system function and to devise a mathematical expression—an algorithm—that defines the operation of each process. Where several hypotheses exist to explain a particular process, alternative expressions are needed, and the model should be run to compare the results of using these alternatives. In all cases, a qualitative concept of how scientists think a process operates lies behind each quantitative aspect of the model, which must take the form of a mathematical expression. In modeling a complex system like the physical climate system, the system must be subdivided into a number of subsystems, each of which may involve numerous processes that operate within that subsystem. The physical climate system, for example, has three major subsystems involving the land, atmosphere, and oceans. The dynamics or circulation of the oceans and atmosphere are especially

important parts of these subsystems. Moisture and energy exchanges between the land and the atmosphere are also of major importance. Biogeochemical cycles affect all of these subsystems. Modeling such a complicated system is a challenging job.

Construction of climate models must take into account the time frame and the size of the area under consideration. Some processes that influence climate operate over long periods of time. For example, the shifting of plates causes the latitude of areas to change over millions of years. Two hundred and eighty million years ago, India and Australia were near the South Pole. At that time, both were partially covered by ice sheets. As they moved northward, the climate changed, as did the vegetation and other features. Both those changes operated on a totally different time scale from the climate changes associated with a major volcanic eruption like that at Mount Pinatubo. In preparing a model, time under consideration must be taken into account. Modelers must also give careful attention to the size of the area covered by the model.

In making models of the entire Earth, modelers must determine how far into the atmosphere to extend the model and how to subdivide the surface. In one model used at the National Center for Atmospheric Research, nine layers stacked to an altitude of 30 kilometers (19 miles) represent the atmosphere, and grid points located 4.5 degrees apart define the surface. At this spacing, grid points are many miles apart. Placing the points closer to obtain better spatial resolution greatly increases the number of calculations required and the need for more closely spaced observational data. To overcome the disadvantages of the more widely spaced grid points, modelers try to find statistical relationships between variables that can be resolved at the grid spacing (e.g., temperature and humidity) and other variables that cannot be resolved at the grid spacing (e.g., clouds). For example, although clouds are much smaller than the grid spacing, a statistical relationship exists among the cloud cover and average temperature and humidity over an area. By determining the temperature and humidity, meteorologists can estimate the cloud cover.

Modelers design tests to provide evidence of the validity of a model for its intended uses. The most convincing test is one in which the model attempts to reproduce results about an aspect of the environment for which relatively complete observational data already exist. Figure 13.12 provides an example of this. In this case, the model program predicts the distribution of pollen from spruce trees in North America at various times during the past 18,000 years. This distribution is well known from samples collected from lake-bed deposits.

Depending on their complexity and the data input, models may predict a great variety of relationships. Modeling conditions from the past is especially informative because it is possible to check the validity of the model with ice cores and marine sediment. In using models to predict the future, the model is no better than the data put into it. In addition, an understanding of how processes operate in the natural system and the degree to which the processes have been correctly translated into mathematical expressions determine how accurate the model will be. Models help improve understanding of the results of climatic changes arising from human activities, as well as those caused by natural processes.

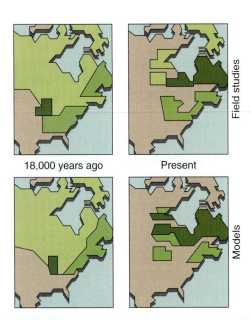

18,000 years ago   Present

Field studies

Models

Figure 13.12   Distribution of Spruce Pollen in North America 18,000 Years Ago and Today. The top drawings shows pollen counts from actual observations. The lower drawings are model simulations. The pollen percentages are indicated by shade. Dark tint indicates high pollen percentages; light tint indicates low pollen percentages.

CAUSES OF LONG-TERM CLIMATE CHANGE

Solar Radiation

The Sun influences Earth's physical climate system through solar radiation and the influx of particles in the solar wind. The amount of solar radiation reaching the troposphere and surface of the Earth is the single most important factor governing Earth's climate. As described in chapter 11, the amount of solar radiation reaching Earth's surface varies with latitude. The amount of radiant energy leaving the Sun also affects the amount that reaches Earth. One of the first astronomers, E. Walter Maunder, observed in the 1600s that the minimum amount of ultraviolet radiation produced by the Sun coincides with minimum sunspot activity, and he suggested that reduced solar output caused the Little Ice Age. Recent studies have shown that few sunspots formed during the Little Ice Age. This may indicate that output of solar radiation was low at that time.

Is the output of solar radiation essentially constant? At the present time, 1 square centimeter placed perpendicular to incoming sunlight outside Earth's atmosphere receives approximately 1.94 calories of heat energy per minute. This value is often described as the *solar constant*. During the 1980s, measurements of solar output from spacecraft confirmed that the solar "constant" is not really constant but that total radiation output varies over the eleven-year sunspot cycle, with increased radiation coinciding with maximum sunspot activity. During times of sunspot activity,

ultraviolet radiation increases, and ozone in the stratosphere is heated. Connections like this between the stratosphere and the troposphere could influence surface temperature. Although observations have not yet covered enough time, some think longer-term variations in total solar output occur and that they may be an important factor in long-term climate changes, such as those associated with the advances and retreats of ice sheets.

Solar radiation and particles reaching the atmosphere affect processes including chemical reactions, production of carbon-14, and formation of ions that take place in the upper atmosphere and the magnetosphere. One study suggested that a decrease in the number of sunspots between 1654 and 1715 caused world temperatures to fall and the size of snow-covered areas to increase. More recently, meteorologists found a strong correlation between the strength of sunspot activity and changes in circulation in the stratosphere. As with variations in solar output, it is too early to evaluate this theory, but in recent years, climatologists have found increasingly convincing evidence that another theory first advanced by Milutin Milankovitch provides insight into long-term climatic change.

Milankovitch Theory of Climatic Change

In 1842, soon after Louis Agassiz recognized the glacial origin of many of the unconsolidated materials at the foot of the Alps, Joseph Adhemar advanced the hypothesis that variations in Earth's orbit cause glaciers to advance and recede in high latitudes and midlatitudes. Eighty years later, the Yugoslav astronomer Milutin Milankovitch explained how slight changes in the position of the Earth relative to the Sun might have the effect of cooling the high latitudes (above 60°). Milankovitch pointed to three components of Earth's motion that affect the amount of solar radiation reaching the surface (figure 13.13).

First, the shape of Earth's orbit around the Sun varies over a period of about 100,000 years. This path changes from an ellipse to nearly a circle, and the distance of Earth from the Sun changes. Because the elliptical orbit brings Earth closer to the Sun for several months, more sunlight reaches Earth when its orbit is elliptical than when it is circular. The second orbital variation is related to the tilt of Earth's axis relative to the plane in which it circles the Sun. The tilt of Earth's axis with respect to a perpendicular to the plane of the orbit changes from 21.5° to 24.5° every 41,000 years. As the angle of tilt changes, the amount of radiation striking polar regions changes. A similar effect results from the third variation caused by wobbling of the Earth's axis. The Earth wobbles like a spinning top. This wobble, called **precession,** repeats every 23,000 years. When the tilt of the axis is greatest, the polar regions receive less solar radiation during winter than they do at other times. As a result, the polar regions become colder during these times.

Milankovitch estimated that the combined effects of the three orbital variations might cause the amount of solar radiation reaching high latitudes to vary by as much as 20%. These changes in orbit cause a systematic variation in the amount of solar radiation reaching Earth and make it possible to predict variations in temperature over time. An especially close correlation exists between the eccentricity of Earth's orbit and temperature variations as indicated by oxygen isotope ratios (figure 13.14). In recent years, comparisons of climatic data of many types with the variations in the amount of solar radiation reaching Earth as predicted by Milankovitch have confirmed the theory. For example, oxygen-isotope-ratio data collected from the shells of foraminifera vary as predicted by the Milankovitch cycle. During the Pleistocene,

the volume of ice on Earth appears to have peaked about every 100,000 years, with surges corresponding to the 23,000- and 41,000-year intervals. In determining climate, the Milankovitch cycle must operate in conjunction with other parts of Earth's systems, especially plate tectonics, which causes land areas to move, and with circulation in the oceans.

SUDDEN AND IRREGULAR CLIMATE CHANGES

As more detailed and precise determinations of past climates have emerged, it has become clear that not all fluctuations in climate match the smooth curves of the Milankovitch cycle. Much of these data have come from careful analysis of the dust content and oxygen-isotope ratios in ice cores taken from the Greenland and Antarctic ice caps (figure 13.15), from cores of marine and lake sediments, and from pollen and tree-ring studies. At least twenty-four abrupt climate shifts have taken place during and following the last major glacial advance. In addition to the irregular short-term variations indicated by these data, the abrupt changes in temperature, which amount to as much as 5° to 10°C (9° to 18°F), are most impressive and worrisome. The data suggest that these changes occurred over a period of only a few years. A change in average temperature of a few degrees centigrade over a period of years or even decades would have devastating consequences. A drop in temperatures of that magnitude would plunge much of the northern parts of North America and Eurasia into extreme cold, and it would change patterns of rainfall, drought, and flooding. Sediment cores reveal that just before each of the last six of these cold snaps, huge numbers of icebergs broke loose from the Arctic and floated south into the Atlantic where the ice melted and dropped sediment trapped in the ice to the sea floor. Rapid breakup of the ice in the Arctic would have the effect of releasing the icebergs. Hugh quantities of fresh water from the melting ice would flood into the North Atlantic where it would float on top of the salt water. The lower density of the fresh water would slow or shut down sinking of water in the North Atlantic and bring the oceanic conveyor belt to a halt, with worldwide climatic impacts related to oceanic circulation.

Eccentricity

The shape of the Earth's orbit varies, from more circular to more elliptical, putting Earth closer to or farther from the Sun.

Cycle duration:
100,000 years

Tilt

Earth's axis of rotation is tilted. The degree of inclination varies above and below 23 degrees.

41,000 years

Precession

Like a spinning top as it slows down, the axis of rotation wobbles, changing the Earth's orientation to the Sun.

19,000 and **23,000** years

Figure 13.13   Motions of the Earth Affecting the Amount of Solar Radiation That Reaches Earth's Surface.

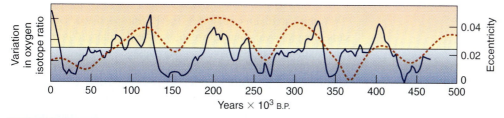

Figure 13.14   Correlation Between the Eccentricity of Earth's Orbit and Temperature Variations. The smooth red line depicts the eccentricity (a measure of how circular the orbit is) of Earth's orbit. The irregular blue curve depicts variations in temperature inferred from studies of the oxygen isotope ratio.

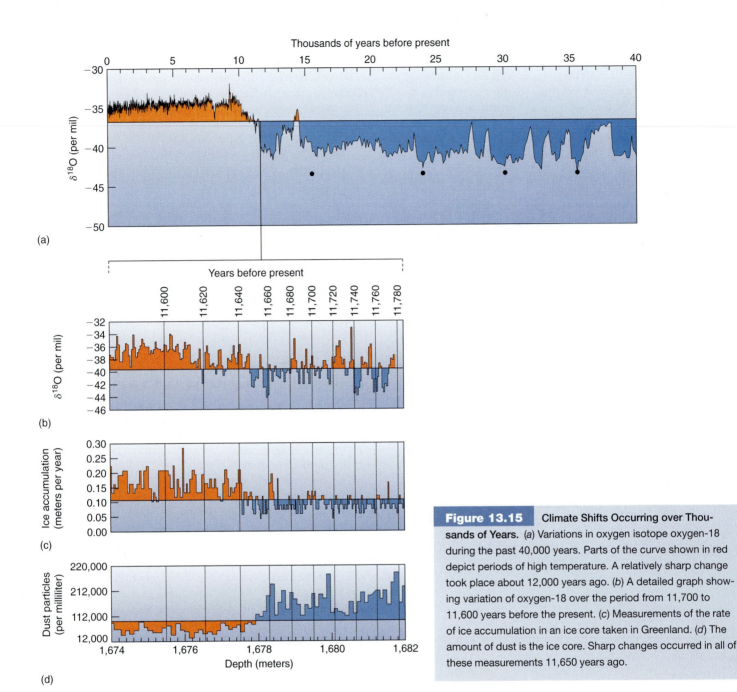

Figure 13.15 Climate Shifts Occurring over Thousands of Years. (*a*) Variations in oxygen isotope oxygen-18 during the past 40,000 years. Parts of the curve shown in red depict periods of high temperature. A relatively sharp change took place about 12,000 years ago. (*b*) A detailed graph showing variation of oxygen-18 over the period from 11,700 to 11,600 years before the present. (*c*) Measurements of the rate of ice accumulation in an ice core taken in Greenland. (*d*) The amount of dust is the ice core. Sharp changes occurred in all of these measurements 11,650 years ago.

THE ROLE OF OCEANIC CIRCULATION

Connections between circulation in the oceans and atmosphere are unmistakable, but the operation of the mechanisms is still uncertain. The most obvious of these connections between El Niño and climate are described in chapter 9. During El Niño, warm surface waters from the western part of the equatorial Pacific move eastward, causing dramatic changes in the climate along the western coast of South America. These changes extend to North America and may affect climate throughout much of Earth. Although El Niño events usually last no more than a year or two, from 1976 to

1986, this circulation pattern in the Pacific persisted. During this same time interval, unusual weather patterns emerged in many far-removed areas. Five severe freezes damaged the citrus groves in Florida during this period. Usually, such freezes occur about once each decade. Unusually high waves repeatedly struck the coast of Southern California. For nearly five years after 1976, the large low-pressure cell over the Gulf of Alaska and the Aleutians became much more intense than usual and shifted eastward. As a result, most of Alaska received more warm air than normal, and the central Pacific became cooler than usual. In this instance, it is obvious that changes in one part of the Earth's climate system caused changes in other parts of the system. However, the exact sequence and cause-and-effect relationships remain uncertain. El Niño events provide evidence of climate changes operating on a scale of a few years to a decade.

Additional evidence of a connection between ocean circulation and relatively short-term climate was discovered in events that took place in the North Atlantic Ocean between 1968 and the early 1980s. During that period oceanographers found a large body of unusually cold water with abnormally low salinity north of Iceland. This water mass moved around the North Atlantic for nearly fifteen years before disappearing in the early 1980s. During these years, sinking of the surface waters in the North Atlantic that form the North Atlantic deep-water mass slowed and may have stopped. About the same time, the quantity of water moving in the Gulf Stream decreased. These changes in the ocean coincided with a cooling trend in northern Europe. It seems likely that the low-salt content of the water caused lower density, which impeded the sinking of the water. In turn, the presence of this mass of cooler water interfered with the northward movement of water at the surface in the Gulf Stream. As less warm water moved in the Gulf Stream to the northeastern Atlantic, air temperature dropped, and the climate of northwestern Europe turned cooler for most of a decade. The coincidence of this slowing down of oceanic circulation with cooling in the North Atlantic region reinforces the idea that deep-water circulation and climate are closely related.

Connections Among Climate, Carbon Dioxide, and Ocean Circulation

Evidence from ice cores collected in Antarctica and Greenland confirms that most of the major climatic changes during the late Pleistocene occurred at the same time in the Northern and Southern Hemispheres. Furthermore, the global drops in temperature coincided in time with pronounced decreases in atmospheric carbon dioxide content (see figure 13.10). The oceans provide the most likely place to store large quantities of carbon dioxide.

When high concentrations of a gas are present in the air above the sea surface, some of that gas tends to move into the water. Wave action facilitates the diffusion of gases across this water-air

interface. Photosynthesis by plants helps regulate the concentrations of carbon dioxide in the water. When plants that have taken in carbon dioxide die and sink to the sea floor, that carbon dioxide becomes part of the sediment. An increase in the number of plants in the upper layers of the ocean may result in a depletion of carbon dioxide in the seawater, which would cause more carbon dioxide to move from the atmosphere into the ocean. The ultimate results could be depletion of atmospheric carbon dioxide, a reduction in heat generated by the greenhouse effect, a lowering of global temperature, and increased glaciation.

Short-Term Oscillations in the Ocean-Atmosphere System

Recent studies are revealing how remarkably close atmospheric and oceanic circulation are connected. The exceptionally strong El Niño of 1998 and the La Niña that followed in 1999 provided a unique opportunity to document these connections. Warming of waters in the Indian Ocean appears to precede the onset of El Niño in the Pacific. The warming of the tropical eastern Pacific at the height of the El Niño altered weather patterns worldwide. As winds that blow from east to west in the tropics decrease, the rise of cold water from depth along the South American coast slows, allowing the waters along the coast to warm. After months of warming, the wind shifts begin to affect the tropical Atlantic, and the ocean there becomes warmer. This alters the weather in Africa. The patterns reverse as La Niña sets in. This type of oscillation affects the high latitudes in the Northern Hemisphere as well as the tropics.

Oscillations in oceanic and atmosphere circulation in the Atlantic and North Pacific recur with a frequency of two to fifteen years. Bands of alternating cool and warm water that cross the Atlantic appear related to changes in the atmospheric pressure between the Azores and Iceland. Variation in the strength of stratospheric winds that circle the Arctic influence circulation and weather at the ground surface in the North Atlantic. Changes in the North Atlantic appear to coincide with changes in the North Pacific, which have been named the Pacific Decadal Oscillation (figure 13.16), marked by eastward shifts of the atmospheric low-pressure

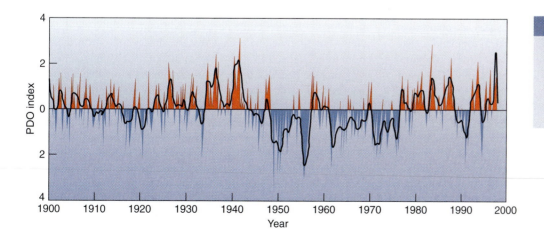

Figure 13.16 The Pacific Decadal Oscillation's Ocean-Temperature Patterns. Since 1900, fluctuations in the temperature of the North Pacific have taken place with an average period of about ten years.

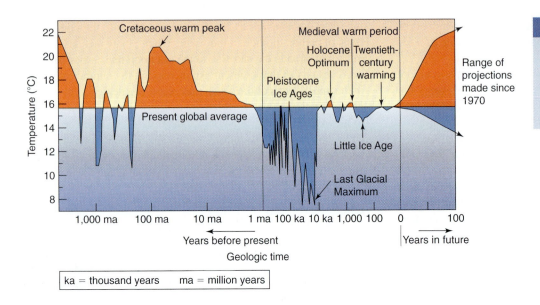

Figure 13.17 The Temperature Record of the Earth over Geologic Time. Note the use of a logarithmic scale and the prominent breaks that occur at the present and at one million years before the present.

center off the Aleutian Islands. These changes have coincided with El Niños.

PREDICTING FUTURE CLIMATE

Everyone who has studied past climates agrees that climatic conditions have changed dramatically over time (figure 13.17). Some of those changes appear to have been slow, incremental variations that have persisted for millions of years. Within the period for which more data are available, and especially the last few thousand years, significant global changes in temperature have taken place within decades. The record of variation in global temperature for the past thousand years (see figure 13.6), assembled by the Intergovernmental Panel on Climate Change, depicts a gradual drop in temperature averages until the last century when temperature increased about 0.5°C (0.9°F). During the first 900 years of the millennium, the variations are thought to represent natural conditions caused by such things as variations in solar radiation, volcanism, and the Milankovitch cycles. The last century shows rises in temperature that are far above natural variability. Many climatologists conclude that effects of increased greenhouse gases and changes in land use, notably the conversion of forestland to cropland, cause these rises. Since the industrial revolution, carbon dioxide, methane, nitrous oxide, chlorofluorocarbons, and CFC-12, all greenhouse gases, have increased in the atmosphere. Projections of a doubling of the carbon dioxide content of the atmosphere suggest even higher temperatures in the range of 1° to 3.5°C (1.8° to 6.3°F) by the year 2100. Changes of these amounts would alter the mix of

plant species in land ecosystems and alter the geographic distribution of major vegetation types, changing regions of savannas, forests, and tundra. These increases in temperature would affect the biogeochemical cycling of carbon, nitrogen, phosphorus, and sulfur in Earth systems and would change the types of trees growing in much of the world's forestland. Unfortunately, modeling of the effects of such temperature increases do not agree on how the changes would affect specific regions.

Studies are still in the early stages of deciphering the causes of climatic change with enough precision to make reliable long-term predictions, but it seems clear that long-term climate for the whole Earth depends on the amount of solar radiation reaching the surface as predicted by Milankovitch. According to the Milankovitch curve (see figure 13.14), the present is close to the predicted peak of an interglacial period. If world temperatures follow the pattern repeated so many times during the Pleistocene, world temperature averages will decline, reaching a minimum level in about 60,000 years. Examination of the rapid and extreme fluctuations in air temperatures indicated by ice cores from the Greenland ice cap (see figure 13.15) or the decadal oscillations indicated from ocean temperature over the last hundred years (see figure 13.16) suggests that drastic changes could occur in the future. From these data, it is clear that sharp climate changes are superimposed on the long-term trends related to solar insolation. An increasing body of data supports the idea that these short-term changes involve interaction between the atmosphere and oceans. Any substantial change in the temperature or circulation of one of these fluids affects the other. For this reason, special attention to recent changes in the greenhouse-gas content of the atmosphere is warranted.

SUMMARY POINTS

1. Long-term averages of weather conditions, referred to as climate, vary locally and on regional scales. Climatic variability is so great that it is difficult to identify trends and to determine the cause of variations. Despite this difficulty, most observers find that evidence global temperatures are rising is convincing. Recent years are among the warmest in recorded history.

2. Numerous natural processes connect climate to the biosphere and the surface of the land, oceans, and ice sheets. Major changes in any of these have direct effects on the others.

3. The mean temperature of Earth's atmosphere is determined primarily by compilation of temperature measurements from all areas of the surface. Changes in temperature of the surface layers in the oceans also provide an indication of temperature. Because neither atmosphere nor ocean is static and because both experience seasonal changes, it is difficult to measure the average temperature of the entire Earth's surface. However, most evidence indicates that the global mean temperature has been rising for the past century.

4. Climatic changes have a number of causes. The addition of carbon dioxide and other greenhouse gases to the atmosphere, shifts in the position and size of oceans and land areas, and variations in the motions of Earth contribute to changing climate. Change in the amount of solar radiation reaching Earth may also be a factor; however, it is not established that such changes have persisted in Earth history.

5. Regional climates are classified based mainly on mean values, ranges, and seasonal variations in temperature and precipitation.

6. Evidence of major climatic shifts abounds in the rock record, but "ice ages" are rare. Plants and fossils of marine invertebrate animals provide excellent indications of climate changes. The present Ice Age began about 1.5 to 2 million years ago. During this time, ice sheets have repeatedly advanced and retreated over large areas of the Northern Hemisphere.

7. Oxygen isotopes provide an independent indication of temperature. The ratio of heavy to normal oxygen isotopes in ice and shells of organisms is related to the temperature of the air and water at the time the ice and shells formed.

8. Some sedimentary rocks also provide an indication of past climate. Salt and gypsum deposits indicate high rates of evaporation. Dust deposits indicate wind direction and speed.

9. The quantity of data involved in defining the climate and the complexity of the relationships among the variables that determine climate make computer modeling a promising method for studying climate. Once models are designed to reproduce present climatic conditions, variables are modified to examine how changes in the variables might affect future climate.

10. Many of the changes observed in Earth climate over the last million years appear to be closely related to variations in the motions of Earth. These variations, first described by Milankovitch, involve changes in the shape of the orbit of Earth around the Sun, changes in the tilt of the axis, and a wobble of the axis.

11. Through studies of El Niño events, climatologists have discovered close connections between the circulation of the oceans and climate. Studies of ice cores reveal a strong correlation between atmospheric carbon dioxide levels and global temperature. This discovery increases concern about human additions of carbon dioxide to the atmosphere.

REVIEW QUESTIONS

1. How do volcanic eruptions affect climate?
2. How is the heat reaching Earth (from incoming solar radiation) in equatorial latitudes distributed to other parts of Earth?
3. On what bases are climates classified?
4. What types of evidence provide clues to past climates?
5. Explain how oxygen isotopes are used to define ancient temperatures?
6. What types of problems do climatologists encounter that limit the use of computers to predict future climate?
7. What variations in Earth motions are considered in the Milankovitch theory?
8. What connection exists between past atmospheric carbon dioxide concentrations and temperatures?
9. How is climate related to oceanic circulation?
10. How has global climate changed during the Pleistocene epoch?
11. How does geologic evidence link climate with carbon dioxide content of the atmosphere?
12. Why is it difficult to determine changes in global mean temperature?
13. What methods are used to determine changes in global mean temperature?
14. How do sedimentary rocks contain a record of past climates? Which rock types are the best climatic indicators?
15. Explain how changes in the deep circulation of the ocean can affect climate.

THINKING CRITICALLY

1. What would be the effect of a mean annual temperature increase of several degrees Celsius on the area in which you live or where you are in school?
2. What steps could the government take to regulate global temperature? Would action by the government of a single country be sufficient?
3. What knowledge is needed to develop policies that could regulate global temperature?

KEY WORDS

aerosols	global climate
climate	precession

Introduction to Earth's Land Environments

Rainfall, mountains, glaciated valleys, streams, and ground water create conditions favorable for the development of a temperate rain forest in New Zealand.

Chapter Guide

This chapter has two primary objectives. It emphasizes the close connections among ecology, climate, and geology, and it provides an introduction to the character of the land surface and the agents that cause changes in the landscape. Subsequent chapters examine the operation of these agents in more detail.

The physical climate, biological, and plate tectonic systems interact, causing the environmental niches on Earth's surface. Processes in the plate tectonic system produce mountain belts, volcanoes, and rock structure. The physical climate system processes shape these features and produce the land surface. These processes include weathering, formation of soil, and changes at the surface caused by erosion, transportation, and deposition of materials. Mass wasting, streams, glaciers, wind, and ground water on land and waves and currents in the coastal environment shape the land surface. The plate tectonic and physical climate systems provide a framework within which organisms live and modify the environment. These three systems combined define Earth environments. If one system changes an environmental niche, its inhabitants must also change.

INTRODUCTION

Today's landscape is unique and continually changing. It is far different from what it was in the past and will be in the future. Because most changes occur so slowly, it is difficult to imagine the extent of the changes that led to the present state or the changes that lie ahead (figure 14.1). The geologic record provides evidence of the magnitude of long-term changes. Four hundred million years ago, a shallow sea covered most of the eastern and central parts of the United States. It was a time before the rise of any of the majestic mountains that are now such distinctive parts of the continents.

During the peak of the last interglacial, about 120,000 years ago, ice disappeared from the north polar region and retreated from much of Antarctica. Plants and animals that now inhabit tem-

perate regions lived near the poles. Sea level rose and covered the sites that are now large cities along the coasts. As the climate grew cold again, ice sheets developed and advanced southward. The peak of this last glaciation occurred about 18,000 years ago, and at that time, ice covered most of Canada and extended into the northern United States to the present Missouri and Ohio rivers. Cool climates shifted toward the equator as the environments between the polar regions and the equator were compressed into a narrow belt. Sea level dropped until the peninsula of Florida was nearly twice as wide as it is today. The site of New Orleans, Louisiana, was inland from the sea, and 100-meter- (328-foot-) high bluffs lined the edges of the lower Mississippi River valley.

The processes responsible for past changes, such as those previously described, continue to modify the environments inhabited by

(a)

(b)

(c)

(d)

Figure 14.1 Terrestrial Environments. (*a*) The Beartooth Plateau has a high mountain environment. (*b*) The coast of Maine is largely rocky and dominated by wave action. (*c*) Swamp environment of the Dismal Swamp in North Carolina. (*d*) Bare rock or rock with a thin cover of stream deposits characterize many deserts.

Table 14.1	Earth Environments Defined Mainly Based on Landforms and Geologic Characteristics

I. Terrestrial Environments

(Note: Streams, lakes, swamps, caves, and other localized features exist within much larger crustal divisions, described in chapter 3.)

Continental cratons

(portions of the crust that have been stable for hundreds of millions of years)

    Continental shields

    High plains

    Interior lowlands

Mountains

Coastal plains

Ice sheets and polar regions

Hot deserts

II. Transitional Environments

Beach and barrier islands

Deltas

Estuaries

Tidal flats

Marine swamps

III. Marine Environments

Continental shelves

Continental slopes and rises

Oceanic ridges

Abyssal plains

Deep-sea trenches

Ecologists, soil scientists, climatologists, geologists, and geographers are all involved in describing, classifying, and analyzing Earth's surface. Each group has a different background, and each focuses on different aspects of the environment. Consequently, each group uses different criteria to distinguish and classify environments. Nevertheless, the basic factors that determine the characteristics of each niche are closely related. Efforts to recognize the combined effects of climate, soil, biogeochemical processes, and geology lead to identification of ecosystems and biomes.

Ecosystems and Biomes

Ecology is the study of the relationships between organisms and their environment. Ecologists study communities of flora and fauna, patterns of life, natural cycles, and the relationships of organisms to each other. Broadly defined, an **ecosystem** consists of the environment and the living organisms that exist in and affect it. **Biomes** are the largest subdivision of terrestrial ecosystems (figure 14.2 and see figure 8.19). Biomes include the community of plants and animals that characterize a particular large natural region such as tropical rain forests. Many ecosystems, including streams, soils, and trees, are present within such a biome. Change in one component of an ecosystem causes changes in other components and in the operation of the system as a whole. For this reason, many people are concerned about changes caused by human activities, such as cutting down rain forests, addition of greenhouse gases to the atmosphere, and the destruction of wetlands.

Ecosystems may be as small as a tidal pool or as large as the Antarctic. The larger units can always be subdivided into many smaller units. Mountain lakes, rocky seashores, or streams that exist in specific physical environments and contain certain plants and animals that interact with one another are types of ecosystems. Rarely are the boundaries between ecosystems sharp. Usually they overlap or are transitional. Such a transition is present around the edge of a marine swamp or on a mountain where the vegetation gradually changes with elevation. These transitions make defining the geographic boundaries of ecosystems difficult.

Although geologists have not developed a formal classification of environments comparable to the biome model illustrated in figure 14.2 and table 14.2, close connections exist between biomes and geologic provinces, based on landforms and bedrock. Just as ecologists identify many localized ecosystems within each biome, geologists recognize localized environments, that is, those in various types of lakes, streams, swamps, and caves that lie within larger divisions characterized by the shape of the land and referred to as **physiographic provinces** (figure 14.3). Many environments defined according to geologic criteria differ in both geology and ecology. For example, the ecologic and geologic characters of lakes in glacial valleys, deserts, and volcanic craters are quite different from one another.

Climate

Climate is a critical factor in determining the geography of the biomes identified in table 14.3. Variation in the amount of solar

plants and animals today (table 14.1). Internal sources of energy are responsible for mountain building, volcanic activity, and for most of the large-scale features, such as oceanic ridges on the sea floor. These elevated portions of the crust rise despite the ever-present pull of gravity. In response to changes in the weight of different parts of the lithosphere, isostatic adjustments (described in chapter 3) cause large areas to rise and fall. Gravity holds the atmosphere against the oceans and solid Earth, and gravity operates everywhere on the surface. As a result, streams, glaciers, ground water, and anything else that is loose at the surface tend to move downslope. Most of the remaining energy that drives surficial processes comes from the Sun. Solar energy drives the physical climate system, including the hydrologic cycle and most biological processes. Climate, soil, the structure and composition of the crust, and plant populations provide the primary basis for defining terrestrial environments.

DEFINING EARTH'S LAND ENVIRONMENTS

On Earth, the term **environment** usually refers to the conditions and influences that affect life and living organisms. Environment is also used in a more general way to describe the physical, chemical, and biological conditions that exist at a place or over a region.

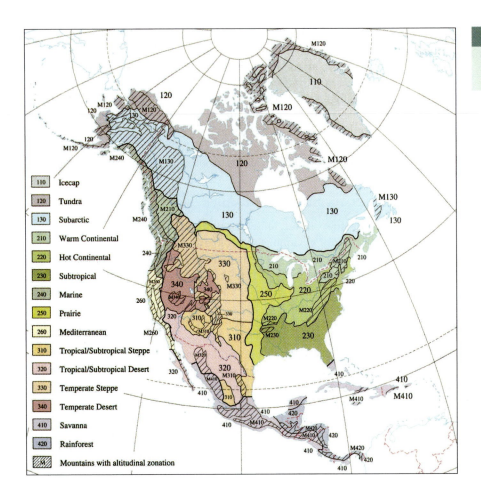

110	Icecap
120	Tundra
130	Subarctic
210	Warm Continental
220	Hot Continental
230	Subtropical
240	Marine
250	Prairie
260	Mediterranean
310	Tropical/Subtropical Steppe
320	Tropical/Subtropical Desert
330	Temperate Steppe
340	Temperate Desert
410	Savanna
420	Rainforest
M	Mountains with altitudinal zonation

radiation reaching Earth's surface is one of the primary causes of the differences between the tropics and polar regions. Climate classifications are based on long-term averages of temperature and precipitation (see figure 13.1). In making finer distinctions among climates, topography, elevation, and other characteristics that are peculiar to a particular locality come into play. For example, the local climate, also called **microclimate,** on south-facing slopes, which receive more hours of sunlight, differs from that on north-facing or shaded slopes. This difference influences the types of trees and other plants that will grow. Localized climatic conditions are common. For example, large bodies of water moderate the climate along coasts and beside lakes. In urban areas, buildings and pavement hold heat. The excess heat and reduced vegetation cause localized climates of cities to differ from those of nearby rural areas. Differences in vegetative cover, such as those between a pine forest and cultivated land, also affect these climates. Thus, microclimates reflect influences of topography, soil, vegetative cover, and the average amount of solar radiation and precipitation reaching the ground at a particular latitude.

Cratons, mountain systems, and coastal plains cover such large areas that they frequently extend across climatic zones. In these cases, climate has more influence than the geologic setting has on the environment. Nevertheless, each physiographic province contains certain characteristic ground-water conditions, streams,

Table 14.2 Terrestrial Biomes Identified by the International Geosphere-Biosphere Programme (IGBP)

Ecologists used soil properties and climate coupled with plant physiology and dominance to develop the global model of biomes shown in figure 14.2. The subdivisions in this model are natural geographic regions characterized by particular plant communities such as those found in tropical rain forests or savannas. Some of these biomes and their subdivisions show a close correlation with physiographic divisions.

Tropical

Tropical dry forest and savanna	Tropical rain forest
Tropical seasonal forest	

Midlatitude

Shrub	Warm evergreen and mixed forest
Hot desert	
Warm grass and shrub	Temperature deciduous forest

High Latitude and Mountain

Cool mixed forest	Cold deciduous forest
Cold mixed forest	Tundra
Cool conifer forest	Semidesert
Cool grass and shrub	Ice and polar desert

Table 14.3	Major Climates and Biomes That Occupy the Same Geographic Locations

An asterisk indicates the closest correspondence between these biomes and climates. (See chapter 13 for more information about climates.)

Climatic Classification (by Köppen)	IGBP Biome Classification
A. Tropical moist climates	Tropical rain forest* and tropical seasonal forest*
B. Dry climates	Hot desert,* tropical dry forests, and savanna*
	Warm grass and shrub
C. Moist midlatitude climates	Warm evergreen mixed forest* with mild winters and temperate deciduous forest*
	Cold deciduous forest,* cool conifer forest with severe winters, taiga (northern conifer forest),* and mixed forest
D. Polar climates	Ice,* polar desert,* semidesert,* and tundra*

IGBP = International Geosphere-Biosphere Programme

topographic relief, bedrock structure and composition, and soils. All of these are critical components of the environment.

Geologic Influences

The composition and structure of bedrock and **geomorphic agents,** (see page 309), such as mass wasting, streams, glaciers, the wind, waves, ocean currents, and ground water, form a framework within which climate and organisms shape natural terrestrial environments. Differences in the geologic framework are responsible for the contrast between many ecosystems. For example, central Canada, which is underlain by granitic rocks, differs from the interior lowlands of Ohio, Indiana, and Kentucky where the bedrock is limestone. Areas where glaciers actively eroded the landscape have different environments than those shaped by streams or the wind.

As they shape terrestrial environments, geologic processes, climate, and organisms interact. For example, elevation of the land influences local climatic conditions. At the bottom of Mount Mitchell, the highest peak in the Appalachian Mountains, vegetation is typical of the piedmont region of North Carolina. Close to the top of the mountain, vegetation more typical of Canada, such as spruce trees, cover the mountain. Climate, the slope of the land, and composition of the bedrock determine what type of soil develops. The composition of the soil and climate determine which plants will grow. The vegetation determines which animals will populate the area, and the plants and animals contribute to the breakdown and decomposition of the bedrock.

Soil

Soils are products of the interaction of the geologic, climatic, and biological aspects of the environment. Soils directly affect the development of ecosystems, including those favorable for human habitation. From the beginning of agricultural societies, soil and soil conditions have greatly impacted on human activities and have determined where the food needs of large human populations can be met.

Biogeochemical Processes

Water, oxygen, carbon, nitrogen, and trace amounts of other elements are essential for life processes. Their availability in any particular ecosystem determines which plants and animals prosper and which ones barely survive. In terrestrial environments, plants derive these elements from the soil and the atmosphere. In the oceans, plants obtain these essential materials from seawater. Animals obtain oxygen from the atmosphere, but they obtain most essential elements by consuming plants or other animals.

In the oceans, many of the essential elements needed by marine organisms reach the ocean only after being freed from rocks on land and carried into the ocean by streams. Others reach the marine environment by moving directly across the interface between seawater and the atmosphere. Organisms obtain these elements directly from the seawater. On land, essential elements and compounds move between plants and animals and the atmosphere or soil. An example is the movement of carbon dioxide and oxygen between plants and the atmosphere. This close connection is illustrated in figure 14.4, which shows changes that take place in the areas covered by vegetation each year as the seasons change. Alterations in the quantity of vegetation coincide with fluctuations in the amount of carbon dioxide in the atmosphere over these areas. During the Northern Hemisphere's summer, the quantity of carbon dioxide in the atmosphere falls because plants are abundant in the north and they absorb carbon dioxide. During winter, plants die off, and atmospheric carbon dioxide rises as plant matter decays. Because survival of plants is so closely related to the environment in which they live, ecologists use plants to identify biomes (see figure 14.2).

THE GEOLOGIC FRAMEWORK OF LAND ENVIRONMENTS

Through processes involving erosion, transportation, and deposition of materials, each of the geomorphic agents, such as glaciers or streams, causes change in terrestrial and transitional environments. Where these agents operate for a long time, certain characteristic landforms evolve. Streams produce gullies and V-shaped valleys as they erode the land surface and produce sediment that is later deposited on flood plains and deltas. Glacial erosion in high mountains produces sharp peaks and causes valleys to assume U-shaped profiles. Generally, several of these geomorphic agents operate in every terrestrial and transitional environment. Downslope movement of material, called **mass wasting,** is present everywhere but is much more effective on steep slopes. Climate alone is responsible for two extreme environments, hot and cold deserts. Extremely low amounts of rain fall in both of these environments. Glaciers and ice sheets are common features of cold deserts, and heat, dry air, wind, and occasional rainstorms shape hot deserts. These extreme

Coastal Mountains

Cascade Mountains

Columbia Plateau

Northern Rocky Mountains

Middle Rocky Mountains

Black Hills

Great Plains

Basin and Range

Sierra Nevada

Colorado Plateau

Southern Rocky Mountains

Great Plains

Figure 14.3 **Physiographic Diagram of the Contiguous United States.** Compare this figure with the maps of biomes (figure 14.2), climate (figure 13.3), areas of subsidence caused by withdrawal of ground water (figure 16.15), and the map of ground-water provinces (figure 19.22).

Canadian Shield

Northern Appalachian Mountains

Interior Lowlands

Appalachian Mountains

Ozark
Plateau

Ouachita Mountains

Atlantic Coastal Plain

Gulf Coastal Plain

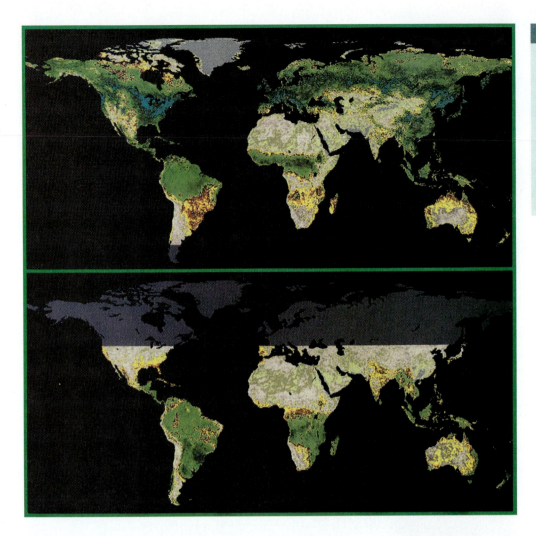

Figure 14.4 Satellite Photographs Enhanced with Color Illustrate the Vegetation Cover. The top image depicts summer conditions in the Northern Hemisphere. Winter conditions in the Southern Hemisphere are shown in both images. The darker green areas have the most lush vegetation. Areas with sparse vegetation are yellow and red. Areas with low vegetation are tan.

environments occur in a great variety of crustal divisions, including cratons, mountains, and coastal plains (table 14.4). The close connection between the shape of the land surface and the geomorphic agents makes landforms useful indicators of geologic and climatic influences on the environment.

The Continually Changing Landscape

Some changes in the landscape are rapid and dramatic. In 1964, a new volcano named Surtsey appeared off the coast of Iceland. In 1959, a huge landslide blocked the Madison River Canyon in Montana. More often, however, the changes are so slow that they are almost imperceptible. Comparison of recent photographs with those taken of the same place a hundred years ago may reveal these changes. The changes often appear negligible, but analysis of maps and landscape drawings made many years earlier gives a clear impression of the retreat of glaciers and changes that have taken place in the course of rivers. Satellite images document relatively rapid changes that occur during floods and hurricanes. Factors that influence the changes and the way they progress over time include

- weathering and erosion
- base level (elevation of the area above sea level)

Table 14.4 Major Divisions of Continents Based on Physiography and Bedrock Structure

North American examples are cited to the right.

Continental Division	North American Examples
Cratons	
Shields	Canadian Shield
Plains and steppes	Great Plains
	Interior Lowlands
Mountain systems	Appalachian and Ouachita mountains
	Cordilleran Mountain system
Coastal plains	Atlantic and Gulf coastal plains

- tectonic setting
- composition and structure of the bedrock
- geomorphic agents operating in the area
- length of time the geomorphic agents have been operating in the area

Weathering and Erosion Weathering processes cause rocks exposed at the surface to disintegrate and decompose. Erosion, which includes all processes that pick up and move materials from one place to another, starts with gravity pulling these weakened materials downslope. Other agents, especially streams, move the weathering products to lower elevations. The combined effects of removal of materials at the surface, called **degradation** (figure 14.5*a*), and the accumulation of sedimentary deposits, called **aggradation,** are referred to collectively as **gradation** (figure 14.5). As a result of gradation, soil and rocks removed from one place settle and build up in another place. The long-term effect of geomorphic agents is to lower the level of the land and move the material to the sea. For example, most of the sediment that weathering and erosion produce in Montana moves downslope into the headwaters of the Missouri River. The river picks up more sediment from tributaries along the way. Eventually, this load moves downstream into the Mississippi River, and ultimately, the solid portions settle and build the delta where the Mississippi enters the Gulf of Mexico. In a similar way, glaciers pick up material from high mountains, move it to lower elevations, and deposit it where the ice melts. As with streams, glaciers remove material from higher to lower elevations.

Base Levels John Wesley Powell, the one-armed Civil War veteran who made the first trip by boat through the Grand Canyon, recognized that rivers cannot reduce the surface of the land below the sea. Powell was impressed by how abruptly the drop in elevation of the Colorado River changes where it flows out of the Grand Canyon. In the narrow canyon, water moves through rapids with high velocity. As the river emerges from the canyon, the slope of the stream decreases, rapids disappear, and the river flows across a region of low relief that lies near sea level. Powell recognized that the river had cut the canyon, and he reasoned that it could not cut the canyon deeper as it approached sea level because streams have lower velocity and lose their power to erode as the slope of the stream decreases. Powell referred to sea level as the **ultimate base level.**

Lakes and enclosed or landlocked depressions may serve as temporary base levels. For example, all the streams around the Great Salt Lake in Utah flow into the lake. No streams flow out of that basin into the ocean. Geomorphic agents cannot lower the level of the area that drains into the lake as long as the lake remains. Eventually temporary base levels disappear, and the sea becomes the level below which streams, wind, and glaciers cannot reduce the land surface. If all the dynamic processes that elevate the Earth's surface ceased, the land surface would eventually be eroded to a nearly flat plane close to sea level. Weathering is the first step in the lowering of the land surface.

Continued over millions of years, degradation changes high mountains into rolling hills. Eventually, the Himalayan Mountains, with peaks close to 10 kilometers (6 miles), may resemble the modern Appalachian Mountains, which has few peaks over 1.2 kilometers (0.75 mile) high. The quantity of sediment removed from an area by major streams makes it possible to estimate the rate at which erosion reduces the elevation of an area. Using the rate at which sediment is removed from continents, it is possible to calculate how long it would take for erosion to remove every bit of soil, sediment, and rock above sea level. The answer is 20 to 40 million years. Enough time has elapsed over Earth's 4.6-billion-year his-

(a)

(b)

Figure 14.5 Effects of Gradation. (*a*) Severely eroded landscape. The bedrock here in Spain is an easily eroded silt and claystone. Running water is responsible for the formation of this severely eroded, "badlands," type of topography. (*b*) The amount of sediment in the stream channel is more than the stream can move. Sediments eroded from the Rocky Mountains and transported by streams cover large areas in the Great Plains of the United States.

tory for processes of erosion to lower its surface to the level of the oceans many times over. Other processes have more than counterbalanced the destructive tendency of degradation on Earth's surface, however. Today, the average surface elevation of the land is about 600 meters (1,800 feet) above sea level.

Tectonic Setting The land surface retains its elevation because of constructive processes such as igneous activity, mountain building, and isostatic uplift of the crust, which were discussed in chapters 3, 5, and 7. The edges of colliding plates thicken as they press against one another. The pressure deforms the margins of the continents, folding sedimentary layers and stacking sections of the continental margins where thrust faults cut through the rocks. Ultimately, uplift, folding, and faulting produce a mountain system such as the Himalayas, where vast areas rise thousands of meters above sea level.

Once part of the continental crust increases in thickness, it tends to stay elevated even though degradation and erosion occur rapidly on uplifted mountainous portions. The level of the land surface at any given time represents a balance between the forces of upheaval in the crust—volcanism, igneous activity, tectonism, isostasy—and the forces of erosion and degradation. The dynamic processes acting inside Earth counteract surficial processes that cause erosion and tend to lower the surface.

The landscape in stable regions evolves more slowly than the landscape in areas of active volcanism or mountain building. Volcanoes may develop rapidly and change shape during a single eruption. Where faulting is active, furrows and escarpments in the landscape may mark the location of displacements in the crust. Cliffs and furrows in the landscape mark portions of the San Andreas fault in California. Such displacements disrupt the drainage and offset the course of streams (figure 14.6).

Figure 14.7 Differential Erosion. Rocks that are resistant to weathering and erosion form cliffs. Those that are easily eroded form lower slopes. The ridges that are prominent on this vertical image are resistant.

Figure 14.6 Aerial View of the San Andreas Fault, an Active Fault in California. Note that displacement on the fault has offset the stream that flows across it.

Composition and Structure of the Bedrock Because some rocks weather and erode much more rapidly than others, rocks such as granite and quartz-rich sedimentary rocks that resist weathering and erosion stand higher in the topography where moisture is abundant than do less-resistant rocks such as limestone or shale. In arid climates, limestones may form ridges. Such differences in resistance produce **differential erosion** (figure 14.7). Over time, weathering and erosion remove the less-resistant materials, leaving resistant materials as more-prominent features in the landscape. As weathering and erosion etch out weak materials, the structure of the rock becomes more apparent.

At the time of deposition, most sedimentary rocks are more or less horizontal. Crustal deformation may modify the structure of these layers, causing them to fracture, fold, or break and slip along faults. Eventually, the weak parts of the rock are eroded away, leaving an imprint of the shape of structural features, such as folds, fractures, and faults (figure 14.8). The effects of structure and composition are especially prominent in the southwestern states where horizontal strata are common and in the Appalachian and Canadian Rocky mountains. Long valleys and ridges have been etched out of folded strata of different resistances to weathering and erosion in the Appalachian and Canadian Rocky mountains.

Geomorphic Agents

Climate affects weathering processes and determines which geomorphic agents have the most influence. Warm temperatures and high moisture increase the effectiveness of chemical-weathering processes. Temperature also determines whether precipitation falls in the form of rain or snow. Thus, climate affects the type of soil that forms, soil moisture, chemical weathering, and the vegetation. Climate also determines whether glaciers, streams, or wind prevail as significant geomorphic agents. Glaciers exist only in high mountains or polar regions. Effects of streams and ground water are more obvious in humid climates than in arid regions, but streams affect most areas at least part of the time. Wind is most effective where it is strong and persistent and where vegetation is scarce. These conditions prevail in deserts and to a lesser extent along shorelines. Over long periods, it is the relative magnitude of the impacts of the geomorphic agents that produces landscape

Figure 14.8 Weathering and Erosion in the Atlas Mountains. Differences in the resistance of rocks to erosion and the structure of the bedrock are responsible for the landscape in this part of the Atlas Mountains, Morocco. The ridges are composed of a layer of rock that is resistant to erosion in the arid climate. Folding of that layer is responsible for the sinuous pattern.

characteristics. Climatic conditions determine which geomorphic agents are most active and which plants and animals are most likely to inhabit the environment.

USING CLIMATE AND PLANTS TO DEFINE ENVIRONMENTS

Climate, plant and animal populations, and geologic factors that produce the landscape provide a good basis for defining land environments. Many of the biomes recognized by the International Geosphere-Biosphere Programme (see figure 14.2)—polar regions, boreal forests, midlatitude forests, savannas, tropical forests, high mountains, and hot deserts—correspond closely to the major climatic divisions defined in chapter 13 (see table 13.1 and figure 13.2). The match is not perfect and boundaries between both climatic zones and biomes are transitional. Combinations of climates, biomes, and physiographic provinces produce extraordinarily large numbers of diverse environmental niches. Nevertheless, a number of distinctive and well-defined environments stand out, and these cover large areas of the land surface. The following sections examine each of these with emphasis on North America.

Polar Regions

Ice covers large parts of the polar regions—Antarctica, Greenland, Iceland, the Arctic Islands, and adjacent areas (see figures 13.1b, 20.2, and 20.3). The marginal areas have no ice part of the year. Most of these areas are **tundra**—the name given to the treeless plains of the Arctic. Tundra covers vast stretches in the northern parts of Canada, Alaska, Europe, and Asia. Long, cold winters and short, cool summers characterize these areas. Plants in this region have short growing seasons. As a result of warming in recent years, these plants are blooming earlier than usual.

Permanently frozen ground, called **permafrost,** makes this harsh environment inaccessible to plants that have deep roots. Of plants, only algae and bacteria manage to survive on the ice. Lichens, grasses, sedges, and some small willows do grow in the soil in regions with permafrost, cold temperatures, and limited sunlight. Animal life fares much better in the tundra than on ice, and insects are especially abundant. The huge numbers of mosquitoes and black flies impress even the short-term visitor. Large herds of caribou roam the tundra, migrating as the seasons change. Reindeer, musk oxen, bears, foxes, wolves, and arctic hares accompany the caribou on their migration. During the summer months, some species of migratory birds nest in the tundra. Farther south, near the margin of the permanently frozen ground, the climate moderates, and a much more diverse and abundant fauna survives the harsh winter. Penguins in Antarctica, polar bears in the Arctic, and some birds spend part of their time on the ice and inhabit the rocky areas near the ice during the warm summer months. Most of the birds migrate farther south during the winter, and many of the animals hibernate during the coldest part of the winter.

Figure 14.9 A Taiga Forest. This taiga forest, composed mainly of spruce and fir trees, is located in the mountains north of Yellowstone Park. Similar forests cover vast areas farther north in Canada.

Boreal (High-Latitude) Forests

South of the tundra, winters are also extremely cold, but they are short compared with arctic winters. Summers vary in length. Some are warm; others are cool and wet. When the ice sheets retreated from these areas, they left a low-lying, poorly drained rocky surface with thin soil. Some evergreen trees prosper under these conditions. Fir, spruce, and pine dominate the forests of northern Canada, northern Europe, and parts of Asia, but in northern Siberia, most of the trees are larches. Ecologists refer to these high-latitude forests as **taiga forests** (figure 14.9). Broadleaf deciduous trees, such as oaks and elms, which are so abundant farther south (figure 14.10), are rare in the taiga forests. The wet, mossy, and boggy ground is an ideal habitat for mosquitoes, and the bogs also make an ideal home for moose, mink, and beavers.

Midlatitude Forests

In the midlatitudes, global atmospheric circulation causes the eastern sides of continents to have cold winters and warm summers, and precipitation is evenly distributed throughout the year. These are ideal conditions for conifers and a great variety of deciduous trees, such as oaks, hickory, maples, elms, and sweet gums. Seasonal changes, especially the loss of leaves in the fall and new growth in the spring, are prominent features of the temperate deciduous forests (see figure 14.10). Most of these forests lie south of the areas that were glaciated during the last glacial advance that culminated about 18,000 years ago. Undoubtedly, the advancing ice disrupted similar forests that grew farther north during the last interglacial period.

With the changing seasons, numerous migratory birds come and go as they move from summer habitats in Canada to winter feeding grounds in temperate or tropical climates. These birds as well as other animals help disperse seeds of trees and shrubs that produce their fruit in the autumn and early winter. Large numbers of animal species, including bears, foxes, deer, rabbits, squirrels, skunks, and groundhogs, also populate these forests.

Savannas

The grass-covered plains that define savannas cover large areas in both tropical and temperate regions. Small areas of trees that prosper along rivers and in places where ground water rises dot the

Figure 14.10 A Deciduous Forest. This deciduous forest is in the Blue Ridge Mountains, part of the Central Appalachian Mountains of Virginia and contains a great variety of deciduous trees, including oak, poplar, and maples, as well as some conifers.

Figure 14.11 A Savanna. Grasses cover the flat ground surface of the Great Plains where the moisture permits. The drier portions of the plains support scrub plants, including cactus and tumbleweed.

surface. Most savannas are in midlatitudes, but some—such as the Great Plains in North America (figure 14.11) and the steppes of Russia—extend far north, where seasonal changes in temperature are extreme. Other savannas extend into the tropics. In savannas, most rain falls in one season—either winter or summer. Because of this concentration of rainfall, evaporation exceeds precipitation most of the year. Consequently, plants live a marginal existence, especially where the land becomes dry and subject to fire.

In the tropical savannas, where temperatures remain high year round, the amount of precipitation and its seasonal distribution vary, and these determine which plants grow best. Some tropical regions, especially large areas in the interior of Africa south of the Sahara Desert, receive little rain, and most of it falls in just a few months. Most trees do not survive long dry spells, but cacti and thorn trees that can hold water for long periods prosper under these conditions. Forests populated by these plants grow in parts of Africa and Mexico. In Africa, large populations of grazing mammals reduce the number of some species of trees and create an environment in which other plants prosper.

Before people began to affect the savanna environment, lack of water, fires resulting from lightning, and huge herds of grazing animals, such as buffalo, killed most shrubs and trees before they could become established. This repeated killing of shrubs and trees maintained grass as the primary plant in the savannas. One- to 2-meter-high stands of grass covered much of the eastern part of the U.S. Great Plains. This grass grew in a thick sod that held moisture and formed the rich soil now used extensively for cultivation of corn. Shorter grasses grew on the drier high plains farther north and west. Since the mid 1800s, farmers have cultivated large parts of the American plains using water drawn from relatively shallow water-bearing layers called *aquifers*. Excessive pumping from these aquifers is depleting the water supply and producing serious problems, which threaten the future productiv-

ity of the region. Today, farmers grow wheat and ranchers graze cattle on this section of the plains. Little moisture manages to cross the Rocky Mountains, so the grass cover in the plains is scattered, thin, and fragile. Nevertheless, ranchers use some of this land for grazing.

Tropical Forests

In the tropics, average temperature remains in the mid-to-high 26°C (80°F) range year round, but the amount of rainfall and its distribution during the year varies from one area to another. Where rain falls throughout the year, both deciduous and evergreen forests prosper. Where rainfall is seasonal, except for trees growing along rivers, the deciduous trees lose their leaves during the long, hot dry season. Heavy precipitation, often more than 254 centimeters per year (100 inches per year), gives rise to exceedingly dense, diverse forests. Where rainfall is heavy, exceptionally deep soil forms. Ironically, the heavy rainfall that fuels such abundant plant growth also removes soluble minerals from near the surface, leaving a soil that is virtually devoid of mineral nutrients. Plants also remove remaining nutrients from the upper part of the soil. Only the thick vegetative cover contains the mineral nutrients needed for plant growth. Young plants obtain the nutrients they need from decaying plants on the forest floor (figure 14.12). Plants literally grow out of the decaying remains of other plants. Soils in areas of the tropics that have lower amounts of rainfall retain larger amounts of mineral nutrients and are more valuable for agricultural development.

In many parts of the tropics, farmers cut or burn forest cover to expose the soil so they can plant crops or graze cattle. In 1997, many such fires got out of control in Indonesia and raged for months. The slash-and-burn approach to farming is unfortunately self-perpetuating. The soil contains only enough nutrients to support crops or grazing for a few years, but once the nutrients are exhausted, crops fail, and the plant cover diminishes. After that, the soil erodes quickly. To survive, farmers generally abandon the eroded land and continue the process of cutting forest for crops or grazing elsewhere.

Plant and animal diversity reaches a maximum in the tropical rain forests. Plant diversity is so great that ecologists have not yet identified all the species present. Some of the rare plants in tropical forests have medicinal properties that make them particularly valuable to humans. Loss of potentially important medicines is one of the reasons people are concerned about the destruction of these forests. In addition, the soil erodes easily and recovers slowly from the effects of clear-cutting. Other concerns include loss of biodiversity, loss of cultural diversity, erosion, and loss of a valuable place for removal of carbon dioxide from the atmosphere.

High Mountains

Although high mountains are present even in tropical latitudes, environments in high mountains throughout the world resemble those in the polar regions. Because temperature drops with increasing elevation, at some level, the temperature is too low for trees to grow. Above this elevation, called the **timberline,** snow,

Figure 14.12 A Rain Forest. Rain forests like this one in Washington support a diverse assemblage of plants, including many varieties of ferns and mosses.

ice, and tundra create a high-altitude equivalent of the polar environment (figure 14.13). In Antarctica, snow and ice completely cover most of the mountains (see figure 13.1*b*). **Snow fields** and valley glaciers remain in the high parts of the Alps, the Himalayas, the Cordilleran Mountains of Canada and Alaska, and in the New Zealand Alps. However, most of the glaciers in these mountains have been retreating rapidly since the early 1900s. The altitude of the timberline varies even within mountain systems, reflecting the effects of the climate of the surrounding regions. Consequently, the timberline is much lower in the colder Alaskan part of the Cordilleran Mountain system than it is in the Rocky Mountains or

the northern part of the Andes. The effects of glacial advances during the ice ages and especially over the past 20,000 years remain. During these advances, ice eroded the mountain tops, and glaciers scoured the valley sides and floors, leaving bare rock exposed over much of the high mountain landscape. Few plants other than lichens and mosses survive on such surfaces. Tundra covers the plateaus and valley floors at high altitudes where soil is present or has started to form.

Because air masses rise as they move over mountains, the higher elevations receive more rainfall and snowfall than surrounding areas. Moisture-laden clouds often cover mountains, even where

Figure 14.13 High-Mountain Environment. High mountains like this in Glacier National Park, Montana, have plants that normally grow much farther north.

the amount of precipitation is low. The large quantity of precipitation and steep mountain slopes produce streams of much higher velocity than those in lower lands. For these reasons, the ecosystems of streams that flow rapidly down high mountain slopes in a series of cascades and waterfalls differ from those of streams that slowly meander across nearly flat coastal plains.

On high mountains, plant and animal communities gradually change as the elevation changes. The deciduous forests that cover the lower parts of mountain ranges in temperate climates give way to evergreen forests at higher elevations. Finally, all trees disappear, and plant communities consist mainly of mosses, lichens, and small shrubs. Only a few species of animals make the highest parts of these mountains their home. Mountain goats, marmots, and bears roam the peaks of many mountains during summer months, but where the winters are severe, most animals hibernate or descend to lower altitudes.

The environment on and around volcanoes may change suddenly and dramatically depending on the amount of activity. Some, such as Stromboli in Italy, remain sporadically active over long periods. For others, such as Mount Pinatubo or Mount St. Helens, violent eruptions punctuate long periods of dormancy, during which the environment may resemble that of other, nonvolcanic mountains.

Earthquakes occasionally shake these ranges and cause massive landslides. Otherwise, crustal deformation causes change slowly.

Hot Deserts

Most hot deserts, areas where evaporation exceeds precipitation, lie between latitudes 20° and 40° north and south of the equator (figure 14.14). In the Northern Hemisphere, the largest deserts lie in a great swath that crosses north Africa, Arabia, Iran, and China, but it also includes northern Mexico and the southwest United States. In the Southern Hemisphere, deserts cover large areas in southern Africa, South America, and Australia. In these latitudes where most deserts form, extremely dry air from high altitude sinks to the ground. As the dry air descends, its temperature rises, and it absorbs moisture from surrounding air. This makes the desert air very dry. Because they are so far removed from sources of moisture, the middle of the Sahara Desert in Africa and the interior of Australia are especially dry. Frontal systems in the winter or thunderstorms in summer bring some rain to most deserts, but hot deserts rarely receive sustained rains.

Although arid regions throughout the world have sparse rainfall, streams play an important role in shaping the landscape.

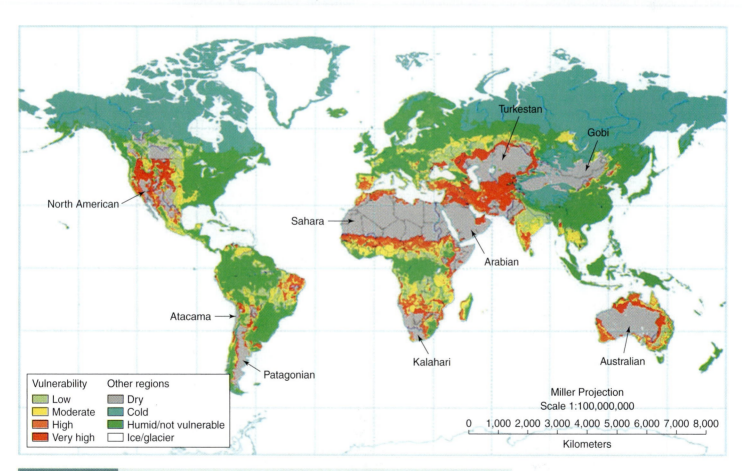

Figure 14.14 Deserts of the World and Areas That Are Vulnerable to Climatic Changes That Would Turn Them into Deserts.

When rains come, they drench the desert for short periods of time during which streams rise and flow rapidly. As in humid regions, stream action in deserts operates in conjunction with weathering and downslope movements to shape the landscape. In humid regions, vegetation promotes infiltration of water into the soil and holds the soil in place. In deserts where vegetation is sparse, most precipitation runs off on the ground surface, causing flash floods and mudflows.

Most desert streams flow only part of the time in steep-sided gullies. For a few hours after a rainstorm, the streams swell with water. During this time, streams move large amounts of the disintegrating and decomposing rock and sediment that cover much of the desert ground surface. However, the water level in the stream falls quickly after the storm is over. The large quantities of rock and sediment brought into the stream during rainstorms clog the stream channels. Unable to carry the sediments any further, the stream resumes a shallow, braided pattern. Eventually, this water filtrates into the ground or evaporates, leaving the channel dry.

Most of the southwestern part of the United States is arid or semiarid. It contains landforms that are representative of many hot desert regions around the world. These include deep canyons cut by the rivers as at Canyonlands National Park (figure 14.15a), broad valleys filled with deposits carried into the basin from adjacent mountains, unusual temporary lakes, and salt flats such as the one shown in Death Valley (figure 14.15b). The deserts of the southwestern states contain few large areas of sand accumulation. Sand rarely covers more than 30% of a desert region (figure 14.16). Instead, bare rock and thin soil-cover are more characteristic of deserts than are sand dunes (see figure 14.1d). Nevertheless, vast expanses of drifting sand that blanket parts of the Sahara and Arabian deserts justify the name "sand seas."

Except for the driest areas, deserts support a surprisingly large and diverse assemblage of plants and animals. The animal populations may not be apparent during the day when temperatures are high, but as the cool of evening descends, the desert animals become active and more obvious. Lizards and snakes are common in the desert. Large populations of rodents and insects disperse fruits and pollinate many of the plants. Like the animals, only plants that can tolerate the extreme temperatures and dry conditions live in hot deserts. Succulent plants that store large quantities of water are particularly well-suited for the desert climate (figure 14.17). These plants do not have leaves; instead, they carry out photosynthesis through their stems. If the soil contains sufficient nutrients, succulent plants abound in deserts, but few plants of any type survive in some deserts, such as the Simpson Desert in Australia, that have nutrient-poor soils.

Although hot deserts contain many species of animals and plants, the total amount of organic material (see figure 14.4) available in the environment is far less than that in many other environments. Consequently, hot deserts, like the cold polar deserts, play a less significant role in Earth's biogeochemical cycles than do biomes in temperate and tropical regions. Although climate and biomes are closely related, both are influenced by the geologic framework within which they operate.

WETLANDS

Any land that is wet part of each year is a *wetland*. The term has no technical definition other than that given to it by law and government regulations. Some land remains wet most of the time (figure 14.18). For example, with each high tide, salt water

(a)

(b)

Figure 14.15 Representative Desert Landforms. Many of the landforms found in deserts, especially in the desert regions of North America, are one of the types shown here. The landscape shown in (a) is the desert region of Canyonlands National Park, Utah. Prominent sandstone layers have been stripped over overlying sedimentary rocks, producing flat surfaces. (b) This image from Death Valley was taken across a lake in which salts are forming. The gradual slopes in the background are formed by sediment washed out of the mountains into the basin.

inundates the marshes on the surface of the Mississippi and San Joaquin deltas and along the edges of the Chesapeake Bay and other estuaries. Water covers much of these areas for at least part of every day. Swamps, marshes, and bogs along river bottoms, low-lying areas along shores of lakes and the ocean, and land near the terminus of glaciers may remain wet most of the time. Other wetlands hold water for short periods. For example, depressions, known as the *prairie potholes,* located in the northern plains are lowlands created by glaciers as they advanced over the region about 18,000 years ago. Today, only after heavy rains or during the spring thaw does water stand in these depressions. Occasionally, water covers portions of the flood plains of most major rivers, but they may remain dry for extended periods.

At the time of the Revolutionary War, more than 200 million acres of land qualified as wetland. Until recently, most people viewed wetlands as wastelands that should be "reclaimed" if possible. In the mid 1800s, this perspective led Congress to pass the Swamp Land Acts. These acts established a national policy of draining and filling wetlands, reclaiming the land for productive uses such as agriculture and development. Initially, the rate of reclamation was slow, but once large earth-moving equipment became available, the rate increased dramatically. In some areas such as the Florida Everglades, the ecological character of vast tracks of land changed as people modified the natural drainage systems. In recent years, people have drained or filled an estimated 300,000 acres each year. The U.S. Fish and Wildlife agency reports that the country lost close to 10,400 square kilometers (4,000 square miles) of wetlands between 1975 and 1985. These losses include wetlands drained for agricultural, commercial, and housing development; lands contaminated by chemical waste disposal; lands drained by water-development projects; and lands lost as a result of rising sea level.

Recognition of the role wetlands play in the environment and the value of leaving the land undeveloped has come slowly. Wetlands along flood plains reduce flood crests downstream by providing storage for water during floods. Wetlands facilitate the infiltration of water into the ground, providing an important way of recharging aquifers in such areas as the prairie-pothole country. As water infiltrates through wetlands, plants absorb and filter pollutants that would otherwise degrade lakes, rivers, reservoirs, or aquifers. Along coasts, wetlands stabilize shorelines and provide buffers for the effects of hurricanes and wave action.

Wetlands are especially important because of their biological productivity and their production of oxygen. Wetlands are second

Figure 14.17 Tall Saguaro Cactus Plants in the Desert near Tucson, Arizona.

(a)

(b)

Figure 14.18 Wetlands. (a) Cypress trees grow in this swamp located close to the Mississippi River in Arkansas. (b) Water stands in this mountain valley, except during winter months when it is frozen.

only to rain forests as sources of atmospheric oxygen. Globally, they provide habitat for more than 150 species of birds and 200 types of fish. More than a third of all nonmigratory birds live in wetlands, and at least half of all migratory birds depend on wetlands for temporary habitat. In addition, wetlands along the coast are the primary spawning, nursery, and feeding grounds for more than half of the fin and shellfish. The future of the fishing industry depends heavily on maintaining the existence and health of wetland environments.

Having recognized the problems arising from the Swamp Lands Acts, the government reversed its policies concerning wetlands. The Clean Water Act of 1970 prohibits the filling of wetlands without a permit issued by the U.S. Army Corps of Engineers, but the act does not regulate the draining of wetlands. The Environmental Protection Agency formulates the guidelines for the Army Engineers and can overrule their decisions. The Agriculture and Conservation Service (formerly the Soil Conservation Service) and the U.S. Fish and Wildlife Service also regulate wetlands. In 1989, the government issued a manual for identifying wetlands over which the government exercises jurisdiction. The manual defines the vegetation, hydrologic conditions, and soil types that qualify for federal regulatory protection. Under the definitions in this manual, large areas of marginally wet land qualify for wetland designation. Despite the Army Corps of Engineers' record of denying only about 1% of the requests to fill wetlands, developers have complained that the wetland definition includes too much marginally wet land. The debate and adjustments in the regulations will undoubtedly continue.

SUMMARY POINTS

1. Geologic processes, climate, and biological organisms combine to shape terrestrial environments.

2. Bedrock composition and structure, geologic processes that originate deep inside Earth, and geomorphic agents, such as streams, glaciers, ground water, wind, and mass movements, form a framework within which terrestrial environments evolve.

3. Biomes and ecosystems can be closely correlated to geomorphic (physiographic) regions and can sometimes be defined by geologic criteria.

4. Major crustal elements, such as cratons, mountain belts, and coastal plains, have distinctive characteristics that influence the environment. However, the boundaries of these elements do not coincide with climatic divisions or biomes.

5. In any given location, climate affects the development of soil and the types of plants that survive to reproduce. A relatively close connection exists between climate and biome divisions. Most boundaries between biomes and climate zones are transitional.

6. Distinctive environments characterized by climate, plant life, and fauna include polar regions, boreal forests, midlatitude deciduous forests, savannas, tropical forests, high mountains, and hot deserts.

7. The landscape is continuously changing, and factors that influence the changes include weathering and erosion, tectonic setting, composition and structure of the bedrock, and geomorphic agents.

8. Sea level is the ultimate base level below which geomorphic agents cannot reduce the land surface.

9. Geomorphic agents act on Earth materials in different ways, producing different types of landforms. The geomorphic agents are downslope movement of materials (mass wasting), streams, glaciers, wind, ground water, waves, and currents.

10. Geomorphic agents—streams, ground water, glaciers, wind, and mass movement—shape the land surface. Gradation refers to the erosion, transportation, and deposition of materials on the service. Degradation is the lowering of the surface; aggradation is the building up of the surface. Degradation begins with weathering and erosion of soil and loose materials on bedrock.

11. Volcanic and tectonic activity causes the surface to remain high in places despite the long-term effects of erosion by geomorphic agents.

12. Each geomorphic agent erodes, transports, and deposits materials in unique ways, and each agent leaves a clear indication of its action in the shape of the land and in the deposits formed. We can determine which agents operated by the shape of the land and features of its deposits.

13. In addition to the operation of geomorphic agents, the character of the bedrock, its structure, tectonic setting, and the length of time the geomorphic agents have operated affect the landform. Ultimately, erosion lowers the land surface to sea level, which is the base level at which geomorphic agents affecting the land cease to do so.

REVIEW QUESTIONS

1. How does the geologic (including the topographic) setting of an area influence the climate of that area?
2. How does the climate of an area influence the ecology of that area?
3. How do boreal forests differ from midlatitudes and tropical forests?
4. What climatic conditions give rise to hot deserts?
5. What adaptations must plants and animals make to survive in hot deserts?
6. Why are climatologists concerned about the cutting down of rain forests?
7. How does the character of the soil influence the biomes of an area?
8. What processes cause the level of land to rise or fall relative to the sea?
9. How does the structure of bedrock affect the shape of the land surface? Give examples.
10. What are geomorphic agents?
11. What is meant by base level? What might form a temporary base level?
12. How is the shape of the land related to the plate tectonic and climatic Earth systems?

THINKING CRITICALLY

1. How do systems that operate deep inside Earth affect the biological communities that live on the surface?
2. How does climate influence the biomes in different ecosystems? How could changes in climate affect the environment of polar regions? Of tropical forests? Of high mountains? Of hot deserts?
3. Why might one expect to find different ecological niches on the north and south sides of a ridge?
4. Discuss the response one might expect to occur in the geography of biomes as a result of changes in Earth's climate system, such as an increase in average global temperature.
5. How does climate affect the weathering process and the effects of geomorphic agents?
6. In the area where you live, how does climate effect the evolution of the landscape?

KEY WORDS

aggradation	mass wasting
biome	microclimate
degradation	permafrost
differential erosion	physiographic provinces
ecology	snow fields
ecosystem	taiga forests
environment	timberline
geomorphic agents	tundra
gradation	ultimate base level

Weathering and Soil Development

Chapter Guide

The interaction of water, atmospheric gases, and organisms with the solid materials of Earth's crust, causing them to physically disintegrate and chemically decompose, lead to the development of soil. Such weathering processes as the freezing of water and thermal expansion and contraction of materials involve purely physical reactions. These processes break down solids and produce fragments of a smaller size. Other weathering processes are primarily chemical or biochemical in nature and result in the alteration of the composition of affected materials. Still other weathering processes are a direct result of the action of organisms. These range from the burrowing of worms in the soil to the trampling of the ground surface by herds of animals.

Weathering depends on the climate to which rocks are exposed and on the composition of the materials. Weathering makes most materials susceptible to erosion, which involves movement of the materials from where they originate. Concentrations of some economically important mineral resources occur as a result of weathering processes. The operation of these processes and the resulting production of soil are the primary focus of this chapter.

INTRODUCTION

WEATHERING

MINERAL RESOURCES FORMED BY WEATHERING

SOILS

Solution of gypsum by rainwater has formed this sculptured surface near St. George, Utah.

INTRODUCTION

As late as the sixteenth century, some students of Earth believed that rocks and minerals were alive—that they were simply lower forms of life than plants and much lower than animals. This belief was founded on the observation that rocks appear to be in various stages of life—youth, maturity, and old age. The same rock that appears solid, strong, fresh, and youthful in one outcrop might be recognized elsewhere as a decaying mass, obviously dead, and starting to disintegrate. Rocks are not living substances. These philosophers were merely observing the results of weathering.

Evidence of weathering is present in most rock exposures. Differences in the color of the inside and outside of broken rocks are usually caused by weathering (figure 15.1). Effects of weathering and attempts to delay the processes are visible every day. The construction of sidewalks and concrete highways in separate sections helps prevent them from cracking when temperature falls and concrete contracts. Houses in moist climates must be painted to prevent decay of the wood. Oxygen will attack a knife left outside in the rain and change the bright, shiny metal into rust. Similar processes act on the rocks and minerals in the crust and convert them into loose, granular materials called **regolith.** Some scientists define **soil** in the same way, but most geologists and soil scientists restrict the term soil to surficial materials containing or capable of supporting plants. Weathering processes are also responsible for the concentration of some economically important elements such as aluminum in deposits suitable for mining.

Figure 15.1 Weathering Rinds. Alteration of the rock usually changes the color of the rock and modifies the composition of the minerals in the weathering rind. Solutions move along fractures as they have in this outcrop in the Ouachita Mountains of Arkansas.

WEATHERING

The decay and decomposition of materials exposed at Earth's surface occur where rocks and minerals come into contact with the atmosphere, water, and organisms. Weathering involves chemical, mechanical, and biological processes. Mechanical processes, including the freezing of water, growth of crystals, unloading caused by removal of materials on the surface, and heating and cooling, cause physical disintegration. Chemical processes, such as oxidation and reactions of materials with water and carbon dioxide, cause decay or chemical breakdown, also called chemical decomposition. Biological processes may manifest themselves in both mechanical and chemical processes. The burrowing of organisms and root growth are examples of the mechanical effects of organisms. Bacterial action and the alteration of the composition of

water through organic processes exemplify biochemical processes. The relative importance of these processes depends largely on the composition of the material being weathered and the climatic conditions under which weathering takes place.

Mechanical weathering dominates in areas subject to temperature extremes and especially where precipitation is low. Chemical weathering is most effective where moisture and temperature are uniformly high. The effectiveness of biological processes depends on the numbers and types of organisms present. Over most of Earth, mechanical, chemical, and biological processes combine to destroy the coherence of rocks and sediment and bring about chemical changes in materials on or near the ground surface.

Mechanical Weathering

Under such circumstances as extremely cold climatic conditions, mechanical disintegration is the principal type of weathering. **Mechanical weathering** is most apparent where cold, variable temperatures and water are both present.

Freezing Water The freezing of water breaks down rocks because water, unlike other liquids, expands when it crystallizes. As it freezes, water increases in volume by about 9%. When water expands, pressure develops inside any container confining it. In an enclosed vessel, freezing water exerts as much as 2,100 kilograms per square centimeter (30,000 pounds per square inch or 2,100 newtons per square meter) on the vessel. This amount of pressure far exceeds the strength of even the hardest and most durable rocks.

Fractures are present in most rock outcrops (figure 15.2). If water in a fracture freezes, the pressure exerted by the expanding water spreads the fracture apart and extends it in depth. This is especially effective where freezing and thawing occur repeatedly as temperature rises above freezing during the day and falls below freezing at night. New fractures open and expose more of the rock to mechanical and chemical weathering. The effects of freezing and thawing accumulate over time. Repeated freezing and thawing, day after day, year after year, make this process highly effective in breaking down rocks and causing them to disintegrate (figure 15.3*b*).

Frost Action In areas where temperature drops below freezing in winter months, ice forms in the soil (see figure 15.3). As water freezes, it expands and forces the soil or rocks above it upward.

This process, called **frost heaving,** loosens the soil and allows air to pass into it. This improves the soil's fertility, but heaving and loosening the soil on steep slopes may also cause the soil to move downslope or expose it to erosion by running water. Once ice starts to crystallize in the soil, the intergranular pore spaces act like capillary tubes and pull water to the surface of the ice layer. Heat released as the water changes state from water to ice keeps a thin film of water on the ice. As long as this film is present, water continues moving to the surface of the ice.

Frost heaving is a problem on farmland where moisture trapped beneath rocks buried in the soil freezes and forces rock fragments upward to the surface. By promoting the formation and growth of potholes, it also takes a heavy toll on roads. The ice lifts the rock by forming in the small capillary-sized opening under the rock. When large numbers of rock fragments are present in the top layer of soil, a rock field may form at the surface. These rocks frequently form polygonal-shaped patterns. The center of each polygon is the point of greatest upward bulging, and the edges of the polygons are correspondingly lower. Where the ground-surface slope increases, rock polygons tend to merge and form stripes. At high latitudes, the ground remains partially frozen throughout the year. This permanently frozen ground is called **permafrost.** In these areas, layers of ice often form in the soil.

Crystal Growth In arid and semiarid climates, crystals of soluble minerals such as gypsum, calcite, and halite (salt) form in the soil. These minerals crystallize where water, that has soaked into the

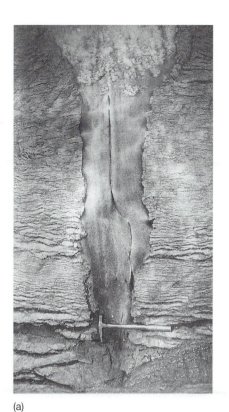

(a)

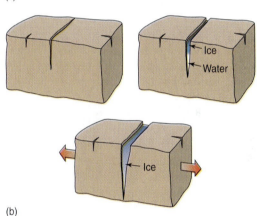

(b)

Figure 15.3 Weathering Effects of Ice. (*a*) In cold regions, ice forms as veins or sheets in soil. (*b*) Freezing of water in cracks in bedrock wedges the rock apart and may extend fractures into the solid rock.

Figure 15.2 Fluids Move Along Fractures. Most rock outcrops have fractures formed where the brittle rock was at least slightly deformed. The fractures in this outcrop provide easy access for fluids that move into the rock mass.

ground after a rain, dissolves soluble compounds in the soil. When evaporation begins, **capillary action** pulls water back toward the surface. As the water evaporates, elements in solution precipitate and form crystals or crystalline aggregates. Growing crystals can exert great pressure on the soil that surrounds and confines them, and they may break up the soil. The process of crystal growth may ultimately lead to the formation of a layer of the precipitated minerals. A layered or nodular mass of calcium carbonate called **caliche** forms in this way. Caliche occurs in the soils in deserts and

Figure 15.4 Sheeting Developed in the Granite Quarry at Mount Airy, North Carolina.

semiarid regions where calcium-rich dust settles or where rocks that contain calcium carbonate lie beneath the soil. Also formed in this way are some deposits of sulfates, phosphates, nitrates, gypsum, alum, epsom salts, and saltpeter.

Unloading Several geomorphic agents, especially glaciers and streams, remove soil and rock from the ground surface. Through construction of open-pit mines, quarries, and highways, humans also remove large quantities of near-surface materials. As unloading takes place, pressure on the underlying rock decreases. The reduction in pressure allows fractures, called **sheeting,** to develop parallel to the ground surface, and layers of the fractured rock may pop up (figure 15.4). This phenomenon also occurs where natural forces caused by unloading split the rock. In Yosemite Valley, glaciers have cut a deep valley into the granitic rocks that underlie much of the western side of the Sierra Nevada. Beautifully developed sheeting has formed on the top and side of Half Dome, one of the most famous of the glaciated mountains in Yosemite. Glaciers removed the soil and gouged the deep valley, cutting nearly vertical cliff faces into the solid granite (figure 15.5).

Heating and Cooling Early in human history, people learned to use fires as a way of breaking down rocks. Primitive quarrying operations used fire to crack rocks. Fire is rarely used for this purpose today, but rocks

Figure 15.5 Sheeting Developed on Half Dome in Yosemite National Park, California.

exposed to the high temperatures associated with wild fires still disintegrate. Exposure to the much lower temperature of the Sun's heat also affects rocks. Daily temperature variations of 20°C or more may occur in desert regions where rapid cooling at night follows high daytime temperatures. Most solids expand when heated and contract when they cool. Even though the amount of expansion and contraction is usually small, repeated expansion and contraction can, over a long period, lead to mechanical breakdown of rocks. Because some minerals expand and contract more than others, gradual breakdown is likely to take place between the grains. Once cracks form around grains, solutions penetrate and cause chemical decomposition.

Animals and Plants Animals and plants are responsible for many forms of weathering. Worms and other burrowing organisms produce holes in the soil and loosen it, providing passageways for atmospheric gases and solutions. Bacteria play an important role in breaking down organic compounds and liberating humic acids. Several marine invertebrates, especially gastropods and echinoids, contribute to the weathering of exposed rocks along shorelines, particularly in the zone between high and low tide. These animals bore into the rock, producing holes or pits in which they live. Eventually, individual holes coalesce as the animals bore more and more holes into the rock surface.

Over the last century, human activities have caused far more extensive modification of the surface of Earth than any natural process. The extent to which people are disturbing the ground surface is apparent from the air. The effects of bulldozers working on construction sites are most dramatic, but farming and grazing also disturb the soil over vast areas. Some crop plants remove particular chemical constituents from the soil and upset the natural chemical balance. Whether natural or cultivated, plants and plant remains are important in the development of soil. These activities accelerate the rate of weathering, expose the deeper layers of soil to erosion, and affect the capacity of the soil to hold water.

The Role of Plants in Rock Weathering

Biochemical processes are among the most important ways plants affect rock weathering. Some plants remove unusually large quantities of certain elements from the soil, and they use these in their plant structure. For example, horsetails, corn, palm trees, bamboo, reeds, and other grasses remove large amounts of silica from the soil. Silicon dioxide comprises nearly 77% of the inorganic ash of giant reeds and 70% of the ash of laurel and palms. In a tropical rain forest,

The Role of Plants in Rock Weathering. Tree roots extend into cracks in otherwise solid rock. As the tree grows, the roots exert pressure on the walls of the crack. This pine is growing on a rock that has fallen from the walls of the Grand Canyon.

Chemical Weathering

Many of the minerals exposed in rocks and sediments at Earth's surface formed as igneous rocks at high temperature. During **chemical weathering,** chemical reactions alter these minerals, producing new minerals that are stable in the presence of water and at the low temperatures that prevail on Earth's surface. A number of elements and compounds in the atmosphere react with minerals. The most common of these are water, carbon dioxide, and oxygen, but both organic and inorganic acids form naturally, and these are highly reactive with many minerals. Chemists describe the strength of acids, the **pH,** based on the concentration of hydrogen ions (H^+) in the solution. The pH of alkaline solutions ranges from 8 to 14. Acid solutions have a pH that ranges from 0 to 6; a neutral solution has a pH of 7. The effectiveness of these acids and bases in altering minerals depends on the size of the individual crystals or mineral fragments, temperature, and composition.

Texture, structure, and composition cause rocks to vary in susceptibility to chemical weathering. Because chemical reactions take place on surfaces, fine-grained rocks or intensively fractured rocks break down more readily than those that are coarse grained and impervious. Fractures, bedding planes, fine openings of capillary size along boundaries between individual minerals, and pore spaces all provide surfaces along which solutions can move. In most rocks, chemical weathering concentrates along bedding planes or fractures. Where bedding and fractures or intersecting fracture systems break rock masses into blocks, weathering is concentrated on the corners of the blocks (figure 15.7). Eventually, these weathering effects reduce the block to a spherical or oval-shaped form (figure 15.7a). This process can lead to the formation of spherical masses of rock, a process called **spheroidal weathering** (figure 15.7b).

Relative resistance of minerals to chemical weathering is the same as their position in Bowen's Reaction Series (see chapter 2). Of the common minerals in igneous rocks, olivine and pyroxene weather most rapidly, and mica and quartz are highly resistant (table 15.1). Quartz is almost insoluble in rainwater and is unaffected by carbon dioxide and other constituents of the atmosphere. Consequently, rocks composed largely of quartz decompose slowly. This property of quartz makes it especially useful for the production of glass, which is composed of melted quartz. Although quartz is highly resistant to chemical weathering, many other minerals including other silicates are less resistant. After a few months of exposure to the atmosphere in hot humid regions, most common minerals exhibit changes in color that are a sign of chemical reactions. Over longer periods of exposure to the atmosphere, outcrops that contain layers of different composition usually exhibit signs of **differential weathering.** The more-resistant layers stand out, and the less-resistant layers recede into the outcrop (see figure 14.8).

General chemical principles apply to reactions involving minerals. Almost all chemical reactions take place more rapidly at high temperature than at low temperature. Because chemical reactions take place on the surface of solid particles, chemical weathering affects small particles more quickly than large particles. For example, powdered sugar dissolves faster than granular sugar because the

plants may remove almost 450 kilograms (half a ton) of silica per 4,000 square meters (an acre) each year. It is not surprising that most soils found in the tropics are deficient in silica. The ash returns to the ground surface when a plant dies and decays, so not all of the silica removed by plants is lost, but the silica-rich ash is vulnerable to erosion. It can be blown away by strong winds, and during heavy rains, surface runoff quickly washes away any ash that lies on the surface of the ground.

Plant canopies and roots also affect weathering processes. The canopy of branches and leaves intercepts light rainfall, shields the ground from sunlight, and tends to slow weathering processes. Roots facilitate weathering by root growth and by drawing water and elements from the soil into the plant. As the roots form, they force their way into the soil and occasionally into fractures in solid rock. As they grow, roots enlarge the openings through which they pass (figure 15.6). When the plant dies, the root cavity becomes a conduit for water that hastens the weathering of the rock through chemical processes. Where the system of roots becomes interwoven, a mat of root-matter forms. This mat protects the underlying soil from erosion. Plants play important roles in the formation of soils and in protecting the soil from erosion. Root systems even affect the weathering of rocks exposed along the shore where algae extend fine roots into the rock to which they attach, loosening grains and altering the chemistry of the solutions around the roots.

The action of bacteria on decaying organic matter may produce organic acids known as *humic acids.* These are present in ground water in hot, humid regions where vegetation is abundant, and they may increase the acidity of the ground water enough to make it a highly effective chemical weathering agent.

(a)

(b)

(c)

Figure 15.7 Spheroidal Weathering. (a) Weathering along fractures has produced this unusual pattern in a sandstone. (b) Spheroidal weathering in granitic rocks. (c) Weathering of granite near Pikes Peak in Colorado produced these rounded masses.

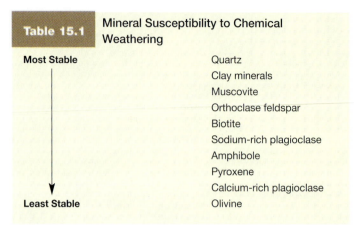

Table 15.1	Mineral Susceptibility to Chemical Weathering
Most Stable ↓	Quartz
	Clay minerals
	Muscovite
	Orthoclase feldspar
	Biotite
	Sodium-rich plagioclase
	Amphibole
	Pyroxene
	Calcium-rich plagioclase
Least Stable	Olivine

surface area of the smaller particles is greater than it is in the same volume of coarser sugar. The same is true of minerals. Small fragments of the most common of all minerals, feldspar, weather much more rapidly than larger pieces. Almost all chemical-weathering processes cause expansion of minerals involved in the reactions. As the altered minerals grow in size, they exert pressure and cause mechanical breakdown of the surrounding materials.

Four processes cause most chemical weathering:

- oxidation
- hydration
- hydrolysis (including solution)
- carbonation

Oxidation The process by which oxygen combines with other elements or compounds is called **oxidation.** Oxygen has a particular affinity for sulfur, manganese, and iron compounds, and these are among the most commonly oxidized materials. Because iron is especially susceptible to oxidation, minerals that contain iron, such as the pyroxenes, hornblende, and olivine, oxidize quickly. The iron in these minerals reacts with oxygen and changes to ferric oxide, the mineral hematite, or to ferric hydroxide, the mineral limonite (common rust). Water facilitates oxidation and may enter into the chemical reactions, as indicated in the chemical reaction below. Iron sulfide, the mineral pyrite, also reacts with oxygen and water to form hematite.

Iron sulfide + oxygen + water ⇌ hematite + sulfate + hydrogen

The sulfate and hydrogen formed through this process produce sulfuric acid, an especially strong acid that reacts with and alters other minerals.

Color changes accompany most oxidation reactions. Yellow, orange, and brown stains so commonly seen on rocks result from oxidation of iron. Green and black soils and rock containing iron change to red, yellow, or brown as limonite forms (figure 15.8). Consequently, soils in warm, moist climates commonly exhibit red, yellow, or brown colors. Reduction, the process by which oxygen is freed from its compounds, is not common in weathering, but it may occur in sediments deposited in stagnant bodies of water

Figure 15.8 Iron-Stained Rocks and Soil Exposed in a Gully That Developed on a Steep Slope. Iron and manganese stains often form in soil or on rock faces where minerals that contain iron are present.

Figure 15.9 Decaying Feldspar in Granitic Rock. Once the feldspar is converted to clay, the quartz in the rock is liberated.

where oxygen is scarce. Where reduction affects iron compounds, a greenish color forms.

Hydration Combination of water with other compounds, a process known as **hydration,** is exemplified by the mineral anhydrite (calcium sulfate), which can absorb water and become the mineral gypsum, hydrous calcium sulfate. Gypsum, a major building material used to make drywall, forms as layers of sedimentary rock deposited under arid conditions.

Hydrolysis Chemical reaction between water and other compounds is called **hydrolysis.** These reactions, especially when combined with the effects of carbon dioxide, are extremely effective in weathering common rocks and minerals. Hydrolysis is important because the process is involved in chemical weathering of silicate minerals, the most abundant minerals in igneous and metamorphic rocks. The following formula describes the chemical weathering of K-feldspar (orthoclase) (figure 15.9):

K-feldspar + water ⇌ potassium + clay + silicic acid

Weathering of feldspars is the main source of clay, which is a stable mineral at surface temperatures and tends to accumulate in the soil. The silica set free in the preceding chemical reaction becomes part of a silicic acid (H_4SiO_4) solution. This solution moves through the soil and into bedrock as part of ground water. Plant roots absorb this form of silica and use it to manufacture cellulose.

Figure 15.10 Petrified Wood. The cell structure of the log is clearly visible in this specimen located in the Petrified Forest in Arizona.

If the concentration of silicic acid is high enough, silica may come out of the solution as colloidal (see chapter 1) or amorphous silica. It may recrystallize into small quartz crystals that line open fractures or cavities in the rock, or it may fill in spaces in buried wood, creating petrified wood (figure 15.10). Opal and chalcedony are two other forms of recrystallized silica.

the valuable components. In other instances, acid produced during weathering leaches the valuable constituents from part of the rock and brings them together elsewhere. The iron deposits in the Great Lakes region, bauxite in Arkansas, nickel in New Caledonia, phosphates in Florida, zinc in Virginia and Tennessee, manganese in India, the Gold Coast, Brazil, and Egypt, and copper deposits in western states are all examples of important deposits concentrated by weathering processes.

The production of bauxite from a red, clay-rich soil called **laterite,** which has a tendency to harden into a brick-like solid, is an example of how weathering may create an economically important mineral deposit. Laterite forms in tropical regions where highly variable rainfall causes large fluctuations ranging from fractions of an inch to many feet in the level of the water table. As silica leaches out of the soil, hydrous aluminum oxides concentrate (figure 15.13). Further chemical alteration of laterites leads to the formation of the aluminum ore **bauxite.** The deposits in Jamaica and in parts of the Gulf Coast, notably Arkansas, are sources of the bauxite used in North America to produce aluminum.

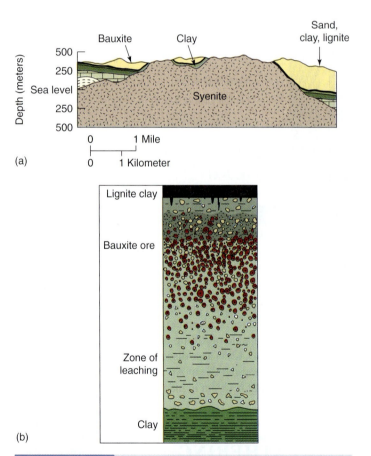

(a)

(b)

Figure 15.13 **Bauxite Deposits.** (a) Deposits of bauxite in Arkansas, the largest in the United States, as a thin layer on top of a large igneous intrusion containing a high percentage of aluminum-rich feldspar. (b) The bauxite is covered with organic-rich clay and underlain by a layer of clay that rests on the igneous rocks from which it weathered.

Nickel deposits in New Caledonia, the world's largest source of nickel, also formed as a result of weathering processes. The nickel is in peridotite, a magnesium- and iron-rich rock derived from the mantle. Chemical reactions between hot water and the peridotite change the original rock into serpentine. In the tropics, these masses of peridotite weather to a lateritic soil. As the peridotite decomposes, silicate minerals break down, and hydrous silicates of magnesium and nickel form in the laterite.

Chemical weathering by ground water causes oxidation of some ore minerals, such as pyrite. Acids produced by this oxidation dissolve and react with other minerals. As the host rock that contains ore elements weathers, the valuable elements leach out of the host rock and move downward in solution. If these metal-bearing solutions encounter water or rocks with which they can react, such as limestone, the metals may precipitate. As this process continues over time, the deep deposit is enriched in the metals leached from above. Pyrite (iron sulfide) and chalcopyrite (copper, iron sulfide) oxidize to form limonite and sulfuric acid. Iron oxide (limonite) forms where this process takes place, producing a surface cap called **gossan** on the ground above the ore body. An enriched zone of the metal develops below the gossan. Copper and silver deposits are also enriched in this manner. Prospectors who looked for the yellow limonite deposits in the soil found many of the major copper deposits in the southwestern United States.

Weathering processes also played an important role in the formation of certain types of banded ironstone, a significant source of iron ores in many parts of the world, including the Great Lakes area of North America. These rocks contain thin alternating bands of chert (silicon dioxide) and a variety of iron-bearing minerals. All of these formations are of Precambrian age and occur in shield areas. Most geologists agree that the iron in the formations originated as a result of deep weathering of continental crust and was deposited as mineral precipitates. The absence of clay minerals in the deposits and the fact that all are Precambrian in age has promoted debates that continue about the conditions under which the weathering occurred that released the iron from rocks. Some argue that these deposits could form only when little or no free oxygen existed in the atmosphere.

SOILS

Products formed during the decay, decomposition, and disintegration of rock combine with organic materials to make up soil. Most soils contain fluids including water, air, organisms, and decaying organic matter. **Residual soils** form from the bedrock materials that lie under them. However, soil may develop on the deposits left by streams, glaciers, or the wind, and soil itself may be transported from one place to another. Consequently, soils formed in one place on one type of bedrock may lie on bedrock to which they are unrelated. The top layer of most soils contain organic matter, such as roots, leaves, and worms. These materials provide important sources of nutrients, and they hold soil moisture, which facilitates plant growth. Water that infiltrates the soil picks up ions dissolved carbon dioxide and carbonic acid are part of the carbonate-silicate biogeochemical cycle (figure 15.11). This process plays an important role in the removal of carbon dioxide from the atmosphere. For this reason, chemical weathering of silicates helps regulate Earth's atmospheric temperature. Higher temperature should speed up weathering thereby reducing temperature.

One of the most obvious effects of carbonic acid is evident on buildings with copper roofs. After exposure to the atmosphere for a few months, these roofs turn green as the copper reacts with rainwater to form a copper carbonate that has the same deep green color as the copper carbonate mineral malachite.

Carbonic acid in rainwater is also responsible for the solution of carbonate minerals (e.g., calcite). Solution of calcite (and the rocks limestone and marble formed primarily of calcite) by rainwater involves several steps. First, water and carbon dioxide in rainwater combine to form carbonic acid. Second, the carbonic acid breaks down to form hydrogen and bicarbonate ions. Third, calcite combines with hydrogen ions to form bicarbonate and calcium ions. The effect of these reactions is that the solid calcite reacts with carbonic acid to produce calcium and bicarbonate ions that are in solution. As long as the water moves, fresh water replaces the solution that contains the calcium and bicarbonate ions, and dissolution continues. If the solution becomes saturated, calcium carbonate deposits form. Evaporation of the water or an increase in the temperature may cause saturation of the solution.

Acid Rain

Because of the reaction between pure water and carbon dioxide, described in the "Carbonation" section, all rainwater is slightly acidic. Gases emitted during volcanic eruptions also produce natural acids. For example, sulfur dioxide reacts with water to produce the highly reactive H_2SO_4 that has a pH of 1. The oxidation of ozone produces sulfur trioxide, which also combines with water to produce sulfuric acid.

$SO_2 + O_3 \rightleftharpoons SO_3$ sulfur dioxide plus ozone produces sulfur trioxide

$SO_3 + H_2O \rightleftharpoons H_2SO_4$ sulfur trioxide plus water produces sulfuric acid

However, acidity of rain has increased significantly as a result of air pollution. The burning of coal, which contains sulfur, and the burning of other hydrocarbons are responsible for much of this pollution. In addition to the sulfur produced from coal-fired electric generators, automobiles and some industrial processes produce nitrogen oxides (NO_x) that return to Earth's surface as particulate matter or after combining with rainwater to produce strong acids.

After the discovery that some lakes and many streams around the world, especially those in Scandinavia and the Adirondack Mountains, had become so acidic that they no longer supported many types of aquatic life, concern about the increasing acidity of rainwater became a global environmental issue. The pH of rainwater over large areas of eastern North America and Europe, where the bedrock and soil contain little lime that is capable of neutralizing acid, had dropped into the range of 4 to 5. This level

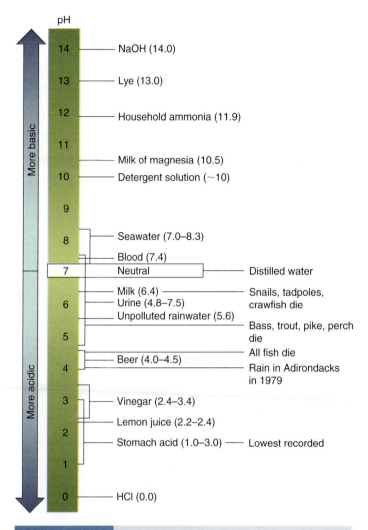

Figure 15.12 pH Scale Showing Levels Tolerated by Certain Plants and Aquatic Animals.

is sufficient to kill many fish and most shellfish. In areas closer to industries that produce acid emissions, rainwater may be even more acidic and capable of killing plants as well as most aquatic organisms (figure 15.12). These discoveries prompted Congress to pass and later amend the Clean Air Act. Despite this effort, acid rain remains a major problem in many parts of the world.

MINERAL RESOURCES FORMED BY WEATHERING

Metals are common in Earth's crust, but many of the most important metal-bearing minerals are disseminated in host rocks and cannot be extracted for a profit unless they have been concentrated by some natural process. Chemical reactions occurring during weathering may accomplish this concentration. In some cases, weathering removes the undesirable parts of the rock and leaves the valuable components. In other instances, acid produced during weathering leaches the valuable constituents from part of the rock and brings them together elsewhere. The iron deposits in the Great Lakes region, bauxite in Arkansas, nickel in New Caledonia, phosphates in Florida, zinc in Virginia and Tennessee, manganese in India, the Gold Coast, Brazil, and Egypt, and copper deposits in western states are all examples of important deposits concentrated by weathering processes.

The production of bauxite from a red, clay-rich soil called **laterite,** which has a tendency to harden into a brick-like solid, is an example of how weathering may create an economically important mineral deposit. Laterite forms in tropical regions where highly variable rainfall causes large fluctuations ranging from fractions of an inch to many feet in the level of the water table. As silica leaches out of the soil, hydrous aluminum oxides concentrate (figure 15.13). Further chemical alteration of laterites leads to the formation of the aluminum ore **bauxite.** The deposits in Jamaica and in parts of the Gulf Coast, notably Arkansas, are sources of the bauxite used in North America to produce aluminum.

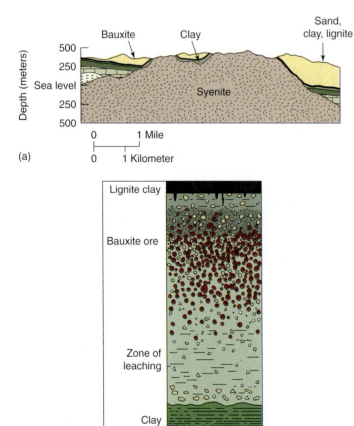

(a)

(b)

Figure 15.13 **Bauxite Deposits.** (a) Deposits of bauxite in Arkansas, the largest in the United States, as a thin layer on top of a large igneous intrusion containing a high percentage of aluminum-rich feldspar. (b) The bauxite is covered with organic-rich clay and underlain by a layer of clay that rests on the igneous rocks from which it weathered.

Nickel deposits in New Caledonia, the world's largest source of nickel, also formed as a result of weathering processes. The nickel is in peridotite, a magnesium- and iron-rich rock derived from the mantle. Chemical reactions between hot water and the peridotite change the original rock into serpentine. In the tropics, these masses of peridotite weather to a lateritic soil. As the peridotite decomposes, silicate minerals break down, and hydrous silicates of magnesium and nickel form in the laterite.

Chemical weathering by ground water causes oxidation of some ore minerals, such as pyrite. Acids produced by this oxidation dissolve and react with other minerals. As the host rock that contains ore elements weathers, the valuable elements leach out of the host rock and move downward in solution. If these metal-bearing solutions encounter water or rocks with which they can react, such as limestone, the metals may precipitate. As this process continues over time, the deep deposit is enriched in the metals leached from above. Pyrite (iron sulfide) and chalcopyrite (copper, iron sulfide) oxidize to form limonite and sulfuric acid. Iron oxide (limonite) forms where this process takes place, producing a surface cap called **gossan** on the ground above the ore body. An enriched zone of the metal develops below the gossan. Copper and silver deposits are also enriched in this manner. Prospectors who looked for the yellow limonite deposits in the soil found many of the major copper deposits in the southwestern United States.

Weathering processes also played an important role in the formation of certain types of banded ironstone, a significant source of iron ores in many parts of the world, including the Great Lakes area of North America. These rocks contain thin alternating bands of chert (silicon dioxide) and a variety of iron-bearing minerals. All of these formations are of Precambrian age and occur in shield areas. Most geologists agree that the iron in the formations originated as a result of deep weathering of continental crust and was deposited as mineral precipitates. The absence of clay minerals in the deposits and the fact that all are Precambrian in age has promoted debates that continue about the conditions under which the weathering occurred that released the iron from rocks. Some argue that these deposits could form only when little or no free oxygen existed in the atmosphere.

SOILS

Products formed during the decay, decomposition, and disintegration of rock combine with organic materials to make up soil. Most soils contain fluids including water, air, organisms, and decaying organic matter. **Residual soils** form from the bedrock materials that lie under them. However, soil may develop on the deposits left by streams, glaciers, or the wind, and soil itself may be transported from one place to another. Consequently, soils formed in one place on one type of bedrock may lie on bedrock to which they are unrelated. The top layer of most soils contain organic matter, such as roots, leaves, and worms. These materials provide important sources of nutrients, and they hold soil moisture, which facilitates plant growth. Water that infiltrates the soil picks up ions dissolved

Figure 15.8 Iron-Stained Rocks and Soil Exposed in a Gully That Developed on a Steep Slope. Iron and manganese stains often form in soil or on rock faces where minerals that contain iron are present.

Figure 15.9 Decaying Feldspar in Granitic Rock. Once the feldspar is converted to clay, the quartz in the rock is liberated.

where oxygen is scarce. Where reduction affects iron compounds, a greenish color forms.

Hydration Combination of water with other compounds, a process known as **hydration,** is exemplified by the mineral anhydrite (calcium sulfate), which can absorb water and become the mineral gypsum, hydrous calcium sulfate. Gypsum, a major building material used to make drywall, forms as layers of sedimentary rock deposited under arid conditions.

Hydrolysis Chemical reaction between water and other compounds is called **hydrolysis.** These reactions, especially when combined with the effects of carbon dioxide, are extremely effective in weathering common rocks and minerals. Hydrolysis is important because the process is involved in chemical weathering of silicate minerals, the most abundant minerals in igneous and metamorphic rocks. The following formula describes the chemical weathering of K-feldspar (orthoclase) (figure 15.9):

$$\text{K-feldspar} + \text{water} \rightleftharpoons \text{potassium} + \text{clay} + \text{silicic acid}$$

Weathering of feldspars is the main source of clay, which is a stable mineral at surface temperatures and tends to accumulate in the soil. The silica set free in the preceding chemical reaction becomes part of a silicic acid (H_4SiO_4) solution. This solution moves through the soil and into bedrock as part of ground water. Plant roots absorb this form of silica and use it to manufacture cellulose.

Figure 15.10 Petrified Wood. The cell structure of the log is clearly visible in this specimen located in the Petrified Forest in Arizona.

If the concentration of silicic acid is high enough, silica may come out of the solution as colloidal (see chapter 1) or amorphous silica. It may recrystallize into small quartz crystals that line open fractures or cavities in the rock, or it may fill in spaces in buried wood, creating petrified wood (figure 15.10). Opal and chalcedony are two other forms of recrystallized silica.

Chapter Fifteen Weathering and Soil Development **325**

Minerals differ greatly in solubility. Most minerals, with the exception of calcite, dissolve more rapidly in hot water than in cold water. Most silicate minerals dissolve slowly in pure water. In contrast, salt and gypsum dissolve rapidly in water. The opening photograph for this chapter shows gypsum pockmarked by solution effects in the semiarid region of southwestern Utah. Consequently, salt and gypsum are rarely found in the soils of humid regions because water moving through the soil removes them. Over a long period, the solubility of minerals in the bedrock affects the development of the landscape. This happens where carbonate minerals, such as calcite, which dissolves in water containing carbon dioxide, are part of the bedrock. In regions of high rainfall, belts underlain by carbonate mineral usually become valleys. Belts of limestone lie beneath many of the valleys in the Appalachian Mountains. In contrast, limestone forms ridges in arid and semiarid regions like those in the western United States.

Carbonation Carbonation is the process in which carbon dioxide combines with oxides of calcium, magnesium, sodium, and potassium to form carbonates of these metals. For example, calcium oxide may combine with carbon dioxide to become calcium carbonate—the mineral calcite.

Carbon dioxide in the atmosphere combines with water to form a weak acid, called **carbonic acid** (H_2CO_3).

$$CO_2 + 2H_2O \rightleftharpoons H_2CO_3 \rightleftharpoons H^+ + HCO_3^-$$
carbon dioxide + water   carbonic acid   hydrogen ion + bicarbonate ion

Consequently, normal rainwater is a weak acid. It has a pH of about 5.6 and is the most common solvent acting on Earth's crust. It reacts with the most abundant silicate minerals in the crust. For example, water containing carbon dioxide reacts with feldspar to produce clay, silica, and a solution containing ions. Reactions involving the breakdown of silicate minerals by reactions with

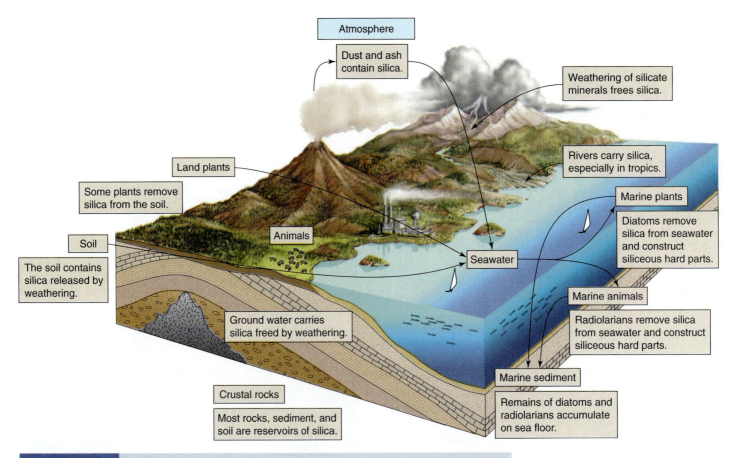

Figure 15.11 Biogeochemical Cycle Involving Silica. Carbon dioxide released from volcanic activity or other processes into the atmosphere combines with water and becomes a weak acid that reacts with many silicate minerals. Part of this silica moves in ground water and is incorporated in petrified wood or deposited as crystals in fractures. Streams carry these weathering products into the ocean where diatoms (algae) and radiolarians (single-celled animals) remove the silica and use it to construct hard parts. When the organisms die, the siliceous remains sink to the sea floor and become part of the sediment. Silica also enters the atmosphere where it is part of the dust particles blown from deserts or volcanic dust. Hot waters that come to the surface at geysers or hot springs may bring silica back to the surface where it is deposited.

from minerals at the surface and deposits these ions at lower levels in the soil. However, if excessive amounts of moisture pass through the soil, leaching of valuable elements may leave the soil deficient in elements needed for plant growth.

Geologists, engineers, and agronomists are interested in different aspects of soils. To geologists, soil is primarily a product of weathering. It develops through processes involving the interaction of rocks, organisms, and components of the atmosphere. Engineers are primarily concerned with the strength and structural characteristics of surficial materials. They study how soil behaves as a foundation material and whether it will swell when wet or otherwise affect a particular construction project. Engineers may refer to all unconsolidated surficial materials, including stream alluvium, lake-bed deposits, windblown sand, and weathering products, as soil. To agronomists, soil is the medium in which agricultural products grow. A specialized field called *soil science,* or *pedology,* is concerned with the study of the origin, use, and conservation of soils.

Soil Horizons

Although some soils are uniform from top to bottom, many, especially residual soils, contain three well-defined layers called the **A, B,** and **C horizons** (figure 15.14). An additional horizon, labeled **O horizon,** is used to designate dead leaves and animals that lie on the surface. A horizon may be missing in some soils; others may have subhorizons of the three main horizons, but in all cases, unweathered parent material, consisting of solid rock or sediment, lies beneath the soil. The lowest soil horizon, the C horizon, contains weathered parent material, loose and partly decayed rock from which the overlying soil originates. The B horizon, also known as **subsoil,** consists of finer material than that in the C horizon. Some of the materials from the overlying soil accumulate in the B horizon. For example, it may have a high clay content that has accumulated from clay-rich sediment or rocks containing minerals that break down to form clay. Consequently, soil in the B horizon is harder when dry and stickier when wet than the soil in either of the other horizons. In humid and tropical climates, the subsoil contains high levels of iron and aluminum hydroxide, but calcium, gypsum, and other soluble materials are removed by infiltrating water. In arid climates where little water is present, these soluble constituents may accumulate in this layer.

The top layer of soil, known as the **topsoil,** or A horizon, is the site of maximum organic activity. Processes that involve removal of soluble constituents in water are most active here. From the top down through an idealized A horizon, one encounters leaves and loose, largely undecomposed organic debris, matted and partially decomposed organic material, a mixture of organic and mineral matter, and finally, a transition into clay-rich subsoil.

Soil Chemistry

The minerals quartz, feldspar, micas, pyroxene, amphibole, and olivine comprise most igneous and metamorphic rocks. Of these, quartz does not decompose rapidly. The others break down into clay

(a)

(b)

Figure 15.14 **Soil Horizons.** (*a*) Soil horizons A, B, and C are clearly shown in this cross section of soil. (*b*) The soil formed by weathering of a limestone that contains clay impurities is very thin in this exposure. Roots from plants on the surface extend down into the partially decayed rock.

minerals, iron compounds, silicon dioxide, and other ions in solution. Because the minerals in sediments and sedimentary rocks ultimately come from the weathering and breakdown of igneous and metamorphic rocks, these sediments and sedimentary rocks contain many of the same minerals as the crystalline rocks. Sedimentary rocks form the underlying rock beneath soil over nearly three-quarters of the land surface. The most common sediments are clay, sand (particularly quartz sand), lime mud, and mixtures of these sediments or their sedimentary rock equivalents—shale, sandstone, and limestone. Of over 100 elements, soils generally contain large quantities of only a few. Many more elements are usually present in trace amounts.

Clay, the most common weathering product of the minerals in igneous and metamorphic rocks, is present in most sediments and sedimentary rocks. Of the various minerals in soils, clay minerals are particularly significant. Clay colloids (see chapter 1) contain a number of elements needed by plants. Although the colloids have a negative electrical charge that attracts and holds positive ions to their surface, the bonds help keep them from being leached by water. Nevertheless, because colloids are so small, they readily enter into chemical reactions in soil and supply plants with the nutrients needed for growth. These nutrients come from the clay colloids, fine mineral fragments, and organic matter in the soil. Clay also holds soil moisture, which is the main medium involved in transferring nutrients from minerals to plants.

Soil moisture picks up ions produced as rock fragments decompose (see figure 1.16). Some of the ions remain in the soil moisture solution; other ions (all positive ions) adhere to the surface of negatively charged clay colloids. Hydrogen, calcium, sodium, and smaller quantities of aluminum, magnesium, and potassium remain attached to the clay. These ions move between the soil moisture and solids or clay surfaces. As plants remove ions for use as food, the minerals, organic matter, and clay release more of those ions. Soils differ in their ability to exchange ions and maintain a supply of the ions needed by plants. Their capacities to do this determine the soil's desirability as a medium in which plants may grow.

Soils containing large quantities of hydrogen and aluminum are acidic. Those with calcium and magnesium are neutral, and soils that contain sodium are alkaline. Although most plants thrive in soils with a pH between 6 and 7, each type of plant grows best in soils of a particular acidity. If soil is acidic, it is necessary to apply lime to neutralize the acid for crops that require neutral pH. Conifers and many shrubby plants prefer slightly acidic soils. Many other plants prefer slightly alkaline conditions. When plants die, the organic acids formed contribute to the weathering processes that produce the soil.

In uncultivated areas, most of the ions plants remove while they are alive are returned to the soil when the plants die and the organic matter decays. In most soils, the total supply of plant nutrients exceeds the needs of resident plants. However, when farmers harvest crops and remove the plant matter from the land, the soil loses these constituents. Similar losses happen where farmers use slash-and-burn techniques to clear forested land for agricultural use.

Because most of the physical, chemical, and biological processes involved in the formation of soil take place slowly, the passage of time is important in the development of soils. However, soils are easily degraded. In a matter of minutes or hours, a flood or a bulldozer can remove soil that formed over many decades or centuries. Once removed, soils do not regenerate quickly. For this reason, loss of soil is an especially serious problem.

Soil Types

A number of different types of soil may develop from the same parent rock. The slope of the ground and especially the climate (see chapter 13) determine the type of soil that develops (figure 15.15).

The soil classification currently used by the U.S. Department of Agriculture is shown in table 15.2. The classes listed are called orders (figure 15.16). All of the order names end with *sol,* meaning *soil.* Each order includes suborders defined in a book published by the U.S. Department of Agriculture entitled *Soil Taxonomy.* Two classifications of soil types are widely used. Listed in table 15.3 are long-established soil types with definitions from the U.S. Department of Agriculture yearbook. Note the relationship between soil thickness and environmental factors such as precipitation, evaporation, and temperature shown in figure 15.17.

Soil Degradation

Loss and degradation of soil are not new phenomena. In earlier times, farmers simply left degraded soil behind and moved to areas that offered better soil, but in the twentieth century, the area under cultivation expanded. Few areas of unused soil remain, and soil degradation has become an increasingly serious problem. Since 1950, the world population has more than doubled. This rapid growth has placed great pressure on the world's soil resources. Intensive agricultural development, deforestation, overgrazing, and human neglect have caused serious degradation of soils. In arid

Table 15.2	Data Bank: Soil Classification

Alfisols: Soils with gray to brown surface horizons and high supplies of exchangeable cations (positively charged ions) other than hydrogen. They usually form in moist areas that are usually dry during the summer. The subsurface horizons are composed of clay accumulations.

Andisols: Soils that have volcanic ash as a parent material.

Aridisols: Soils with soil horizons that have low organic content. These soils form in areas that are dry at least half of the year.

Entisols: Young soils that do not have well-developed soil horizons.

Gelisols: Soils formed on frozen ground.

Histosols: Soils with high organic content.

Inceptisols: Moist soils with soil horizons formed from parent materials, but they lack horizons in which clay or other substances have accumulated.

Mollisols: Soils with so much organic content that the surface horizon is nearly black.

Oxisols: Soils that have an oxic horizon.

Spodosols: Soils that have accumulations of amorphous materials in subsurface horizons.

Ultisols: Soils that are usually moist, have layers of clay accumulation, and low supply of exchangeable cations.

Vertisols: Soils that contain large amounts of swelling clays. Wide cracks develop in these soils during dry seasons.

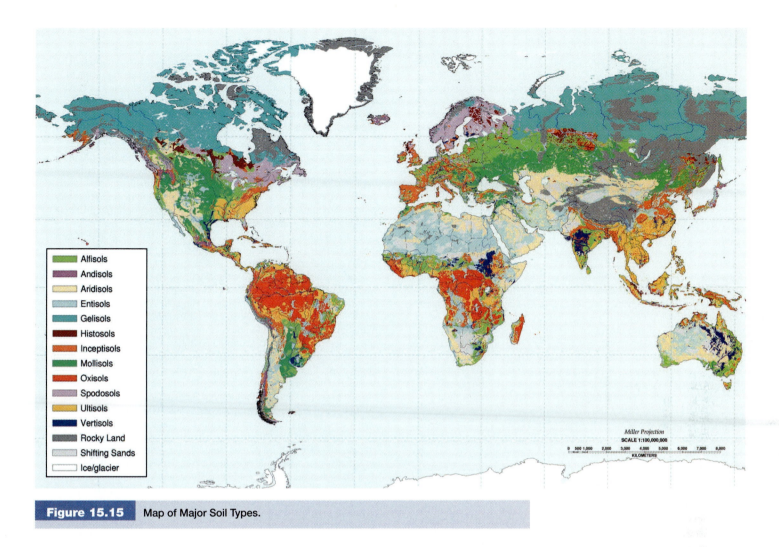

	Alfisols
	Andisols
	Aridisols
	Entisols
	Gelisols
	Histosols
	Inceptisols
	Mollisols
	Oxisols
	Spodosols
	Ultisols
	Vertisols
	Rocky Land
	Shifting Sands
	Ice/glacier

Miller Projection
SCALE 1:100,000,000

0   500 1,000   2,000   3,000   4,000   5,000   6,000   7,000   8,000
KILOMETERS

Figure 15.15   Map of Major Soil Types.

regions, these neglects can transform the area into a desert. The World Resources Institute reports that an area the size of China and India combined has suffered moderate to extreme soil degradation. Erosion by water and wind removes soil and causes long-term losses that are difficult to repair. Degradation also occurs as chemical and physical alterations that affect soil fertility and its usefulness. However, it is much easier to reverse chemical and physical degradation than it is to replace soil.

Causes of soil degradation vary considerably from region to region. For the world as a whole, overgrazing followed by erosion is the leading cause of soil degradation. This practice accounts for about 35% of the total. Close behind overgrazing are deforestation and poor agricultural practices, such as excessive or insufficient use of fertilizers and cultivation on steep slopes or in arid regions. Each of these problems accounts for about 30% of the total. Stripping land of vegetation for fuel causes approximately 7% of soil degradation. Industrialization, including use of land for waste disposal and land severely affected by airborne pollutants, accounts for 1%.

Water Erosion   The U.S. Natural Resources Conservation Service (formerly the Soil Conservation Service) estimated that in 1994,

erosion removed about 5.7 tons of soil per acre on land in cultivation. Steep slopes and loss of vegetation expose soil to this type of degradation. In its initial stage, the easily recognized problem of erosion appears as small widely spaced rills—miniature valleys—cut by water as it drains off the land. As erosion continues, the rills deepen and eventually form deep gullies. Initially, the gullies are widely spaced, but gradually they expand, and eventually they merge to form a topography called **badlands** that is inaccessible for cultivation (see figure 14.5a). However, shallow erosion and the flow of sheets of water across the ground surface remove most of the nutrient-rich topsoil long before badlands form. Where water erosion is especially severe, it may eventually remove all of the soil down to the level of bedrock. On steep slopes, this takes place rapidly once gullies form.

Wind Erosion   In arid and semiarid regions, wind and water account for removal of soil. Low levels of soil moisture and loss of vegetation make soil particularly susceptible to wind erosion. Overgrazing, deforestation, and poor agricultural practices also expose soil to wind erosion. Wind erosion was a major problem during the dust-bowl conditions that devastated parts of Oklahoma and Texas in the 1930s. Soil losses were especially great at that

Table 15.3 Data Bank: Types of Soil

Desert Soil: Desert soil is a zonal group that has light-colored surface soils and usually is underlain by calcareous (e.g., limestone) material and frequently by hard layers. Desert soil develops under extremely scanty scrub vegetation in warm to cool arid climates.

Large portions of some deserts have no soil cover. In the absence of moisture, wind erosion produces large expanses of bare rock, sand dunes, and lag gravels that consist of boulders, cobbles, or pebbles left as a residual cover on deserts when the wind has blown finer sediment away. Where soil is present in a desert, it is usually thin, loose, and brittle. The low rainfall prevents leaching of soil and favors accumulation of salts, especially in low areas of desert basins. As water close to the surface evaporates, salt and calcium carbonate crystallize in the soil, creating a layer called **hard pan.**

Tundra soil: Tundra soil is present on high mountains at all latitudes, but the largest areas of tundra lie at high latitude in the Northern Hemisphere. Tundra forms where cold weather prevails most of the year. In the Arctic, permanently frozen ground, the permafrost, is present within a meter of the surface. In addition to the permafrost, soil at the surface freezes in winter. As freezing takes place, water crystallizes out in layers of irregular size, disrupting the soil and forcing it up into small mounds. A dense growth of moss, sedges, and flowering plants usually covers these mounds. In summer, when ice in upper portions of the soil melts, swamps and lakes form. Low slopes, permafrost, and glacial deposits interfere with surface drainage. Though moisture is abundant on the ground, low air temperature reduces the rate of evaporation, slows chemical reactions, and inhibits the growth of most plants. Consequently, the soils are thin.

Podzol Soil: A zonal group having surface organic mats and thin layers of organic materials that lie over gray leached horizons are called podzol soils. Dark brown layers composed of colloids and soluble salts form a lower zone. These layers develop under coniferous or mixed forests or under vegetation consisting of low evergreen shrubs having whorls of needle-like leaves. All of these plants grow in a cool-temperature, moist climate.

Podzol soils are the traditional agricultural soils of Europe and North America. They are also present in the central and eastern United States, which has a humid climate and large areas of hardwood forest cover. One of the most common varieties of podzol soils has a gray-brown color, is crumbly, and is composed of silty loam, a mixture of clay, silt, sand, and organic matter. In wooded areas, it has a matted layer of decaying leaves near the surface. The A horizon of podzol soil supports abundant plant growth and animal activity. The B horizon is partially leached. It is a light gray color and is richer in clay, coarser in texture, and darker in color than the A layer. The A and B normally have an aggregate thickness of about 1 meter (3 feet). The C horizon contains blocky clay. Soil profiles above rocks as different as limestone, shale, and glacial deposits share similar characteristics, despite the different compositions of the parent rock. The soil is acidic and requires liming, but it supports a wide variety of plants, including grains, legumes, root crops, vegetables, and fruits.

Chernozem Soils: The zonal group of chernozem soils has a deep and dark-colored surface horizon that is rich in organic matter. The top layer grades into lighter-colored soil below. At 0.5 to 1.3 meters (1.5 to 4 feet), calcium carbonate accumulates in layers. Chernozem soil develops under tall and mixed grasses in a temperate to cool subhumid (semiarid) climate.

These soils develop in climates where moisture is not abundant or consistent. A large expanse of this soil group underlies the plains of North America—a major wheat-producing area. The low level of soil moisture reduces leaching of the soil. Because ground water does not remove ions from the soil, a layer of lime usually accumulates in the subsoil, and it does not need fertilizer.

Latosol soils: The zonal latosol soils are characterized by deep weathering and abundant hydrous-oxide minerals. They develop under tropical forest where rainfall is abundant. Areas with tropical and subtropical climates have a greater variety of soils than all other climates combined. This is due in part to the combinations of high temperature, variable amounts and distribution of rainfall, and topography. Tropical climates generally have heavy rainfall. Portions of the tropics have up to 10 meters (33 feet) of rainfall per year. In some areas, rain falls throughout the year, while others have hot, humid periods of high rainfall separated by warm dry spells.

Soils in this climatic belt are usually thick. The conditions under which they form are ideal for chemical weathering. Plant matter is abundant, and it quickly decomposes into soil. Decomposing minerals release plant nutrients rapidly, but the high rainfall soon washes these nutrients out of the soil. Common farming practices in some areas consist of clear-cutting the forest and then cultivating the land for four or five years until its fertility decreases. When the land no longer produces good crops, the farmer abandons it, allowing it to grow into forest again, and cuts another section of forest. It takes several decades for the soil to regain fertility after farming. Because leaching of nutrients out of the soil is so extensive, soils are richest where new mineral supplies that provide the needed nutrients can continually rejuvenate them. These mineral supplies come from volcanic ash, stream deposits, or continued erosion of soil parent material. Generally, these mineral nutrients build up at a slow but continuous rate. However, fires that burn near-surface organic matter during dry seasons interrupt the buildup of nutrients and leave the ground exposed to erosion and runoff. The runoff following heavy rains removes the nutrients from the plant ash before they return to the soil.

Laterite: Laterite is a special type of clay-rich soil that has a tendency to harden into a brick-like solid. Laterite generally develops where rainfall is highly variable. In these places, the level of the water table fluctuates from within half a meter of the ground surface to depths of many meters. Ground water in these places leaches silica from the soil and causes hydrous aluminum oxides to concentrate. These soils contain so much aluminum that they are potential sources of the metal. Alteration of laterites leads to the formation of bauxite, the principal ore of aluminum.

Alluvial Soils: Alluvium is a material transported and deposited by streams. Large streams flowing on low slopes usually have wide valleys. During floods, these streams deposit alluvium on the flat areas next to the stream. Soils that develop on this transported alluvium are often rich and provide valuable farmland, although this land is susceptible to flooding.

Mountain Soils: A great variety of soil types may cover mountain slopes. In many mountainous areas, several types of soil form close together. Because climatic conditions in mountains depend on elevation, the soil often varies from the lower slopes to the top of a mountain. On slopes, mountain soils contain fragments of broken rock in the process of slow transport down the mountainsides.

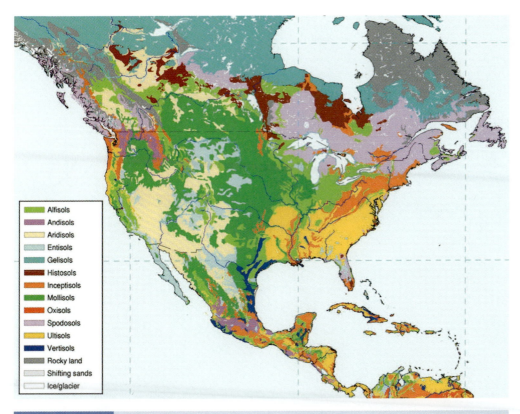

Alfisols
Andisols
Aridisols
Entisols
Gelisols
Histosols
Inceptisols
Mollisols
Oxisols
Spodosols
Ultisols
Vertisols
Rocky land
Shifting sands
Ice/glacier

Figure 15.16 Map of Large Parts of North America Showing the Distribution of Soil Classes as Defined by the Recent Soil-Classification System.

time because farmers replanted their fields after drought conditions caused their crops to fail. Repeated plowing of the land caused loss of soil moisture and left the dry soil exposed to the effects of wind erosion. Strong winds blew the finest soil particles away and piled sand-sized particles in dunes.

Chemical Degradation Changes in soil chemistry often produce serious damage to soil. Human activities, especially increases in salt content of soil and excessive use of fertilizers or pesticides, cause most chemical soil degradation. The salt content of soil may increase as the result of several practices. Excessive pumping of ground water in coastal areas causes seawater to intrude into the ground-water reservoirs. When farmers or homeowners use this brackish water for irrigation, the salt content of the soil increases. Water used for irrigation picks up salts near the surface as it infiltrates back into the ground. Gradually, the salt content of the water increases as the same water passes repeatedly through the irrigation system. Pesticides, industrial pollutants, and acid rain also contribute to chemical deterioration of soil. In addition, poor agricultural practices may remove nutrients from the soil.

Physical Degradation Any change in the structure of soil that impairs its use for agricultural purposes results in physical degradation. Vehicles and trampling by cattle cause compaction of

clay-rich soils. As the clay compacts, it hardens, decreasing the infiltration of water into the soil. As infiltration decreases, more water runs off on the surface, and water erosion increases.

Geography of Soil Degradation

Overgrazing and stripping of land for firewood are damaging to soils in arid and semiarid regions. These problems affect vast areas in northern Africa, areas around the Mediterranean, and portions of North America and Australia. In Europe, intensive industrialization and urbanization, accompanied by problems related to waste disposal and pollutants, affect large areas. In the United States, the Natural Resource Conservation Service estimates that nearly 25% of the cropland is eroding so rapidly that it cannot be sustained for agriculture. Problems with salinization affect areas in the arid southwestern states, the rich soils of the Great Plains have lost some of their productivity, and water erosion remains a problem where marginal farming is practiced. The areas least affected by degradation lie in the high latitudes of North America and Eurasia.

Soil Reclamation

Time is an important element in the formation of soils. Most of the processes involving chemical and biochemical reactions with minerals in the source material occur slowly. Consequently, soils

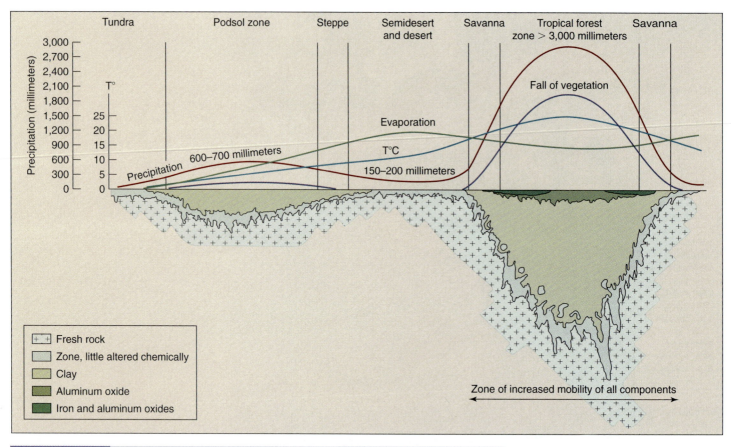

Figure 15.17 Relationships Between Major Soil Types and Climatic Conditions Responsible for Their Development.

Legend:
- Fresh rock
- Zone, little altered chemically
- Clay
- Aluminum oxide
- Iron and aluminum oxides

Zone of increased mobility of all components

evolve. In some areas where weathering has continued for long periods, the soil may extend to depths of more than 33 meters (100 feet). The long time needed for soil formation makes it especially important that productive soils be protected and preserved. The greatest protection is currently given to soils used for growth of organic produce. As more people purchase food products that are free of all types of toxins, interest has grown in protection of soils from pesticides, herbicides, pollutants of all types, and even chemical fertilizers. The effort is directed toward producing "organic soils" suitable for growth of organic produce. This recent trend in specialized farming has limited application compared with the more general efforts of government agencies to reduce soil losses.

In the 1930s, the U.S. government created the Soil Conservation Service, now known as the Natural Resources Conservation Service, within the Department of Agriculture. The primary objectives of this agency are to save soils and improve crop production by encouraging farmers to grow crops that are well suited to the types of soils present in a given area. Programs developed by the Natural Resource Conservation Service have reversed much of the damage caused earlier when farmers cultivated a piece of land until crop yield began to fail or gullies made it impossible to plow

the land. Erosion control measures, such as filling gullies, planting trees on badly eroded slopes, planting lines of trees or shrubs to reduce wind erosion, and use of contour plowing, have all been highly successful. In contour plowing, the farmer plows parallel to contours rather than up and down slopes. The furrows that are parallel to contours catch runoff and increase infiltration, whereas furrows perpendicular to contours provide ready-made channels for runoff and greatly increase soil erosion.

Soil reclamation is now required for most mining operations, especially those involving strip mining in which topsoil is removed. Generally, these regulations stipulate that the land be restored as nearly as possible to its original configuration and that topsoil be set aside before mining and used to cover the mined area after the mining is complete. Efforts are also made to reduce surface runoff by use of barriers such as bales of hay or plastic sheeting that prevent soil from washing away. Such barriers are used on most construction sites as well as around mines and other areas where the soil cover is removed.

Although many problems remain, the methods needed to reduce soil loss and degradation are now well known and widely applied, especially in North America and Western Europe where government agencies are dedicated to solving these problems.

SUMMARY POINTS

1. Weathering refers to the decay and disintegration of rocks where they come in contact with the atmosphere, water, and living organisms. Mechanical weathering processes, such as expansion and contraction caused by solar radiation, crystal growth, unloading, and freezing and thawing, cause rocks to disintegrate. These processes break down the whole rock. Chemical weathering processes, such as oxidation, carbonation, dissolution, and hydrolysis, which act on surfaces, cause rocks to decompose. Because chemical processes act on surfaces, smaller particles are more prone to chemical weathering than larger particles.
2. Rocks and minerals differ greatly in their susceptibility to weathering and erosion.
3. Soil is a product of weathering. Some soils have distinct zonal structure, and others have no zones. In representative zonal soils, the upper A horizon contains organic matter; the B horizon is leached and rich in clay; the C horizon consists of decaying fragments of the bedrock. Among the factors that influence the development of soil are composition of the parent rock, climatic conditions, and slope.
4. Clay plays an especially important role in the value of soil for plant growth. The clay holds and releases nutrients to the roots of plants.
5. Soil degradation is an important environmental problem. Soil erosion, chemical degradation caused by changes in soil chemistry, and physical degradation, such as the compaction of soil by traffic and cattle, may severely damage soils.

REVIEW QUESTIONS

1. What common rocks are most susceptible to weathering by dissolution?
2. How does acid rain form?
3. What bedrock conditions help neutralize effects of acid rain in streams?
4. Where is freezing and thawing most effective as a weathering agent?
5. Why does clay play such an important role in soil chemistry?
6. What are the main causes of soil degradation?
7. How do the soils formed in humid climates differ from those formed in arid climates?
8. What is the role of freezing water in weathering processes?
9. Based on susceptibility to weathering, what minerals would you expect to find in beach sand?

THINKING CRITICALLY

1. How might weathering processes in the continental North America change if Earth's climate experienced a shift toward cooler temperatures? Toward warmer temperatures?
2. Explain why limestone strata commonly form cliffs in arid climates but most cliffs in humid climates are composed of quartz minerals.
3. What steps should individuals take to protect soil on land they own?
4. What steps can a community take to help protect soil?

KEY WORDS

A horizon	hydrolysis
badlands	laterite
bauxite	mechanical weathering
B horizon	O horizon
caliche	oxidation
capillary action	permafrost
carbonation	pH
carbonic acid	regolith
chemical weathering	residual soils
C horizon	sheeting
differential weathering	soil
frost heaving	spheroidal weathering
gossan	subsoil
hard pan	topsoil
hydration	

chapter sixteen

Mass Wasting— The Work of Gravity

Chapter Guide

The characteristics of Earth's surface, on land, beneath the sea, and inside Earth, change continuously because gravity exerts a downward pull on everything (figure 16.1). This pull affects all geomorphic agents—streams, glaciers, ground water, wind, and waves. It causes water in streams and valley glaciers to move downslope; it makes ice sheets spread laterally; it causes water in the ground to percolate downward; it causes sand, ash and dust particles in the air and various solids of all sorts in the oceans to settle to Earth's surface; and it causes seismic sea waves (tsunami) to travel far away from their place of origin. The long-term effect of this downslope movement is to transfer material from high to lower elevations.

In the early stages of Earth history, gravity caused the separation of the high-density materials in the core to separate from the lighter materials in the mantle and lithosphere. Gravity plays an important role in the spreading of the sea floor. It causes lithospheric slabs to sink in subduction zones, causes huge masses of sediment on the sea floor to slide downslope, and it exerts a downslope pull on mountains even as they rise.

This chapter discusses the mechanics of mass wasting, famous cases that changed Earth's appearance, and ways of avoiding and preventing mass wasting.

INTRODUCTION

MATERIALS AND MECHANISMS

FAMOUS CASES OF MASS WASTING

SUBSIDENCE AND COLLAPSE

AVOIDING AND PREVENTING MASS WASTING

Cultivation on steep slopes in Guatemala led to the development of downslope movement of soil.

INTRODUCTION

Mass wasting is an integral part of Earth systems. Volcanic activity and mountain building, parts of the plate tectonic system, form mountains and elevate sections of the crust, creating the slopes involved in mass wasting. Glaciers and streams cut into volcanic materials, and the uplifted crustal rocks accentuate steep slopes. Earthquakes shake Earth's surface, triggering many downslope movements on land and in the oceans. Heavy rains associated with hurricanes and frontal systems saturate surficial materials, making them more likely to move. Freezing and thawing free fragments of massive rock exposures and other weathering processes create the clay and other fine materials involved in many forms of mass wasting. Examples will make the close connections between mass wasting and other parts of Earth systems more obvious.

The downslope movement of materials on Earth's surface takes many forms. In the winter of 1997, unusually heavy rains drenched the slopes of the mountains along the west coast of North America. As water soaked into the ground, filling pore spaces in unconsolidated sediment, large masses of soil and sediment slumped off steep hillsides. In Seattle, some of these masses buried houses, causing tragic loss of lives. Unfortunately, the downslope movement of water-saturated surficial materials is not uncommon. During volcanic eruptions, like that of Nevado del Ruiz, described in chapter 7, melting snows, heavy rains, or steam that causes torrential rains combine with the ash. The resulting slurry of mud flows down the sides of the volcano, following stream valleys. The eruption of Mount St. Helens in 1980 caused similar problems—problems that threaten most areas near active volcanoes. Oceanographers have discovered huge submarine landslides around the Hawaiian Islands. Even larger crustal blocks appear to have been involved in submarine slides that slipped off of rising mountains. The huge mass of material that now makes up the Taconic Mountains of Vermont and New York slipped off the flank of high mountains that were rising farther east in New England.

In all of these instances, slope instability took place naturally, and little could be done to prevent it. People must learn to live with and avoid many types of slope instabilities; however, modifications of Earth's surface by humans cause marginally stable slopes to become unstable. As our use of land has intensified, it has become increasingly difficult to avoid developing areas where potential natural instability exists—areas where, for example, floods, wave action, or earthquakes may trigger mass wasting. Construction of building foundations, dams, reservoirs, bridge abutments, tunnels, highways, and canals often induces slope failure (figure 16.2). Costly and sometimes even disastrous results follow where the dangers of potential mass wasting go unrecognized or ignored. By understanding the mechanisms that cause instability, it is easy to avoid much of this damage. Because slope stability is critically important in many construction projects, it is one of the subjects in which civil engineers and geologists share an interest.

Figure 16.1 Massive Blocks of Sandstone Topple off the Edge of a Cliff at Muleys Point, Utah.

MATERIALS AND MECHANISMS

The force of gravitational attraction drives mass wasting. Although this force acts uniformly on everything at Earth's surface, some materials are unstable and move downslope, subside, or collapse, while others remain stable. The main factors that determine whether a slope will remain stable or not are the angle of slope of the ground surface; the type of material and the presence of water, ice, compressed gases, or steam in the material; the existence of planes, such as bedding surfaces, faults, or fractures on which sliding may occur; and the amount of vegetative ground cover.

A few general terms are used to describe types of mass wasting. For example, the term **avalanche** is applied to rapid falling, sliding, or flowing of large masses of snow, ice, soil, debris, or mixtures of these materials under the force of gravity. However, most terms used to describe types of mass wasting have more precise meanings, and many follow a system suggested by D. J. Varnes in a report for the National Academy of Sciences (1978). In this system, the name of the type of material involved in the movement is usually combined with the mechanism of movement. Rock slides and mudflows are examples. Varnes also defined the rate of movement for mass wasting (table 16.1).

Defining Materials In describing materials involved in mass wasting, scientists and engineers use rather precise definitions of common materials. **Rock** is solid, unbroken material. **Earth** is a general term applied to soil, weathered or disintegrated rocks, and loosely consolidated sediments. The general term **debris** applies to mixtures of rock, soil, plant matter, and mud. **Soil** is composed of the products of rock disintegration and decomposition, modified by geologic agents and capable of supporting plant life. **Mud** is a liquid or semi-liquid mixture of water and finer-sized particles of earth and soil.

(a)

(b)

Figure 16.2 Slope Failure. (a) During construction of Interstate Highway 81, the toe of an old landslide deposit was undercut. The material continued to move until a large part of the landslide materials had been removed. (b) Slopes undercut in unconsolidated sediment or soil during construction often fail. This one is an example of slumping.

The Nature of Materials That Become Unstable

Some rocks are inherently strong and stable even in almost vertical cliffs. For example, the walls that glaciers cut in granitic rocks in Yosemite Valley, California, are nearly vertical for many meters (hundreds of feet). Slabs of rock that separate from the rock mass along fractures occasionally fall from these walls as they did in 1996, but the walls have remained nearly vertical for thousands of years. These walls remain stable because the minerals in the gran-

ite are only slightly weathered and they form an interlocking network of crystals that gives the rock mass great strength and stability.

Even sedimentary rocks may stand in steep cliffs where natural cement holds the grains together or the rock consists of particles that interlock. The bonds between the minerals of such rocks may be strong in all directions through the rock mass, but even massive rocks usually contain planes or zones of weakness on which movements may occur. Fractures, faults, layers containing strongly aligned minerals (especially common in some metamorphic rocks), or layers composed of materials that have low resistance to sliding may occur in bodies of otherwise massive rocks. The physical properties of sedimentary rocks are generally uniform within layers, but these properties vary greatly from one layer to another. Some rocks, such as limestone or gypsum, are soluble. Other rocks, such as sandstone or conglomerate, have open, interconnected pore spaces in which water can accumulate. Still other rocks, such as shale, may expand or become plastic when wet. These inherently weak layers yield when the rock is subjected to movements such as earthquakes, the buildup of fluids in the rock, or oversteepening of slopes.

Most unconsolidated sediments are more prone to mass wasting than compact and cemented rocks, but sediments differ from one another greatly in their physical properties and stability. Sediments composed of angular fragments in a mixture of sizes tend to be more stable on steep slopes than those containing rounded and uniform fragments. For example, windblown dust called **loess** (pronounced *luss*), which covers large areas of the Mississippi Valley, consists of angular particles that hold together so firmly that this unconsolidated material will stand in a nearly vertical road cut (see chapter 17). In many sedimentary rocks, grains are held together by natural cements such as calcium carbonate, silica, or clay. In general, these are more stable than uncemented materials just as dry sediment is more stable than wet sediment, and fragments with rough surfaces are more stable than those that are smooth. As will be evident in following sections, the mechanics of slope failure depends on the type of material (figure 16.3).

Mechanisms of Movement

Mass wasting is most likely to occur where the ground surface slopes at a steep angle and where vegetation does not hold the soil in place. Most solids remain stable where the ground slopes are less than about 35°. However, some materials, especially those that are unconsolidated and contain a large amount of water or ice, spread by flowing or yielding as a plastic material where they are not restrained by obstacles such as valley walls or mountains. Most mass wasting falls into one of four mechanisms of movement: **falls, slides, slumps,** and **flows.**

Falls

Falls may involve rocks, soil, or debris. Most rockfalls, such as the one at Yosemite in 1996, involve relatively small masses.

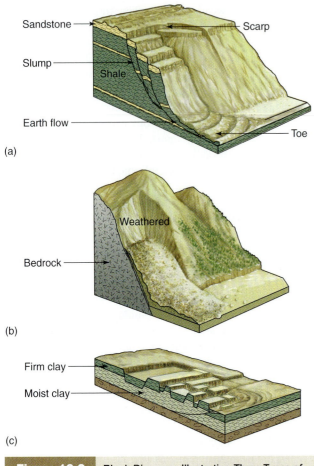

(a)

(b)

(c)

Sandstone — Scarp
Slump —
Shale
Earth flow —
— Toe

Weathered
Bedrock —

Firm clay —
Moist clay —

Figure 16.3 Block Diagrams Illustrating Three Types of **Slope Failure.** (a) Slumps, (b) debris avalanche, and (c) lateral slip on wet clay layer.

Rockfalls are extremely rapid movement of solid materials through the air. Falls generally entail rolling and bouncing, as well as free fall. Almost all falls occur on cliffs or very steep slopes, like the one that killed seven hikers at the Sacred Falls on Oahu, Hawaii. Deep, nearly vertical gulches are common in volcanic rocks.

Steep slopes are common in glaciated valleys, in places where streams have cut deep notches, along escarpments formed by active faults, along wave-cut cliffs such as those described in chapter 10, in quarries, along road cuts, and in places where cliffs are retreating. Soil falls occur where streams undercut the banks of a river.

Downslope Movement of Dry Fragmental Materials

Physical disintegration of most rock types produces fragments. Most dry fragmental materials on steep slopes exhibit similar behavior regardless of whether the fragments are sand or blocks of rock. These fall or roll downslope and accumulate as piles or sheets. Children learn about the behavior of sand while playing in sandpiles. For example, they learn that sand can stand in a cone-shaped pile. The slope of these piles is nearly constant for a given type of sand. That slope is known as the **angle of repose** (figure 16.4). All sediment has a natural angle of repose, usually about 30° to 35°. In most unconsolidated sediment, steeper slopes are unstable. Attempts to increase the slope by adding more to the top of the pile or by removing part of the base of the pile invariably fail. Fragments roll and slide down the slope until the slope reaches the angle of repose again. Dry, angular blocks and mixtures of fragments of different sizes have higher angles of repose than rounded, wet, particles of uniform size. Materials with rough or angular surfaces like those dislodged from cliffs by freezing and thawing have high angles of repose and are more resistant to slope failure than those with low friction resistance.

Once a fragment breaks free, it falls, rolls, or slides downslope, dislodging others as it goes. Finally, it comes to rest on a pile of similar fragments that accumulate at the base of the slope. This type of fragmental material, called **talus,** forms cone-shaped masses where the localized source is a narrow notch in a cliff or sheets where the rocks fall from a cliff face (see figure 16.4).

If the slope of the valley floor is sufficient, converging piles of talus may coalesce and form a tongue-shaped projection down the valley called **rock glaciers** (figure 16.5). These lobes of talus move, but the movement is extremely slow. Repeatedly checking the position of the end of the lobe over a period of years will reveal the movement. Despite their slow movement, some rock glaciers extend many miles down valley.

Most rock glaciers occur in high mountains where valley slopes are steep, where freezing and thawing occur repeatedly, and where both ice and water may become incorporated within the masses of freshly broken rock. As the proportion of ice increases, the

Table 16.1	Rates of Mass Wasting	
Extremely rapid	10 feet/second or more	(3 meters/second or more)
Very rapid	0.01–10 feet/second	(0.003–3 meters/second)
Rapid	5 feet/day–5 feet/month	(1.5 meter/day–1.5 meter/month)
Moderate	5 feet/month–5 feet/day	(1.5 meter/month–1.5 meter/month)
Slow	5 feet/year–5 feet/month	(1.5 meter/year–1.5 meter/month)
Very slow	1 foot/5 years–5 feet/month	(0.06 meter/year–1.5 meter/month)
Extremely slow	Less than 1 foot in 5 years	(less than 0.06 meter/year)

Source: after Varnes, 1958

Figure 16.4 **The Angle of Repose.** Fragments of rock are loosened by freezing and thawing on the steep slopes of the Beartooth Mountains, Montana. The loose blocks accumulate at large cone-shaped or sheet-like piles at the base of the slopes. The inclination of the surface of the cones is the angle of repose for this material.

movement of that mass changes from that typical of dry fragmental materials to movement characteristic of glaciers (see chapter 20).

Slides and Slumps

The importance of inclined planes on the movement of materials is obvious in the case of the sliding of a block on a sloping surface. The plane could be a fracture surface, a fault, a bedding plane, or the surface of the ground. To observe the effects, place a block on a board and gradually lift one end of the board until the block slides downslope. The block slides when the component of force (caused by gravity) acting down the slope overcomes the frictional resistance of the block to sliding.

The weight (W) of anything on Earth is the force (F) caused by the action of gravitational attraction on the mass (m) of the object ($W = F = mg$), where g is the acceleration of falling objects caused by gravity. Although the weight of the block on this plane tends to pull it straight down, the plane makes vertical movement

impossible. Instead, the block behaves as it would if the two forces—one parallel to the board and one perpendicular to the board—acted on it (figure 16.6). Friction between the block and the board resist the downslope movement of the block. If the block is not moving, frictional resistance to movement along the contact between the plane and the block counterbalances the component of force due to the weight of the block directed down the plane. If the slope increases or if the frictional forces between the plane and the block decrease, the block becomes unstable and slides down the plane. Friction depends on the nature of the surfaces, the presence of lubricants, and the weight of the block. The slope required for sliding is steep when the friction is great, but as friction decreases, movement occurs on lower slopes.

The same principles apply to the downslope movement of rocks, soil, or debris on a solid unyielding base. Slope failure by sliding of masses of rock on inclined planes occurs most frequently where steeply inclined strata or fractures slope in the same direction as the ground surface. Removal of material from the base of slopes by streams, glaciers, or during highway construction can destabilize the slope. Even massive rocks may slide where materials that have low frictional resistance to sliding, such as shale, salt, or gypsum, occur beneath the sliding material. The presence of water in the zone of movement also facilitates sliding. This is especially the case where clay is present because wet clay becomes slick—it has less frictional resistance to shearing.

Landslide is a general name applied to downslope movement of many different types of materials, usually soil, rock, or artificial fill. Landslides differ from other slides because the land surface rather than an internal plane (e.g., a fracture or bedding plane) is the plane along which movement occurs. Where large blocks remain intact in a slide, the name **block-glide** applies. Large blocks of material underlain by slick clay may slide on very low slopes. This is especially likely where the material is subjected to vibrations such as those that accompany earthquakes (see section entitled "Mass Wasting Triggered by the 1964 Alaskan Earthquake," page 346).

Slump is a special type of slide in which the plane of failure is a curved surface resembling the shape of a spoon (see figure 16.3). During movement, the slump mass rotates downward and outward. A curved scar forms at the top of the mass, and material in the moving mass bulges outward at the base of the slump, referred to as the toe. Sliding occurs on a curved plane. Material at the toe of the slump flows. Once the plane of failure develops, it is difficult to control slumps. Removal of the base of the slope destabilizes the mass, and the crack at the top of the slump allows water to enter the slump. The water destabilizes the mass by lubricating the plane on which slip occurs, by reducing the frictional resistance of clay minerals, and by adding weight to the slump mass.

Flows—Effects of Water, Ice, Compressed Gases, and Steam

Although water is usually present in the ground, an excess of water may cause the ground to become unstable. A number of different mechanisms may contribute to this instability, but the lubricating effect of water is perhaps most important. When water accumulates

(a)

(b)

FAMOUS CASES OF MASS WASTING

Landslide in Madison Canyon, Montana

The Madison River, one of the major tributaries of the Missouri River, has its headwaters in the area west of Yellowstone Park (figure 16.11). A large earthen dam constructed on the river created a large reservoir, called Hebgen Lake. Downstream from the dam, the river flows through a 400-meter-deep, steep-sided canyon cut in the Madison Mountains. At about midnight on August 17, 1959, a severe earthquake centered near Hebgen Lake shook the region (see chapter 6 for details). Surface waves rocked the water in the

reservoir, shook the soil on the slopes around the reservoir, and cracked the highways and ground surface. The vibrations dislodged bricks from chimneys over 150 kilometers (93 miles) away from the epicenter. Displacement along a fault on the north side of the steep, narrow canyon of the Madison River set the ground in motion. The vibrations triggered movements in decayed soils and rock debris on the south side of the canyon, and in the chaos that followed, rocks on the south side of the canyon collapsed. Approximately 27 million cubic meters (a trillion cubic feet) of rock, soil, and trees slid, fell, and flowed into the canyon, leaving a scar from the top of the mountain to the river. Debris engulfed the valley, the highway, and a campground where twenty-eight campers died beneath the mixture of rock and soil. When the mass

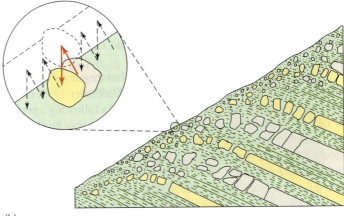

(a)

(b)

Figure 16.10   Soil Creep. (a) The ripple-like features on this slope illustrate soil creep. Cattle walking along the hillside are largely responsible for these ripples. (b) During freezing and thawing, particles in the upper few inches of soil follow the path shown in the enlarged section.

Figure 16.4 The Angle of Repose. Fragments of rock are loosened by freezing and thawing on the steep slopes of the Beartooth Mountains, Montana. The loose blocks accumulate at large cone-shaped or sheet-like piles at the base of the slopes. The inclination of the surface of the cones is the angle of repose for this material.

movement of that mass changes from that typical of dry fragmental materials to movement characteristic of glaciers (see chapter 20).

Slides and Slumps

The importance of inclined planes on the movement of materials is obvious in the case of the sliding of a block on a sloping surface. The plane could be a fracture surface, a fault, a bedding plane, or the surface of the ground. To observe the effects, place a block on a board and gradually lift one end of the board until the block slides downslope. The block slides when the component of force (caused by gravity) acting down the slope overcomes the frictional resistance of the block to sliding.

The weight (W) of anything on Earth is the force (F) caused by the action of gravitational attraction on the mass (m) of the object ($W = F = mg$), where g is the acceleration of falling objects caused by gravity. Although the weight of the block on this plane tends to pull it straight down, the plane makes vertical movement

impossible. Instead, the block behaves as it would if the two forces—one parallel to the board and one perpendicular to the board—acted on it (figure 16.6). Friction between the block and the board resist the downslope movement of the block. If the block is not moving, frictional resistance to movement along the contact between the plane and the block counterbalances the component of force due to the weight of the block directed down the plane. If the slope increases or if the frictional forces between the plane and the block decrease, the block becomes unstable and slides down the plane. Friction depends on the nature of the surfaces, the presence of lubricants, and the weight of the block. The slope required for sliding is steep when the friction is great, but as friction decreases, movement occurs on lower slopes.

The same principles apply to the downslope movement of rocks, soil, or debris on a solid unyielding base. Slope failure by sliding of masses of rock on inclined planes occurs most frequently where steeply inclined strata or fractures slope in the same direction as the ground surface. Removal of material from the base of slopes by streams, glaciers, or during highway construction can destabilize the slope. Even massive rocks may slide where materials that have low frictional resistance to sliding, such as shale, salt, or gypsum, occur beneath the sliding material. The presence of water in the zone of movement also facilitates sliding. This is especially the case where clay is present because wet clay becomes slick—it has less frictional resistance to shearing.

Landslide is a general name applied to downslope movement of many different types of materials, usually soil, rock, or artificial fill. Landslides differ from other slides because the land surface rather than an internal plane (e.g., a fracture or bedding plane) is the plane along which movement occurs. Where large blocks remain intact in a slide, the name **block-glide** applies. Large blocks of material underlain by slick clay may slide on very low slopes. This is especially likely where the material is subjected to vibrations such as those that accompany earthquakes (see section entitled "Mass Wasting Triggered by the 1964 Alaskan Earthquake," page 346).

Slump is a special type of slide in which the plane of failure is a curved surface resembling the shape of a spoon (see figure 16.3). During movement, the slump mass rotates downward and outward. A curved scar forms at the top of the mass, and material in the moving mass bulges outward at the base of the slump, referred to as the toe. Sliding occurs on a curved plane. Material at the toe of the slump flows. Once the plane of failure develops, it is difficult to control slumps. Removal of the base of the slope destabilizes the mass, and the crack at the top of the slump allows water to enter the slump. The water destabilizes the mass by lubricating the plane on which slip occurs, by reducing the frictional resistance of clay minerals, and by adding weight to the slump mass.

Flows—Effects of Water, Ice, Compressed Gases, and Steam

Although water is usually present in the ground, an excess of water may cause the ground to become unstable. A number of different mechanisms may contribute to this instability, but the lubricating effect of water is perhaps most important. When water accumulates

Figure 16.5 A Rock Glacier in the Mountains of Colorado.

in surficial materials, it fills the pore spaces and adds weight to the mass. If the water is trapped in the mass, it is bouyant. At the same time, water reduces friction between the particles of the materials and increases the tendency of the whole mass to flow.

Flow of unconsolidated materials resembles flow seen in liquids. The movements of streams, glaciers, and masses of unconsolidated rock are less distinct than one might think. Generally, the term *flow* applies to masses of continuously moving unconsolidated material containing water or ice. Streams generally carry solid material, especially clay, in suspension. As the clay content increases, at some point, so much clay is present that it seems more appropriate to call the watery mass a **mudflow** rather than a muddy stream (figure 16.7). Even though ice is solid, under pressure it will move continuously in a type of flowage described in chapter 20. Glaciers carry large amounts of rock fragments and other debris. In some places, the quantity of rock in the flowing mass is much greater than the amount of ice, and the name rock glacier is appropriate.

An intermittent water supply, lack of vegetation, and abundant unconsolidated rock debris contribute to the formation of mudflows. In the semiarid regions of the southwestern United States where surficial materials remain dry most of the year and become saturated during the heavy rains, mudflows are common. Often they follow dry periods during which vegetation is lost through fires. As water saturates the ground, soil and other unconsolidated materials exposed on the surface turn into a slurry of mud that flows down steep slopes and exits through channels and gullies. Velocities of these mudflows may approach those of streams, as much as a meter (several feet) per second. Although mudflows are not dry, many do not contain free water. The water is held in

the clay. Some mudflows are thick enough to carry boulders in suspension.

Solifluction is downslope movement of soils, rock debris, and other water-saturated fragments over a frozen base. This type of flowage occurs most frequently in regions where the ground freezes in the winter and partially thaws during the summer. When thawing occurs, meltwater saturates the upper layers of the soil, producing a watery layer of soil on top of a solidly frozen base. Water lubricates the

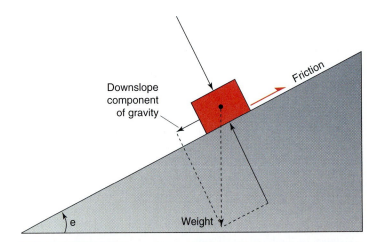

Figure 16.6 Mechanics of Sliding. A block on an inclined plane will slide when the component of force caused by gravity and directed down the inclined plane is greater than the friction holding the block in place.

Figure 16.7 A Mudflow Swept Through This House Leaving 3 to 5 Feet of Mud.

interface between the frozen and thawed layers, allowing the upper layers to move over the lower frozen ones. The frictional resistance to movement is so low that soil may flow even on low slopes. The moving mass of soil and debris takes sheet-like, lobe-shaped, or tongue-like forms as it follows the slope of the ground to lower elevations.

Earth flows involve the movement of soil and weathered rock over a well-defined basal shear surface that is roughly parallel to the ground surface. Rotation of the moving material like that found in slumps is not present. A disastrous earth flow took place where the Gros Ventre River cuts through the mountains east of Kelly, Wyoming. From 1909 until 1911, earth flowed slowly down inclined layers of shale, clay, and thin sandstones toward the Gros Ventre River. Movement began in layers of clay and was most rapid when snow melted in the spring. The moving mass left a scar up the mountain side as sliding concentrated on the slippery clay base (figure 16.8). The mass gradually formed a natural dam across the river, and water accumulated behind the dam. In 1925, the river topped the dam and rapidly cut through the uncompacted sediment. The reservoir drained rapidly and caused a flood that destroyed the town of Kelly.

Debris flows are similar to earth flows, but they involve movement of mixtures of rock plus sediment. In December 1999, disastrous debris flows struck the coast of Venezuela following a period of unusually heavy rain. Caraballeda, a city that was built along the coast on a fan-shaped mass of river deposits formed at the mouth of a river that drains a large region of high mountains, suffered severe damage (figure 16.9a). Debris flows carrying soil, river alluvium, and even boulders 3 meters (10 feet) across swept through the city, destroying houses and severely damaging high-rise buildings (figure 16.9b).

Plastic Behavior

Plastics are materials that appear to be solid but deform continuously if subjected to pressure above a certain level. This yielding takes the form of a solid flow—like that of a piece of silly putty or a pile of clay. Many natural materials exhibit **plastic behavior.** Clay and mixtures of sediment containing a high percentage of clay exhibit plastic behavior when they are wet; beds of salt are plastic even when they are dry. Clay occurs as an unconsolidated sediment deposited in lakes and on the sea floor; it forms where shales absorb water; it is a residue left as limestone dissolves; and it is a common weathering product of many rocks. Because clay is present in most soil and is part of many unconsolidated sediments, mass wasting often involves the plastic behavior of clay. If clay is present, water increases the plasticity of the sediment. Water seeping down along fractures or through pore spaces into claystone or mudstone may induce plastic flow. The failure of one of the largest earthen dams ever built, the Fort Peck Dam in Montana, resulted from the swelling and flowage of a layer of clay under the dam. As the reservoir filled, water seeped into the clay layers beneath the dam, causing them to swell and flow. The clay began to flow out from under the dam, removing support for the dam. A large section of the dam subsided; then it split apart, and the reservoir emptied. The presence of water clearly affected the stability of the clay beneath the dam.

Creep is the name given to imperceptibly slow movements that involve internal rearrangement of particles in the mass by rotation, sliding, or plastic yielding. Creep of soil on steep slopes is probably the most widespread of all mass wasting. Tilted telephone poles, stone walls, and ripple-like soil surfaces on steep slopes where cattle walk across the slope are evidence of creep (figure 16.10a). Freezing of water in soil facilitates creep. When water freezes in soil, it loosens the soil and heaves it upward. If heaving occurs on a slope, the soil moves upward perpendicular to the slope, but when the ice melts, gravity pulls the soil vertically down, and the soil creeps downslope. As a result, each cycle of freezing and thawing moves the soil slightly downslope. This effect is less pronounced where plants with deep roots are present.

Figure 16.8 The Gros Ventre Earthflow in Wyoming.

Figure 16.9 A Debris Flow. (*a*) View of Caraballeda, Venezuela before a debris flow destroyed part of the city. (*b*) Close view of buildings damaged by the debris flow.

(a)

(b)

FAMOUS CASES OF MASS WASTING

Landslide in Madison Canyon, Montana

The Madison River, one of the major tributaries of the Missouri River, has its headwaters in the area west of Yellowstone Park (figure 16.11). A large earthen dam constructed on the river created a large reservoir, called Hebgen Lake. Downstream from the dam, the river flows through a 400-meter-deep, steep-sided canyon cut in the Madison Mountains. At about midnight on August 17, 1959, a severe earthquake centered near Hebgen Lake shook the region (see chapter 6 for details). Surface waves rocked the water in the reservoir, shook the soil on the slopes around the reservoir, and cracked the highways and ground surface. The vibrations dislodged bricks from chimneys over 150 kilometers (93 miles) away from the epicenter. Displacement along a fault on the north side of the steep, narrow canyon of the Madison River set the ground in motion. The vibrations triggered movements in decayed soils and rock debris on the south side of the canyon, and in the chaos that followed, rocks on the south side of the canyon collapsed. Approximately 27 million cubic meters (a trillion cubic feet) of rock, soil, and trees slid, fell, and flowed into the canyon, leaving a scar from the top of the mountain to the river. Debris engulfed the valley, the highway, and a campground where twenty-eight campers died beneath the mixture of rock and soil. When the mass

(a)

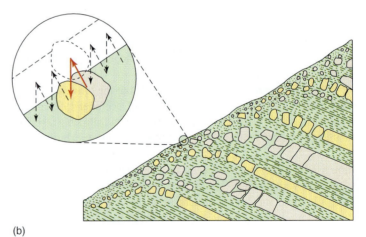

(b)

Figure 16.10 Soil Creep. (*a*) The ripple-like features on this slope illustrate soil creep. Cattle walking along the hillside are largely responsible for these ripples. (*b*) During freezing and thawing, particles in the upper few inches of soil follow the path shown in the enlarged section.

(a)

Figure 16.11 Landslide in Madison River Canyon, Montana. (*a*) View of the landslide in the Madison River Canyon. (*b*) Cross-sectional view of the slide area showing profiles of the land surface before and after the slide. (*c*) Aerial view of slump scars at the top of the slide.

(b)

(c)

settled, it had covered an area of over half a square kilometer (about 135 acres) with a layer of rock nearly 45 meters (148 feet) thick. The mass moved extremely fast—estimated at 96 kilometers per hour (60 miles per hour)—forcing air in the valley aside and creating a gale-force wind. When the slide hit the river, it splashed water, leaving mud in trees high on the sides of the valley and sending a wave of water from the reservoir up the river. Momentum of the moving mass carried this wave of debris 122 meters (400 feet) up the opposite side of the valley (see figure 16.11).

The potential for mass wasting in the Madison Canyon had existed for many thousands of years. As the river deepened its canyon, the slopes of the canyon sides became too steep to remain stable. A massive, steeply dipping layer of dolomite acting like a retaining wall held deeply weathered gneisses and schists on the upper slopes. Because carbonate rocks such as dolomite resist weathering in semiarid climates, the dolomite had decayed less than the rock upslope from it. All of these rocks—the gneiss, schist, and dolomite—inclined toward the river on the south side of the valley and contained fractures. When the earthquake caused the rock to vibrate, blocks of the dolomite became dislodged, releasing the vast quantity of decayed rock that it had held in place.

Slides, Slumps, and Flowage in the Panama Canal

Before construction of the Panama Canal in the early 1900s, it was necessary for ships traveling from the Atlantic to the Pacific oceans to go all the way around South America and through the dangerous waters of the Straits of Magellan. The economic benefits of a shortcut across land at the shortest distance between the Atlantic and Pacific in Panama were obvious. In 1905, construction on the canal project finally began. The canal was planned so that it ran through the narrowest portion of land with little regard for the types of soils and rock formations in the area. Problems encountered during the excavation almost terminated the project. Costly earth movements plagued the construction of the canal almost from the start of the project. Cutting the sides of the canal too steeply caused slumps and slides (figure 16.12). Slumping of the sides of the canal was particularly severe where the canal cut a layer of rock known as the Cucaracha (cockroach) Formation, a sediment partially composed of a highly plastic clay.

Instability worsened when the contractor lowered the slopes of the side of the canal. Ordinarily, lower slopes are less prone to failure. However, the material removed from the side of the canal

was put on the ground at the top of the cut. This additional weight on the already unstable material accelerated the movements. The situation worsened when the contractor dumped material dredged from the canal (a mixture of water and clay) at the upper edge of the cut. The weight of this mass compressed the clay in the underlying formations, squeezing the plastic clay in the Cucaracha and causing it to flow. It moved laterally in the layer until it reached the canal, where pressure from the overlying material was less. There, the clay bulged upward, forming lumps of clay in the canal. Water in the dredged material facilitated this plastic behavior. The company that originally contracted to build the canal went bankrupt as a result of these problems. At times, the mass wasting was so extensive and costly to remove that doubts were raised about the feasibility of completing the canal. However, construction continued, and finally, the workers removed almost all the clay layer beneath the canal. At that point, the sides of the canal stabilized, and it held.

Mass Wasting Triggered by the 1964 Alaskan Earthquake

Anchorage, Alaska, lies on a glacial outwash plain that slopes from the base of the high, rugged Chugach Mountains down to the sea. Sediments deposited by melting glaciers and in lakes form a terrace at the foot of the mountains. This terrace drops off abruptly along a series of bluffs about 10 to 20 meters (33 to 66 feet) above sea level. A flat-lying layer of clay, the Bootlegger Cove clay, lies beneath the portion of the terrace on which Anchorage was built. This layer is gray clay, silt, and mud deposited by melting glaciers in lagoons during the ice ages. It ranges in thickness from about 30 to 75 meters (98 to 246 feet) and lies only one to two meters below the land surface. When Anchorage was built, people took little notice of the clay and were generally unaware that unconsolidated

clay is a particularly poor soil for foundations. Wet clay can behave like a liquid under certain conditions.

When an earthquake shook the south central region of Alaska on Good Friday, 1964, the shaking of the earth caused the clay under Anchorage to fail. In some places, it slid toward the waters of Cook Inlet; in other places, it turned into a liquid-like material and flowed toward the waterfront. Great landslide cracks opened in the soil, and some houses moved along with the clay toward the sea as much as 150 meters (492 feet). About 200 acres of the terrace surface broke into a series of blocks approximately a meter across. These blocks were tilted at various weird angles to form a chaos of the bluish gray silt and clay, brownish peat and muskeg, black soil, and road pavement. This earthquake set off slope failures that cost over a hundred lives and hundreds of millions of dollars in property damage. Movements in the unconsolidated sediment beneath the city of Anchorage caused most of this damage (figure 16.13).

The Sherman Landslide

The Sherman landslide was another of many movements triggered in surficial materials by the 1964 Alaska earthquake. This landslide is of special interest because the debris that covered a large area of the Sherman Glacier traveled on a layer of trapped and compressed air (figure 16.14).

When the earthquake occurred, a massive slab of sandstone and shale on the mountainside above the glacier disintegrated. These rocks had stood as a steeply dipping, highly fractured rock mass on the side of a steep ridge. A mass of loose rock about 450 meters (1,480 feet) long, 300 meters (980 feet) wide, and 150 meters (490 feet) deep plunged down the 40° slope of the mountainside. The debris slid up to 5 kilometers (3 miles) across the snow-covered surface of the Sherman Glacier and spread out in

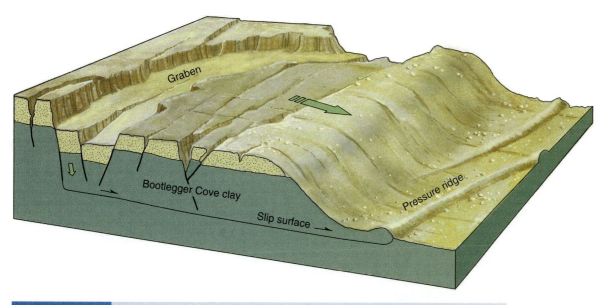

Figure 16.13
Movement of Ground Surface Caused by the 1964 Alaskan Earthquake.
Schematic diagram of the development of a graben during the subsurface movement of clay.

an irregular lobe 1 to 3 kilometers (0.6 to 2 miles) wide with a re-markably uniform thickness of 3 to 6 meters (10 to 20 feet). The debris traveled with sufficient momentum to climb 25 meters (82 feet) up the opposite valley wall, 5 kilometers (3 miles) from its source.

One of the most remarkable things about this slide is that it did not gouge a track across the snow. Instead, the slide mass spread uniformly over the surface of the glacier. The surface of the debris reflected the shape of the top surface of the underlying ice. Only at the outer edge of the slide, where debris formed a rim, did the slide plow into the glacier surface. Ice and snow compose much of this

Figure 16.14   Aerial View of the Sherman Landslide.

rim, which is up to 15 meters (50 feet) high, is 15 to 150 meters (50 to 490 feet) wide, and is characterized by a chaotically hummocky, lumpy topography. Crevasses through the debris exposed a sharp contact between the lower boundary of the debris and its contact with the underlying ice. Angular boulders of the slide material lie in contact with undisturbed, flat-lying layers of ice or snow. Had the debris been in contact with the ice while it was moving, the debris in the slide would have disrupted and gouged the top layer of snow.

The low temperatures in the area rule out the possibility that water and mud lubricated the Sherman landslide; however, dry snow or a mixture of snow and air may have lubricated it. Compressed air, trapped below rapidly falling rock as it reached a break in slope, may have formed a lubricating layer on which the slide rode. If air were a primary lubricant, then the loss of air at the leading edge would account for the pronounced increase in friction there and for the snow subsequently plowing into ridges. The high velocity of the slide mass (about 160 kilometers per hour, 100 miles per hour) had to require some type of lubrication.

Among the most unusual features of the slide are small cones of debris on top of large boulders, as though the debris dropped onto them from the air. All of this evidence indicates that the slide was airborne on top of a layer of compressed air trapped beneath the slide mass. When the air escaped, the debris dropped onto the snow surface.

SUBSIDENCE AND COLLAPSE

Even areas where the ground surface is flat may fail as a result of sudden or gradual loss of underlying support. Subsidence and collapse of the surface are examples of these vertically directed movements.

Subsidence is the gradual lowering of the surface of the ground. Generally, subsidence follows compaction or development of a void in the rocks below the ground surface. In areas where ground water has dissolved soluble rocks, voids are created that cannot support the overlying rock (discussed in chapter 19). During volcanic eruptions, large volumes of lava and pyroclastic materials move from beneath the volcano to the surface, leaving voids (see figure 7.13). Many calderas develop as the volcano subsides or collapses into these voids. Subsidence is also a serious problem in areas where underground mining removes support for the surface; subsidence over coal mines has destroyed property in a number of Pennsylvania towns. Where unconsolidated or semiconsolidated sediment is present, subsidence often occurs during or following the removal of fluids (water or oil and gas) from the ground (table 16.2 and figure 16.15). Subsidence caused by withdrawal of ground water is widespread and affects many urban areas, including Los Angeles, California; Houston, Texas (see table 16.2); and New Orleans, Lousiana. In an oval-shaped area around Houston, subsidence affects the region from the coast at Galveston to nearly 48 kilometers (30 miles) northwest of Houston. Small faults have been activated, and many house foundations have been damaged. The maximum subsidence of about 3 meters (10 feet) is located about 24 kilometers (15 miles) east of Houston. Restrictions on pumping of ground water has reduced the problem since the 1980s.

Compaction of unconsolidated sediments also contributes to subsidence. This continues to take place in the sediments in the coastal plain and beneath the continental shelf along the northern margin of the Gulf of Mexico. Subsidence also takes place in the Gulf as a result of the lateral movements of deeply buried clay and salt deposits. Such yielding of clay contributed to the problems encountered during the construction of the Panama Canal. Finally, tectonic activity and isostatic adjustments of the Earth's crust may cause subsidence over large regions (see chapters 3 and 5).

Collapse differs from subsidence in that the movement happens suddenly and usually with failure and displacement along breaks. The material may disintegrate, breaking into small pieces because of the collapse. Collapse also occurs in areas where water moves beneath the ground surface, dissolves limestone, gypsum, or other soluble rocks, and creates a cavity into which the surface collapses (see discussion of sinkholes in chapter 19). Similar collapses may occur where mining has removed rocks beneath the surface or where lava has flowed, removing support for the overlying rock and soil.

Table 16.2	Magnitude of Land Subsidence Due to Ground-Water Withdrawal in the United States
The Amount of Subsidence Is Indicated in Square Kilometers Affected	
Savannah, Ga.	330
New Orleans, La.	150
Houston-Galveston, Tex.	12,000
Stanfield, Ariz.	700
Eloy, Ariz.	1,000
Las Vegas Valley, Nev.	300
Sacramento Valley, Calif.	500
Santa Clara Valley, Calif.	650
Los Banos-Kettleman City, Calif.	6,200
Tulare-Wasco, Calif.	3,700
Arvin-Maricopa, Calif.	1,800

Source: From UNESCO

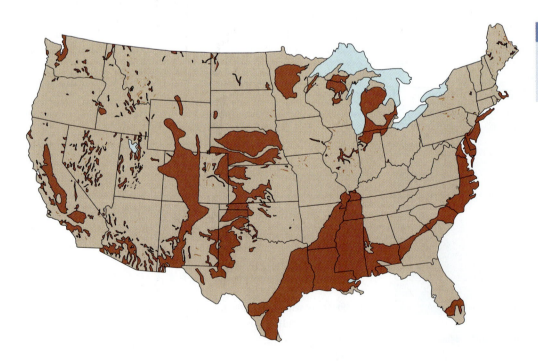

Figure 16.15 Areas of Subsidence. All of the red areas have had subsidence caused by the withdrawal of water from the ground.

Subsidence in the Los Angeles Basin

The city of Los Angeles, California, lies within a large basin filled with geologically young, semiconsolidated sediment. Mountains composed of harder and older rocks rise along the landward sides. The whole region is involved in the crustal movements related to the northwestward slip of the Pacific Plate. During the Cenozoic, subsidence and development of basins containing thousands of meters of sediment have accompanied faulting in southern California. Numerous porous layers of sediment are present in these basins, and some of these contain large volumes of petroleum. Oil and gas production accompanies development of these fields, and removal of ground water has accompanied the rapid growth in population and urbanization of the area.

The Wilmington oil field was among the first major discoveries in the Los Angeles Basin. Oil production in the field, located in the harbor area of Long Beach, started in the 1920s. As withdrawal of oil from the pore spaces continued, the unconsolidated sediment compacted, and the ground surface over the center of the basin began to subside. After twenty years of oil production, the center of the area of subsidence had dropped about 5 meters (16 feet), and subsidence of 1 meter (3.3 feet) or more extended nearly 2 kilometers (1.25 miles) from the center of the depression. As the area subsided, some points on the surface shifted horizontally as much as a meter (figure 16.16).

In 1951, the Baldwin Hills Reservoir was constructed about 11 kilometers (7 miles) southwest of the city of Los Angeles. The reservoir occupies a ravine on the north slopes of Baldwin Hills. A 46-meter- (150-foot-) high earthen dam created a reservoir with a surface area of about 83,000 square meters (20 acres) and

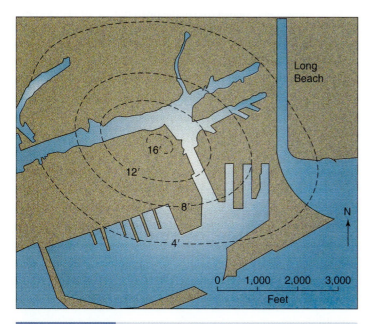

Figure 16.16 Map of the Wilmington Oil Field in Long Beach, California, Showing Subsidence from 1937 to 1951.

contained about 1 million cubic meters (900 acre-feet) of water. (An acre-foot contains about 43,560 cubic feet.) In December 1963, the dam failed, and the water swept through an urban area. Five people died in the flood, and property losses amounted to about $15 million. The dam failed when a crack in the bottom of the reservoir extended through one abutment of the dam. As water flowed through the narrow passage in the easily eroded earth, the split enlarged, and water escaped rapidly through the breach, flooding the area downstream from the dam.

The foundation of the Baldwin Hills Dam consisted of semiconsolidated sands and silts. Repeated surveys in the region revealed that the region around the dam had subsided. By 1962, an area west of the dam had subsided about 3 meters (10 feet), and long cracks in the ground had opened near the dam. Two processes—active tectonic deformation involving subsidence of the basins and horizontal displacements along the faults along the coast of California—probably contributed to the formation and movement on the cracks. In addition, the dam is on the edge of the Inglewood oil field. Large quantities of petroleum have been withdrawn from this field so subsidence caused by compaction probably contributed to the failure.

AVOIDING AND PREVENTING MASS WASTING

Human activities are involved in many incidences of slope failure. The mass wasting encountered during construction of the Panama Canal, the problems with subsidence in the Los Angeles Basin, and numerous smaller slope failures seen along roads and construction sites are examples. Human errors have been a factor in many slope failures, large and small; however, it would certainly be wrong to think that most large construction projects result in failures. Potential problems involved in large earthen-work projects are much better understood today than they were decades ago. Once the problems are recognized, engineers can design projects in ways that will avoid problems.

Generally, it is much cheaper to avoid using areas where slopes are unstable than it is to design and construct works to prevent mass wasting. However, in urban areas, high land values encourage use of open space for development. As the amount of open space decreases, the option of avoiding potentially unstable areas is increasingly difficult. Most problems arise where hazards go unrecognized until it is too late or where development occurs in the hope that preventative measures will overcome extreme hazards. The cost of the site surveys is minor compared with the potential losses or costs of repairing damage after mass wasting have taken place.

It is easy to recognize many signs of potential instability. In general, steep slopes are more likely to be unstable than are lower slopes, and slopes underlain by thick soil, unconsolidated, or semiconsolidated material are more likely to be unstable than are solid bedrock foundations. However, even solid bedrock can be potentially unstable if the rock is fractured or contains fault zones or minerals such as micas that are aligned within the rock. Evidence of past mass wasting should serve as warning of future movements. Boulder-strewn slopes beneath cliffs clearly indicate that the cliff is actively eroding. Many ancient slumps or landslide deposits are marginally stable and are easily destabilized by removing the toe of the mass.

Soil maps and maps showing areas of potentially hazardous soil and slope conditions are available for some areas. The Department of Agriculture, the U.S. Geological Survey, local planning offices, or state geologic surveys generally maintain files of this type of information. If construction on a potentially unstable area is necessary, a detailed survey of the type and extent of the unstable material is essential. Only then can engineers make plans that will ensure future stability. In some instances, the best plan involves complete removal of the unstable material. If the amount of material involved is not too great, retaining walls may be useful. Usually, it is advisable to drain unstable areas and to divert surface drainage around them.

Stabilizing Slopes

Vegetation is especially effective in stabilizing steep slopes in regolith that includes soil, rock fragments, and weathered rock. The roots of plants help hold loose materials together, and they may extend deeply enough into the ground to hold surficial materials in place on the underlying rock. In general, tree-covered slopes are more stable than bare slopes, but if movement occurs at a level that is deeper than the root system, the plants move along with the other materials. Plants also remove water from the soil, reducing the amount of water available to infiltrate to deeper levels, thereby making movements at depth less likely. However, if water reaches deep into the soil or gets into strata that contain clay, it reduces sliding friction and may make the clay much more plastic. Plants have little effect where mass wasting occurs on a deep, steeply inclined plane in the bedrock.

Not all plants are suitable for slope stabilization. A plant known as kudzu was imported from Japan to stabilize road cuts in many parts of the United States. The plant has large leaves and forms an extensive system of vines that grow rapidly during the summer months. The plant was widely used and became well established before highway departments recognized that the plant is difficult to control. The vines will grow into and form canopies over trees, eventually killing them.

The remedies for stabilizing slopes after they fail is often expensive, and continued failures are not uncommon. Retaining walls are often built in attempts to maintain steep slopes. These may be satisfactory solutions if the amount of material behind the wall is not great and if water is drained from behind the wall. If large amounts of loose material are behind the wall and if the plane of movement lies beneath the retaining wall, the wall is likely to become part of the moving mass. Methods used to stop slope failure include reducing the slope, diverting water away from the area that has failed, draining water from the unstable material, construction of retaining walls, and application of riprap to the slope.

Where failure results from excavations in unconsolidated materials that create slopes in excess of 30°, reducing the slope of the cuts is usually the initial attempt to stop additional slope failure. Some highway departments cover the reduced slopes with a layer of impervious plastic sheeting that stops water from entering the slope. This is covered with a layer of large crushed stones. This design dries out the unstable material, increases the removal of rainwater from the slope, and applies weight that inhibits lateral flow of material in the cut.

Removal of the toe of landslides or talus slopes usually destabilizes the slope and induces continued movement. Retaining walls are often used to stop the advance of fragmental materials. The wall has the effect of reducing the slope above the top of the wall. Slumps and slides are especially difficult to control. Cutting the toe of these masses rarely succeeds in stopping movement (see figure 16.2). Diverting water away from cracks that usually form at the top of slumps and slides is often an effective way to stop movement. Diverting water away from unstable soil, sand, silt, or debris that is flowing is also an effective way of controlling these forms of mass movement.

Mass movements play a role in the operation of all geomorphic agents. In the case of streams, most of the material carried by streams gets into the stream from the banks or slopes adjacent to the streams. Much of the load carried on top of glaciers in valleys also comes from slopes. Downslope movements are responsible for the development of the downwind side of sand dunes. Subsidence and collapse are involved in the development of landscapes shaped as a result of solution by ground water, and mass movement is an important process in the sliding and slumping of water-saturated sediment on steep slopes in oceans. Mass movement affects all parts of Earth's surface and is often difficult to control.

SUMMARY POINTS

1. The force of gravity pulls everything on Earth toward the center of the Earth. As a result, all materials near the surface are prone to downslope movement.
2. Water and mixtures of water and solids have low resistance to internal deformation and flow down any slope.
3. Wet clay and ice exhibit plastic behavior. They flow when overlying material reaches a certain weight. They flow downslope, and if the pressure on them is sufficient, they even flow upslope.
4. When the support beneath them is removed, solids, even large rigid blocks, may move downslope by falling, sliding, or rolling. They may also move much more slowly by creeping or solifluction.
5. Slopes become unstable and subject to failure where they become steeper than the angle of repose of the material or where the underlying support is removed. Undercutting may occur naturally as a result of streams, wind, glaciers, waves and

currents, or of solution by ground water. Frequently, undercutting results from human activities, especially highway, canal, and foundation construction.

6. The type of material that moves (e.g., rock, soil, debris, and mud) and the way movement occurs (e.g., fall, slide, slump, subsidence, collapse, creep, or flow) provide a good way of describing mass wasting. The rate of movement also serves as a basis for classification.

7. Mass wasting is one of the most widespread hazards to people. A large proportion of modern mass wasting are induced by human activities, such as construction projects that undercut slopes; removal of water, oil, or gas from underground; mining; and piling loose material in high, oversteepened piles in landfills and mine dumps.

8. Many of the problems arising from mass wasting can be avoided by careful geologic evaluation of potential construction sites. Steep slopes are most likely to be unstable, but depending on the type of materials present in the subsurface, other areas may contain hazards. The cost of an initial study is generally a fraction of the cost of remedial action after mass wasting has taken place.

9. How do plastics differ from liquids? Elastic solids?
10. How do elastic solids, fluids, and plastic Earth materials differ in the way they react to the pull of gravity?
11. Under what conditions does the surface of the ground subside?

THINKING CRITICALLY

1. What features should one look for in trying to determine whether a particular area is potentially unstable?
2. How is mass wasting related to plate tectonic and climatic Earth systems?
3. How might communities built in dangerous locations protect themselves from potential disasters? What actions can individuals take?
4. What steps could have been taken to avoid the Madison landslide disaster?
5. How is it possible for rocky debris to move across ice and snow without disturbing the surface of the ice?

REVIEW QUESTIONS

1. Why do blocks on inclined planes slip more easily as the angle of repose of the plane increases?
2. In what ways does the behavior of inclined layers of wet clay differ from the behavior of layers of dry sandstone?
3. What factors determine the angle of repose of loose material?
4. What are the effects of water on slope stability?
5. What conditions cause collapse of the ground surface?
6. What conditions cause the ground to subside?
7. What would you look for to determine whether the slopes are creeping? Slumping?
8. What caused so much damage in Anchorage during the 1964 earthquake?

KEY WORDS

angle of repose
avalanche
block-glide
collapse
creep
debris
debris flow
earth
earth flow
fall
flow
landslide

loess
mud
mudflow
plastic behavior
rock
rock glacier
slide
slump
soil
solifluction
subsidence
talus

chapter seventeen

The Role of the Wind

Winds that blew across this area millions of years ago created sand dunes that are beautifully preserved in the Colorado Plateau.

Chapter Guide

Wind, or the movement of air across Earth's surface, controls heat balance, spreads pollen and seeds, generates floods, modifies ecosystems, causes waves and currents in oceans, carries and deposits sediment, and even spreads contaminants around the world. Wind is one geomorphic agent that can touch us all no matter where we live. As a geomorphic agent, wind helps shape the coastlines and the land surface, especially in arid and semiarid regions. Wind generates waves and surface currents that cut cliffs where land rises abruptly from the sea. Where the land rises gradually from sea level, waves and currents move sediment along the shore and build beaches.

High-altitude winds carry and distribute ash blown into the atmosphere during volcanic eruptions. On land, wind picks up dust from the desert and from plains near ice sheets and transports it great distances. Sand dunes dominate the land surface in parts of many deserts, but wind has an impact everywhere fine materials are present on the surface, and winds are persistent. Winds also affect ecosystems in less obvious ways. They distribute pollen and seeds; they facilitate evaporation; and where winds persist, plants must be able to withstand the pressure.

The power of the wind is most apparent where such storms as hurricanes, thunderstorms, or tornadoes strike. A single storm may blow down houses, uproot trees, bring floods, and cause surges of water that break open passageways across offshore islands.

Most winds are expressions of the global pattern of atmospheric circulation which influences Earth's heat balance and its climate. This chapter discusses the role of wind, how it acts as a geomorphic agent and its ecological impact.

INTRODUCTION

Strong, persistent winds blow across the Great Plains of North America. The air is especially dry during the summer and fall months when the wind sweeps out of the northwest. In most years, thunderstorms develop over the Plains when strong convection currents carry moist air so high that the moisture condenses and falls back to Earth. Sometimes, cool air from the north collides with warm, moist air masses that develop over the Gulf of Mexico. This collision usually happens over the southern part of the Great Plains states of Oklahoma, Kansas, and the Texas panhandle. The result is the formation of huge thunderstorms that bring short-lived relief from the dry heat of summer. During the early 1930s, the storms failed to materialize. Droughts are not unusual features of the southern Great Plains, but in the 1930s, the region was caught in the grip of the most extended drought in American history, lasting for most of a decade. Studies of tree rings, which are narrow during dry years, confirm the historic severity of the dust-bowl years. Another drought of similar severity occurred in the sixteenth century. The earliest European settlements in North America were attempted during that drought, and all of them failed, as did Sir Walter Raleigh's English colony on Roanoke Island in North Carolina.

As dry winds continued to blow in the 1930s, the usual sporadic precipitation did not fall. From 1930 until 1936, the amount of precipitation dropped across almost all of the contiguous states with the exception of Maine and Vermont, but the effects were most devastating in the Great Plains, from Montana and the Dakotas to the Gulf Coastal Plain.

Moisture in clay minerals and pore spaces holds the soil in place, but gradually, plants draw this moisture up into their vascular system, and transpiration and evaporation take their toll. When the supply of soil moisture is exhausted, the ground cracks open and the soil begins to crumble. Eventually, the soil becomes a powder easily moved by the wind. After the first year of the drought of the 1930s, a few dust storms blew across portions of the Great Plains (figure 17.1). In April of 1933, 179 dust storms struck the Plains, and in November of that year, a huge storm carried dust from the Plains into the eastern states from Georgia to New York. In May of 1934, dust picked up in Montana and Wyoming settled from Chicago, where it fell like a heavy snowfall, all the way to the Atlantic Ocean. Some estimate that 12 million tons of dust fell in Chicago. From 1935 to 1938, unusually severe wind stripped topsoil from large parts of a four-state area (figure 17.2).

A woman who lived in Kansas described the experience of being in one of these dust storms as follows:

All we could do about it was just sit in our dusty chairs, gaze at each other through the fog that filled the room and watch that fog settle slowly and silently, covering everything—including ourselves—in a thick, brownish gray blanket. When we opened the door swirling

(a)

(b)

Figure 17.1 Wind Storms. (a) In the 1930s, dust storms swept across the High Plains of the United States, blowing soil away and devastating the economy. (b) Strong, cold winds blow across the landscape in high latitudes. Here, ice crystals have grown horizontally on the downwind side of this marker.

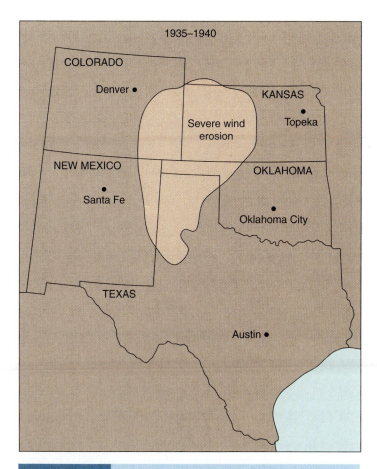

Figure 17.2 **Location of Severe Wind Erosion.** During the 1930s, the most severe wind erosion took place in the four-state area indicated on this map.

whirlwinds of silt beat against us unmercifully. . . . The door and windows were all shut tight, yet those tiny particles seemed to seep through the very walls. It got into cupboards and clothes; our faces were as dirty as if we had rolled in the dirt; our hair was gray and stiff and dirt ground between our teeth. (As reported in the *Kansas City Times*.)

THE ROLE OF WIND IN EARTH SYSTEMS

This introductory account of the Dust Bowl emphasizes some of the connections between climatic conditions, wind, and ecology. Relationships among the wind and other parts of Earth systems have been described in earlier chapters. The critical importance of winds involved in atmospheric circulation was emphasized in chapter 12. Atmospheric circulation helps maintain a global heat balance between equatorial and polar regions. The change in wind directions that occurs when the land surface of Asia becomes warm in the summer causes moist air from the Indian Ocean to move inland. The resulting monsoons are responsible for the heavy rains that drench northern India in the fall. Settling of cool, dry air from high in the atmosphere causes dry air to move from central Asia toward the ocean. Similar mechanisms cause the smaller-scale sea air movements, called **land breezes,** which blow across the seashore at night as the land becomes cool. During the day, **sea breezes** blow from the cool water in the ocean toward the warmer land where hot air is rising and reducing pressure at ground level.

The intense winds associated with hurricanes and tornadoes described in chapter 12 have devastating effects on the ecosystems located in coastal areas. Connections between wind and other Earth systems are seen in many other natural processes. The far-reaching impacts related to volcanic eruptions and winds that are part of atmospheric circulation described in chapter 7 provide evidence of the connection between plate tectonic systems and the climate system. Winds that blow off of ice sheets carry dust that settles far away from the ice margin. Sediments deposited from the wind provide valuable clues to past climates, and the role of wind in generating waves and currents was discussed in chapter 9.

In all cases, air moves in response to variations in atmospheric pressure. Many such variations result from differences in near-surface temperature. Cold air is associated with high pressure because cold air has a higher density than warm air. Atmospheric pressure drops as the air temperature rises. Winds give expression to the movement of air from areas of high pressure to those where the pressure is lower. The resulting wind has the following far-reaching impacts on the surface environment:

- Wind is a factor in controlling heat balance
- Wind spreads pollen and seeds
- Wind-generated storms cause floods and modify ecosystems
- Winds cause waves to form in large bodies of water
- Wind is a factor in the formation of surface currents in the oceans (see chapter 9)
- Wind carries and deposits sediment
- Wind spreads atmospheric contaminants around the world
- Wind is a geomorphic agent

Because the relationships between wind and other systems have or will be examined elsewhere, this chapter concentrates on the ways wind alters the landscape and ecosystems. These geomorphic effects are most obvious in hot dry deserts, described in chapter 14, and in coastal areas, described in chapter 10.

WIND AS A GEOMORPHIC AGENT

Wind shapes the surface of the land where it is strong enough to move sand, silt, or dust that are present on the ground surface. These conditions exist in deserts where the wind is strong, the amount of precipitation is low, and fine-size weathering products are abundant. Wind is also an effective geomorphic agent in coastal areas where waves and currents replenish beaches with sand. In areas near ice sheets where streams flowing from beneath the ice carry and deposit fine sediment ground up under the moving ice, strong

Figure 17.3 Wind Blows down Death Valley, Removing Dust from the Alluvial Deposits in the Floor of the Valley.

winds blow off the ice, pick up the fine sediment, and carry it away. Winds also shape the landscape near volcanoes where eruptions can project huge quantities of dust and ash high into the atmosphere. The coarser fragments fall close to the volcano on its downwind side, but the finer materials may travel great distances.

Potential sources of windblown material are available wherever small particles of soil or sediment lie on the ground surface. Alluvial fans, dried-up stream deposits, and outcrops of sandstone provide sources of sand in hot deserts (figure 17.3). Trees and other types of vegetative cover, as well as soil moisture, which acts as a weak adhesive, hold soil and sediment in place. However, in arid regions, soil moisture is low, plants are few, and weathering products are exposed to wind erosion.

In semiarid regions, precipitation is sufficient to encourage farmers to cultivate soil, but excessive cultivation exposes the soil to wind erosion. The problem is especially serious when the amount of precipitation is inadequate to support the types of crops being grown. Continued cultivation aggravated the problems associated with the Dust Bowl in the 1930s. For decades before the drought, farmers had cultivated the soils of the Great Plains with few problems. Then the amount of rainfall suddenly decreased and droughts began. During the early dry years, plowing continued, accelerating the evaporation of soil moisture and loosening the soil. As the drought continued, plants died; new plants failed to grow; and the persistent winds removed the fine, dry unprotected soil and carried it away in huge dust storms. If farmers had not continued to cultivate dry soil, the amount of soil removed by wind would have been less severe.

Problems similar to those encountered during the Dust Bowl in North America arise as a result of overgrazing dry land. In portions of northern Africa, scrub growth covers lands that were formerly used for grazing. Where herds of cattle have grazed until they have stripped all of the plant cover, the scrub-covered areas change into barren, hot deserts of the types described in chapter 14. Aerial photographs show straight lines that separate barren desert land from scrub-covered land. The lines mark the position of fences.

Continued global warming will aggravate this situation and accelerate **desertification,** which affects large areas around the world (see figure 14.14).

Wind Erosion

The wind is highly selective in the size of materials it moves. As the wind velocity increases, progressively larger particles move. Because the size of the grains that wind moves is so sensitive to wind velocity, windblown sediments are uniform in grain size when wind velocity remains nearly constant. Most of the load moved by the wind consists of small particles: granules (2 to 4 millimeters), sand (1/16 to 2 millimeters), silt (1/256 to 1/16 millimeters), and dust (less than 1/256 millimeters). Occasionally, high winds roll pebbles; they may even bounce along the ground surface, but the wind rarely lifts a pebble off the ground. Even boulders may roll as the wind dislodges supporting sediment from the downwind side.

Fragments of particular sizes begin to move when the wind reaches a critical velocity for that particular particle size. At first, the particles roll along the surface and bounce into the air when they hit one another. Once bouncing begins, the process accelerates as the wind speed increases. The wind propels grains that bounce into the air and imparts more energy to them. As the grains hit the surface again, the impacts dislodge and cause other grains to bounce into the air. This sets up a chain reaction called **saltation** (figure 17.4). Soon, bouncing grains fill the air in a zone that extends from a few centimeters (an inch or 2) to as much as a meter (several feet) above the ground surface, depending on wind velocity.

Ripples cover the surface of most windblown sand deposits (figure 17.5). Like the ripples seen in streams and in shallow water at the beach, currents cause ripples to form. Long asymmetric ridges form more or less perpendicular to the wind direction. The larger grains tend to roll up the windward side of the small ridge and slide down the downwind side of the ridge. This quickly produces a ripple that is asymmetric in shape. The downwind side is steeper than the windward side of the ripple. The size and orientation of the ripples change quickly in response to variations in the wind velocity or direction.

The wind lifts dust particles into the air in a different way, and they move at a much higher altitude than sand or silt. Dust particles are thin and flat, and they have a large surface area relative to their weight. Consequently, slight updrafts can pick up and carry dust high in the air. Turbulence in the air and small whirlwind disturbances, known as dust devils, lift dust into the air. **Deflation** is the name given to the removal of particles from the surface. Once deflation begins, the particles play an important role in wind erosion.

Combinations of deflation, abrasion, and impact of sand cause most erosion accomplished by the wind. Sand, silt, and dust carried by the wind are tools of wind erosion. Long-continued sandblasting by wind-driven sand polishes and abrades any hard material it strikes. When the grains hit a rock surface or soil, the momentum of the impact may crack or dislodge granular material. These impacts are capable of weakening the cement of a rock and

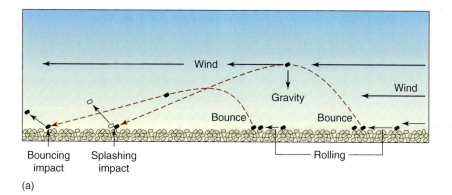

Figure 17.4 Grains of Sand in Saltation.
(a) When strong winds blow across a sand surface, some of the sand rolls along the surface, and some bounces into the air and comes back to the surface where impacts dislodge more sand. (b) Most sand movement occurs within a meter of the surface.

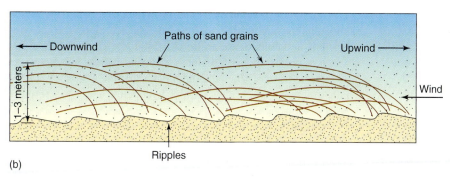

freeing other sand grains. Sandblasting can frost glass, clean stone buildings, and even cut letters in polished stones as hard as granite. Where sand and silt have blown over rocks for a long time, the rocks become worn from the abrasion, achieving smooth, polished surfaces. Such stones, called **ventifacts,** are common features in deserts and a sure sign of wind erosion. In deserts, wind-blown sand may cut notches in rocks near the ground. This type of undercutting produces forms called **table rocks** or **pedestal rocks.** Most of these form where a thin column of a more easily eroded rock supports a larger, more resistant rock (figure 17.6).

During the occasional heavy rains, streams carry mixtures of sediment across the gently sloping surfaces of deserts. The wind selectively removes the finer particles, leaving behind coarse particles that remain as a thin layer of gravel, called **lag gravel** (figure 17.7). The fragments in the lag gravel may form a tight mosaic cover, called **desert pavement.** Once formed, these pavements protect the underlying surface from wind erosion.

In areas of strong wind, where scattered vegetation holds the soil in place, the wind blows away pockets of unprotected sediment, leaving **blowouts.** These broad, shallow depressions are

Figure 17.5 Ripple Marks. These marks are present on the surface of most sand dunes.

Figure 17.6 Pedestal Rock. This pedestal rock gives the name to Mexican Hat, Utah. A layer of rock easily eroded by wind action lies beneath the slab at the top.

Figure 17.7 Lag Gravel. As the wind progressively removes fine materials, rocks that are too large or heavy for the wind to move remain as lag gravel. The fine sediment cracked as it dried, producing the polygonal pattern characteristic of mud cracks.

distinctive features along many coasts where grasses and trees partially cover sand deposits inland from the shoreline.

Deposition from the Wind

As wind velocity drops below the minimum wind velocity required to move particles of a given size, grains of that size stop moving and settle. Consequently, wind does an excellent job of sorting particles according to their size, shape, or weight.

Blankets of Windblown Sediment

Some windblown deposits cover large areas with relatively thin layers of sediment. Generally, these sheets thin toward their margins and away from the source area. Some deposits of volcanic ash are of this type. Deposits of ash lie downwind of active volcanoes. Because the volcano is a point source, the ash deposit converges toward the vent and usually thins out downwind, leaving a lens-shaped mass of ash. Blankets of pumice cover large areas around the volcanic center in the Jemez Mountains of New Mexico and the central portion of the North Island of New Zealand.

Much finer ash and dust deposits appear in cores taken in the ocean. Some of these fine deposits originated from distant volcanoes, but most of them came from desert regions (figure 17.8). The extremely fine dust blown from deserts can be caught up in global circulation and circle the Earth. Dust even appears in ice cores taken from the ice sheet in Greenland. Most of this dust traveled in wind currents that followed a path from the Gobi Desert in Mongolia.

One of the most distinctive wind-laid deposits consists of fine silt with minor amounts of clay and fine sand known as **loess**. These sediments are usually homogeneous, nonstratified, and porous. Most deposits are buff to light yellow in color. Many of the loess deposits in the United States formed at the end of the last major advance of ice sheets in North America. At that time, loess deposits covered a large part of the central United States, and remnants of the loess remain throughout the Mississippi Valley. These deposits buried grasses and continental invertebrates such as gastropods. Fossils of the grasses remain in upright positions, indicating that the dust settled slowly over the fields. Because many of the fragments of silt are angular in shape, loess deposits stand up and can form steep cliffs.

Not all loess originates in the same way. The sediment may originate from the surface of deserts, alluvial valleys, and the sediment deposits from glacier meltwater. Many of the angular fragments are thought to originate from rocks ground up by glaciers. The fragments travel in meltwater from the glaciers into the region immediately beyond the ice margin. The ice chills the air over glaciers, and cold, dense air flows off the ice. This cold air blows across plains composed of fine sediments carried by meltwater away from the ice. When the sediment is dry, the wind picks up the finer fraction and carries it away. Eventually, it settles, forming layers of loess.

Thick and extensive deposits of loess cover about 6.6% of the surface area of China. Much of this is dust blown off the surface of the Gobi Desert. The thickest accumulations are present in the valley of the Yellow River where deposits 200 to 300 meters (600 to 1,000 feet) thick are present. The loess creates large, flat surfaces. Because it is easily eroded, the edges of the loess plateaus are severely eroded. These deposits contain many fossils of Pleistocene animals, including some of the oldest human fossil remains. Among these is the skull of Lanatian Man, which is about 750,000 years old.

Other than loess and volcanic ash, sand and silt make up most windblown deposits. The terms sand and silt apply to sediments of a certain grain size, but so much sand and silt consist of the mineral quartz that it is easy to think that all sand is quartz sand. Quartz is the most common constituent of sand because it is hard and insoluble. Quartz is one of the few common minerals that can survive both chemical and mechanical weathering processes. However, some sand and silt deposits contain minerals other than quartz. For example, sand on some beaches of Italy contains the mineral olivine derived from volcanic rocks. Because olivine weathers rapidly, these sands must be young. Sand-sized pieces of volcanic ash compose the sand on many Hawaiian beaches, and the White Sands National Monument in New Mexico contains sand composed of gypsum, a mineral that dissolves quickly in humid climates. Even sand-sized particles of ice form dunes in parts of the Antarctic. Regardless of their composition, if the sand particles remain intact and survive weathering, all sand may form dunes.

Sand Dunes

Sand dunes form wherever strong winds, either constant or intermittent, act on large quantities of sand-sized particles (figure 17.9). In hot deserts, most of the sand comes from decaying outcrops of sandstone or from intermittent streams. Most other sand dunes lie on the back shore of beaches and are replenished by sand washed

Figure 17.8 Wind Blown Dust Produces Streaks Across the Landscape of Northern Chad in the Sahara Desert of Northern Africa. Dust and sand is funneled between two more resistant landform features—the Tibesti Mountains and the Ennedi Plateau. Both of these features are dark in color on this oblique photo. Strong winds blowing across deserts carry dust great distances. In some deserts the dust settles, forming blanket-like deposits of dust known as loess.

Figure 17.9 Sand Dunes of the Sand Dunes National Park in Colorado. The wind blows from left to right in this area. The slopes on the windward side are low, but steep slopes mark the downwind side of the dunes.

up on the beach. Regardless of its source, sand moves downwind until wind decreases or something interferes with the sand movement. Usually, this happens where obstructions block the movement of the sand and initiate the formation of a pile. As the pile grows, it may eventually move away from the obstruction.

Sand dunes may start as streamlined piles of sand with nearly equal slope on both the up- and downwind side, but as the dune grows larger, eddies form on the downwind side, and the dune begins to assume an asymmetrical shape. In a typical cross section drawn along a line through the center of the dune and in the direction of the prevailing wind, laminations in the dune have the form shown in figure 17.10. On the downwind side and just beyond the crest, wind velocity drops, and saltation of sand stops. The sand on this slope becomes steeper and starts to slump and slide. Eventually, this slope assumes the angle of repose, about 35°. Sand from the windward side continually blows to the crest, causing it to become oversteepened and triggering slumps and slides (figure 17.11). As long as the wind blows, sand continually moves from the windward side to the lee (downwind) side of the dune. The movement of sand that produces sand dunes leaves internal features in the dunes that remain as long as the deposit survives. Laminae inclined to the surface the dune moved across, called **cross-bedding,** are the most prominent of these features (see figure 17.10). Slight variations in the size, composition, or texture of the sand leave an impression of the inclined surfaces on the windward and lee sides of the dune. As the dunes move downwind, they often move across older dune deposits, leaving the imprint of a stack of cross-bedded sand. Cross-bedding is also created in streams that carry sand. The two resemble one another, but cross-bedded sand moved by the wind is usually much more uniform in size than that found in stream deposits. In addition, sand dunes tend to be much larger than deposits in streams.

Figure 17.10 A Cross Section of Ancient Sand Dunes Exposed in Utah.

The position of sand dunes shifts downwind. How rapidly this movement occurs depends on the persistence of the wind and its velocity. Some individual dunes remain stationary for long periods; others move quite rapidly. In a few instances, dunes have moved as much as 70 meters (230 feet) per year.

Figure 17.11 Sand Dune Slumps and Slides. As sand piles up on the downwind side of a dune, the sand slides downward forming the slip face of the dune.

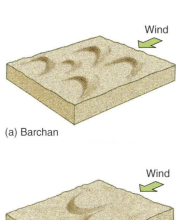

Wind

Barchan dunes, with the isolated forms shown here, develop where the supply of sand is low. The convex side of the dune faces the wind, and the wings point downwind.

(a) Barchan

Wind

Transverse dunes form long ridges oriented nearly perpendicular to the prevailing wind direction. The slip face is downwind.

(b) Transverse

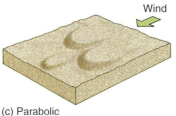

Wind

Parabolic dunes have a shape that resembles that of barchan dunes, but the wings point upwind.

(c) Parabolic

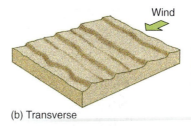

Probable wind directions

Star dunes may develop where the wind changes direction.

(d) Star

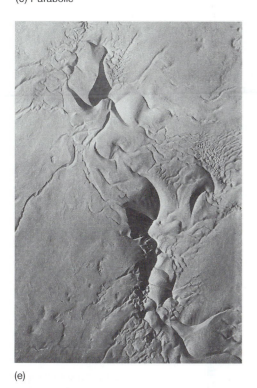

(e)

(f)

Figure 17.12 **Types of Sand Dunes.** (*a*) Barchan dunes. (*b*) Transverse dunes. (*c*) Parabolic dunes. (*d* and *e*) Star dunes. (*f*) Linear dunes, also called longitudinal dunes or seifs.

Although most sand dunes have distinct cross sections of the type just described, three types of sand deposits—sand sheets, sand stringers, and dome dunes—have no distinct slip face. **Sand sheets** are broad, flat deposits; **sand stringers** are long, narrow stripes of sand; and **dome dunes** are circular or elliptical mounds with no identifiable slip face.

The supply of sand, wind velocity, constancy of the wind direction, and the amount and distribution of vegetative cover determine the shape a sand dune will assume. If the amount of sand is limited

and the wind blows from one direction almost all of the time, dunes develop simple shapes like those shown in figure 17.12*a*, *b*, and *c*. However, where the quantity of sand is great and where large dunes come together, individual dunes tend to lose their distinctive form and may even become part of vast reaches of sand known as **sand seas.** In the Sudan, a large sheet of sand moving from west to east buried the channel formerly occupied by the Nile River. The first evidence of this buried channel came from a study of satellite images of the area (figure 17.13). Today, the Nile flows in a channel

(a)   (b)

Figure 17.13 Two Images of a Part of Sudan. (a) This view shows the surface as it might appear in a photograph. (b) This view, taken with a special type of radar system, shows recorded radiation that penetrated the sand to reveal a former channel of the Nile River that is now buried under sand.

cut around the edge of the sand. Compound dunes consisting of several different types of dunes superimposed on one another lend a feeling of infinite variety to the seas of sand.

Ancient Sand Deserts of the American Southwest

Vast deserts filled with giant sand dunes occupied large parts of the Colorado Plateau during the later part of the Paleozoic and again during the Mesozoic. The remains of the older sand dunes are exposed in cliffs in the Grand Canyon where the Coconino sandstone forms a white cliff about one hundred meters (several hundred feet) thick near the top of the canyon wall (figure 17.14). Other exposures of the cross-bedded windblown deposits appear at Canyon De Chelly and at the Natural Bridges National Monument. Exposures of these sand deposits are scattered over a large region, but younger sedimentary rocks bury them in many places.

In the early part of the Mesozoic, shallow bodies of marine waters, enclosed basins in which evaporates accumulated, and large areas of low-lying land prevailed in the Colorado Plateau region. Gradually, the seas disappeared, and the region became a vast low-lying land crossed by streams carrying sand from mountains that lay to the northeast in Colorado. By the middle Mesozoic, the region had become arid, and sand dunes covered most of the plateau region (figure 17.15). In some areas, such as Zion National Park, the sand accumulated to thicknesses of more than 330 meters (1,000 feet) (figure 17.16). The dunes would be impressive by comparison with modern dunes such as those on the Outer Banks of North Carolina or the great dune fields in Arabia and Central Asia.

ECOLOGICAL EFFECTS OF WIND

Tornadoes and hurricanes have devastating effects on the environment. Hurricane-force winds can move enough sand across barrier islands to breach the island and produce new inlets connecting the open ocean with the bays that lie behind the islands. In this way, the local environment is transformed in a matter of hours from that of a beach to a shallow marine environment characterized by strong tidal currents. The strong winds associated with hurricanes pile water up along the coast and produce flooding where the land near the coast is low. Hurricanes drive floodwaters inland, covering large areas along the Atlantic coast and Louisiana. Large deltas, such as the Mississippi River Delta and the delta of the Ganges River, are especially susceptible to flooding during hurricanes because the tops of deltas are close to sea level.

Heavy rain generally accompanies the wind associated with hurricanes, but strong and persistent winds unrelated to storms blow across many regions, such as the Great Plains of North America. In these areas, especially those located in the latitudes 20° to 30°, the wind promotes evaporation. Frequently, the potential evaporation far exceeds the amount of rainfall. Therefore, the wind is a major factor controlling the climate and, through climate, the ecology of those regions.

Another effect was discovered in the carefully controlled environment created in the glass-enclosed experimental facility known as the Biosphere, which Columbia University now operates near Tucson, Arizona. In some of their early studies of forests in the Biosphere, scientists were surprised to find that some of the trees growing in the Biosphere lacked the strength to remain upright. The missing element in their environment was wind. Movement of the trees caused by wind affects the cell structure of the trees and produces a stronger plant. In more recent experiments at the Biosphere, fans produced enough air movement to make the plants sway. The effects of wind on trees are obvious in high mountains and along coasts where strong winds are persistent. Where wind is strong, trees assume asymmetric shapes and are shorter; where winds are extremely strong, the trees rise a meter or less above ground level. The wind is a limiting factor of their growth but also produces a sturdy tree.

Figure 17.14 Coconino Sandstone. The white cliff in the foreground and in the distance is the Coconino Sandstone, a layer of sand that formed as sand dunes in the Paleozoic. The tops of the sand dunes were cut off as the environment changed.

Lack of water is one reason the plants that inhabit deserts are distinctly different from those found in other environments. Desert plants not only survive on less water from the ground, they lose moisture through transpiration in areas where evaporation and wind quickly carry the moisture away from the plant.

Many effects of the wind on ecosystems are subtler than those associated with strong storms. The winds associated with strong storm systems carry spores and pollen hundreds or even thousands of miles. This is one of the most effective mechanisms involved in the spread of plants and bacteria. Images from space reveal that dust blown off the desert regions of northern Africa is carried by the wind across the Atlantic Ocean. Records of the amount of dust reaching Barbados Island in the eastern part of the Caribbean reveal that much larger quantities of dust reached the Caribbean during the 1970s and 1980s than in earlier years. Scientists estimate that hundreds of millions of tons of African dust cross the Atlantic each year. The peak quantities of dust coincide with the onset of serious problems on Caribbean reefs, including devastation of one of the sea urchins that inhabit the reef. Later, the branching coral known as *Acropora* also experienced high mortality, and coral reefs are severely damaged, causing bleaching of the coral. In the Mediterranean, a specific bacterium causes bleaching of some coral species. This coincidence suggests that the wind may have transported fungi, bacteria, and other microorganisms to the Caribbean from soils in Africa and that these organisms are responsible for severely damaging the ecosystem.

Wind affects animal as well as plant populations. Strong winds associated with storms may blow small animals downwind, and even moderate winds transport insects and move them to new locations where they establish themselves. Birds are also affected by winds. Although most birds remain quiet in protected places on windy days, storms can blow seagulls hundreds of miles inland.

The role of wind as a factor in climate and ecology is not as obvious as that of temperature or water, but it affects flora, fauna, and the landscape in which they live.

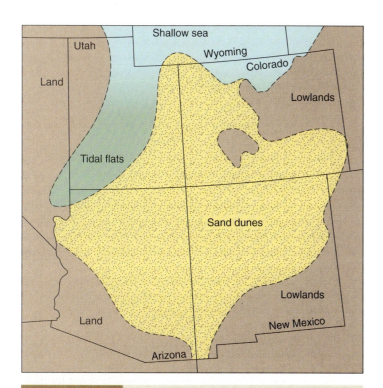

Figure 17.15 Ancient Sand Desert. Much of the Colorado Plateau region was covered by a sea of sand dunes in the middle of the Jurassic period, about 165 million years ago.

REDUCING ADVERSE EFFECTS OF THE WIND

The Dust Bowl in the 1930s devastated agriculture over a vast region of the central United States. Even today, large areas in parts of Africa and Asia are damaged by wind erosion, and in some areas,

Figure 17.16 Zion National Park. Crossbeds of the Jurassic sand dunes are beautifully exposed in the cliffs at Zion National Park.

the advance of sand dunes poses serious problems. What can be done to mitigate these problems? The task is especially difficult because climate, over which humans have no control, determines surface temperature, evaporation, and wind. Attempts to cultivate land repeatedly during the extended dry periods in the 1930s aggravated the problem by reducing soil moisture and exposing more soil to wind erosion. Overgrazing during dry periods reduces the plant cover that holds soil in place and commonly initiates wind erosion in arid regions. Retaining a plant cover is especially important as a way of reducing erosion. This can be facilitated by spreading plants that resist extremely dry conditions. Sea oats are used along many beaches to help hold sand dunes in place. Use of rows of trees planted perpendicular to the prevailing wind direction can also reduce wind erosion.

SUMMARY POINTS

1. Differential heating of the ground surface causes lateral variation in air pressure. Air moves from areas of high pressure (cold) to areas of low pressure (warm).
2. Some winds, such as movement of air from ocean to land during the day—a breeze called a sea breeze—and from land to ocean at night—called a land breeze—are due to differential heating of land and water.
3. Large-scale winds develop as a result of global atmospheric circulation, described in chapter 12. The strongest winds form in connection with storms such as thunderstorms, hurricanes, and tornadoes.
4. The wind can pick up fine dust from the surface and transport it great distances. Volcanic dust that reaches the stratosphere may travel around the globe for years. Larger particles such as sand rarely rise more than tens of centimeters off the surface. Sand moves mainly by the process of saltation.
5. Sand deposits often assume dune shapes. The shape of dunes depends on the supply of sand, the strength of the wind, and the persistence of the wind direction.
6. In addition to its importance as a geomorphic agent in arid and semiarid regions, the wind plays a role in shaping the ecology of desert and semiarid regions. In these regions, wind is a significant factor in promoting evaporation.
7. Spores and pollen of many plants are spread by the wind.
8. Wind, as part of the global circulation system, is a major determinant of climate.
9. The impact of windblown sand produces a sand-blasting effect on exposed rocks.
10. Wind erosion and deposition cause many landforms, such as desert pavements, pedestal rocks, and hollows in cliff faces found in deserts.

REVIEW QUESTIONS

1. Where is wind an important factor in shaping the environment?
2. How does material from the ground get into the upper levels of the atmosphere?
3. What evidence suggests that loess is deposited from the wind rather than from water?
4. What is the basis for classifying sand dunes?

5. What conditions cause isolated barchan dunes (see figure 17.12) to form?
6. Describe the internal structure of sand dunes, and explain how they form.
7. What conditions cause dust bowls?

THINKING CRITICALLY

1. Both wind and water are fluids. In what ways do their roles in Earth's systems resemble one another, and how do they differ?
2. Where is the wind an important factor in shaping the environment?
3. In what ways does the wind affect the environment where you live or go to school?

KEY WORDS

barchan dunes	parabolic dunes
blowouts	pedestal rocks
cross-bedding	saltation
deflation	sand seas
desertification	sand sheets
desert pavement	sand stringers
dome dunes	sea breeze
lag gravel	star dunes
land breeze	table rocks
linear dunes (seifs)	transverse dunes
loess	ventifacts

Streams

Waterfalls along the steep slopes of a glaciated valley near Milford Sound, New Zealand.

Chapter Guide

Water is as important to Earth systems as blood is to the human body. Blood flows throughout the human body, nourishing it, influencing its growth, and eliminating waste. Similarly, water is essential to many Earth processes, particularly weathering, erosion, climate, and the movement and availability of nutrients within soil. This chapter deals with the processes that operate in streams, the ways the surficial runoff of water in streams shapes the landscape, even in arid regions, and the great impact streams have when the amount of precipitation exceeds their normal capacity and water floods adjacent lowlands. Chapter 19 explores the movement and storage of water underground. It will explain the close connection between streams and ground water, why streams continue to flow long after surficial water has drained away, and the importance of water stored below ground level as a water supply. Chapter 20 examines water stored as snow and ice and the effects that changes in the ice-water balance have on sea level. Water plays other important roles in Earth systems. It is essential for survival of organisms, moves critically important nutrients into the soil, and change in the state of water is one of the main determinants of climate.

INTRODUCTION

The Hydrologic Cycle

Great reservoirs of water reside in the oceans, on land, in streams and lakes, in the soil, and underground (table 18.1 and see figure I.7). The hydrologic cycle is a convenient way of looking at the global transfers of water. Because oceans occupy nearly 70% of Earth's surface, most precipitation falls directly into the ocean. Nearly 59% of the water that falls on land returns to the atmosphere through evaporation, transpiration by plants directly, and sublimation from ice and snow. The remainder runs off on the surface through streams and lakes, infiltrates into the ground, becomes part of soil moisture or ground water, or becomes part of glaciers where it may remain tied up as ice for thousands of years. Some surface water infiltrates into the soil where clay and organic matter absorb a portion of it. The remaining water moves into the underlying sediment or rock.

Most water on Earth remains within an essentially closed system. Because the total quantity of water is relatively constant, movement of water from one reservoir to another is part of a process of recycling. This system of water transfer and return constitutes the hydrologic cycle (see figure I.7). This cycle involves transfer of water between Earth's interior and the surface and among the surface reservoirs. Although water is continually recycling through natural processes, water resides in one or the other of the natural reservoirs. The oceans, stream channels, natural lakes, glaciers, the atmosphere, soil and organisms, and Earth's crust are the main reservoirs. Table 18.1 indicates the quantity of water presently held in each of these. Artificial reservoirs, such as tanks, pipelines, ponds, and bodies of water behind dams, contain a small amount of the total.

Water moves between Earth's interior and the atmosphere and oceans through tectonic and related volcanic processes (figure 18.1). Although the rates of tectonic recycling are difficult to determine, the paths it follows are clear. Volcanic activity is the most obvious manifestation of the movement of water from inside Earth into the atmosphere. Steam, the most abundant of the volcanic gases, originates from magma and from the rocks through

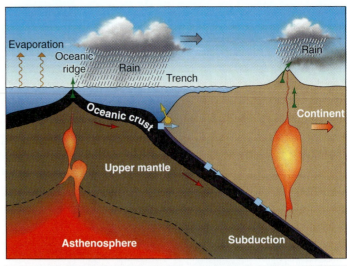

Figure 18.1 Tectonic and Volcanic Systems are Involved in the Hydrologic Cycle. Water vapor that rises from volcanoes into the atmosphere joins water evaporated from the sea surface. Most of this water returns to the oceans as precipitation, and some of it saturates sediments deposited on the seafloor. As these sediments move into subduction zone, fluids in the sediment migrate into the crust beneath volcanic arcs where magma is generated. These fluids in the magma are the main source of the gases emitted by volcanoes.

which the magma and associated gases pass as they move to the surface. Water from the surface finds its way back into Earth's interior in subduction zones where portions of the lithosphere containing water-saturated sediments and volcanic rocks sink into the mantle. Water in this sinking slab of lithosphere facilitates metamorphism and melting of rocks in the subduction zone, and some of it becomes part of the magma generated in the subduction zones.

The oceans are by far the largest of the surface-water reservoirs. Water in the deeper parts of the ocean basin moves so slowly that it takes over a thousand years for it to complete the journey through the ocean basins (see chapter 9). Water from the ocean enters the atmosphere as a result of evaporation. Solar radiation raises the water temperature and promotes evaporation, especially in low latitudes. A large proportion of the solar energy reaching Earth's surface is involved in this process. Water vapor rises into the atmosphere where an amount of water equal in volume to that found in the upper few centimeters of the ocean resides in the form of clouds. Atmospheric water vapor becomes involved in atmospheric circulation and eventually returns to Earth's surface as precipitation in the form of rain, snow, or ice. As rain or meltwater collects on the surface, it flows downslope, leading to the development of streams.

Table 18.1	World's Estimated Water Supply	
Location	**Volume (cubic miles)**	**Percentage of Total Water**
World ocean	317 million	97.2
Surface water		
Freshwater lakes	30,000	0.009
Saline lakes and inland seas	25,000	0.008
Stream channels	300	0.0001
Subsurface water	2,016,000	0.625
Ice caps and glaciers	7,000,000	2.15
Atmosphere	3,100	0.001

Source: USGS

FORMATION OF STREAMS

Images of the surface of other planets reveal how different Earth's surface would appear if streams did not exist. In the absence of streams, weathering would continue, but only the wind would move material from one place to another. Most landforms would reflect the structure of the underlying rock. Nothing would etch out rock layers of different resistance to weathering and erosion. Plains would mark areas where strata are flat-lying; dome-shaped land surfaces would reflect domal uplifts. Volcanoes and mountains would stand high, but no streams or glaciers would cut into these high areas. Stream valleys and deltas would not exist. Because streams transport loose materials from high to lower elevation, the land surface on Earth changes continuously. Envision this process as starting with a newly created land surface. Such surfaces form where major volcanic eruptions blanket large areas with volcanic ash and where sea level drops (or land rises), exposing the continental shelf to erosion. In such situations, it would be possible to observe the formation of new streams.

During a rain, water infiltrates into the soil. As open spaces in the soil fill, water runs off on the surface downslope. Runoff starts as a thin sheet of water on the ground surface. Gradually, the water coalesces into channels that progressively form larger and larger streams. Small streams become the tributaries of larger streams that carry the water into lakes and ultimately to the ocean or into areas where the water evaporates or soaks into the ground. A network of small, barely noticeable drainage channels that are dry most of the time form on the ground (figure 18.2). Only the larger channels appear as distinct valleys with recognizable stream channels. Only a few of these contain water all the time. These are **permanent streams.**

Most streams have evolved over thousands, in some cases, millions, of years. The permanent streams and their tributaries provide a network through which water runs off of the ground surface. The **surface-drainage basin** of a river is the area from which surface runoff flows into that river. Major rivers, such as the Amazon in South America or the Mississippi in the United States, contain water that collects from huge areas through extensive systems of tributaries (figure 18.3).

Each tributary of major streams has its own drainage basin. In humid regions, streams continue to flow long after all the water on the surface and in the soil has drained off. Ground water feeds these streams and maintains their flow. Much of this water rises in springs seeping out of cracks in the bedrock or along bedding planes. Water may also rise into stream channels beneath the waterline. In deserts,

Figure 18.2 Aerial Photograph of a Small Drainage Basin in Colorado.

Figure 18.3 Drainage System of the Mississippi River.

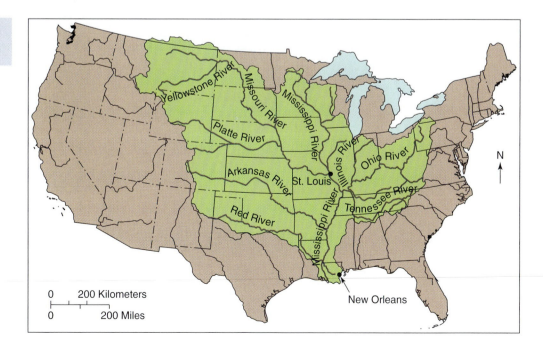

stream channels are usually dry. Streams flow for a short time after a heavy rain, but after the rain stops, the water soaks into the ground quickly.

Only part of the total rainfall runs off on the surface. Runoff is much greater on steep slopes than it is on low slopes; it is greater on rock than on soil; and runoff is greater where plant cover is sparse than it is in heavily vegetated areas. Of these factors, the slope of the land and the character of the soil remain unchanged throughout the year, but the amount and type of vegetation and the presence of ice and snow varies considerably with the season.

PRINCIPLES GOVERNING STREAM ACTION

Scientists characterize streams by their length, the fall in elevation along the stream, the quantity of water passing through the stream, the shape and pattern of the stream channel, the velocity and type of flow of water in the channel, and the load of the stream. Changes in one of these factors affect the others. These relationships help us understand how streams function.

The Stream and Its Channel

Stream channels and the valleys in which streams flow reveal much about the geologic history of the region and about the processes of erosion, transportation, and deposition that are taking place today. Compare the stream patterns, the character of the channels, and the shapes of the valleys depicted in the chapter opening photograph and those in figures 18.4 and 18.5. The stream in the opening photograph cascades down a steep mountain side. The stream in figure 18.4a flows through a straight valley cut in solid rock; figure 18.4b depicts a stream so filled with gravel that its channel is poorly defined. The Colorado River at Lees Ferry (see figure 18.4c) has steep cliffs on either side, and a few sandbars are present along the edge. The Waitake River in New Zealand (see figure 18.5), has many channels that weave a braided pattern. Understanding what causes these differences is an important objective of this chapter. The first step is defining the characteristics of streams and their channels.

The width and depth of a stream vary as the quantity of water in the stream changes. As the water level in a stream rises above the banks, it covers part of the larger valley in which the stream flows (figure 18.6). By multiplying the area of the cross section of a

(a)

(b)

(c)

Figure 18.4 Stream Channels. A number of factors, including the quantity of water and sediment carried by the stream and the stream's velocity, determine the character of the stream channel. (a) The channel of this stream in the Catskill Mountains follows a system of joints in the bedrock. (b) This stream contains so much sediment that no distinct channel has formed. (c) Sandbars have formed in the low-velocity parts of this channel that is confined to the cliff side of the valley, the Colorado River near Lee's Ferry.

stream by the average velocity of water passing through that section, one can determine the quantity of water passing through that section, a quantity known as **discharge** of the stream. Because tributaries bring more water into a stream, the discharge changes from one section of a stream to another, and in humid regions, it increases downstream.

Variation in discharge causes the water level to fluctuate and results in erosion or deposition of loose, unconsolidated sediment, called **alluvium,** in the stream channel. The shape of the channel changes as the quantity of sediment in the bottom of the channel varies. In some streams, water flows over bedrock in channels essentially devoid of alluvium (see figure 18.4*a*). In others that carry heavy loads of sand, silt, or gravel, the stream flows in several shallow, interwoven, and rapidly changing channels that form a **braided** pattern (see figure 18.5).

During floods, water overflows the channel bank, and streams erode the valley beyond the channel. As discharge decreases, the stream deposits alluvium across the valley floor. In this fashion, a nearly flat surface called the **flood plain** (see figure 18.6) forms at a level slightly above the stream banks. Streams that flow in the bottom of canyons or across bedrock surfaces generally have narrow and poorly defined flood plains. Streams such as the Mississippi River that flow across wide expanses of unconsolidated sediment have wide flood plains.

Channel size and shape vary from the **head** of the stream where the stream starts to its **mouth** where the stream flows into another stream, a lake, or the ocean. Except in arid climates, channels increase in size downstream as the area being drained and the discharge increases. Usually, the channel also becomes deeper and assumes a shape that allows the increased discharge to move more efficiently.

The slope of the stream channel—called the **stream gradient**—the shape of the channel, the frictional resistance it offers to flow,

and the discharge determine the velocity of the stream. To calculate the gradient of a stream, divide the difference in elevation between two points along the stream by the distance between the points. Generally, gradient is expressed in meters per kilometer or feet per mile. For most streams, the gradient is steepest near the head of the stream and lower downstream. A graphical plot of elevation against distance along a stream channel provides a more precise picture of the way the gradient varies from the head to the mouth of the stream. Such plots are called **longitudinal profiles** (figure 18.7). The stream gradient is one of the factors that determines the way water flows in the stream.

Stream Flow Water in a stream exhibits one of three flow patterns—turbulent, laminar, or shooting flow (figure 18.8). In mountains, where water flows in narrow channels filled with boulders and rock debris, the water jumps, boils, and foams as it pours over rocks and rushes down steep slopes. Where obstructions interfere with the smooth flow of water, the water becomes **turbulent.** Eddy currents, whirlpools, and boiling movements characterize this type of flow. Water is extremely turbulent where it cascades down a rough stream channel. On medium to low slopes, the surface of the water is smooth in some places, slightly undulating in others, and may exhibit a mild, slow eddying or boiling motion.

Where the stream moves slowly through a smooth-sided channel or in places where the velocity increases rapidly, water may move as though it flows in fine layers that slip over one another, called **laminar flow,** following smooth, streamlined paths. True laminar flow is extremely rare in streams. Where the stream gradient and velocity increase so much that the water level drops, stream flow, referred to as **shooting flow,** is nearly laminar (see figure 18.8). This usually occurs at the lip of a waterfall or where water flows over the top of a dam.

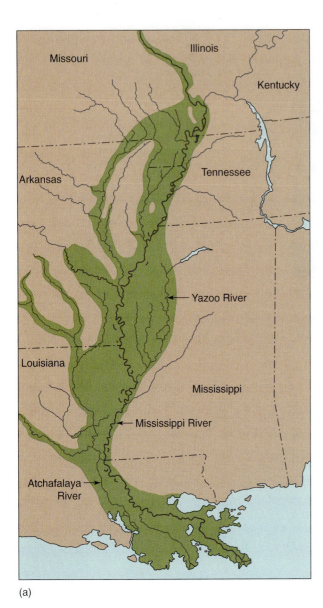

(a)

Figure 18.6 Flood Plain. The flood plain of a river is the area beside the stream that is covered by water when the channel overflows during floods. (a) The flood plain of the lower Mississippi. (b) Sketch of a flood plain.

Velocity Distribution If the stream flows in a straight, symmetrical channel, the greatest velocity lies near the middle and about one-third of the way down from the surface (figure 18.9). On curves, the greatest velocity shifts toward deeper water, on the outside of a curve. Generally, the maximum turbulence lies just beneath and to either side of the maximum velocity.

Stream Erosion

Water that moves as a thin layer on the ground surface—called *sheet wash*—downslope movements of surficial materials, and stream action cause most erosion in drainage basins. After heavy rain, sheet wash moves fine materials. Streams pick up and move the materials supplied to them by sheet wash and mass wasting, and under certain conditions they also erode bedrock under and beside the channel. Because wet clay is highly cohesive, clays are more difficult for streams to lift into the water than silt- or sand-sized particles. For this reason, channels in clay are much smoother than those composed of larger-sized particles. Consequently, the flow is less turbulent, and less erosion results. A much higher stream velocity is required to entrain clay than silt, but once suspended in the stream, clay remains in suspension longer than other particles (figure 18.10).

Within the stream channel, the impacts of water and suspended sediment, solution, and the bursting of bubbles account for most erosion. The pressure of the flowing water may dislodge rocks in the bottom of the channel and in the stream banks. Once loose, rocks bounce along in the channel in the process of saltation, described in chapter 17. They hit and dislodge other material in the bed of the stream. The impacts of the bouncing rocks fracture and break solid rock. In swift and highly turbulent water, the part of this load, known as the **bed load,** bounces along the bottom. Movement of bed load concentrates erosion on the bottom of the channel and causes the stream to cut downward. As a result, some streams flow in narrow, deep gorges like those of the Colorado River and its tributaries at the Grand Canyon, Arizona (see figure 18.4c). Where downcutting is rapid, erosion of the sides of the channel may be negligible.

Streams that flow in winding or sinuous courses erode laterally. On each curve, water shifts toward the outside of the bend where higher velocity and turbulence concentrate erosion. As a result, the channel becomes deeper, and the bank becomes steeper on the outside of the curve. As the slope of the bank increases, it becomes unstable. Sections of it slump into the stream. The stream picks up and removes this material, and the stream channel shifts laterally toward the outside of the curve.

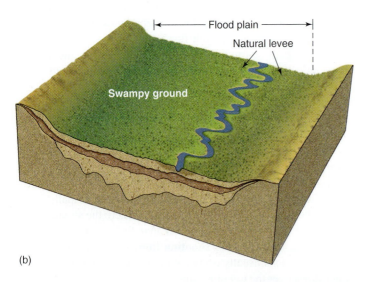

(b)

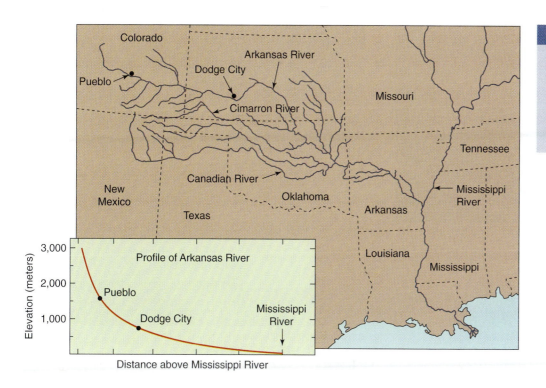

Where streams flow across bedrock, abrasion, impact, and dissolution erode bedrock. Rounded depressions, called **potholes,** form in such streams (figure 18.11). These usually contain a few rocks or coarse sediment. When the velocity of the stream is high, rocks in the hole swirl around and abrade the walls of the hole. Dissolved and suspended products wash up and out of the hole. Some potholes formed in this way are several meters deep.

Where water becomes charged with air bubbles, as it does in rapids and at the base of waterfalls, collapse of air bubbles causes erosion. Bubbles implode as the water pressure on the submerged bubble increases. Collapse of this sort creates a strong shock wave capable of causing rapid erosion even in hard materials. The erosive power of bursting bubbles is so great that engineers design dam spillways to minimize this effect.

Stream Load

The material transported by a stream is its **load.** As previously discussed, parts of the load consist of solids that roll or bounce along the bottom of the channel, the bed load (see figures 18.4b and 18.5). Fine particles make up the **suspended load,** and dissolved materials form the **solution load.**

Stream banks and channels provide a ready source of load. Ordinarily, alluvium left from earlier periods of high water partially fills the channel. Most of this alluvium moves only during periods of high water. With higher discharge, the stream can cut its channel deeper by removing alluvium from the bottom of the channel. As the stream removes alluvium or undercuts, the banks fail by slumping or sliding into the stream. This process is particularly important

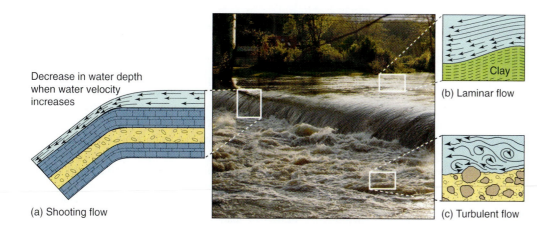

Decrease in water depth when water velocity increases

(a) Shooting flow

(b) Laminar flow

(c) Turbulent flow

Figure 18.8 Stream Flow. Arrows indicate the paths that water follows in (a) shooting, (b) laminar, and (c) turbulent flow. Water at the base of a waterfall is extremely turbulent.

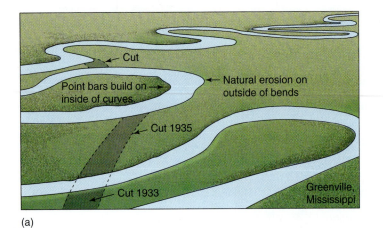

Cut

Point bars build on inside of curves

← Natural erosion on outside of bends

Cut 1935

Cut 1933

Greenville, Mississippi

(a)

Figure 18.9 Stream Velocity. (a) Sketch of the Mississippi River as viewed northward from over Greenville, Mississippi. Several cuts have been made at necks to straighten and shorten the stream. (b) The water in a stream that follows a sinuous or meandering course has a higher velocity on the outside of curves. Around these bends, erosion is rapid, and sediment shifts to the inside of the curves where sandbars form.

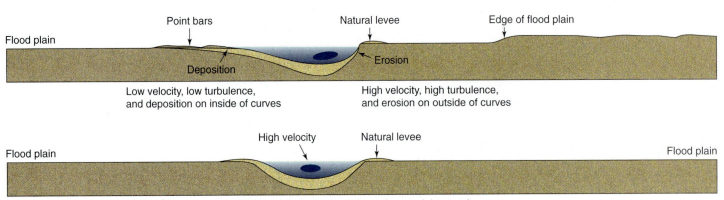

Point bars

Natural levee

Edge of flood plain

Flood plain

Deposition

Erosion

Low velocity, low turbulence, and deposition on inside of curves

High velocity, high turbulence, and erosion on outside of curves

High velocity

Natural levee

Flood plain

Flood plain

High velocity in center of channel on straight stretches

(b)

along streams that flow through easily eroded unconsolidated sediments, such as flood-plain alluvium, glacial deposits, lake-bed deposits, or unconsolidated marine deposits.

Although streams obtain most of their load from the channel and valley slopes, they may also receive load from other more remote sources such as glaciers or the wind. Meltwater from glaciers carries clay and silt ground up as the glaciers scrape across bedrock. The wind carries volcanic ash and may move sand and dust great distances. Windblown sand accounts for much of the load of streams located in deserts where strong winds sweep the loose sediment from the desert floor.

When heavy surface runoff sweeps large quantities of soil or sediment into streams, suspended matter makes the water appear muddy. Most suspended load consists of silt and clay, but in turbulent water, sand-sized particles may also remain suspended. The water buoys up everything under water, reducing the effective weight of submerged objects by an amount equal to the weight of the displaced water.

The load carried in solution consists mainly of elements and ions released by weathering processes. Streams flowing through regions underlain by bedrock composed of limestone, dolomite, or salt beds may carry much of their load in solution. Where rainfall is abundant, most of this dissolved load comes from ground water that finds its way back to the surface through springs. Solution load composes a small part of the total load of most streams, but in arid

regions, where large quantities of soluble minerals are present in the soil, and in humid regions, where streams may be sluggish and chemical weathering dissolves more material, a large part of the load is carried in solution. In arid regions, these soluble materials accumulate until a heavy rain occurs. In humid regions, soluble materials are continually flushed out in the flow.

Most streams carry some of their load by each of these means, but for any particular part of a stream, one type of load may be far greater than others. For example, the lower Mississippi River carries most of its load in suspension and solution and only a small part as bed load.

The **competence** of a stream is the largest particle the stream is capable of moving. Experimental evidence indicates that the largest size that will roll along a gently sloping stream bottom varies as the sixth power of the velocity of the water. Even a slight increase in velocity causes a great increase in the size and weight of rocks the stream can move. For example, a stream that can move a 0.45-kilogram (1-pound) rock at a velocity of 25 centimeters per second (10 inches per second) will be able to roll a boulder weighing 29 kilograms (64 pounds) at a velocity of 50 centimeters per second (20 inches per second). **Capacity** is the total amount of load a stream can transport under the existing conditions of discharge and velocity. Sometimes, insufficient load prevents the stream from filling its capacity, or the stream may not be able to pick up or move the load that is present. Nevertheless, it is by

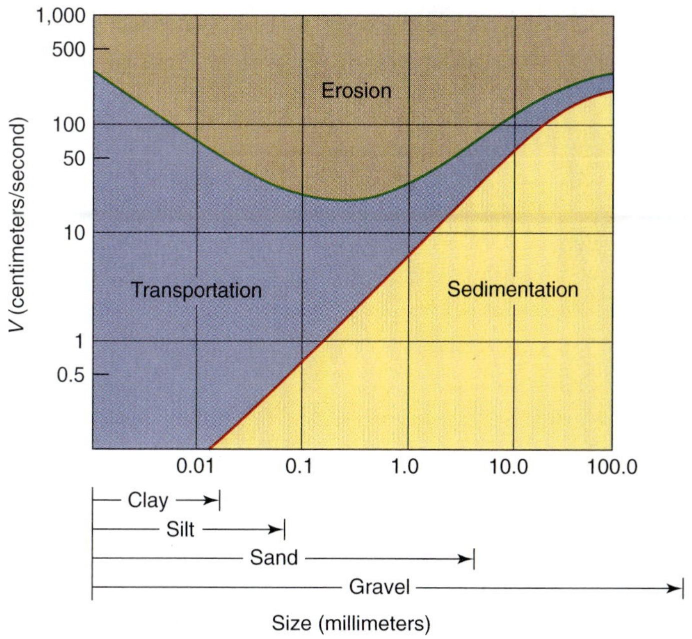

Figure 18.10 **Stream Erosion.** Stream velocity determines the size of materials a stream can move. At high velocities, streams can pick up (erode) a great range of sizes. At low velocities, streams can carry clay, but they deposit most sizes.

moving material along the stream channel that streams accomplish erosion.

Stream Deposition

Because velocity and discharge determine the ability of the stream to move its load, a decrease in either velocity or discharge causes deposition of part of the load. Velocity is especially sensitive to a reduction in the stream's gradient. Consequently, deposits accumulate on deltas where streams lose their velocity as they enter lakes or the ocean (see deltas in chapter 10) and in places where streams issue from mountains onto lower slopes. Large fan-shaped stream deposits, called **alluvial fans** (figure 18.12), form at these sites, and extensive deposits fill some basins in arid regions (see discussion later in this chapter).

The thickness of the alluvium in a stream channel varies with each change in discharge and velocity. Thick channel fill may be left undisturbed for long periods between major floods, but in channels cut into bedrock, thin alluvial deposits are transient features. The increased velocity and discharge associated with floods increase stream capacity. As floodwater spills over the stream banks and spreads across the flood plain, shallow depths and obstructions, especially trees and plants, slow movement of the broad, thin layer of water on the flood plain. When the floodwater subsides, the velocity of the water on the flood plain drops, and its capacity to move sediment decreases. The suspended sediment settles, leaving a thin layer of new sediment across the flood plain. Sometimes, a wedge-shaped deposit, called a **natural levee,** forms along the edge of the channel (figure 18.13). The levee contains coarse sediment deposited near the channel. The finer sediment that settles out on the flood plain produces rich agricultural land.

In humid regions, streams eventually carry most of their load into the sea where the deposits form deltas, but in dry regions, where water quickly evaporates or infiltrates into stream alluvium, the load available to the stream frequently exceeds the stream's capacity. The channel fills with alluvium, and the stream, unable to move its load, takes on a braided form. Over time, the stream shifts back and forth over the surface, eventually constructing the land counterpart of a delta (see figure 18.12).

Interaction of Factors Affecting Stream Behavior

Velocity, water depth, channel width, and the capacity of streams to carry load are interdependent (figure 18.14). Because a change in one of these affects the others, they are called **dependent variables.** Several other variables are independent of changes in

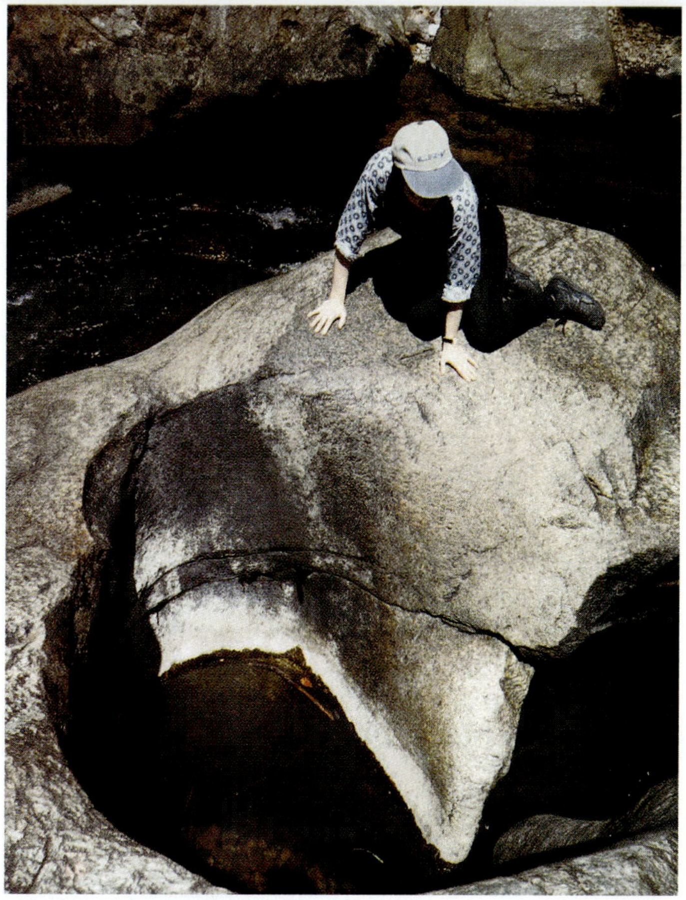

Figure 18.11 **Potholes.** Where streams have excess capacity, they may remove all sediment from the bottom of the channel and cut the channel. At this locality in Virginia, small eddies have gradually cut potholes into solid granite.

the stream. For example, streams do not control the amount of rainfall (i.e., discharge), the elevation of the regional base level, tectonic or volcanic activity in the drainage basin, or the amount of load supplied to the stream.

Changes in stream systems operate in both long and short time frames. Some variables change rapidly. The sudden increase in discharge after a thunderstorm causes an immediate increase in velocity, transporting power, and the cross-sectional area of the stream channel. These changes quickly work their way downstream, temporarily changing the character of the stream. Other changes like those related to climate or tectonic activity bring about slow changes in streams and their valleys. These changes may continue for hundreds of thousands or even millions of years, and they alter the landscape.

Over a period of years, many streams achieve a condition in which any change in one of the controlling variables sets in motion changes in other variables that counteract the initial change. In such streams, known as **graded streams,** the slope of the stream changes to provide the velocity required to transport the load supplied from the drainage basin at the prevailing discharge and chan-

nel conditions. For example, if load increases in the upper reaches of a stream, the gradient in that part of the stream increases as the upper end of the stream channel fills with more sediment than it can move. The steeper slope causes the velocity of the stream to increase, and this increases the capacity of the stream. As capacity increases, the stream erodes its channel and moves more bed load, lowering the gradient and, in turn, the velocity. This model of stream behavior allows us to predict the series of changes in other variables. All streams exhibit these types of adjustments, and many approach the graded stream model closely.

HOW STREAMS SHAPE THE LANDSCAPE

Continued over many thousands of years, erosion and deposition by streams has shaped much of today's landscape. Where deposition of sediment has been occurring long enough to bury the underlying bedrock, the shape of the deposits controls the landforms that develop. Alluvial fans form where streams carrying large quantities

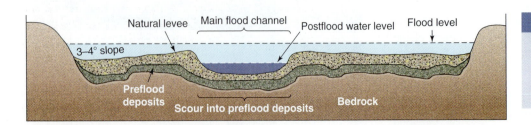

of sediment flow out of high mountains into low-lying areas. Deposition results as the gradient of the stream decreases, and the stream lacks the capacity to carry its load. Over time, streams on alluvial fans shift position, spreading their load first in one direc-

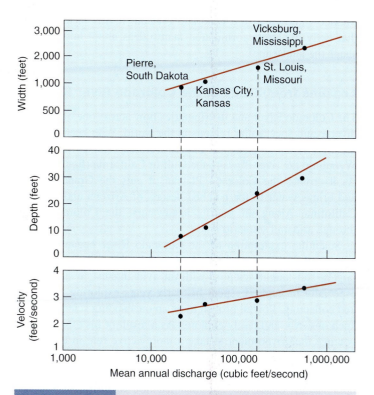

Figure 18.14 Interaction of Factors Affecting Stream **Behavior.** Variation in width, depth, and velocity with increases in discharge along the Mississippi River at Pierre, South Dakota; Kansas City, Kansas; St. Louis, Missouri; and Vicksburg, Mississippi. Note that width, depth, and velocity increase at all of the localities as discharge increases.

tion, then another, building the fan shape (see figure 18.12). Such fans are especially prominent features of the landscape in arid regions where many streams flow intermittently, but they are also present in humid regions. Deltas, discussed in chapter 10, are partially submerged counterparts of alluvial fans that develop where heavily loaded streams flow into bodies of water.

Alluvial fans may overlap and coalesce along mountain fronts, and where deposition happens for a long time beyond mountain fronts, alluvial plains may form. Large alluvial plains are present along the western edge of the Great Plains of North America. Here, streams such as the Arkansas River, the Missouri River, and the Platt River, which have been flowing out of the Rocky Mountains for many millions of years, have laid a sheet of alluvium that in some places extends over a hundred miles into the Great Plains. The streams that construct alluvial fans and alluvial plains usually flow in a braided pattern, a sure sign of their excess load. Most such streams have moderately steep gradients, but as stream gradients decrease, streams change character.

Features of Streams with Low Gradients

Stream patterns consisting of large sinuous loops, called **meanders,** are one of the most characteristic features of streams that have low gradients (figure 18.15). These streams usually flow over unconsolidated sediment and erode laterally rather than cutting downward. By eroding laterally, the curves become accentuated and eventually form loops. The resistance of the banks determines how rapidly they erode and how wide the floodplain becomes. The lower Mississippi River south of Cairo, Illinois, is one of the best-known meandering streams, but meanders may appear, even in high mountain valleys where the valley floor is nearly flat. Meandering streams are present on the bottom of many valleys that were shaped by glaciers.

In experiments, streams that flow over uniform material quickly form curves even if they start in a straight channel. As soon as a

Figure 18.15 Features Associated with **Meandering Streams.** This aerial photograph of meandering streams shows oxbow lakes, an impending cutoff at a neck, point bars, and the edge of the meander belt.

Figure 18.18 Air View of a Water Gap Formed Where a River Cuts Across a Ridge.

Stream Piracy The origin of valleys developed by stream action but now dry was one of the major insights of G. K. Gilbert, who worked with Powell in his explorations of the West. Seeing dry valleys shaped like a river valley and containing river gravel, Gilbert concluded that another stream had diverted the stream that originally cut the valley. He called this process stream **piracy.** Piracy occurs when a stream with a channel below the level of another stream cuts into the valley of the higher stream and diverts the flow of water. Because it is flowing on a steeper gradient, the head of the pirate stream erodes its valley more rapidly than the stream it cuts off. Pirate streams grow by headward erosion into the drainage basin of the adjacent stream and eventually intersect the main channel of the captured stream.

Stream piracy accomplished by headward erosion and stream capture is one explanation of how streams cut through resistant ridges, forming passes called **water gaps** (figure 18.18 and see figure 18.16). Some water gaps develop along faults, fracture zones, or other lines of structural weakness that are easily cut by the headward erosion of the pirate stream.

Entrenched Streams Along most of its path to the sea, the Colorado River and its tributaries, the Green and San Juan rivers, flow through a succession of steep-walled canyons (see figure 18.4c). In some places, the courses of these streams meander through canyons cut thousands of feet deep into the nearly flat plateau surface. Such streams in both humid and arid regions originally had low gradients and some meandered across flat surfaces. As uplift of the region occurred, the streams began to cut downward and gradually cut the deep canyons they now occupy. The term **entrenched** applies to such rivers (figure 18.19 and see figure 18.4c).

How Streams Shape Landscapes in Arid and Semiarid Regions

Streams in combination with mass wasting create the features of arid landscapes. Rains fall throughout the year in most humid regions; heavy rains of short duration are typical of drier climates. Even in mountainous areas where glaciers have carved much of the landscape, streams modify those forms as soon as the glaciers waste away.

Most streams flow through valleys eroded by stream action. Exceptions are found where streams flow along depressions that crustal movements initially caused. Most such depressions are fault zones or down-dropped crustal blocks—grabens—built by extension of the crust (see figures 5.28 and 6.17). Even in structural depressions, streams quickly modify the landscape. More often, streams flow in valleys cut by the streams over millions of years as the landscape evolved. During this evolutionary process, streams and their tributaries develop drainage patterns related to the character of the underlying bedrock as previously described.

Figure 18.19 Entrenched Stream. The San Juan River flows in a meandering pattern in southern Utah.

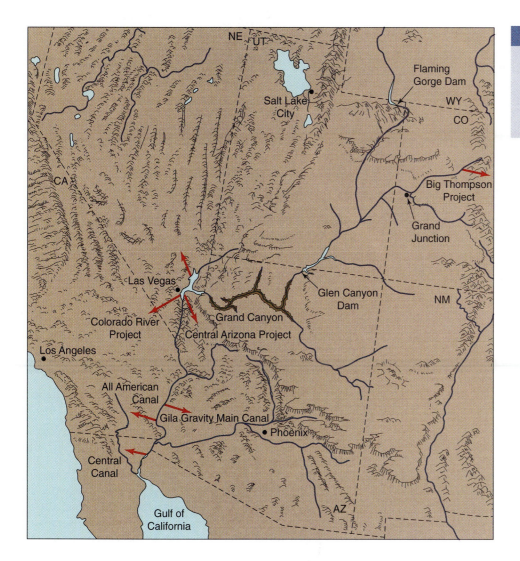

Figure 18.17 Drainage System of the Colorado River. The drainage system cuts across the Colorado Plateau. The Basin and Range Province lies to the south and west of the plateau. The topography of this region is shown on figure 14.3.

evolution of the landscape. Powell was the first to recognize and explain the concept of **base level,** described in chapter 14. The Colorado River and its tributaries flow well above their base level throughout much of their length. Powell also saw that the streams in the Colorado River basin differed in their relation to the structure of the bedrock on which they flow. He proposed a classification of these relations that is still in use. It contains four categories of streams: consequent, superimposed, antecedent, and subsequent.

Consequent streams flow down the initial slope of a newly formed surface. Streams flowing off volcanoes or down recently emerged portions of the continental shelf are examples. In the Colorado Plateau, Powell saw streams flowing down the bedding surface of tilted strata.

Superimposed streams are those that have cut down to their present courses through an overlying cover. For example, a stream that initially flows across horizontal strata might, as it cuts down, encounter and cut through an unconformity into crystalline rocks or deformed layers that lie beneath sediment. The course of the stream in the rocks beneath the unconformity might then be deter-

mined by its earlier course, established when the stream flowed on the overlying cover.

Antecedent streams persist in their course even as the bedrock beneath them deforms by folding, uplifting, or faulting. Such streams cut their channels down rapidly enough to maintain a slope across a rising structure. Powell thought that the course of the Green River, where it cuts across the Uinta Mountains in northern Utah, originated in this way. He concluded that the Green River flowed south into the Colorado River before the Uinta uplift took place and that the Green River cut its channel downward fast enough to maintain its course across the uplift even as it rose. In contrast, many of the smaller streams that flow off the Uinta Mountains developed after uplift of the mountains had taken place. These small streams are consequent streams.

Subsequent streams follow courses dictated by the resistance of the underlying bedrock. These streams flow in valleys underlain by rock that is less resistant to erosion. This is true of the minor streams that flow parallel courses to the ridges as seen in figure 18.16. Other subsequent streams may follow fault lines or fracture patterns, creating a rectangular drainage pattern.

Figure 18.18 Air View of a Water Gap Formed Where a River Cuts Across a Ridge.

Stream Piracy The origin of valleys developed by stream action but now dry was one of the major insights of G. K. Gilbert, who worked with Powell in his explorations of the West. Seeing dry valleys shaped like a river valley and containing river gravel, Gilbert concluded that another stream had diverted the stream that originally cut the valley. He called this process stream **piracy.** Piracy occurs when a stream with a channel below the level of another stream cuts into the valley of the higher stream and diverts the flow of water. Because it is flowing on a steeper gradient, the head of the pirate stream erodes its valley more rapidly than the stream it cuts off. Pirate streams grow by headward erosion into the drainage basin of the adjacent stream and eventually intersect the main channel of the captured stream.

Stream piracy accomplished by headward erosion and stream capture is one explanation of how streams cut through resistant ridges, forming passes called **water gaps** (figure 18.18 and see figure 18.16). Some water gaps develop along faults, fracture zones, or other lines of structural weakness that are easily cut by the headward erosion of the pirate stream.

Entrenched Streams Along most of its path to the sea, the Colorado River and its tributaries, the Green and San Juan rivers, flow through a succession of steep-walled canyons (see figure 18.4c). In some places, the courses of these streams meander through canyons cut thousands of feet deep into the nearly flat plateau surface. Such streams in both humid and arid regions originally had low gradients and some meandered across flat surfaces. As uplift of the region occurred, the streams began to cut downward and gradually cut the deep canyons they now occupy. The term **entrenched** applies to such rivers (figure 18.19 and see figure 18.4c).

How Streams Shape Landscapes in Arid and Semiarid Regions

Streams in combination with mass wasting create the features of arid landscapes. Rains fall throughout the year in most humid regions; heavy rains of short duration are typical of drier climates. Even in mountainous areas where glaciers have carved much of the landscape, streams modify those forms as soon as the glaciers waste away.

Most streams flow through valleys eroded by stream action. Exceptions are found where streams flow along depressions that crustal movements initially caused. Most such depressions are fault zones or down-dropped crustal blocks—grabens—built by extension of the crust (see figures 5.28 and 6.17). Even in structural depressions, streams quickly modify the landscape. More often, streams flow in valleys cut by the streams over millions of years as the landscape evolved. During this evolutionary process, streams and their tributaries develop drainage patterns related to the character of the underlying bedrock as previously described.

Figure 18.19 Entrenched Stream. The San Juan River flows in a meandering pattern in southern Utah.

of sediment flow out of high mountains into low-lying areas. Deposition results as the gradient of the stream decreases, and the stream lacks the capacity to carry its load. Over time, streams on alluvial fans shift position, spreading their load first in one direc-

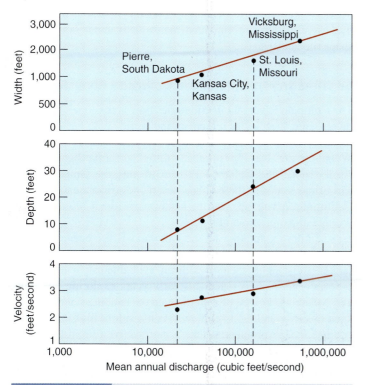

Figure 18.14   Interaction of Factors Affecting Stream Behavior. Variation in width, depth, and velocity with increases in discharge along the Mississippi River at Pierre, South Dakota; Kansas City, Kansas; St. Louis, Missouri; and Vicksburg, Mississippi. Note that width, depth, and velocity increase at all of the localities as discharge increases.

tion, then another, building the fan shape (see figure 18.12). Such fans are especially prominent features of the landscape in arid regions where many streams flow intermittently, but they are also present in humid regions. Deltas, discussed in chapter 10, are partially submerged counterparts of alluvial fans that develop where heavily loaded streams flow into bodies of water.

Alluvial fans may overlap and coalesce along mountain fronts, and where deposition happens for a long time beyond mountain fronts, alluvial plains may form. Large alluvial plains are present along the western edge of the Great Plains of North America. Here, streams such as the Arkansas River, the Missouri River, and the Platt River, which have been flowing out of the Rocky Mountains for many millions of years, have laid a sheet of alluvium that in some places extends over a hundred miles into the Great Plains. The streams that construct alluvial fans and alluvial plains usually flow in a braided pattern, a sure sign of their excess load. Most such streams have moderately steep gradients, but as stream gradients decrease, streams change character.

Features of Streams with Low Gradients

Stream patterns consisting of large sinuous loops, called **meanders,** are one of the most characteristic features of streams that have low gradients (figure 18.15). These streams usually flow over unconsolidated sediment and erode laterally rather than cutting downward. By eroding laterally, the curves become accentuated and eventually form loops. The resistance of the banks determines how rapidly they erode and how wide the floodplain becomes. The lower Mississippi River south of Cairo, Illinois, is one of the best-known meandering streams, but meanders may appear, even in high mountain valleys where the valley floor is nearly flat. Meandering streams are present on the bottom of many valleys that were shaped by glaciers.

In experiments, streams that flow over uniform material quickly form curves even if they start in a straight channel. As soon as a

Figure 18.15   Features Associated with Meandering Streams. This aerial photograph of meandering streams shows oxbow lakes, an impending cutoff at a neck, point bars, and the edge of the meander belt.

curve forms, the maximum velocity shifts toward the outside of the curve, accelerating erosion there. The surface of the water in the stream takes on a slight slope toward the inside of the curve. This directs the flow back across the channel and against the opposite bank, and the conditions favoring a sinuous path begin. Deposits of sediment called **point bars** (see figure 18.15) form on the inside of the curve, and as the process continues, the curves become accentuated, and eventually, meanders form. As a meander grows, it may come back on itself, cutting off sections and leaving behind curved lakes called *oxbow lakes* (see figure 18.15).

Effects of Bedrock Structure

Where streams have eroded into bedrock, the structure and composition of the bedrock and the elevation of the area above base level control the formation of the landscape even if the bedrock is horizontal. **Dendritic** stream patterns that resemble the branches of a tree usually form in regions underlain by flat-lying sediment or sedimentary rocks that are easily eroded and more or less uniform in composition (see figure 18.2). However, if the bedrock varies in its resistance to erosion, streams tend to follow paths of least resistance in the most easily eroded material. After erosion continues for a long time, the stream pattern is likely to reflect the structure of the bedrock. Thus, streams around domes may form concentric patterns. If weaknesses have formed along fractures and faults, the streams are likely to have straight channels resulting in drainage patterns that may have the shape of a **trellis.** Otherwise, streams tend to form in easily eroded materials, such as shale or unconsolidated sediment. In humid climates, streams often flow in valleys underlain by limestones that are soluble, thus weathering and eroding more rapidly than many other rock types. In arid climates, limestones may form cliffs. In mountain belts that have been eroded for long periods, the stream patterns may reveal the presence of folds or inclined beds (figure 18.16). Especially good examples of these drainage features are found in portions of Utah, Arizona, Colorado, and New Mexico—known as the Colorado Plateau. It was in this region that early explorers formulated many ideas about how streams shape the landscape.

Lessons from Study of the Colorado River Basin

The Colorado River and its tributaries flow across one of the most diverse and spectacular regions in the world. With headwaters in the high mountain peaks of western Colorado and as far north as Wyoming, the river flows to the Gulf of California (figure 18.17). As it crosses the Colorado Plateau, it cuts the Grand Canyon and brings water to the parched deserts of Arizona and southern California. Along all but its headwaters, the Colorado flows through arid and semiarid lands.

Scientific exploration of the Colorado River began with the famous expeditions of John Wesley Powell in 1867. Powell, a largely self-taught scientist with an avid interest in nature and especially in geology, spent his early years exploring, teaching, and collecting. Powell proved to be an outstanding scientist and a keen observer and hardy explorer. As a result of his efforts to encourage the government to take a more active role in the exploration and development of its western lands, he was appointed the first director of the U.S. Geological Survey in 1879. Powell's observations about the Colorado led him to formulate theories about the relation of streams to the land they drain. Many of his ideas are among the basic principles used to understand the

Figure 18.16 Stream Patterns May Identify Bedrock Structures. (a) Variation in the resistance of rock to erosion produces stream patterns that may reveal the structure. (b) In this image taken over the Susquehanna River in Pennsylvania, the ridges are composed of sandstone. Note that the river cuts across the ridges.

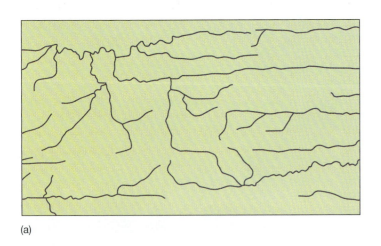

(a)

(b)

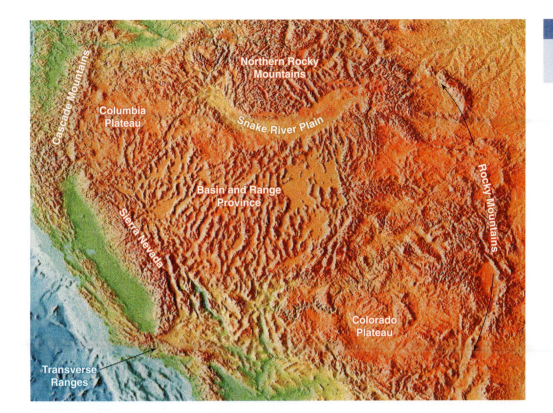

Cascade Mountains

Columbia Plateau

Northern Rocky Mountains

Snake River Plain

Rocky Mountains

Basin and Range Province

Sierra Nevada

Colorado Plateau

Transverse Ranges

Streams in Arid Regions of the American Southwest

Most of the arid region in western North America lies within the Basin and Range Province and the Colorado Plateau (figure 18.20). The Sierra Nevada and the Transverse Ranges of southern California border these provinces on the west. Vast young lava flows of the Columbia River Plateau and the Cascade Mountains lie to the north and northwest, and the Rocky Mountains form the eastern border. This arid region extends southward into the Sierra Madre, a rugged field of volcanic rocks in Mexico (see figure 14.3).

Broad, flat to gently sloping basins separated by block-like mountains characterize the Basin and Range Province (see figures 18.12 and 18.20). Some of these mountains are 3 to 4 kilometers (9,800 to 13,000 feet) high and hold snow on their peaks well into the summer months. Most of the basins and ranges are elongated in a north-south direction.

Structurally complex rocks, deformed during the Paleozoic and Mesozoic, form the bedrock of the mountains in the Basin and Range Province. Younger block faulting affected the region in the Cenozoic. The boundary between many of the ranges that rose and basins that dropped during the most recent deformation is composed of high-angle faults, some of which are still active. Streams do not flow out of these basins, and few permanent streams exist in the region. The bottoms of these basins define the regional base level. The floor of Death Valley, one of the regional base levels, is about 85 meters (280 feet) below sea level. Those that flow from the mountains into adjacent basins disappear as the water evaporates or disappears into the ground.

The Colorado Plateau is strikingly different from the regions around it. In the Colorado Plateau, nearly flat-lying sedimentary strata cover most of the region. Horizontal terraces develop where strata are flat (figure 18.21). In other areas, broad gentle warps of strata separate one terrace surface from another. In some places, high-angle faults offset the strata. Broad arches or flexures are also common in the plateau region, but the rocks in the plateau are much less deformed than those in the Basin and Range Province to the west or in the Rocky Mountains to the north and east. The Colorado Plateau region rose as a gigantic block during the deformation that elevated the Rocky Mountains. Today, the plateau surface stands about 1.8 to 2.5 kilometers (6,000 to 8,000 feet) above sea level.

Plains Produced by Erosion Extensive flat or gently sloping plains characterize the landscape of both the Basin and Range and the Colorado Plateau, but these plains did not develop in the same way. Most of the plains in the Colorado Plateau formed as a result of the stripping of subhorizontal layers (see figures 14.15*a* and 18.21). A cliff face marks where erosion is now removing the upper layer. Plains built in this way are flat because underlying rock layers are nearly flat. The plain expands as erosion progressively removes the face of the cliff. Many such cliffs are composed of sandstone. Shales or mudstones usually lie beneath the cliffs. The cliff face slowly retreats as blocks of the cliff-forming rock layer fall away. In some places, the cliff face slumps. At others, the base of the cliff is undercut as less-resistant material washes away.

In the Basin and Range Province, gently sloping surfaces, called **pediments,** rise from the centers of the basins and cut

Figure 18.21 Flat-Laying Rocks in the Colorado Plateau.

across outcrops of different rock types. After heavy rains, water washes across the pediment surface in shallow channels, leaving a thin veneer of sediment as the runoff decreases. The slope of the pediment is approximately the slope needed by runoff to move the products of weathering and erosion downslope across the pediment. Active erosion takes place at the upper edge of the pediment, where mass wasting and running water erode cliffs along the mountain front. Debris eroded from the cliffs accumulates at their bases and on the upper edge of the pediment, producing a steeper slope there. Over time, surface wash gradually removes this accumulation of debris, and the cliff retreats. Cliff retreat and the ensuing growth of the pediment are common in arid regions. If the process of slope retreat continues long enough, the high ground forming the cliff disappears, leaving a gently sloping plane called a **pediplane.**

Plains Produced by Deposition Plane-sloping surfaces also form as a result of deposition of sediment. Alluvial fans, described earlier, are common features in arid regions (see figure 18.12). Depositional plains also form farther away from the mountain front, typically at the foot of the pediment where alluvial fans coalesce and in the central part of the basin below the pediment. This depositional surface, called a **bajada,** forms where the sediment carried across the pediment settles out of the streams as the water evaporates or infiltrates into the ground. These sediments form the thick deposits in the basins of the Basin and Range Province.

Shallow lakes, called **playa lakes,** form in the center of the basin after heavy rainfall. The water in these lakes gradually evaporates, leaving flat areas of fine sediment and salt deposits (figure 18.22). These deposits may contain economically important concentrations of some salts and are mined for borax and other salts. The high salt content of the lakes makes them an inhospitable environment for most organisms, but some bacteria survive under even these harsh conditions.

HUMAN IMPACTS ON THE HYDROLOGIC CYCLE, AND VICE VERSA

Surface-Water Supply

For centuries, humans living in the humid regions of the world viewed the water supply as virtually inexhaustible. Streams served both as domestic water supply and as a medium for waste disposal. The quantity of water was sufficient to dilute the waste so it posed no health problems. As population increased, problems with both the quantity of water and water quality emerged. In North America, the supply problem in semiarid regions of the Southwest were obvious. Pollution problems first arose in the areas of highest population density and greatest development. Today, water-pollution

Figure 18.22 **A Playa Lake in Death Valley.** The close-up shows salt deposits in the part of the lake that has dried.

problems are worldwide and affect most communities. Water is essential for fisheries, for the growth of forests and other types of vegetation, and for wildlife. Water generates electricity in hydroelectric power plants. In rivers, water provides for navigation, recreation, and aesthetic needs.

Without water, human activities cease. Where water is available, the quantity and quality of water determine the range of sustainable human activities. Generally, human use involves removal of water, at least temporarily, from natural systems, and in some degree, pollution of used water. Humans withdraw most of the water they use from the ground or from streams. Globally, domestic and municipal uses account for about 8% of the water removed from natural systems. Agricultural activities, mainly irrigation, consume nearly 69% of withdrawn water; industries consume another 23% in manufacturing processes, power generation, and for cooling. Where water is scarce or of poor quality, individuals, communities, and governments face hard choices. Disputes over the rights to water and to its uses lead to large-scale water management programs and not infrequently to legal action.

Competition for water resources has grown as population and industrialization have increased. According to the World Resources Institute, the amount of water withdrawn from natural systems has increased 35-fold in the past three centuries. The amount of water consumed by humans has been increasing 4 to 8% each year with the highest rate of use being in the more-developed countries. In North America, humans take approximately 1,700 cubic kilometers (60 million cubic feet) of water from the ground and streams each year. Comparable figures for other areas are about 726 cubic kilometers (25 million cubic feet) per year for Europe, about 526 cubic kilometers (18.4 million cubic feet) per year for Asia, about 476 cubic kilometers (16.6 million cubic feet) per year for South America, and about 244 cubic kilometers (8.5 million cubic feet) per year for Africa. The quantities are sufficient to create water shortages in some areas.

Critical water shortages currently exist over large regions of East and West Africa, in northern China, in the Middle East, and in North Africa. Less severe shortages are present in many other parts of the world. The options for alleviation of these shortages are to limit population growth and development; to reuse water; or in areas close to the ocean, to turn to desalination. None of these alternatives can be easily implemented. Custom and religion make limitation of population growth difficult; contamination limits the reuse of water; and desalination is expensive.

Surface-Water Supply in Arid Regions

It may be hard to believe that the sea of brine at Great Salt Lake was once a huge body of fresh water. The water came from the nearby mountains and from farther north as sheets of ice melted and the meltwater flowed into the enclosed basin that holds the Great Salt Lake. Winds blowing across the lake set up waves that cut notches where they crashed into shore around this freshwater lake. Deltas, bars, wave-cut notches, and other shoreline features are still visible on the mountainsides near Salt Lake City. More than 10,000 years have passed since the lake reached its maximum extent. During this time, streams flowing into the lake have carried the salts dissolved from rocks over which they flow. Hot, dry air evaporated water from the lake surface, leaving a lake saturated with sodium chloride, and a region that depends on wells and runoff from nearby mountains for fresh water.

The irregular supply of surface water in arid regions encourages people to find ways to store water. Generally, they have stored it behind dams in reservoirs where the high rate of evaporation takes a toll. Losses to evaporation are minimized by constructing high dams with deep reservoirs and smaller surface areas. The dams and their reservoirs provide water for irrigation, for municipal needs, and for the generation of electricity, but they also destroy natural rivers. This has given rise to continuing controversies over not only whether or not to build dams, but also who should get the impounded water. The history of the development of the water supply on the Colorado River illustrates these issues.

The Colorado River Basin

The Colorado is the only river that flows across the Basin and Range. Despite its 2,300-kilometer (1,450-mile) length and a drainage basin larger than France, the Colorado often carries little or no water where it enters the Gulf of California. Evaporation rates are exceedingly high throughout most of the drainage basin, but water use and modification of the drainage by humans are the primary reasons for the drying of the Colorado River Delta. As a result of farming, the water that reaches the delta is also salty and laced with pesticides.

The earliest visitors from the eastern states to the arid southwest found the region so desolate they concluded that people would never inhabit the region or use it for agricultural purposes, but John Wesley Powell recognized the important role water would play in the future of the region. However, it is unlikely that even Powell could have envisioned the extent to which humans would eventually modify the drainage system. Today, water from the Colorado irrigates more than 8 million square kilometers (2 million acres) of rich, highly productive farmland. More than thirty dams and hydroelectric plants intercept the flow. Major aqueducts move water out of the basin to help provide water for development in other areas. The Big Thompson Project takes water to the eastern slopes of the Colorado Front Range. The Central Arizona Project takes water to Phoenix and Tucson. The Colorado River Aqueduct and the All-American Canal move huge quantities of water into southern California (see figure 18.17).

The first major diversion of water from the Colorado River was the Grand Ditch, a trench and pipeline built in the early 1900s that carries water over the Rocky Mountains at an elevation above 3,300 meters (10,000 feet). This system provides water for Fort Collins, other towns along the mountain front, and for irrigation of sugar beets cultivated in the western edge of the Great Plains. Today, the Big Thompson Project moves water through a 6-meter (20-feet) wide, 21-kilometer (13-mile) long tunnel under the crest of the Rocky Mountains to sustain development in central Colorado.

By 1922, many people in the states traversed by the Colorado River recognized the future importance of the water. The seven states in the drainage basin drew up the Colorado River Compact. Under the terms of this agreement, the states divided the drainage basin into upper and lower basins. The agreement provides for the withdrawal of 7.5 million **acre-feet** per year of water (1 acre-foot is the quantity of water in a reservoir that is 1 acre in size and filled to a depth of 1 foot. One acre-foot = about 40,500 cubic meters) from each basin. In 1944, a separate pact with Mexico promised 1.5 million acre-feet per year of water for use in Mexico. The total water allocated under these agreements is 16.5 million acre-feet per year. Unfortunately, since about 1930, the total water supply in the river has averaged closer to 14 million acre-feet per year. Of that, about 2 million acre-feet escape through evaporation. Before 1990, the water supply in the southern part of the basin was adequate to meet demands. For many years, the supply of water in the river exceeded demands, but in recent years, the desire for water has increased. It is likely that disputes over the use of Colorado River water will continue both within states and between states. For many years, California took more than its share, but with opening of the Central Arizona Project, Arizona demanded and got its allocated share. Southern California had to reduce its consumption of water from the Colorado and sought a water supply from other sources in northern California. For the first time, water use in the southern basins equaled or exceeded the total supply.

Within states, much of the debate concerns the division of water between agricultural and domestic interest and on the retention of water rights for future development. In Colorado, movement of water from the western slopes, where population density is low, to the eastern slopes, where growth is rapid, is an issue. Once a state commits the water originating in one area to another area, the source region loses its right to future use of that water. Development in one area today prevents future development in another.

Throughout many parts of the arid southwest, people are debating the appropriate uses of water. In Los Angeles, local ordinances designed to conserve water are in effect. Water patrols issue citations to homeowners for washing sidewalks or watering the lawn in the middle of the day, but not far away in the desert at Palm Springs, golf courses surround artificial lakes with fountains and boat docks, and hotels provide lagoons for their guests.

Depletion of the Aral Sea

In 1960, the Aral Sea, which is located on the border between Uzbekistan and Kazakhstan, covered 67,000 square kilometers (26,000 square miles) and was the fourth largest lake in the world after the Caspian Sea, Lake Superior, and Lake Victoria. Between 1960 and 1987, the lake lost 40% of its surface area and 66% of its volume (figure 18.23). The average depth dropped from 16 to 9 meters (50 to 29 feet), and the salinity increased to 27 grams per liter. The lake changed more over that twenty-seven-year period than it had over the past 1,300 years! Human actions are the primary cause.

The Aral Sea lies at the bottom of a vast area of internal drainage. Like the Great Salt Lake of Utah, no streams drain out of the Aral Sea. Consequently, the water level rises and falls as the amount of water reaching the lake changes. Two major rivers, the Amu Darya and the Syr Darya, that flow northward out of the Himalayan Mountain chain provide most of the water reaching the Aral Sea. During the Pleistocene, the quantity of water reaching the lake fluctuated as glaciers grew and receded in response to changes in climatic conditions. Late in the Pleistocene, the lake level rose toward its modern level. Between 1910, when accurate observations of water level began, and 1960, the level of the lake changed less than a meter. Then, the effects of human activities in the basin began to have significant impact.

Withdrawal of water from the major tributaries for irrigation and diversion of water from the Amu Darya account for the most significant loss of water. Much of this water irrigates cotton crops in the valleys of the major tributaries of the lake. As the lake

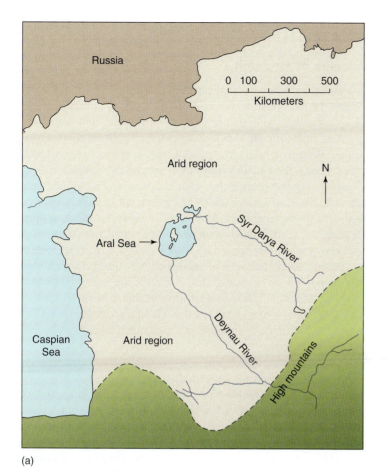

(a)

(b)

Figure 18.23 Depletion of the Aral Sea. (*a*) The Aral Sea is located in an arid region north of the Himalayas and east of the Caspian Sea. (*b*) The rapid retreat of the shoreline of the Aral sea from 1960 to 1997.

shrinks, towns and cities that formerly stood on the lakeshore become landlocked. The major fishing port of Muynak is now over 32 kilometers (20 miles) from the sea. To maintain jobs, the government brings frozen fish to the cannery. Large areas around the lake have become salt marshes. Reduction in the size of the lake caused the local climate to change and resulted in more extreme temperatures in both summers and winters. Sand dunes now threaten some villages, and salts blown by windstorms from the salt flats around the lake pose health problems for the more than 35 million people affected by the changing lake.

The future of the Aral Sea is uncertain. Unless the governments allow more water to reach the lake, shrinkage of the lake will continue, and salinity of the sea will increase. More water can reach the sea only if cotton farmers withdraw less water for irrigation, or if the government constructs canals or pipelines to divert water from other streams into the basin. The choices are especially difficult because a reduction in irrigation would certainly have an adverse impact on the agricultural productivity along the tributaries, and demand for water throughout the basin already exceeds supply.

The Nile River

Waters of the Nile River are essential to the economic well-being of Egypt. As the Greek historian Herodotus expressed it, "Egypt is the gift of the Nile." The headwaters of the Nile extend from the large lakes in Kenya, Uganda, and Tanzania near the equator, across the Libyan and Nubian deserts, to the shores of the Mediterranean Sea at 30° north latitude (figure 18.24). For millions of years, the river carried sediment from its drainage basin into the Mediterranean where one of the largest and most agriculturally productive deltas in the world developed. Each century, floods along the Nile added a few centimeters of topsoil to the flood plain and to the Nile Delta. In 1965, this natural system was drastically altered by construction of a new dam upstream of the older Aswan Dam. The new dam raises the water level 67 meters above the former level of the river. During the wet season, water accumulates in the reservoir. During the dry season, water is released to irrigate large tracts of land in the deserts. The high level of the dam makes it possible to produce cheap electricity. The dam has also created unanticipated

Figure 18.24   Nile River. This stream flows out of mountainous regions across an arid region and forms the Nile delta in the southeastern corner of the Mediterranean Sea.

environmental problems in a country with rapid population growth.

By interrupting the flow of water in the river, the dam has reduced flooding and the freshwater supply to the delta. The annual supply of topsoil to the surface and front of the delta has been reduced. Consequently, the delta is less productive, and waves and currents are eroding the delta front (see figure 10.33). The floodwaters carry nutrients needed by plants on the delta as well as sediment. Formerly, these nutrients flowed into the sea where they supported a large population of microscopic-sized plants called *phytoplankton*. The reduction in the numbers of these plants, the base of the food chain in the sea, has adversely affected fish populations.

A large canal that will carry 13.6 billion liters (3.6 billion gallons) of water per day out of the Aswan Reservoir, named Lake Nasser, is under construction. By 2017, this canal, the New Valley Canal, is projected to extend 300 kilometers (186 miles) out into the desert and make it possible to create millions of acres of farmland in what is now barren desert.

Both the Aswan Dam and the New Valley Canal have been controversial projects since their inception. Proponents point to the value of the new farmland, cheap electricity, and flood control. Opponents note the high rate of water loss to evaporation, erosion and loss of water on the delta, and the loss of land in the reservoir itself. As with most such major changes in water management, the projects have both positive and negative impacts, and the long-term balance between these is not clear.

FLOODING

Causes of Flooding

Streams flood when discharge exceeds the capacity of the channel (figure 18.25). Most floods result from excessive rainfall, such as that associated with thunderstorms, hurricanes, and stationary weather fronts, or the rapid melting of heavy snow cover in the spring. Because a high proportion of rainfall runs off in areas with steep slopes, the stream valleys in mountains are subject to flash floods during and immediately after heavy rains. Flat ground near streams, especially where underlain by impervious soils or shallow bedrock, floods frequently. Many such areas lack well-developed drainage systems needed to move surface runoff out of the area.

Low permeability caused by frozen ground, high clay content in the soil, or the presence of solid rock near the surface reduces the proportion of rainwater that infiltrates the soil. Plant cover plays an important role in several ways. Plants slow surface runoff and delay the rise of water in streams. Plants also reduce runoff because the root systems improve infiltration and take up water that the plant transpires.

Human activities, including construction of flood-control projects, have increased the magnitude and frequency of floods on some rivers. Roads, parking lots, and buildings cover large parts of some drainage basins in urban areas. In these places, impervious pavement replaces vegetative cover. All of these activities tend to seal the ground surface, increase the proportion of rainfall that runs off, reduce the natural storage areas for floodwater, and change channel characteristics so water moves more rapidly through some portions of the channel than others.

Methods Used to Control Floods

Although natural stream channels have adequate cross-sectional areas to contain the waters that normally flow along the stream,

Figure 18.25   The James River at Richmond, Virginia, During a Flood in 1969.

during floods, water in a natural system spreads over the flood plain. Where human development occurs along streams, problems arise because people build cities next to the stream channel and cultivate the nutrient-rich soil deposited on the flood plain during floods. Huge investments of public funds have been made in flood-control projects. Most of the time, these projects have worked. They have provided protection for communities along rivers, furnished cheap electric energy for whole regions, and made it possible to cultivate some of the most fertile land in the world. The first major efforts to control a river focused on the Mississippi River and its tributaries.

Before the construction of levees along the Mississippi River, floodwaters covered large sections of the flood plain almost every year. As farmers discovered that the soils on the flood plain are exceptionally fertile, they cleared the land and began cultivating cotton on the flood plain of the lower Mississippi (south of its junction with the Ohio River). The need for better drainage and flood protection was quickly realized. Before levees were constructed, farmers built houses on stilts to keep them dry during the wet season. It soon became apparent that long sections of levees would be required to solve the flooding problem. With the growth of towns and cities on flood plains, the amount of damage to buildings and roads during floods steadily increased. Political pressure to undertake projects designed to protect the property on the flood plains followed. Finally, the federal government became involved in efforts to protect the flood plains. The Corps of Engineers was assigned the project. At first, engineers designed flood-control projects, such as a levee around a town, to provide local protection. Later, as engineering capabilities increased, the Corps began to undertake more ambitious projects. These included construction of large dams with hydroelectric capabilities on major rivers, lining long sections of the lower Mississippi with levees, and changing channel characteristics by straightening river channels or dredging sections to greater depth.

The primary objective of flood control is to confine discharge to the stream channel along as much of the stream as possible. A number of simple but costly techniques can accomplish this goal. Artificial levees can raise the edge of the channel. Dams can control peak discharge in the stream by holding water. By closing spillways, water that would normally pass downstream goes into storage. Changing channel characteristics (e.g., straightened or deepened) can move water more rapidly along a section of the stream. By increasing infiltration, runoff is reduced or slowed. Most of these techniques involve major projects that are expensive, require continual maintenance, and drastically change the character of the streams. Each has its own set of advantages, limitations, and problems.

Artificial Levees Low earthen dams, called **artificial levees,** now parallel the channels of the lower parts of many of the major rivers in the United States. Levees contain the stream in its channel, but they also reduce the size of wetlands and prevent the storage of water on the flood plain during times of flood. During floods, levees and floodwalls raise the water level in the protected portion of the stream and increase the velocity of the passage of water through that area. This technique may cause flooding farther downstream where levees are not present. Levees and flood walls may also prevent drainage of small tributaries into the main stream. If the tributaries to a stream flood, the levee may act as a dam, impounding water on the "wrong" side of the levee. Flood walls may contain gates to regulate the flow of tributaries into a main stream, but high cost restricts their use. Because construction costs increase with height, it is undesirable to build levees higher than is necessary. Engineers generally design levees for a flood level expected about once every hundred years. If the river rises above that level, as it did in 1997 along the Red River in North Dakota, water flows over and destroys part of the levee and floods the land protected by the levee.

Dams Dams are useful for power generation, water storage, and recreation, as well as for flood control (figure 18.26). Because

Figure 18.26 Glen Canyon Dam in Northern Arizona. The Glen Canyon dam has been one of the most controversial of all dams. The reservoir fills a canyon said to be as beautiful as the Grand Canyon.

hydroelectric power is clean, dams are attractive sources of energy, but they have serious drawbacks and limitations. Streams flowing into reservoirs deposit their loads, eventually fill the reservoir with silt, and destroy its capacity for water storage. As a reservoir fills with silt, its usefulness decreases, and dredging or abandonment becomes necessary. Large dams are often controversial. As reservoirs fill, they displace populations, destroy habitat, and flood good agricultural land, archeological sites, and burial grounds. John Muir tried unsuccessfully to stop construction of a dam in Hetch Hetchy Canyon, a canyon similar to Yosemite Valley. Controversy over the Glen Canyon Dam in Arizona centered on flooding of one of the most beautiful canyons in that region, and plans for additional dams in the Grand Canyon are discussed from time to time.

The availability of suitable dam sites limits the usefulness of dams in many areas. A good dam site must meet special foundation requirements. The foundation must be composed of stable rock capable of withstanding the pressure exerted against the dam when the reservoir fills with water. The reservoir should be watertight, and in arid regions, the reservoir should have a small surface area relative to the volume of water in storage. Otherwise, evaporation causes excessive loss of water. The cost of the project also includes the value of the land and developments on the land submerged in the reservoir. Dams already exist on many of the best dam sites in the United States. Other uses of the sites or areas that would lie within the reservoir preclude construction on many potentially good sites. For example, both Yosemite Valley and the Grand Canyon contain excellent sites for dams, but their value for uses other than reservoir sites is considered more important.

Dams are expensive to build, and when water levels drop, large mudflats form around the edge of the reservoirs. Despite these drawbacks, dams are an important and effective means of flood control. If the reservoir is large enough, a dam can regulate the flow downstream.

The cost-effectiveness of constructing a few large dams versus many smaller dams on the small tributaries of major streams is an important question in flood-control decisions. Small dams are cheaper to build; they help reduce the buildup of sediment in large reservoirs; and they improve infiltration. The main drawback to building small dams is that they are less effective than large dams in reducing the effects of major floods.

Construction on unprotected flood plains causes most of the losses associated with flooding. In 1968, the federal government took steps to reduce these losses by initiating a program of flood insurance that covers buildings constructed on flood plains. Under provisions of the program, insurance is available in participating communities, but to qualify, a community must agree to follow certain procedures in the management of flood-prone areas. They must map and identify areas that are likely to be flooded during a hundred-year period and restrict development allowed in that area. This is accomplished by flood-plain zoning ordinances that regulate development. The effect has been to make people and communities much more aware of flood hazards and to limit the amount of new construction placed on unprotected flood plains.

Channel Modification Changing the shape of a stream channel can help reduce flood damage locally. Channel modifications lower the water level by increasing the velocity of the stream. Channelization usually involves removing curves and reducing the roughness of the channel. In extreme instances, a concrete-lined channel or a culvert replaces the natural stream channel (figure 18.27). Channelization reduces flood levels along the channel, but it increases flood levels downstream where the velocity returns to normal.

Figure 18.27 Channel Modification. Streams in many urban areas are placed in concrete-lined channels like this one in France.

Flood Prediction

By using computer models, hydrologists can predict flood levels along major rivers. Gauging stations provide much of the information used in these models. By knowing the shape of the channel and the water level (also called the stage of the river), it is possible to calculate the discharge. The width, water level, average velocity, and discharge at any place on a stream relate to one another in a systematic way that can be expressed by mathematical equations. By making observations at various places along a stream system over a long time, it is possible to predict the way these stream characteristics will change downstream as the discharge at points upstream and on major tributaries varies.

By collecting data from many gauging stations located throughout a drainage basin, hydrologists use both mathematical (computer) and physical models to predict changes in the water level at points downstream. These models predict how long it will take for a flood crest to move downriver and what the flood stage (level of water in the stream) will be at various places. These models are accurate enough to provide flood warnings on major streams hours and sometimes days before a flood crest arrives. One result of using these predictions has been a dramatic decrease in the number of deaths caused by floods. If the prediction comes early enough, it is possible to raise levees downstream to avert flooding before the flood crest arrives.

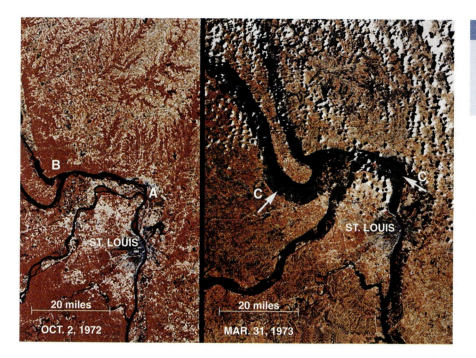

Figure 18.28 Mississippi River Flooding. Satellite images showing the area near St. Louis, Missouri. The left image shows the region in the fall of 1972. The right image shows the same area during the flood of 1973.

20 miles

20 miles

OCT. 2, 1972

MAR. 31, 1973

Flooding on the Mississippi River

Efforts to control flooding and improve navigation on the Mississippi River that began before the Civil War continue to the present. Downstream from Cairo, Illinois, levees provide flood protection for the flood plain all the way to New Orleans. Upstream on the Mississippi, Missouri, and Ohio rivers, systems of dams and nearly 3,200 kilometers (2,000 miles) of levees provide a measure of protection, but the levees are not continuous. Despite the billions of dollars spent to build levees and dams, dredge channels, drain the flood plain, and divert the river, floods in 1973 and 1993 destroyed property over huge areas. In 1973, one of the worst floods on record covered 12 million acres of land, damaged 30,000 homes, and produced a record of 77 days of flooding at St. Louis, Missouri, 63 days at Memphis, Tennessee, and 88 days at Vicksburg, Mississippi. Parts of ten states flooded, 50,000 people had to be evacuated, and estimates of property damage exceeded $400 million (figure 18.28). In 1993 and 1997, flood levels exceeded those of 1973. In 1993, 23 million acres of the upper Mississippi basin flooded; fifty people died; property damage exceeded $10 billion; and floodwaters damaged or destroyed over 70,000 homes.

Similar conditions led to both of these floods. The Bermuda high-pressure system located over the western Atlantic and mid-Atlantic states blocked passage of storms to the east. Although these conditions are normal, an unusual amount of warm, moist air from the Gulf of Mexico flowed northward over the Midwest until it encountered cold arctic air from Canada. This triggered exceptionally persistent and large amounts of rainfall. In 1993, parts of the Midwest received more than a meter (39 inches) of rain during the late spring and summer months.

An unusually wet fall in 1972 preceded the floods of 1973. Reservoir levels were high, and most streams in the central and southern part of the drainage basin were already well above their normal stages. Unusually high precipitation continued in February and March. Much of the snow accumulation in the northern part of the basin began to melt, while the ground remained frozen and unable to absorb the meltwater. Flooding in the upper Mississippi River started in early March, and as heavy rains continued, flooding extended south. By April, the entire mainstream of the Mississippi south of Cairo, Illinois, was above flood stage. Gradually, streams north of Cairo exceeded their flood stages, and water spread onto the flood plain. As rains continued, floodwater rose on both sides of some levees, trapping water on the wrong side of the levee. Satellite photographs document the rise and fall of the waters.

Although unusually heavy precipitation over a period of months caused the flooding in both 1973 and 1993, many hydrologists concluded that levees and channelized drainage, intended to keep the stream in its channel, had contributed to the flood damage. Where the levees hold, they remove natural storage capacity on the flood plain for floodwaters. This shifts the water farther downstream where it raises the flood level. In both of these floods, rains were so heavy and so widespread that levees prevented some water from returning to the main channel. In 1993, flood levels rose so high that many of the levees on the upper Mississippi failed, and in places, engineers destroyed some levees to help reduce flood levels in more highly developed areas. The flood wall at St. Louis, built to withstand a flood crest 17 meters (about 56 feet) above normal, withstood floodwaters that came within two-thirds of a meter (about 2 feet) of the top of the wall.

The 1993 floods renewed debate about several flood-control measures under construction on the flood plain, use of federally funded flood insurance, and restoration of wetlands. The discussion centered on whether to move whole towns out of the flood plain or construct levees to protect them. Many mortgage companies require people who purchase homes to buy flood insurance. The federal government subsidizes flood insurance for people living on flood plains, but many do not buy it. Some opponents of the program say that residents in flood-prone areas will not buy insurance because they know the government will help pay the cost of flood damage. Proponents of flood insurance advocate providing no help for those who refuse to buy insurance. Another federally funded program, the Wetland Reserve Program, designed to protect land in the flood plain from development, originally established a goal of restoring a million acres of wetlands by 1995. That program is still far short of the goal.

Floods Caused by Hurricanes

Heavy rains always accompany hurricanes. These storms form over warm water in the ocean, and it is the rise of moisture-laden air that maintains the unstable atmospheric condition that causes the hurricane (see chapter 12). Hurricanes are so large that they carry huge amounts of moisture with them as they cross the coast and move inland. The effects of hurricanes may extend hundreds of miles inland as they did in the case of Hurricanes Camille and Agnes, described in chapter 12.

Development and Urbanization as Causes of Flooding

Some agricultural and forestry practices increase flood potential. Cultivation of land results in the destruction of natural plant cover and increases the potential for runoff and erosion. If heavy rains fall while the land is bare, more water runs off on the surface and increases the amount of sheet wash and its effectiveness in moving the loose soil. The additional runoff quickly increases the discharge of streams and can lead to flooding. Deforestation and destruction of forest by fire have effects similar to those caused by pavement. Because forests grow on slopes that are too steep for agricultural use, the potential for erosion and increased runoff is even greater than it is on most cultivated land.

The impacts of urbanization are more complex. Three severe drainage problems accompany urbanization: streams flood more frequently; flood stages are higher; and sediment load increases during construction phases. Water runs off paved surfaces and roofs more quickly than from natural ground surfaces. Large areas of pavement in the Houston, Texas, area contributed to the damage done by tropical storm Allison in June 2001. The flooding that resulted from the high rainfall and surface runoff made this one of the most expensive storms in American history. The most effective ways to resolve these problems are to increase infiltration artificially and to design channels or storm sewers that have the capacity to carry the increased discharge. In practice, neither solution is ordinarily implemented until floods demonstrate the magnitude of the problem.

Erosion at construction sites increases the amount of silt that gets into streams. Maintenance or quick replacement of plant cover and use of screens can effectively reduce the amount of silt that reaches streams. In many places, building codes and zoning ordinances now require that contractors take steps to reduce erosion and silting. These requirements may include the following:

1. Minimum disturbance of the plant cover
2. Use of temporary vegetation and mulch
3. Construction of sediment basins on sites
4. Planting of permanent cover at the end of construction
5. Selection of sites that have the least erosion potential

All of these steps are highly effective in reducing the problems of erosion and silting.

SURFACE-WATER–GROUND-WATER CONNECTION

The concept of the hydrologic cycle makes it clear that surface drainage and water that seeps into and moves underground are closely connected. In humid regions, ground water feeds streams. Otherwise, streams would cease to flow shortly after rainfall stops. In arid regions, so much water infiltrates into the ground from stream channels that the streams dry up. Chapter 19 examines the ground-water reservoir, the way water moves through the ground, how the interchange of water between the surface and ground takes place, and ways of analyzing these connections.

SUMMARY POINTS

1. The action of streams represents a balance between the many factors that affect stream flow and deposition. Climate, slope, and the presence of ground cover all help determine the amount of water that enters a stream from runoff or underground sources. Permanent streams flow in well-defined channels whose form is the result of the interaction of several factors, including stream discharge, the amounts and types of alluvium in the channel, the nature of the underlying bedrock, and the stream gradient. These variables give rise to channels ranging from broad, low valleys carrying braided streams to steep-sided V-shaped canyons.

2. Water in streams may exhibit smooth laminar flow in unobstructed paths, mildly turbulent streaming flow, or highly turbulent flow where the channel is rough and stream gradients are steep.

3. Streams carry loads of silt, sand, rock, and dissolved material, largely provided to them by sheet wash and mass movement. The load is carried in solution, in suspension, or as part of the bed load. The climate of the stream's drainage area and the velocity and turbulence of the flow in the stream determine how much of the load moves in each of these ways.

4. The combination of moving water and its load erodes the stream channel, cutting it downward toward base level or laterally. Lateral erosion is most pronounced where streams flow on low gradients.

5. Where the velocity or volume of flow drops, so does the stream's capacity to carry load. Deposition takes place where the gradient decreases suddenly; where a stream overflows its banks, forming flood plains; or where it enters a lake, a stream with less capacity, or the ocean.

6. The drainage patterns formed by streams in an area reflect the topography and bedrock of the region. Most common are dendritic patterns formed where the underlying bedrock is uniform in character, concentric or radial patterns formed around domes, and straight-line or rectangular patterns formed where fractures and faults in bedrock define stream channels. Meandering patterns form where streams flow on low slopes through unconsolidated materials.

7. Streams flood when discharge exceeds the amount a stream can carry within its banks. The abnormally high discharge generally results from an increase in runoff caused by heavy rains combined with limited infiltration. Steep slopes, impervious surfaces of rock or pavement, frozen ground, and the absence of plant cover contribute to stream flooding. Because human activity has tended to increase the amount of impervious surface and to strip away the protective plant cover, it has increased the severity of flooding along many rivers. As a result of improved understanding of the factors in flooding, many communities require construction practices aimed at increasing infiltration to avert flooding. Artificial levees, dams, and holding reservoirs are among the artificial structures used to contain floodwaters and regulate river discharge.

REVIEW QUESTIONS

1. What are some of the factors that shape the depth of a stream's channel?
2. How does the load carried by a stream get into the channel?
3. What characteristics of a stream indicate that it is eroding laterally? Vertically?
4. What are the effects of an increase in stream discharge on the velocity, capacity, depth, width, and gradient of a stream?
5. Under what conditions do streams entrench their channels?
6. Why do so many floods occur in the spring?
7. Why does urbanization cause streams to flood?
8. What arguments are made for and against the construction of dams?
9. What alternatives exist for flood control other than dam construction?

10. Explain how hydrologists predict flood stage downstream.
11. Approximately how much water does the atmosphere contain?
12. What problems are involved in trying to determine whether water is being added to the surface reservoirs from Earth's interior?
13. List the major uses humans make of water resources.
14. How is water involved in biogeochemical systems?
15. What is the role of water in shaping Earth's surface?

THINKING CRITICALLY

1. How is it possible that measures taken to control floods may actually be causing higher water levels during floods?
2. If you were on the planning commission for a small city, what type of information would you want to see to determine whether growth of the city might cause flooding?
3. In what ways are streams affected by processes related to Earth's plate tectonic system?
4. How does the movement of water among the various reservoirs differ in arid and humid climatic regions?
5. In what ways would surface streams respond to a change in average global temperature?
6. How is water related to Earth's plate tectonic and climatic systems?
7. The power of streams to pick up alluvium and erode their channels critically affects what human activities?

KEY WORDS

acre-feet	longitudinal profiles
alluvial fan	meanders
alluvium	mouth
antecedent streams	natural levee
artificial levee	pediment
bajada	pediplane
base level	permanent streams
bed load	piracy
braided	playa lake
capacity	point bars
competence	potholes
consequent streams	shooting flow
dams	solution load
dendritic	stream gradient
dependent variable	subsequent streams
discharge	superimposed streams
entrenched streams	surface-drainage basin
flood plain	suspended load
graded streams	trellis
head	turbulent flow
laminar flow	water gaps
load	

chapter nineteen

Ground Water

Chapter Guide

Part of the precipitation that falls in any area sinks into the ground and adds moisture to the soil and other unconsolidated material at the ground surface that make up the regolith. As this water percolates deeper into the sediment or rock beneath the regolith, it becomes ground water. Where this water goes after leaving the surface is unclear, but by using dyes and observation wells, it is possible to trace its movement. In most places, it moves slowly through interconnected pore spaces between grains; in others, it flows rapidly through subsurface passageways created by solution of the bedrock. Where the effects of solution are pronounced, the surface of the ground sinks, and much of the surface drainage diverts into underground passageways. In such areas, large holes may suddenly open in the ground, swallowing streams, cars, or whole buildings. This chapter examines ground water as part of the hydrologic cycle, the ways ground water affects the land surface, the occurrence of ground water, and some of the problems people encounter in developing water for human uses.

Water vapor formed by heat at depth forces water to the surface at Old Faithful in Yellowstone National Park. Although hot water is present in many places, geysers of this type are rare.

INTRODUCTION

Once water infiltrates into the ground, it is out of sight, but the processes that occur underground shape the surface. One morning, following a night of heavy rain, a family walked out the front door of their house in rural Virginia to discover a circular hole about 5 meters (16 feet) wide and 4 meters (13 feet) deep. The ground had collapsed, exposing the foundation of the house and their chimney. Although it is unusual for a hole of this type to develop under or adjacent to a house, sudden collapse or subsidence of the ground is not as rare as we might think—not if you live in an area underlain by limestone. Water that seeps into the ground dissolves limestone and creates voids. This solution process operates slowly, but eventually, it removes the support needed to maintain a stable surface. Even in areas underlain by insoluble rock, water that moves through the soil supplies plants with the nutrients they require, facilitates the chemical reactions involved in chemical weathering, and as the water percolates deeper, it passes through voids in the rock and accumulates in the ground. Some ground water comes back to the surface at springs and feed streams and lakes (figure 19.1). The water stored in the ground provides the water many communities need for domestic, industrial, and agricultural use. Many states depend on ground water for more than half of their total water use (table 19.1). Sustained economic development requires replenishment of this water supply. Otherwise, it is necessary to import water or recover fresh water from the sea.

This water infiltrates through sandstones at the surface and becomes concentrated in fractures and along bedding planes.

THE ROLE OF GROUND WATER IN EARTH SYSTEMS

Most of the processes involving ground water occur below the ground surface and out of sight. If one considers what would happen if water did not penetrate the surface, the important roles it plays in Earth systems become obvious. Ground water is essential for the survival of plants and animals that live or find nourishment in the soil. Without ground water, Earth's surface would become a barren desert. Water in the ground is critical for processes of chemical weathering. As it percolates through the soil and bedrock, water picks up the ions freed by weathering and carries them into streams, lakes, and ultimately, many of them find their way into the oceans. Without ground water, lakes and oceans would contain few salts. The impact on marine organisms that rely on these nutrients would be dramatic. Many depend on these nutrients. The supplies of silica needed for the construction of hard parts of diatoms and radiolarians and the calcium carbonate required by most marine invertebrates would decline drastically. Clearly, water that gets into the ground is not only essential for human consumption, it plays critical roles in terrestrial and marine ecosystems.

THE ORIGIN OF WATER IN THE GROUND

Most ground water is rainwater, also known as **meteoric water,** that infiltrates from the surface; however, sediment may retain water trapped during deposition, and water also comes from magmas, the initial source of most water. On average, enough rainwater falls each year to cover the land areas of the world to a depth of 74 centimeters (29 inches). Depending on the climate, the slope of the ground surface, and the amount and type of plant cover, this water evaporates, plants remove it from the ground and transpire it into the atmosphere, it runs off on the surface, or it infiltrates into pore spaces in the regolith and rocks.

Water derived from magmas, called **juvenile water,** contains a great variety of elements found in igneous rocks. The acids present in it make it unsuitable for human use. Most water trapped in sediments at the time of deposition, called **connate water,** is marine water. The salts also make it unsuitable for human consumption or for use in irrigation. Although connate water is commonly encountered in deep oil wells, it is rarely present at shallow depths because circulating meteoric water flushes it out of pore spaces.

INFILTRATION OF WATER INTO THE GROUND

Gravity pulls water on the surface downward through interconnected pore spaces. As it moves through the regolith, some water adheres to regolith particles. Clay and organic materials absorb some of the water as it percolates downward, but if sufficient water

Table 19.1 States That Produce More Than a Billion Gallons of Ground Water Each Day

The numbers indicate the millions of gallons used in each state in 1980.

	Amount Produced for States That Withdraw a Billion or More Gallons Each Day	Percentage of Total Water Supply for States That Produce More Than Half of Their Supply from Ground Water
California	21,000	49%
Texas	8,000	73%
Nebraska	7,200	72%
Idaho	6,300	—
Kansas	5,600	87%
Arizona	4,200	61%
Arkansas	4,000	79%
Florida	3,800	62%
Colorado	2,800	—
Louisiana	1,800	—
New Mexico	1,800	49%
Mississippi	1,500	73%
Indiana	1,300	—
Georgia	1,200	—
Oregon	1,100	—
Pennsylvania	1,000	—
Utah	1,000	—
Iowa	Less than 1 billion	70%
Minnesota	Less than 1 billion	49%
Vermont	Less than 1 billion	49%

Source: U.S. Water Resources Council

is present, it moves through the regolith. Porosity and permeability determine how quickly this movement occurs. **Porosity** is a measure of the amount of empty space in an otherwise solid material. It is expressed as a percentage of the total volume of the material. **Permeability** is a measure of the ease with which water can move through a material and results from interconnections between pore spaces (figure 19.2). A simple measurement of permeability, called a **perk test** (a measure of percolation), consists of digging a hole, filling it with water, and measuring the amount of time required for the water to seep into the regolith. A more precise method for determining permeability involves measuring the rate of flow of water through a material under carefully controlled conditions.

If porous and permeable rock or sediment lies beneath the regolith, water moves into it. In the upper part of the regolith, air and percolating water mix in a zone called the **zone of aeration** (figure 19.3). As the water moves through the zone of aeration and into the underlying material, it reaches a depth below which water fills all the pore spaces. The top surface of this **zone of saturation,** called the **water table,** may be in the regolith shortly after a rain, but it is usually in the underlying sediment or rock (see figure 19.3). Despite its name, the water table is not flat. Both its level and shape change as the supply of water from above and the movement of water below vary. Because the character of rock or sediment layer determines porosity and permeability, some layers contain and

yield much more water than others do. Such water-bearing layers are called **aquifers.** If sediment or rock with low permeability is present, the water either accumulates above it or moves laterally downslope on top of the impervious material until it encounters a more permeable material or reaches the ground surface.

STORAGE OF WATER UNDERGROUND

Most ground water occupies spaces between grains in rock and regolith, open fractures, and small solution cavities (see figure 19.3). In granular materials such as sandstones and conglomerates, the amount of intergranular pore space depends on the shape, packing, the amount of cement that fills pore spaces, and the degree of sorting of the grains. Smaller grains may reduce porosity by filling spaces between the larger grains. Texture also affects porosity. Maximum porosity results where every grain is directly above the center of another. Porosity is less when grains are strongly aligned than it is when grain alignment is random. Intergranular pore spaces are present in materials composed of grains, but all rock types break, and most contain fracture porosity.

Because most rocks are brittle, even slight warping of the layers is sufficient to generate fractures. If the fractures are open, they

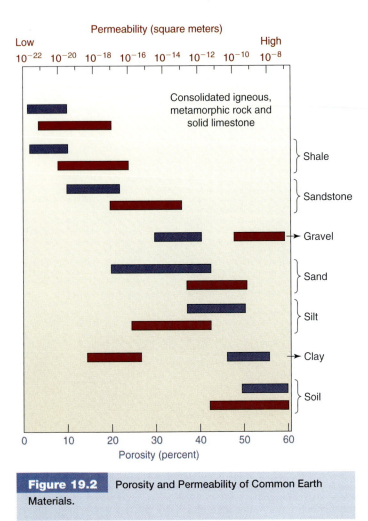

Permeability (square meters)

Low 　　　　　　　　　　　　　　　　 High
10^{-22} 10^{-20} 10^{-18} 10^{-16} 10^{-14} 10^{-12} 10^{-10} 10^{-8}

Consolidated igneous, metamorphic rock and solid limestone

Shale

Sandstone

Gravel

Sand

Silt

Clay

Soil

0　　10　　20　　30　　40　　50　　60
Porosity (percent)

Figure 19.2 Porosity and Permeability of Common Earth Materials.

provide storage space for water and avenues along which water can move. In most igneous and metamorphic rocks, fractures are the only type of porosity present. In these rocks, the number and spacing of fractures and the size of the open spaces between fracture surfaces determine the porosity.

In many soluble rocks, such as limestone and dolomite, solution cavities contain most of the water. These cavities form along bedding surfaces or fractures, but they may be irregular in shape and size. In some caves, subterranean lakes and channels hold water, but most solution pores and passageways are small.

Such substances as calcium carbonate or silica carried in ground water and deposited as calcite or quartz in pore spaces may reduce or even eliminate porosity. As the dissolved mineral matter precipitates, solids fill the cavities, intergranular pores, or fractures. Human activities also may destroy the porosity of water-bearing sediments. If pumping removes water too quickly or completely, sediments, especially those containing a high percentage of clay, may compact under their own weight and become consolidated. Generally, this process is not reversible; so, rapid withdrawal of water may drastically reduce the storage capacity of some sediments.

Clay, the main mineral constituent of shale, has high porosity, but although clay absorbs water, the pore spaces are poorly

interconnected, so clay has low permeability. For this reason, layers of claystone or shale generally form barriers to the movement of water underground.

THE SHAPE OF SATURATED ZONES IN THE GROUND

The Water Table

Porosity and permeability control the movement of water, determine the shape of the water table, and the distribution of water-saturated zones in the ground. If interconnected pore spaces provide a continuously connected hydraulic system through which water may move freely from the surface, hydrologists describe the ground water as **unconfined** (figure 19.4). Unconfined conditions exist where thick, granular, porous, and permeable sediments or sedimentary rocks extend from the ground surface down to a lower material that is impervious or of low permeability. The water table makes up the upper surface of the saturated zone. Generally, the shape of the water table is a somewhat suppressed version of the topography. Maps showing contour lines that indicate the elevation of the top of the zone of saturation depict the shape of the water table. Where the distribution of permeability in the

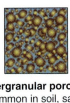

Intergranular porosity
Common in soil, sand, gravel, sandstone, and other granular materials

Solution cavities
Common in limestone, gypsum, and salt

Fracture porosity
Common in all consolidated rocks

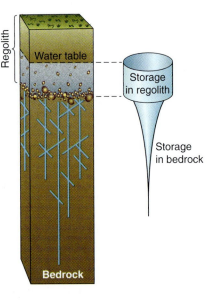

Figure 19.3 Water Infiltration into the Ground. Water infiltrates through regolith pulled downward by gravity. The water table is the upper surface of the zone of saturation. Water is stored in the spaces between grains, in fractures, lava tubes, and in solution cavities.

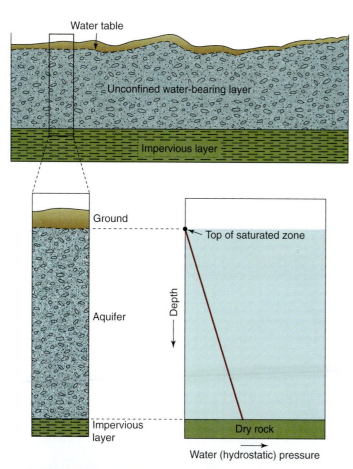

Figure 19.4 **Unconfined Ground Water.** In the case of unconfined ground-water conditions, the water table conforms roughly to the shape of the topography. In humid climates, the water table intersects the ground surface along most streams and lakes. Hydrostatic pressure in the pore spaces increases with depth below the level of the zone of saturation.

subsurface materials causes water to move into layers or zones partially surrounded or overlain by barriers, the ground water is **confined** (figure 19.5).

In humid climates, the water table is deeper under hills and usually comes to the ground surface at the edges of lakes and along streams, where water from beneath the water table moves into lakes and streams. During dry spells, water from ground-water reservoirs continues to come to the surface. This water rises to the surface in springs or in the channel of streams. This infusion of ground water makes it possible for streams to flow long after runoff ceases. In arid or semiarid regions and during droughts in humid regions, the water table may drop below the level of streams and lakes. Water then infiltrates from the lakes or streams and percolates down towards the water table. When these conditions persist, lakes and streams may dry up entirely.

A localized impervious barrier may prevent water from moving down to the regional water table and cause a localized near-surface zone of saturation, a **perched water table,** to form above the

regional water table. Generally, water escapes from a perched water table through springs located where the impervious layer intersects the ground surface.

Springs

Springs are places where water issues from the ground. In unconfined ground-water systems, springs form where the water table intersects the land surface. Water usually comes to the surface along streams or at the edge of lakes—often at low places in the topography. Although springs develop under many conditions, they usually form where porous and permeable zones along bedding planes, fracture zones, or faults crop out in a valley (figure 19.6). In confined ground-water systems under hydrostatic pressure, **artesian conditions** may be present in which springs may develop at high elevations. Most artesian springs appear where fractures or faults break through a confined aquifer. Artesian water may also find its way to the surface in places where the water-bearing layer turns up and is exposed at the ground surface.

In limestone, even though the rocks may be impervious, water may come to the ground surface through cavities formed by solution along fractures, beds, or faults. Similar conditions exist in lava flows. Although lava is impervious, blocky lava, cooling cracks, and lava tubes provide passages for water. Springs commonly issue from the base of lava flows, especially where less permeable rock lies beneath them. Of the largest springs in North America, those with flows exceeding 3 cubic meters per second (100 cubic feet per second), thirty-eight are in volcanic rocks, twenty-four are in limestone, and three are along faults.

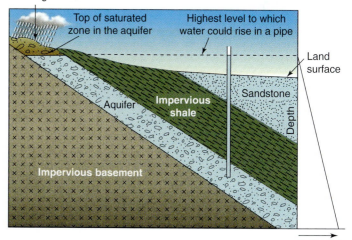

Figure 19.5 **Confined Ground Water.** Where water exists in an aquifer confined above by an impervious layer, hydrostatic pressure at any depth in the aquifer is determined by the difference in elevation of the top of the saturation zone and that depth. Water in a well will rise almost to the level of the top of the zone of saturation.

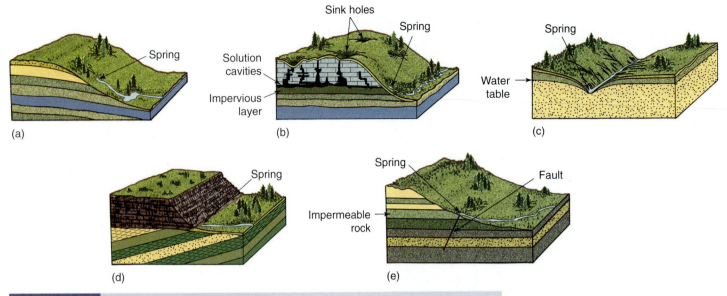

(a)

(b)

(c)

(d)

(e)

Figure 19.6 **Five Conditions That Give Rise to Springs.** (*a*) A spring forms where a layer containing water in fractures intercepts the ground surface. (*b*) A spring forms where a layer containing solution cavities intercepts the ground surface. (*c*) A spring forms in the bottom of a valley where the water table reaches the ground surface. (*d*) A spring forms at the base of a pile of porous and permeable lava flows. (*e*) A spring forms where water moves up along a fault that cuts through a confined aquifer.

The largest spring in the United States is Silver Springs, Florida. It is one of many springs in the Ocala limestone, a formation that underlies much of the Florida peninsula. Solution cavities give this limestone extremely high porosity and permeability. Because impervious sediment lies over most of the Ocala limestone, water in many parts of the Ocala is under artesian pressure (figure 19.7). The pool at Silver Springs is about 80 meters (262 feet) in diameter and 12 meters (39 feet) deep; the main opening into the cavity, a hole 20 meters (66 feet) wide and 4 meters (13 feet) high, is clearly visible from the surface. The water in this spring is famous for its clarity. The discharge from the spring, 2 million cubic meters per day, is sufficient to form a river. The water is unusually warm, averaging about 23°C (72°F).

Spring Deposits

Water in springs and streams that flow through limestone contains large quantities of calcium carbonate. As evaporation takes place near waterfalls or cascades, the water becomes saturated with calcium carbonate, and precipitation occurs. Such deposits form a loose, porous material called **tufa** or a compact rock known as **travertine.** Tufa frequently contains organic matter such as leaves and twigs on which spray falls. Most such deposits are irregular in shape, but some grow into beautiful terraced features like those at Mammoth Hot Springs in Yellowstone Park (figure 19.8). Firmly cemented tufa is a good building stone. The porous, rough texture makes it highly effective for soundproofing.

Deposits called **siliceous sinter** develop where ground water issues from hot volcanic rocks. Sometimes the silica forms a

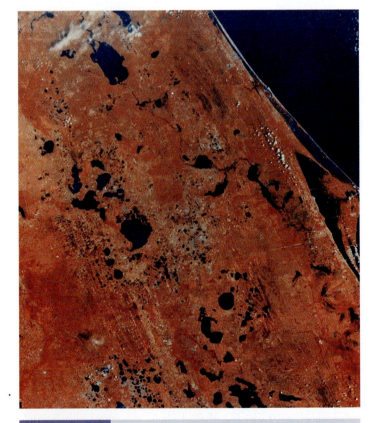

Figure 19.7 Ground-Water Conditions in Florida Where Large Springs Have Formed in Solution Cavities in the Ocala Limestone.

Figure 19.8 Spring Deposits in Yellowstone National Park, Wyoming. Terrace-like deposits form around many springs that carry large amounts of silica or calcium carbonate in solution.

gel-like substance in the hot springs. Where the hot water accumulates in shallow pools, siliceous deposits may build terraces resembling those of many calcium-carbonate deposits.

Contacts Between Fresh Water and Salt Water

Along the coast, fresh water percolates into the ground, but salt water moves into pore spaces under the ocean. Because of its lower density, fresh water floats on salt water. Consequently, where ground water is unconfined, a wedge-shaped zone saturated with salt water lies beneath the fresh water and extends landward from the shoreline (figure 19.9). A similar situation may exist in a confined aquifer that contains or comes in contact with salt water. Along the contact, salt water mixes with and contaminates the fresh water, making it unsuitable for drinking, for irrigation, and for many other uses. This condition is a serious problem on many islands, along much of the coast of eastern North America, and in the continental interior where connate water is present at relatively shallow depths. Where an aquifer contains both salt water and fresh water, removal of fresh water through wells at a rate in excess of the rate at which fresh water enters the aquifer will cause the level of the salt water to rise in the aquifer. If this condition persists, saline water will contaminate water being drawn into the well. Examples of this condition will be examined later.

Hydrostatic Pressure

The behavior of slow-moving water in the ground closely approximates that of static water in a pipe. In a static system, such as a water line connecting houses at different elevations with a water tank, water can rise to the level of the water in the tank (figure 19.10). Houses at lower elevations have higher water pressure than houses at higher elevations. The pressure at any depth in a

body of water is a function of the water depth. Pressure equals the density of the water (ρ) multiplied by the water depth (h) and the gravitational acceleration (g) ($P = \rho gh$). Pressure in water, called **hydrostatic pressure,** at any point in a porous and permeable material is equal to the weight of the overlying water. The pressure is equal at all points located at the same level throughout a body of rock regardless of the shape of the body. **Hydrostatic head** at a point in a water-saturated body is the height of the column of water supported by the pressure at that point.

Artesian Conditions

Artesian conditions exist where hydrostatic pressure in a ground-water system is sufficient to raise water above the level of the water table. The level of water in a well drilled into the aquifer will reveal the presence of artesian conditions. In some places, the pressure in the ground-water system is high enough to bring water to the ground surface.

These conditions always develop in confined aquifers. Water enters the porous layer, the aquifer, at the ground surface or through

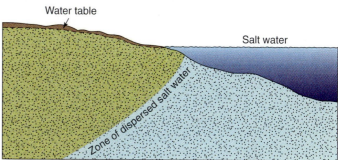

Unconfined conditions

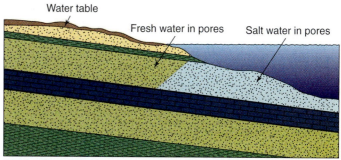

Confined conditions

Figure 19.9 Contacts Between Fresh Water and Salt Water. Where the ocean comes in contact with freshwater-bearing sediment, a zone of fresh water and salt water mix, and salt water is dispersed into the sediment containing fresh water. If ground water is unconfined, the salt water–freshwater contact occurs at a depth equal to about 18 times the elevation of the water table above sea level. If the water is confined, the salt water moves up through porous and permeable layers. Pumping of fresh water induces rise of the saltwater–freshwater contact.

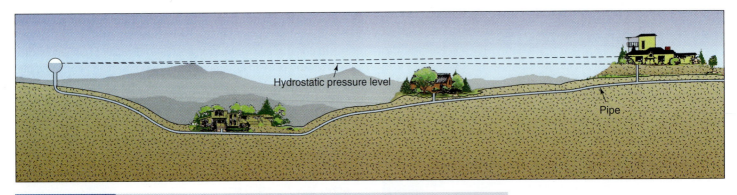

Figure 19.10 Hydrostatic Pressure. Water will flow through interconnected conduits of any shape. As a result of hydrostatic pressure, water rises almost to the level of the water in the tank.

a connection with other water-bearing rocks in which water moves freely (figure 19.11 and see figure 19.5). **Recharge** of the aquifer occurs through these connections. As water moves into the aquifer, the available pore spaces fill, and hydrostatic pressure increases. If a well penetrates the aquifer at an elevation below the level of saturation, water can flow from the well under hydrostatic pressure.

Artesian conditions exist in extensive aquifers, such as the Dakota sandstone, a porous layer of rock that lies under the Great Plains. Where the Dakota sandstone rises to the ground surface along the Rocky Mountain front in Colorado, water infiltrates and flows through the sandstone (see figure 19.11). The water flows down away from the mountain front under the plains. At lower elevations in the plains, wells drilled into the aquifer produce water for domestic water supplies and for irrigation. Some of the deep layers of sandstone contain salty water that was trapped in the sand when it was deposited in a marine environment. Smaller lens-shaped bodies of sand or sandstone, such as buried river channels and alluvial fans, may also contain artesian water.

Unconfined water lies above confined water in artesian aquifers. If the recharge area of an artesian aquifer is higher than the local ground level, the level to which the artesian water could rise can be higher than the local water table. Because of this pressure, water from a deep well could rise to the surface under pressure, but water from a shallow well would require pumping.

MOVEMENT OF GROUND WATER

Hydrologists trace the movement of ground water by injecting dyes or radioactive tracers into water wells. Such studies prove that ground water usually moves slowly. In rocks with low porosity and permeability, water may move no more than 1 to 10 meters per year. More rapid movement occurs in porous rocks such as coarse alluvial gravels with rates of several hundred meters per day. Even higher rates exist where water flows through solution cavities as underground streams.

In 1856, Henry Darcy described the flow of water in unconfined sand and in a tilted confined aquifer. He demonstrated that the velocity of water moving down the surface of the water table in unconfined water-bearing materials is equal to the product of the permeability (k) and the slope of the surface (X/Y), where X is the change in elevation per unit of horizontal distance (Y). The

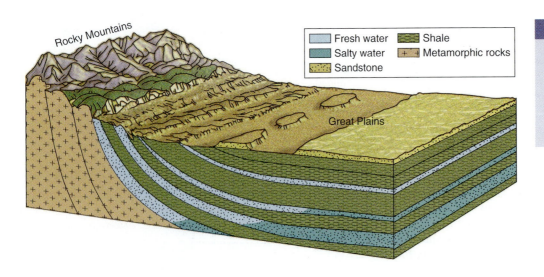

Fresh water
Salty water
Sandstone
Shale
Metamorphic rocks

Rocky Mountains

Great Plains

Figure 19.11 Aquifer. Along the Rocky Mountain front, aquifers, such as the Dakota sandstone, have been tilted up along the mountain front and exposed by erosion. Water is confined in aquifers such as the Dakota sandstone.

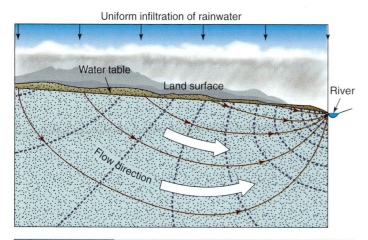

Uniform infiltration of rainwater

Water table

Land surface

River

Flow direction

Figure 19.12   A Flow Net. In this cross section, two sets of lines drawn so they cross each other at right angles form a flow net. One set of lines depicts the direction of flow, and the second shows the progress of water moving in the zone of saturation. In this hypothetical model, the aquifer is uniform in porosity and permeability.

velocity of movement of confined water in an aquifer equals the product of the permeability and the hydraulic gradient, the difference in hydrostatic head (delta H) between two points divided by the length (L) of the path of flow between the points. An equation known as Darcy's law describes these relations.

$$V = kH/L$$

Hydrogeologists depict the movement of water through the ground with **flow nets** (figure 19.12). The first of two sets of lines, one with arrows, indicates the direction water moves; the second set of lines, drawn perpendicular to the first set, connect points that move like a wave front. Flow nets are hypothetical constructions based on certain assumptions about the porosity, permeability, and recharge of the water-bearing layer. The actual flow path is determined by injecting dyes or radioactive tracers in selected wells and using observation wells to trace the direction and velocity of flow. The circulation of ground water usually brings water back to the surface in streams, lakes, or the ocean, or as springs on the ground surface.

GEYSERS

Geysers are conduits through which steam and hot water under pressure rise to the ground surface. When a geyser erupts, steam and boiling hot water may rise high into the air. All geysers develop in areas underlain at shallow depths by magmas or still-hot igneous rock. In some areas, naturally occurring hot water is an important source of energy. Although volcanically active areas are numerous, only a few contain geysers. Most notable are the geysers at Yellowstone Park in Wyoming, in Iceland, and on the North Island of New Zealand. Since its discovery in 1870, Old

Faithful Geyser in Yellowstone National Park regularly spurted 38 to 45 cubic meters (1,300 to 1,600 cubic feet) of steam and water to an average height of 40 meters (131 feet) (rarely as much as 70 meters (230 feet) about once every hour). Old Faithful in eruption is shown in the opening photograph for this chapter. Over the last two decades, the interval between eruptions has varied and is often less than an hour. Although no other geysers are this regular in their activity, processes similar to those at Old Faithful cause all of them.

Following an eruption at Old Faithful, most of the water returns to the ground through the vent or nearby holes. A period of quiet follows during which only a few whiffs of steam rise from the vent. After about forty minutes, steam again rises, and hot water appears in the vent. Next, water and steam flow from the vent in a series of small outpourings. Some of this runs off, but most goes back into the hole. A few minutes later, boiling hot water rises and erupts to a height of several meters above the vent. After several minutes of these minor eruptions, the major eruption occurs, and a column of water and steam spurts upward for several minutes before it dies down, and the cycle repeats.

The processes acting below the surface also follow a regular sequence. The ejected water comes from an irregularly shaped network of fractures and cavities dissolved by water that circulates through the volcanic rocks. Heat is supplied by huge masses of igneous rocks and perhaps some magma that lie at a depth of about 1.6 kilometers (a mile) below the surface. After an eruption, the water percolates back into the subsurface fractures and solution channels and refills them. The temperature of the water, cooled by its ascent into the air, quickly rises again. Because the heat in the rock increases with depth, water in the lower parts of the network becomes hot quicker than that closer to the surface. Water at the surface boils when it reaches 100°C (212°F) (slightly less at Yellowstone because atmospheric pressure at high elevations reduces the boiling point). When the system of interconnected passages is full of water, the weight of the column of water increases the pressure and the boiling point at depth. Consequently, the temperature of deeper water must rise to a higher temperature than 100°C before it will begin to boil. Eventually, somewhere in the underground network, the water temperature reaches the boiling point, and some water changes into steam. The expansion of the water as it changes into steam exerts sufficient pressure to force the overlying water out at the top of the vent. A small amount of water comes out at first, then more, until reduced pressure in the network permits large quantities of water to convert suddenly into steam. The main part of the eruption starts with this explosive generation of steam at depth.

The water that rises to the surface in geysers is heavily laden with elements dissolved from the rocks in the feeder system. At the surface, the sharp drop in temperature leads to precipitation of this mineral matter in the form of a rock called *geyserite*. Eventually, the precipitated geyserite becomes a mound around the opening (figure 19.13). These localized deposits are but one of many features in the landscape created by the action of ground water. Pools of clear, bluish water that are only slightly below the boiling point are among these features found in Yellowstone National Park (figure 19.14).

Geothermal Energy

Heat within earth and hot ground water are sources of heat energy, called **geothermal energy,** that are used for domestic heating in many places and for generation of electricity in a few locations. Temperature normally increases with depth at a rate of about 1°C per 30 meters (about 2°F per 100 feet) of depth. The rate of increase with depth is much greater in areas where bodies of magma are near the surface and where large intrusions are at, or buried close to, the surface. Geothermal energy is readily available in New Zealand, Yellowstone Park, and Iceland where geysers are present. The largest development of natural steam is located in California at a site known as The Geysers, which is situated near the Long Valley volcanic center. Here and in New Zealand, steam is used to generate electricity. The center at The Geysers supplies a large proportion of the electricity used in San Francisco. The gey-

sers on the North Island of New Zealand supply much of the electricity used at Auckland. In New Zealand and in Iceland, hot water from the geyser basins is piped directly into homes and businesses.

Heat in the ground is being used increasingly in the contiguous United States for domestic heating. The temperature in the ground at depths of 1 or 2 meters (a few feet) is relatively stable at about 10°C (50°F) throughout the year. Water or other fluids circulated through a system of underground pipes, or in wells, may be used with heat pumps for heating in winter months and cooling in summer months.

Although the potential for use of geothermal energy is significant for domestic purposes, geothermal energy supplies less than a tenth of a percent of total energy used in the world.

HOW GROUND WATER SHAPES THE LANDSCAPE

Rainwater containing carbon dioxide from the atmosphere is responsible for the solution of most limestone. Water combines with carbon dioxide to form carbonic acid, a weak acid in which limestone slowly dissolves. Solution stops once the water becomes saturated with calcium carbonate. For solution to continue, unsaturated water must continue to flow over or through the limestone. Recent measurements of rates of limestone solution indicate that the elevation of the surface of a region underlain by limestone could decrease between 0.005 millimeter and 0.5 millimeter each year. At this rate, the elevation of the land might decrease between 5 to 500 meters (16 to 1,600 feet) over a million years. Given enough time, solution can change the shape of the landscape.

Karst Topography

The name **karst topography** applies to areas having landforms shaped by the solution effects of ground water. Where such rocks as salt, gypsum, and limestone lie near the ground surface, dissolution produces a distinctive landscape. Caves, enclosed surface depressions, stream valleys that end suddenly where a stream disappears into the ground, natural bridges, and tunnels characterize this type of landscape (figure 19.15).

Salt and gypsum dissolve rapidly, but neither is responsible for most karst topography in humid regions. In contrast, carbonate rocks originally deposited on shallow continental shelves have been uplifted and are at or near the surface over vast areas and cause most karst regions. The Florida peninsula, parts of the southeastern Atlantic coastal plain, large areas in central and western Texas, the Appalachian Mountains and plateau, the interior lowlands of Illinois, Kentucky, and Ohio, the Ozark dome, and the Carlsbad Plateau region of New Mexico are the largest areas of karst topography in the United States. Relatively young, semiconsolidated limestones underlie Florida, Texas, and the Coastal Plain. Much older, Paleozoic age, recrystallized limestones lie beneath most other karst regions of the United States.

Because long-continued movement of water through limestone is critical for the development of karst topography, solution effects are greatest where permeability and porosity of the limestone allow the water to move freely. Limestones, especially younger ones like those composed of broken shell fragments and other semiconsolidated carbonates, are usually porous and allow passage of water. Older limestones are fine-grained, compact, and recrystallized. They have little or no intergranular porosity, so most of the water moving through them passes along fractures and bedding planes.

Sinkholes and Diversion of Surface Drainage Enclosed topographic depressions called **sinkholes** (figure 19.16) are the most characteristic feature of karst topography. Some form as depressions where the top surface of limestone dissolves and water drains off through underlying fractures or pore spaces. Others form as water infiltrates and dissolves the soluble rock beneath the ground surface. Then, the ground surface may subside slowly into the resulting cavities, or the roof may collapse suddenly without warning as fracture-bound blocks drop from the ceiling (figure 19.17). Once a depression begins to form by any of these processes, water on the ground surface flows into the depression, and the solution of underground limestone accelerates as more and more water moves into the depression and passes into the dissolution passageways beneath the sinkhole. Streams flowing across karst terranes may disappear into sinkholes. Most such streams reemerge elsewhere as surface streams, but the water in others may reemerge as springs along streams, under lakes, or in the ocean.

Over a long time, sinkholes grow in size and in number. The resulting karst landscape gradually changes as more and more of the surface water disappears into the underground drainage system. Where streams disappear into sinkholes, the stream abruptly ends in a **swallow hole.** In other places, materials normally carried away by the surface stream accumulate on the dry, abandoned valley floor. If the stream returns to the surface, its underground passage becomes a natural tunnel. Remnants of such tunnels form natural bridges. Where the solution of bedrock has continued for many hundreds of thousands or millions of years in areas of thick, soluble bedrock, only remnants of the limestone—mound-like or tower-like hills—may rise above relatively flat river valleys.

Caves Naturally constructed underground cavities or voids are called caves. Most form as a result of solution of limestone, gypsum, or salt, but they also exist where hot liquid lava flows out

Figure 19.15 Karst Topography Example. Natural Bridge, Virginia, is a remnant of a tunnel formed by a subsurface stream that connected a field of sinkholes with an outlet in the James River.

beneath a solid crust, leaving a tube-shaped underground passageway (figure 19.18).

Above the water table, caves in limestone, gypsum, and salt grow by dissolution effects and stream erosion. Subterranean streams flow through channels in many caves. Some are streams diverted from the surface into sinkholes. Other streams originate within the system of underground passageways as water collects. The processes of stream erosion in these underground streams are identical to those seen on the surface. Potholes, even meanders, exist in caves. More often, fractures and bedding surfaces are the prime avenues for movement of water through the rock. Consequently, many caves have straight, narrow passageways or become lateral extensions within the plane of more soluble layers.

Several lines of evidence support the idea that some caves develop below the water table. Crystals that apparently grew underwater cover the walls and ceiling of some caves. Stream erosion above the water table cannot explain the internal network of blind passageways and branching tubes, pockets, and smooth, rounded cavities present on the roofs of some caves. In such caves, the cavities have a three-dimensional, sponge-like pattern with blind pockets and dead-end tubes rather than a stream-like pattern. A two-phase history provides the best explanation for the origin of most caves. The first

Figure 19.16 Sinkhole. The roof of a solution cavity collapsed following heavy rain.

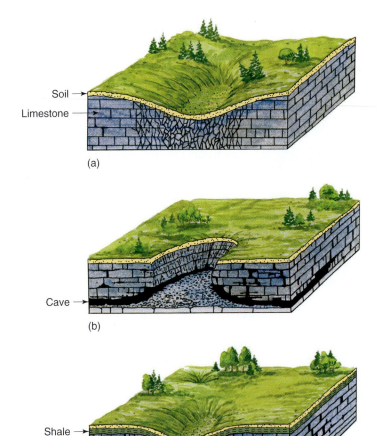

Soil →
Limestone →

(a)

Cave →

(b)

Shale →

(c)

(a)

(b)

phase consists of dissolution of the soluble rock by slowly circulating water moving along cracks and bedding planes beneath the water table. The second phase involves erosion by water moving along passageways after the zone of saturation drops below the level of the cave and most water drains out of it. The position of the cave relative to the water table can change as streams cut their channels to lower levels. Changes in the regional base level can also have this effect. If the water table drops below the cave, water drains out. Conversely, if the water table rises above the level of a cave, it is flooded.

Cave Deposits Usually, cave deposits have the same composition as the surrounding rock. In limestone caverns, the deposits are calcite, and caves in beds of gypsum contain gypsum deposits. Ground water dissolves the soluble rock as it moves from the surface into the cave. The cave deposits grow as water evaporates or loses carbon dioxide and the solution becomes saturated with calcium carbonate or, in the case of gypsum deposits, with calcium sulfate. In active caves supplied with large quantities of water, cave deposits grow rapidly—as much as a centimeter a year. In dry caves, such as the famous cavern at Carlsbad, New Mexico, deposition is slow and may cease altogether.

Moisture seeps into caverns and drips from the ceiling and sides of the passageways to form pools and streams on the floor. Places where water drips from the ceiling become sites of deposits shaped like thin straws. These hanging deposits, called **stalactites,** start as a drop of water hangs from the ceiling. If the cave air is dry, or the solution loses CO_2 from the drop, a thin ring of calcite or gypsum forms. Eventually, the ring extends to form a straw. If the straw becomes filled, water moves as a thin film over the surface of the straw, and the deposit slowly grows layer on layer until a large tapering cone forms (figure 19.19).

Water that drops from the end of a stalactite falls to the floor, where more calcite or gypsum precipitates. Slowly, a stump-shaped deposit, called a **stalagmite,** rises from the floor. Many stalagmites have a small saucer-shaped top into which drops of water fall from the ceiling. Overflow from this saucer runs down the sides of the stalagmite, building it out. Eventually, stalactites and stalagmites grow together to form columns.

Other formations grow on the sides and floor of caverns. As droplets of water on the walls evaporate, small, rounded, knob-like projections form on the walls. Cascade-like deposits, called **flow stone,** develop on the walls and floor of caverns where water flows

(a)

(b)

Figure 19.19 Cave Deposits. (a) Common cave formations include stalactites, stalagmites, and columns. (b) Drops of water hang from the base of straw-like stalactites. As the water evaporates and loses carbon dioxide, calcium carbonate is deposited at the base of the straw.

over a surface. Rims of calcite or gypsum form around the edge of the pools. Over time, the flow and rim accumulations create terraced deposits.

ENVIRONMENTAL PROBLEMS IN KARST REGIONS

Pollutants get into ground water from trash dumps, from leaking sewer lines, injection wells used to dispose of waste, pesticides sprayed on cropland, or from leaks in septic tanks, drain fields, and cesspools. The slow movement of water through the regolith and rocks can remove some pollutants. Fine-grained sediment traps small particulate matter, and chemical reactions, especially between clay minerals and pollutants, can remove some elements and help clarify and purify ground water. Nevertheless, transfer of pollutants into drinking water by ground water is a serious environmental problem in every part of the world (figure 19.20).

Polluted ground water is usually identified when someone's water supply changes color or taste or when serious illnesses are traced to the drinking water. All three of these indicators occur frequently. In some instances, toxins in ground water have caused numerous deaths. Many of these have been traced to improper disposal of toxic industrial chemicals. An infamous case involved the results of chemical waste buried in the site of an abandoned trench designed by W. T. Love to be a canal near Niagara Falls, New York. In the 1920s, the dump was used until the trench was filled. In 1953, the trench was covered with a clay seal. In following years, homes, swimming pools, and playing fields were built over and around the trench. During unusually heavy rains, the ground became saturated, barrels of waste rose to the surface, and polluted water spilled out of the trench into ground water. Birth defects and damage to chromosomes among residents were eventually traced to the toxins in the dump and over 850 families were moved from the area. The events at Love Canal alerted the public to ground-water contamination. In 1981, Congress passed a bill establishing the Superfund, initially $1.6 billion to clean up hazardous waste sites. In subsequent years, the Environmental Protection Agency has identified hundreds of sites that need to be cleaned up, and both state and federal laws have reduced the spread of toxins.

Once a point or area where pollutants are entering the ground has been identified, the direction and rate of movement of ground water determines how large an area is or will be affected. Removing or reducing the threat of the pollutants may involve a variety of techniques, ranging from pumping out contaminated water to actually digging and removing the contaminated ground. Many of the techniques are expensive, and in some instances, it is not feasible to remove all of the pollutants because they have spread too far.

The direction and rate of movement of water in karst areas are much less predictable than the movement of water through homogeneous rock or sediment with intergranular porosity. Any pollutant that gets into the ground may move rapidly through the network of solution cavities and cracks. In karst areas, pollutants located anywhere in the drainage basin of a sinkhole will likely contaminate

Figure 19.20 Ground-Water Pollutants. Sinkholes are often used as trash dumps. Some contain toxic household products that enter the ground-water supply.

ground water. The problems are especially common in rural areas where septic systems with drain fields are used for sewage disposal. In many areas, outhouses are still widely used.

Many individuals and some communities have used sinkholes as waste-disposal sites. Before the potential for water pollution became well known, many communities used sinkholes for garbage dumps. Because the depressions are unsuitable for cultivation or other uses, many farmers and other landowners discard all types of waste in them. These uses of sinkholes are especially harmful because surface waters draining into sinkholes go directly into the ground-water system. Water can move rapidly through underground solution cavities and streams, quickly carrying toxic waste or bacteria great distances. Consequently, in karst areas, it is necessary to take special care constructing landfills to ensure that any pollutants are retained within the fill and do not leak into the subsurface.

Sinkholes also pose serious hazards for construction (figure 19.21). Because sinkholes contain a mixture of moist and unconsolidated soil, weathering products, and blocks of rock, they make unstable sites for foundations. Not only is the material in the sinkhole likely to subside as it compacts under the weight of a building, continued subsidence or even collapse is possible. In some places, rapid withdrawal of ground water may induce a sinkhole to collapse. In 1981, a large sinkhole appeared at Winter Park, Florida (see figure 19.21a). It expanded quickly, consuming houses and even parked cars. Although the process is usually less dramatic, collapse or subsidence of sinkholes occurs throughout most karst regions.

Problems with foundation stability and failure of reservoirs to hold water also arise in regions of karst. Many examples exist, such as the Hales Bar Dam in Tennessee, where problems were apparent only after construction of the dam was complete. The reservoir would not hold water. Water disappeared into sinkholes

and bypassed the dam, flowing through solution cavities that opened farther downstream. Before the reservoir was sealed, the cost of closing the sinkholes exceeded the original cost of the dam.

GROUND-WATER SUPPLY

Development of Ground-Water Supplies

Some people think that most ground water flows in underground streams similar to the streams on the surface. To them, locating a water supply means finding where those streams are. In a few places, mainly in karst topography, water does flow in underground

(a)

(b)

Figure 19.21 Sinkholes as Construction Hazards. (a) A large and active sinkhole formed at Winter Park, Florida, in 1981. (b) Most of a house disappears into a sinkhole in Virginia.

streams, but such streams are rare. In most places, ground water fills intergranular pore spaces or open fractures. The manner in which these pore spaces are distributed underground depends on the types of sediment or rocks that are present and on the structural configuration of those rocks. Figure 19.22 illustrates regions in which ground water occurrence is similar. Examples of the subsurface conditions of several of these ground-water provinces are discussed in following sections of this chapter.

Hydrogeologists require detailed knowledge of the subsurface in the area in which they plan to explore for water. They use local topographic and geologic maps and cross sections. Topographic maps locate streams, springs, and sinkholes. Aerial photographs may reveal the presence, orientation, and location of major fracture systems. Geologic maps illustrate the surface distribution of rock bodies of various types, the orientation of rock units, the location of major fault zones, and other structural features. Geologists draw cross sections from these maps to project these rock units into the subsurface. Cross sections containing subsurface data from nearby wells provide information about the depth to water-bearing units. Various types of geophysical surveys can provide additional precise information about the location and depth of water. Once the probability of success in finding water and favorable locations is complete, drilling a well follows.

After drilling a short distance in the ground, a casing, a pipe that is larger than the pipe through which water will flow, is placed in the hole. The casing keeps surface water and regolith from slumping into the hole. Once the well is completed, a pump test is conducted to determine the quantity of flow that can be expected and the rate at which the water level in the well is drawn down. In an unconfined aquifer with intergranular porosity, the water table is almost flat before pumping begins. After pumping starts, the water table around the well responds by dropping. The upper surface of the water table assumes the form of a cone, called a **cone of depression.** The cone grows larger until the rate at which water flowing toward the well is sufficient to replace the water being withdrawn.

Each water well has a cone of depression around it. If wells are placed close together, the cones of depression will overlap, resulting in a lowering of the water table between the wells. Clearly, wells can be too close together. If many wells located close together remove water from the same aquifer, the water table may be drawn down over a large region.

Safe yield is the amount of water that can be withdrawn from a ground-water reservoir without producing undesired effects. These include depletion or pollution of the water supply; damage to the aquifer by destroying its porosity and permeability, causing salt-water intrusion; or subsidence and cracking of the ground surface. Most undesirable effects happen as a result of long continued **overdraft,** a condition that reduces the quantity of water in the ground faster than recharge takes place.

Pumping wells may draw water from a considerable distance. In wells located close to the coast or in an area where saline, connate water is present, salt water may flow toward the well and contaminate the water supply. The effects may take the form of an advancing saltwater wedge, such as the one under Long Island that moves progressively inland as a result of overdrafts. Where salt water lies beneath fresh water, pumping may cause the salt water

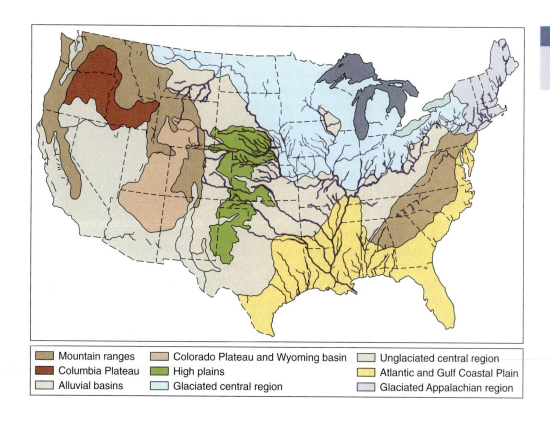

Figure 19.22   Ground Water Regions of the Conterminous United States.

▨ Mountain ranges	▨ Colorado Plateau and Wyoming basin	▨ Unglaciated central region
▨ Columbia Plateau	▨ High plains	▨ Atlantic and Gulf Coastal Plain
▨ Alluvial basins	▨ Glaciated central region	▨ Glaciated Appalachian region

to rise in an inverted cone shape beneath a well. Overdrafts can cause long-term damage to aquifers and have serious impact on the economic welfare of communities. Because surface water and ground water are so closely connected, it is useful to analyze the water supply.

Water Budgets

The surface drainage basin may coincide with natural boundaries separating bodies of ground water. In the Basin and Range Province (see figure 19.22), mountain barriers between ground-water basins hinder and may even prevent water movement underground from one basin to another. However, aquifers may also extend across surface drainage divides. In the Atlantic and Gulf coastal plains, some aquifers underlie large parts of several major surface drainage basins (see figure 19.22).

A **ground-water basin** is an area with natural geologic boundaries that contains the areas of recharge and discharge of water within an aquifer. In analyzing the water supply in a drainage basin, hydrologists try to account for all the water entering, leaving, or stored in the basin. They accomplish this by formulating a **water budget** for the area. The water budget contains an inventory of the water in the area plus water that is moving into and out of the area. At any given time, streams, lakes, the ground and artificial storage compounds such as reservoirs contain water. Water that originates as precipitation or meltwater from glaciers continually comes into and goes out of basins through streams and underground flow. Pipelines and canals may move additional large quantities of water in or out of the basin. Evaporation and transpiration, called **evapotranspiration,** account for much of the loss. Consumption of water and artificial movements of water, such as pumping of ground water into streams or artificial recharge of ground water, are locally important.

Precipitation, evaporation, transpiration, infiltration, and runoff are important factors in the water budget. The amount of surface runoff is approximately equal to the amount of precipitation minus amounts lost by infiltration, evaporation, and transpiration. Rain and snow gauges provide a measure of precipitation. The amount of evaporation from open pans indicates probable losses from lakes and stream surfaces. The level of water in the ground is the most direct indication of the amount of water that is infiltrating into the ground. Stream gauge data also provide the information needed for calculation of runoff.

The U.S. Geological Survey estimates the water budget for the entire country. According to these estimates, the average rainfall over the contiguous United States is about 75 centimeters (30 inches) each year. Of that, about 30% runs off in streams to the ocean, and about 60% goes back into the atmosphere through evapotranspiration. In recent years, the quantity of water evaporated into the atmosphere has increased. In part, this increase is related to higher air temperatures. The increased supply of water in the atmosphere has, in turn, led to increased precipitation, and in some years, to greatly increased snowfall. A large supply of water is present in the ground in most regions, but the way that water is stored varies greatly from one area to another.

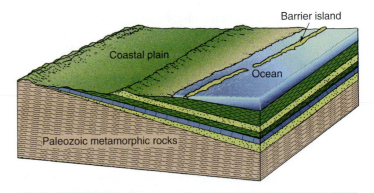

Figure 19.23 Ground Water in Coastal Areas. Cross section across a coastal area like that in the Atlantic coastal plain where aquifers are underlain by an impervious layer (see figure 19.9).

Ground-Water Provinces

Rock structure, composition, and the types of surficial deposits determine the occurrence of ground water. These conditions may be uniform over large regions, or water may occur in a variety of geologic settings within a single area. In all cases, the potential for continued use of water supplies depends on recharge of the ground water.

Recharge occurs most readily where water accumulates near the surface. Stream and glacial deposits contain much of this water. Large water supplies exist in the alluvial deposits in the flood plains along the courses of major streams such as the Mississippi, Ohio, and Missouri rivers (see figure 18.3). Other important supplies are in alluvial deposits, such as the fan-shaped deposits in the Basin and Range (see figure 18.12) or the shoestring-shaped sand and gravel deposits in the Great Plains. Extensive sheets of semiconsolidated beach sands also contain water in the coastal plains and in some inland basins. Water-bearing unconsolidated glacial sediments, discussed in chapter 20, exist over vast areas of the north central United States that were occupied by ice sheets during the Pleistocene (see figure 19.22). In all of these areas, water occupies intergranular pore spaces in silts, sands, and gravels.

Ground Water in Coastal Areas

The ground-water conditions on Long Island, New York, are typical of those found throughout much of the Atlantic and Gulf coastal plains (figure 19.23 and see figure 19.22). The history of development is well documented, and many of the problems encountered on Long Island are present in other coastal areas. Sediments underlain by impervious crystalline rocks are present in the coastal plains. The crystalline rocks crop out in Manhattan and Connecticut, but farther south and around the Gulf, these rocks are a kilometer or more (thousands of feet) deep. Semiconsolidated marine sediments inclined toward the ocean lie on top of these crystalline rocks. In New York, the sediments crop out along the shore of Long Island Sound and reach a depth of 600 meters (about 2,000 feet) along the southeastern coast of Long Island where they continue eastward beneath the continental shelf. Long Island differs from the coastal areas farther south because glacial deposits formed at the edge of the last great ice sheet accumulated on Long Island. These deposits of sand and gravel form ridges that extend the length of Long Island and cover older coastal plain sediments.

Both the semiconsolidated sediments and glacial deposits contain porous and permeable zones. Precipitation infiltrates the glacial deposits and sand, forming a water table in the glacial deposits that roughly conforms to the configuration of the land surface. Before Long Island became heavily populated, the long-term average ground-water recharge equaled discharge. The average infiltration of rainfall on Long Island, about 1.5 million liters per square kilometer, passed into and through the glacial deposits. Eventually, the water reached the sea. Part of it infiltrated into the underlying sediments where it filled pore spaces in three layers separated from one another by shales of low permeability. The permeable layers contained fresh water under Long Island and salt water from the ocean in their seaward continuations beneath the ocean and Long Island Sound.

Initially, the people on Long Island obtained their water supply from shallow wells dug into the glacial deposits. Soon, disposal of sewage into these deposits created a pollution problem, and they had to abandon the shallow wells. To find more water, they drilled wells through the glacial deposits into the underlying semiconsolidated sediments.

Population growth and water usage for industrial purposes increased most rapidly at the western end of the island. In this area, large-scale ground-water withdrawal from deeper aquifers caused the freshwater-saltwater contact in the aquifers to move landward. Intrusion of the salt water accelerated as the construction of storm sewage systems diverted ever-increasing amounts of runoff overland into the ocean. Construction of buildings, streets, and parking lots further reduced infiltration of rainwater into the ground. By the late 1930s, large-scale contamination of the aquifers by salt water led to abandonment of this supply of drinking water.

Today, the western part of Long Island obtains its water from the New York City supply, which comes from surface reservoirs in the Catskill Mountains. Changes in water management have slowed, and in some places even reversed, the saltwater intrusion. These include laws requiring that water removed from the ground for air-conditioning be pumped back into the ground. Other efforts to increase recharge involve construction of leaky ponds that trap surface runoff and direct it back into the ground.

Ground Water in the Basin and Range

Average annual rainfall in the Basin and Range is from 10 to 15 centimeters (4 to 6 inches). Impervious rocks in the mountains isolate water in the deep basins. The whole region has poorly developed internal surface drainage. Little water moves in or out of basins on the surface or underground. Alluvial fans contain most of the ground water in the Basin and Range Province. These fans consist of unconsolidated mixtures of boulders, gravels, sand, silt, and clay. Long string-like deposits of well-sorted, highly porous, and permeable gravel accumulate in the stream valleys on modern alluvial fans. Interstream areas contain relatively impervious deposits from mudflows and sheet wash. At the outer edge of the fans, coarser materials give way to less permeable fine-grained sands, silts, and clay interbedded with modern lake-bed deposits.

During parts of the last 2 million years, ice covered the mountains in this region. Increased precipitation around the ice and some melting of ice produced lakes that filled some of the basins and saturated the alluvial deposits. Layers of clay, silt, and fine sand of low permeability settled in the lakes. These deposits, which are 300 meters thick, lie on top of older, more compacted and cemented alluvial-fan and lake-bed deposits.

Today, surface water moves toward the lower parts of the basins in intermittent streams. Only in the mountains is rainfall sufficient to provide the excess needed for infiltration. As water flows out of the mountains around the edge of the basins, it seeps down into the aquifers and becomes confined under layers of clay or other impervious beds. The porous and permeable beds in the unconsolidated alluvial deposits make up the principal ground-water reservoirs in these basins (figure 19.24 and see figure 18.12). Most of

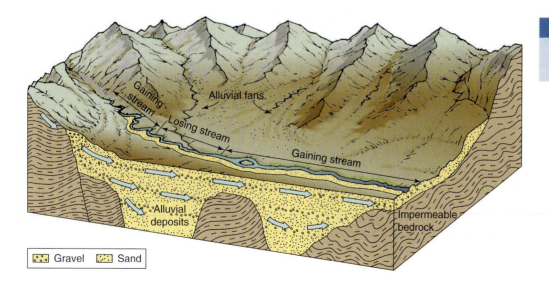

Figure 19.24 Subsurface Geologic Conditions in the Alluvial Basins of the Western United States.

Gaining stream

Alluvial fans

Losing stream

Gaining stream

Alluvial deposits

Impermeable bedrock

Gravel   Sand

the aquifers leak water into other beds, producing a more or less continuous hydraulic system. Where these discontinuous water-bearing layers are interbedded with impervious beds, zones develop that contain water under artesian pressure. Young, high-angle faults located along the edge of the basins cut and displace many of these alluvial deposits including some of the artesian layers, and in some places, the artesian water rises along these faults, producing springs at the surface along the fault trace.

Las Vegas, Nevada, lies in one of these basins. The ground-water conditions in this basin are representative of those found throughout much of the Basin and Range. Demand for water is highest in and around the city of Las Vegas. Here, excessive pumping reduces artesian pressure in the aquifers and removes more water than is naturally replenished. As a result of the rapid withdrawal of water, the ground has slowly subsided. The weight of the water stored in nearby Lake Mead Reservoir contributes to this subsidence.

Withdrawal of water causes fine-grained sediments to compact. This destroys their porosity. Only expensive techniques such as injection of water at high pressure can restore it, and in some places, even these techniques fail. Compaction following withdrawal of ground water is not confined to the Basin and Range. Compaction also occurs in the San Joaquin Valley of California, and it may be destroying the capacity of aquifers to store water in the High Plains and even in the Atlantic and Gulf coastal plain, where water is abundant.

Water Crisis in the Great Plains

The High Plains states, an eight-state area extending from Texas to South Dakota (figure 19.25), produce nearly 12% of the nation's cotton, corn, and wheat. Much of the land is irrigated with water pumped from approximately 150,000 wells drilled into the Ogallala Formation. Pumping is lowering water levels in the aquifer so rapidly that some hydrogeologists estimate that within a few years, the Ogallala will contain only enough water to irrigate less than half the area irrigated in 1990.

The sands, silts, and lenses of gravel that compose the water-bearing part of the Ogallala originated as a result of erosion of the Rocky Mountains (see figure 19.11). Streams flowing out of the mountains carried and deposited the sediment as an extensive, almost sheet-like layer that is up to 200 meters (656 feet) thick in places.

Generally, thin deposits of more recent alluvium, sand dunes, glacial deposits, or windblown silt cover the Ogallala. Surface water infiltrates into the Ogallala through these thin covers where they are porous and permeable. Unfortunately, rainfall in the plains and recharge from streams are not adequate to replace the water that is being removed. Nor does all the water removed for irrigation purposes infiltrate into the ground. Much of that water evaporates from the ground surface or in the air where irrigation involves spraying. Plants also transpire part of the soil moisture back into the atmosphere.

The Ogallala is not a uniform layer of sediment. The water-bearing beds vary in porosity, permeability, and thickness, and natural recharge varies from place to place. The quantity of water

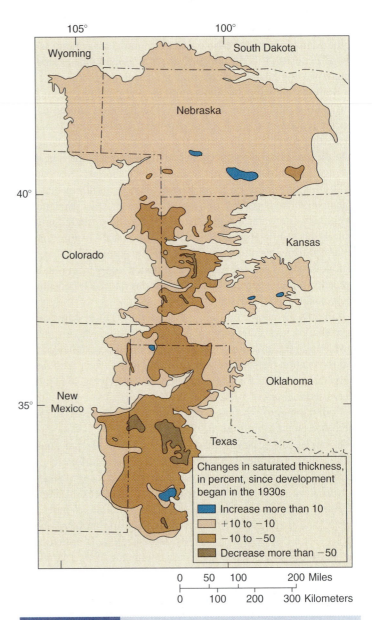

Figure 19.25 The Thickness of the Saturated Zone in the Aquifer Under the High Plains Between the 1930s and the 1980s. The potential reservoir of water has decreased more than 50 meters in parts of northern Texas, and most of the area has experienced a drop in water level.

stored in the ground also varies. Most wells reach the water table at depths of 20 to 150 meters (65 to 490 feet), but as overdrafts continue, this depth and the cost of pumping water to the surface increases. In some areas, the cost of pumping may make the water economically unavailable long before the supply gives out.

The outlook for the future is bleak. Water in the aquifer is being "mined"—drawn down 10 to 15 times faster than it is being replenished by recharge. Water that accumulated in the Ogallala over thousands of years is being withdrawn in a century. The cost of pumping is increasing, and the cost of importing water by canals or pipelines would be exceedingly high. Some hydrologists predict that, eventually, vast areas in the region will be removed from agricultural production and given over to capturing water so the remaining areas underlain by the Ogallala can be recharged at a rate close to the rate at which water is withdrawn.

Avoiding Problems in Development of Ground-Water Supplies

Both natural conditions and human activities may lead to problems of contamination and inadequate water supply. If the nature of the water occurrence and potential sources of pollution are known, many of these problems can be avoided. The following are among the most common of these problems.

1. Long-term overdrafts (when pumping exceeds recharge) lead to exhaustion of the water supply. After the water supply reaches exhaustion, the rate of water withdrawal cannot exceed the rate of recharge.
2. When the rate of pumping of water from an aquifer exceeds the rate of recharge, damage to the aquifer by compaction and loss of porosity is likely.

3. Where salt water is present in an aquifer or in adjacent rocks, withdrawal of fresh water faster than recharge replaces it may cause salt water to contaminate the freshwater supply.
4. By placing wells too close together, the cones of depression of the water table around the wells will intersect and increase the lift required to raise water to the surface.
5. Excessive pumping of water from karst areas may induce subsidence or collapse of the ground surface.
6. Sewage and chemical or radioactive waste disposal in water-bearing rock or sediment contaminates ground water and will move with it through the aquifer (figure 19.26). Impurities may remain in the water in such an aquifer for many years. Radioactive waste may damage an aquifer permanently.
7. The salt content of ground water increases where water infiltrates salt-encrusted regolith in arid regions. For this reason, recycling of ground water by irrigation tends to increase its salinity (see figure 19.26).

Replenishing Aquifers

In areas of low rainfall, people depend on ground water for domestic and industrial use. As development occurs in these areas,

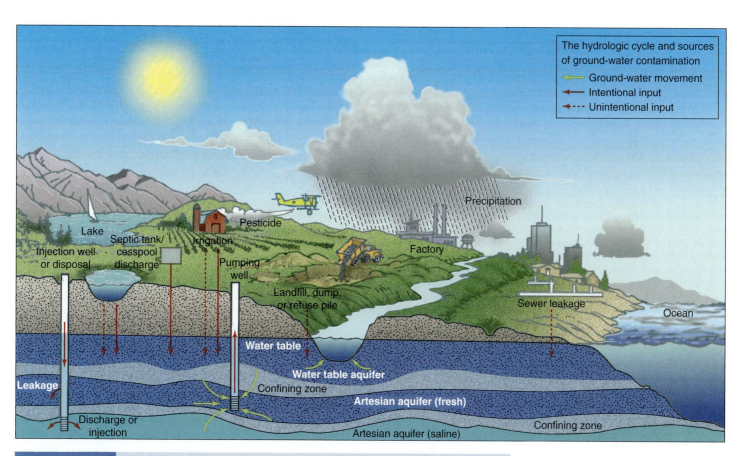

Figure 19.26 Sources of Ground-Water Pollution. Pollution sources include leakage of pollutants from sewage systems, landfills, dumps, septic tanks and cesspools, and from wells used to inject fluids.

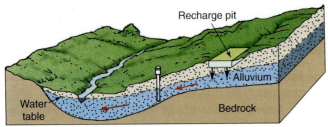

Recharge pit

Alluvium

Water table

Bedrock

Artificial recharge of aquifers can be accomplished by constructing pits filled with porous and permeable rock in the valley of stream tributaries.

(a)

Figure 19.27 Ways of Increasing Recharge of Aquifers.
(a) Use of recharge pits and wells. (b) Pumping from a well located near a stream so infiltration from the streams is increased. (c) Use of canals to increase infiltration of water from a stream.

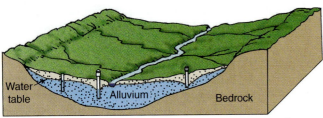

Water table

Alluvium

Bedrock

Infiltration along a stream can be increased artificially by pumping water from wells near the stream. In this way, the heavy runoff from urban areas can be used to recharge ground-water supplies.

(b)

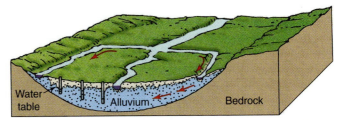

Water table

Alluvium

Bedrock

Canals can be used to recirculate streamflow across porous and permeable alluvium, increasing infiltration.

(c)

the rate of production of ground water approaches and may exceed the rate of natural recharge. At some point, the water budget must balance. The government may ensure this by restricting the withdrawal of ground water, by importing water, or by increasing the rate and amount of recharge (figure 19.27).

Generally, the rate of recharge of an aquifer increases as water levels drop in the natural area of recharge. By keeping the water table lower, infiltration increases and less water runs off on the surface. Wells located near major streams can draw down the water table and encourage infiltration from streams. In some places, pits or basins constructed in tributary valleys of major streams increase infiltration. Small dams constructed in such a way that water leaks into the regolith serve the same purpose. Reusing water can help solve the problem of inadequate water supply. For example, water withdrawn for air-conditioning is suitable for other uses; we can pump it back into the ground through wells or allow it to infiltrate from holding ponds. In all of these methods, the basic idea is to increase the amount of water that infiltrates into the ground and reduce the amount of the water budget that moves out of the area through surface runoff.

SUMMARY POINTS

1. Underground water originates from three sources: meteoric (derived from precipitation), connate (trapped in sedimentary rock), and magmatic (coming from magma or lava). The water accumulates within pore spaces, solution cavities, and open fractures in the regolith and bedrock.

2. All regolith and rocks contain voids that water can occupy. Porosity is the percentage of the total volume that is void. Permeability is the capacity of a material to transmit fluid. Materials with high porosity are not necessarily permeable.

3. Most near-surface water is meteoric water that has infiltrated regolith and rock. Below some depth, water saturates all the pore spaces. The upper surface of this saturated zone, the water table, tends to follow the surface topography, but in a subdued form. Generally, it is deeper under hills and closer to the surface in low areas.

4. The term unconfined applies to ground water that can move freely through rock and regolith. Impermeable layers situated above a porous, saturated zone confine water and restrict its movement. Impervious layers may create perched water tables or cause artesian conditions if the area of recharge is above most of the aquifer.

5. Where the water table intersects the land surfaces, ground water may issue from springs. If the ground water has percolated through limestone, calcium-carbonate deposits of tufa and travertine may develop around the springs or along streams that carry high concentrations of these ions.

6. Long-continued solution of bedrock by ground water may lead to the formation of karst topography, which is characterized by caves, swallow holes in streams, blind-stream valleys, natural bridges, and tunnels. Sinkholes form when the surface

subsides or collapses into these solution cavities. Within caves, stalactites, stalagmites, and columns develop as ground water, saturated in calcium carbonate or calcium sulfate, drips from the ceiling of the cave.

7. A water budget is an account of the amount of water entering an area, usually a drainage basin, and the amount that is leaving that area. Serious problems will eventually arise in areas where the uses and losses of water exceed the influx of new supplies.

8. Ground water is a critically important natural resource used for domestic, industrial, and agricultural purposes. Many communities depend entirely on ground water. In some areas, these supplies are endangered by saltwater contamination, excessive pumping, and improper disposal of wastes.

9. Based on the way ground water occurs, the country can be subdivided into ground-water provinces. These are closely related to the types of rocks in the subsurface and their structure.

10. Potentially serious water-supply problems exist in a number of ground-water provinces. In some arid and semiarid regions, the quantity of water used exceeds the supply stored in the ground. Aquifers are not adequately recharged in the Great Plains and in the Basin and Range provinces. Along the Atlantic and Gulf coasts, excessive pumping has caused saltwater intrusions in a number of localities.

11. Ground water can be replenished by construction of small dams and by recharge of aquifers through wells.

REVIEW QUESTIONS

1. What conditions determine how rapidly water infiltrates the regolith?
2. Other than by infiltration, how does water get into the subsurface rocks?
3. What rocks have both low permeability and high porosity?
4. Under what conditions is the top of the water table a smooth surface?
5. Under what circumstances can removal of water cause subsurface water to become polluted?
6. How does the movement of water in an unconfined aquifer differ from that in a confined aquifer?
7. How do sinkholes come about?

8. Describe the steps that precede the eruptions of a geyser.
9. Why are some sinkholes unstable while others are stable?
10. Why are all sinkholes potentially hazardous?
11. Where would you look in your area for springs?
12. Why is the depth to the zone of saturation so uniform in some areas and so irregular in others?

THINKING CRITICALLY

1. How would an increase in precipitation affect the supply of ground water in your region? What would be the effect of an increase in average temperature?
2. How would a change in sea level affect ground-water conditions in the region where you live?
3. Would ground-water conditions be affected in the region where you live if the region were elevated by crustal deformation?
4. Do processes that come about as part of the plate tectonic system affect the porosity of rocks?

KEY WORDS

aquifer	perk test
artesian conditions	permeability
cone of depression	porosity
confined ground water	recharge
connate water	safe yield
evapotranspiration	siliceous sinter
flow nets	sinkhole
flow stone	stalactite
geothermal energy	stalagmite
geysers	swallow hole
ground-water basin	travertine
hydrostatic head	tufa
hydrostatic pressure	unconfined ground water
juvenile water	water budget
karst topography	water table
meteoric water	zone of aeration
overdraft	zone of saturation
perched water table	

The Role of Ice in Earth Systems

Glaciers have shaped much of the landscape of high mountains throughout the world. This glacier is located in the Parque Nacional los Glaciares in Argentina.

Chapter Guide

Ice has shaped the landscape at high latitudes and in high mountains throughout the world. It affects the level of the sea, the climate, and the growth of plants and animals. In the past, ice covered much of North America, Europe, and Asia. Even today, one entire continent is covered by ice.

About 2 million years ago, global temperature began to drop, and for the past million and a half years, Earth's climate has oscillated between periods of glacial advance and warmer interglacial intervals. Large accumulations of ice have controlled many aspects of Earth's environments. This chapter examines the formation of glacier ice, the ways ice and deposits from ice shape the topography, and the effects that ice has had on the regions beyond the ice margin. Additional impacts of the ice ages are discussed in more detail in other chapters. Ice sheets affect the heat balance and climate on Earth (chapter 11). Ice influences the circulation in the oceans (chapter 9). Ice affects the formation of air masses and local winds (chapter 12). Glaciers change ecosystems near the ice. As the quantity of water stored as ice on land increases, sea level drops (chapter 10). This chapter reviews and provides additional information about some of the important connections between glaciers and other parts of Earth systems. It also focuses attention on the dramatic effects glaciers have on the shape of the land surface.

INTRODUCTION

Today, almost 1.5% of all of the water on Earth is ice, and it covers about 10% of the land surface. Glaciers occupy the valleys of many mountains in high latitudes (figure 20.1), and some glaciers like those on the east African volcano Kilimanjaro (elevation 5,960 meters [19,575 feet]) exist in equatorial latitudes. A thick sheet of ice covers most of one entire continent; Antarctica (figure 20.2) and other sheets cover large parts of Greenland, Iceland, and islands in the Arctic Ocean (figure 20.3). Eighteen thousand years ago, ice extended as far south as the Ohio and Missouri rivers in North America, and it covered Great Britain and much of northern Asia. This last glacial advance was but one of a number of glaciations that affected North America. Four of these episodes of glaciation are named for one of the states in which the deposits occur. The glacial advance known as the Wisconsin glaciation is the most recent of many Pleistocene glacial advances separated from one another by interglacial episodes similar to the present. Most geologists think that ice sheets will expand again in the future.

Glaciers have played a major role both in shaping the landscape and in controlling the evolution of modern environments all over Earth. At the peak of the last glacial advance, ice covered 30% of Earth's land surface. Sea level stood 100 to 130 meters (330 to 430 feet) lower than it is today, climatic zones were compressed toward the equator, and the distribution of fauna and flora over much of Earth was quite different from that of today. Deposits from the ice sheets, sediments on the sea floor and in lakes, tree rings, gases trapped in the ice in Greenland and Antarctica, and erosional effects of this glacial advance left a well-documented record of the immense importance of glaciation in shaping the face of Earth.

THE CLIMATIC EFFECTS OF ICE ON EARTH

If the amount of solar radiation that strikes Earth, especially at high latitudes, decreases over a long period, cold, polar climates result. Snow accumulates and glaciers form as a result of precipitation at air temperatures below freezing. Although glaciers are an effect rather than a primary cause of cold climatic conditions, once snow and ice cover large areas of the surface, they reflect solar radiation, decrease the amount of solar radiation absorbed by the surface, and reduce temperature further. Large ice sheets influence the heat balance on Earth and the patterns of atmospheric circulation, precipitation, and water supply far beyond the margins of the ice. At the peak of the last glacial maximum, ice covered all of Canada and extended into the northern states. A wide belt along the margin of the ice was a polar desert or covered by tundra (figures 20.4 and 20.5). Forest covered the southern and central Appalachians and the interior lowland region. Dramatic changes occurred in the southwestern United States where the region that is arid today was wet and fertile during the glacial maximum. The tropics shifted toward the equator.

As sheets of ice grow, they maintain the low temperature of the air over them. The size of the area with cold air grows with the ice cover. These larger air masses compress all other climatic zones toward the equator and modify the patterns of atmospheric circulation. As the movement of air across Earth's surface changes, the distribution of clouds, the amount of cloud cover, and the amount and distribution of precipitation also change. A record of the changes that accompanied the growth and decay of ice sheets during the ice ages (the last 1.5 to 2 million years) remains trapped in

Figure 20.1 Valley Glacier **Located in Antarctica.** The snowfield from which the glacier moves is visible in the distance.

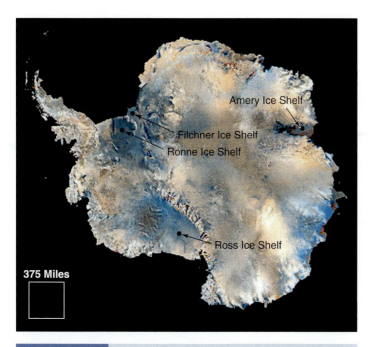

the ice in Greenland and Antarctica and in sediments deposited both in the ocean and in lakes.

THE INTERACTION BETWEEN ICE AND THE SEA

The growth of ice sheets on land areas has two main effects on the oceans. First, the climatic conditions that cause sheets of ice to form also reduce the temperature of the surface layers of the oceans, and any decrease in surface-water temperature directly affects the deeper circulation of the oceans. Second, any large change in the quantity of ice on land is accompanied by a corresponding change in the quantity of water in the oceans.

The Ice-Water Balance and Sea Level

Changes in the relative amounts of ice and water affect sea level throughout the oceans. At the height of the last major ice advance, sea level stood between 100 and 130 meters (330 and 426 feet) lower than it does today. As the ice melted, sea level rose, rapidly at first, but the rise has been proceeding at slower rates for the last 5,000 years. If all the ice presently stored on land melted, sea level would rise by an amount conservatively estimated as being between 25 and 60 meters (80 and 200 feet). Changes in sea level of this magnitude have taken place repeatedly over the last 1.5 million years. Past changes in sea level had pronounced effects on the evolution of continental shelves and coastal landforms (see chapter 10).

It is difficult to imagine the impact a 25- to 60-meter (80 to 200 feet) rise in sea level would have on the overall shape and character of the coasts. In North America, the entire Florida peninsula would lie under water, as would most of the Atlantic and Gulf coastal plains. Boston, Massachusetts; New York City; Washington D.C.; Charleston, South Carolina; Los Angeles and San Francisco, California; Philadelphia, Pennsylvania; Houston, Texas; New Orleans, Louisiana; and Seattle, Washington would lie within embayments along the newly created coast. Seas occupied these areas during some of the interglacial periods. A drop in sea level of 100 meters (330 feet) would drastically reduce the size of the Gulf of Mexico, double the size of Florida, and land would extend far to the east of Newfoundland (see figure 20.5). Obviously, major changes in the geography of the oceans accompany the rise and fall of sea level. Equally dramatic changes take place in the location and strength of currents. Even circulation in the deeper parts of the ocean is affected by changes in sea level and accompanying changes in the temperature of the surface layers of the ocean.

The Role of Sea Ice in Oceanic Circulation

The temperature and salinity of seawater drive the deep circulation in the oceans. As water temperature drops, its density increases, and it sinks, pushing underlying water ahead of it. This cold, high-density water initiates movement of the conveyor-belt-like circulation described in chapter 16. Seawater acquires its cold temperature at high latitudes where average air temperatures are far below freezing during the winter and somewhat below freezing even in summer months.

Because of its salt content, seawater does not freeze until the temperature is about −2°C (28°F). As the water freezes, salt is partially excluded from the ice crystals, increasing the salinity and density of the surrounding water. During winter months, the area

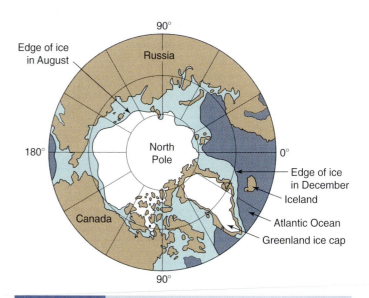

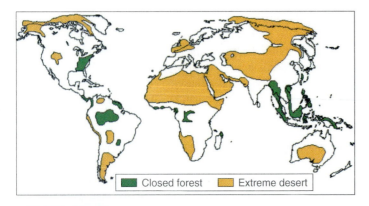

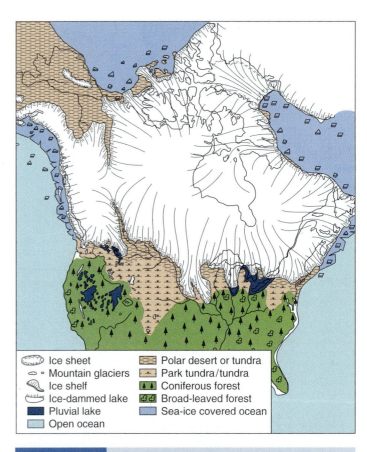

Figure 20.4 Last Glacial Maximum. During the last glacial maximum, about 18,000 years ago, most forests were located in equatorial regions, and vast cold deserts surrounded the ice sheets in the Northern Hemisphere.

covered by **sea ice,** frozen seawater, gradually expands, and during warm summer months, sea ice retreats. Satellite images track this annual advance and retreat of the sea ice (see figure 4.2). While the ice insulates the underlying seawater from the extremely cold air temperature, it keeps the temperature of water in contact with it close to freezing. In winter, sea ice covers the entire Arctic Ocean and extends far south through the passages between Labrador, Greenland, Spitsbergen, and Norway. Floating masses of ice cover most of the sea as far south as Newfoundland.

In spring, more solar radiation reaches the surface at high latitudes, and air temperature rises. The surface and edges of the ice shelves and glaciers begin to melt and break apart, setting huge icebergs adrift, free to move with surface currents. One such iceberg sank the *Titanic*. Because the salt content of sea ice is less than that of seawater, melting of sea ice reduces the salinity of seawater at the same time that the surface temperature of the water rises. Consequently, if melting of sea ice continues for a long time, the density of seawater decreases, reducing the rate at which the water sinks. With less water sinking, the deep circulation in the oceans slows, and less deep water returns to the surface in areas of upwelling. The conveyor belt slows, and climatic effects associated with it around the world are affected (see chapter 16).

GLACIERS

Glaciers are large, perennial masses of ice formed on land by compaction and recrystallization of snow. Glaciologists classify them according to their geography and temperature. **Valley glaciers** may coalesce along a mountain front, forming **piedmont glaciers.** Ice sheets, also known as **continental glaciers** because they may cover large parts of continents, are not confined in valleys.

Based on the temperature of the ice, glaciologists classify glaciers as either temperate (warm) or polar (cold) glaciers. The temperature of the ice affects the movement of the glacier and its ability to erode the rock beneath the ice. Warm glaciers move more

rapidly and shape the land more quickly than do cold glaciers. Surface temperatures of all glaciers vary with the season, but once the upper few inches are below freezing, conduction of heat through ice is negligible. For this reason, the temperature in the interior of a glacier tends to remain constant throughout the year. Ice in a warm glacier may have uniform temperature only slightly below freezing from top to bottom. Percolation and refreezing of meltwater warm the deeper portions of the ice. Heat (80 calories per gram) is liberated when water freezes, and this heat raises the temperature of the surrounding ice. The warm ice yields to stress and recrystallizes, making it flow more readily than cold ice. The temperature of cold glaciers like those in Greenland and Antarctica ranges between $-10°$ and $-20°C$ ($14°$ to $-4°F$) at the surface; they are only slightly warmer at depth.

Polar Ice Masses

At present, most of the land and water area covered by ice lies in polar latitudes. The Antarctic ice sheet contains about 88% of the world's ice. This mass of ice covers about 13.5 million square

Ice sheet
Mountain glaciers
Ice shelf
Ice-dammed lake
Pluvial lake
Open ocean
Polar desert or tundra
Park tundra/tundra
Coniferous forest
Broad-leaved forest
Sea-ice covered ocean

Figure 20.5 Map of North America at One Stage During the Last Glacial Advance. The map depicts the ice sheet, regions of tundra, and location of coniferous and deciduous forest. Note the position of the shoreline around North America at a time when sea level was 100 meters lower than it is today.

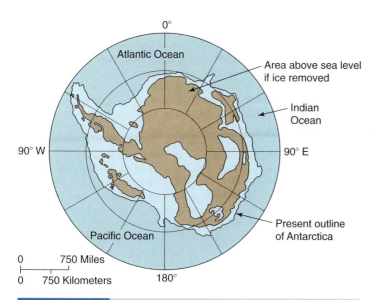

Figure 20.6 Antarctic Ice Sheet. Much of Antarctica is covered by sea ice that is part of the Ross, Ronne, and Pine Island shelves. If all of the ice melted, the land of Antarctica would be much smaller and would include a number of islands.

(a)

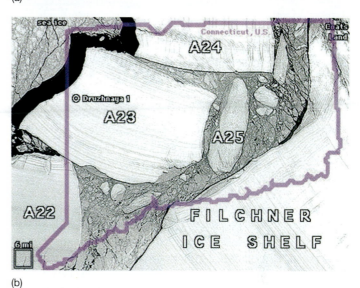

(b)

Figure 20.7 Filchner Ice Shelf. (a) Perspective sketch of Filchner ice shelf. (b) Satellite image of part of the Filchner ice shelf showing pack ice and large icebergs. The outline of the state of Connecticut illustrates the size of the huge iceberg that broke off of the shelf in 1986. Other large icebergs have continued to drift away from the Filchner and other ice shelves in Antarctica.

kilometers (5.3 million square miles). The ice in Antarctica is more than 4 kilometers (2.5 miles) thick in the thickest part. Mountain ranges, including one 1.57 kilometers (5,150 feet) high and more than 160 kilometers (100 miles) long, lie buried beneath this ice. These mountains are close to the height of the Appalachian Mountains. If the ice melted rapidly, a large part of Antarctica would be under water, and the shape of the continent would resemble that shown in figure 20.6. The lowest point of Antarctica would be about 1.44 kilometers (0.9 mile) below sea level, and the eastern part of the continent would be a high plateau standing about 1.9 kilometers (1.2 miles) above sea level. The shape of the continent would change rapidly as the areas formerly covered by ice began to rise due to isostatic rebound—a process described in chapter 3.

Glacier ice from the Antarctic ice sheet flows into the sea, producing ice shelves that are unsupported by land. In places, the ice shelves on the Ross and Filchner seas are about 1 kilometer (3,280 feet) thick. In addition to the ice shelves, wind-driven floating pieces of ice pack together. **Pack ice** (figure 20.7b) covers most of the surface of the Antarctic Ocean. At the end of winter, pack ice reaches its greatest extent and covers most of the ocean south of latitude 65° S.

THE ROLE OF GLACIERS IN SHAPING THE MODERN LANDSCAPE

Growth and Decay of Glaciers

The only condition necessary for the growth and expansion of glaciers is that for an extended period, the amount of snow and ice accumulation must exceed the amount of ice lost through melting and sublimation. Increased snow and sleet and temperatures low enough to preserve the ice throughout the year are more than is necessary for glaciers to grow. The extremely low temperatures encountered in Antarctica today are unnecessary for the formation and growth of continental glaciers. Because air that is close to freezing can hold more moisture than air at extremely low temperatures, glaciers grow more rapidly when temperatures are near freezing. For this reason, a decrease of a few degrees in the average annual temperature would favor growth of larger masses of ice.

Measurement of the supply and loss of ice and snow over such a vast area as Greenland or Antarctica is difficult. In 2000, airborne measurements indicated that the elevation of the ice surface in parts of the Greenland ice sheet above 2 kilometers (1.24 miles)

have a nearly balanced budget, but the regions of lower elevation are thinning. The net loss over the Greenland ice sheet is enough to raise sea level about 0.13 millimeter (0.03 inch) per year.

The West Antarctica ice sheet covers nearly a quarter of Antarctica (see figure 20.6). Impressive changes have been taking place in the ice sheet during the past few decades since detailed studies started. Mud obtained from a hole that penetrated the ice sheet contained evidence that at some time in the last 1.3 million years, the ice sheet had disintegrated and wasted until little remained. By 1998, studies of radar images of the ice sheet that were first obtained in the early 1990s revealed that the line between the part of the ice sheet that rests on land and the part that floats on water is shifting at a rate of more than a kilometer each year. This suggests that the ice sheet is losing mass by melting at its base. In March of 1999, a super iceberg 293 kilometers (183 miles) long by 37 kilometers (23 miles) wide broke off the Ross ice shelf. The surface of this ice mass is about the size of the state of Connecticut, and the submerged part is nearly ten times larger. Within two months, the iceberg had split and drifted 32 kilometers (20 miles) from shore. Other huge icebergs have broken off the ice shelves in Antarctica for many years. Massive icebergs broke off of the Filchner ice shelf in 1989 (see figure 20.7). The movements of these massive ice blocks that break off Antarctica are largely unpredictable. They contain huge amounts of water. This one could supply the water needs of Los Angeles for several hundred years. Only a small part of the ice in an iceberg (figure 20.8) is present above water.

Despite the dramatic events on the Ross shelf ice, scientists studying this area estimate that the West Antarctic ice sheet has a projected life span of 4,000 to 7,000 years and that melting of the ice sheet over the past 11,000 years has raised sea level about 0.8 millimeter per year. At present rates of wastage, melting of the ice sheet may be raising sea level about 1.3 millimeters per year.

Coastal areas, which receive the most precipitation, have the highest rate of ice accumulation in Antarctica (more than 1 meter in thickness per year). Most losses result from the breaking away of ice from the ice sheet where it protrudes into the ocean, from sublimation—the process by which ice evaporates without going into a liquid state—and from melting at the base of the ice sheets. The mean annual temperature of $-30°C$ ($-22°F$) keeps melting at the surface to a minimum.

Formation of Glacial Ice and Movement of Ice Sheets

For any area, the lowest altitude at which snow persists year-round fluctuates, reflecting changes in temperature and the amount of snowfall. Above the snowline, snow remains from year to year as snowfields. During glacial advances, snow covers vast regions such as central Canada where elevations are low. In mountainous regions, however, the snowfields may lie at high elevations just below the summit levels of high peaks, which may protrude through the snowfields and glaciers (figure 20.9).

As snow accumulates, snowflakes recrystallize, forming a granular ice called **firn,** or **neve.** The weight of overlying snow compacts the granular ice together into a solid mass of ice. If the temperature rises above freezing, the snow near the surface melts, and water percolates deeper into the accumulation where it freezes again and forms ice.

A typical glacier has a brittle upper crust about 60 meters (200 feet) thick. At deeper levels, the ice flows. Fractures and crevasses form in the brittle upper portion of the glacier as it moves (figure 20.10). These crevasses, which are often a meter (3.28 feet) wide, open and close in response to shearing, compression, and extension of the ice as the lower ice flows over irregularities and around obstructions in its path. When the pressure on ice is equal to the weight of about 70 meters (230 feet) of ice, it begins to deform continuously. This deformation involves slip or gliding between layers of molecules within individual ice crystals and movement along shear planes that cut through individual crystals (figure 20.11). Recrystallization accompanies both mechanisms. In effect, glacier ice undergoes a type of solid flowage. Once flow starts, the ice moves in the direction of the least pressure. That direction is generally downslope for valley glaciers and out around the lower margin of ice sheets. The ice at the bottom of an accumulation over 70 meters (230 feet) high would begin to spread laterally, even on a flat surface.

The ice sheets that covered Canada 18,000 years ago were over a thousand meters (several thousands of feet) thick. These are comparable to the ice sheets in the Antarctic today. Driven by the pressure created by their own weight, these sheets slowly spread. As in Antarctica today, the pressure deep in the ice sheets that once covered Canada, Scandinavia, and Siberia was sufficient to make the ice spread over topographic irregularities hundreds of meters high.

Movement of Valley Glaciers Stream valleys are the courses of least resistance for ice to follow as it begins to flow, especially in mountains. Tongues of ice begin their slow movement down the valleys as valley glaciers. How far the glaciers move depends on the accumulation of snow, the temperature, the shape of the surface over which the ice moves, and the rate at which melting and sublimation occur at the end of the glacier. The rate of advance increases as more snow accumulates in the snowfield and as the ice thickness increases. The rate of advance also increases down steeper slopes and across smoothed and streamlined topography.

Several techniques make it possible to measure the movement of glaciers. The oldest method involved observing changes in the positions of stakes driven into the ice along a straight line across the surface of the glacier. Today, glaciologists determine glacier movements by establishing stations on the surface of the ice and tracking their change of positions with global-positioning systems. These instruments receive signals from satellites and can determine the location of points within a few centimeters. Satellite images taken periodically also reveal changes in the position of the ice margin. These studies reveal that ice at the center of the glacier moves much faster than that at the sides, where friction along the contact between ice and the rock retards movement (figure 20.12). On the surface of some glaciers, patterns in the ice reveal the movement in the ice. The arrangement of crevasses, the internal structure of the glacier ice, and the form taken by debris on the glacier surface also indicate movement in glaciers. To detect ice

Figure 20.9 Aerial Photograph Across the Harding Ice Field in Kenai Fiords National Monument in Alaska Showing a Few Mountain Peaks That Stand Above the Snow.

movement beneath the surface, glaciologists may drill holes in the glacier and measure the progressive deviation of the hole from its initial vertical orientation.

Some glaciers move so slowly that trees and other plants grow in the debris on top of them; others move at rates as high as 6 to 8 meters (20 to 26 feet) per day and, in exceptional cases, over 30 meters (100 feet) per day. Hassanabad Glacier in the Karakoram Mountains in Kashmir once advanced over 3 kilometers (two miles) in 2.5 months—about 140 meters (460 feet) per day. Such extremely rapid movements of glaciers, called **surges,** rarely continue for long periods.

Glaciers and ice sheets leave records that make it possible to trace the direction of their movement long after the ice is gone. One of these is the presence of isolated pieces, and occasionally huge blocks, of rock, called **erratics,** that are carried by glaciers and deposited when the ice melts. If the place of origin of the erratic is known, the erratic provides an indication of the direction of movement of the ice. Erratics may be transported far from where they originated. Some erratics found in South America are composed of rocks that crop out in Africa. This provided one of the early lines of geologic evidence that these two continents were once connected and that the theory of continental drift is valid.

Figure 20.10 Glacial Crevasses. Crevasses form in glaciers where the base of the ice moves over irregularities in the valley floor or where the ice down-valley from the crevasses moves at a high rate. For scale, note the group of people standing on the ice surface.

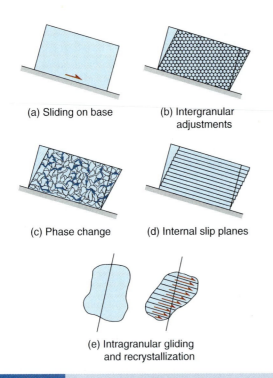

(a) Sliding on base   (b) Intergranular
                                  adjustments

(c) Phase change   (d) Internal slip planes

(e) Intragranular gliding
    and recrystallization

Figure 20.11 Mechanisms of Glacier Movement. (a) Glaciers may slide over their base; (b) the grains of firn or ice may move relative to one another; (c) some of the ice may melt and refreeze at lower levels; (d and e) the ice may move by shearing along slip planes.

main valley. Waterfalls form where the postglacial streams flow from the hanging valley into the main valley.

Rocks frozen into the ice cause most of the erosion accomplished by glaciers at the contact between glacier ice and the underlying rock. Meltwater seeps into fractures in the bedrock. When the water freezes, it expands, wedging blocks of rock out of place and onto or into the glacier ice. As the glacier moves, the blocks of rock, frozen into the glacier, move with it. Deep within a glacier, the weight of the overlying ice squeezes ice into cracks in the bedrock. Then, blocks of the rock are dislodged as the moving glacier pulls away. Glaciologists call the processes by which a glacier lifts blocks of bedrock out of place **glacial plucking,** or **quarrying.** Because plucking occurs rapidly at the head of valley glaciers, they enlarge the head of the glacier and create huge amphitheater-shaped features called **cirques.** Eventually, the divide between glacial valleys is reduced to a sharp, narrow ridge, known as an **arête.** As parts of the drainage divide are cut to lower elevation, passes called **cols** are created, and finally, isolated steep-sided peaks called **horns** are all that remain of the former drainage divides.

The rocks frozen into the base of glaciers abrade and polish the surface of the underlying bedrock as the glacier moves. This abrasion carves striations (lines) and grooves into the bedrock (see figure 20.13). The grooves remain long after the ice melts away, and they reveal the direction the ice was last moving.

Valley glaciers that flow into the sea cut the floor of the valley below sea level, producing a **fiord** (figure 20.14). Some fiords are more than a thousand meters deep. This depth is due partly to the

The bottom of most erratics bears striations and grooves formed while the block, frozen into the glacier, scraped over bedrock beneath the glacier. These scratches and grooves also occur on bedrock surfaces and indicate the direction the ice was moving and one of the ways glaciers accomplish erosion.

Erosion by Valley Glaciers

Many of the distinctive features found in high mountains are products of erosion by glaciers that carved glacial landforms in former stream valleys (figure 20.13). Many valley glaciers have been retreating for several thousand years. As the glaciers recede, sediments carried by the glacier settle to the valley floor, streams reoccupy the floor of the glaciated valleys, and lakes, called **tarns,** develop in depressions and behind natural dams formed by sediment. Nevertheless, the landforms shaped by glacial erosion remain apparent. As they move from high elevations in mountains, glaciers smooth and streamline the valleys. They straighten the valley and change the common V-shaped profile of stream valleys to large **U-shaped valleys** that offer less resistance to the movement of ice. Tribuary glaciers cannot cut the floor of their valleys as deeply as the main glaciers. Thus, when the ice melts, the tributary valleys, called **hanging valleys,** may remain high above the

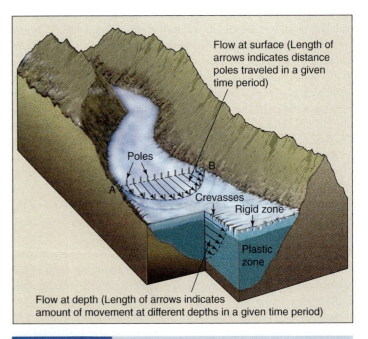

Flow at surface (Length of arrows indicates distance poles traveled in a given time period)

Poles
A   B
Crevasses
Rigid zone
Plastic zone

Flow at depth (Length of arrows indicates amount of movement at different depths in a given time period)

Figure 20.12 Block Diagram of a Valley Glacier Showing the Variation in the Rate of Ice Movement in Different Parts of the Glacier.

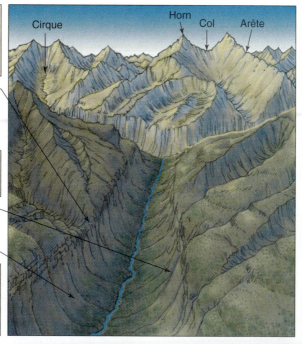

Hanging Valleys When the glaciers recede, the valleys of the tributary glaciers open into the large valley of the main glacier high above the valley floor, forming hanging valleys.

Hanging valley —

Truncated spurs are steep rock walls formed along the sides of the valley between tributaries to the main glacier.

Truncated spurs —

U-shaped valley —

U-shaped Valley Valley glaciers streamline the valleys through which they flow. Gradually, glacial erosion transforms the typical v-shaped stream valley profile into a U-shape.

Cirque Horn Col Arête

Cirques are amphitheater-shaped features hollowed out near mountain peaks by erosion at the head of valley glaciers. Rock walls produced by glacial quarrying action become precipitous before curving onto the floor of the valley.

Horns are the sharp, isolated peaks formed where several cirques join; erosion of the cirque walls produces a sharp isolated peak.

Cols are narrow, sharp-edged passes formed where the cirques intersect and lower the ridge by headward erosion.

Arêtes are sharp ridges formed along drainage divides and on the ridges that separate adjacent glacial valleys.

(a)

(b)

(c)

Figure 20.13 **Valley-Glacial Erosion.** (a) Landforms developed as a result of glacial erosion in valleys that have been occupied by glaciers. (b) Movement of the glacier along this rock wall polished the surface and caused scratches that indicate the direction of movement of the ice. (c) Photograph of horns, including the Matterhorn, that stand in the high, internal part of the Swiss and French alps.

rise in sea level since the last major glacial advance and partly to the glacier's cutting its valley below sea level. If the ice moves fast enough, the glacier is able to carve its valley below sea level before the ice melts in the seawater or breaks away to form icebergs. Because ice floats on seawater, glaciers are usually unable to cut a valley far offshore. The coasts of Norway, Greenland, British Columbia, Patagonia, and southwest New Zealand are famous for their spectacular fiords.

Erosion by Ice Sheets Like valley glaciers, ice sheets also carry most of their load near the base of the ice. Once the sheet has become thick enough to flow over flat areas, its top surface is usually above all but the highest mountain peaks. Consequently, rocks and debris rarely fall on the top of the sheet. The basal load is effective in shaping the land over which the ice moves. Ice sheets flatten, smooth, and streamline preexisting topography, removing many irregularities, such as sharp ridges, and producing streamlined hills called **roches moutonnées.** Many of these hills have smoothed and streamlined surfaces on the side from which the ice came. A steeper and more angular side produced by glacial quarrying faces away from the incoming ice.

Figure 20.14   A Fiord. This photograph depicts a hanging valley that once fed glacier ice into the much larger glacier that carved the fiord in the foreground.

How Glaciers Obtain and Transport Loads

Like bulldozers, the front edge of advancing glaciers scrapes the ground and pushes soil and unconsolidated sediment ahead of the advancing ice. As the quantity of debris along the margin increases, the ice usually overrides this debris and drags big blocks of rock under the ice. Eventually, much of the debris under the advancing glacier freezes into the ice. This load may include all of the rock types that crop out in the terrain over which the glacier has moved. Soil, alluvium, and any other unconsolidated materials encountered by the moving ice are incorporated into it, along with solid rock plucked from bedrock and materials produced by the grinding and abrasion of bedrock. Much of this load moves with the ice until the ice wastes away. Valley glaciers also carry substantial quantities of load on top of the ice. This portion of the load comes mainly from the sides of the valley.

Removal of rock from the sides of the valleys by glaciers steepens the slopes and accentuates peaks. As a glacier undercuts the sides of its valley, the slope of the valley walls increases, and the slopes become unstable. Slumps and landslides, as well as talus, snow, and debris avalanches, dump great quantities of debris onto the top surface of the glacier. Most of this load lies on the surface of the ice where it forms ridges along both sides of valley glaciers (figure 20.15). Where two valley glaciers converge, the ridges of debris come together to form a single ridge that lies near the center

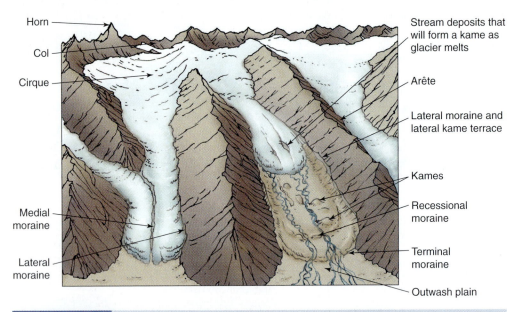

Figure 20.15   Deposition of Sediment from Valley Glaciers. Sediment carried by glaciers is deposited at the end of the glacier, along the margins of the ice, and across the floor of the valley. The terminal moraine is located at the most advanced position of the ice. Recessional moraines form where the end of the ice remains stationary long enough for an accumulation to form. Lateral moraines and terrace deposits of sediments called kame terraces form at the former edge of the ice. Ridges of sediment may also develop in the center of the valley where lateral deposits came together where tributary glaciers came together. Small hills of sediment, called kames, are left where sediment had accumulated in depressions on the surface of the glacier.

Cirque

Medial moraine

Lateral moraine

Figure 20.16 Aerial View of the Yenta Glacier in the Alaska Range. The dark strips of debris formed on top of the glacier where tributary glaciers merge with the main glacier will become medial moraines when the ice melts.

of the glacier (figure 20.16). After the glacier has wasted, a ridge of debris is left in the middle of the valley.

Some debris may remain on top of the glacier all the way from the head to the terminus of a valley glacier, but much of it slowly works its way toward the bottom of the ice. Debris on top of the glacier falls into crevasses that open as the ice moves over topographic irregularities. This debris increases the load frozen into the bottom of valley glaciers.

Glacial Deposits

All glacial deposits are called **glacial drift.** While the ice is moving, glaciers continue to carry most of their load to the end of the ice where it accumulates on the ground. The resulting heterogeneous mixture of material, called **till,** contains blocks of bedrock

plucked by the glacier and mixed with sand, gravel, and silt. Till may also contain some poorly stratified sediment, but the deposits contain a variety of different sizes of material. Some till forms as the glacier presses, crushes, and plasters glacial debris transported near the base of the ice to the surface under the glacier. However, most till forms where debris transported within or on the ice sinks to the ground as the glacier wastes away. The resulting deposit is loose, shows no signs of pressure effects, and is called **ablation till.**

Glaciers waste away through the combined effects of melting and sublimation—change of state directly from solid to gas. As the ice melts, the load carried by the glacier drops to the ground and leaves distinctive features identified in figures 20.15 and 20.17. When a glacier stops advancing, rapid deposition occurs at the terminus of the glacier, where the forward movement of ice is

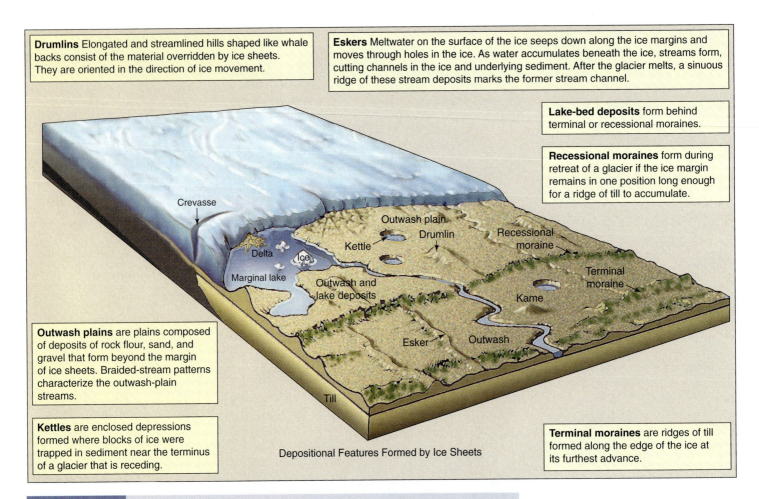

Drumlins Elongated and streamlined hills shaped like whale backs consist of the material overridden by ice sheets. They are oriented in the direction of ice movement.

Eskers Meltwater on the surface of the ice seeps down along the ice margins and moves through holes in the ice. As water accumulates beneath the ice, streams form, cutting channels in the ice and underlying sediment. After the glacier melts, a sinuous ridge of these stream deposits marks the former stream channel.

Lake-bed deposits form behind terminal or recessional moraines.

Recessional moraines form during retreat of a glacier if the ice margin remains in one position long enough for a ridge of till to accumulate.

Outwash plains are plains composed of deposits of rock flour, sand, and gravel that form beyond the margin of ice sheets. Braided-stream patterns characterize the outwash-plain streams.

Kettles are enclosed depressions formed where blocks of ice were trapped in sediment near the terminus of a glacier that is receding.

Terminal moraines are ridges of till formed along the edge of the ice at its furthest advance.

Crevasse

Outwash plain
Drumlin
Recessional moraine
Kettle
Delta
Ice
Marginal lake
Outwash and lake deposits
Terminal moraine
Kame
Esker
Outwash
Till

Depositional Features Formed by Ice Sheets

Figure 20.17 Depositional Glacial Features. These features include recessional and terminal moraines, kettles, outwash plains, eskers, and lake-bed deposits formed by deposition of sediment from an ice sheet. Drumlins result from erosion by the ice sheet. Ice sheets also tend to streamline the topography, remove soil, and scour the bedrock over which they move.

balanced by wastage. This balance between the loss of ice and the advance of more ice determines the position of the terminal edge of the glacier. The terminus may move forward even though ice is continually melting along this leading edge, but if the amount of melting and sublimation is equal to the amount of ice moving to the end of the glacier, the terminus remains fixed in one position. If the terminus remains fixed in position, a ridge of material deposited from the melting ice accumulates, forming a deposit. When the rate at which the ice melts and sublimates exceeds the rate of ice movement, the terminus of the glacier recedes, leaving behind the variety of deposits shown in figures 20.15 and 20.17. These deposits include **moraines,** distinct accumulations composed mainly of unsorted, unstratified glacial till. Many moraines are ridges formed at the terminus of valley glaciers and ice sheets, at places where the glacier paused during recession, along the side of the valley glaciers, or in the middle of the valley. These are defined in figures 20.15 and 20.17.

Streams issue from beneath most glaciers, and they also form along the ice margin when the ice is melting. Deposits from these waters, which consist mainly of sand and silt, are known as **glaciofluvial drift.** The water in streams of glacier meltwater has a characteristic milky color caused by powdery, silt-sized material, called **rock flour,** that is carried in suspension. This silt forms as rocks frozen into the base of the ice grind against the bedrock beneath the base of the ice. Unlike most sediments deposited from streams, these sediments may contain fragments of minerals such as hornblende or feldspar that ordinarily decompose in water. They survive in glacial deposits because they have not been subjected to chemical weathering. The presence of fresh little-decomposed fragments helps distinguish sediments of glacial origin from sediments derived from weathered rock. Streams may carry rock flour and other coarser sediment far beyond the margin of the ice, where the silt gradually settles in lakes or in alluvial deposits to become part of what is called **outwash** sediment (see figure 20.17).

BEYOND THE ICE MARGIN

The environmental impacts of continental-sized ice sheets extend far beyond the edge of the ice. The climatic conditions that create ice also produce permanently frozen ground and change the environment so drastically that plants and animals must either adapt to the cold or move to warmer climates. Melting ice and precipitation produces streams that flow away from the glaciers, carrying large quantities of fine-grained sediment. Where the quantity of sediment exceeds the capacity of the streams, sediment settles out and forms vast deposits beyond the ice margin. Plains formed in this way are called **outwash plains.** Strong winds blowing off the ice pick up the finest portion of this sediment and spread it far from the ice. *Loess,* discussed in chapter 17, is the most widespread of these deposits. Another widespread but quite different environment, frozen ground, covers vast areas at high latitudes.

Permafrost

Today, nearly 20% of the land area of the world has permanently frozen ground, called **permafrost** (figure 20.18). This includes all of Antarctica and most of the land above latitude 50° N, including 85% of Alaska. Continuous, thick layers of permafrost underlie some areas. Four hundred meters (1,300 feet) of permafrost exist near Barrow, Alaska, and frozen ground extends to depths of 600 meters (2,000 feet) on Melville Island, Northwest Territories, Canada. Farther south, permafrost is thin and discontinuous or occurs as isolated masses.

Above the Arctic Circle, large areas of thick, permanently frozen ground develop where the air temperature rarely rises above freezing. Most areas inhabited by people have seasonal temperature variations, with temperature rising above freezing in summer. When the top of the frozen ground thaws, water saturates the ground. In poorly drained areas, the soil turns into a mud-like mass that flows downslope. Because the ground tends to subside where the ice melts, serious foundation problems exist in areas of permafrost. American armed forces encountered these foundation problems when they first began to construct buildings on permafrost. Heat from the buildings melted the ice under the building, and the structure subsided. Differential subsidence also produces problems for railroads, highways, buildings, pipelines, and airstrips. In some areas, even removal of vegetation that acts as a thermal cover is sufficient to trigger melting.

In areas of permafrost, polygonal-shaped patterns form when the soil contracts during the cold months (figure 20.19). In polar regions, wedges of ice that have filled deep cracks usually underlie the edges of these polygons. The repeated freezing and thawing of this ice gradually forces rocks in the soil to the surface where they form polygonal-shaped rock patterns.

Unfrozen water may be present below permafrost, within the permafrost, or above the frozen ground. Where water is present below the ground surface, small, dome-shaped hills 40 to 100 meters (130 to 300 feet) in diameter and sometimes over 33 meters (100 feet) high, called **pingos,** may form. These mounds develop

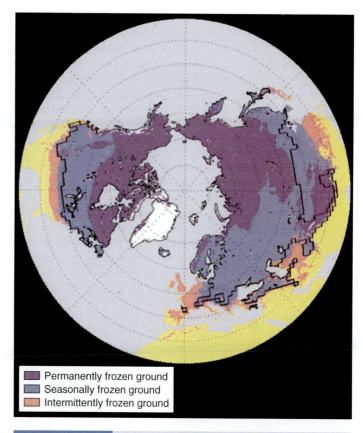

Permanently frozen ground
Seasonally frozen ground
Intermittently frozen ground

Figure 20.18 Regions of Modern Permafrost and the Extent of Ground That Is Seasonally Frozen or Intermittently Frozen at the Present Time.

as lens-shaped masses of ice grow within or under permafrost. Pingos can originate in several ways. Some form where ground water moves from high ground into and between layers of frozen ground; others form on the sites of former lakes. The depth to which the ground freezes in winter is usually much greater beneath a stream or a lake where water saturates the ground. As the temperature drops, frozen ground surrounds water-saturated sediments under the lake. Gradually, the ground around the lake freezes progressively closer to the center of the lake, and water moves toward former deep spots in the lake where it freezes, expands, and forces the lake sediments upward. If this process continues, all of the lake water will freeze, and the pingo will have a solid ice core. Most of the lakes that give rise to pingos are small bodies of water, many of which form in moraines. Glaciers may produce much larger lakes and cause dramatic changes in stream drainage near the ice.

Drainage Disruption and Glacial Lakes

Thousands of lakes, great and small, mark the landscape of areas that large ice sheets formerly covered. These areas have sluggish drainage because deposits dropped from the ice as it melted form a

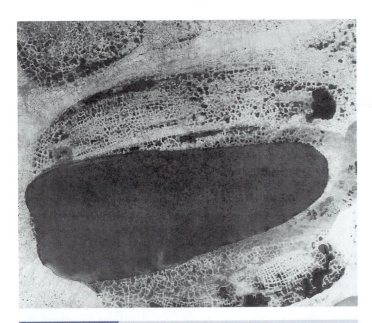

the front edge of the ice to move up the valleys that drained toward the north. The ice gouged depressions in the valleys before it rose high enough to move across the top of the Catskills. Striations scratched on the bedrock still indicate the direction of movement, and erratics picked up by the glacier much farther north remain where they were dropped as the ice wasted.

Before glaciation, a river known as the Teays River was the main stream south of the Great Lakes. Today, most of the channel of this once-mighty river lies buried beneath glacial deposits. The Teays had its headwaters in the Appalachian highlands of North Carolina. It followed the present-day course of the New River in Virginia and West Virginia, but in southern Ohio, its course departed from modern drainage paths, and it flowed north across Ohio, Indiana, and Illinois. There it flowed into an arm of the Gulf of Mexico that extended up the Mississippi Valley into southern Illinois. Since the last glaciation of the region, isostatic rebound of the crust has caused this area once covered by ice to rise. This uplift tilted moraines and lake deposits and redirected drainage to their present courses.

blanket of irregular thickness. Toward the later stages of wastage, sediments surround and encase large chunks of ice that eventually melt, leaving depressions called **kettles.** Where the water table is high, some of these depressions contain water. Lakes such as Lake Agassiz in Canada, which originally covered a larger area than the present Great Lakes, formed as the retreating ice left natural dams in the form of **recessional moraines.**

During the Pleistocene epoch, ice repeatedly covered parts of the northern United States, including the area occupied by the Great Lakes. Before the first glacial advance, the region was upland of low relief located on the edge of the Canadian Shield (see chapters 3 and 5). Because it contains rocks that are slightly less resistant to erosion than surrounding areas, the area now occupied by Lake Superior was already topographically low, perhaps a broad valley. As the ice sheet formed and pushed south, ice stripped the soils and gouged out less-resistant rocks. The weight of the ice in central Canada caused the lithosphere to subside and tilted the area of the Great Lakes slightly northward toward the advancing ice. The ice sheets melted during each interglacial period, and lakes formed in the depressions and behind dams of moraine (figure 20.20). Each time, the lithosphere rebounded, just as it is doing today, and the lakes gradually drained. These older lakes occupied the same general areas as the modern lakes, but they had somewhat different shapes. The advance and retreat of ice sheets and the concurrent tilting of the lithosphere had dramatic effects on the drainage of areas around the ice.

The Finger Lakes of New York are also products of glacial erosion. As ice sheets moved south from Canada, they encountered the Catskill Mountains. Weight of the high parts of the ice sheet forced

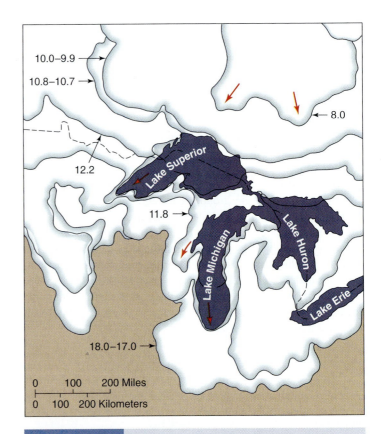

Glacial Lakes Beyond the Ice Margin

Even during the advance of ice sheets, increased precipitation and water released from melting ice during summer months accumulated in lakes beyond the margin of the ice. Many of these lakes formed in regions such as the western United States, northern Africa and the Kalahari, Arabia, and the Atacama and Patagonian deserts in South America. In the western United States, many of these lakes formed in depressions created by crustal deformation that had dropped large blocks of the crust. Large areas around the lakes were moist, perhaps fertile, and well-watered during times of glacial advances.

The Great Salt Lake is a remnant of one of the lakes that formed beyond the ice margin. The original lake is known as Lake Bonneville; it covered about 80,000 square kilometers (31,000 square miles) in Utah, Nevada, and southern Idaho. A second somewhat smaller lake, named Lake Missoula, formed behind an ice dam and occupied valleys in western Montana.

Lake Bonneville arose west of the Wasatch Mountains in the Basin and Range Province (see figure 14.3). Streams in this region flow into enclosed basins where water accumulates in temporary lakes until it evaporates. As the last major advance of ice sheets in North America ended, the reduced rate of evaporation, increased precipitation, and melting ice fed rivers flowing into both Lake Bonneville and Lake Missoula. Since the lakes had no natural outlet, a huge amount of water accumulated in the enclosed basin.

Approximately 14,500 years ago, the level of Lake Bonneville reached the level of the drainage divide between it and the Snake River in Idaho (figure 20.21). When this happened, water flowed over the divide and down the steep slope at the divide and into the Snake River. Water flowing over the outlet quickly cut into the underlying sediment and rock and created a deep gorge located at Red Rock Pass in Idaho. In the ensuing flood, about 4,750 cubic kilometers (1,135 cubic miles) of water rushed through the gorge and down the Snake and Columbia rivers to the Pacific Ocean. This was a flood of such gigantic proportions that features in the landscape formed during the flood have survived the more normal erosion that has taken place since. Among these features are abandoned channels suitable for much larger streams than the modern Snake River, abandoned giant waterfalls, and boulders that moved along these channels. Some of these boulders are 10 meters (33 feet) in diameter—far too large for the modern Snake or Columbia rivers to move.

Another flood resulted from the catastrophic drainage of glacial Lake Missoula. This lake developed behind a natural dam created by large ice lobes. Glacial Lake Missoula found its outlet in northern Idaho. As water topped the ice lobes, the rushing water quickly destroyed the dam. During the flood that followed its release, water rushed westward down steep slopes into eastern Washington (figure 20.22). This portion of Washington is part of the Columbia River Plateau, an area underlain by lava flows that had spread over the region and filled depressions millions of years earlier. Ultimately, the lava flows created a vast flat surface. As the Missoula floodwaters reached this surface, they rushed across the top of these flows, cutting channels into the basalt. No single channel contained all of the water. Instead, a broad network of channels developed in the basalt as the water scoured first one, then another, section of eastern Washington. When the waters receded, they left an unusual topography characterized by large abandoned channels cut in bedrock, large stream valleys in which small streams flow, and oversized waterfalls over which little water flows. Geologists refer to this region as the **channeled scablands** (figure 20.23). It is a monument to the power of running water and its effectiveness as an agent that shapes the land surface. Ice sheets of continental dimensions also have great impact on the ecology of the regions beyond the margin of the ice.

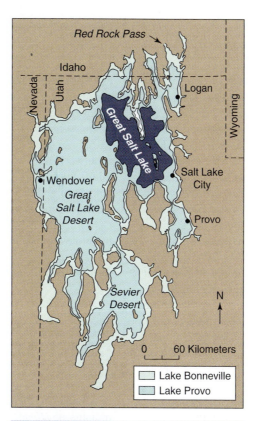

Figure 20.21 Map Showing the Maximum Extent of Lakes Bonneville, Provo, and the Great Salt Lake.

Effects of the Ice Ages on Plants and Animals

Temperature and moisture conditions are among the most important factors that determine plant diversity and where different species of plants thrive. Tropical rain forests contain the greatest diversity of plants on Earth. In general, plant diversity decreases from the equator to the poles. Trees disappear from the landscape in the far north, where lack of sunlight and low temperatures create an inhospitable environment. A similar effect accounts for the changes in flora with elevation. At low elevation, a great variety of plants are generally present, but the number declines with elevation. Eventually, the temperature is unfavorable for tree growth.

Plants provide a guide to local climates. Paleobotanists use this principle to reconstruct past climatic conditions by using plant fossils and pollen. The wind distributes pollen over the area immediately surrounding plants, and some of that pollen falls into ponds and lakes where it is buried along with sediment. By studying pollen in sediment in glacial lakes, scientists can determine what plants lived nearby. Cores

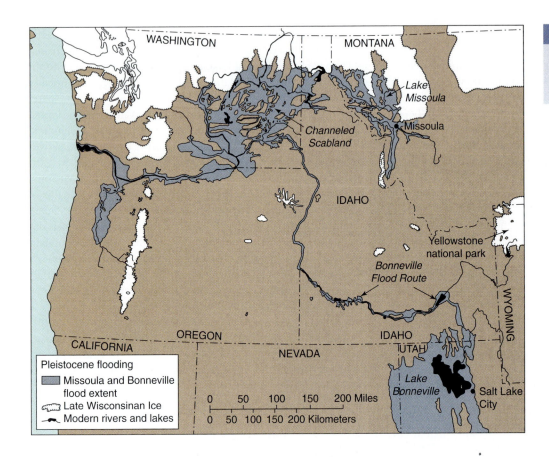

Figure 20.22 Areas of the Northwest Affected by the Floods Following the Drainage of Glacial Lake Missoula and Lake Bonneville.

Map labels:

WASHINGTON
MONTANA
Lake Missoula
Missoula
Channeled Scabland
IDAHO
Yellowstone national park
Bonneville Flood Route
WYOMING
OREGON
CALIFORNIA
IDAHO
UTAH
NEVADA
Lake Bonneville
Salt Lake City

Pleistocene flooding
Missoula and Bonneville flood extent
Late Wisconsinan Ice
Modern rivers and lakes

0   50   100   150   200 Miles
0   50 100 150 200 Kilometers

taken from these lake sediments provide a record of changes that have taken place in the plant populations around the lakes. The pollen at different levels in these cores reveals the shifting geographic distribution of different species of plants during the ice ages. In lakes that freeze over during the cold months, ice reduces the sediment supply, and a thin, dark layer of fine sediments settles to the lake bottoms. In warmer months, the surface of the lakes thaws, and streams bring in coarse sediment along with the pollen. The two new layers of sediment deposited each year form **varves** (see figure 4.22). By counting the varves in modern lakes, scientists can determine the age of each layer. This type of chronology provides even greater precision in dating than does carbon-14 dating. Based on pollen analysis of lake-bottom cores, it is evident that certain plants multiply in response to changes in climate. For example, as temperature drops, hemlock, pine, and spruce forests cover larger areas. If moisture levels rise at the same time, beeches also prosper. A change of only 1°C (1.8°F) in average summer temperature over a long period is enough to have a pronounced effect on vegetation. Consequently, changes in vegetation are sensitive guides to changing climatic conditions.

Although it is clear that animals also move in response to a changing climate, their movements are more difficult to trace than the shifting of plants. Animals, particularly land animals, move rapidly and leave few traces. Fish and marine mammals also move long distances. Marine organisms that live on the bottom or are free-floating provide more valuable clues to water tempera-

ture. Water temperature in the range of 4° to 7°C (39° to 45°F) is optimal for cod to multiply. Below 2°C (36°F), their populations begin to drop. Records of the cod fisheries of countries in the North Atlantic tell us the successes and failures of fishing for cod. During the period from 1625 to 1629 and again in 1675, the cod fisheries in Iceland failed for lack of cod. These years coincide with severe cold spells in Iceland and northern Europe. Water temperature also affects many species of animals in the ocean, including microorganisms such as protozoans. Oceanographers have long used the types of protozoans present in cores to determine the approximate water temperature of the ocean. From protozoans to humans, changing climates affect the geographic distribution of life.

THE EFFECTS OF CONTINENTAL GLACIATION ON HUMAN HISTORY

Sediments of Pleistocene age contain the oldest known fossil remains of *Homo sapiens*. These people were hunters. At times of glacial advances, when ice covered much of the Northern Hemisphere, these early humans remained in low latitudes, but as the ice retreated northward, they followed the herds of wild animals on which they depended. We know of their movements because

they left behind weapons made of the sharpened stones used in hunting.

By the end of the last glacial advance, sea level had dropped far below its present level, leaving bridges of land that connected many of the islands of the southwest Pacific with southeast Asia. Other land bridges connected Asia with North America across the Bering Strait. By that time, early human hunters had made their way into Mongolia and northern Siberia. Sometime while these land bridges remained above water, hunters began to cross from Asia into North America. Some anthropologists think this migration began between 11,000 and 12,000 years ago; however, a few archeological sites appear to be much older. People who may have followed the same path during an earlier glacial advance inhabited these sites. From the Bering Strait, the trail of migration extended across the mountains of Alaska and down along the eastern side of the Canadian Rocky Mountains. In the Great Plains, the immigrants found huge herds of animals. It is not clear why the immigrants didn't stop where they found this great food supply. Perhaps it was too cold. Some anthropologists think the hunters had to keep moving south to find even more game. These early times in human history are poorly documented, but more-recent written histories confirm what is known as the Little Ice Age and the dramatic effects climatic change can have on humans.

A vivid impression of climatic change comes from records compiled from the thirteenth to the nineteenth century. Iceland had been settled between A.D. 870 and 930, long before the thirteenth century. During this early part of its history, Icelanders grew grain crops, made fishing trips to Greenland, and maintained commerce with Norway. During these times, vineyards in Great Britain prospered 480 kilometers (300 miles) north of their present limits, but after A.D. 1200, sea ice and storms began to interfere with passages

between Greenland, Iceland, and Norway. After 1347, voyages between Norway and Iceland stopped for many years. As the winters became colder and longer, grain no longer ripened in Iceland, and fish began to migrate away from the area. Later in the 1300s, a succession of disastrous harvests caused repeated famines in England and Iceland. Glaciers in mountainous areas in Europe advanced, streams flooded, and landslides and avalanches triggered by more snow affected the mountains. Later in the fourteenth century, climatic conditions moderated in northern Europe; this period of moderation in the climate was a pause between severe climatic conditions.

From about 1550 until 1800, extremely cold conditions periodically returned to Europe and Iceland. Glaciers advanced from 1750 to 1760 and again about a hundred years later, just before the American Civil War began. Then, suddenly in the twentieth century, the quantity of sea ice in the North Atlantic near Iceland began to decrease, and glaciers throughout the region and in most other parts of the world began a retreat that has continued to the present.

Rapid retreat of valley glaciers that extend into the ocean along the coast of Alaska is an example of their widespread disappearance. During the past 400 years, most of the fiords of Alaska were occupied by glaciers that extended to coastal positions. Rapid calving—blocks breaking off the end of a glacier—has caused glaciers to retreat dramatically. Despite flow rates of 30 meters (100 feet) per day, the Columbia Glacier has retreated 33 kilometers (about 21 miles) in twenty years.

With one continent covered by ice and the Arctic Ocean frozen over each winter, it is reasonable to argue that the present is part of the Ice Age. The future of this ice is highly uncertain. However, it is now clear that climate and the growth and decay of ice on Earth are closely related to processes that take place in the oceans.

WHAT CAUSES ICE AGES?

Based on the geologic record, the presence of sheets of ice big enough to cover continents is rare. Although an ice sheet has covered the eastern part of Antarctica for over 30 million years, modern ice sheets began to form in the Northern Hemisphere less than 2 million years ago. The time span between this most recent ice age and the previous major glaciation is about 200 million years. Another 200-million-year gap separates the late Paleozoic ice sheets from those that were present in the early part of the Paleozoic. The conditions needed for formation of ice sheets include several factors. First, land areas must be located near the polar regions. High levels of precipitation of snow are needed on these land areas, and temperature must be low enough for precipitation to fall as snow rather than rain. Although many ideas have been considered, the currently favored explanation for decreasing global temperatures sufficiently is that suggested by James Croll in 1875 and used by the Yugoslavian geophysicist Milutin Milankovitch to explain the glacial-interglacial cycle (see chapter 13 for discussion of the Milankovitch cycles). He pointed out that the amount of solar radiation reaching Earth's surface varies from time to time as a result of several periodic changes in the inclination of the axis of rotation, the shape of Earth's orbit around the Sun, and a wobble of Earth's axis. Conditions favorable for colder temperatures peak when all three of these effects combine.

SUMMARY POINTS

1. During periods of glaciation, large masses of ice accumulate as valley glaciers or as large ice sheets. Modern valley glaciers form in high mountains at all latitudes, but ice sheets exist only in the polar regions. The Antarctic ice sheet is about 13 million square kilometers in area and probably about 4,000 meters thick. This sheet contains most of the ice on Earth. Where they meet the sea, ice shelves and pack ice extend out into the water. In the Northern Hemisphere, ice sheets exist on Greenland and Iceland, but the entire Arctic Ocean freezes during the winter months.

2. Glaciers grow and advance when the amount of ice formed from precipitation exceeds the losses caused by melting, sublimation, and the formation of icebergs. In the Arctic, sea ice is becoming thinner.

3. Snow that persists from one year to the next gradually compacts and recrystallizes under the weight of water. At first, snow changes into firn. As the firn compacts and recrystallizes, it changes into the solid ice of the main body of the glacier. Slippage and shear within crystals and recrystallization of the ice cause glaciers to flow. Valley glaciers flow downhill through existing stream beds. Ice sheets flow outward under the pressure of the overlying mass on the lowest layers. The upper part of a glacier is under less pressure and is also colder. Consequently, it is more brittle than the base and may crack to form crevasses as the ice mass flows.

4. Because of friction with the valley floor and walls, a valley glacier flows most rapidly near the center of its surface. Along the sides and bottom, it picks up and carries blocks of bedrock, abrading both the valley and its rock load as the glacier moves forward. The striations formed on bedrock indicate the direction of flow.

5. Valley glaciers produce steep U-shaped valleys with cirques at their heads. After the glacier melts, a small lake, a tarn, remains in the cirque. Truncated spurs and hanging valleys form where tributary glaciers enter larger glaciers. Sharp ridges called arêtes form between tributaries, and where several glaciers head around a single peak, they form a horn.

6. Glacial deposits are of two kinds: glaciofluvial deposits (deposited from water) and till (material carried by the ice). Till is a poorly sorted mixture of soil, stream deposits, and bedrock left under the glacier or dropped as the ice melts. These deposits form moraines. Lateral moraines, consisting of debris that falls from steepened valley walls, form along the sides of valley glaciers (see figure 20.15). Where two valley glaciers flow together, the lateral moraines may come together, producing a medial moraine (see figure 20.16).

7. Streams of meltwater under glaciers form **eskers** (see figure 20.17). Where streams leave glaciers, they may form alluvial fans. Small hills form as sediment that accumulated in depressions on the ice settles to the ground.

8. Although ice ages are relatively rare and relatively short-lived events, geologists have found evidence of several periods of glaciation. Many of the most prominent topographic features on Earth formed during the most recent Ice Age, which involved several major advances of ice in the Northern Hemisphere. The most recent of these advances, the Wisconsin glaciation, culminated about 18,000 years ago. This glaciation left behind the Great Lakes and drastically altered the drainage pattern for the continent.

REVIEW QUESTIONS

1. In what ways does the presence of large masses of ice tend to affect climate?
2. How does the cause of movement of ice in an ice sheet differ from that in a valley glacier?
3. What distinctive landforms result from erosion by valley glaciers?
4. What distinctive landforms result from erosion by ice sheets?

5. How does glacial till deposit differ from stream deposits?
6. How does the windblown silt deposited far from ice sheets originate?
7. Briefly outline the origin of the Great Lakes.
8. How did the channeled scablands form?
9. What are land bridges, and what role have they played in human history?

THINKING CRITICALLY

1. What would be the environmental impacts of rapid melting of the ice sheets? How might the area where you live be affected by this event?
2. Can humans control advancing ice sheets? If so, how? What types of costs would be involved?
3. Many mountains contain a mixture of landforms resulting from both glacial and stream erosion. How can one distinguish landforms caused by glaciers from those caused by streams?
4. Discuss the impacts of glacial advance and retreat on human history. Consider how drastic temperature changes would affect everyday life. How might climate have influenced the outcome of historic events, such as World War I?
5. Discuss how the landscape in the area where you live, or other areas that you know, has been affected by glacial ice. What clues does the landscape provide?

KEY WORDS

ablation till	moraine
arête	neve
channeled scablands	outwash
cirque	outwash plains
col	pack ice
continental glacier	permafrost
drumlins	piedmont glacier
erratics	pingos
eskers	quarrying
fiord	recessional moraine
firn	roches moutonnées
glacial drift	rock flour
glacial plucking	sea ice
glaciofluvial drift	surge
hanging valley	tarns
horn	terminal moraine
kames	till
kettle	truncated spurs
lake-bed deposits	U-shaped valley
lateral moraine	valley glacier
medial moraine	varve

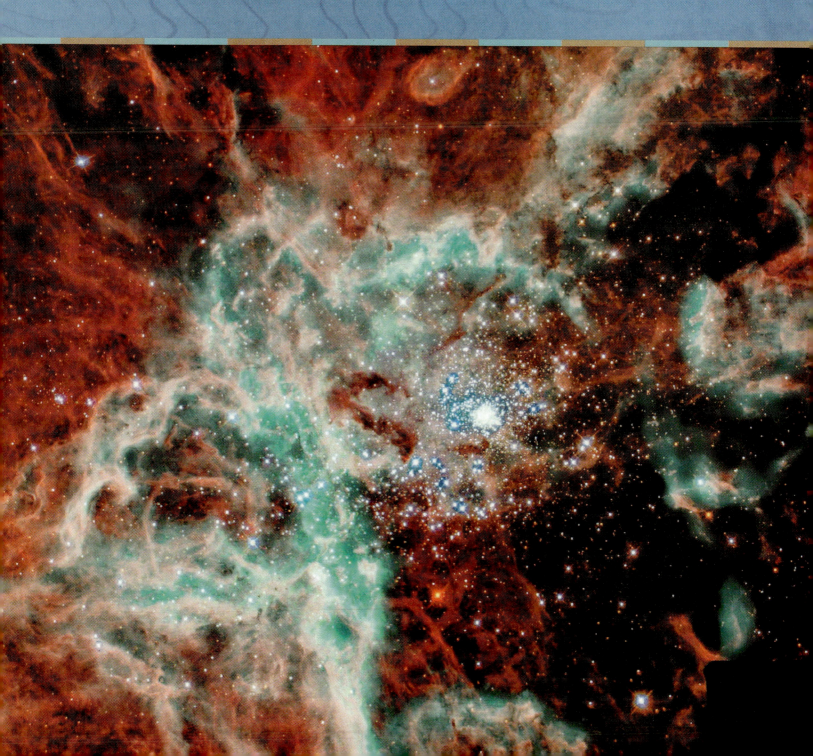

unit four The Solar System and Its Place in the Universe

chapter twenty-one

Earth's Neighbors in Space—The Solar System

Chapter Guide

The idea that Earth revolves around the Sun was suggested as early as 350 B.C., but during the Middle Ages, the idea that Earth is the center of the Universe prevailed. In the fifteenth century, Nicholas Copernicus laid the groundwork for modern ideas about the configuration of the solar system. Giant steps followed the development of telescopes, and finally in the twentieth century, spacecrafts made it possible to collect a few samples and obtain extraordinary images of the planets and the smaller bodies, the satellites or moons, that revolve around them. This chapter explores some of the remarkable discoveries about the Sun, the Moon, and other neighbors of Earth. Visits to the Moon have helped clarify many questions about the earliest history of Earth. Data from other planets have contributed to understanding the origin of the solar system.

An artistic view of the Solar System.

INTRODUCTION

Earth became a planet before the oldest rocks known on Earth originated about 4 billion years ago. Thus, it is necessary to seek clues to the major events that took place before the rock record formed elsewhere in the solar system. The Moon, other planets, or even other systems of stars are sources of this information. It is possible to make some deductions based on the earliest part of the geologic record. Although the evidence is sparse, this intriguing part of Earth history has attracted considerable attention from those who continue to piece together this dim portion of Earth's past.

Scientific evidence suggests that the planets originated between 4.5 and 5 billion years ago from a cloud of gases surrounding the Sun. Evidence of the earliest parts of Earth's history comes from the Moon and other planets. Tectonic activity combined with the effects of geomorphic agents, such as running water, wind, and glaciers, have changed the surface of Earth so drastically that few, if any, features survive from Earth's early history. Extraterrestrial material bombarded Earth, shattering its thin crust and contributing to volcanic activity. As the atmosphere evolved from volcanic gases, rains fell, the oceans developed, and a sedimentary veneer spread over Earth's surface. Growth of this sedimentary cover continues today, as does generation of molten materials and alteration of the crust. Because these processes are still in operation, it is possible to observe them in action and to use what is known about Earth as a way of interpreting the diverse features found on other planets in the solar system and their satellites. The Sun is central to this system and to an understanding of the origin of Earth. Evidence for the origin of the planets must be compatible with knowledge of the Sun.

THE SUN

The Sun is an immense sphere of hydrogen and helium gases held together by gravity. The temperature and pressure in the interior of stars is so high that atoms break down, forming clouds of incandescent plasma composed of subatomic particles. Like other stars, the Sun has an extremely hot, dense center surrounded by a less-dense outer region, the Sun's atmosphere. In recent years, new information about the Sun has come from a stationary satellite known as the SOHO (Solar and Heliospheric Observatory), which operates far outside Earth's atmosphere where observations are unaffected by the Earth's atmosphere.

The Sun's Atmosphere

The vast envelope of gases, primarily hydrogen and helium (table 21.1), that compose the Sun's atmosphere consists of a series of layers, the outer atmosphere called the **corona,** an inner zone called the **chromosphere,** and the **photosphere,** which forms the visible surface of the Sun (figure 21.1). Gases in the corona are so thin that they are visible from Earth only during total solar eclipses when the Moon blocks out light from the main body of the Sun. Corona temperatures reach over 1,000,000°C (1,800,000°F). Gases in the chromosphere, Sun's inner atmosphere, reach temperatures

in the range of 5,000° to 35,000°C (9,000° to 63,000°F). Large flares of hot gases often extend into the chromosphere and produce the solar wind. The next lower layer, the photosphere, is the source of most of the visible light that comes to Earth. Large, cooler dark spots appear in the photosphere. These are associated with huge magnetic storms that are larger than Earth. The number of sunspots follows an eleven-year cycle. The peaks of these cycles are associated with magnetic storms on Earth and interference with radio signals that pass through the ionosphere. The temperature in the photosphere, about 6,000°C (11,000°F), is still thousands of times lower than the temperature in the Sun's deep interior.

The great temperature difference between the surface and the interior of the Sun causes convection, a slow boiling motion in the gases that accounts for the granular appearance of the Sun's surface (figure 21.2). Great bubbles of hot gases rise toward the surface (figure 21.3). These convection cells lie below the photosphere and extend deep into the Sun's interior. They carry heat outward from the central part of the Sun and may play a role in generating the Sun's magnetic field. Gases flow steadily outward from the center of each cell to the cell boundaries. At greater depth, Sun's interior includes a core in which hydrogen is packed so tightly and the temperature, estimated at about 17,000,000°C (30,000,000°F), is so high that collisions between protons and nuclei produce nuclear fusion reactions in which helium is created. As this energy rises outward from the core, it radiates into the zone of convection currents and, ultimately, away from the Sun.

From Earth, astronomers observe the photosphere, a relatively thin 15,000-kilometer (9,300-mile) layer. During solar eclipses, huge, arch-shaped protrusions called **prominences** are visible. They leave dark spots on the surface of the Sun and extend far out from the surface, at times penetrating through the corona (see figure 21.2). These prominences resemble gigantic atomic explosions that would obliterate Earth if they occurred nearby. They provide a supply of charged particles that come to Earth as the solar wind. These particles enter the upper reaches of Earth's atmosphere and are trapped by Earth's magnetic field. As they

Table 21.1	The Most Abundant Elements in the Sun and Earth

Figures for the Sun are percentage by weight of the Sun. Figures for the Earth are percentage by weight of the Earth.

Element	Sun	Earth
Hydrogen	92	Trace
Helium	7.8	Trace
Oxygen	0.8	30.0
Carbon	0.2	Trace
Iron	0.2	35
Nitrogen	0.1	Trace
Neon	0.1	Trace
Silicon	0.07	15
Magnesium	0.05	13
Sulfur	0.04	2.0
Nickel	Trace	2.4

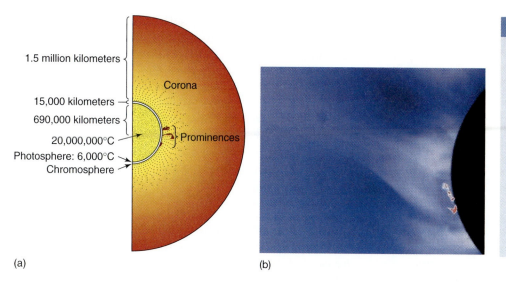

(a)

1.5 million kilometers

15,000 kilometers

690,000 kilometers

20,000,000°C

Photosphere: 6,000°C

Chromosphere

Corona

Prominences

(b)

Figure 21.1 The Sun's Atmosphere. (a) The solar atmosphere consists of three main divisions: the corona, the chromosphere, and the photosphere. (b) Solar flares rise out of the corona. The Sun's corona is a thin gaseous layer that is usually visible only when the moon covers the photosphere during total eclipses. Parts of the corona extend over 1 million kilometers (600,000 miles) above the photosphere. Even in its denser parts, the corona is only about 1 million-millionth of the density of the Earth's atmosphere at sea level.

descend to lower levels, they interact with nitrogen and oxygen ions in the ionosphere and cause auroral displays.

The level of activity on the Sun is of great importance for Earth's environment. The Sun's radiation is the primary source of energy for the physical climate system and for biogeochemical cycles that shape Earth's surface. Until recent years, measurements of solar radiation seemed to indicate that the amount of solar radiation reaching Earth has remained nearly constant. The amount of solar energy striking 1 square meter at the top of Earth's atmosphere is known as the *solar constant*. Improved instrumentation and more careful studies have revealed variations in the solar

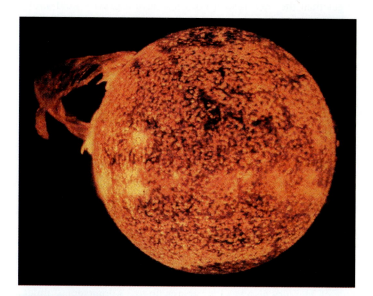

Figure 21.2 Sun Prominences. High temperatures and turbulent motion of gases near the surface of the Sun cause convection cells to form in the Sun's atmosphere. This image was photographed from space using light emitted by ionized helium. The large loop is a prominence rising far out into the Sun's corona. Sunspots are present at both ends of these loops.

constant and in the amount of ultraviolet radiation reaching Earth. The length of time covered by these observations is so short that it is still difficult to evaluate how important the variations may be in changing Earth's climate.

Composition of the Sun

Hydrogen is by far the most abundant element in the Sun. It is about ten times more plentiful than the next most abundant element, helium, which astronomers found in the corona even before chemists discovered it on Earth. These two elements combined constitute well over 99% of the material in the Sun (see table 21.1). Neither free hydrogen nor helium is especially abundant on Earth or on its close neighbors, Mercury, Venus, and Mars. In part, this is because these elements are gases at extremely low temperatures. Helium is still a gas at −269°C (−454°F) and is so light that it escapes from Earth's gravity field. In addition, helium is inert and does not combine with other elements to form compounds. If, as seems likely, the planets and Sun all originated from a single mass of material, why do the planets differ so much from the Sun in composition, especially in the relative amounts of hydrogen and helium?

THE PLANETS

In seeking to explain the origin of the planets, planetary scientists attempt to determine their size, mass, composition, their distance from the Sun, and their motions. The planets closest to the Sun—Mercury, Venus, Earth, and Mars, also known as the terrestrial planets—are much smaller both in mass and in radius than are the giant outer planets, Jupiter, Saturn, Uranus, and Neptune, known as the Jovian planets. Pluto is much smaller than the other planets. It is located a great distance from the Sun, and its orbit around the Sun is quite different.

The spacing of the planets is systematic. In general, each planet is about twice as far from the Sun as the next closer planet, but a gap containing smaller bodies, called **asteroids,** occurs between the

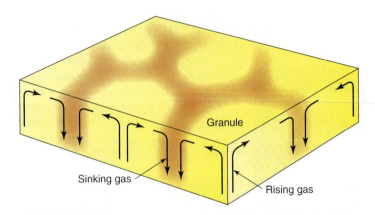

Granule

Sinking gas

Rising gas

terrestrial and Jovian planets. The rotation of planets about their axes is probably inherited from the early stages of the solar system's origin. Although the orbits of Pluto and Mercury are more elliptical, all of the other planets revolve in nearly circular orbits about the Sun and in the same direction that the Sun rotates on its axis. Most of the planets, with the exception of Venus, Uranus, and Pluto, rotate in the same direction about their axes. The arrangement and movement of the planetary satellites resemble those of the planets around the Sun, and most of these satellites lie in or close to the plane in which Earth circles the Sun (figures 21.4 and 21.5).

The terrestrial planets have high density, and presumably they are composed of materials similar to those in Earth. The much larger, less-dense Jovian planets contain large quantities of hydrogen, helium, ammonia, and methane, some of which occur as ices. The planetary atmospheres of the two groups also differ. The terrestrial planets have atmospheres composed of gases, such as nitrogen, oxygen, carbon dioxide, and water vapor. Atmospheres of the Jovian planets contain hydrogen compounds, such as ammonia, and hydrocarbons, such as methane. These compounds and helium are rare in the atmospheres of terrestrial planets.

Water vapor was present on Earth early in its history, but the temperature was too high for water to exist as a liquid. Later, as the temperature dropped, water condensed from the hot gases emitted by volcanoes, and water may have come to Earth as ice in comets that entered the atmosphere and melted. Other planets evolved in different ways. Planets farther away from the Sun than Earth were too cold for water or water vapor to exist. Water on those planets is tied up in ice. Of the four inner planets, Mercury, like the Moon, lacks sufficient mass to hold water vapor in its atmosphere. Venus developed a thick atmosphere composed largely of carbon dioxide and is much too hot for liquid water. Mars, the planet most like Earth, has an ice cap that grows and decays seasonally, and water may rise to the surface at times, but the temperature-pressure conditions do not allow water to remain on the surface for long.

In summary, the bodies in the solar system have a disc-shaped arrangement of Sun, planets, and satellites, in which rotation and revolution are in one direction (see figure 21.4). This arrangement constrains scientific theories about how the solar system originated. Other constraints come from an understanding of the materials that make up the Earth and other planets as well as the bits of matter that come to Earth from space—meteorites.

THE TERRESTRIAL PLANETS

Mercury

Mercury is the closest planet to the Sun. From Earth, Mercury always appears close to the Sun. It is best seen when it rises in the east just before the Sun or when it sets in the west, just after sunset. These observations led the ancient Greeks to believe that Mercury was two different planets. Mercury, named for the Roman messenger god who traveled from Earth to the heavens at great speed, completes its orbit around the Sun faster than any other planet making the trip in eighty-eight days. Its rotation, once every fifty-nine days, is very slow by Earth's twenty-four-hour standard.

Proximity to the Sun ensures that the surface temperature on the sunny side of Mercury is high, reaching a peak of nearly 370°C (700°F) at midday. The nightside has a low temperature of −180°C (−292°F). Both poles of Mercury may be sites of ice caps, and ice may also exist inside deep impact craters on the surface. If Mercury had an atmosphere, circulation in the atmosphere would regulate these temperature extremes. However, Mercury is only slightly larger than the Moon, so it does not have a gravity field strong enough to hold an atmosphere. The weak gravity field and the high level of solar radiation removed any atmosphere that may have existed on Mercury long ago.

Mercury resembles Earth in several ways. Both are in the same part of the solar system and probably developed in the same way. Although Mercury is the smallest planet, it is almost as dense as Earth. The composition of Mercury also resembles that of Earth, and it has a core composed of iron and a mantle of silicate minerals (figure 21.6). Despite these similarities in composition and internal structure, the surface of Mercury looks much more like that of the Moon than of Earth. Both are pockmarked by impact craters formed early in their history (see figure 21.6a). In the absence of an atmosphere, the surface of Mercury has changed little since the rate of impacts slowed down about 3 billion years ago. The perfectly preserved craters also provide evidence that little tectonic activity has taken place after the craters formed. The largest of these impact features is about 1,300 kilometers (800 miles) in diameter. The surface of Mercury also has long scarps up to 3 kilometers (2 miles) high and 500 kilometers (300 miles) long that appear to be faults.

Venus

As viewed from Earth, a dense, highly reflective cloud cover enshrouds Venus, making the planet appear bright. At twilight, a few hours before the Sun rises or after it sets, Venus appears

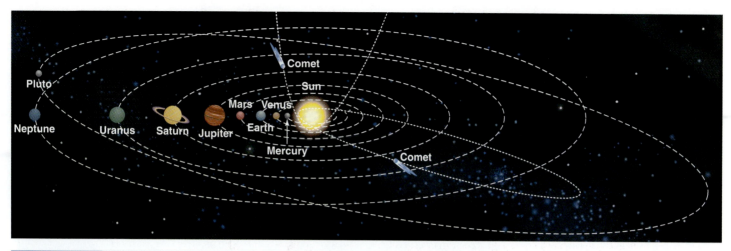

Figure 21.4 Schematic Drawing of the Solar System Showing the Orbits of the Planets Around the Sun.

fifteen times brighter than the brightest star. For this reason, it is called the morning or evening star. Venus is of special interest because it is Earth's closest neighbor and because it is similar in size, mass, and density to Earth. However, its 243-day period of rotation is much longer than Earth's twenty-four hours, and it rotates in the opposite direction from Earth and most other planets. Despite its proximity, nothing is directly visible on its surface because a dense cloud cover entirely obscures it. Consequently, intense interest in getting a closer look developed, and it was the first planet to be explored by rockets and spacecraft. *Mariner II* passed within 39,000 kilometers of Venus in 1962; it was followed in 1967 by the first of several Russian spacecraft that parachuted into the atmosphere and landed on the surface. None of these craft survived more than a few hours in the hot atmosphere on Venus, but in 1979, an American atmospheric probe and an orbiter circled Venus for eight months. Radar images obtained from the spacecraft *Magellan* in the 1990s provide more detailed information about the surface of Venus than is available about most of the ocean basins, which cover nearly 70% of the surface of Earth (figure 21.7).

Vast areas of the surface are flat, but impact craters are present. A trough-like depression, 120 kilometers (74 miles) long and with relief of 3 kilometers (4.8 miles), resembles the valleys of East Africa where grabens formed as the crust of eastern Africa pulled apart. A mountain nearly 250 kilometers (155 miles) in diameter with a central depression nearly 85 kilometers (53 miles) in diameter near its top is of volcanic origin. A number of elongated and nearly parallel mountain ranges with relief of 100 meters (330 feet) to 2 kilometers (1.24 miles) resemble mountain systems like the Alpine-Himalayan chain on Earth. Volcanoes, faults, and mountain ranges on Venus provide proof that geologic processes are active on the planet, but unlike on Earth, these features do not appear to result from plate tectonics.

Although Venus is more like Earth in size than any other planet, it does not have a hospitable environment for humans. The atmosphere of Venus is 97% carbon dioxide and 1 to 3% nitrogen. The carbon dioxide absorbs radiation reradiated from the surface and heats the atmosphere—producing the greenhouse effect—which is responsible for high temperatures between 400° and 500°C (700° and 900°F). A haze near the top of the cloud cover appears to be composed of sulfuric acid. In 1995, the *Hubble* space telescope revealed that the amount of sulfur dioxide in the atmosphere has declined by a factor of 10 since 1978 when the *Pioneer Venus* orbiter recorded its presence. A gigantic volcanic eruption that took place before the earlier observations probably blew the sulfur dioxide into the atmosphere. In the intervening years, the sulfur dioxide presumably settled out. The lack of water and oxygen and high atmospheric pressure, ninety times those on Earth, rule out the possibility that types of life familiar to us will be found on Venus.

Mars

No planet has attracted more popular attention than Mars. This interest stems in part from the long-held hope that some form of life, perhaps even a Martian civilization, might account for the canal-like patterns reported by early astronomers who viewed the planet with telescopes. Modern telescopes demonstrated that the markings thought to be canals are natural features, but Mars has emerged from its obscure image as a planet with a highly varied and interesting surface shaped by processes similar to some of those acting on Earth. Ice caps, deserts, volcanoes, and landforms created by gigantic floods and crustal deformation are among these landforms.

The existence of polar ice caps on Mars was recognized through telescopic observations that revealed white regions near the poles that grow and shrink with a regular seasonal cycle. Water vapor is present in the atmosphere near the poles. Both the amount of carbon dioxide and water vapor in the Martian atmosphere also show seasonal variations; so ices of water and carbon dioxide form the ice sheets. Seasonal changes of color patterns in equatorial

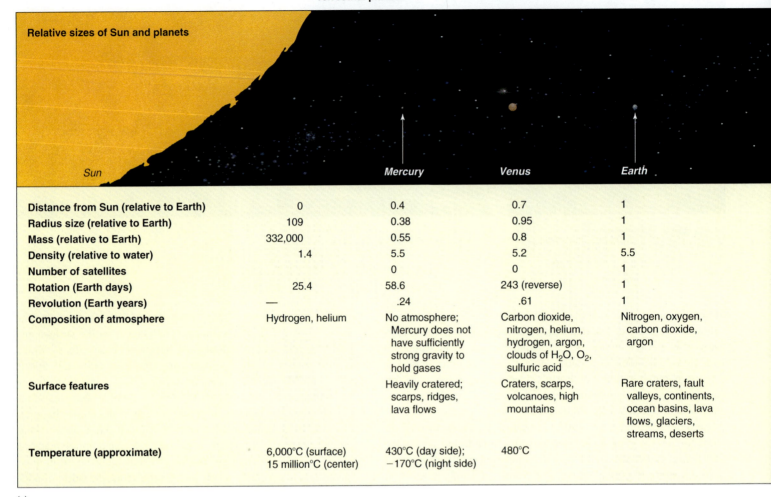

Relative sizes of Sun and planets

	Sun	Mercury	Venus	Earth
Distance from Sun (relative to Earth)	0	0.4	0.7	1
Radius size (relative to Earth)	109	0.38	0.95	1
Mass (relative to Earth)	332,000	0.55	0.8	1
Density (relative to water)	1.4	5.5	5.2	5.5
Number of satellites		0	0	1
Rotation (Earth days)	25.4	58.6	243 (reverse)	1
Revolution (Earth years)	—	.24	.61	1
Composition of atmosphere	Hydrogen, helium	No atmosphere; Mercury does not have sufficiently strong gravity to hold gases	Carbon dioxide, nitrogen, helium, hydrogen, argon, clouds of H_2O, O_2, sulfuric acid	Nitrogen, oxygen, carbon dioxide, argon
Surface features		Heavily cratered; scarps, ridges, lava flows	Craters, scarps, volcanoes, high mountains	Rare craters, fault valleys, continents, ocean basins, lava flows, glaciers, streams, deserts
Temperature (approximate)	6,000°C (surface) 15 million°C (center)	430°C (day side); −170°C (night side)	480°C	

(a)

Figure 21.5 **The Two Groups of Planets.** The planets fall into two groups, (a) an inner group of small high-density planets called terrestrial planets and (b) an outer group of much larger (with the exception of Pluto) giant planets referred to as Jovian planets. A belt of asteroids lies between these two groups of planets.

regions suggested plant growth to some viewers. These darkened areas were initially described as dark green in color, which encouraged the view that these were plants that were growing and spreading south as the ice caps melted in summer and provided an abundance of water. Closer examination of the color changes revealed that they were brown rather than green. They appear to be areas of windblown deposits rather than plants. An orange region near the equator appears to be desert-like; yellowish clouds appear over the desert, and similar clouds covered and obscured the entire surface of the planet during the *Mariner* flyby in 1972. Results of the *Viking* landing mission revealed a desert consisting of undulating topography strewn with rocks of varied size, color, and texture covered with a thin layer of fine dust. Numerous dunes and other forms closely resembling windblown deposits on Earth are present (figure 21.8). Wind action is clearly an important agent that has shaped the surface of Mars.

A giant volcano, Olympus Mons (figure 21.9), stands out as one of the most imposing features not only of the Martian landscape but of the entire solar system. It is about 600 kilometers (370 miles) in diameter and over 19 kilometers (12 miles) high, exceeding in size even the largest volcano in Hawaii, Mauna Loa. Impact craters of all sizes appear on Mars. Vast basins, some of which are almost circular, mark impact sites of large asteroids. Following the impacts, thin but extensive lava flows flooded the impact craters. Younger craters dot both the plains and surrounding regions. Patterns found in the basin floor of the larger craters and on the plains resemble polygonal structures seen on Earth in soils at high latitudes where the ground is permanently frozen, suggesting that the ground on Mars is frozen. Volcanic activity may have caused frozen ground to melt, resulting in what appear to be large river channels. The presence of so much ice on Mars is not surprising because it is much farther away from the Sun than is Earth. This

Jovian planets

	Mars	Asteroid belt	Jupiter	Saturn	Uranus	Neptune	Pluto
							100
	1.5		5.2	9.5	19.2	30	39.4
	0.53		11.23	9.41	4.06	3.88	0.2
	0.11		318	95	15	17	0.002
	3.9		1.4	0.7	1.3	1.6	—
	2		16	18	15	8	1
	1		0.4	0.4	0.45 (reverse)	0.6	6.4
	1.88		11.86	29.46	84	164.8	284.4
	Carbon dioxide, argon		Hydrocarbons, ethane, acetylene	Largely nitrogen	Methane	Methane	Unknown; probably methane and water
	Craters, volcanoes, stream valleys, fault scarps, lava flows		Obscured by clouds	Obscured by clouds	Unknown	Unknown	Unknown
	−50°C (average)		−130°C	−185°C	−200°C	−215°C	−230°C

(b)

ice, as well as that from the polar ice caps, could provide meltwater sufficient to account for other surface markings, which look like stream channels. *Viking* photographs reveal well-developed braid-like channel forms, but none reveal any streams containing water at the present time (see figure 4.24). Liquid water cannot exist on Mars under present temperature and pressure conditions (refer to the discussion of states of water in chapter 1).

Movements within the crust of Mars are suggested by the large valleys that developed as a result of subsidence of large blocks of the Martian crust. The largest of these valleys, Valles Marineris, is nearly 7 kilometers (4 miles) deep and 240 kilometers (150 miles) wide.

Mars has an atmosphere, but its clouds, unlike those surrounding Venus, are not thick and do not generally obscure the entire surface of the planet. With only 1/100 of the density of Earth's atmosphere, the Martian atmosphere is composed largely of carbon dioxide with

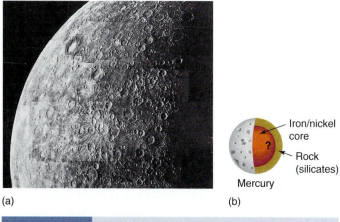

(a)   (b)

Figure 21.6 Mercury. (*a*) Impact craters produce a pock-marked appearance on Mercury. (*b*) The internal structure of Mercury consists of a metallic core surrounded by a silicate mantle.

Iron/nickel core

Rock (silicates)

Mercury

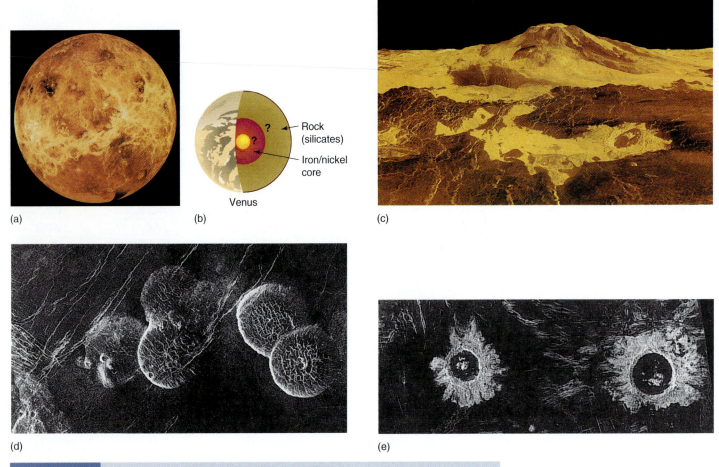

Maat Mons volcano (Vertical scale exaggerated)

(a)
(b)
Rock
(silicates)

Iron/nickel
core

Venus

(c)

(d)
(e)

Figure 21.7 **Venus.** (*a*) The surface of Venus, as revealed by radar imagery, is highly varied. (*b*) The internal structure of Venus resembles that of Earth. (*c*) The volcano Maat Mons on Venus is flanked by lava flows that show as light areas in this image. (*d*) The circular areas on these images are thought to result from rising magma below the crust. As the magma approaches the surface the crust fractures. (*e*) The bright line features in this image are thought to be faults.

the inert gas argon making up the second largest part (possibly 20%). Nitrogen, the most abundant constituent of Earth's atmosphere, is present in small quantities, and minor constituents of Martian atmosphere include oxygen, carbon monoxide, and water vapor. The small amount of oxygen may originate from the breakdown of carbon dioxide of water vapor by ultraviolet radiation. The absence of plants on Mars eliminates the possibility that photosynthesis is a source of oxygen. Clearly, the Martian atmosphere is quite different from Earth's, and this suggests that the two atmospheres evolved in different ways. The atmosphere is probably thin because the planet is less massive and of lower density than Earth and therefore cannot hold gases to its surface as effectively as does Earth.

Life on Mars? Early viewers with telescopes described canals on Mars, which were supposedly built by Martians to move water from polar ice caps to the desert regions in the middle latitudes.

Unfortunately, this romantic notion had to be abandoned in view of observations first made with larger telescopes, later supported by photographs taken during the *Mariner* flyby in 1972, and finally confirmed by the *Viking* landing mission to Mars in 1976 that failed to reveal any sign of life. Hope that life ever existed on Mars faded until 1996 when a panel of scientists at NASA announced the discovery of minute fossil-like forms in a meteorite collected in Antarctica. The isotopic composition of the meteorite is unlike that of materials found on the Moon or in most other meteorites, but the ratio of the nitrogen isotopes and rare gases in this meteorite is the same as that found in Martian rocks. The discovery has proven to be almost as controversial as the idea that canals are present on Mars. Most studies of the rock have concluded that the fossil-like form is not a fossil, but the argument continues and may not be resolved until humans reach the surface of Mars and collect more samples for investigation.

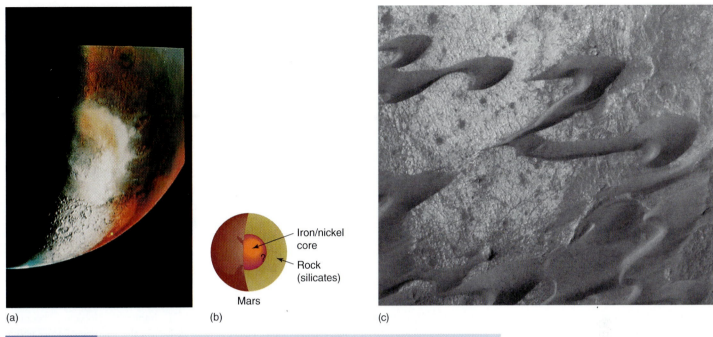

(a) (b) Mars (c)

Iron/nickel core

Rock (silicates)

Figure 21.8 **Mars.** (*a*) The ice fields and impact craters on Mars are clear on this image of the south polar region. (*b*) The internal structure of Mars resembles that of Earth, with an iron/nickel core surrounded by a mantle composed of silicate minerals. (*c*) These sand dunes on Mars resemble the barchan dunes found on Earth. The wind responsible for these dunes is blowing from upper right to lower left. Most of the Martian surface in middle latitudes is covered by rock fragments, sand, and dust.

THE JOVIAN PLANETS

The four giant planets, known as the Jovian planets, have much in common. The external portions of all are gases. They all have satellites, some of which are close in size to the terrestrial planets. All four have ring systems, and all four hold atmospheres that, above the tops of the clouds, are composed of hydrogen and helium or their compounds. In addition, all of the giants have magnetic fields. The magnetic fields trap particles coming to the planets from the Sun, creating radiation belts similar to those on Earth.

The closest look at the Jovian planets came from the unmanned *Voyager II* spacecraft, which left Earth in August of 1977. Twelve years later, after passing all of the giant planets, it came close to Neptune, the most distant of the giant planets. The giant planets are so far away from the Sun that temperature is so extremely low that substances that are liquid on Earth occur as ices on all of the giant planets (figure 21.10). Information received from *Voyager* and more recently from *Galileo,* a spacecraft that began to orbit Jupiter in December of 1995, has provided incredibly sharp images of the giant planets and their satellites.

Figure 21.9 **The Gigantic Martian Volcano, Olympus Mons.** The volcano is about 600 kilometers (370 miles) in diameter and rises nearly 25 kilometers (15 miles) above its surroundings. The surface around the volcano exhibits patterned ground caused by fractures and faults in the planet's crust.

Jupiter

Jupiter is the largest planet in the solar system. Its composition may be close to that of the cloud of gases from which the solar system formed, making it of special interest to scientists. Spectroscopic studies of light reflected from its clouds and direct samples from the *Galileo* probe show that its upper atmosphere is composed primarily of hydrogen and helium. Many anticipated that the abundance of hydrogen and helium would be approximately the same relative abundance as in the Sun. However, somewhat less helium is present in the atmospheres of Jupiter and Saturn than originally expected. This proves that these giants have changed since they first formed. In addition to hydrogen and helium, ammonia, water vapor, and methane are present in the atmosphere. The abundance of hydrogen and helium accounts for Jupiter's low density—a property it shares with the more distant planets Saturn and Neptune. Despite its low density (1.3 grams per cubic centimeter; density of water is 1 gram per cubic centimeter), it contains nearly two and a half times the combined mass of all the other planets, and its volume is about 40% greater than that of all the other planets combined.

The atmosphere is organized into zones that run parallel to the equator. These prominent zones on the exposed part of Jupiter's atmosphere contain huge oval storms such as the Great Red Spot (figure 21.11*a*) that are moving in the planetary circulation. The Great Red Spot, which is larger than Earth, is one of a number of distinctive and remarkable color markings on the surface of Jupiter. The red spot was not reported from 1713 until 1831, but it has been present ever since. Smaller clouds have been seen swirling around in a counterclockwise current within the Great Red Spot. Radical changes have taken place in some of the smaller spots since 1989. A new spot emerged in the Northern Hemisphere, and a large spot found in 1989 in the Southern Hemisphere disappeared between 1989 and 1995. Some of these spots are thought to be clear areas in the methane-based atmosphere of the planet. These "windows" make it possible to see to deeper levels in the atmosphere where ammonium hydrosulfide cloud layers are present. Four belts of clouds of varying coloration, shades of reddish brown and greens, lie parallel with the equator of the planet and are persistent. These streams of rapidly moving clouds are similar to the high-altitude zonal flow pattern above Earth. The most plausible explanation for the colors on Jupiter is that they are organic compounds produced as a result of electrical discharges in a mixture of methane and ammonia gases. Experiments confirm that this is a possibility, and lightning is a frequent feature of the Jovian atmosphere.

Jupiter has several faint rings composed mainly of tiny particles of dust. These rings extend 128,000 kilometers (80,000 miles) from its surface. Astronomers have recently concluded that Jupiter's rings are caused by interplanetary meteoroids smashing into Jupiter's four small, inner moons. The impacts knock off dust particles that remain in the same orbits as the moons. Jupiter has a short period of rotation, approximately ten hours; the surface near the equator moves at a rate of nearly 45,000 kilometers/hour (28,000 miles per hour). As a result, the planet is distinctly flattened at the poles.

The composition of the interior of the planet is unknown, but in view of its size, the pressures should be great enough to compress hydrogen into a liquid metallic state (figure 21.11*b*). The transition to liquid metal is estimated to occur at a depth of about one-third of Jupiter's radius, and a small rocky core may be present.

The remarkable sets of images of the sixteen Jovian satellites reveal an unexpected variety of surface features. Some of these features resemble landforms on Earth. Images showing volcanic eruptions on the satellite Io prove that active volcanoes are not confined to Earth. *Voyager 1* and 2 recorded nine eruptions, and more than 100 volcanoes are active on Io, including one that produces lava fountains that extend half a mile above the surface. These eruptions are unlike those on Earth in several respects. The erupted materials contain great quantities of sulfur dioxide, and the eruptions extend far above the surface, some reaching 250 kilometers. Although most volcanism on Earth is related to plate tectonics, tidal effects cause volcanism on Io. The gravitational attraction between Io and Jupiter produces tides in much the same way gravitational attraction between Earth and the Moon causes tides in the oceans on Earth, as described in chapter 9. However, because Jupiter is so massive, the tidal bulge on the surface of Io is about 100 meters (330 feet) high.

Extremes in surface topography occur on Jovian satellites. Callisto possesses one of the most densely cratered surfaces in the solar system. In contrast, Europa has one of the smoothest surfaces known. Its surface resembles a cracked egg with the cracks filled by ice. These surface features support the idea that an ocean lies below the icy crust. Ganymede is the largest satellite in the solar system. Its surface shows both cratered and grooved terrain. Surface temperatures on all these satellites are low, in the range of −100° to −200°C (−148° to −328°F). Consequently, little liquid is expected on the surface of any of them, but temperature increases with depth, and liquid is presumably present as the temperature rises.

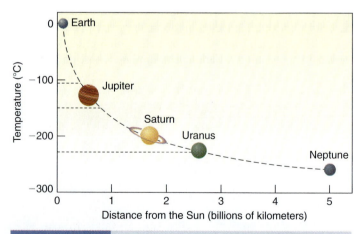

Figure 21.10 **Extremely Low Temperatures of the Giant Planets.** The vertical scale on this graph shows the change of temperature with distance away from the Sun.

(a)

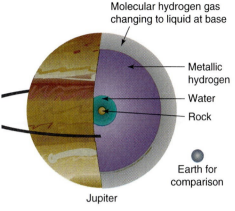

Molecular hydrogen gas changing to liquid at base

Metallic hydrogen

Water

Rock

Earth for comparison

Jupiter

(b)

Figure 21.11 Jupiter. (a) The Great Red Spot in Jupiter's atmosphere is one of the most distinctive features on the planet. Note the swirling pattern in clouds near the spot. These patterns reveal the atmospheric circulation on the planet. (b) Unlike the inner planets, the internal portions of Jupiter are thought to consist of rock surrounded by water and metallic hydrogen. The outer part of the planet is hydrogen gas that changes to liquid hydrogen at depth.

Saturn

Like Jupiter, the main body of the planet Saturn consists of hydrogen. It is surrounded by a hydrogen-rich atmosphere that contains huge masses of clouds that swirl around in storms on its surface. A powerful equatorial jet stream moves at 500 meters per second (1,500 feet per second), and near the top of Saturn's atmosphere, winds are fierce with speeds of 1,800 kilometers per hour (1,100 miles per hour). A system of well-developed rings composed of icy particles, some the size of houses, gives Saturn the most distinctive appearance of any planet (figure 21.12). Saturn requires nearly thirty years to complete its orbit around the Sun. Because its axis is inclined to the plane in which it circles the Sun, the rings are viewed at different angles. These rings circle the planet in the plane of its equator and have complex structure and change over time. Gravitational attraction between the materials in the rings and the large satellites probably accounts for many of the unusual features seen in these rings. The smaller particles may also interact with the magnetic field of Saturn, creating patterns of spokes and tightly wound spiral formations. In recent years, a huge white cloud, probably composed of ammonia ice crystals, has risen from near Saturn's equator and has spread to cover large areas along the equator.

Saturn has seventeen satellites, but only the largest, Titan, has a dense atmosphere, composed mainly of nitrogen and methane. Its surface temperature is about −179°C (−290°F). Satellites generally lack the gravitational pull needed to hold gases. Even gases that escaped through volcanoes on the Moon drifted off into space long ago. However, Titan has sufficient mass to hold an atmosphere that is denser than Earth's atmosphere. Although a thick haze obscures the surface of Titan, the composition of this haze suggests that Titan's surface may contain lakes composed of a hydrocarbon known as ethane. The surface of many of Saturn's satellites are pockmarked by impact features, and some exhibit topography that suggests the presence of faults and other signs of tectonic activity (figure 21.13).

Uranus and Neptune

These two planets share many features. Both have upper atmospheres composed of hydrogen and helium, but significant amounts of the light gas hydrogen have apparently been lost from them. Large spots caused by gigantic storm systems are the most distinctive features on the face of both planets. *Voyager* clocked wind velocities on the surface of Uranus that are several times greater than Earth's most intense hurricanes. The atmospheres of Uranus and

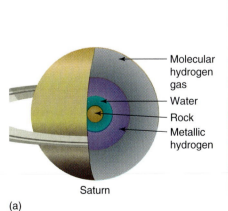

Molecular hydrogen gas

Water

Rock

Metallic hydrogen

Saturn

(a)

(b)

Figure 21.12 Saturn. (a) The internal structure of Saturn resembles that of Jupiter. (b) Details of some of the rings of Saturn reveal satellites.

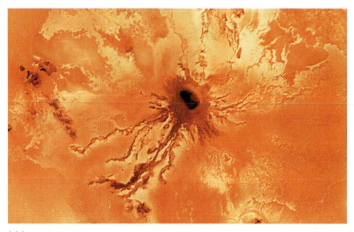

(a) Io

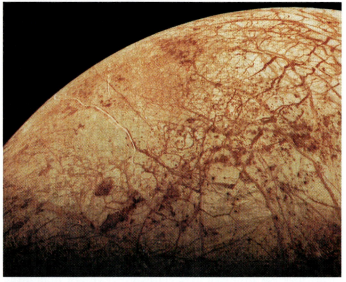

(b) Europa

Figure 21.13 Satellites of the Jovian Planets Exhibit a Great Variety of Form and Surface Features. (a) Lava flows spread out from the central vent of volcano on Io. Volcanoes on Io are active. Clouds of gas have been seen rising from their vents. (b) The icy crust of Europa is extensively cracked.

Neptune contain much more methane than do those of Jupiter and Saturn. The high clouds consist of ices of methane. Like the other giants, Uranus and Neptune have rings, but they are thin, narrow, and dark in color. These rings consist of dust particles and in the case of Neptune, larger pieces of matter. These rings have gaps with more matter being concentrated in certain parts of the ring. It is likely that satellites hold the matter in this configuration. Planetary scientists think much of the matter in these rings comes from the breakup of small moons.

Both Uranus and Neptune have satellites. Impact craters and faults are present on the icy surfaces of some satellites. Others are free of such features, which were probably later erased by melting of the ice and formation of younger smooth surfaces of frozen liquids.

Neptune was not discovered until astronomers were trying to understand changes they observed in the orbit of Uranus. These unusual motions could be explained if Uranus were affected by another planet. Later, the position of this hypothetical planet was calculated, and Neptune was found. Although Neptune receives very little sunlight, it has an active atmosphere with several large spots that resemble Jupiter's Great Red Spot. Measurements by *Voyager* revealed winds that blow with velocities over 1,900 kilometers per hour (1,200 miles per hour), the highest-speed winds found on any planet.

Voyager revealed the presence of eighteen satellites around Neptune. The largest of these, Triton, is especially interesting. Active geyser-like eruptions spew clouds of nitrogen gas and dark dust over 1.6 kilometers (a mile) into the atmosphere. Triton has an extremely thin, nitrogen-rich atmosphere that is the source of a nitrogen-based frost that covers most of the satellite's surface. Triton has unusually high density and revolves around Neptune in the opposite direction from Neptune's orbit around the Sun, which suggests that it may be a captured object rather than an original satellite.

Uranus shares one unusual feature with Pluto. Their axes of rotation lie in the plane of the ecliptic rather than tilted at a high angle to this plane.

Pluto

Pluto, the outermost and smallest of the planets, is about two-thirds the size of our Moon. It is one of about sixty objects that lie near the outer edge of the solar system. Unlike most other planets, Pluto's orbit does not lie in the plane of the ecliptic. This has led some to think it is a body captured by Neptune. Pluto's highly elongated orbit takes it inside Neptune's orbit for a twenty-one-year period during each revolution. The Hubble telescope reveals a dozen bright and dark regions on the surface. With surface temperatures of about −230°C (−382°F), snows composed of nitrogen, carbon monoxide, and methane probably cover the light regions. Differences in brightness may be due to alteration of methane ice by radiation that turns it dark. Much of the planet may be composed of a mixture of rock and ices composed of methane and water. Pluto has one satellite, Charon, which is nearly half the size of Pluto. At the extremely low temperatures, the surfaces of both Charon and Pluto are frozen, but surprisingly and for unknown reasons, Pluto is covered by methane ice, but Charon is covered by water ice.

COMETS, ASTEROIDS, AND METEORITES

Comets

Comets are bodies composed of gases and ices that often have a long tail and follow an orbit around the Sun (figure 21.14). Most comets have a nucleus composed of water ice, carbon-bearing and

Figure 21.14 Comet Hale-Bopp.

silicaceous material. Some scientists think the carbon materials found in comets may be the original source of organic materials on Earth. The atmosphere around the nucleus, called the **coma,** is composed of the gases produced as the solid nucleus goes into a gaseous state. Although the nucleus is only a few miles in diameter, tails composed of extremely thin gases can extend away from the nucleus for more than a million kilometers. Solar radiation illuminates comets. Some of the most spectacular comets appear in the night sky as a ghostly pale glow. The material in the comet tail is so thin that it is affected by the solar wind and is blown away from the Sun. Earth often passes through these tails without any adverse impacts.

Asteroids

Asteroids are small bodies that circle the Sun. About 20,000 such bodies have been identified, but many thousands more are located between Mars and Jupiter. This zone, referred to as the *asteroid belt,* consists of materials that failed to form a planet. Although the asteroids are widely separated, a spacecraft almost came into contact with one. The largest asteroid, Ceres, is over 1,000 kilometers (620 miles) in diameter. The asteroids are thought to have intense collisional histories and are the major sources of the meteorites that reach Earth. Thousands of asteroids move in orbits that cross Earth's orbit. Impacts of asteroids may have caused the extinction events that took place on Earth at the end of the Paleozoic and Mesozoic eras. Because many of these asteroids are composed mainly of iron, some scientists have already discussed the possibility of bringing one into orbit around Earth to establish mining operations.

Meteoroids and Meteorites

Meteoroids are bodies of rock and metal, smaller than planets or asteroids, that move in space (figure 21.15). **Meteorites** are meteoroids that fall through the atmosphere and strike Earth. As meteoroids pass through the atmosphere at speeds of 10 to 15 kilometers per second (6 to 9 miles per second), friction within the atmosphere heats them and causes them to glow or vaporize, producing the familiar bright fiery trails called shooting stars, or *meteors.* Meteorites become hot enough to glow at approximately 150 kilometers (90 miles) above Earth's surface. Small particles, many of them no larger than peas, vaporize completely before they reach the surface. Larger meteorites have surfaces sculptured by this heating.

The meteorites found on Earth can be divided into several different classes according to their composition. Most belong to one of two groups. The first group includes the iron meteorites, called **siderites.** These contain about 90% iron and 10% nickel. The second group, known as stony meteorites, or **chondrites,** consists mainly of aggregates of the minerals olivine and pyroxene. Some of the minerals in chondrites contain water locked up as part of the crystal structure. This is a possible source of the water found on Earth. Some meteorites also contain extremely small quantities of carbon and hydrocarbons including several other basic building blocks of organic molecules. Other meteorites, carbonaceous chondrites, contain a considerable amount of carbon (see figure 21.15). Still other meteorites contain mixtures of iron, nickel, olivine, and pyroxene. Although chondrites resemble rocks and are

(a)

(b)

Figure 21.15 Meteorites. (a) This sample consists of olivine crystals in a matrix of iron-nickel metal. The sample is 16 centimeters across. (b) This chondritic meteorite belongs to a class referred to as carbonaceous chondrite. The darker gray material is a fine-grained matrix in which calcium-aluminum-rich inclusions are present.

(a)

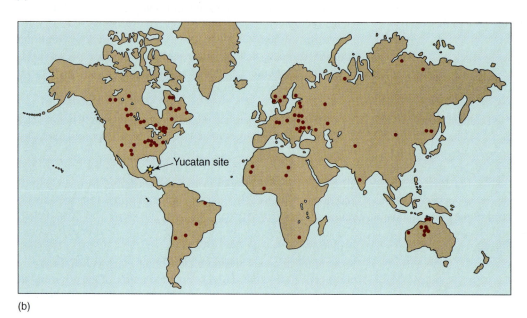

Yucatan site

(b)

less likely to be identified as meteorites, they are probably more abundant than siderites. Nevertheless, the largest known meteorite, measuring nearly 4 meters (13 feet) in diameter and weighing about 50,000 kilograms (about 55 tons), is a siderite that fell in southwest Africa. Most stony meteorites are small by comparison, the largest of them weighing a little over 900 kilograms (2,000 pounds).

When a large meteoroid hits the surface of Earth, the impact forms a crater. Sometimes the meteorite explodes, excavating a crater that is much larger than the meteorite itself. Huge impact craters give the Moon and several other planets, as well as many of their satellites, characteristic pockmarked surfaces. The famous Meteor Crater of Arizona (figure 21.16a), over 1.3 kilometers (.75 mile) wide and 200 meters (660 feet) deep, is an example of an explosion crater. The meteorite that formed this crater struck relatively horizontal sedimentary rocks, shattering them and folding the layers around the edges of the crater into an upright

position, similar to the patterns produced by near-surface underground explosions. This impact fused the sand grains in the bottom of the meteor crater, forming an unusual variety of silica that occurs at impact sites where sand is subjected to extreme pressure. The discovery of this type of silica in sedimentary rocks formed at the end of the Mesozoic era lends support to the idea that meteoroid impacts caused the extinction of the dinosaurs. Impact craters exposed at the surface of the ground, such as the Meteor Crater in central Arizona, were discovered early. More recently, drilling operations have revealed much larger craters buried along the northwestern margin of the Yucatan Peninsula in Mexico and under the southern part of the Chesapeake Bay (see figure 4.27).

Meteorites also provide evidence about the composition of Earth's interior. Earth and meteorites appear to contain similar materials. The density of Earth's core is approximately that of a nickel-iron alloy, and the outer mantle of Earth consists largely of olivine and pyroxene. The density of the other terrestrial planets suggests that they may also consist of these materials. Based on these observations, many believe that meteorites are pieces of asteroids that never became incorporated into a planet.

Regardless of their origin, in all the meteorites tested, the minerals crystallized about 4.5 to 4.7 billion years ago. These ages agree with the 4.6-billion-year age of the Moon, suggesting that the meteorites and the planets formed during the same cosmic event. Armed with this background of knowledge about the Sun, the planets, and meteorites, it is possible to examine some ideas concerning the origin of planets.

THE ORIGIN OF PLANETS

Most theories concerning the formation of planets are evolutionary theories in which the planets and Sun developed from a single mass of materials undergoing a progressive evolution. According to evolutionary theories, planets should orbit around other stars in the universe.

Early Ideas About the Origin of Planets

One of the early attempts to describe the origin of the solar system appears in the writing of the French philosopher René Descartes (1596–1650). He envisioned the solar system developing from a vast cloud of gas within which eddy-like vortices swirled. He proposed that the Sun condensed at the center of one of these large gaseous vortices, and that the planets condensed at smaller vortices radially distributed around the Sun.

The German philosopher Immanuel Kant (1755) and the French mathematician the Marquis de Laplace (1796) independently developed similar evolutionary models for the solar system. Both conceived the system as initially consisting of a large, flattened rotating cloud of gases and particles surrounding the Sun. Such bodies are now called solar *nebula*. They pointed out that gravity would cause contraction in the denser portions of this cloud of gases and that the contracting mass would rotate even more rapidly, just as a spinning ice skater does by pulling arms or legs closer to the body. This rapid rotation caused the contracting mass to flatten eventually into a disc-shaped body. Kant envisioned the formation of a number of secondary concentrations of condensing matter within the nebula, which would become planets, but Laplace concluded that rings of matter would remain around the outer edge of the contracting nebula as its rate of rotation increased. Later, the matter in these rings would condense and coalesce into the planets and their satellites. Early nebular hypotheses postulated that the nebula rotated faster as it condensed. If this were true, the Sun should rotate much faster than it does.

Growth of the Solar System from a Cloud of Gases and Dust

Once it was recognized that the same elements exist in the Sun, planets, and other bodies in the solar system, most scientists concluded that the Sun and planets formed from a single cloud of gases and dust. It follows that the present configuration of planets

is a natural outgrowth of the evolution of that material. During this evolution, the inner planets lost most of the hydrogen and helium, which are so abundant on the Sun. How this cloud separated into the bodies now found in the solar system is the major question. The Sun contains most of the original mass of that cloud of matter; just how this concentration occurred is not clear. Similar central concentrations of matter, surrounded by disc-shaped clouds of gases and dust are familiar sights to astronomers who study star formation in other parts of the universe (figure 21.17).

If the Sun and planets originated from a single mass of materials in interstellar space, the materials in that cloud probably started out being uniform in composition with most of the matter concentrated near the center. If these ideas are correct, the theory must explain what events in the evolution of the nebula ultimately caused the marked contrast between the terrestrial planets and the more-distant giant planets. Some differences between the terrestrial and Jovian planets are quite apparent. Because the terrestrial planets are closer to the Sun, they receive more solar radiation.

If the nebula hypothesis is to explain the origin of the solar system as it is presently understood, the hypothesis must account for the fact that the Sun now contains a much higher percentage of

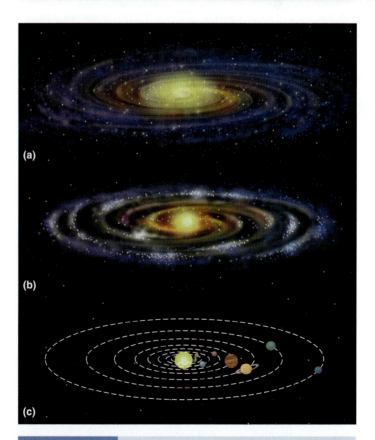

(a)

(b)

(c)

Figure 21.17 Schematic Drawing of the Evolution of the Solar System. (a) Matter is distributed over a vast region with a concentration near the center. (b) As accretion takes place, the matter is gradually swept into the larger bodies. (c) In its final stage, the matter is highly concentrated in planets.

light elements, such as hydrogen and helium, than do the planets. If the nebula from which the solar system originated initially had the same composition as the Sun, each planet developed from a cloud of matter that had the same proportions of the elements found in the Sun. For each planet, it is possible to calculate how much of each element would be needed to re-establish the proportions of the elements found in the Sun. For example, the terrestrial planets, which are highly deficient in hydrogen, helium, methane, and ammonia, would require large amounts of these elements to balance the huge proportions of silicates and heavy metal they now contain. To account for the quantity of silicates and heavy metals present in the terrestrial planets, a cloud 500 times the present mass of these planets would be necessary. In contrast, Jupiter could have formed from a cloud only five times its present mass.

Astrophysical evidence indicates that nebulae are initially cool and diffuse clouds of matter with only a few atoms per cubic centimeter. However, as the matter in them reacts to its gravitational force, they begin to contract. The density of matter and temperature increase tremendously during this process. Nebulae are exceedingly hot when stars form; after that, they cool slowly.

Temperature, pressure, and the materials of the nebula determine how the nebula evolves. In its early stages, the solar nebula was hot enough to be entirely gaseous. When contraction ceased, the nebula began to cool. When temperatures in the solar nebula dropped to 1,660°C (3,020°F), iron condensed into small droplets. Condensation from gases generally occurs on some particle that serves as a condensation nucleus. This is true of water vapor that condenses on fine dust particles. If no nuclei for condensation are present, temperatures may drop well past the normal condensation point without any droplets forming. Under these supercooled conditions, once condensation begins, it is rapid. The structure of many iron-rich meteorites suggests that they crystallized rapidly into solids. When temperatures dropped to 1,400°C (2,550°F), silicates also condensed, perhaps as coatings on the iron. Iron and silicates concentrated in the inner part of the nebula where temperatures were high and where the density of matter was greatest. Meteorites probably originated relatively close to the Sun, probably at the distances now occupied by the terrestrial planets. Water and hydrocarbons require very low temperatures to condense. As a result, they initially condensed at great distances from the Sun.

Once the process of condensation began, the solar system consisted of a hot, gaseous star, the Sun, surrounded by clouds of gases containing dust particles that formed as the gases condensed. Cooling of these gases quickened as hot condensates developed, blocking out sunlight in the more distant reaches of the nebula. Far away from the Sun, temperatures were low enough (−70°C) to allow crystallization of ices, not only of water, but also of methane and ammonia. It was from these ices that the Jovian planets evolved. Presumably, the terrestrial planets do not contain these ices because solar radiation drove off most of these constituents before the planets formed, or the ices changed to gases and drifted off.

After liquids condensed and solid particles of silicates, metals, and ices formed, growth of larger masses occurred as the particles came together, a process known as **accretion** (see figure 21.17*b*). At this stage, the small grains of ices and silicates moved within great cloud-like cells of rotating turbulent gases, rather like wind-driven snow. These cells may have resembled the satellite images of hurricanes. The entire mass of gases and particles moved rapidly, but because they moved in the same direction, the relative velocity between particles was not great. This low relative velocity is important because growth by accretion requires solid particles to stick together. High-velocity collisions would break the particles.

Accretion of grains of silicates and metals probably started in the inner portion of the solar system while the particles were still hot enough to weld together on contact. The internal structure of some meteorites was hot enough to recrystallize (see figure 21.15*b*). At a greater distance, where the temperatures were lower, the ices could stick together as wet snowflakes do. In either case, the particles slowly grew larger. Estimates of how fast this process might proceed vary considerably from a slow growth rate of .001 centimeter per year, which would produce a body 10 meters (33 feet) in diameter in 1 million years, to rapid growth rates that could produce a Moon-sized body in little over 100,000 years. As these larger bodies, called **planetesimals,** developed, they moved toward and concentrated in a plane (see figure 21.17*c*). Initially, no Sun existed. It was not a star that was fusing hydrogen to helium and producing radiation. Condensation and density increases proceeded until the pressure in the center of the nebula was high enough for temperature to reach the level needed to produce fusion and begin the life of the new star, the Sun. Eventually, the Sun and planetesimals contained most of the material that had previously made up the nebula. At this stage, the nebula began to clear, and growth of bodies accelerated. Once sunlight penetrated the cloud of dust and

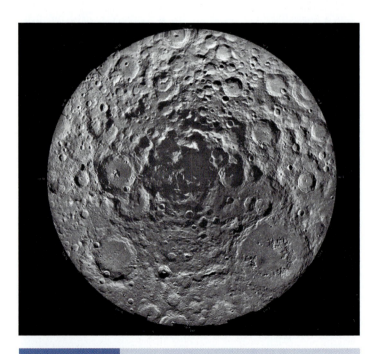

Figure 21.18 Mosaic Image of the Moon. This image gives an exceptional view of the cratered surface of the south polar region of the Moon.

particles, solar radiation and gas flowing out of the Sun's atmosphere drove off the remaining light elements, such as hydrogen and helium, and ices on the innermost bodies evaporated. Gradually, the terrestrial planets lost their lighter and more volatile constituents, which concentrated farther out in the solar system.

The next critical phase in the evolution of the system occurred when the planetesimals—now freed of the turbulent clouds that spawned them—grew large enough for their gravitational fields to attract matter. Planets above a critical size would sweep up the smaller bodies moving in nearby orbits. Growth at this stage took place rapidly. These larger bodies soon dominated the orbital zone in which they moved. Initially, many smaller planets fell into the dominant ones, and the resulting impact created the large craters present on Mercury, Venus, Mars, and the Moon (figure 21.18). This growth process took place early in the history of the solar system. According to this theory, most of the Martian and lunar craters formed within a few hundred million years of one another. Lunar studies already indicate that many of the lunar craters formed during the same stage in the Moon's history. Presumably, Earth's surface once bore similar markings as its most distinctive features. The impact of one or more large planetesimals is probably responsible for the tilt of Earth's axis to the plane of the ecliptic—the plane in which Earth revolves around the Sun.

The impact of two planetesimals could either lead to the fusion of the two into a single body—as appears to be the usual condition—or, if the energy of the impact is sufficient, one or both of the bodies might break up. It is especially interesting here to note that the belt of asteroids located between the terrestrial and Jovian planets lies in an orbit predicted by astronomers for a planet. The giant planet Jupiter almost certainly influenced the accretions of planetesimals in this orbit. They may have collided at velocities too great for fusion and instead broke up. Jupiter may also have disrupted the orbits and ejected planetesimals from the solar system that came close but not close enough to fall into it. It has also been suggested that some of its satellites were ejected from the planet as a result of tidal effects.

Astronomers and planetary scientists have offered several explanations for the origin of satellites such as the Moon. Some favor the idea that larger bodies moving in a nearby orbit captured smaller planetesimals destined to become satellites. This might have taken place during the accretion of the planets or after the planets formed. Other scientists favor the idea that satellites were fragments of planets created during violent impacts (figure 21.19). Still other scientists argue that accretionary processes created both planets and satellites at the same time.

Evidence from Earth

The Earth provides evidence that bears directly on the origin of the solar system. From studies of seismic waves generated during earthquakes, seismologists conclude that Earth has a nickel-iron core covered by a silicate mantle and a highly differentiated outer lithosphere about 100 kilometers (60 miles) thick. If Earth grew by accretion of planetesimals, as the modern nebular hypothesis suggests, it must have passed through a later stage when it was

(a)

(b)

Figure 21.19 A Hypothesis of the Moon's Formation. (*a*) A lone giant impact on Earth early in its history may have led to the formation of the Moon. (*b*) A portrayal of Earth soon after the giant impact hypothesized to have led to the formation of the Moon. The young Moon is still sweeping up the debris splashed into orbit. The Mars-size impact would have melted Earth throughout.

sufficiently molten to allow the heavy nickel-iron compounds to separate from the lighter silicates—a process called **differentiation.** While Earth was molten, the heavier solids sank toward the center of the Earth. Differentiation would explain why more dense matter occurs in the core than in the mantle of Earth (figure 21.20).

An alternate theory is that the core developed during condensation in the nebula. According to this hypothesis, initial accretion of solid materials occurred while nebular temperatures near Earth were about 1,200°C and nickel-iron was condensing. As Earth grew, it became hotter as a result of the impact of material falling into it. The nickel-iron condensate of which Earth initially consisted formed a molten surface layer. As the nebula cooled further, olivine and pyroxene started to condense and fall into the growing Earth, but liquids composed of metals did not mix with silicate melts. Because the metals are heavier, a sharp boundary developed between core and mantle. During this high-temperature stage at Earth's surface, volatile materials evaporated but were held to Earth by gravity and became part of a primitive atmosphere; the lightest constituents, hydrogen and helium, escaped to space. Only when the temperatures on Earth began to drop did many of the remaining light constituents become part of the outermost layers of Earth. At that time, Earth had a metallic core covered by a silicate mantle, a lithosphere composed of rocks made up of elements that are lighter than those in the mantle, and a primitive atmosphere.

According to both of these theories, Earth was molten at an early stage in its history. If this is correct, all rocks on Earth ultimately come from rock melts and from the materials that crystallized near the surface during this earliest and most obscure stage in Earth's history. As conditions on Earth changed, and especially as the atmosphere, oceans, and living organisms originated and evolved, the primary igneous rocks weathered and eroded, sediments formed, and the history now preserved in Earth's crust began.

The accretion of the materials that compose Earth ended hundreds of millions of years before the oldest-known Earth rocks formed. Geologists have not found rocks older than about 4.2 billion years, and these oldest rocks were recrystallized, apparently by high temperatures. Some geologists think that Earth underwent a major thermal event, during which the minerals containing radioactive elements used to determine ages recrystallized, making it impossible to find any older dates. If they are right, it will be impossible to reconstruct the history of the time before this strong heating occurred. Fortunately, rocks older than 4.2 billion years exist on the Moon, and they provide evidence used to infer conditions during the early history of the solar system, including Earth.

The Moon: Its Implications for Early Earth History

Studies of the Moon rocks that the Apollo astronauts brought back to Earth and of lunar landforms have enabled geologists to outline the major events in its history. Lunar history has four major stages:

Accretion

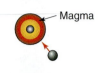
Planetesimal

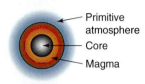
Magma

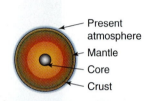
Primitive atmosphere
Core
Magma

Present atmosphere
Mantle
Core
Crust

Figure 21.20

Stages of Accretion. The Earth starts accreting material. When the infalling velocity exceeds about 2 kilometers per second, at approximately lunar size, planetesimals devolatilize upon impact, producing in succession a primitive atmosphere and an ocean of magma that will slowly cool and crystallize to form a crust.

1. formation and early melting
2. formation of a lunar crust
3. formation of the lunar seas, the maria
4. formation of new impact craters, minor flows, and some volcanoes

Initial Formation and Early Melting How the Moon originated is still uncertain, despite the tremendous efforts directed to answering this question. Several possibilities are under investigation. According to one, the Moon was blown out of Earth during a major impact event (see figure 21.19). Others think that the Moon and Earth formed at about the same time, but the Moon was never part of Earth itself. A third possibility is that the Moon formed separately from Earth, essentially as a separate planet, and was later captured by Earth's gravitational attraction as the two passed near each other.

Although the correct theory is still unknown, a number of geologic observations constrain these theories. First, the oldest lunar samples dated radiometrically at 4.6 billion years old provide an approximate age for the Moon. This figure is close to the age of most meteorites, and it is the estimated age of Earth and the solar system. Second, a considerable amount of evidence indicates that the Moon originated in a relatively short time. Current estimates of the time involved range from 200 million years down to a few hundred years. Third, the Moon almost certainly either formed as a hot body or became hot enough to melt rock at an early stage in its history. The evidence for this conclusion is that the oldest rocks found on the Moon have low gas content, suggesting that the high temperature drove gases from the lunar crust. Finally, Moon rocks bear a strong resemblance to iron- and magnesium-rich igneous rocks found on Earth. This similarity supports the view that either Earth and Moon originated close to one another or they were a single body that was split apart by collision with a Mars-sized body at an early stage.

Formation of the Lunar Crust The second stage of the Moon's history is also obscure—not because lunar rocks of this age are rare but because they crop out in the most rugged and least-explored portion of the Moon, the highlands. The highlands consist of primitive lunar crust, characterized by rough, mountainous, cratered terrain that developed during the early bombardment of the Moon as large masses fell into it. This old crust consists of basalt, gabbro, and anorthosite (a rock composed mainly of plagioclase feldspar). Earlier in its history, falling material bombarded the primitive lunar crust, creating the closely spaced, sometimes overlapping craters shown in figures 21.18 and 21.21. In 1996, ice was reported in the debris in some of these craters. This may be an important resource in future extended exploration of the Moon's surface. The

Figure 21.21 The Moon's Terrain. Oblique lighting on the Moon's surface emphasizes the complex cratered terrain of the lunar highlands. The oldest rocks on the Moon are in the highland regions. The extent of cratering on the Moon is evident in this photograph. These craters formed as a result of the impact of large meteorites during the early history of the solar system. The large plain areas, such as the one in the lower part of this picture, are vast lava flows erupted through the thin solid crust that had formed. Some of the most prominent craters lie in these lunar plains called seas, or maria.

large lunar seas, called the **maria** (singular form, *mare*), formed in this 3.5 to 4.5 billion-year-old, severely battered crust as large bodies fell into the Moon, creating huge impact basins.

Formation of the Lunar Seas The formation of the maria—the vast seas of iron-rich basaltic lava flows—was one of the most dramatic events in lunar history. These lavas filled large impact craters and covered parts of the earlier topography (see figure 21.21). They provide a valuable benchmark by which planetary scientists distinguish older landforms and rock structures from those that formed after the mare basalts solidified. These lavas are so vast and continuous that they must mark a major event in which large volumes of the outer part of the Moon partially melted. Because the flows on the Moon were hot and fluid, they spread evenly over immense areas, filling the largest of the previously formed impact craters. Dating of lava flows from the Sea of Tranquillity indicates that these events took place about 3.2 to 3.7 billion years ago.

Post-Mare History Most of the major features of the lunar landscape had formed by the time the last of the mare lava flows filled in the remaining low areas. However, some meteorites continued to fall into the Moon, producing the youngest and many of the most perfectly shaped impact craters (figure 21.22). Some impact-induced volcanism accompanied these last major impacts, and faults developed in and around the mare. Although the infall of large fragments became rare, many small meteorites have continued to fall onto the Moon just as they have on Earth, but the re-

sulting effects on the two landscapes are dramatically different. Few scars or traces of meteorites impacts remain on Earth because most of the impacts occurred many millions of years ago. Weathering processes that decompose rocks and erosion have long since altered the early impact craters on Earth beyond recognition. In contrast, on the Moon, impacts crushed surface rock, resulting in a veneer of dust and fragments. This covering has remained barren and almost unchanged since its creation because the Moon has no atmosphere, streams, glaciers, or oceans that act upon it. However, if ice remains on the Moon, at least small quantities of water may have been present at some point in the Moon's history. Following what most geologists believe was a similar early history, the Moon and Earth evolved in strikingly different ways.

Based on what is known about the Moon, it seems likely that early in its history meteorites bombarded Earth's surface as heavily as they did the Moon's surface. Molten and crushed rock probably covered the surface of Earth as it does the surface of the Moon today. Clouds of hot gases rose from the thin crust and especially from craters and cracks that the impacts caused. These breaks through the crust allowed the molten rock below the surface to flow out as hot lava. On the Moon, gases produced in this way drifted off into space because the Moon's gravity was too weak to hold them. Some gases escaped from Earth, too, but because of its mass, Earth's gravity field held other gases, which must have accumulated to form a primitive atmosphere. The initial atmosphere was enriched with hydrogen and helium and later with gases similar in composition to the gases that escape today from

volcanoes like those in Hawaii (see figure 7.5*a*). When the temperature of this early atmosphere cooled enough for water to condense, rains fell on the hot crust, eventually accumulating as large bodies of water. Even at this early stage, Earth's evolution differed from that of the Moon, where all gases escaped and no rains fell. The role of the atmosphere in the evolution of Earth's surface clearly caused many of the differences now seen today between Earth and the Moon.

SUMMARY POINTS

1. Much of the evidence used to deduce the origin of Earth and the solar system comes from studies of the Sun, meteorites, and the other planets. The Sun is a large, extremely hot cloud of incandescent gases and plasma composed of dissociated electrons and atomic nuclei. Its atmosphere has three parts: the photosphere, the chromosphere, and the corona. The Sun consists almost entirely of the two lightest elements, hydrogen and helium; all other elements combined make up less than 1% of its mass.

2. According to their composition and physical characteristics as well as their locations, the planets fall into two classes. The small dense terrestrial planets—Mercury, Venus, Earth, and Mars—travel in the orbits nearer the Sun. The giant Jovian planets—Jupiter, Saturn, Uranus, and Neptune—have different compositions and atmospheres. Pluto is much smaller and denser than the other Jovian planets, and it is not classified with either the terrestrial or Jovian planets. Some planetary scientists think it may have been a satellite of one of the Jovian planets.

3. The Sun and planets form a disc-shaped array in which most planets revolve around the Sun and rotate on their axes in the same direction.

4. Most of the meteoroids that fall to Earth are about the same age, 4.3 to 4.7 billion years, also the best estimate of the age of Earth and Moon. Metallic meteorites (siderites) have the same composition as Earth's core, and the stony meteorites (chondrites) contain the same minerals that dominate Earth's mantle. This evidence suggests that similar processes created both Earth and the meteorites. According to one widely held theory of the origin of the solar system, a gaseous nebula gradually evolved into the Sun and planets. As the nebula cooled, first metals and then lighter silicates condensed, forming liquid droplets and hot solid particles. These grew by accretion into planetesimals. Some of these planetesimals became large enough to attract matter by gravity, sweeping up and accreting smaller bodies. In its early history, Earth was a hot body composed mainly of nickel-iron condensate from the nebula. As the nebula cooled, silicates accumulated as an outer mantle covering the metallic core or the mantle and core separated as the higher-density iron matter settled to the center.

5. According to evidence from lunar rocks, the Moon is about the same age as meteorites and consists of material much like the iron- and magnesium-rich igneous rock on Earth. The Moon's surface shows scars of many of the impacts of falling meteoroids and accretionary planetesimals. Some of the impact craters later filled with lava flows of basaltic composition

similar to those found in Hawaii. The flat areas caused by these flows contain younger craters and a soil-like veneer of dust and fragments that remains virtually unchanged today.

REVIEW QUESTIONS

1. What evidence supports the idea that the planets originated from a rotating disc-shaped mass of matter?
2. How do the terrestrial and Jovian planets differ in composition?
3. Why do planetologists think impact craters, similar to those on the Moon, originally marked the Earth's surface?
4. Why do ancient impact craters remain on the Moon's surface while those on Earth have largely disappeared?
5. What evidence suggests that Earth was once molten?
6. What processes are thought to have caused the differences in composition of terrestrial and Jovian planets?
7. Why are the planets located close to the Sun smaller and denser than the Jovian planets?
8. In what ways does Earth differ from the Moon and other planets?
9. What do present motions of the planets reveal about the probable origin of Earth and the solar system?

THINKING CRITICALLY

1. What environmental conditions found only on Earth make it more suitable for life than conditions on other planets in the solar system?
2. What effects, if any, do other planets in the solar system have on Earth?
3. Explain how Earth might have formed through accretion processes.

KEY WORDS

accretion
asteroids
chondrite
chromosphere
coma
comet
corona
differentiation

maria
meteorite
meteoroid
photosphere
planetesimal
prominences
siderite

chapter twenty-two

Beyond the Solar System

Chapter Guide

What is Earth's place in the universe? This final chapter examines the scientific evidence that bears on this question. What types of objects can we observe in the night sky? How big are they? What is their composition? Are they moving, and, if so, in what direction? Does a pattern to this movement exist? Does it hold answers to the past history of the universe. Is it possible to get some idea about the time of origin of the universe? Recent advances in the use of telescopes hold answers to these questions.

Images obtained by the Hubble space telescope are revealing new wonders of the universe. Here, huge clouds are illuminated by ultraviolet radiation from a star cluster at 30 Doradus.

INTRODUCTION

Scientific models are the products of our imagination. They are constructed to help us visualize what we cannot see. The model itself does not exist in nature. The creation of the model is simply an effort to provide a better description of nature. If the model, along with any mathematical formulation, truly reflects reality, it eventually will be validated experimentally. The greatest value of a model, however, lies in its ability to predict unknown behavior . . . When a model fails to agree with reality, we must modify it or discard it in favor of a better idea.
F.M. Miller (1984, p. 163 *Chemistry,* McGraw-Hill)

Although earthbound telescopes like those in Hawaii and South and North America have long probed space for clues about the shape, dimensions, and evolution of the universe, incredible new and dramatic views of distant space reached Earth in 1993 when the Hubble telescope began to produce sharp images of the celestial sphere (figure 22.1). The Hubble telescope orbits 592 kilometers (367 miles) above Earth's surface, far beyond the haze and lights that interfere with Earth-based telescopes. With a 238-centimeter (94-inch) mirror, Hubble has produced clear, sharp images of objects that have not been previously seen. Some of these objects lie at the margin of what astronomers refer to as the fringe of space, or the "edge of the universe." Although its viewpoint has made it possible for the Hubble telescope to produce sensational images, analysis of the light coming to Earth from objects in space has suggested answers to many questions about the universe. In all such studies, the first step in analyzing the objects in the sky is knowing where they are relative to Earth.

LOCATING OBJECTS IN SPACE

To an observer on Earth, the sky resembles a sphere, called the **celestial sphere,** on which stars appear fixed in the nighttime sky. Of course, this is an illusion. Actually, the stars are located at various distances from Earth and appear projected against a dark, infinitely distant background. To describe the position of an object as seen projected against this background, one needs a set frame of reference (see appendix E for a discussion of how to locate objects in the sky).

ANALYZING THE COMPOSITION OF STARS

Most of what is known about the composition of stars and clouds of gases in the universe is derived from the study of their spectra. The visible part of the spectrum, previously discussed in chapter 11, may appear as a **continuous spectrum,** showing variation in colors and corresponding wavelengths that range from the violet color of short wavelength radiation to the dark red produced by long wavelength radiation (figure 22.2). If radiation passes through cool gases, selective parts of the spectrum may be absorbed. This leaves dark lines in the otherwise continuous band of radiation and produces a **dark line absorption spectrum.** By emission of radiation, hot gases may produce specific lines of color, called an **emission spectrum.** Both the emission and dark line absorption spectra are indicative of the composition of the materials responsible for the lines produced (see figure 22.2).

The spectrum of the Sun is a continuous spectrum with dark absorption lines. This spectrum has wavelengths most of which range from 300 to 1,000 nanometers (1 nanometer = 10^{-7} centimeter). See the discussion of the spectrum in chapter 11. Radiation from a variety of elements in the interior of the Sun produces the bright part of the solar spectrum. It may be viewed as resulting from the excitation of atoms, which causes them to emit radiant energy. The absorption lines provide evidence that certain elements are present. Most of these lines are produced in the photosphere. For this reason, the spectrum can be used for chemical analysis.

As early as 1924, the spectra of over 200,000 stars had been obtained and catalogued. From this compilation, it was soon discovered that almost all of these spectra could be arranged in an orderly sequence. This sequence is broken into seven classes labeled O, B, A, F, G, K, and M. These classes reflect temperature differences. Hot stars have spectra that differ from cool stars. By noting which lines are present, it is possible to determine a star's temperature.

Class O stars are the hottest. In these, the lines of ionized helium, oxygen, and nitrogen are present (their electrons have been stripped), and hydrogen is present. Then, in sequence, the ionized lines disappear, and hydrogen lines increase in brightness (classes B and A); then hydrogen lines weaken, and lines of metals become intense (classes F, G, and K); finally, in class M, lines of titanium

Figure 22.1 A View of the Deep Field of the Universe Photographed by the Hubble Telescope.

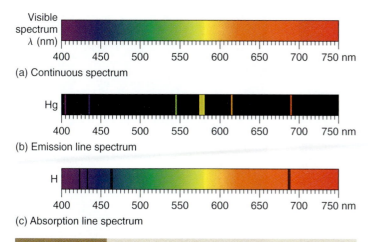

Visible spectrum
λ (nm)

400 450 500 550 600 650 700 750 nm

(a) Continuous spectrum

Hg

400 450 500 550 600 650 700 750 nm

(b) Emission line spectrum

H

400 450 500 550 600 650 700 750 nm

(c) Absorption line spectrum

Figure 22.2 Spectra Used in Analysis of Stars. Three types of spectra are used to analyze the light coming from stars. (a) The visible part of the continuous spectrum that reaches Earth from the Sun extends from violet to deep red. (b) Hot gases of specific elements emit certain lines of the spectrum. (c) Cool gases also absorb specific lines in the continuous spectrum.

oxide are prominent. In color, these classes of stars vary from blue (the hottest stars) to yellow and finally red.

If light from a star passes through a cooler gas, the atoms of the cooler gas absorb some of the light. The particular wavelengths absorbed are exactly those that the gas would radiate if it were heated. The spectrum produced in this way has a bright background with dark gaps referred to as dark line spectra. Elements also differ in their power to absorb their particular wavelengths. This absorption depends on temperature. Thus, spectral analysis provides an indication of chemical composition, temperature, and density of the hot gases that produce the spectral lines and, simultaneously, of the composition of cooler outer atmospheres of the star. Analysis of spectra is a critical tool in the study of individual stars and in the analysis of large groups of stars so far away that they look like single-point sources of light to the eye.

VAST GROUPS OF STARS

With a small telescope you can see that many of the glowing objects in the night sky are not individual stars like the Sun. Instead, some of the objects that appear to the naked eye as single stars look like fuzzy patches with small telescopes. Larger telescopes reveal that these are actually vastly greater accumulations of stars and clouds of dust and gases, called **galaxies.** Based on images from the Hubble telescope, billions of galaxies exist, each of which has a mass billions of times greater than that of the solar system.

Galaxies

Galaxies contain billions of stars, and the clouds of gases and dust from which stars form. Old stars are most prominent in the central portions of galaxies; however, they are accompanied by much younger stars that are called **giants,** because they are large and very bright. The giants account for the brightness of the central portion of galaxies.

A great variety of types of galaxies have been identified and classified. The classification devised by Edwin Hubble distinguishes three main types of galaxies on the basis of their shapes: elliptical, spiral, and irregular. Some spiral galaxies have a central straight, bar-like core. The elliptical galaxies have a smooth and featureless appearance. Stars appear in random patches in irregular galaxies. These three main sequences account for over 90% of all known galaxies.

Most galaxies seem to fit into a pattern of continuously varying characteristics. Astronomers are still unsure about what creates these different forms. Rotation probably plays a role because most flattened galaxies rotate more rapidly than the globular-shaped galaxies.

The Milky Way—Earth's Galaxy

The distribution of visible stars in the sky is not random. The greatest concentration falls in a belt known as the Milky Way, which consists of our solar system, an estimated 400 billion other stars, and vast clouds of gases and dust. The Milky Way is similar in shape to Andromeda (figure 22.3a). Both have the form of a disc with globe-shaped centers and spiral arms. Distances used by astronomers are so great they are often described in terms of the distance light would travel at the rate of 298,000 kilometers per second (186,000 miles per second) in a year. Andromeda is about 2.2 million light years away. The Milky Way is about 100,000 light years in diameter. Stars are concentrated in the central part of the disc and form a flattened ellipsoid bulge that is about a third of the diameter. The solar system is located about 30,000 light years from the center.

Movement of the Solar System in the Milky Way Galaxy

The Sun is not stationary in the Milky Way. It orbits within the disc of our galaxy much as Earth orbits the Sun. Astronomers use a simple method to detect this motion. Assume that the Sun, with its satellites, is moving toward a particular point in the sky. Then the stars ahead, if they have a random motion, should appear to move out from that point, while stars in the opposite direction should appear to be closing in behind. These movements are just what is observed. The speed of the Sun's motion can be approximated by noting the average speed with which those stars near the Sun's path straight ahead are approaching the Sun. The value is checked by obtaining the average speed of recession of the stars directly behind the Sun. The results indicate that the Sun is moving with a speed of about 19 kilometers (12 miles) per second in relation to nearby stars.

Stellar Objects

The objects we see in the night sky differ greatly in brightness, in distance from Earth, and in their spectra. Some of the bright

(a)

(b)

(c)

objects visible in the sky at night are single stars like the Sun. A few are planets and satellites of planets like the Moon. Although all of these planets and satellites reflect light, none produce light, and none outside the solar system are bright objects in the sky. Despite this, evidence that planets are present, and that they revolve around other stars like the Sun, is mounting. Many astronomers think that planets similar to those in the solar system circle numerous stars.

Some of the objects that appear to be single stars are actually great clusters of stars, two stars that rotate around one another, or clouds of dust and gases. These clouds of dust and gases, called **nebulae,** are widespread both inside and outside the Milky Way. Some of these clouds are dark; others that are located close to stars are bright. The clouds vary greatly in shape (figure 22.4). Though the gas density in them is less than a very good vacuum on Earth, they are estimated to contain from 5 to 20% of the total mass of the Milky Way. The dust is generally thought to be composed of tiny flecks of iron, silicon, carbon, and their compounds, with coatings of ice or other frozen gases.

The spectra of stars may be taken through even the denser portions of nebulae. Gases present give absorption lines in the spectra from the stars. These spectra reveal that nebulae are composed of sodium, calcium, and hydrocarbons. Hydrogen has also been de-

tected by means of radioastronomy. Under certain conditions of temperature, stars in the nebulae not only illuminate it but also heat the gas so it produces a bright line spectrum. The lines produced in this way indicate the presence of hydrogen, helium, nitrogen, and oxygen. The composition of nebulae is strikingly similar to that of stars. This and other evidence points to the probability that the stars might have formed from nebulae. The name **planetary nebulae** is given to certain clouds that are small and more or less round. Some of them resemble smoke rings. These clouds are slowly expanding and are thought to be gases ejected from a dying star.

Modern telescopes have revealed that the variety of stellar objects is much greater than was recognized a few decades ago. These objects include great accumulations of stars known as **globular clusters.** Many of the oldest stars in the Milky Way are located in these huge massive clusters of stars. Some stars, known as **cepheid variable stars,** vary in the amount of light they emit. Some change from being bright to dim in one to sixty days. Other stellar objects are invisible.

Black holes are invisible centers of extremely high mass. They were recognized by the emission of exceptionally strong pulses of radiation in the X-ray part of the spectrum from gases that surround the black hole. This radiation effect was identified long before the idea that black holes are associated with collapsing centers of extremely high mass (figure 22.5). After this collapse, mass is concentrated in such high density that the gravitational attraction of the mass prevents the escape of light. The Hubble telescope has found black holes at the center of most galaxies. Clouds of gases swirl around the holes. The speed of this cloud movement is a measure of the mass in the hole. An incredible mass equivalent to 3 to 5 billion Suns is present in a typical galactic black hole. As matter enters the hole, it is heated to very high temperatures, and the inner parts of the disc around the black hole eject extremely high volumes of radiation from bodies known as quasars.

Quasars, also known as quasi-stellar objects, are highly energetic and extremely far away. They were first recognized by studies of visible light in which the red wavelengths were shifted much farther than had been noticed before. They are so far away that telescopes on Earth had revealed little about them. In this case, the Hubble telescope revealed that most quasars are the small but luminous cores of very remote galaxies. Radio waves have detected lobes of matter that extend out from quasars. One popular interpretation is that supermassive black holes located near the center of an active galaxy produce quasars according to the process described in the description of black holes.

Luminosity

Luminosity, the total quantity of light emitted by a star (or galaxy), is especially important, because a number of stellar characteristics and measurements, such as stellar distances, mass, and radii, are related to it. In the time of Ptolemy, a scale of six magnitudes of brightness was established, which provided a rough way of classifying stars according to their apparent visual brightness. The twenty brightest stars were called *stars of the first magnitude.*

The apparent magnitude of a star depends on its distance as well as its actual luminosity. It is known from experiments that the apparent brightness of a point source of light is inversely proportional to the square of its distance. Therefore, if the apparent magnitude of a star and its distance from Earth are known, it is possible to calculate the apparent brightness that star would have if it were at some other distance. It is possible to compare stars by calculating their luminosity at some standard distance. The distance selected is 32.6 light years. At this distance, the direction to the star changes by 1 second of arc as Earth makes its journey around the Sun. The Sun is a star of the fifth magnitude and thus much less bright than many of its companions in the Milky Way.

Gas and dust: The 6-trillion-mile-high fingers, made of dust and hydrogen gas, turn blue when they are hit by ultraviolet radiation

Star eggs: Clumps of hydrogen inside the pillars called Evaporating Gaseous Globules, or EGGs, eventually hatch into stars

Figure 22.4 Gas and Dust Clouds in Which Stars Are Being Formed. These finger-shaped masses extend for trillions of miles. Photograph taken by the Hubble telescope.

Hertzsprung-Russell (H-R) Diagram Figure 22.6 shows the **Hertzsprung-Russell (H-R) diagram,** a plot of luminosity against spectral types for some 7,000 stars located close to the Sun. One of the interesting features of this plot is that the distribution is nonrandom. A main sequence of stars falls on the curve across the center of the diagram. In addition, a concentration of points is located above the main sequence (giants and supergiants), and a few isolated stars lie below the main sequence (white dwarfs—hot, but small stars with exceedingly high density). This diagram has proved to be one of the mainsprings from which astronomy has advanced in modern times. It provides a simple, pictorial way to summarize stellar properties such as the relationship between temperature and luminosity. The hottest stars are blue in color; the coolest stars are red. On the H-R plot, the hottest and more massive stars are located near the top of the main sequence. The H-R plot also reveals clear distinctions between main-sequence, red giant, and white dwarf stars.

Detecting Movements of Stars

As viewed from Earth, stars appear to be fixed on the celestial sphere. However, several techniques described in the following

Figure 22.5 A Black Hole in Galaxy NGC 4261. A cloud of gases surrounds the hole. These gases swirl around the hole as they are drawn in.

section, "Determining Stellar Distances," prove that stars are located at varying distances from Earth, and they provide a way of determining distances to stars. In 1718, Halley first noted that stars are not fixed in position. He found that several of the bright stars were no longer in positions assigned to them by Ptolemy in the second century A.D. The motion of a star can be seen in two ways. The first of these, called **proper motion,** is an apparent shift of position with respect to other stars on the celestial sphere. The second is **radial speed,** the speed by which the star appears to approach or recede from Earth.

Proper motions are determined by comparing two images of the sky made several years apart. If a star has moved (transverse to the viewer's line of sight) between the times the two images were taken, they will not coincide.

Shifts in spectral lines of the star's light are used to determine radial velocities. This shift is known as the **Doppler effect** (figure 22.7). This effect is generally familiar because it occurs in sound as well as in light. The rise in pitch of an approaching train and the sudden drop in pitch of a train whistle as it passes are examples of this effect. Visualize the phenomena as the apparent shortening of the wavelength of a sound coming from the source of that sound as it approaches and the lengthening of the wavelength as the source recedes from us. Light waves behave the

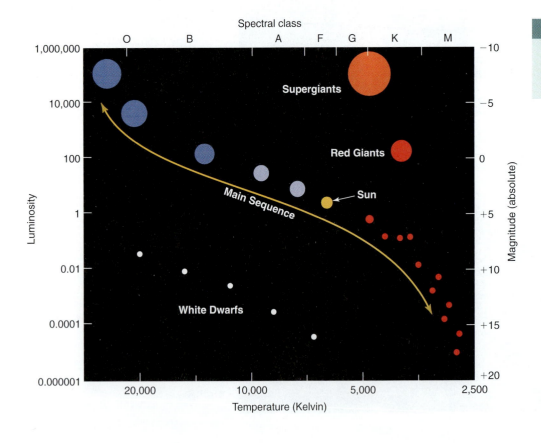

Figure 22.6 The Hertzsprung-Russell (H-R) Diagram. The H-R diagram plots the absolute magnitudes of stars against spectrum types.

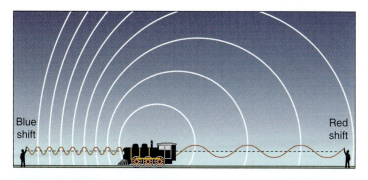

Figure 22.7 Doppler Effect. The wavelength of the sounds from a car or train that is moving away from you are longer, and the sound has a lower pitch. Sounds from an approaching car have shorter wavelengths and have a higher pitch.

type, and distance for thousands of stars. By plotting the apparent magnitude and spectral type of a star, its distance can be read from the H-R plot. If a star's true brightness is known by measuring how bright it looks—its apparent brightness—it is possible to calculate its distance. For example, based on a star's spectrum class, the H-R diagram provides an indication of its luminosity. Next, the apparent brightness of the star is measured. From its apparent brightness and luminosity, its distance is determined by using the inverse square law. This law states that the apparent brightness of a body decreases inversely as the square of its distance.

Cepheid variables can also be used to determine distance. Cepheid variables are yellow supergiants that change in brightness periodically. The periods range from about one day to as long as sixty days. What makes Cepheids so useful is that their period is

same way. If a body, such as a star, is rapidly receding away, lines in its spectrum will be displaced toward the red end (longer wavelengths). The amount of the shift is proportional to the speed of the light source. By comparing the spectrum of a source that is not moving, the star's radial velocity can be measured. The radial velocities range widely and exceed more than 96 kilometers per second (60 miles per second). Earth is traveling at a rate of approximately 29 kilometers per second (18 miles per second) around the Sun, so the observed value must be corrected for Earth's speed in its orbit around the Sun.

Determining Stellar Distances

Astronomers use many methods to determine the distances from Earth to stars and galaxies. However, some methods work only for nearby stars while other methods require information about the star that may be difficult to determine. For example, for nearby stars, astronomers use **trigonometric** or **parallax methods** as a measure of distance (figure 22.8). Careful observation of a nearby star demonstrates that as Earth orbits the Sun, the star appears to change position with respect to background stars that lie at greater distances. If the direction to the star is measured at various points in Earth's orbit, the star's position will be found to vary. Figure 22.8 shows how the position change can reveal the star's distance. The diameter of Earth's orbit and the angle between Earth, the star, and the Sun can be measured. The distance to the star is then the side of a triangle. Distances to stars are so great in relation to the size of Earth's orbit that only relatively close stars can be measured in this way. This method can be used for distances less than 3,000 light years. The size of Earth's orbit limits the use of parallax. Most stars in our galaxy are so far away that the parallax is too slight to be measured accurately.

Using Luminosity as a Measure of Distance Luminosity is used to determine the distance to stars that are too far away for use of the parallax method. This method involves use of the H-R plot, which illustrates the relationship of apparent magnitude, spectral

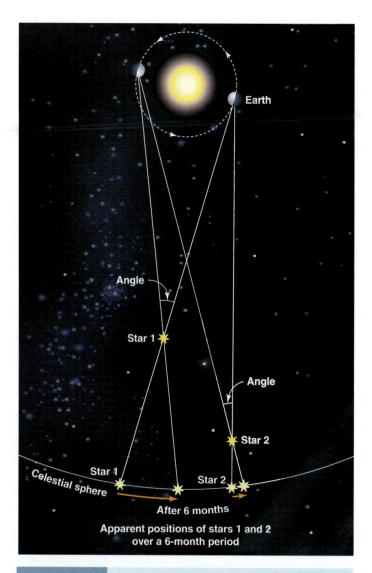

Figure 22.8 The Parallax Effect. Note how the position of the two stars against the celestial sphere changes with the position of Earth in its orbit. Also note that the closer star shows greater displacement than the more-distant star.

related to their luminosity. The slower the star pulsates, the more luminous it is. Observations allow astronomers to calibrate this relation so that a measurement of period gives luminosity. Once luminosity is known, we can again use the inverse square law to find the distance.

The period of the change in brightness of Cepheids has been plotted against their apparent magnitude. Once it was possible to determine the distance to one Cepheid by one of the previously discussed methods, the distance to all others could be calculated if their periods and apparent magnitudes could be determined. Fortunately, the distances calculated in this way can be checked for a large number of Cepheids by determining their distances through luminosity and the Hertzsprung-Russell diagram. Cepheids are bright enough to allow measurements throughout the Milky Way and even as far as a few close galaxies.

Each of the above methods is used to determine successively greater distances than the preceding one, and each is based on the assumption of the accuracy of the information obtained by its predecessor.

Stellar Evolution

Stars originate as the gases and dust in nebulae are drawn together by gravitational attraction. As the density grows, the temperature of the compressed gases rises until the components in the core of the star reach temperatures at which nuclear reactions begin. The most important of these reactions fuses hydrogen into helium. Stars contain enough hydrogen that they can "burn" for millions of years, but eventually, the quantity of hydrogen decreases, and finally, hydrogen is depleted. A complex reaction then occurs in the star, causing it to expand and become a red giant. At this time, radiation from the core drives off the outer layers of the star. These form a cloud of gas surrounding what was once the core. The core originally rich in hydrogen is now mainly carbon and oxygen. In massive stars, the gravitational pull of material toward the core is great enough to overcome the remaining outward pressure in the star. At this point, the star may collapse or may even explode.

The sky contains many stars that are in these last stages of stellar life. White dwarf stars are among these. These are hot stars in which the nuclear fuels have been expended. The density of white dwarfs is exceedingly high. If the Sun collapsed to form a white dwarf, its diameter would be 1/100 of its present value. Because the stars are so small, they are dim. Because the dwarfs have such high density, they may attract material from a companion star orbiting around it. The material may include hydrogen, which forms a new shell on the old hydrogen-depleted core. Under these conditions, the hydrogen is extremely unstable and may explode (figure 22.9). If explosions occur, a brilliant burst of light comes from the star, and stars that were previously invisible suddenly become bright. These outbursts are called *novae,* a word that means "new star."

Stars have been known to appear suddenly where no stars had previously been known to exist. Actually, these are now known to be faint stars that suddenly become bright in a burst of energy that may increase their apparent brightness by 10,000. In some instances, more than one explosion occurs. After this outburst, novae

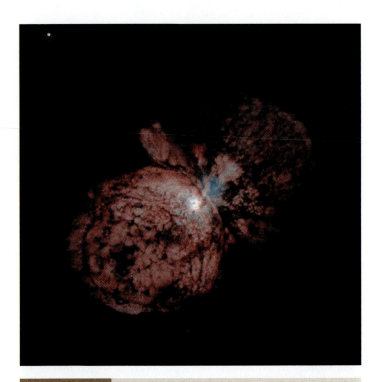

Figure 22.9   An Exploding Star. Eta Carinae turns brighter after its explosion. The lobes that extend out away from the star contain dust that makes them reddish. The vein-shaped stripes that extend out from the star may be lanes of dust or condensation.

rapidly return to their original dim state. The largest nova, called **supernova,** follows one of several courses. High-mass stars use all their nuclear fuel and form a core of iron. These cores collapse, and the supernova explodes. Low-mass stars turn into white dwarfs. If the dwarfs are part of a binary system, they may be able to attract enough mass to collapse and explode.

It has been estimated that several, perhaps as many as two dozen, such explosions occur each year within the Milky Way. After these outbursts, a cloud of gases is usually left. The explosion of a supernova produced the Crab Nebula, a cloud of gases that appeared about the year A.D. 1000, and has been expanding ever since (figure 22.10).

ORIGIN AND EVOLUTION OF THE UNIVERSE

One of the ultimate questions of science is "How did the universe originate?" Most astronomers currently favor a theory for the origin of the universe that postulates that all matter in the universe was concentrated in a relatively small space about 13 to 14 billion years ago. Extremely high pressures and temperatures within that initial cloud of matter were too high for elements to exist; matter consisted instead of a cloud of subatomic particles including protons and electrons. From this state, the matter of the universe began to expand. As the universe expanded, it cooled, and gravity drew the gas into clumps that eventually became galaxies. The Milky Way is one of these. Within these clumps, gravity continued to draw material into yet smaller clumps that became stars. The Sun is one such star.

Evolutionary Theory

The idea that the universe originated in one "central" place in space; that all matter was there at one time; and that since that time, this matter has been expanding and evolving is supported by many observations. As the largest telescopes scan space, they detect what appears to be a rather homogeneous distribution of galaxies. However, these galaxies are in motion. Studies of the spectrum of light from each galaxy reveal that the lines of the spectrum are shifted toward the red end. This is interpreted to mean that the galaxies are moving away from us. It is further observed that as the distance to a galaxy increases, the speed of its movement away from us increases as well. This prompts the generalization that the farther away a galaxy is, the greater its speed. Some of these velocities are close to the speed of light. The simplest way to envision the expansion and the observed effect of all galaxies moving away from Earth is to imagine that all of the galaxies are represented by dots on a balloon. As the balloon expands, the dots all move away from one another and out from a central region.

Most astronomers and astrophysicists believe the evidence currently available supports the idea that the universe originated about 13 to 14 billion years ago and that it formed after a gigantic explosion, the "Big Bang." Immediately after this event, all of the matter in the universe was a hot soup composed of the particles from which atoms would quickly evolve. These subatomic particles included electrons and their positive counterparts (positrons) plus particles known as photons (light), neutrinos, and antineutrinos, which have very little mass. After the explosion, this mass of matter and radiation, at a temperature on the order of 100 million degrees, expanded rapidly and has continued to expand to the present. Within a fraction of a second after the explosion, as matter separated and cooled, neutrons and protons began to form, and they quickly combined to form stable atomic nuclei. Hydrogen and helium were the first elements created in this cloud of matter. Once the universe cooled and galaxies and stars formed, other, heavier elements developed by the fusion in stars. Fusion of helium produced the nucleus of carbon atoms that, in turn, may fuse and break down into such elements as oxygen, magnesium, sodium, and neon. Oxygen atoms can fuse to form sulfur and phosphorus.

Microwave radiation from space has made it possible to study conditions when the universe was very young and before stars and galaxies formed. The picture obtained with microwave telescopes shows only irregular gas clouds. There were no stars or galaxies to see when the radiation was born just 300,000 years after the "Big Bang." Two billion years later, these clumps had become much more distinct and are recognized on images obtained with the Hubble telescope as galaxies. Images of objects that existed between 10 and 11 billion years ago depict a universe with many more galaxies than now. These images indicate much more interaction between galaxies than once thought. The new Hubble images indicate that collisions between galaxies were more frequent in the past and that causes their shapes to change. Small disc systems merge into objects that are nearly spherical.

In the future, more observations will be added to the remarkable record the Hubble telescope revealed. A new spectrograph will be installed on Hubble in a few years, and work progresses on the next generation of space telescopes. Scientists are hopeful that it will show images of the formation of new galaxies. A new telescope sensitive to infrared radiation will soon be capable of receiving radiation that penetrates the clouds of dust and of recording images of stars being formed. Clearly, exciting times lie ahead.

Figure 22.10   The Crab Nebula.

SUMMARY POINTS

1. The composition of stars and clouds of gases and dust is determined by use of spectral analysis, which includes both bright lines and dark lines. The dark lines represent parts of the spectrum that are absorbed by gases. On the basis of their spectrum, stars are classified into one of seven classes. These classes provide an indication of the temperature of the body.

2. Galaxies are huge accumulations of stars and clouds of gases and dust. Many of these accumulations have a spiral or spherical shape; others are irregular in shape. The solar system is located in a galaxy known as the Milky Way, which is a large rotating disc-shaped body.

3. New stars, called novae, originate within the clouds of gases and dust, called nebulae, that exist in all galaxies. In addition to nebulae, galaxies also contain globular clusters of stars, Cepheid variable stars, and black holes formed by such extreme accumulations of matter that light cannot escape. Bursts of x-radiation, derived from quasars, come from hot gases falling into black holes.

4. Parallax and the luminosity of a star are used to determine the distance to a star. The distance to galaxies is often found from the period of Cepheid variables.

5. By comparing the position of a star against the celestial sphere at different times of the year as Earth moves around the Sun, the movement of stars relative to Earth can be determined.

6. The shift of spectral is used to determine the speed with which stars or galaxies move away from Earth. This shift is a result of the Doppler effect.

7. As viewed from Earth, most galaxies appear to be moving away from the same point in space. The time needed for all of them to return to this point and the initial extremely high temperature and density is about 13 to 14 billion years.

REVIEW QUESTIONS

1. Where in the sky are the celestial poles and the celestial equator (see appendix E)?
2. How can you describe the location of a star by using its right ascension and declination (see appendix E)?
3. What is the source of most information about the composition of celestial objects, including the Sun?
4. What does the color of a star tell us about the physical condition of the star?
5. Where is Earth located in the universe?
6. What is a black hole?
7. What are quasars, and how are they related to black holes?
8. What important information has been obtained by use of the Hertzsprung-Russell diagram?
9. What is the Doppler effect, and how is it used to determine the movement of stars?
10. Outline modern ideas about the early history of the Universe.

THINKING CRITICALLY

1. Compare the application of the various methods used to determine the distance to stars.
2. What information about the universe is obtained by use of telescopes, X rays, and radio waves?
3. Why do so many astronomers think other planets exist in the universe?

KEY WORDS

black hole	Hertzsprung-Russell
celestial sphere	(H-R) diagram
cepheid variable star	luminosity
continuous spectrum	nebulae
dark line absorption	parallax method
spectrum	planetary nebulae
Doppler effect	proper motion
emission spectrum	quasar
galaxy	radial speed
giants	supernova
globular cluster	trigonometric method

Units and Conversions

Both English and Système International d'Unités (SI units) are used in this text. Although the English units are more familiar to students in the United States, the SI units have the advantage of being based on the decimal system. Computations are much easier with SI units. Tables A.1 through A.6 show some common conversion factors.

METRIC PREFIXES

mega	10^6	milli	10^{-3}
kilo	10^3	micro	10^{-6}
centi	10^{-2}	nano	10^{-9}

Table A.1   Conversion Factors for Distances

English	Metric	Metric	English
1 inch	= 2.54 centimeters	1 centimeter	= 0.39 inch
1 foot	= 0.305 meter	1 meter	= 39.36 inches = 3.28 feet
1 fathom	= 6 feet = 1.83 meters	1 kilometer	= 0.62 mile
1 mile	= 1.61 kilometers		

Table A.2   Conversion Factors for Areas

1 square centimeter = 0.155 square inch	1 square inch = 6.45 square centimeter
1 square meter = 10.8 square feet	1 square foot = 0.093 square meter
1 square kilometer = 0.386 square mile	1 square mile = 2.59 square kilometers
	1 acre = 43,560 square feet
	1 square mile = 640 acres

Table A.3   Conversion Factors for Volume

1 cubic centimeter = 0.06 cubic inch	1 cubic inch = 16.39 cubic centimeters
1 liter = 1.06 quarts	1 cubic foot = 7.48 gallons = 0.028 cubic meter
1 cubic meter = 35.3 cubic feet	1 quart = 0.95 liter
1 cubic kilometer = 0.24 cubic mile	1 gallon = 0.13 cubic feet = 3.78 liters
1 acre foot = 43,560 cubic feet = 1,234 cubic meters	

Units of temperature

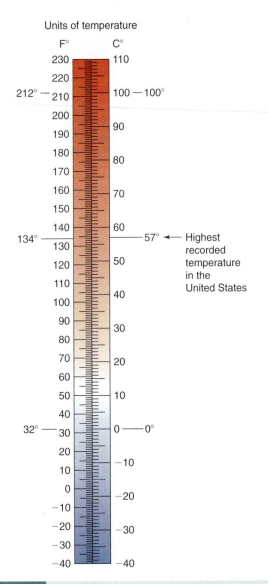

134° — .... — 57° ← Highest recorded temperature in the United States

Figure A.1 Conversion Chart for Temperature.

Table A.4	**Conversion Factors for Mass**		
1 gram	= 0.035 ounce (avoirdupois)	1 ounce	= 31.3 grams
1 kilogram	= 2.205 pounds	1 pound	= 0.45 kilogram
1 metric ton	= 2,205 pounds		

Table A.5	**Conversion of Pressure**
1 pascal (Pa)	= 1 newton (N) per square meter
	= 1 kilogram per meter × seconds
1 pascal (Pa)	= 1.013×10^{-5} bars
1 bar	= 1 atmosphere = 10^5 pascals
1 bar	= 1,000 millibars (mb)
1 inch of mercury (for air pressure)	= 33.863 millibars (mb)

Table A.6	**Conversion of Temperatures**

See figure A.1.

$°C = (°F − 32)/1.8$

$°F = (1.8 \times °C) + 32$

$°K = °C + 273$

Minerals

PHYSICAL PROPERTIES

All materials exhibit a range of physical properties including such characteristics as density (mass per unit volume), optical properties (the way they transmit and reflect light), electrical properties, and behavior in magnetic fields. Of the many possibilities, geologists generally use physical properties that can be determined quickly and without the use of laboratory equipment as a way of identifying minerals in the field. For this reason, physical properties—hardness, luster, color, color of the powder of a mineral, the way a mineral breaks, and its density—are most frequently used in mineral identification.

Hardness refers to the ability of one mineral to scratch another. A scale of ten minerals, called **Mohs' hardness scale,** table B.1, is used as a basis for hardness comparisons. For the first nine minerals, the scale is nearly linear (that is, a mineral of hardness 5 is almost five times as hard as a mineral of hardness 1, but diamond, which has a hardness of 10, is several times harder than corundum, which has a hardness of 9).

Luster is the appearance of a mineral's surface in reflected light. Geologists classify all minerals as having the luster of metals (such as copper or pyrite) or nonmetals (such as quartz or calcite). Nonmetallic lusters may be further subdivided into classes, such as glassy (vitreous), resinous (waxy), silky, or dull.

Color can be used to help identify most metallic minerals, but color is unreliable in identifying many nonmetallic minerals because the color may be caused by small amounts of impurities rather than the main components of the mineral itself. Minerals, such as quartz or calcite, may occur in many different colors.

Streak, the color of the powder of a mineral, is more diagnostic than the color of the mineral. Although the color of many nonmetallic minerals varies, most of them have a white or colorless streak. In contrast, for metallic minerals, both the color of the mineral and the color of its streak are characteristic.

All minerals will break. Some minerals tend to break along smooth planes and are said to exhibit **cleavage.** The way in which a mineral breaks is governed by the internal arrangement of the atoms and by the strength of the bonds between atoms and atomic groupings in the mineral. If the bonds are consistently weak in certain directions, the mineral will break across the weak bonds, producing smooth planes called cleavages. Among minerals that exhibit cleavage, differences exist in the number of cleavages, the angles between cleavages, and smoothness of the cleavage surfaces.

Minerals that do not have differences in bond strength do not break along cleavages. They may break into splinters; fracture smoothly and evenly; or break to form rough, irregular surfaces. A few minerals, such as quartz, break with a smooth, curved fracture known as a conchoidal (shell-like) fracture.

Density of a material is the mass per unit of volume. Although this physical property is not frequently used for hand specimen identification, density can be measured in the laboratory and is a valuable property for mineral identification. (Table B.2.)

CRYSTALS AND CRYSTAL FORMS

Because the internal arrangement of atoms in minerals is systematic, minerals tend to grow into regular geometric shapes. The regular arrangement of atoms is known as a *lattice.* The lattice determines the shape or shapes of crystals a particular mineral will have if the mineral is free to grow. The crystal shape may be used as a means of identification. There are fourteen types of lattices, but some of these produce the same form. For example, three different arrangements of atoms produce an external cubic form.

Unfortunately, the space into which minerals generally grow is constrained by other minerals that are growing at the same time. Thus, well-formed crystals grow under only special circumstances. Many of the most perfect crystals grow into water-filled spaces such as cavities in lava flows or along open cracks in rocks. Generally, the external shape is governed by the space in which the mineral grows, but if growth is unrestricted, the mineral may grow into a regular geometrical shape called a **crystal** (see figures 1.10 and 2.1).

Table B.1	Mohs' Hardness Scale	
1. Talc		6. Feldspar (a file)
2. Gypsum	(fingernail)	7. Quartz
3. Calcite	(copper penny)	8. Topaz
4. Fluorite		9. Corundum
5. Apatite	(steel knife, plate glass)	10. Diamond

Table B.2 Mineral Identification Chart for Select Minerals

Minerals selected for this list are common rock-forming minerals and a few minerals of economic importance.

Metallic Luster	Hardness	Color	Distinctive Features
Graphite, C	1	Gray	Greasy feel
Native copper, Cu	2.5	Copper red	Ductile; familiar color
Galena, PbS	2.5	Gray	Color; cubic crystals; high density
Native gold, Au	2.5	Gold	Familiar color
Chalcopyrite, $CuFeS_2$	3	Brass yellow	Slightly different in color and hardness from pyrite
Hematite, Fe_2O_3	6	Red or black	Red streak; one variety, specularite has steel-gray color and is metallic; other varieties occur as compact masses or oolites
Magnetite, Fe_3O_4	6	Black	High density; magnetic
Pyrite, FeS_2	6	Brass yellow	Resembles gold; is called "fool's gold"
Nonmetallic Luster			
Talc, $Mg_3(OH)_2Si_4O_{10}$	1	Variable	Pearly luster; greasy feel
Gypsum, $CaSO_4$	2	White	Some forms are fibrous with silky luster; massive forms are soft
Clays, complex silicate	2	Usually white	Greasy feel; will not effervesce in acid
Sulfur, S	2	Yellow	Distinctive color; no cleavage
Halite, NaCl	2.5	Colorless	Cubic crystals and cleavage; taste
Mica group, silicates	2.5	Black, pink, or white	Single perfect cleavage
Serpentine, complex	2.5	Greenish or black	Greasy feel; distinctive luster
Calcite, $CaCO_3$	3	Variable	Perfect rhombohedral cleavage (3); effervescence
Sphalerite, ZnS	3.5	Brownish yellow	Resinous luster
Dolomite, $(Ca,Mg)CO_3$	3.5	Variable	Looks like calcite; does not effervesce in dilute acid
Bauxite, Al hydroxides	Variable	Variable	Distinctive texture; round pea-shaped concretions
Fluorite, CaF_2	4	Variable	Four perfect octahedral cleavages
Limonite, iron oxide	5	Rust	Color
	5	Black	Hexagonal crystals; imperfect cleavage
Hornblende, complex silicate	5.5	Black	Long prismatic crystals; two cleavages not at right angles
Augite, complex silicate	5.5	Greenish black	Two cleavages at right angles
Feldspar group	6	Variable	
Orthoclase, $KAlSi_3O_8$		Pink	
Plagioclase, $(Na,Ca)AlSi_3O_8$		White to grayish	Striations; one variety has a play of colors
Garnet, complex silicate	6.5	Variable	Deep wine-red color is distinctive of one variety
Olivine, $(Mg,Fe)_2SiO_4$	6.5	Variable	Usually greenish
Quartz, SiO_2	7	Variable	Hexagonal crystals; hardness; great variety of forms
Diamond, C	10	Variable (colorless)	Hardness

HOW TO IDENTIFY MINERALS AND ROCKS

As you become more familiar with minerals and rocks, you will recognize some types immediately. For others, it will be important to know how to distinguish two types that resemble one another. Do not try to memorize all the physical properties of all minerals. Learn the distinguishing characteristics. A good initial approach is outlined as follows.

Step 1 Does the mineral have metallic or nonmetallic luster?
Step 2 Is the hardness greater or less than glass (H = 5.5)?
Step 3 Does the mineral exhibit cleavage? If yes, how many, and what are the angles between them?

Step 4 What minerals does this one resemble?
Step 5 Are distinguishing characteristics present (e.g., color, striations, play of colors)?

COMMON SILICATE MINERALS

Several silicate minerals are especially important because they occur so frequently as rock-forming minerals.

Quartz Quartz (silicon dioxide [also silica], SiO_2) crystallizes from cooling magma; it forms in metamorphic rocks at high temperature, and it may also crystallize from water circulating near the surface of the ground. Hard and durable, quartz is nearly insoluble in water, and it remains intact while other minerals in rocks containing quartz chemically break down. Thus, the quartz becomes a major component of sediments such as sand. It is the main mineral in most sand, and it occurs mixed with clay and other products of rock weathering and disintegration in muds and other sediment.

Quartz, along with calcite, occurs in a greater variety of forms than other minerals (see figure 2.1). Silica develops as a result of crystallization of silica-rich magma, chemical precipitation, and organic activity. Some animals build their hard skeletal parts out of silica, and occasionally, silica deposited from water that passes through the sediment replaces the shells of invertebrate animals and the woody matter of petrified wood. Silica also occurs as veins in all types of rocks and as cement in sediments composed of fragments. Sometimes, quartz crystals line cracks in rocks where silica has chemically precipitated from water moving through the crack. Most large crystals of quartz form in this way. Hot waters at some springs deposit silica, and nodules composed of silica, known as *flint,* or *chert,* occur in marine sediments where quartz formed while sediment changed into rock. Chert is extremely fine-grained, containing aggregates of crystals so small they cannot be seen with a microscope.

Feldspar Group Feldspars are the most abundant minerals in the Earth's crust. Most feldspars form at high temperatures and are especially prominent as components of many igneous and metamorphic rocks (see figure 2.7). Unlike quartz, in water—such as ordinary rain, which contains carbon dioxide—feldspars slowly decompose through chemical alteration to form clay minerals. For this reason, feldspars weather and may be absent from sediments derived from rocks that contain feldspars. Like quartz, the feldspars are silicates; however, unlike quartz, all feldspars do not have the same composition. Feldspars contain aluminum as well as silicon, bound together in a three-dimensional network of silicon-oxygen and aluminum-oxygen tetrahedra. Positive ions of potassium (K), sodium (Na), and/or calcium (Ca) bind the tetrahedra together.

Some feldspars, referred to as potassium feldspars or orthoclase, contain potassium ($KAlSi_3O_8$). The other main group of feldspars, **plagioclase feldspars,** contain sodium or calcium or both sodium and calcium together. Sodium and calcium ions can substitute for one another in the crystal structure. The amount of sodium and calcium in members of the plagioclase feldspar group varies continuously from all sodium ($NaAlSi_3O_8$) to all calcium ($CaAl_2Si_2O_8$).

Ferromagnesian Minerals—Olivine, Pyroxene, and Amphibole

Minerals that contain large amounts of iron and magnesium constitute the **ferromagnesian minerals.** These minerals crystallize only at high temperatures from magmas or in metamorphic rocks. Because they decompose rapidly when exposed to the atmosphere, they occur less commonly than quartz in sediment.

Olivine Group Olivine is the main component of Earth's mantle, and it is also the principal component of stony meteorites, which are probably fragments of a disintegrated planet. Olivine frequently occurs in igneous rocks that contain large quantities of iron and magnesium, but it does not generally occur in quartz-rich rocks that have a high silica content. Magnesium and iron can substitute for one another in the olivine crystal lattice. Consequently, members of the olivine group may contain both magnesium and iron. In these minerals, the atoms in the olivine crystal lattice form a relatively open pattern. Consequently, when olivine is subjected to high pressure deep inside the Earth, the atomic configuration collapses into a more dense packing. This higher-density mineral, called spinel, has the same composition as olivine but a different atomic structure.

Pyroxene Group Minerals of the pyroxene group occur in igneous rocks that crystallize from magmas rich in magnesium and iron. The pyroxenes are silicate minerals in which the silicon-oxygen tetrahedra join to form single chains. Varying amounts of magnesium, iron, and calcium hold these chains together. Lithium, sodium, and aluminum may also be present. These metal ions join chains of silicon-oxygen tetrahedra that share oxygen atoms on two corners. The general formula for the pyroxenes is $Ca(Mg,Fe)Si_2O_6$.

Amphibole Group Because the lattice structure of the amphiboles allows easy substitution of one ion for another, a wide range of chemical compositions occurs. In amphiboles, calcium, sodium, potassium, manganese, magnesium, iron, aluminum, and some other less common ions hold double chains of silicon-oxygen and aluminum-oxygen tetrahedra together.

Layer Lattice Silicates—Micas and Clay

A number of silicate minerals including mica and clay are characterized by a layer-type crystal lattice. These layers are of two types. One consists of silicon-oxygen tetrahedra bound together in sheets by sharing of oxygens. A second type of layer consists of octahedral arrangements of hydroxyl (OH) ions and magnesium or aluminum ions. These sheets are joined together in various combinations. Often they are linked by ions of calcium, sodium, iron, or potassium.

The micas are common constituents of all rock types. The group includes black mica, or **biotite;** white mica, or **muscovite,** both of which occur in igneous and metamorphic rocks; and glauconite (also called greensand), which occurs as rounded, fine-grained aggregates in marine sediments. Because the bonds between layers in the micas are weak, the minerals of this group have a sheet-like structure referred to as perfect cleavage—the most distinctive physical property of the mica group. Cleavage is the tendency of minerals to break in certain planes. These planes are weak because the strength of the bonds holding atoms together varies. In the case of mica, bonds within the sheets are much stronger than the bonds between sheets of atoms.

Geologists use the term *clay* as a size designation to describe sedimentary particles that are less than 2 microns (1/512 millimeters) in diameter. When used in this way, the name clay may be applied to any mineral of that size in a sediment. But the term **clay**

mineral applies to a group of silicates similar in structure to the micas. In these minerals, the silica-oxygen tetrahedra form sheets held together by metal ions. Aggregates of clay minerals have distinctive physical properties. Many clay minerals can absorb water, and when they are wet, they become soft and pliable.

Clay minerals are mainly the result of chemical weathering of other silicate minerals such as feldspar and the ferromagnesian minerals. They are components of soils and many sedimentary rocks, especially shale.

CARBONATE MINERALS

The carbonate ion (CO_3^{-2}) (see figure 2.8) is the essential structural unit of minerals in the carbonate group. Although it occurs compounded with manganese, iron, barium, strontium, and other positive ions, by far the most abundant carbonates are those of calcium, $CaCO_3$ (which has the mineral name *calcite*), and a combination of calcium and magnesium, $CaMg(CO_3)_2$ (the mineral *dolomite*). Because these two minerals are so abundant in the rocks found at the Earth's surface, special attention is focused on them here.

Calcite Calcite and quartz probably occur in a greater variety of forms than any other mineral (see figures 2.6 and 2.8). Most of the calcium in the oceans is derived from weathering of calcium-bearing minerals like plagioclase in igneous rocks. Ions produced by weathering are carried into the sea by water. Under certain conditions, these ions combine to form calcium carbonate, the mineral calcite, which is the primary mineral of which the rock limestone is composed. Limestones develop in both marine and nonmarine waters. Many limestones now exposed on land developed in the shallow waters of the continental shelf. Over geologic time, some of these shelf areas have been uplifted above sea level.

Many marine organisms use calcium carbonate derived from seawater to build shells and other hard parts that eventually become constituents of sediment. For example, the chalk cliffs of Dover are composed of billions of minute shells of single-celled plants and animals. Calcium carbonate may also be chemically precipitated from seawater as it evaporates. Limestone is made up of compacted and consolidated deposits of calcium carbonate. When limestone is subjected to heat and pressure, a metamorphic rock called *marble* is formed. Because calcium carbonate is soluble in rainwater, rocks containing calcium carbonate, such as limestones or marbles, may be dissolved. The calcium carbonate moves as ions in streams or water underground until it is either carried back to the ocean or deposited as calcite elsewhere in the crust. Calcite formed through chemical precipitation occurs in veins usually in limestone and around springs, collects in cavities, and forms coatings on plants where a fine mist of water is produced. Most of the deposits found in caves result from chemical precipitation of calcite from evaporating water.

The structure of calcite resembles that of sodium chloride described earlier, except that the crystal lattice has a rhombohedral shape (see figure 2.4) rather than a cube. Calcite has a hardness of 3, is nonmetallic, varies greatly in color, and may also be nearly clear. It has three cleavages, none of which intersect at right angles, and it breaks along these cleavages to form nearly perfectly shaped rhombohedrons. If dilute hydrochloric acid is placed on calcite, it fizzes as carbon dioxide is released. Its solubility in water is greatly increased if carbon dioxide is present. Unlike most other minerals, calcite becomes more soluble at lower temperatures.

Rock Identification

Most rocks are aggregates of minerals. They form as products of igneous activity, metamorphism, or consolidation of sediments. Although they form under very different conditions, rocks of different types often resemble one another. All rocks are classified according to texture and composition. Identification generally follows three steps.

Step 1 Examine the texture. Grain boundaries of igneous and metamorphic rocks generally form a mosaic pattern. Many sedimentary rocks consist of fragments of various sizes.

Step 2 Learn how to identify grains of minerals. All rocks are classified according to texture and composition.

Step 3 Learn how to distinguish samples that resemble one another.

Texture refers to the size, shape, and geometric relationships of the component particles or crystals that compose the rock. Most rock textures belong to one of two major textural classes—fragmental or crystalline (see figures 2.6 and 2.19).

Many sedimentary rocks and volcanic rocks consist of fragments. The general terms **clastic** (**pyroclastic** in the case of volcanic rocks) or **fragmental** apply to such materials. Clastic materials are further subdivided on the basis of the size and shape of the fragments. The size categories and names associated with them are clay (less than 1/256 millimeter), silt (1/256 to 1/16 millimeter), sand (1/16 to 2 millimeters), granule (2 to 4 millimeters), pebble (4 to 64 millimeters), cobble (64 to 256 millimeters), and boulder (more than 256 millimeters). The sizes indicated are the diameter of spheres, but perfectly spherical fragments are rare. The roundness or angularity of particles is an indication of shape, but some particles are best described as *platy* (e.g., flakes of mica) or *prismatic* (e.g., elongate columnar shapes).

Most igneous intrusive and metamorphic rocks and some sedimentary rocks have **crystalline textures** (see figures 1.12 and 2.2). In these rocks, mineral components of the rock have grown together during crystallization. As with fragmental materials, the size of the crystals is a useful way of distinguishing these textures. General terms such as fine-grained, medium-grained, or coarse-grained may apply to crystalline rocks.

Geologists commonly depict classifications of rock types in chart form. The general format of these charts (see figure 2.13) shows variations in composition plotted against textural varieties. Gradual transitions in both composition and texture are common. Although the distinctions between rocks that lie within these transitions are arbitrarily defined, for many samples, it is possible to assign the correct name without difficulty.

IGNEOUS ROCKS

Geologists apply several special terms to igneous rock textures. If individual minerals in igneous rocks are too small to be distinguish without magnification, the texture **aphanitic** applies, and if individual minerals can be distinguished in igneous rocks, the texture is **phaneritic.** If needed, exact measurements of crystal sizes provide greater precision.

In general, grain size of igneous rocks is an indication of the rate of cooling and crystallization. If cooling takes place so rapidly that crystals do not grow, **glass** (an amorphous material—one lacking internal crystalline structure) develops. Rapid crystallization produces fine-grained textures; slow cooling produces coarse-grained textures. In some instances, slow crystallization that produces coarse crystals (called phenocrysts) precedes rapid crystallization that produces fine crystals or glass. The mixed crystal sizes produces a texture known as **porphyry.**

Because liquid surrounds the first minerals that crystallize in a magma, these crystals may grow to form nearly perfectly shaped crystals. Minerals that crystallize later must occupy the spaces between earlier crystals and cannot grow into their natural crystal shape. Thus, it is possible to determine the order in which minerals in igneous rocks crystallize.

Textures of igneous rocks range from glass to coarse-grained and porphyritic. For purposes of identifying samples in hand specimens, one can separate specimens into five categories—glass, porphyry, vesicular, fine-grained (aphanitic), and coarse-grained (phaneritic). The crystals in a porphyry rarely exceed a few centimeters in diameter, and most coarse-grained plutonic rocks contain crystals less than a centimeter in diameter. Composition is generally estimated on the basis of the presence and abundance of a few common minerals—quartz, biotite, potash (orthoclase) feldspar, plagioclase feldspar, hornblende, pyroxene, and olivine. The composition of each of the major igneous rocks varies through a range. Thus, the name *granite* applies to rocks that contain quartz, potash feldspar, and hornblende, but the amount of each of these varies from one granite to another. Note that quartz and olivine do not ordinarily occur in the same rock, and it is unusual for pyroxene and quartz to occur together. Note also that each of the coarse-grained plutonic rocks has a fine-grained equivalent. These fine-grained rocks crystallized quickly either as extrusives or shallow intrusive rocks. Obsidian—natural glass—may have the chemical composition of any of the igneous rocks.

COMMON IGNEOUS ROCKS

Coarse-Grained Intrusive (Plutonic) Rocks

- **Granite** is a coarse- to medium-grained equigranular rock composed of potassium feldspars (orthoclase and microcline) and quartz, with some biotite mica and hornblende, usually present in small amounts. Other minerals may be present but in small amounts. These include black specks of magnetite, honey yellow crystals of sphene, red grains of garnet, and other less common minerals.
- **Migmatites** are rocks of mixed nature containing metamorphic and igneous-appearing rock. Sometimes the term is restricted to injection of granite into schist along the layering of the schist.
- **Graphic granite** is a rock of granitic composition that has a texture consisting of an intergrowth of orthoclase and quartz in such a way that cross sections resemble hieroglyphics.
- **Granite pegmatite** is a coarse-grained rock with granitic composition. It generally contains microcline, quartz, and mica (biotite or muscovite). A great variety of rare minerals may be present. Pegmatites are usually found near the borders of large plutons composed of rocks of finer-grain size. Water-rich solutions may have been present in sufficient volume to permit migration of ions to develop large crystals.
- **Syenite** is a coarse- to medium-grained equigranular rock composed primarily of orthoclase. Some hornblende, biotite, and pyroxene are present. Either a small amount of quartz or nepheline is usually present, depending on the degree of silica saturation.
- **Granodiorite** is a medium- to coarse-grained plutonic rock that contains quartz, calcic plagioclase, and orthoclase as light-colored constituents and biotite, hornblende, or pyroxene as mafic constituents. It contains at least twice as much sodic plagioclase as orthoclase.
- **Diorite** is a coarse- to medium-grained rock composed of intermediate plagioclase and biotite, hornblende, or pyroxene. Small quantities of quartz may be present. The amount of dark minerals ranges from 12 to 36% of the total.
- **Gabbro** is an equigranular rock composed of coarse- to medium-grained pyroxene, hornblende, and biotite. The amount of these may exceed the amount of feldspar, which is calcic plagioclase. The mineral olivine is usually present, and quartz is absent. Most gabbros are dark because dark minerals predominate. It is not uncommon to find gabbro pegmatites. Like the granite pegmatites, these are simply extremely coarse-grained equivalents of the gabbro, probably formed from residual patches of water-rich magma.
- **Peridotite** is a dark, coarse-grained, equigranular rock containing a large quantity of olivine and/or pyroxene or hornblende, but no quartz or feldspar.

- **Dunite** is a rock composed almost completely of the mineral olivine; accessory pyroxene and chromite may be present.
- **Pyroxenite** is a dark, coarsely crystalline rock composed mostly of pyroxenes.
- **Hornblendite** is an igneous rock of coarse texture and nearly equigranular fabric composed mostly of hornblende.

Fine-Grained Igneous Rocks

Note: Most fine-grained rocks are formed by solidification of lava extruded on the ground surface or by rapid cooling of magma around the edges of plutons or in thin, shallow bodies.

- **Felsite (rhyolite)** is a general term applied to igneous rocks of fine-grained texture and light color, indicating a granitic composition. If large crystals are present in a felsitic groundmass, it is called a *felsite porphyry*. Colors of felsites may range from white, gray, pink, yellow, brown, to purple and light green. When phenocrysts of quartz are present, the rock is termed a *quartz felsite* or, more commonly, a *quartz porphyry*.
- **Basalt** is a dark-colored, fine-grained rock composed primarily of plagioclase and pyroxene, with or without olivine. Basalts are dark because they contain large percentages of pyroxene, biotite, olivine, or other dark minerals. It is often an extrusive igneous rock and may contain cavities formed by bubbles of gas.
- **Diabase (dolerite)** is a rock of basaltic composition, consisting mainly of labradorite (a plagioclase feldspar) and pyroxene, with a texture characterized by euhedral or subhedral crystals of plagioclase embedded in a very fine-grained matrix of pyroxene.

Glasses

- **Obsidian** is a solid, natural glass containing no crystals. The name has in recent years been somewhat restricted to glasses having a very low water content as contrasted with pitchstones. Most obsidian is black, but it also occurs in green and brown colors. It breaks with a conchoidal (curved) fracture. Natural glasses frequently contain round or spherical bodies composed of incipient crystals of feldspar and silica. These may be white, gray, or red in color and often have a radiating, fibrous internal structure. These bodies are called **spherulites** and develop where parts of the magma cool more slowly than the surrounding glass. Also found in the natural glasses are the collapsed remains of glass bubbles, called **lithophysae.** They form as gases escape from the glass, the bubbles break, and the hot glass flows back down but does not completely melt again.
- **Pitchstone** is a variety of obsidian with a luster like that of a resin instead of glass. Pitchstone differs in composition from obsidian in that it contains 5% or more of water, whereas obsidian contains less than 1%. Pitchstone may be black, gray, red, brown, or green in color.

SEDIMENTARY ROCKS

Composition, indicated by the minerals present in the rock, and texture provide the basis for identification and classification of sedimentary rocks. Most sedimentary rocks belong in one of two major groups. The first group, clastic sedimentary rocks, consists of fragments. The second group includes materials that are so fine-grained that the individual constituents cannot be seen. Some of this second group are clastic sedimentary rocks composed of exceedingly small fragments or clay particles. Others in this second group have crystalline textures formed by chemical or biochemical precipitation. Figures 2.19, 2.20, and 2.21 illustrate some of the most common sedimentary rocks.

COMMON SEDIMENTARY ROCKS

Clastic Sedimentary Rocks

- **Conglomerate** is a rock consisting of rounded rock fragments generally of mixed sizes in excess of 2 millimeters. Most conglomerates develop as stream gravel and beach deposits where fragments become rounded as they move.
- **Breccia** is a rock consisting of angular rock fragments generally of mixed sizes in excess of 2 millimeters. Most breccias form along faults or where rock masses collapse, as in caves or along cliff faces.
- **Sandstone** is a general name for clastic rocks composed of fragments that fall in the size range of 2 millimeters to 1/16 millimeter. Some sandstones are almost pure quartz, but sandstone can be of any composition. Sandstones originate mainly from stream and beach deposits.
- **Siltstone** is a general name for clastic rocks composed of fragments that occur in the size range of 1/16 millimeter to 1/256 millimeter in diameter. Most siltstones are stream deposits.
- **Arkose** consists of sand-size particles composed in large part of feldspar and rock fragments. Arkose forms as stream deposits or on beaches where burial occurs rapidly before the feldspar breaks down as a result of chemical weathering.
- **Graywacke** consists of sand-sized particles including enough dark minerals to produce a grayish color. Graywacke is generally a stream or marine deposit containing fragments derived from volcanics.
- **Shale** consists of particles less than 1/256 millimeter in diameter. Clay is the predominant constituent, but quartz and limestone may be present. Shale forms from clays deposited in streams, lakes, or in marine environments. Because clay may remain in suspension, it may occur in sedimentary deposits formed far from the source of the clay.
- **Fragmental limestone** is composed of fragments of fossils or other limestone. Most fossiliferous limestones develop in shallow marine water, commonly on or near reefs.
- **Coquina** consists of a mixture of shell fragments and other fine-grained sediment. Most coquina deposits develop in shallow marine water.

Nonclastic Sedimentary Rocks

- **Oolitic limestone** is composed of small, spherical masses of calcium carbonate that has been precipitated in agitated water (as in the Bahama Banks region where waves wash over shallow banks).
- **Chert** (also known as **flint**) is a precipitate composed of silicon dioxide. Chert commonly occurs as nodules or thin layers in marine limestones. It also occurs in layers especially associated with bedded ironstone.
- **Micrite limestone** forms from recrystallized lime mud. These muds may occur in both shallow and deep marine or nonmarine waters.
- **Dolomite** is a fine-grained carbonate composed of calcium-magnesium carbonate. Dolomite develops today in a few localities in marine water. Dolomites formed long ago may be altered limestones.
- **Salt (halite)** is a fine- to coarse-grained precipitate of sodium chloride, generally found in dried lakes. Salt now develops in the Great Salt Lake in Utah.
- **Gypsum** is a fine- to coarse-grained precipitate of calcium sulfate, generally found in dried lakes.

Procedures for Identifying Sedimentary Rocks

1. Is the rock composed of fragments, visible crystals, or grains?
2. Does the sample contain quartz (H = 7)?
3. Identify fragments that are visible and measure the sizes of the fragments.
4. Locate the coarse clastic sample on the identification chart.
5. Test the sample with (1 drop) HCl. Rocks that effervesce are, or contain, calcium carbonate.
6. Check the distinguishing characteristics of the various types of limestones.
7. Very fine-grained sedimentary rocks that break in slabs and are gray, brown, or black may be shale.

See tables C.1 and C.2 for a summary of sedimentary rocks.

METAMORPHIC ROCKS

Common Metamorphic Rocks

- **Slates** are fine-grained, perfectly foliated rocks characterized by a tendency to break along nearly perfectly parallel foliations, called **slaty cleavage.** Slate is derived from shale. In the process of metamorphism, the clay minerals have been altered to small but aligned mica minerals. Individual minerals are too small to be identified in hand specimens.
- **Phyllites** are fine-grained rocks characterized by lustrous, silky sheen, caused by light reflected from the chlorite and muscovite micas of which they are composed. Quartz and plagioclase feldspar are often present. The grain size of phyllites is larger than that of slates but finer than that of schists. Phyllites are usually greenish or red and may show the initial stages of segregation of some mineral constituents into layers.
- **Schists** are foliated rocks of medium to coarse crystalline texture. Unlike phyllites, most of the mineral constituents of schists are easily identified by the parallel or nearly parallel

Table C.1 Clastic Sedimentary Rocks

Breccia

Angular fragments in fine matrix; fragments more than 2 millimeters

Conglomerate

Rounded fragments in fine matrix; fragments more than 2 millimeters

Graywacke

Fragments less than 2 millimeters; gray or brown; poorly sorted; quartz, feldspar, dark minerals

Arkose

Fragments less than 2 millimeters; various colors (usually reddish); contains abundant feldspar

Sandstone

Fragments less than 2 millimeters; usually composed mainly of quartz

Siltstone

Fragments less than 1/16 millimeter (very fine-grained); often composed of quartz

Shale

Exceedingly fine particles of clay size (less than 1/256 millimeter); color varies (black, tan, and gray are common); may contain calcium carbonate as an impurity; breaks into thin slabs (has rock cleavage)

Bioclastic Limestone and Coquina

Bioclastic limestone that may be compact; effervesces; coquina composed of loose shell fragments

Table C.2 Nonclastic Sedimentary Rocks

Micrite Limestone

Aphanitic in texture; commonly gray and dense; effervesces in HCl

Chert (flint)

Aphanitic in texture; commonly black, white, or pink; hardness of 7; will not react with acid

Chalk

White and very fine-grained; composed of protozoan shell fragments; will effervesce in HCl

Oolitic Limestone

Looks like fish-egg masses; commonly white or grayish; may look like sandstone; will effervesce

Dolomite

Looks very much like limestone, but may have a slightly granular appearance; gray to blue-gray is common color; will not effervesce in acid (but powder will)

Salt

Has all the same properties as halite but is more massive; resembles gypsum but has characteristic taste

Gypsum

Has properties of the mineral gypsum

Anhydrite

Similar to gypsum but is generally massive and granular in form; may be somewhat harder than gypsum and often contains impurities

---

alignment of micaceous minerals. The most common minerals in schists are quartz, feldspars, and micas. If one of the constituents makes up 50% or more of the rock, its name is attached as a modifier (e.g., mica schist, quartz schist, or hornblende schist). If no constituent comprises 50%, the names of the two most abundant constituents are used (e.g., garnetiferous-mica schist).

- **Gneisses** are metamorphic rocks exhibiting compositional layering that produces a banded appearance. Gneisses may be medium- to coarse-grained. Most contain quartz and feldspar interlayered with thin layers rich in hornblende or mica. The quartz, feldspar, and other constituents usually have interlocking boundaries. The layering is thought to develop as a result of segregation of mineral constituents during metamorphism. In some cases, it corresponds to original bedding of sedimentary rocks.

- **Marbles** are coarse-grained rocks formed by metamorphism of calcite, limestone, or dolomite. Impurities in the rock may impart a foliation, but generally, the texture consists of an interlocking mosaic growth of calcite or dolomite crystals.

- **Quartzites,** metamorphic equivalents of quartz sandstones, are composed of about 80% or more quartz. Tightly cemented quartz sandstones may resemble metaquartzites in that both often break across sand grains. However, the grain boundaries of metaquartzites have usually grown together and become interlocked, and the rock exhibits foliation.

- **Hornfels** are fine-grained, nonfoliated, dense, usually dark rocks formed near the contacts of igneous intrusions. The minerals form a mosaic of interlocking grains. Hornfelses commonly break into splintery fragments that have translucent edges, like horn.

- **Amphibolites** are composed of plagioclase and hornblende and biotite. Prismatic hornblende crystals may be aligned. Quartz-free amphibolite is usually derived from basalt or iron- and magnesium-rich tuffs. The presence of quartz suggests that the amphibolite may be derived from sedimentary rocks.

Topographic and Geologic Maps

TOPOGRAPHIC MAPS

A topographic map is a way of accurately representing the configuration of the land on a plane surface. Generalized impressions of landforms may be obtained by use of shading or hachure lines, but a topographic map is constructed so a close approximation of the elevation of any point on the map is possible. This is accomplished by use of contour lines that connect points of equal elevation. These lines are drawn at set, but arbitrarily determined, intervals. The contour interval might be 1 meter (3.3 feet) in a relatively flat area but 100 meters (330 feet) in a high mountainous region. An interval is selected to provide an optimum amount of information without having the lines merge. Normally, a heavier line is used to mark the elevation of every fifth contour.

Generally, points on the map that lie between contours have elevations intermediate between the elevations of the lines to either side. This generalization fails along the top of ridges or in the floor of a valley where the elevation of intermediate points may be above or below that of two contours of identical elevation. Contours may be closed, as they are around a hill that is entirely shown on the map, or they may run off the margin of the map. A single map may have more than one contour of the same elevation. For example, two hilltops may each be ringed by contours of the same elevation. Contours never cross one another unless they are superimposed over a vertical cliff or along an overhanging ledge. Where contours go around enclosed depressions, the contour lines have short hachures on the downslope side of the line.

The U.S. Geological Survey, the Coast and Geodetic Survey, and some states and private firms produce topographic maps. Government agencies produce maps at a variety of scales. Older maps were made at a scale of 1 inch to each mile. These are known as fifteen-minute maps because they cover areas fifteen minutes longitude wide and fifteen minutes latitude high. They have a scale of 1:62,500. One inch on the map is equivalent to 62,500 inches horizontally across the mapped area. More recent maps are produced at scales of 1:24,000, 1:50,000, 1:100,000, and 1:250,000.

In addition to contour lines, topographic quadrangle maps provide the following information:

1. Scale in miles, feet, and kilometers.
2. Cultural features such as houses, churches, cemeteries, county lines, park boundaries, and state boundaries.
3. Physical features such as rivers, lakes, marshes, levees, beaches, and mountain peaks.
4. The name of the map area and the names of those that adjoin at corners and along the sides are often printed at map margins.
5. Exact elevations of points that have been determined by ground surveying. These include bench marks (BM with elevations given) and triangulation stations (a small triangle with elevation), and exact elevations may be given for peaks and other prominent points.

Geologists frequently use topographic maps for interpreting landforms, for studying drainage systems, and especially as base maps on which other data are recorded. Topographic maps are also of great value in surveying, highway planning and construction, and in regional and city planning.

GEOLOGIC MAPS

Geologic maps show interpretations of the areal distribution of one or more of the following: (1) units of rock, called formations; (2) rock types; or (3) rocks of certain ages. Some geologic maps show the distribution of surficial materials, such as alluvium, sand dunes, or terrace deposits. Others show bedrock that may be covered by surficial deposits. Often, bedrock maps also include some surficial deposits. Most detailed maps show rock units, called *formations,* that geologists can recognize and map. The rock types and fossils it contains identify a rock unit. The rock units distinguish igneous, metamorphic, and sedimentary rocks from one another.

Lines, called *contacts,* are drawn to outline the area in which a map unit is known or inferred to be present. Many geologic maps use colors to help distinguish the various units whether they be formations, rock types, or rocks of different ages. Geologic maps usually show the location of faults and may also indicate where folds are located. A legend on the map contains key symbols used on the map. These symbols help identify the map unit, the type of fault, and the orientation of beds of sedimentary rock.

Maps that cover large areas, such as a state or country, may group all rock units of a single age together. Usually these maps separate igneous intrusions from sedimentary rocks of the same age. Geologic maps drawn at scales of 1:24,000, 1:50,000, 1:62,500, or 1:100,000 are generally drawn on a topographic map base. This makes it possible to locate the map unit contacts precisely. Maps on topographic bases show contours, cultural features, streams, and latitude and longitude.

Star Charts

Star charts (figure E.1) identify the most obvious stars in the sky and give the names of the constellations. Observers on Earth's surface can use the coordinate systems shown in figures E.2 and E.3 to locate objects in the sky. The observer needs to be familiar with the terminology used in these systems. As the observer looks toward the sky, all objects in space appear as points on a spherical surface called the **celestial sphere** (see figure E.2). **Zenith** is the point on the celestial sphere vertically overhead at any given moment. **Nadir** is the point on the celestial sphere opposite the zenith. A **great circle** on Earth or on the celestial sphere is a circle formed where a plane that passes through the Earth's center intersects the sphere. The **celestial horizon** is the great circle on the celestial sphere halfway between

Figure E.1 Star Charts. (a) The night sky in spring. (b) The night sky in winter.

(a)

zenith and nadir. **Vertical circles** are great circles that pass through the zenith and nadir. The **celestial meridian** for any observer is the great circle that passes through the zenith of the point on which the observer is standing; often it is used as a reference plane.

To describe a point on the celestial sphere with the above coordinates, azimuth and altitude are measured. The **azimuth** of a star is the angle from the north or south point of the horizon to the intersection with the horizon of the vertical circle that passes through the star. **Altitude** is the angle measured from the horizon to the star in the vertical circle that passes through the star. While the system using azimuths and altitudes is valuable for navigation and surveying, the position of stars is constantly changing because of Earth's rotation.

The second coordinate system is independent of Earth's rotation. It uses coordinates based on the position of the projection of Earth's poles on the celestial sphere. These two points, known as the **celestial poles,** remain essentially still. The polestar, Polaris, is close to the projection of the north celestial pole on the celestial sphere. The

celestial equator is the great circle on the celestial sphere halfway between the north and south celestial poles (see figure E.3).

Hour circles are great circles that pass through the north and south celestial poles. In this system, north is toward the north celestial pole; west is the direction of the apparent daily movement caused by Earth's rotation. The reference hour circle in this system is the one that passes through the point where the Sun's path crosses the celestial equator at the start of spring. In this system, the two measurements that are equivalent to longitude and latitude are called, respectively, right ascension and declination. **Right ascension** is the angular distance measured eastward in the plane of the celestial equator between the vernal equinox and the hour circle that passes through the star. **Declination** is the angular distance of the star north or south of the celestial equator (0° to 90°) measured in the hour circle passing through the star.

Hour angle is another measure frequently used with declination to define the position of a star. The hour angle is the angle between the observer's celestial meridian (the great circle that passes through

(b)

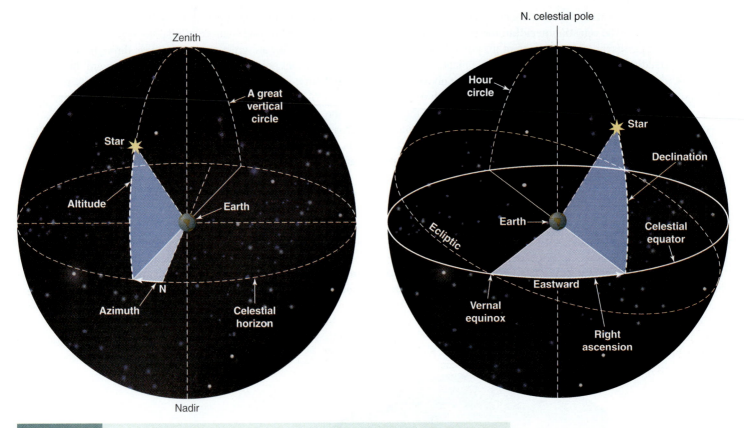

Figure E.2 Coordinate Systems Used to Locate Planets, Stars, and Galaxies in the Celestial Sphere.

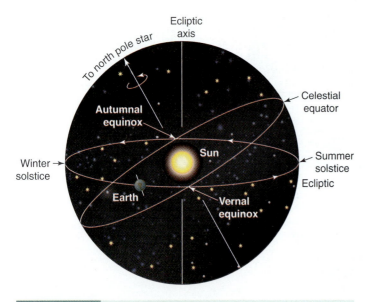

Figure E.3 Schematic Illustration Showing the Geometry of the Celestial Sphere, Equator, Ecliptic, Ecliptic Axes, and the Equinoxes.

north, south, and the zenith of the observer) and the hour circle of the star measured westward in the celestial equator. Unlike right ascension, which is nearly constant, the hour angle of a given star to a given observer changes with the hour of the day. All angles measuring right ascension, declination, or hour angle have a common apex, the center of the plane of the celestial equator, the point through which the axis between the north and south celestial poles passes.

These systems of coordinates make it possible for professional and amateur astronomers to track the movement of the planets in the solar system, the stars that lie far beyond, and to share that information.

appendix F
The Periodic Table of Elements

The periodic table of elements grew out of work by Dmitri Mendeleev in 1868. He formulated what he called the Periodic Law:

> If all the elements be arranged in the order of their atomic weights, a periodic repetition of properties is obtained. This is expressed by the law of periodicity; the properties of the elements, as well as the forms and properties of their compounds, are in periodic dependence.

The table contains an array of rectangles, one for each element, arranged in rows and columns. Each column is designated as a group. The elements in each group exhibit similar behavior. For example, the inert gases occur in a group at the right of the table. In general, the column number corresponds to the number of electrons in the outer orbit of the element. This is reflected in certain properties of the atom. The rows in the table are called periods or series. Each series contains one new set of electronic shells. All elements in group 1 have one electron in their outer orbit; all elements in series 2 have two orbits, and so on. Iron and nickel are transition elements that essentially belong to group 8. A full description of these elements, as well as a fuller description of the periodic table, can be found in introductory chemistry texts. Basic data about the elements are given in figure F.1.

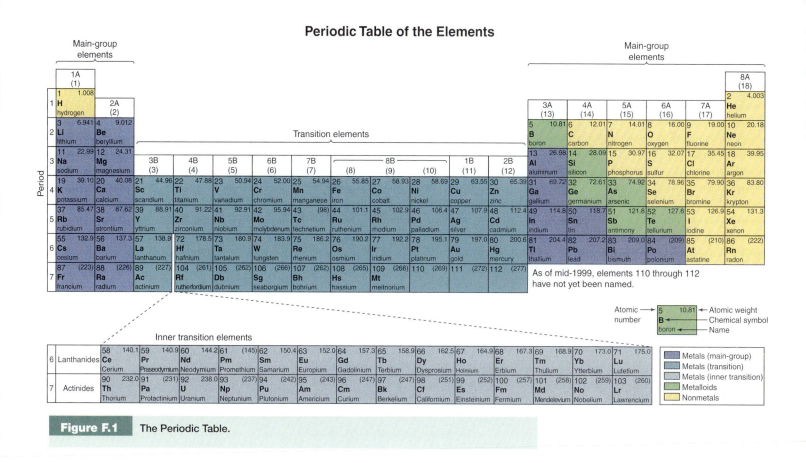

Figure F.1 The Periodic Table.

Glossary

The *Glossary of Geology and Related Sciences,* published by the American Geological Institute, has been used as the authority for geologic definitions in this book and its glossary. Students should refer to the A.G.I. *Glossary* for words not defined here, for various meanings of a word, and for the original sources of words.

aa the Hawaiian term for blocky lava. The flows have rough, jagged, and blocky surfaces.

ablation the combined processes, particularly melting and sublimation, by which a glacier wastes.

ablation till glacial till deposits formed from the debris that is transported within or on the ice and gradually lowered as ablation causes the glacier to shrink.

absolute humidity the amount of water vapor in the air expressed as the mass of water per unit volume of air.

absolute time time expressed in years. In geology, absolute time or age is determined from tree rings, varves, or radiometric dating.

abyssal plain flat area in the deep ocean having low relief; irregular in shape and found in a great range of sizes up to widths of several hundred kilometers.

accumulator plant plant that accumulates and stores certain elements or compounds.

acid rain this term is applied to rainwater that has a pH lower than ordinary rainwater as a result of the addition of atmospheric pollutants such as sulfuric or nitric acids.

adhesion the molecular attraction between adjacent surfaces.

adiabatic processes as used in meteorology, adiabatic processes involve no exchange of heat energy between a parcel of air and the surrounding air.

aftershocks earthquakes that follow a high-magnitude earthquake.

aggradation the process of building up a depositional surface.

alluvial referring to materials (sand, gravel, etc.) transported by and deposited from streams.

alluvial fan a fan-shaped deposit of stream alluvium laid down where a change in stream gradient occurs.

alluvium material transported by streams and deposited in stream valleys.

altitude the vertical distance of an object or a plane directly above Earth's surface.

amorphous solid a solid that does not possess a crystalline structure; for example, glass.

amphibolites metamorphic rocks that contain a high percentage of the mineral hornblende.

andesite a volcanic rock intermediate in composition between basalt and rhyolite. Andesitic lavas are usually found above subduction zones.

andesite line the line around the northern and western Pacific Ocean that separates volcanoes erupting basalt from those erupting andesite or other sialic volcanic products.

angle of repose the steepest slope on which a uniform, uncemented, granular material can maintain stability.

angular unconformity an erosion surface separating folded or tilted rocks below from less deformed rocks above.

anomaly (gravity) a gravity measurement that is greater (positive) or less (negative) than the measurement that would be predicted for a locality based on an idealized model of the Earth.

antecedent a drainage that predates the origin of structural features, such as ridges or anticlines, across which the streams flow. The streams maintain their courses while the structure develops in the underlying rock.

anthracite coal that is dark black in color, dull to brilliant, and even submetallic in luster. It is hard; burns with a short, blue flame; and emits little odor.

anticline an upfold in which the limbs dip away from the fold axis. Beds exposed in the core are older than those exposed on the limbs after erosion.

anticlinorium a fold system located on an uplifted area.

aphanitic a textural description in which individual crystals in a rock are too small to be seen or identified by the unaided eye.

aquifer a water-bearing strata or rock body.

arches a large-scale structural feature that has anticlinal form.

arc-trench gap the horizontal distance in an island arc between the axis of the deep-sea trench and the center of volcanic activity.

arête sharp ridge produced by glacial erosion.

argillaceous containing clay.

arkose a sandstone made largely of a mixture of quartz and feldspar fragments.

arroyo a dry stream channel; commonly found in the desert regions of the southwestern United States.

artesian aquifer a water-bearing rock unit that contains water under pressure.

artesian conditions situation in which ground water is under sufficient pressure to rise above the zone of saturation.

artificial levee a levee constructed along rivers.

ash a designation in size (1/16 to 2 millimeters in diameter) given to fragmental eruptives.

ash flow a hot mixture of volcanic ash and gases that travels down the flanks of a volcano or along the ground surface. The solid materials are generally unsorted and may include dust, pumice, scoria, or blocks.

assimilation the digestion or incorporation of materials originally in the wall rock of a magma chamber into the magma.

asthenosphere a zone in the Earth some tens of kilometers deep in which the rock is plastic and where any stresses are removed by flowage.

asymmetrical fold a fold in which the limbs are not symmetrical about the axial surface.

Atlantic-type margin continental margins of the type found around most of the Atlantic Ocean; characterized by absence of volcanic, seismic, and tectonic activity.

atoll a ring-shaped island with coral reefs that enclose a lagoon.

atom the smallest particle of an element that retains the chemical nature of the element. Atoms are neutral, spherical, and have a positively charged central nucleus surrounded by one or more negatively charged electrons.

atomic number the number of protons in the nucleus of an atom.

auroras light emissions caused by interactions between gases in the ionosphere and charged particles coming mainly from the Sun.

axial plane an imaginary surface that approximately bisects a fold into symmetrical halves.

axis an imaginary line formed by the intersection of the axial surface of a fold with a bedding surface.

azimuth the angle measured clockwise on a horizontal plane between the line to an object and the line formed directed toward true north or true south.

back arc basins the basins located behind island arcs.

backshore the part of the shore that is level or sloping landward, covered by water only during storms.

bajada the gentle, sloping ground surface located between a pediment and a playa or flats in arid regions; often restricted to surface formed by deposition.

bar a sand or gravel deposit submerged at shallow depth.

barchan dune a curved sand dune whose convex side faces the wind and whose wings point downward.

barrier beach a beach deposit formed offshore, such as those along much of the eastern coast of North America between Long Island and southern Texas.

barrier reef the name given to a coral reef that runs parallel to the shore of an island or continent, separated from it by a lagoon.

basalt a dark, fine-grained rock composed primarily of plagioclase and pyroxene, with or without olivine.

base level an elevation below which streams cannot reduce the level of the land by erosion. Sea level is referred to as the ultimate base level.

basin (structural) a circular or elliptical downwarp or structural depression, with younger beds in the center.

batholith a large pluton of intrusive rock, many square kilometers in areal extent.

bauxite the principal ore of aluminum, composed of hydrated alumina ($Al_2O_3 \cdot 2H_2O$).

bay-mouth beach a beach located across the mouth of a bay.

bayou type of stream characterized by sluggish drainage and located on the flood plain of a large river.

beach the gently sloping shore of a body of water, particularly the sea, which is washed over by waves or tides.

beach drifting movement of sediment along the shore by combined effects of incoming waves and return flow from the beach.

bed load sediment transported at the bottom of stream channels. The bed load moves mainly by rolling or saltation. The materials are too heavy to become suspended in the water.

beheaded stream a stream that has lost part of its drainage system as a result of stream piracy.

Benioff zone see *seismic shear zone.*

benthic community the plants and animals that live on the sea floor.

benthos forms of marine life that live on the sea floor. The term is also applied to the ocean bottom.

biogeochemical cycles the cycling of chemical constituents that pass through geologic and biologic systems.

biome the community of plants and animals that characterize a particular natural region.

biotite a member of the mica group of minerals in which iron and magnesium are important constituents. The mica is dark in color.

black smoker hydrothermal vents located on oceanic ridges that emit water blackened by sulfide precipitates.

blind valley valley that leads into a hillside or gradually loses the characteristics of a valley as water from its streams is lost to subsurface channels.

block glide a form of mass wasting in which large blocks of rock remain intact as they slide along bedding, fractures, or faults without rotation. Many block glides occur on low slopes.

blowouts a general term applied to depressions formed by wind erosion.

body wave seismic wave propagated throughout a three-dimensional continuum and not related to a boundary surface.

bog see *swamp.*

bomb fragment (32 millimeters in diameter or larger) of material erupted from a volcano.

boreal forest forest that grows just south of the tundra.

bottom-set beds the subhorizontal layers of sediment deposited in front of a delta.

Bouguer effect the attraction that the slablike mass of material between the geoid or sea level and the elevation at which a gravity reading is taken has on the value of gravity at that elevation.

Bouguer gravity anomaly any residue when the theoretical value of gravity for a given latitude calculated by the international formula is subtracted from the observed gravity at a station corrected for free-air effect, Bouguer effect, and topography.

braided stream a type of stream pattern in which the stream consists of a number of small channels that cross back and forth over one another, producing a braided appearance.

breaker a wave breaking on the shore, formed when the velocity of the top part of the wave begins to exceed the forward velocity of the wave as a whole; the top outruns the rest of the wave and spills over in front of it.

breccia a fragmental rock containing angular, instead of rounded, pieces.

brittle failure rupture or breakage characteristic of brittle substance, preceded by little flow or plastic deformation.

buffering resistance to changes in pH when a small amount of a strong acid or strong base is added to a solution.

calcareous containing calcium carbonate.

calcite the mineral composed of calcium carbonate.

caldera a large enclosed or partially enclosed depression caused by collapse or explosion of a volcano.

caliche surficial materials cemented together by soluble salts. It occurs as layers of calcareous material deposited at the surface or at shallow depth in the soil.

calorie the quantity of energy required to raise the temperature of one gram of water by one degree centigrade.

calving a process in which masses of ice split off to form icebergs where an ice sheet extends into the sea.

capacity a measure of the ability of a stream to transport suspended and bed load.

capillarity the action in which a fluid is drawn up in small interstices or tubes as a result of surface tension. (syn., capillary action)

carbonaceous containing carbon.

carbonation process by which carbon dioxide is added to oxides of calcium, magnesium, sodium, and potassium to form carbonates of these metals.

carbonic acid a weak acid formed by a combination of carbon dioxide and water.

catastrophism the concept that violent, short-lived events have greatly modified the Earth's crust, the organisms on Earth, or both of these.

cave a naturally formed underground cavity or void.

celestial equator the great circle on the celestial sphere formed by the projection of Earth's equator.

celestial horizon a circle that lies on the celestial sphere. Its position is defined by the plane that passes through the Earth and perpendicular to the line directed toward zenith or nadir.

celestial meridian half of a great circle on the celestial sphere that passes through the zenith of a given place and ends at the celestial poles.

celestial poles points on the celestial sphere formed where the projection of Earth's axis of rotation penetrates the celestial sphere.

celestial sphere an imaginary sphere of infinite radius on which all heavenly bodies except the Earth appear projected. Earth lies at the center of the celestial sphere.

Cenozoic era the most recent era into which geologic time is divided. It extends from the end of the Mesozoic era, about 70 million years ago, to the present.

central vent eruption a volcano that erupts from a single vent.

chalk a fossiliferous limestone composed of the shells of protozoans and particularly fossil globigerina. It is white and soft.

chemical weathering the breakdown of rocks and minerals at Earth's surface as a result of chemical processes.

chert (flint) a dense, hard rock composed of SiO_2, with color ranging from white through gray to black.

chondrite see *stony meteorite*.

chute a narrow passage of water between an island and the main bank of the stream; commonly found on the inside of curves along meandering streams.

cinder cone a cone-shaped hill formed from the accumulation of cinders and ash around a volcanic vent.

cirque an amphitheater-shaped depression formed by glacial quarrying or plucking where the glacier starts its movement down a valley.

cirrus clouds high clouds composed of ice crystals that have a wispy or filament-like appearance.

clastic sedimentary and volcanic materials that are composed of fragments.

clay mineral the term *clay* is applied to rocks or minerals composed of extremely small fragments with diameters less than 1/256 millimeters. Clay minerals are finely crystalline or hydrous silicates that contain aluminum. Many clay minerals also contain calcium, sodium, potassium, iron, and magnesium. Most of these minerals form through weathering of silicate minerals, such as feldspars, pyroxenes, and amphiboles.

cleavage (1) mineral: the property exhibited by some minerals of breaking along definite smooth planes; a manifestation of the internal orderly arrangement of atoms in a mineral. (2) rock: the property exhibited by some rocks of breaking along definite, smooth, subparallel, planes; caused by closely spaced fractures or by alignment of platy minerals.

climate the long-range averages of weather in a region.

closed system as used in chemistry, a closed system is one in which no matter is transferred into or out of the system.

coal vegetable matter that has been altered both physically and chemically through geologic processes to a black, rock-like substance.

coast a broad zone directly landward from the shore.

coccolith minute calcareous plates formed on some marine flagellate organisms.

cohesion the resistance a rock offers to shear deformation not caused by friction between particles in the rock.

col pass formed where two cirques converge, cutting into the same wall and thus lowering it below the level of the remainder of the summit area.

collapse sudden failure or breakdown of material as when the roof of a cave falls into the cave.

colloid a particular size of matter in the range of 10^{-5} to 10^{-7} centimeters in diameter.

colluvium a mixture of soil, stream deposits, and talus.

compaction process by which material is pressed together (for example, as a result of the weight of sediment deposited on a given layer).

competence the largest-sized particle (measured by diameter) a stream can move.

composite cone volcanic cone composed of both cinders and lava flows; such cones are circular in plan, and the sides are concave upward in profile.

compound a substance composed of two or more elements combined in fixed proportions.

compressibility an elastic modulus, a constant characteristic of a given material, indicating how much the volume of a material changes in response to applied hydrostatic pressure.

compressional seismic waves seismic waves that are transmitted as a result of the compressions and expansions of the medium through which they move. (syn., P waves or longitudinal waves) See *dilatational wave*.

compressive stress a stress applied (as in a vise) in such a way that material is forced together.

conchoidal fracture a broken surface that is curved or shell-shaped, as exhibited by glass.

concordant injections intrusions emplaced along planes of stratification, layering, or schistosity so that the borders of the intrusion are parallel to preexisting layers.

condensation the process in which a substance changes from gas to liquid phase.

cone of depression as water is withdrawn from a well, the water table around the well takes on the shape of a cone. The pointed end of the cone is located at the bottom of the well pipe.

confined ground water ground-water conditions in which the water in an aquifer is confined to that aquifer by an overlying impervious cover, which prevents its upward movement.

confining pressure term applied to nondirected pressure due usually to depth of burial in rocks.

conglomerate a coarse-grained sedimentary rock composed of rounded fragments that are more than 2 millimeters in diameter.

connate water marine or fresh water trapped in sediments when they are deposited on lake or sea bottoms.

consequent streams a stream that flows in the same direction as the inclination of the slope on which it originally formed.

contact metamorphic aureoles concentric zones of alteration around plutons due to the effects of heat and chemically active fluids in the magma on the country rock.

contact metamorphism the alteration of country rock around an intrusion near its contact.

continental accretion the theory that continents have grown by incorporation of mountain belts around their margins.

continental drift the theory that continents have moved relative to one another.

continental margin the zone of transition from crustal structure of the continents to that of the ocean basins.

continental platform the area around a continental shield that is covered by sedimentary rock. The term is also used as a synonym for continental shelf.

continental rise the gently sloping surface located at the base of the continental slope.

continental shelf a shallow, gently sloping surface of low local relief that extends from the shoreline to the shelf break, where the seaward gradient sharply increases to greater than 1:40.

continental shield the area of basement rock exposed in a craton.

continental slope the relatively steep portion of the sea floor that occurs at the seaward border of the continental shelf.

convection a general term applied to density-driven movement of liquids and other materials that flow. The term is applied to movements in the atmosphere, oceans, and in the core and mantle of Earth. Differences in density are usually caused by differences in the heating of the materials.

convergent plate boundary in plate tectonics theory, the boundary between the plates that are moving toward one another.

conveyor-belt model a general term applied to the pattern of movement of deep oceanic currents.

coordination number number of anions in contact with a cation in any given packing arrangement.

coquina sedimentary rocks composed of a loose aggregation of shell fragments cemented together.

core the central portion of the Earth's interior, bounded above by the mantle and thought to be composed of iron.

core-mantle system the processes that take place within Earth's core and mantle.

Coriolis effect the deflection of any object in motion on a rotating sphere.

country rock the rock in which an igneous intrusion is emplaced.

covalent compound a compound that consists of atoms bonded together by shared electron pairs.

crater a depression around a volcanic vent shaped like an inverted cone.

craton the stable portion of each continent; the shields and the surrounding areas.

creep (1) rock: extremely slow deformation of materials subjected to stress below the elastic limit. (2) soil: the imperceptibly slow movement of soil downslope.

crest the highest part of a fold form; also, the top of a waveform.

cross-bedding successive, systematic, internal bedding that is inclined to the principal surface of accumulation.

crosscutting relationship a term for the intersection of two features such as a dike or fault and a layer. The relationship is used to determine the relative age of the two features. The dike is younger than the layers it cuts across.

crust that part of the Earth above the M-discontinuity, approximately the outer 10 kilometers in ocean basins and the outer 20 to 40 kilometers under continents.

crystal a homogeneous body, the surfaces of which have grown into smooth planes as a result of the internal orderly arrangement of atoms.

crystalline solids solids that have an orderly internal arrangement of their atoms, molecules, or ions. These solids may have a well-defined shape as a result of their internal structure.

crystalline textures rock textures characterized by minerals that have grown together, forming a mosaic-like pattern.

crystallization separation of the components of a mixture through differences in solubility in which a component comes out of the solution as a solid. The solid may have the form of crystals.

cumulonimbus clouds cumulus clouds that produce precipitation. Generally, the name is applied to the large, high clouds associated with thunderstorms.

cumulus cloud a cloud that has a base within 2 kilometers of the surface. These clouds have fluffy shapes resembling cotton.

Curie point the temperature at which a material loses its spontaneous magnetization; about 570°C for ferromagnetic minerals.

cutoff the short channel formed where a stream cuts through the neck of a meander loop.

daughter isotope a nuclide formed by disintegration of a radioactive parent.

debris a general term applied to mixtures of rock, soil, plant matter, and mud.

declination (magnetic) the horizontal angle at a locality between the direction to magnetic north and true (geographic) north.

décollement "ungluing" of a bed or zone during which beds above slip over those below. Usually, the layers are composed of salt, gypsum, clay, or other weak rock.

deep-sea trench long, narrow trough-like depressions in the floor of the ocean. The deepest of these, the Mariana trench, is over 11.6 kilometers (38,000 feet) deep.

deflation blowing away of materials.

degassing emission of gases from Earth's interior. The term is usually used in relation to the idea that the atmosphere originated as a result of volcanic emissions.

degradation the general lowering of the land surface by erosion.

delta deposit of alluvial material, sometimes triangular, formed at the mouth of a stream.

dendritic having a branch-like outline.

denudation laying bare of the surface of the Earth; now applied to reduction of the level of the land by processes of erosion.

depth of compensation the depth in Earth at which the weight of different parts of the lithosphere is equal.

desertification the transformation of an area into desert.

desert pavement a tight mosaic cover formed by lag gravels that protects underlying material from wind erosion.

dew point the temperature at which atmospheric water vapor begins to condense into dew as the air cools.

diabase a fine-grained igneous rock composed largely of pyroxene and plagioclase feldspar.

diastrophism deformation of the Earth's crust; the processes by which the crust is deformed and the results of this deformation.

diatom simple siliceous plant of microscopic size with one of many different shapes; may resemble a rod, sphere, or circular disc.

diatom ooze a deposit consisting of siliceous remains of diatoms.

differential erosion erosion that occurs at different rates as a result of differences in the resistance of materials.

differential weathering differences in the weathering effects caused by differences in the resistance of rock and minerals to weathering processes.

dike an injected body with parallel or subparallel walls that is narrow relative to its lateral extent and discordant with respect to preexisting layers.

dike swarm term used when many dikes, often of similar trend or orientation, occur together.

dilatational wave (*or* **primary, P, compressional,** *or* **longitudinal wave**) seismic wave propagated like sound waves with movements in the medium through which it passes parallel to the direction of propagation.

diorite a coarse- to medium-grained rock composed of intermediate plagioclase and biotite, hornblende, or pyroxene.

dip the angle between a plane and the horizon, measured in a vertical plane at right angles to the strike.

dip-slip fault a fault in which the direction of displacement is up or down the dip of the fault.

directed pressure application of a pressure, stress, or force in such a way that it has direction.

discharge a measure of the quantity of water passing through a cross-sectional area of a stream per unit of time.

disconformity a break in a succession of sedimentary layers caused by a period of erosion or nondeposition. Layers above the break are parallel to those below.

discordant intrusion intrusion injected across planes of stratification or schistosity.

disharmonic fold fold form whose geometry varies in an irregular way with depth.

displacement the straight-line distance after movement between two points that were originally adjacent on opposite sides of a fault.

dissolution a process by which mineral and rock materials pass into solution. During the process, solid material dissolves, changing into ions in a medium, usually water.

distributary branch of a stream, usually at the mouth of a river on a delta, formed by the breaking up of a single stream into smaller streams.

diurnal tides a tide with only one high tide and one low tide each day.

divergent plate boundary in plate tectonics, a boundary between plates that are moving away from one another.

doldrums the belt of light and variable winds that occur near the equator.

dolerite synonym of *diabase*.

doline see *sinkhole*.

dolomite a sugary-textured, dense rock that does not effervesce in dilute acid unless it is powdered. Dolomite is a calcium magnesium carbonate.

dome a circular or elliptical uplift; older beds are exposed in the center after erosion.

dome dunes sand dunes that have a domal shape.

drainage basin the area drained by a stream.

drainage divide an imaginary line separating the area drained by one stream from the areas drained by adjacent streams.

drift glacial deposit.

drumlin a streamlined hill shaped somewhat like a whaleback formed by a glacier and generally composed of glacial till.

dune accumulation of sand that is mobile and independent of obstructions.

dunite a rock composed almost completely of the mineral olivine.

dynamic metamorphism metamorphism produced when deformation is significant in the transformation of rock. The term is applied when effects of directed stress are important in metamorphic processes.

earth a general term describing disintegrated rocks and loosely consolidated sediments.

ebb tide a tide that is falling or the period between high tide and the next low water.

ecliptic the plane in which Earth revolves around the Sun.

eclogite metamorphic rock composed primarily of a red garnet and a green pyroxene, having a bulk chemical composition that could only have formed from rocks of basaltic or gabbroic composition.

ecology the study of the relationships between organisms and between organisms and their environment.

ecosystem a community, together with the physical and chemical factors with which it interacts.

Ediacaran fauna late Precambrian soft-bodied faunas that resemble a famous fauna found in Australia.

effluent water that moves from beneath the water table into lakes and streams.

elastic limit the limit of stress beyond which a rock cannot fully recover its original shape after a deforming force is removed.

elasticoviscous material behavior characterized by a combination of elastic and viscous properties.

elastic rebound theory the concept applied to earthquakes caused by elastic response of crustal rocks to applied strain. Elastic rebound occurs when failure (the earthquake) occurs and built up strain energy is released.

electron one of the fundamental particles of which atoms are composed. The mass is 9.107×10^{-28} grams and is assigned a relative negative electrical charge of one unit, which equals 1.602×10^{-19} coulombs.

element an atom or group of atoms that is stable and cannot be broken down by ordinary chemical methods.

entrenched meander the meandering course of a river that flows in a deep valley.

entrenched stream a stream that flows in a channel cut below the general level of the surrounding land and often into bedrock. Cliffs may occur along one or both banks.

environment as used in biology, the environment consists of the external factors and conditions that influence organisms or a community. Geologists apply the term to describe the physical, chemical, and biological conditions that characterize a certain type of geomorphic area, such as a lake or flood plain.

environmental lapse rate the rate of change of temperature with elevation in the atmosphere.

epeirogenic a type of regional deformation characterized by broad uplifting or downwarping of a large area.

epicenter point of the ground directly over the focus of an earthquake where motion is initiated.

epoch a division of geologic time. Geologic periods are often divided into epochs on the basis of unconformities.

era the largest division of the geologic time scale. Time since the Precambrian is subdivided into the Paleozoic, Mesozoic, and Cenozoic eras.

erosion in broad context, includes the effects of all processes that loosen and remove earth materials: weathering, solution, corrosion, and transportation. Many geologists prefer to restrict use of the term to processes involving removal of material, such as by streams, glaciers, wind, waves, and ground water.

erratic a rock carried by a glacier and laid down on a different type of rock.

esker sinuous ridge composed of till and glaciofluvial deposits and thought to originate in stream channels and sometimes under glaciers.

estuary the mouth of a river that empties into the ocean. Seawater mixes with fresh water in the estuary, and the estuary is affected by tides.

eugeosyncline many geosynclines can be subdivided into two major parts, miogeosyncline and eugeosyncline. The eugeosyncline is located generally on the oceanward side of the miogeosyncline and is usually characterized by greater mobility and the presence of volcanic products and graywacke as well as other sediments.

evapotranspiration loss of water from the land caused by a combination of evaporation and transpiration.

exfoliation scaling on exposed rocks.

extension stretching or pulling apart.

extrusion igneous rock bodies and lava that crystallized or solidified on the surface of the Earth.

fault a shear zone along which displacement has occurred.

fault block a mass bounded by faults on at least two sides.

fault block mountain mountain formed as a result of the displacement of sections of the Earth's crust along steeply inclined faults.

fault plane the zone along which movement occurs when faulting takes place.

fault scarp an escarpment marking the position of displacements where one side moved up in relation to the other.

feldspar one of a group of silicate minerals, including potash feldspars (orthoclase and microcline) and the plagioclase feldspars, which contain Ca, Na, or both Ca and Na.

felsite a general term applied to igneous rocks of fine-grained texture and light color, indicating a granitic composition.

ferromagnesian minerals minerals such as pyroxene and hornblende that contain a high proportion of iron and magnesium.

ferruginous containing iron.

festoon cross-bedding a type of cross-bedding formed when channel-filling takes place. Beds have a characteristic curve (concave upward) appearance in cross section.

fiord glaciated valley whose floor is submerged below sea level.

firn compacted, granular snow.

fissure eruption volcanic eruptions that occur along cracks, which may be fracture zones or faults.

flagellate a microscopic plant with one of a variety of shapes, ranging from that of a balloon with a string attached to that of a pot or fancy mask; all have whip-like projections that move.

flat-pebble conglomerate a conglomerate or breccia layer formed where a newly deposited sediment layer is broken by wave action, producing platy pebbles.

flexural fold a fold formed where flow or slip is restricted by the layer boundaries. The layering exercises an active control on the deformation; the resulting folds represent a true bending of layers.

flint a synonym for *chert*.

flood plain the flat area along a stream that is occasionally flooded.

flood tide the rising, or flow, of water toward the shore as the tide approaches the coast.

flow as used in hydrology, refers to the movement of water. Structural geologists apply the term to rock deformation that is not instantly recoverable without loss of cohesion. As used with reference to mass movement, flow is the movement of unconsolidated materials that exhibit plastic or semifluid behavior.

flow casts (or load casts) rolls, lobate ridges, or other raised features that may be produced in sand or silt as a result of the flow of sediment in an underlying bed, usually composed of some soft sediment.

flow nets maps showing an orthogonal set of lines in which one set depicts the direction of ground-water movement and the second set represents the front of movement.

flowstone calcium-carbonate deposits formed from water in caves, often taking the shape of a cascade-like mass.

focal depth the depth below the Earth's surface at which an earthquake occurs.

focus the point at which an earthquake originates or where motion is initiated.

fog a ground-level cloud.

fold bend or distortion of rock masses; most easily recognized when bedding is deformed.

foliation a secondary rock fabric in which platy minerals or shear surfaces are in parallel alignment. Schistosity and rock cleavage are types of foliation.

footwall the block below a dipping fault.

foraminifers single-celled marine animals with small shells composed of calcite.

foreland the portion of an orogenic belt located on the flanks of the highest and most severely deformed parts of the mountain system.

foreland fold fold formed in a belt located on the flanks of a mountain system's core.

fore-set bed layer of inclined sediment on the front slope of a delta.

foreshore the zone between low tide and the area where the beach either becomes horizontal or slopes landward.

fossil the remains of plants or animals that lived in the geologic past, including tracks and trails left by living organisms.

fossiliferous limestone limestone formed from the shells of marine animals or in which fossils make up a significant part of the rock.

fractional crystallization separation of a magma into two or more phases.

fracture (or joint) break in rock along which displacements have not occurred.

fragmental a sediment or rock composed of pieces.

fragmental limestone a limestone composed of fragments of calcium carbonate.

free-air gravity anomaly a gravity anomaly that remains after the reading has been corrected for elevation above the reference surface (geoid).

fringing reef reef formed along the edge of an island or continent at the shoreline.

front as used in meteorology, a front is the narrow zone separating air masses of different physical properties, especially air masses that have different temperatures and density.

frost action the effects caused by freezing water, particularly as an agent of mechanical weathering.

frost heaving the process in which the soil or rocks are lifted up as a result of the freezing of ice under them.

fumarole a hole or vent through which fumes or vapors issue in an area underlain by volcanic materials.

gabbro rock composed of coarse- to medium-grained plagioclase feldspar, pyroxene, hornblende, and biotite.

gal a unit used to express gravitational acceleration. 1 gal = 1 cm/sec^2.

gamma a unit of measure of magnetic field strength. 1 gamma = 10^{-5} oersted.

gauss (or oersted) a measure of magnetic field strength. 1 gauss = 1 dyne/unit pole.

geodesy the field of geophysics concerned with the shape and dimensions of the Earth.

geoid the shape of a sea-level surface on Earth extended through continents.

geomagnetic time scale a time scale based on the age of reversals in Earth's magnetic field. Radiometric dating is used to determine the age of the reversals.

geomorphic agents the agents, such as running streams, ground water, wind, waves, and mass wasting, that bring about changes in the shape of the Earth's surface.

geomorphology the study of the shape of the Earth's surface.

geophone instrument designed to pick up vibrations of the ground; used in seismic exploration.

geosyncline long, narrow belt of long-term subsidence and sedimentation.

geothermal energy energy obtained from inside Earth.

geothermal gradient the rate of change of temperature with depth in Earth.

geyser the forceful emission of water from beneath the ground as a result of the sudden change of part of the water to steam.

geyserite a chemical sediment composed of SiO_2 formed around the vents of geysers.

glaciofluvial deposit deposit formed from water on, in, or under a glacier, as well as beyond the glacier terminus, by meltwater.

glass a noncrystalline, amorphous, material. Glasses are produced naturally when magma is frozen. The magma cools so fast that crystals do not form.

glassy the texture of amorphous materials; such materials show no crystals and thus are glass-like.

global climate see *climate*.

globigerina single-celled marine animal with a globular snail-like shell.

globigerina ooze calcareous deposits on the sea floor composed of the shells of the protozoan globigerina.

Glossopteris **flora** a group of plants, seed ferns, that grew only in very cold climates during the late Paleozoic era.

gneiss a medium- to coarse-grained metamorphic rock characterized by compositional layering that produces a banded appearance.

Gondwanaland the name of the southern continental mass thought by some to be the source of Africa, India, South America, and Australia.

gossan deposits of hydrated oxide of iron formed in the upper parts of mineral veins and masses containing pyrite as a result of oxidation and leaching.

gouge soft powdery material resulting from grinding and abrasive action along a fault.

graben elongate crustal block, bounded by normal faults on either side, that has moved down in relation to these sides.

gradation a term applied to the combined effects of aggradation and degradation of the Earth's crust; sometimes restricted to effects of stream action.

graded bedding sedimentary structure in which a systematic vertical change occurs in sediment grain-size.

graded stream stream that, over a long period of time, is adjusted so that the slope of the stream channel is sufficient to give the stream the velocity needed to move its load.

gradient (stream) the rate of change of elevation of the stream with distance along the stream (that is, slope).

granite a coarse- to medium-grained equigranular rock composed of potassium feldspars and quartz, with some biotite mica and hornblende usually present in small amounts.

granite pegmatite a very coarse-grained rock with granitic composition.

granitization the metasomatic processes by which granitic rocks are produced from sedimentary rocks.

granodiorite an igneous rock that is intermediate in composition between granite and diorite.

granulation the crushing of rock under such conditions that no visible openings result.

graphic granite a pegmatite composed of quartz and feldspar minerals intergrown to produce a graphic pattern.

gravity separation the process by which the first crystals that form in a magma, if heavier or lighter than the melt, will sink or rise through the melt and become segregated either at the bottom or as a raft of floating minerals near the top of the magma chamber.

graywacke an impure sandstone consisting of quartz and feldspar fragments and small fragments of rocks.

great circle a circle on Earth's surface that lies in a plane that passes through the center of the earth.

greenschist facies metamorphic facies produced by temperatures and pressure conditions that cause the original sedimentary minerals, including clay and zeolite, to be altered. Rocks assume a greenish appearance as a result of the formation of chlorite and sericite.

ground moraine glacial deposits resulting from melting so that debris on the surface, in the ice, and caught near the bottom of the ice is slowly lowered to the ground.

ground water water moving beneath the ground surface.

ground-water basin a subsurface region characterized by well-defined areas of recharge, collection of water, retention, and discharge of water. Some basins have the shape of a basin. Others are areas that have well-defined boundaries with adjacent ground-water basins.

guide fossils (*or* **index fossils**) fossils that are used to determine the age of sedimentary rocks. Generally, these are fossils of animals that existed for a limited time.

guyot submerged flat-topped seamount. Most are thought to be volcanoes that have been eroded by wave action.

gypsum a mineral composed of calcium sulfate. It has a hardness of 2 and is formed mainly as a result of precipitation from seawater.

hail chunks of ice formed in cumulonimbus clouds as a result of repeated precipitation of ice as the chunk is carried up in updrafts. The name is restricted to ice masses that are at least 5 millimeters in diameter.

half-life the constant amount of time required for one-half the mass of a radioactive isotope to break down.

halite a mineral composed of sodium chloride (syn., *salt*).

hanging valley a valley having a floor higher than the valley or shore into which it leads; usually formed by glaciation.

hanging wall the surface or block of rock above an inclined fault zone.

hardness (1) in water, the amount of calcium carbonate and magnesium carbonate in solution. (2) the resistance of mineral to scratching.

hard pan (*or* **caliche**) a hard, impervious layer of soil containing insoluble materials and clay.

head the upper reaches of a stream or glacier.

headland a part of the shore that protrudes into the ocean.

headward erosion progressive extension of the head of a stream as a result of erosion.

heat capacity the quantity of heat required to raise the temperature of a system by one degree at constant pressure and volume.

heterosphere the upper part of the atmosphere in which components are not well mixed.

hogback a sharp ridge formed as a result of the outcrop of a resistant layer.

Holocene epoch a division of geologic time encompassing the last 10,000 to 11,000 years. Also known as recent time.

homocline general name for any layers of sedimentary rocks that have a single, relatively uniform direction of dip (one limb of a large fold may be described as homoclinal).

homosphere the lower part of the atmosphere in which gases are well mixed.

hook a deposit of sediment along a beach that has the shape of a fish hook.

horn sharp peak that projects above the surrounding area and was shaped by glacial erosion.

hornblendite a rock composed largely of hornblende, a common member of the amphibole group.

hornfels fine-grained, nonfoliated, dense, usually dark rock formed near the contacts of igneous intrusions.

horse latitudes the informal name given to the latitudes located along the axis of the subtropical high-pressure belt. These latitudes center on 30° north and south of the equator.

horst an elongate block, bounded by normal faults on either side, that has moved up in relation to the sides.

hot spots in plate tectonics theory, a center of volcanic activity thought to form as a result of the rise of hot materials from deep in the mantle. Hot spots may remain fixed for long periods of time while the overlying plates move.

hour angle the angle between the observer's celestial meridian and the hour circle of the star measured westward in the celestial equator.

hour circles great circles that pass through the north and south celestial poles. The reference hour circle is the one that passes through the point where the Sun's path crosses the celestial equator at the start of spring.

hundred-year flood a flood that may be expected to recur once every hundred years on the basis of past records of flood stages.

hurricane a tropical cyclone in which the winds exceed 74 miles per hour. Most hurricanes are about 500 kilometers (300 miles) in diameter.

hydration absorption and combination of water with other compounds, as when anhydrite absorbs water to become gypsum.

hydraulic gradient the rate of pressure head change per unit of distance at a given place and in a given direction.

hydraulic head the pressure at a point in a body of moving water, expressed in terms of the height of the column of water that can be supported by the pressure.

hydrologic cycle (*or* **water cycle**) the cycle of phenomena through which water passes.

hydrolysis a reaction between water and another compound.

hydrostatic head an expression of the hydrostatic pressure in terms of the height of a column of water that can be supported by the pressure.

hydrostatic level the level to which water will rise in a sealed well.

hydrostatic pressure the pressure exerted at a given point in a body of water at rest by the weight of the overlying body of water.

hydrothermal solution hot, watery solutions commonly associated with volcanic or igneous bodies.

hypothermal deposits the deepest ore deposits associated with metamorphic rocks; usually lenticular in shape and include such minerals as native gold, chalcopyrite, tin, and molybdenite.

hypothesis a tentative theory adopted to explain a set of observations.

iceberg big block of ice that has broken off of a glacier and drifts in the sea.

ice sheet large body of ice not confined to a valley; may reach continental dimensions and bury whole mountain ranges.

igneous a term meaning "born from fire," applied to both volcanic rocks and rocks that crystallize slowly within the Earth's crust.

inclination a general term applied to sloping lines or surfaces. It is the angle between the surface or line and the horizontal plane.

index minerals minerals used to identify the different grades of metamorphism. From low grade to high grade, these minerals are chlorite, biotite, garnet, staurolite, kyanite, and sillimanite.

inertia a property of matter by which it remains at rest or in a uniform state of motion unless it is acted on by an external force.

influent a condition existing when the water table is depressed below the level of stream and lake bottoms and water moves from the lake or stream toward the water table.

inner core the central part of Earth's core. The inner core is a solid composed primarily of iron.

insolation the weathering effect of the Sun's heat on exposed rocks.

intensity a measure of the effects experienced by people at the surface of the ground when earthquakes occur. A standard scale is used.

internal drainage a drainage system with no streams leading out of it.

intertropical convergence zone the equatorial region of low surface air pressure along which the trade winds converge. It separates the northern trade winds from the southern trade winds. (syn., ITCZ)

intrusion an igneous body emplaced and crystallized beneath the surface of the ground.

ion a charged particle that forms from an atom or covalently bonded group of atoms when it gains or loses one or more electrons.

ionic compound a compound that consists of oppositely charged ions.

island arcs arcuate chains of volcanic islands formed over subduction zones.

isobar a line on a map that connects points of equal atmospheric pressure. The lines may represent pressure conditions at ground level or at some other designed level in the atmosphere.

isograd line connecting points of equal metamorphic grade; forms a boundary between zones.

isostasy the principle that the upper parts of the Earth's outer layers are in a state of flotational equilibrium on a lower level.

isostatic anomaly a gravity anomaly that remains after the reading has been corrected for density variations thought to exist down to the level of isostatic compensation.

isostatic equilibrium the condition of the Earth's outer layers when they are in flotational equilibrium, with no tendency to move up or down.

isotopes atoms of an element having the same number of protons in the nucleus as generally found in the element but different numbers of neutrons. Isotopes of an element have the same atomic numbers but different atomic weights.

jet streams a narrow zone of high-velocity air currents that generally flow near the tropopause.

jetty a structure like a pier that is extended into a body of water to influence the currents. Usually, these are built to protect harbors or sections of beach.

joint a crack or break in rock masses across which the rock has lost cohesion; also called a fracture.

juvenile water water derived from magma.

kame glacial deposit laid down in contact with ice. Although used in various ways, the term is often applied to small conical hills formed of stratified drift.

kame terrace terrace built along the side of a glacial valley by load that is dropped by water between the glacier and the adjacent valley wall.

karst topography topography shaped in part by solution and diversion of surface water underground in areas of limestone and dolomite bedrock.

kettle an enclosed depression left in drift as the last remnants of ice melt away.

klippe an isolated block of material separated from underlying rocks by a fault; associated with low-angle thrust faults.

lag gravel loose gravel-sized particles left in arid and semiarid regions as wind removes finer materials.

laminar flow flow in which the fluid behaves as though it were made up of very thin layers that glide over one another; streamline flow.

landslide perceptible downslope movement of earth, rock, soil, and mixtures thereof.

lapilli small fragments (4 to 32 millimeters in diameter) erupted from volcanoes.

lapse rate the rate at which air temperature drops with altitude.

lateral moraine a deposit from a valley glacier composed of debris that accumulated on the glacier surface as a result of normal downslope movement of debris from the sides of the valley.

laterite a type of clay-rich soil that has a tendency to harden into a brick-like solid.

Laurasia the name of the northern continental mass thought by some to be the source of North America, Europe, and Asia.

lava magma that reaches the ground surface.

law of faunal succession principle stating that the fossil fauna of rocks of different age are different and provide a key to identification of rocks of that age.

law of superposition principle stating that in a normal stratified sequence, each stratum was deposited on one of greater age.

lightning an electric discharge that occurs between a cloud and the ground or between two parts of large clouds.

limb the flank or side of a fold.

linear dunes sand dunes that are elongated enough to form a linear feature on the ground.

liquidus a curve (in a two-component system) connecting points in a temperature-composition diagram representing the saturation of the liquid phase.

lithification the process by which unconsolidated sediments become consolidated rocks.

lithophysae glass bubbles found in some lavas.

lithosphere the relatively rigid outer rind of the Earth. The exact limits are not specified, but it is distinguished from a plastic zone below called the asthenosphere.

lithostatic pressure the pressure at a point in the Earth caused by the weight of the overlying body of rock.

Little Ice Age the period extending from A.D. 1200 to 1900 during which global temperatures dropped and glaciers advanced.

load of a stream the material in transport in a stream. Usually the load is of three parts: material moved along the bottom of the stream channel, material in suspension, and material in solution.

loess an unconsolidated, unstratified, and homogeneous sediment composed of small, angular particles derived from a variety of rock types.

longitudinal dune a sand dune that is elongate in the direction of the prevailing wind.

longitudinal profiles a stream profile constructed to illustrate the drop in the elevation of the channel with distance.

longitudinal waves seismic waves in which the local movements in Earth are oriented in the same direction as the direction in which the wave is traveling. (syn., P wave and compressional wave)

longshore current a component of motion in the water directed along the shoreline.

longshore drifting the movement of sand on the shore generally in the direction of the longshore current.

love (L) wave a seismic wave that travels near the surface (within the lithosphere) in which the local motion involves movement that is transverse to the direction of movement and parallel with Earth's surface.

low-velocity zone (*or* **Gutenberg low-velocity zone or channel**) a zone in the upper mantle in which seismic waves move with reduced velocity. The depth of the zone is between 100 and 200 kilometers.

luster appearance taken on by a mineral in reflected light.

mafic rock an igneous rock rich in minerals containing iron and magnesium.

magma (*or* **rock melt**) multicomponent silicate melt that usually contains some solid minerals, liquids at high temperature, and gases under pressure.

magmatic differentiation processes by which a magma becomes altered and yields different rock types during crystallization.

magmatic water water produced by magmatic activity, also called juvenile water.

magnetic anomaly abnormally high or low values for the strength, declination, or inclination of the Earth's magnetic field.

magnetic declination the angular difference between the compass direction to magnetic north and the direction of true north.

magnetic field the area around a magnet within which the influence of the magnet can be detected.

magnetic-field strength the force (in dynes) that would be exerted on a unit pole at a point in a magnetic field.

magnetic induction magnetic effects produced in materials that contain iron-bearing minerals (notably magnetite) by an external field, such as Earth's magnetic field.

magnitude a measure of the amount of energy released at the focus of an earthquake.

main magnetic field the component of Earth's magnetic field that arises from processes that occur in the core. This field varies over time while the field component caused by accumulations of magnetic minerals is permanent.

manganese nodules pebble- to cobble-sized masses precipitated on the sea floor. They are composed of manganese, copper, nickel, and other metals.

mantle the zone in the Earth's interior below the crust and above the core.

marble the metamorphic equivalent of calcite, limestone, or dolomite.

mare the large, flat areas on the Moon formed by extensive outpourings of lava; also known as lunar seas.

marl mixture of shells and shell fragments with muds, clay, particles of calcite or dolomite, or sand; an impure limestone found in a semiconsolidated state.

mass wasting the downslope movement of surficial materials as a result of the effects of gravity.

matter anything that possesses mass and occupies space.

M discontinuity (*or* **Moho**) the major seismic discontinuity used to define the base of the Earth's crust. It occurs at a depth of 20 to 40 kilometers beneath continents and 10 kilometers under oceanic crust.

meandering stream a stream that flows in a strongly curved course consisting of loops in which the stream almost doubles back on itself.

mean sea-level the average height of the sea surface for all stages of the tide over a nineteen-year period.

mechanical concentration concentration of minerals through mechanical processes, such as gravity separation in streams, along coasts, or by wind.

mechanical weathering processes of weathering by physical processes that break down a rock into fragments.

medial moraine ridge of till left in the middle of a valley after a valley glacier melts.

melting the change of state from a solid to a liquid.

mesopause the zone in the atmosphere that separated the mesosphere from the thermosphere. Usually, this zone occurs at an altitude of about 85 kilometers.

mesosphere the layer of the atmosphere that separates the troposphere from the thermosphere.

Mesozoic era the interval of geologic time following the Paleozoic and preceding the Cenozoic.

metamorphic aureole the concentric zones of alteration usually found surrounding igneous intrusions.

metamorphic rock rock formed through the process of metamorphism.

metamorphic zone an area that was exposed to similar metamorphic conditions. Zones are recognized by certain diagnostic minerals.

metamorphism the processes acting beneath the ground by which rock is so extensively altered that it obtains a new texture or a new mineral composition or both.

metasomatism a process of simultaneous solution and deposition that replaces old minerals.

meteoric water all forms of precipitation, including rain, sleet, snow, and hail.

meteorite a body of stone or metal that has fallen to the Earth from space.

mica one of the most distinctive of all mineral groups, exhibiting single, perfect cleavage and having a sheet structure. Included are biotite, muscovite, and other less common varieties.

micaceous (*or* **lamellar, foliated**) descriptive of minerals that consist of thin, flat plates.

micrite limestone a limestone with an extremely fine-grained texture. The individual crystals of the limestone are too small to be visible to the unaided eye.

microclimates climatic conditions restricted to a particular geographic location. The special conditions found over a large city or on a south-facing slope are examples.

microseism small-amplitude seismic disturbance caused by wind, waves, and other earth vibrations and recorded on seismograms.

migmatite rocks of a mixed nature containing metamorphic- and igneous-appearing rock.

mineral a naturally occurring element or compound formed by inorganic natural processes and having generally a definite chemical composition and a characteristic atomic structure, which is expressed in an external crystalline form and in other physical properties.

miogeosyncline a long, relatively narrow belt in which subsidence and accumulation of sediment (notably shallow-water sand, mud, and limestone) occur. See also *eugeosyncline*.

mixed tides tidal conditions during which two high and two low tides of different ranges occur each day.

Moho see *M discontinuity*.

Mohs' hardness scale a standard scale used to measure the resistance of minerals to scratching, which is taken as an indication of their hardness. The scale is based on the relative hardness of 10 minerals. One, talc, is the softest, and diamond is 10 on the scale.

molecule a structure that consists of two or more atoms that are chemically bound together and behave as an independent unit.

monocline a flexure across which the beds maintain a constant direction of dip.

moraine a sedimentary deposit formed along the edges, beneath glaciers or at the end of glaciers from material carried by the glacier.

morass see *swamp*.

mud a mixture of water and the finer particles of earth and soil.

mud cracks cracks that commonly form in mixtures of clay, sand, and silt when the sediment is dried.

mudflow a flow of a mixture of mud and water.

muscovite a member of the mica group of minerals, which contains little or no iron and magnesium. Potassium is present instead of iron or magnesium. The mineral is colorless or slightly yellow to brownish.

nadir a point on the celestial sphere that lies directly below the observer.

nappe a large mass of folded rock that has been moved a long distance from its point of origin.

native elements any element that is uncombined in a nongaseous state in nature. Carbon, sulfur, arsenic, and selenium are examples.

native metals native elements that are metals.

natural bridge rock bridge formed by the diversion of surface streams underground, remnant of a cave, or a natural tunnel.

natural levee a low ridge formed on either side of a stream channel from deposits laid there during a flood.

neap tide low tidal range occurring about the time of quadrature of the Moon.

nebula an interstellar cloud of gas or dust.

nekton pelagic animals that are active swimmers.

net slip the straight-line distance after fault movement between two points that were originally adjacent on opposite sides of the fault.

neutrons an uncharged subatomic particle found in the nucleus of atoms. Its weight is slightly more than a proton.

nonconformity an ancient erosion surface found in rock bodies where the rock below the erosion surface is crystalline rock, either igneous or metamorphic, and that above the erosion surface is sedimentary.

normal fault steeply inclined fault along which the hanging wall has moved downward in relation to the footwall.

nuée ardente a type of volcanic eruption in which incandescent gases and dust are erupted.

nunatak peak or isolated hill that protrudes through a snowfield or glacier.

obduction in plate tectonics, the process by which part of the sea floor is uplifted and faulted onto other crust, sometimes onto continental crust.

oblique-slip fault a fault on which the displacement is oblique across the fault plane.

obsequent stream a stream that flows in the direction opposite that of the dip of the underlying rocks.

obsidian naturally formed glass. Obsidian forms when lava is frozen so quickly that crystallization does not occur.

oersted a magnetic force designated in terms of dynes per unit pole strength. See also *gauss*.

offshore from the position of low tide seaward.

olivine a mineral series varying in composition from forsterite (Mg_2SiO_4) to fayalite (Fe_2SiO_4).

oolite round, egglike body usually having concentric internal layering.

oolitic limestone a limestone composed of small, spherical masses formed by precipitation of calcium carbonate in agitated water.

ooze marine sediment consisting largely of shell fragments of microscopic marine animals.

open system a system in which mass passes into and out of the system.

ore rock or mineral that can be economically mined to produce a metal.

orogenic belt a mobile belt in the Earth's crust that has been subjected to folding and other deformation.

orogeny deformation of the crust; usually mountain-building processes.

outer core the outer part of Earth's core. The outer core is a liquid composed primarily of molten iron.

outwash deposit sediment deposited beyond the ice margin by meltwater produced as an ice sheet or valley glacier wastes.

overdraft the ground-water condition created by removal of more water from an aquifer than is recharged.

overturned fold a fold in which the axial surface is inclined so one limb is over part of the other limb.

oxbow lake a lake formed as a result of the cutoff of a meander.

oxidation the process by which oxygen combines with elements or compounds.

ozone hole as viewed from high in the atmosphere, the ozone hole is a circular-shaped region in the stratosphere in which the quantity of ozone is abnormally low.

pack ice floating pieces of ice driven together.

pahoehoe the Hawaiian name for fluid lava on which a surface forms that appears ropy.

Paleozoic era a division of geologic time covering that interval after the Precambrian and before the Mesozoic.

Pangaea name of the single protocontinent thought by some to be the original source of all continents.

parabolic dunes sand dunes that have a parabolic shape as viewed from above. These dunes have a scoop shape in which the horns point upwind.

parent isotope an isotope that is radioactive and breaks down to form daughter isotopes.

passive fold fold in which flow or slip crosses the layer boundaries, the layers exercising little or no control on the deformation. Layer boundaries serve merely as markers, parts of which are displaced relative to other parts to produce an apparent bending.

peat an accumulation of vegetable matter that is in the initial stages of disintegration and decomposition.

pedestal rocks large rocks balanced on smaller outcrops. Most pedestal rocks are formed by wind erosion.

pediment a gently sloping surface formed by erosion of cliffs or steeper slopes; often a rock surface with a thin veneer of alluvial covering.

pediplane a widespread erosion surface covered in places by a thin veneer of alluvium. The surface is formed by the coalescence of two or more adjacent pediments.

pedology study of the soil, its origin, character, and uses.

pegmatite rock composed of very coarse crystals, usually in the form of a dike or lens, formed in the late stage of crystallization of a magma.

pelagic community the plants and animals that live in the open ocean, not on the bottom or along shorelines.

pelean volcanic eruptions that resemble the eruption of Mount Pelée on the island of Martinique.

peneplain a nearly plane surface formed as a result of erosion by stream action.

perched water table a condition found where a zone of saturation is located above an unsaturated zone and above the regional water table.

peridotite a dark, coarse-grained, equigranular rock containing a large quantity of olivine and/or pyroxene or hornblende, but no quartz or feldspar.

period (1) the time between arrival of adjacent peaks in a wave train. (2) one of the major divisions of time used in the geologic time scale.

perk test a method used to determine the ability of a soil or sediment to transmit fluids.

permafrost permanently frozen ground.

permanent stream stream that flows throughout the year.

permanent thermocline the vertical portion of the zone in which the change of water temperature with depth is constant and does not reflect seasonal changes.

permeability a measure of the ability of a stratum or rock unit to transmit fluids.

phaneritic textural description in which crystals in a rock are large enough to be seen with the unaided eye. Phaneritic textures may be coarse-, medium-, or fine-grained.

phenocryst large crystal in a glassy or finer-grained igneous rock.

photic zone the depth in the ocean through which light penetrates.

photochemical dissociation the breakdown of water vapor by ultraviolet radiation, releasing free oxygen.

photosynthesis the process by which plants make their own food using solar energy.

pH scale measures the acidity of a solution (the concentration of hydrogen ions). Values range from 0 to 14, 7 being neutral, less than 7 being acid, and greater than 7 being alkaline.

phyllite fine-grained schistose rock identified by the lustrous, silky sheen that characterized light reflected from the chlorite and muscovite micas of which it is composed.

physical climate system Earth systems that are driven by the effects of solar radiation. This system includes life and geomorphic processes.

physiographic province a region in which the geologic structure, climate, and resulting landforms are distinctive and different from those of adjacent regions.

phytoplankton marine organisms that float on or near the surface of water bodies. They photosynthesize and produce much of the oxygen in the atmosphere and serve as food producers in aquatic ecosystems.

piedmont glaciers glaciers that coalesce along a mountain front.

piezometric level the static level to which water in an artesian aquifer will rise if it is penetrated by a well.

piezometric surface an imaginary surface that everywhere coincides with the static level of the water in the aquifer.

pillow lava lava characterized by ball- or pillow-shaped masses, formed by eruption of lava underwater.

pitchstone a volcanic glass with a waxy, dull, resinous luster.

placer deposit accumulation of minerals that are heavy, resistant to chemical weathering, hard, malleable, and durable enough to survive the mechanical concentration processes active in streams or in waves.

plagioclase feldspars a group of feldspars composed of various mixtures of sodium and calcium with silicon, aluminum, and oxygen. The end members contain only sodium or calcium, but a continuous range of percentages of these two lie between the end members.

planetesimal a small, planetlike body.

plankton organisms that float on or near the surface of bodies of water.

plastic a material that deforms continuously and permanently once the stress applied to it exceeds a certain amount.

plastic flow deformation characterized by continuous nonreversible strain.

plate tectonic system the name given to processes that are closely related to the movement of lithospheric plates. The system includes most earthquakes, igneous, and volcanic activity.

playa lake (*or* **salina**) a shallow lake located in a basin, usually having internal drainage. Such lakes commonly contain water only intermittently and are generally high in salt content as a result of water evaporation.

Pleistocene epoch a division of geologic time also known as the Ice Ages. The Pleistocene started 2 million years ago and continued until the most recent interglacial stage when ice began to melt.

plucking (*or* **quarrying**) process by which pieces of bedrock are lifted out of place by glaciers.

plunge the angle (measured in a vertical plane) between a line and the horizon.

plunging axis the angle of inclination measured from the horizon when the axis of a fold is not horizontal.

pluton a large igneous intrusion.

point bars sand bars located on the inside of curves in streams.

polar covalent compound compound made of molecules that behave somewhat like small magnets.

polar easterlies winds in the polar regions that blow out of the east.

polar front the front located between polar air masses and air masses in the westerly belt.

polar front jet a jet stream generally located close to the poles and near the polar front.

polar molecules molecules in which the electron distribution is such that the molecule behaves like a small magnet with positive and negative ends.

pole (1) geographic: the ends of the Earth's axis of rotation. These rotational points are the north and south geographic poles. (2) magnetic: any of several points on the surface of the Earth where the lines of magnetic force are vertical. Magnetic poles do not coincide with geographic poles, though the Earth has north and south magnetic poles.

porosity the volume of pore space in a rock.

porphyry an igneous rock composed of large crystals embedded in a fine-grained or glassy groundmass.

pothole oval or round hole formed in the channel of a stream by concentration of abrasion, impact, and eddy current action.

Precambrian a general term applied to the geologic time starting with the origin of Earth and continuing to the beginning of the Cambrian period.

precession the wobble of Earth's axis of rotation.

precipitate the insoluble product of a chemical reaction in which two soluble ionic compounds react.

precipitation water that falls from the atmosphere in the form of ice, snow, hail, sleet or in the liquid state.

primary (P) waves see *dilatational wave.*

process as used in Earth Science, a process is a series of actions or operations that proceed to an end result.

progressive wave wave in which the waveform advances. The wave created by flipping one end of a rope that is tied at the other end is an example.

Proterozoic eon the more recent of two major divisions of Precambrian time.

protons subatomic particles with positive charges that are found in the nucleus of atoms.

pumice a light-colored, porous, glassy, highly vesicular rock.

P wave see *dilatational wave.*

pyroclastic a general term applied to fragmental materials of volcanic origin.

pyroclastic flow a general term applied to flows of ash and other pyroclastic materials from volcanoes.

pyroxene a group of minerals that are single-chain silicates.

pyroxenite an igneous rock that contains a high percentage of pyroxene minerals.

quadrature resultant condition when lines from the Sun to the Earth and from the Moon to the Earth are at right angles.

quarrying see *plucking.*

quartz a common mineral composed of silicon and oxygen, SiO_2.

quartzite the metamorphic equivalent of quartz sandstones.

Quaternary period a division of geologic time including the Pleistocene and Recent time.

radioactive elements nuclides that undergo spontaneous disintegration of the atoms to form other nuclides. The process involves emission of alpha, beta, and other energetic particles.

radioactive isotope isotope that is unstable and breaks down spontaneously by the emission of radiant energy.

radiolarian single-celled marine animal with an internal skeleton composed of radiating siliceous projections; important constituent of sediment in some areas of the ocean.

range variations in level of water surface resulting from tidal influences.

rarefaction movement of Earth in a seismic wave that is directed toward the focus of the earthquake.

Rayleigh wave a seismic wave that travels at or close to the surface and involves retrograde elliptical motion.

rays lines that represent the direction of movement of wave motion.

recessional moraine a ridgelike mass of drift accumulation from debris dropped by the glacier when the margin of the ice remains in one position for any period of time during retreat of the glacier.

recharge a general term applied to the addition of water to an aquifer.

recrystallization reorganization of the elements of the original minerals in a rock that may take place when the rock is subjected to high temperatures and pressure, particularly when water is present.

recumbent fold a fold with a nearly horizontal axial surface.

red clay the second-most extensive deep-sea deposit, brown to reddish in color, and containing films of manganese and manganese nodules.

reef flat the flat part of a reef composed of dead reef materials and commonly strewn with coral fragments and sand. It may be dry at low tide and may have some living coral colonies living in pools.

refraction (1) the bending of water waves as they pass into shallow water (water that is less than half their wavelength). (2) The bending of light rays as they pass through different media. (3) The bending of seismic waves as they pass through materials that have different seismic velocities.

regional metamorphism the name applied to those metamorphic alterations that affect rocks over large areas and are indicative of widespread environmental changes rather than localized deformation or magmatism.

regolith a general term applied to soil and other unconsolidated material that lie above bedrock.

rejuvenation condition whereby the behavior of a stream and its effects on the topography appear to revert to those of a more youthful stage in the erosion cycle.

relative humidity the ratio of the amount of water vapor in the air to the amount that could be in the air if the air were saturated at the same temperature.

remanent magnetization magnetic effects unrelated to the existing external field but due to permanent magnetism of the substance.

residual concentration a process by which some valuable mineral constituents accumulate while undesirable constituents are removed by weathering and erosion.

residual soils soils that were formed in the same location where they now occur.

reverse fault steeply inclined fault along which the hanging wall has moved up in relation to the footwall.

rhyolite a fine-grained igneous rock that has a composition similar to that of granite.

right ascension a coordinate for locating objects in the sky, analogous to longitude on the Earth's surface. Measurements are made in terms of hours and minutes.

ripple mark an undulating surface marking on sediment formed as a result of movement of a granular sediment in the medium of deposition.

roche moutonnée rounded hill that has been streamlined by glacial abrasion and scouring.

rock a consolidated aggregate of minerals; sometimes large masses of a single mineral.

rock cleavage the tendency of rock to break along closely spaced subparallel surfaces.

rock flour a powdery substance composed of fine (silt-sized) particles of rock produced by the grinding, abrasive action of glacier ice moving over bedrock.

rock flowage flow in which material remains solid, yet undergoes continuous deformation. Flowage of this type is accomplished by movements along planes of slip within the component crystals of the rock and by intergranular rotation; usually it is accompanied by recrystallization of minerals.

rock glacier mass of rock fragments that assumes a form similar to a valley glacier as a result of slow downslope movement.

rock structure the configuration of rock bodies, as they are expressed both at and beneath the ground surface.

rotational fault a scissors-like motion involving rotation of the fault blocks relative to one another.

runoff the portion of precipitation that flows off the ground surface in streams.

runup the elevation above mean sea-level reached by a tsunami.

safe yield the amount of water that can be withdrawn from a ground-water basin without producing undesired results.

sag pond depression, often filled with water, located along a fault where material in the rock has weakened as a result of deformation and shearing.

salinity a measure of the salt content of seawater. It is usually expressed as parts per thousand.

salt the common name for the mineral *halite,* which is composed of sodium chloride.

saltation the bouncing movement of materials in transport in streams and in the air. The material is heavy enough to return to the bottom of the channel or to the ground surface, where it hits other rock and may bounce back into suspension while it rolls in the direction of the current.

sand fragments of sediment that fall in the range of 1/16 and 2 millimeters (0.0025 and 0.08 inch).

sand drift sand deposit formed where a gap between obstructions acts as a funnel.

sand seas large areas covered by sand and sand dunes. The distribution of sand dunes may give the surface a form that resembles that of the ocean.

sand shadow sand accumulation located downwind from and sheltered by an obstruction, such as a boulder, grove of bushes, or cliff, that interrupts the streamlined flow of wind and reduces its velocity.

sand sheets large accumulations of sand that lack dune forms.

sandstone any rock composed of fragments between 1/16 to 2 millimeters in size regardless of composition.

sand stringers informal term applied to long, narrow bodies of sand.

saturated with reference to ground water, saturated refers to soil, sediment, or rock in which the pore spaces are filled with water.

saturated solution an ionic solution that contains all of the ions it can hold without causing precipitates to form.

schist strongly foliated metamorphic rock of medium- to coarse-crystalline texture.

schistosity ease of parting in metamorphic rocks, such as schist and gneiss, resulting from the parallel or subparallel alignment of platy minerals.

science the use of observation, experiments, and hypotheses to form a better understanding of natural phenomena.

scoria a vesicular volcanic rock formed by the expansion of gases in lava as it is ejected; compositionally equivalent to basalt.

scour-and-fill small-scale bottom scour or channels that are subsequently filled.

sea caves caves that open into the ocean.

seafloor spreading the theory that the sea floor is growing by means of intrusion and formation of new crust along oceanic ridges.

sea level the mean levels of the sea after tidal and wind effects are removed.

seamount a mountain that rises from the sea floor but does not break the surface of the ocean. Most seamounts are extinct volcanoes.

secondary (S) waves see *shear wave.*

secular change long-term variation that is observed as progressive change in declination, intensity, and inclination of the magnetic field; rates of change vary with time at any given point.

sediment materials that settle out of water, wind, or ice, as well as materials precipitated from solution and deposits of organic origin.

sedimentary facies parts of the sedimentary rock unit reflecting the conditions under which it originated. Often these are characterized by distinctive rock types or primary features.

sedimentary rock hard, solid rock formed when sediments become consolidated.

seif longitudinal dune characterized by a long, sharp ridge crest, one side of which is rounded and the other of which falls abruptly as a slip face.

seismic discontinuities boundaries inside Earth at which seismic wave velocity or the type of wave transmitted changes.

seismic sea wave see *tsunami.*

seismic shear zone (*or* **Benioff zone)** the zone that dips from deep-sea trenches under island arcs or adjacent continental margins; defined by the concentration of earthquake foci within it.

seismic tomography use of seismic observations (differences in travel times of seismic waves) to obtain a cross-sectional view of Earth's interior.

seismic wave the vibration produced by earthquakes or other disturbances that are propagated through Earth's interior as a result of the elastic response of Earth to the disturbance.

seismograph an instrument designed to detect and record seismic waves.

seismology the study of the origin and propagation of wave motion through Earth's interior.

self-exciting dynamo theory the concept that electromagnetic effects generate Earth's magnetic field produced spontaneously in Earth's core.

semidiurnal tides tides that have a period equal to about half of a lunar day.

shadow zone the zone between 103 and 143 degrees distance from an earthquake epicenter in which direct waves do not arrive because of refraction of the Earth's core.

shale consolidated mud and clay, often containing sand or lime in large amounts and having an earthy odor.

shear fractures breaks or fractures in rock caused by shear failure.

shear stress a stress applied in such a way that two adjacent parts of the material slip past one another.

shear wave (*or* **secondary, S,** *or* **transverse wave)** seismic wave that involves a shearing of the material; the wave vibrates perpendicular to the direction in which the wave front is moving.

sheeting a set of joints or fractures that develop nearly parallel to the surface of the ground as a response to the release of confining pressure when the weight of the overlying rock is removed.

shield those areas of Precambrian rock exposed at the surface or shallowly buried that have not been folded or complexly deformed since the end of the Precambrian.

shield volcano the largest type of volcanic cone; it resembles a shield or low, sloping dome in profile.

shooting flow flow of water in which a sudden increase in velocity of flow causes the surface level to drop. Commonly seen at the lip of waterfalls.

shore the zone between mean low tide and the landward edge of wave-transported sand.

shoreline the line along which the sea and the land meet.

sialic rock an igneous rock rich in silicon and aluminum.

siderite a meteorite composed primarily of metal, especially nickel and iron.

silicate minerals mineral that contains SiO_4 tetrahedra. The most common rock-forming minerals belong to this group.

siliceous containing silica.

siliceous sinter hot-spring deposits composed of siliceous material.

sill concordant, sheet-like intrusive body with large lateral extent relative to its thickness.

siltstone a sedimentry rock composed of silt-size fragments.

similar fold fold within which the shape remains the same from one layer to another, but the thickness varies from place to place in each layer.

simple eutectic system a type of behavior of some two-component melts when the components do not form a solid-solution series. These systems are plotted on temperature-composition plots.

sinkhole (*or* **doline**) enclosed depression formed by solution, subsidence, or collapse of the surface into a cave.

slate the most perfectly foliated metamorphic rock, thought to result from dynamic metamorphism of argillaceous material.

slaty cleavage rock cleavage like that of slate in which the micaceous minerals show a strong alignment in the plane of the cleavage.

slickensides striations and grooves that appear where rocks on either side of a fault have moved across one another.

slip face the downwind face of a sand dune formed as a result of slip and mass movement of sand downslope from the crest of the dune.

slump mass downslope movement of material on a curved surface of failure.

smog a mixture of fog and smoke or polluted air that reduces visibility.

snowfield the area above the snow line where snow remains from year to year.

snow line the lower limit of the area of perennial snow.

soil the product of rock disintegration and decomposition by weathering, modified by biological agents; capable of supporting plant life.

solar constant the rate at which solar radiant energy is received outside the atmosphere on a surface normal to the incident radiation. The average is 1.94 gram calories per minute per square centimeter.

solar wind the movement of ionized particles away from the Sun. A portion of these particles are drawn toward Earth and cause auroras as they pass through the ionosphere.

solid-solution series minerals that can undergo solid-melt reactions within certain limits in such a way that the composition and physical properties vary continuously with varying amounts of the components, as exhibited by plagioclase feldspar.

solidus the curve in a temperature-composition diagram (for two-component systems) connecting points above which the solid and liquid are in equilibrium and below which only a solid phase exists.

solifluction soil flowage, particularly that induced by saturation of the soil as a result of partial thawing of the ground.

solubility the maximum amount of an ionic compound dissolvable in a given amount of solvent under a given set of conditions of temperature and pressure.

solution a liquid mixture that has no visible boundaries among its components.

sorting separation of particles according to size, shape, or weight.

spatter cones mounds or tower-like structures built up by successive addition of layers of lava ejected from small openings.

specific gravity the ratio of a mineral's weight to that of an equal volume of water.

spheroidal weathering chemical weathering effect in which the corners of the rock are altered, producing a more or less spherical core often surrounded by shells of altered rock.

spherulites rounded masses composed of crystals or internal radial structure that radiates from a central point.

spit a bar and beach built out from the tip of the headland beach, forming a projection out from the landmass.

spring a place where water issues from beneath the ground surface onto the ground.

spring tide unusually high and low tides that result when the Sun and Moon lie in line with the Earth and the attraction of the two bodies is cumulative.

stacks columnar masses of rock, standing isolated as islands just offshore.

stalactite deposit of calcite that hangs down from the ceiling of a cave.

stalagmite a stump-shaped cave deposit, usually of $CaCO_3$, formed on the floor of a cave.

standing wave wave motion in which the form does not move laterally, but the crest and troughs reverse.

star dunes sand dunes that have the shape of the symbol commonly used to identify stars.

stick-slip movement a type of fault movement characterized by sudden and intermittent movements interspersed by periods when the moving surfaces are stuck together.

stock an igneous intrusion that is less than 40 square miles in surface area. Often these are small intrusions of larger plutonic bodies.

stony meteorite (*or* **chondrite**) meteorite composed of nonmetallic materials, especially olivine.

strain the response of the rock to an applied stress, especially the distortion or change in the volume of the rock.

stratification rock layering.

stratigraphic record general name applied to the historic record contained in sedimentary rocks.

stratosphere the layer of the atmosphere located above the troposphere and below the mesosphere. Temperature increases with altitude in the stratosphere.

stratovolcano a volcano formed by combined eruptions of cinder and lavas. Eruptions of cinder built the cone high near the vent. Eruptions of lava extend the base of the volcano.

stratum (*or* **bed**) layer of sediment or rock of varying color, texture, or composition.

stratus clouds a low, gray cloud layer with a relatively uniform cloud base.

streak the color of the powder of a mineral.

stream gradient the rate of change of channel slope with distance measured along the stream.

streaming flow flow of water moving on medium and low slopes. The water is smooth in some places and exhibits mild, slow eddying in others.

stream piracy the process by which one stream extends in a headward direction, cutting into the valley of a second stream and eventually diverting the flow of the second stream.

striation any linear scratch-like mark. For example, striations are produced where glaciers drag rocks over one another, along fault surface, and where cleavages intersect the surface of a mineral.

strike the compass direction (bearing) of any horizontal line on a plane.

strike-slip fault a fault on which the direction of movement is horizontal along the strike of the fault.

stromatolites a structure produced by sediment trapping, finding, or precipitation as a result of the growth of microorganisms, principally blue-green algae.

strombolian volcanic eruptions that resemble those of Stromboli in Italy. This volcano erupts fountains of basaltic lava from a central crater.

structural terrace a flexure that results in a flat, nearly horizontal surface.

structure form and orientation of parts of rock masses and their relation to the whole. Structural geology is concerned with the origin as well as the geometry of rock masses.

subduction in plate tectonics, the process by which one plate sinks or moves beneath another and into the Earth's mantle where it is assimilated or remelted.

subduction zones places where sections of the lithosphere sink into the mantle.

sublimation the process by which a material passes from a solid to gaseous state without becoming a liquid.

subsequent stream a stream that flows in a channel that is adjusted to underlying rock structure (generally follows weaker rock outcrop belts, faults, and the like).

subsidence rapid sinking or gradual downward settling of Earth's surface.

subsoil the B horizon; normally contains fewer organisms than do the overlying layers, but more than the C horizon. Weathering has reduced all rock to fine material.

subtropical jet a high-altitude (12 to 14 kilometers) jet stream usually found between latitudes 20° and 30°.

superimposed stream a stream that cuts down through a cover and establishes itself in the underlying bedrock. Superimposed streams frequently have courses that bear little relationship to the variations in the resistance to erosion of the bedrock across which they flow.

surface-drainage basin the area drained by a stream and its tributaries.

surface layer of the ocean the part of the ocean located above the permanent thermocline.

surface wash the flow of water following a rain or melting snow or ice that brings about downslope movement of surficial materials at the surface of the ground; also called sheet wash.

surface wave elastic wave that is propagated near the surface of a body.

surge rapid (that is, meters/day) movement of valley glaciers.

suspended load the sediment carried by a stream in suspension.

suspension used in hydrology to refer to material that is moved as solid particles within the water.

suture in plate tectonics, the line or zone within a plate along which two plates have come together in the past.

swallow hole opening in a stream channel leading into caverns and solution cavities.

swamp an area in which the ground is wet, usually having low relief and containing lakes or remnants of lakes.

swell wind-generated wave that has advanced into regions of weaker winds or calm.

syenite a coarse- to medium-grained equigranular igneous rock composed primarily of orthoclase with hornblende or biotite.

symmetrical fold a fold in which the two limbs are nearly mirror images of one another.

syncline a downfold in which the limbs dip toward the axis.

system a related set of natural processes, objects, or features. A set or arrangement of things so related or connected as to form a whole.

taiga forest coniferous forest found in Northern Asia and northern North America.

talus (*or* **scree**) rock fragments that accumulate as a heap or sheet at the base of a steep rock surface.

tarn a small lake in the mountains, often in a depression gouged out by a glacier.

tectonics the study of the formation and deformation of the Earth's crust that result in large-scale structural features.

temperature (thermal) inversion the atmospheric condition in which a layer of warmer air lies above cooler air.

temporary base level a level to which the reduction of topography of an area is restricted for a limited time.

terminal moraine ridge of drift (largely till) formed along the edge of a glacier at the position of the glacier's farthest advance.

terra rosa red and clayey soil found in areas of karst topography.

Tertiary a period of geologic time, including approximately 67 million years from the end of the Mesozoic era to the beginning of the Quaternary.

texture the general physical appearance of rocks. The size, shape, and arrangement of component minerals in sedimentary rocks determine the texture. In igneous and metamorphic rocks, the crystallinity and fabric of the minerals are major determinants of texture.

theory an idea, an hypothesis, that is supported by experimental or observation evidence.

thermals a general term applied to warm masses of air.

thermosphere the upper layer of the atmosphere located above the Mesosphere, starting at about 85 kilometers altitude. Temperature increases rapidly with altitude in this layer.

thrust fault a fault inclined at a low angle along which reverse-type movement has occurred.

thrust sheet a thin mass of rocks carried or moved laterally above a thrust fault.

tidal bore a turbulent wave that moves with high velocity up streams as a result of high tides.

tidal flat the flat area located along some coasts that is periodically covered with water during high tide.

tidal period the time between two high tides.

tidal range the difference in height between a high and a low tide.

tide the periodic rise and fall of sea level.

till unsorted, usually unstratified or poorly stratified, glacial drift.

timberline the elevation above which few trees grow.

tombolo sand bar or beach that connects offshore islands with the mainland or another island.

top-set bed bed composed of flat-lying sediments deposited on top of a delta by streams flowing across the delta, marshes, or low areas on either side of the streams.

topsoil a general term applied to the upper fertile part of the soil.

tornado a small-scale cyclone with very strong winds. At ground level, most tornados are less than 500 meters (1,640 feet) in diameter.

transform fault a fault that terminates sharply at a place where the movement is transformed into a structure of another type.

transverse dune a sand dune oriented with its long axis perpendicular to the prevailing wind.

transverse stream a stream that flows across the structural features of the underlying bedrock.

travel-time curve the plot of travel time versus distance of seismic waves.

travertine compact calcareous cave deposits.

tree rings growth rings seen in cross sections of trees. Because tree growth is more rapid in the summer, wider rings form at that time.

Tropic of Cancer latitude 23.5° south of the equator. It is the southernmost latitude reached by vertical rays from the Sun.

Tropic of Capricorn latitude 23.5° north of the equator. It is the northernmost latitude reached by vertical rays from the Sun.

tropopause the boundary between the troposphere and the stratosphere.

troposphere the portion of the atmosphere that is in contact with the surface. The top of the troposphere occurs where the temperature drop with altitude levels out and then begins to rise.

true north the direction from any point of the surface to the geographic North Pole.

truncated spur ridge between tributaries entering a major glaciated valley, cut back as a result of glacial erosion in the main valley.

tsunami a Japanese term applied to any gravity-wave system formed in the sea as a result of a large-scale, short duration disturbance of the surface.

tufa porous or cellular spring deposit; applied to spongy, fragile deposit with an earthy texture.

tuff ash or dust compacted to form a rock.

tundra a flat or gently sloping treeless area in the arctic or subarctic regions.

turbidity current a mass of water highly charged with material in solution and suspension that flows with turbulent motion down slopes through normal waters.

turbulent marked by wildly irregular motion.

turbulent flow fluid flow in which the paths of particle motion are irregular, with eddies and swirls.

ultimate base level the elevation toward which degradational processes tend to reduce the Earth's surface; the elevation the sea would have if all land areas were removed and the material deposited below water level.

unconfined ground water ground water not restricted by impervious material in its movement.

unconformity a substantial break or gap in the geologic record where rock units are overlain by units that do not ordinarily occur in that position. Many unconformities are erosion surfaces or other breaks in the stratigraphic record.

uniformitarianism the theory that the operation of physical and chemical principles and processes has been unchanged throughout geologic time.

unit magnetic pole one pole exerting a force of 1 dyne on another like pole when the two are 1 centimeter apart in a vacuum.

upwelling the rise of cold and more dense subsurface water toward the surface.

valley glacier glacier confined to a valley, generally on a mountain.

vaporization the process of changing from a liquid to a gas.

varves banded sequences of sediment in which a particular sequence of beds has been deposited repeatedly. The beds are caused by seasonal variations; each sequence represents one year's deposit.

vein a thin, sheet-like intrusion or a filling of a fracture.

vent a hole where the pipe or feeder systems for a volcano break through to the ground surface.

ventifact smooth, polished stone resulting from repeated abrasion of sand and silt blowing over a rock for a long time.

vertical circle any great circle that passes through the zenith.

vesicle a cavity formed where gases are trapped in a lava.

viscosity measure of the resistance a fluid offers to shear or flow; the inverse of fluidity.

volcanic breccias rocks containing angular fragments formed as a result of volcanic eruptions.

volcanic neck rock formed by cooling and crystallization of lava in the pipe-like feeder system of central vent volcanoes.

volcanic rock a general term applied to rocks of volcanic origin. They may include rocks formed in the volcano or in the area around the volcano.

water budget an expression of the total water resources of an area— that introduced into the area and that used or lost by any means.

water gap a pass through a mountain ridge through which a stream flows.

water table the surface that separates a zone where the pore space in the rocks or ground are not completely filled with water (that is, they are unsaturated) from a zone below in which they are saturated.

wave the up-and-down movement on the surface of a body of water.

wave-cut terrace the flat surface produced by wave erosion. These are generally found along coasts characterized by cliffs. The terrace is located on the seaward side of the cliff.

wave-cyclone low-pressure systems and accompanying circulation patterns that develop in wave-like patterns. These are commonly formed along the polar front.

wave height the vertical distance between the top of a wave crest and the bottom of an adjacent trough.

wave length the distance between two crests or two troughs.

wave refraction the bending of waves as they approach the shore obliquely.

weather (1) state of the atmosphere including descriptions of such features as temperature, precipitation, storms, and other meteorological factors. (2) alteration of materials (rocks and minerals) exposed to the atmosphere, moisture, and organisms at the group surface.

weathering changes that occur where rocks and minerals come in contact with the atmosphere, surficial waters, and organic life under conditions normally found at the surface of the Earth.

welded tufts (or **welded pumice** or **ignimbrite**) volcanic ash deposit in which the ash and glassy particles are welded together.

westerly winds a general expression to indicate that the wind is blowing from west to east.

wind gap a gap through a mountain originally formed by water passing through the gap, but now dry.

window the place where, as a result of erosion, the rocks beneath a thrust sheet may be seen surrounded by rocks that make up the thrust sheet.

xenolith literally, "foreign" rock; applied to rock fragments found in an igneous rock (for example, inclusions of wall rock).

yield point the stress level beyond which a material strains continuously with no additional stress.

zenith a point on the celestial sphere that lies directly above the observer.

zone of aeration that part of the soil and rock in which both air and percolating water occur.

zone of saturation that part of the sediment and rock in which pore spaces are filled with water.

zooplankton freely floating animals.

Credits

Photographs

Inside Front Cover: Digital Image by Dr. Peter W. Sloss, NOAA/NESDIS/NGDC
Unit Openers: 1: NASA; 2: © Simon Huber/Stone/Getty Images; 3: © Chad Ehlers/Stone/Getty Images; 4: NASA, N. Walborn (STScI), J. Maiz - Appellaniz (STScI) and R. Barba (La Plata Observatory).

Introduction
Opener: © Howie Garber/Getty Images; I.1: NASA; I.12: NASA.

Chapter 1
Opener: E. W. Spencer; 1.1: E. W. Spencer; 1.6a: © The McGraw-Hill Companies, Inc./Stephen Frisch, photographer; 1.10: © The McGraw-Hill Companies, Inc./Doug Sherman, photographer; 1.13: U.S. Geological Survey/HVO; 1.14a-b: U.S. Park Service; 1.17a-b, 1.19: E. W. Spencer; 1.21: Courtesy of Wayne Carmichael (Wright State University), Mark Schneegurt (Wichita State University), and Cyanosite (www.cyanosite.bio.purdue.edu).

Chapter 2
Opener: E. W. Spencer; 2.1, 2.2: E. W. Spencer; 2.3, 2.4a: U.S. National Museum; 2.4b: U.S. National Museum; 2.4c: U.S. National Museum; 2.6: E. W. Spencer; 2.7a: U.S. National Museum; 2.7b, 2.8b: E. W. Spencer; 2.11: U.S. Geological Survey, HVO; 2.12a-g: E. W. Spencer; 2.15a: U.S. Geological Survey, HVO; 2.15b, 2.17, 2.19a: E. W. Spencer; 2.19b: Courtesy of Leica Microsystems, Inc.; 2.20a-b: E. W. Spencer; 2.21: U.S. Geological Survey; 2.22a-b, 2.25a-e: E. W. Spencer.

Chapter 3
Opener: © Stone/Getty Images; 3.7: NOAA; 3.9: Surface of the Earth/NOAA; 3.10: NOAA; 3.23: E. W. Spencer.

Chapter 4
Opener: © FPG/Gary Randall/Getty Images; 4.2: NASA; 4.5: National Geophysical Data Center; 4.6: E. W. Spencer; 4.7a-e: U.S. National Museum; 4.7f, 4.8a-b, 4.9: E. W. Spencer; 4.12: W. B. Hamilton, U.S. Geological Survey; 4.13d, 4.15a, 4.15b-1, 4.15b-2: E. W. Spencer; 4.17: U.S. National Museum; 4.21: D. J. Miller, U.S. Geological Survey; 4.22: E. D. McKee, U.S. Geological Survey; 4.24: NASA.

Chapter 5
Opener: National Geophysical Data Center; 5.4, 5.13, 5.16, 5.21: NOAA; 5.28b-1, 5.28b-2: E. W. Spencer.

Chapter 6
Opener: © James Balog/Stone/Getty Images; 6.1: © Reuter/Sankei/Shimbun; 6.8: E. W. Spencer; 6.9, 6.14a-b: U.S. Geological Survey, NEIC; 6.21: Charles Mueller, U.S. Geological Survey Circular 1193; 6.22, 6.27: U.S. Geological Survey, NEIC; 6.29, 6.32: U.S. Geological Survey.

Chapter 7
Opener: © G. Brad Lewis/Stone/Getty Images; 7.1a: N. Banks, U.S. Geological Survey; 7.1b: J. Janda, U.S. Geological Survey; 7.5b: U.S. Geological Survey, CVO; 7.5c: U.S. Geological Survey, HVO; 7.6a: E.W. Spencer; 7.6b, 7.7: U.S. Geological Survey; 7.8a: NASA; 7.8c: U.S. Geological Survey, HVO; 7.9a: National Archives; 7.9c, 7.10a, 7.11: U.S. Geological Survey; 7.12a: Dennis Slifer; 7.12b: E. W. Spencer; 7.13a: U.S. Geological Survey, HVO; 7.15b: U.S. Geological Survey, CVO; 7.18a: U.S. Geological Survey, CVO; 7.19: Underwood & Underwood, Library of Congress; 7.20a: NOAA, National Geophysical Data Center; 7.23: NOAA.

Chapter 8
Opener: © Art Wolfe/The Image Bank/Getty Images; 8.1, 8.2a-b: NOAA; 8.3: Courtesy of Paul Johnson/School of Oceanography/University of Washington; 8.4: Copyright © 1995 David T. Sandwell and Walter H. F. Smith; 8.6, 8.9: Courtesy of William Haxby, Lamont-Doherty Observatory; 8.10: NOAA; 8.19: NASA; 8.20a-b: E. W. Spencer; 8.21: NOAA.

Chapter 9
Opener: © Don & Liysa King/The Image Bank/Getty Images; 9.4, 9.8: NOAA, National Geophysical Data Center; 9.14: Courtesy of Steve Nerem, NASA; image processing was done by Denis Leconte, Jet Propulsion Lab; 9.18a,b: NASA.

Chapter 10
Opener: E. W. Spencer; 10.1: U.S. Park Service; 10.5, 10.7, 10.12, 10.14, 10.15, 10.16: E. W. Spencer; 10.22: National Park Service; 10.23: NASA; 10.25, 10.26, 10.28: E. W. Spencer; 10.31, 10.32, 10.33: NASA.

Chapter 11
Opener: © Stephen Studd/Stone/Getty Images; 11.1: E. W. Spencer; 11.3: U.S. Department of Weather Bureau; 11.4: © PhotoDisc, Inc.; 11.5a: E. W. Spencer; 11.5b: National Center for Atmospheric Research; 11.8: NOAA, National Geophysical Data Center; 11.12: NASA; 11.17: NASA, EOS; 11.18: NOAA.

Chapter 12
Opener: © Ralph Wetmore/Stone/Getty Images; 12.1: NASA; 12.3: NOAA; 12.6, 12.7a: E. W. Spencer; 12.9: NASA; 12.12, 12,13: NOAA; 12.16: NOAA; 12.19, 12.20: NOAA Photo Library, OAR/ERL/National Severe Storms Lab (NSSL); 12.21: NASA.

Chapter 13
Opener: © Stephane Cande/FPG/Getty Images; 13.1a: E. W. Spencer; 13.1b: W. B. Hamilton, U.S. Geological Survey; 13.1c: E. W. Spencer; 13.7: Josh Landis/National Science Foundation;

Chapter 14
Opener: E. W. Spencer; 14.1a-d: E. W. Spencer; 14.2: U.S. Geological Survey; 14.3: Courtesy of Ray Sterner, Johns Hopkins University; 14.4: Data provided by NOAA/NESDIS, photographs produced by NASA; 14.5 a-b: E. W. Spencer; 14.6: R. E. Wallace, U.S. Geological Survey; 14.7: NASA; 14.8: U.S. Air Force; 14.9, 14.10, 14.11, 14.12: E. W. Spencer; 14.13: J. Mohlhenrich, Park Service; 14.14: USDA; 14.15 a-b, 14.16, 14.17, 14.18a,b: E. W. Spencer.

Chapter 15
Opener: E. W. Spencer; 15.1, 15.2: E. W. Spencer; 15.3a: U.S. Geological Survey; 15.4: E. W. Spencer; 15.5: N. K. Huber, U.S. Geological Survey; 15.6, 15.7a-c: E. W. Spencer; 15.8: F. O. Jones, U.S. Park Service; 15.9, 15.10: E. W. Spencer; 15.14a: USDA Soil Conservation Service; 15.14b: E. W. Spencer; 15.15: U.S. Department of Agriculture.

Chapter 16
Opener: © Stone/Getty Images; 16.1, 16.2a,b, 16.4: E. W. Spencer; 16.5: U.S. Geological Survey; 16.7: U.S. Forest Service; 16.8: E. W. Spencer; 16.9a,b: U.S. Geological Survey; 16.10a: E. W. Spencer; 16.11a, c: J. R. Stacy, U.S. Geological Survey; 16.12: National Archives; 16.14: T. L. Pewe, U.S. Geological Survey.

Chapter 17
Opener: © Kerrick James/Stone/Getty Images; 17.1a: U.S. Soil Conservation Service, photo by Al Carter; 17.1b, 17.3, 17.5, 17.6, 17.7: E. W. Spencer; 17.8: NASA; 17.9, 17.10: E. W. Spencer; 17.11: E. D. McKee, U.S. Geological Survey; 17.12e: U.S. Air Force; 17.12f: U.S. Geological Survey; 17.13: NASA; 17.14: E. W. Spencer; 17.16: E. W. Spencer.

Chapter 18
Opener: E. W. Spencer; 18.2, 18.4 a-c: E. W. Spencer; 18.5: New Zealand National Publicity Studio O; 18.8, 18.11: E. W. Spencer; 18.12, 18.15, 18.16: J. R. Balsley, U.S. Geological Survey; 18.18: U.S Geological Survey; 18.19: E. W. Spencer; 18.20: NOAA, National Geophysical Data Center; 18.21, 18.22a,b: E. W. Spencer; 18.24: NASA; 18.25: Ken Soper, Virginia Department of Highways & Transportation; 18.26, 18.27: E. W. Spencer; 18.28: NASA.

Chapter 19

Opener: E. W. Spencer; 19.1: E. W. Spencer; 19.7: NASA; 19.8, 19.13, 19.14, 19.15, 19.16a, 19.18a-b: E. W. Spencer; 19.19a: James F. Quinlan, Jr.; 19.19b: National Park Service; 19.20: E. W. Spencer; 19.21a: Rich Deuerling; 19.21b: Courtesy of Dave Hubbard, Virginia Division of Mineral Resources.

Chapter 20

Opener: © Robert van Dir Hirst/Stone/Getty Images; 20.1: W. B. Hamilton, U.S. Geological Survey; 20.2: U.S. Geological Survey; 20.4: Compiled by Jonathan Adams, Environmental Sciences Division, Oak Ridge National Laboratory; 20.7a: U.S. Geological Survey; 20.7b: U.S. Geological Survey; 20.8: NOAA; 20.9: M. Woodbridge Williams, National Park Service; 20.10, 20.13b-c, 20.14: E. W. Spencer; 20.16: U.S. Geological Survey; 20.18: NASA; 20.19: U.S. Air Force; 20.23: E. Soldo, National Park Service.

Chapter 21

Opener: © FPG/Getty Images; 21.1b, 21.2: NASA; 21.6a: JPL/California Institute of Technology/NASA; 21.7b-e, 21.8a, c: NASA/JPL; 21.9: NASA/Goddard Space Flight Center; 21.11a, 21.12b: NASA/JPL; 21.13a: Goddard Space Flight Center, NASA; 21.13b: NASA/JPL; 21.14: Courtesy of Mike Skrutskie, University of Massachusetts; 21.15a,b: Ken Nichols, Institute of Meteorites at the University of New Mexico; 21.16a: Courtesy of David Roddy, Meteor Crater, Northern Arizona; 21.18: NASA; 21.21: NASA; 21.22: NASA Apollo 17 Crew.

Chapter 22

Opener: © Anglo-Australian Observatory, Photograph from UK Schmidt Plates by David Malin; 22.1: Courtesy STScI; 22.3a-c: Courtesy of Anglo Australian Telescope Board; photographs by David Malin; 22.4: Courtesy STScI; 22.5: NASA; 22.9: Chandra/NASA/CXC; 22.10: Jay Gallagher (Univ. of Wisconsin), WIYN, AURA, NOAO, NSF;

Line Art

Introduction

I.6: After Shawn Spencer; I.8: After Shawn Spencer.

Chapter 1

1.2: After M. S. Silberberg, 2000, *Chemistry:* McGraw-Hill; 1.4: After M. S. Silberberg, 2000, *Chemistry,* McGraw-Hill; 1.6: After M. S. Silberberg, 2000, *Chemistry,* McGraw-Hill; 1.8: After M. S. Silberberg, 2000, *Chemistry* McGraw-Hill; 1.9: After Duxbury and Duxbury, *Fundamentals of Oceanography,* McGraw-Hill; 1.15: After M. S. Silberberg, 2000, *Chemistry,* McGraw-Hill; 1.16: Elizabeth Spencer; 1.18: After M. S. Silberberg, 2000, *Chemistry,* McGraw-Hill; 1.20: After G. B. Johnson, 2000, *The Living World:* McGraw-Hill; 1.22–1.25: After Elizabeth Spencer.

Chapter 2

2.10: After Shawn Spencer; 2.16: After D. L. Peck, U.S. Geological Survey professional paper 935-B; 2.23: Figure from *Petrogenesis of Metamorphic Rock* by H. G. Winkler. Reprinted by permission of Springer-Verlag GmbH & Co.KG.

Chapter 3

3.4: After the North American Tectonic Map, 1696 compiled by Philip King; 3.6: After Oberholtzer, 1933; 3.13: After O. M. Phillips, 1968, *The Heart of the Earth,* Freeman, Cooper & Co.; 3.19: After L. J. Burdick and D. V. Helmberger, 1978, *Journal of Geophysical Research;* 3.22: After B. Gutenberg, 1951, *Internal Constitution of the Earth,* Princeton University; 3.24: After F. Albaride and R. D. van der Hilst, 1999. New mantle convection model may reconcile conflicting evidence: EOS; 3.25: Modified from T. Lay, 1989, *EOS: Trans. Am. Geophys. Union* 70, 49: American Geophysical Union; 3.28: After X Song and P. Richards, Lamont-Doherty Earth Observatory; 3.29: From T. T. Arny, *Explorations,* McGraw-Hill.

Chapter 4

4.1: After *Earth System Science—A Closer View,* 1988, NASA Advisory Council; 4.4: After A. H. Sallenger, Jr., et al., USGS; 4.18: From *The Earth's Earliest Biosphere: Its Origin and Evolution* by William Schopf. Copyright © 1984 by Princeton University Press. Reprinted by permission of Princeton University Press; 4.23: After F. L. Wiseman, 1985, *Chemistry in Modern World,* McGraw-Hill.

Chapter 5

5.3: From "The Fit of the Continents around the Atlantic" by E. C. Bullard et al. in *A Symposium of Continental Drift* edited by P. M. S. Blackett et al., 1965, pp. 48–49. Reprinted by permission of The Royal Society; 5.5: After J. G. Sclater, Nat. Acad. Press. 2101 Constitution Ave. NW Washington D.C. After National Research Council, 1993, *Solid-Earth Sciences and Society,* National Academy Press; 5.6: From "The Fit of the Continents around the Atlantic" by E. C. Bullard et al. in *A Symposium of Continental Drift* edited by P. M. S. Blackett et al., 1965, pp. 48–49. Reprinted by permission of The Royal Society. 5.7: After K.C. Condie, 1982, *Plate Tectonics and Crustal Evolution,* Pergamon Press Ltd.; 5.8: From "Magnetic survey off the west coast of North America, 40 N latitude to 52 N latitude" by A. D. Raff and R. G. Mason in *Geological Society of America Bulletin,* 1961. Copyright © 1961 by Geological Society of America. Reproduced with permission of Geological Society of America via Copyright Clearance Center; 5.9: After J. R. Heirtzler, LePichon, and J. G. Brown, 1966, *Magnetic anomalies over the Reykjanes Ridge:* Deep Sea Research, v. 13; 5.17: "Geometry of Benioff zones: Lateral segmentation and downwards bending of the subjected lithosphere" by Bryan L. Isacks and Muawia Barazangi, 1977, in *Island Arcs, Deep Sea Trenches and Back-Arc Basins: Maurice Ewing Series 1* edited by M. Talwani and W. C. Pitman III. American Geophysical Union, Washington, D.C., pp. 99–114; 5.22: After R. D. van der Hilst from *Physics Today,* August 1999. Reprinted by permission of the author; 5.24: M. N. Toksoz and P. Bird, 1977, *Formation and evolution of marginal basins and continental plateaus,* in M. Talwani and W. C. Pitman, III, eds, *Island Arcs, Deep Sea Trenches and Back-Arc Basins:* Maurice Ewing Series 1, American Geophysical Union, Washington, D. C.; 5.29: Figure VIII-38c, "Southern Rocky Mountains, structure-sections through southern Cordilleran Orogen, Geology of Western Canada." Taken from *Geology and Economic Minerals of Canada,* Chapter VIII, by R. A. Price and E. W. Mountjoy. Reproduced with the permission of the Minister of Public Works and Government Services Canada, 2002 and Courtesy of the Geological Survey of Canada; 5.31: From "Implications of plate tectonics for the Cenozoic tectonic evolution of Western North America" by Tanya Atwater in *Geological Society of America Bulletin,* 1970. Copyright © 1970 by Geological Society of America. Reproduced with permission of Geological Society of America via Copyright Clearance Center.

Chapter 6

6.4: After H. F. Reid, 1906, *The elastic-rebound theory of earthquakes:* Bulletin Department of Geology, University of California, No. 6; 6.5: After Alpha, et al., 1989, U.S.G.S. open file report; 6.6: After W. R. Hanson, 1965, U.S.G.S. Professional Paper 542; 6.16: Figure by Roy van Aisdale in *Geotimes,* May 1997, Figure 1, p. 16. Reprinted with permission from *Geotimes.* Copyright the American Geological Institute; 6.18: After U.S.G.S. Professional Paper 1313; 6.19: After U.S.G.S. Professional Paper 1313; 6.20: After U. S. G. S. Circular 1193; 6.23: After U.S.G.S. Circular 1045; 6.24: After U.S.G.S. Circular 1045; 6.25: After U.S.G.S. published in EOS, Nov., 1989; 6.28: After U.S.G.S.

Chapter 7

7.2: After U.S.G.S. map; 7.4: From "Troundnjenite Batholith" by R. R. Compton in *Geological Society of America Bulletin,* 1966. Copyright © 1966 by Geological Society of America. Reproduced with permission of Geological Society of America via Copyright Clearance Center; 7.13b: After U.S.G.S. map; 7.14: Figure by Howel Williams, 1962. "The Ancient Volcanoes of Oregon" (Condon Lectures: Oregon State System of Higher Education, 3/e). Reprinted by permission of Oregon University System; 7.15a: After U.S.G.S. Professional Paper 1250; 7.16: After R. D. M. Verbeek, 1885, *Krakatau 1883;* 7.18b: After U.S.G.S. Professional Paper 11250; 7.20b: From Pinatubo Volcano Observatory, 1992. *Lessons from a Major Eruption: Mount Pinatubo, Philippines;* Earth in Space.

Chapter 8

8.5: After Duxbury and Duxbury, *Fundamentals of Oceanography,* McGraw-Hill; 8.11: From "Geophysical studies in the Gulf of Mexico" by R. G. Martin and J. E. Case in *The Ocean Basins and Margins: The Gulf of Mexico and the Caribbean* by A. E. M. Nairn, 1975, pp. 65–106. Reprinted by permission of Kluwer Academic/Plenum Publishers; 8.12: From "Geometry and evolution of salt structures in a marginal rift basin of the Gulf of Mexico, East Texas" by M. P. A. Jackson and S. J. Seni in *Geological Society of America Bulletin,* 1983. Copyright © 1983 by Geological Society of America. Reproduced with permission of Geological Society of America via Copyright Clearance Center; 8.13: After R. G. Martin, 1978, *Northern and eastern Gulf of Mexico continental margin: stratigraphy and structural framework:* American Association of Petroleum Geologists, Studies in Geology No. 7; 8.15: From "Crustal Rupture and the Initiation of Imbricate Thrusting in the Peru-Chile Trench" by D. A. Prince and L. D. Kulm in *Geological Society of America Bulletin,* 1975. Copyright © 1975 by Geological Society of America. Reproduced with permission of Geological Society of America via Copyright Clearance Center; 8.18: After H. U. Sverdrup, M. W. Johnson, and R. H. Fleming, 1942, *The Oceans: Their Physics, Chemistry and General Biology,* University of Chicago Press; 8.23: After Emery, Tracey, and Ladd, 1954, U.S.G.S. Professional Paper 260.

Chapter 9

9.1: After U.S. Navy; 9.2: After Duxbury and Duxbury,1999, *Fundamentals of Oceanography,* McGraw-Hill; 9.3: From *Oceans* by Sverdrup, Johnson and Fleming. Copyright 1942. Reprinted by permission of Pearson Education, Inc., Upper Saddle River, NJ; 9.6: After Duxbury and Duxbury, 1999, *Fundamentals of Oceanography,* McGraw-Hill; 9.11: After V.W. Ekman, 1902; 9.16: After J. R. Toggweiler,1994, *The Ocean's Overturning Circulation:* Physics Today; 9.17: After Wallace Broecker; 9.22: After U.S. Navy.

Chapter 10

10.6: Data from NOAA; 10.29: After E. W. Scholl and M. Stuvier, 1967, *Recent submergence of southern Florida: A comparison with adjacent coasts and other eustatic data:* Geological Society of America Bulletin, v. 78. Copyright © 1967 by Geological Society of America. Reproduced with permission of Geological Society of America via Copyright Clearance Center; 10.34: From C. R. Kolb and J. R. Van Lopik, *Deltas in their Geologic Framework,* eds. Shirley and Ragsdale, p. 22, 1966. Reprinted by permission of Houston Geological Society.

Chapter 11

11.2: Enlarged section of stratosphere reproduced by permission of F. M. Mimes III; 11.9: After M. S. Silberberg, *Chemistry:* McGraw-Hill; 11.10: After T. T. Arny, *Explorations:* McGraw-Hill; 11.11: After T. T. Arny, *Explorations:* McGraw-Hill; 11.14: Figure from *Physics of Climate* by Peixoto and Oont, 1992. Reprinted by permission of Springer-Verlag GmbH & Co. KG; 11.15: Compiled from data collected by NOAA as shown in EOS v. 80, no. 39; 11.16: From data largely due to C. D. Keeling; diagram originally published in the *SMIC Report* and reproduced here by kind permission of the M.I.T. Press.

Chapter 12

12.4a&b: By permission of Edward Linacre and John Hobbs, the authors of *The Australian Climatic Environment.* John Wiley, 1977; 12.14: After Danielson, Levin, and Abrams, *Meteorology:* McGraw-Hill; 12.17; After NOAA.

Chapter 13

13.2: After W. Köppen, 1931; 13.3: After W. Koppen, 1931; 13.5: by K.E. Trenberth et al. from *Intergovernmental Panel on Climate Change,* 1995. Reprinted by permission; 13.6: Reprinted with permission from *Science,* Vol. 288, April 2000, p. 590 by Mann, et al., American Geophysical Union. Copyright © 2000 American Association for the Advancement of Science; 13.8: Figure by Van de Hammen in *Late Cenozoic Glacial Ages: Concepts of Physical Geology* edited by K. K. Turekian. Copyright © 1971. Reprinted by permission of Yale University Press; 13.10: Figure on p. 429 by J. R. Petit et al., "Climate and Atmospheric History of the Past 420,000 years from the Volstok Ice Core, Antarctica" in *Nature,* Vol. 399, 1997. Reprinted by permission of the author; 13.11: From "Late Pleistocene and Holocene aeolian dust deposition in North China and the northwest Pacific Ocean" by K. Pye and Li-Phing Zhou in *Palaeogeography, Palaeoclimatology, Palaeoecology* 73, no. 1–2 (1989), pp. 11–23. Reprinted by permission of the author; 13.12: After *Earth System Science, A Closer View:* 1988, NASA Advisory Council; 13.13: Data courtesy of Dr. David Hodell, University of Florida. Graphic by Juan Velasco, *The New York Times,* February 16, 1999. Reprinted by permission; 13.14: Based on information contained within "Variations in the earth's orbit: Pacemaker of the Ice Ages" by J. D. Hays et al. in *Science,* 1976, Vol. 194, pp. 1121–1132; 13.15: After K. Taylor, 1999: *American Scientist,* v. 87, p. 325; 13.16: Figure from *Science,* Vol. 281, July 1998, p. 157.

Reprinted by permission of Nate Mantua, University of Washington; 13.17: Figure 1.2, p. 5 of *Climate Process and Change* by Edward Bryant. Copyright © 1997. Reprinted with the permission of Cambridge University Press.

Chapter 15

15.13: Edward Roberts/*American Scientist,* 1999, Vol. 87, p. 325. Reprinted by permission of Scientific Research Society; 15.17: After N. M. Strakhov, 1967, *Principles of Lithogenesis:* Oliver and Boyd Publishers.

Chapter 16

16.3: After D. J. Varnes, 1958, "Landslide Types and Processes" in E. B. Eckel, ed., *Landslides and Engineering Practice;* 16.11b: After J. B. Hadley, 1964, U.S.G.S. Professional Paper 435; 16.13: After W. R. Hansen, 1965, U.S.G.S. Professional Paper 490; 16.15: After Clawges and Price, 1999, U.S.G.S. Open File Report 99–77; 16.16: Data obtained from surveys by the Long Beach Harbor Department and Office of Los Angeles County Surveyor and Engineer.

Chapter 17

17.15: After F. Peterson and C. Turner-Peterson, 1989, International Geological Congress, Field Trip Guidebook T130.

Chapter 18

18.1: Modified after Foster and Smith, 1990; 18.10: After F. Hjulstrom, 1935, "Studies of the morphological activity of rivers as illustrated by the River Fjris": Univ. Uppsala Geological Institute Bulletin, 25; 18.14: After L. B. Leopold and T. Maddock Jr, 1953, U.S.G.S. Professional Paper 252.

Chapter 19

19.6: Modified after O. E. Meinzer, 1923, U.S.G.S. Water Supply Paper 494; 19.11: After U.S.G.S. Circular 1139; 19.22: Modified after a map by the U.S. Department of Agriculture; 19.24: After U.S.G.S. Circular 1139; 19.25: After U.S.G.S.

Chapter 20

20.5: B. G. Anderson and H. W. Born Jr., 1994, *The Ice Age World:* Scandinavian University Press; 20.11: After R. P. Sharp, 1960, *Glaciers:* Condon Lectures; University of Oregon Books; 20.20: After U.S.G.S. map; 20.21: From "Quaternary palaeolakes in the evolution of semidesert basins, with special emphasis on Lake Bonneville and the Great Basin, USA" by Donald R. Currey in *Palaeogeography, Palaeoclimatology, Palaeoecology,* Vol. 73, 1990, pp. 189–214; 20.22: After Victor Baker, University of Arizona.

Chapter 21

21.11b: From NASA; 21.12a: From NASA

Chapter 22

22.2: After M. S. Silberberg, 2000, *Chemistry:* McGraw-Hill; 22.7: After T. T. Arny, 2000, *Explorations:* McGraw-Hill.

Index

Glossopteris flora, 108, 289
Gneisses, 54, 478
Gondwanaland, 108, 179
Gossan, 328
Grabens, 116, 117, 122
Gradation, 307
Gradient, 371
Grand Canyon
  as erosion example, 83, 370, 372
  stratification in, 86, 87
Grand Ditch, 384
Granite
  chemical composition, 45
  formation, 44
  as phosphorus source, 31
  properties, 475–76
  resistance to mass wasting, 339
Granite pegmatite, 476
Granodiorite, 476
Granules, 475
Graphic granite, 476
Grasses, 230, 233, 311
Gravity, 10, 212–14
Graywacke, 477, 478
Grazing, excessive, 331, 332, 356
Great Barrier Reef, 191
Great circles, 480
Great Lakes, 428
Great Plains, 311
Great Red Spot, 446, 447
Great Salt Lake, 383, 429
Greenhouse effect. See also Global warming
  causes, 22
  global warming and, 251–53
  on Venus, 441
Greenhouse gases
  climate changes and, 296
  distribution in atmosphere, 243
  global warming and, 251–52
  on Venus, 441
Greenland ice sheet, 419–20
Green River, 379
Gros Ventre earthflow, 343
Ground-level air circulation, 265
Ground motion in earthquakes, 131
Ground water
  geysers, 401–2
  ground subsidence from withdrawal, 348
  infiltration and storage, 394–96
  influences on landscape, 402–5
  movement, 400–401
  pollution, 405–6
  role in Earth systems, 394
  saturated zones, 396–400
  sources, 393, 394
  supply issues, 406–12
Ground-water basins, 408
Growth rates, 93
Guide fossils, 89, 90
Gulf of Aden, 116
Gulf of California, 116, 118
Gulf of Mexico, 181–82
Gulf of St. Lawrence, 214–15
Gulf Stream, 202–3, 204, 295
Gutenberg, Beno, 68

Guyots, 178
Gypsum, 325, 477, 478
Gyres, 201–2, 206, 207

H

Hadean eon, 89
Hail, 270
Hale-Bopp comet, 6, 449
Hales Bar Dam, 406
Half Dome, 322
Half-life, 94
Halite, 38
Hanging valleys, 422, 423
Hanging walls, 122
Hard-ball model, 18
Harding Ice Field, 421
Hardness of minerals, 36, 471
Hard pan, 333
Hassanabad Glacier, 421
Hawaiian Islands
  atmospheric carbon dioxide measurements,
    252, 290
  as hot spot example, 113–14
  volcanic eruption style, 165, 167
  volcanic gases, 153
Heads of streams, 371
Heat balance, 247–53, 285–86
Heat budget, 250, 251
Heat capacity of water, 24
Heating and cooling, weathering effects, 322
Hebgen Lake, 344
Heiskanen model, 69
Helium, 439, 446
Herjolfsson, Bjarni, 198
Herodotus, 198
Hertzsprung-Russell diagram, 464, 465
Hess, Harry, 110–11
Heterosphere, 243
High-altitude air circulation, 266–67
High latitude biomes, 302
High mountain environments, 311–13
High Plains states, 410
High-pressure cells, 265
High-pressure systems, 262–63
High tides, 214, 273
Himalayan Mountains, 60
Histosols, 330
Holocene epoch, 91–92, 288
Homosphere, 243
Hooks, 230
Horizons, soil, 329
Hornblendite, 476
Hornfels, 478
Horns, 422, 423
Horse latitudes, 265
Horsts, 122
Hot deserts, 313–14
Hot spots
  as guide to plate movements, 113–14
  magma from, 42, 151
  seamounts over, 178
  Yellowstone area, 166–69
Hot springs, 398, 399
Hour angle, 481–82

Hour circles, 481
Hubble telescope, 460, 463, 467
Human activities
  atmospheric damage, 253–56
  carbon dioxide from, 193 (see also Global
    warming)
  in coastal environments, 233–34, 235,
    236, 237–38
  ground-water impact, 405–6, 407, 408–9, 410–12
  impact on marine biodiversity, 188–90
  mass wasting from, 345–46
  sea level changes and, 223
  slope failure and, 338, 339
  soil loss from, 331–32
  surface water impact, 382–86, 389–90
  weathering effects, 322
Humboldt, Alexander, 207
Humic acids, 323
Humid continental climate, 283, 284
Humidity, 243–44
Humid subtropical climate, 282, 284
Hurricane Andrew, 275
Hurricane Camille, 275
Hurricane Hugo, 273, 275
Hurricanes
  destruction from, 260, 273, 274–76
  flooding from, 390
  formation, 273, 284
Hydration, weathering effects, 325
Hydrogen, 439, 446, 447
Hydrologic cycle
  effect of human activities, 382–86
  overview, 7, 8, 41, 368
Hydrolysis, 325–26
Hydrostatic head, 399
Hydrostatic pressure, 397, 399, 400
Hydrothermal vents, 176, 177, 178, 190
Hypotheses, 5

I

Iapetus Ocean, 121
Ice
  isostatic effects on crust, 70–71
  on Mars, 441, 442–43, 445
  melting point, 21
  on Mercury, 440
  molecular structure, 20
  on Pluto, 448
  in solar system origins, 452
  weathering effects, 320–21
Ice ages
  causes, 432
  data supporting, 288–89
  effect on plants and animals, 429–30
  effects on human history, 430–31
Icebergs, 418, 420
Iceland, settlement, 431
Ice sheets. See also Glaciers
  climatic effects, 416–17
  current trends, 286–87
  data from trapped oxygen, 289
  effects on continental margins, 180, 221
  erosion by, 423–24
  formation, 92

interaction with oceans, 417–18
occurrence around the world, 416
Ice shelves, 417, 419
Igneous rocks
formation, 18, 22, 37, 42–47
properties, 475–76
Impact craters
on Earth, 6, 100, 450
on Moon, 454–56
on planets, 442, 443
Incandescent cloud eruptions, 162–65
Inceptisols, 330
Inclination, magnetic, 74
Index fossils, 89, 90
Index of refraction, 94
Indian Ocean, 204, 205, 210–11
Inertia, 132
Inner core, 8, 72, 75
Inorganic compounds, 18
Integrated systems, 4
Interdisciplinary approach, 4, 5
Interglacial intervals, 92, 288
Intergovernmental Panel on Climate Change, 296
Internal divisions of Earth, 67
International Geophysical Year, 254
Intertropical convergence zone, 265
Intrusions, 46, 47, 152
Inverse square law, 465
Inversions, 264
Io, 446, 448
Ionic bonding, 19
Ionic compounds, 19
Ionosphere, 246
Ions, 19, 330
Iridium, 100
Iron, 16, 37
Iron ore, 328
Irregular galaxies, 461, 462
Irrigation
effects on water supplies, 384–85, 410–11
soil degradation from, 332
Island arcs
locations, 61
in North American continent, 119
at subduction zones, 116–17, 119, 120
Islands, barrier, 231–34
Isobars, 260, 261
Isostasy theory, 70–71
Isotopes, 94, 95
Izmit earthquake (1999), 138–39

J

James River, 386
Japanese Islands, seismic activity, 141–42
Jet streams
formation, 267–68
relation to pressure systems, 270
on Saturn, 447
Jetties, 229
JOIDES Resolution, 177
Jovian planets, 439, 440, 445–48
Juan de Fuca Ridge
history, 123, 124
seismic activity, 142–43, 160

Jupiter, 443, 446, 447
Jupiter effect, 145
Juvenile water, 394

K

Kames, 424
Kant, Immanuel, 451
Karakoram Mountains, 421
Karst topography, 402–6
Katmai volcano, 153
Kenai Mountains, 143, 421
Kermadec Islands, 115–16
Kettles, 426, 428
K-feldspar, 325
Kilauea volcano
calderas, 156–58
lava lakes, 22, 46
USGS observatory, 165
Kobe earthquake (1995), 128, 142
Kocaeli, Turkey earthquake (1999), 138–39, 140
Kodiak Island, 143, 144
Krakatau volcano
climatic effects, 286
destruction from, 149
eruption style, 159, 163
tsunami from, 225
Krill, 188
Kudzu, 350
Kyoto Protocol, 253

L

Labrador current, 203–4
Labradorite, 40
Lag gravel, 357, 358
Lahars, 150
Lake-bed deposits, 426
Lake Bonneville, 429, 430
Lake Missoula, 429, 430
Lake Nasser, 386
Lakes, glacial, 427–29
Laminar flow, 371, 373
Lanatian Man, 358
Land breezes, 355
Land bridges, 431
Land environments
defining, 301–3
geologic framework, 303–9
identifying by climate and vegetation, 309–14
overview, 300–301
wetlands, 314–16
Landslides
defined, 341
earthquakes with, 132
famous cases, 344–45, 346–47
sea floor, 181
La Niña, 211, 295
Laplace, Marquis de, 451
Lapse rates, 263
Laramide Orogeny, 123
Las Vegas, Nevada, 410
Lateral moraines, 424, 425
Lateral slip, 340
Laterite, 328, 333

Latitude, influence on Coriolis effect, 205
Latosol soils, 333
Lattices, 471
Laurasia, 108
Lava. *See also* Magma
cave formation, 403, 404
crystallization, 45–46, 154
evidence of ancient flows, 84
magnetic fields in, 109–10
on Moon, 455
sources, 23
at subduction zones, 117
varieties, 43–45, 150–51, 153–55
Lava domes, 154–55, 161, 164
Lava lakes, 22, 45–46, 47, 154
Lava plateaus, 165–69
Law of faunal succession, 89
Laws, in science, 5
Layer lattice silicates, 473–74
Levees
flood control, 387, 389
natural, 237, 375, 376
to preserve channels, 237
Lichens, 93
Life, origins, 98–99. *See also* Animals; Organisms
Lightning, 271
Limestone
formation, 29, 39–41, 50, 474
metamorphism of, 53
solution of, 394, 396, 402–4
weathering of, 326
Limonite, 328
Linear dunes, 361
Liquids
basic properties, 20
occurring on Earth, 18, 23–26
Lithification, 49
Lithophysae, 476
Lithosphere. *See also* Plate tectonic system
crust *versus,* 66
defined, 8
properties, 67–71
rock cycle, 41–42
Little Ice Ages, 82, 292, 431
Living organisms. *See* Organisms
Load, of streams, 373–75
Lodestone, 73
Loess, 339, 358
Loihi, 113
Loma Prieta earthquake (1989), 140–41, 142
Long Island ground water problems, 407, 408–9
Longitudinal profiles, 371, 373
Longitudinal waves, 64, 65
Longshore currents, 224
Longshore drift, 224, 229
Long-term climate changes, 292–93
Long-term processes, 82–83
Los Angeles, subsidence in, 349
Love Canal, 405
Love waves, 64, 131, 134
Low-pressure cells, 265
Low-pressure systems, 261–62, 268–70
Low tides, 214
Luminosity, 463–64, 465
Luster, 36, 471

Photochemical dissociation, 97
Photosphere, 438
Photosynthesis
  influence on atmosphere, 97
  in oxygen cycle, 30
  in photic zone, 198
  products of, 27
pH scale, 323, 330
Phyllites, 477
Physical climate system, 6–7
Physical degradation of soil, 332
Physiographic provinces, 301, 304–5
Phytoplankton, 185
Piedmont glaciers, 418
Pilot leaders, 271
Pingos, 427
Piracy, 380
Piston coring, 177–78, 179
Pitchstone, 476
PKP waves, 72, 74, 75
Plagioclase feldspars, 39, 40, 473
Plains
  abyssal (see Abyssal plains)
  on Colorado Plateau, 381, 382
  outwash, 424, 426, 427
Planetary nebulae, 462
Planetesimals, 452–53
Planets
  basic properties, 439–40
  Earth as, 4
  Jovian, 439, 440, 445–48
  origins, 438, 451–56
  terrestrial, 440–44, 445
Plankton, 185
Planktonic communities, 185–88
Plants. See also Vegetation
  in deserts, 363
  effect of ice ages, 429–30
  flood control role, 386
  photosynthesis, 27, 30
  to stabilize slopes, 364
  variation in growth rates, 93
  water uptake, 24, 25, 27
  weathering effects, 322–23
Plastic behavior, 343
Plate tectonic system
  effect on land environments, 307–8
  effects on sea level, 222
  evolution of theory, 107–10
  measuring movements, 114–15
  mountain building, 118–24
  overview, 6, 9, 10, 105, 106–7
  rock cycle, 41–42
  role in hydrologic cycle, 368
  subduction in, 115–16 (see also Subduction zones)
  support for theory, 110–14
  tectonic cycles, 116–18
  volcanoes and, 169 (see also Volcanoes)
Playa lakes, 382, 383
Pleistocene epoch, 91–92, 288
Pliocene era, 288
Plunges, 121
Pluto, 443, 448
Plutonic rocks, 18, 42
Plutons, 45, 47, 151–52
Podzol soil, 333

Point bars, 377, 378
Polar air masses, 265
Polar climate, 283
Polar easterlies, 265
Polar front, 266
Polar front jet, 267
Polar ice masses, 417, 418–19
Polar molecules, 24
Polar regions
  environmental features, 309
  on Mars, 441
  offshore sediments, 180
  rapid climate change, 286–87
Polar wandering, 75, 109–10
Pole strength, 74
Pollen, 429–30
Pollution
  atmospheric, 32, 254–55, 264
  of estuaries, 236
  impact on marine biodiversity, 189
  water, 382–83, 405–6, 411
Porosity, 395, 396
Porphyry, 46, 47, 475
Positive gravity anomalies, 71
Potash feldspar, 39, 40
Potassium, 39
Potholes, 373, 375
Powell, John Wesley, 307, 378–80
Prairie potholes, 315
Pratt, John, 68–69
Precambrian period, 89–90, 107
Precession, 292
Precipitates, 26
Precipitation
  acid rain, 32, 255
  atmospheric, 244–45
  chemical, 48, 50
  effect on deserts, 313–14
  El Niño effects, 211
  flooding and, 389
  with hurricanes, 275–76
  mineral formation from, 26
  with thunderstorms, 270–71
Prehistoric climate data, 287–90
Pressure
  atmospheric, 260–63
  confining, 52
  conversion factors, 470
  effect on states of matter, 21, 43
  as factor in metamorphism, 52
Prevailing wind patterns, 265–66
Primary waves, 64
Prince William Sound, 143, 144
Processes, defined, 6
Progressive waves, 223
Project SHEBA, 287
Prominences, 438–39
Proper motion, 464
Proteins, 28
Proterozoic eon, 89
Protons, 16
Pumice, 155
P waves
  paths through Earth's interior, 72–73, 74, 75
  properties, 64
  velocity/depth relation, 68

Pyroclastic flows, 158, 162, 165
Pyroclastic materials, 152, 155, 475
Pyroxenes, 39, 46, 473
Pyroxenite, 476

Q

Quarrying, 422
Quartz
  in metamorphic rocks, 53
  properties, 472–73
  resistance to weathering, 323
  sand from, 49, 358
  structure, 36, 38, 40
Quartzites, 478
Quasars, 463
Quaternary period, 91

R

Radial speed, 464, 465
Radioactive decay, 44–45, 94–95
Radioactive elements, 16
Radioactive isotopes, 94–95
Radioactive waste disposal, 189
Radiolarians, 187–88
Radiometric age determination, 94–95
Rain forests
  destruction, 253
  environmental features, 311, 312
  soil type, 333
Rainwater
  acidification, 32, 255, 327
  weathering effects, 326–27
Rapid-wasting disease, 191
Rarefactions, 64, 65
Rayleigh waves, 64, 131, 134
Rays, seismic, 65, 66
Recessional moraines, 424, 426, 428
Recharge
  in Basin and Range, 409–10
  defined, 400
  promoting, 411–12
  sediment type and, 408
Recumbent anticlines, 121
Red clay, 178
Red giants, 464, 466
Red Rock Pass, 429, 430
Red Sea, 116
Red tide, 187
Reef flats, 191
Reefs, 190–93
Reelfoot graben, 137, 138
Reflection of waves, 65–66, 223
Refraction
  index, 94
  of ocean waves, 223, 224, 225
  of seismic waves, 65–66
Regional climates, 281–84
Regional metamorphism, 51–52
Regolith, 320
Reid, H. F., 129
Relative humidity, 243
Remanent magnetic effects, 111–13
Remotely operated vehicles (ROVs), 176, 178

Sheeting, 322
Sheet wash, 372
Sherman landslide, 346–47
Shields, 59
Shield volcanoes, 155, 158
Shooting flow, 371, 373
Shoreline, 220. *See also* Coastal environments; Coasts
Short-term processes, 80–81
Siderites, 72, 449–50
Sierra Nevada Mountains, 283
Silent Spring, 4–5
Silica
    in impact craters, 450
    removal from soil by plants, 323
    weathering products, 325, 326–27
Silicate minerals
    common types, 472–74
    in magma, 43, 154
    overview, 36–39
    weathering, 326–27
Siliceous rock, 50
Siliceous sinter, 398–99
Silicon, 37, 39
Silt, 388, 475
Siltstone, 477, 478
Silver Springs, Florida, 398
Sinkholes, 403, 404, 406
Sinks, for carbon dioxide, 193
Slash-and-burn farming, 311
Slates, 54, 477
Slaty cleavage, 477
Slides
    earthquakes with, 132
    famous cases, 344–45, 346–47
    on sand dunes, 360
    sea floor, 181
    types, 341
Slope
    mass wasting and, 339–41
    ocean surface, 206
    stabilizing, 350
Slumps
    mass wasting, 340, 341, 345–46
    in sand dunes, 360
Smith, William, 88–89
Smog, 32, 254
Snake River, 429, 430
Snake River area lava flows, 166, 168
Snider, A., 108
Snow fields, 312, 420–21
Snow formation, 244, 246
Sodium, 39
Soil. *See also* Mass wasting
    creep, 343, 344
    defined, 320, 338
    degradation, 330–32 (*see also* Erosion)
    effects on ecosystems, 303
    features, 328–30, 333
    formation, 21–22, 41–42
    reclamation, 332–34
Soil chemistry, 329–30
Soil Conservation Service, 334
Soil maps, 350
Soil science, 329
Soil types, 330, 331, 333

Solar and Heliospheric Observatory, 438
Solar constant, 247, 292, 439
Solar radiation
    atmospheric heat balance, 247–53
    effects on marine environments, 184–85
    long-term climate changes and, 292–93
    variations, 82, 439
    volcanic influence, 169–70
Solar system
    comets, asteroids, and meteorites, 443, 448–51
    Jovian planets, 439, 440, 445–48
    origins, 438, 451–56
    Sun's properties, 438–39 (*see also* Sun)
    terrestrial planets, 440–44, 445
Solar wind, 247, 438–39
Solids, basic properties, 20
Solifluction, 342–43
Solution cavities, 396
Solution load, 373, 374
Solutions, 24–26, 47
Solvents, 24, 402
Sorting, 50
South America, continental fit with Africa, 109
Southern Hemisphere, 205, 261
Southern oscillation, 210, 211–12
South magnetic pole, 73
Spacecraft, 441, 442, 445, 446, 448
Space radiation, 95
Spectral analysis, 460–61
Spheroidal weathering, 323, 324
Spherulites, 476
Spinel, 71
Spiral galaxies, 461, 462
Spits, 228, 230
Spodosols, 330
Springs, 397–98
Spring tides, 213
Stacks, 227, 228
Stalactites, 25, 404–5
Stalagmites, 405
Standing waves, 214, 215
Star charts, 480–81
Star dunes, 361
Starfish, 191–92
Stars, 460–66
States of matter, 20–21, 23, 24
Stationary fronts, 268
Steam, 20
Steno, Nickolas, 86
Stick-slip, 131
Stocks, 152
Stony meteorites, 449–50
Storms. *See also* Hurricanes
    destruction from, 260
    dust, 354–55
    impact on beaches, 229, 233–34
    types, 270–76
Strains, 129
Stratification
    as basic feature of sedimentary rock, 50, 51
    of mantle, 71
    of marine environments, 185
    of static air, 260–61
Stratigraphic record, 85
Stratopause, 243, 244

Stratosphere, 242, 244
Stratovolcanoes, 155, 157
Stratus clouds, 245
Streak, 471
Stream gradient, 371
Stream piracy, 380
Streams
    flooding, 386–90
    formation, 369–70
    from glaciers, 426
    impact of human activities, 382–86
    landscape-shaping forces, 376–82
    principles of, 370–76
Striations, glacial, 422, 423
Strike-slip faults
    defined, 113, 114, 122
    earthquakes on, 138–41
Stromatolites, 90
Strombolian eruptions, 165
Subduction zones
    causes, 118
    cross section, 11
    defined, 9, 106
    destruction of oceanic crust, 115–16
    earthquakes in, 135, 141–43, 144
    magma from, 43, 150–51
    origin of theory, 107
    volcanic formations, 116–17
Sublimation, 21
Subsequent streams, 379
Subsidence
    Louisiana coast, 220
    mass wasting, 347–49
    on permafrost, 427
    sinkholes, 403, 404, 406
Subsoil, 329
Subtropical high zone, 265
Subtropical jet, 267
Sudan, 362
Sulfur, 169–70
Sulfuric acid, 324, 327
Sun. *See also* Solar radiation
    effect on tides, 212, 213
    energy from, 6, 9–10
    influence on Earth's magnetic field, 74, 75, 76, 247
    properties, 438–39
    role in photosynthesis, 30
    spectrum, 460
    speed of movement, 461
Sunset Crater, 157
Sunspots
    cycles, 438
    influence on climate, 292
    influence on Earth's magnetic field, 74, 75, 76
    prominences and, 439
Superfund, 405
Supergiants, 464
Superimposed streams, 379
Supernovae, 466
Superposition, 5, 86
Surface-drainage basins, 369–70
Surface layer of ocean, 201, 202
Surface tension of water, 23
Surface waves, 64, 131
Surges, 421